AF500320

Progress in Mathematics
Volume 140

Anthony W. Knapp

Lie Groups Beyond an Introduction

Springer Science+Business Media, LLC

Anthony W. Knapp
Department of Mathematics
State University of New York at Stony Brook
Stony Brook, NY 11794

Library of Congress Cataloging-in-Publication Data

Knapp, Anthony W., 1941-
Lie groups beyond an introduction / Anthony W. Knapp.
p. cm. -- (Progress in mathematics ; v. 140)
Includes bibliographical references (p. -) and index.

DOI 10.1007/978-1-4757-2453-0
1. Lie groups. 2. Lie algebras. 3. Representations of groups.
I. Title. II. Series: Progress in mathematics (Boston, Mass.) ;
vol. 140.
QA387.K567 1996 96-17002
512'.55--dc20 CIP

Printed on acid-free paper
 Birkhäuser 

Typeset by the author in Adobe Times Roman, using MathTime fonts © by the TeXplorators Corporation and Y & Y Inc., Math Symbol fonts designed by the American Mathematical Society and © by Blue Sky Research and Y&Y Inc., and Euler Fraktur fonts designed by Hermann Zapf and © by Blue Sky Research and Y&Y Inc.

9 8 7 6 5 4 3 2 1
www.springer.com/mycopy

To Susan

and

Sarah and Dave and Will

CONTENTS

LIST OF FIGURES

PREFACE

Fifty years ago Claude Chevalley revolutionized Lie theory by publishing his classic *Theory of Lie Groups I*. Before his book Lie theory was a mixture of local and global results. As Chevalley put it, "This limitation was probably necessary as long as general topology was not yet sufficiently well elaborated to provide a solid base for a theory in the large. These days are now passed."

Indeed, they are passed because Chevalley's book changed matters. Chevalley made global Lie groups into the primary objects of study. In his third and fourth chapters he introduced the global notion of analytic subgroup, so that Lie subalgebras corresponded exactly to analytic subgroups. This correspondence is now taken as absolutely standard, and any introduction to general Lie groups has to have it at its core. Nowadays "local Lie groups" are a thing of the past; they arise only at one point in the development, and only until Chevalley's results have been stated and have eliminated the need for the local theory.

But where does the theory go from this point? Fifty years after Chevalley's book, there are clear topics: É. Cartan's completion of W. Killing's work on classifying complex semisimple Lie algebras, the treatment of finite-dimensional representations of complex semisimple Lie algebras and compact Lie groups by Cartan and H. Weyl, the structure theory begun by Cartan for real semisimple Lie algebras and Lie groups, and harmonic analysis in the setting of semisimple groups as begun by Cartan and Weyl.

Since the development of these topics, an infinite-dimensional representation theory that began with the work of Weyl, von Neumann, and Wigner has grown tremendously from contributions by Gelfand, Harish-Chandra, and many other people. In addition, the theory of Lie algebras has gone in new directions, and an extensive theory of algebraic groups has developed. All of these later advances build on the structure theory, representation theory, and analysis begun by Cartan and Weyl.

With one exception all books before this one that go beyond the level of an introduction to Lie theory stick to Lie algebras, or else go in the direction of algebraic groups, or else begin beyond the fundamental "Cartan decomposition" of real semisimple Lie algebras. The one exception is the book Helgason [1962],* with its later edition Helgason [1978]. Helgason's books follow Cartan's differential-geometry approach, developing geometry and Lie groups at the same time by geometric methods.

The present book uses Lie-theoretic methods to continue Lie theory beyond the introductory level, bridging the gap between the theory of complex semisimple Lie algebras and the theory of global real semisimple Lie groups and providing a solid foundation for representation theory. The goal is to understand Lie groups, and Lie algebras are regarded throughout as a tool for understanding Lie groups.

The flavor of the book is both algebraic and analytic. As I said in a preface written in 1984, "Beginning with Cartan and Weyl and lasting even beyond 1960, there was a continual argument among experts about whether the subject should be approached through analysis or through algebra. Some today still take one side or the other. It is clear from history, though, that it is best to use both analysis and algebra; insight comes from each." That statement remains true.

Examples play a key role in this subject. Experts tend to think extensively in terms of examples, using them as a guide to seeing where the theory is headed and to finding theorems. Thus examples properly play a key role in this book. A feature of the presentation is that the point of view—about examples and about the theory—has to evolve as the theory develops. At the beginning one may think about a Lie group of matrices and its Lie algebra in terms of matrix entries, or in terms of conditions on matrices. But soon it should no longer be necessary to work with the actual matrices. By the time one gets to the last two chapters, the point of view is completely different. One has a large stock of examples, but particular features of them are what stand out. These features may be properties of an underlying root system, or relationships among subgroups, or patterns among different groups, but they are far from properties of concrete matrices.

A reader who wants only a limited understanding of the examples and the evolving point of view can just read the text. But a better understanding comes from doing problems, and each chapter contains some in its last section. Some of these are really theorems, some are examples that show the degree to which hypotheses can be stretched, and some are exercises. Hints for solutions, and in many cases complete solutions, appear in a section near the end of the book. The theory in the text never relies on a problem from an earlier chapter, and proofs of theorems in the text are never left as problems at the end of the current chapter.

*A name followed by a bracketed year points to the list of References at the end of the book.

A section called Notes near the end of the book provides historical commentary, gives bibliographical citations, tells about additional results, and serves as a guide to further reading.

The main prerequisite for reading this book is a familiarity with elementary Lie theory, as in Chapter IV of Chevalley [1946] or other sources listed at the end of the Notes for Chapter I. This theory itself requires a modest amount of linear algebra and group theory, some point-set topology, the theory of covering spaces, the theory of smooth manifolds, and some easy facts about topological groups. Except in the case of the theory of involutive distributions, the treatments of this other material in many recent books are more consistent with the present book than is Chevalley's treatment. A little Lebesgue integration plays a role in Chapter IV. In addition, existence and uniqueness of Haar measure on compact Lie groups are needed for Chapter IV; one can take these results on faith or one can know them from differential geometry or from integration theory. Differential forms and more extensive integration theory are used in Chapter VIII. Occasionally some other isolated result from algebra or analysis is needed; references are given in such cases.

Individual chapters in the book usually depend on only some of the earlier chapters. Details of this dependence are given on page xiv.

My own introduction to this subject came from courses by B. Kostant and S. Helgason at M.I.T. in 1965–67, and parts of those courses have heavily influenced parts of the book. Most of the book is based on various courses I taught at Cornell University or SUNY Stony Brook between 1971 and 1995. I am indebted to R. Donley, J. J. Duistermaat, S. Greenleaf, S. Helgason, D. Vogan, and A. Weinstein for help with various aspects of the book and to the Institut Mittag-Leffler for its hospitality during the last period in which the book was written. The typesetting was by AMS-TEX, and the figures were drawn with Mathematica®.

May 1996

PREREQUISITES BY CHAPTER

This book assumes knowledge of a modest amount of linear algebra and group theory, some point-set topology, the theory of covering spaces, the theory of smooth manifolds, and some easy facts about topological groups. The main prerequisite is a familiarity with elementary Lie theory, as in Chapter IV of Chevalley [1946]. The dependence of chapters on earlier chapters, as well as additional prerequisites for particular chapters, are listed here.

CHAPTER I. Tensor products of vector spaces (cf. §1 of Appendix A).

CHAPTER II. Chapter I. Starting in §9: The proof of Proposition 2.96 is deferred to Chapter III, where the result is restated and proved as Proposition 3.29. Starting in §11: Tensor algebra as in §1 of Appendix A.

CHAPTER III. Chapter I, all of Appendix A.

CHAPTER IV. Chapter I, tensor and exterior algebras as in §§1-3 of Appendix A, a small amount of Lebesgue integration, existence of Haar measure for compact groups. The proof of Theorem 4.20 uses the Hilbert-Schmidt Theorem from functional analysis. Starting in §5: Chapter II.

CHAPTER V. Chapters II, III, and IV. The proof of Theoreem 5.62 uses the Hilbert Nullstellensatz.

CHAPTER VI. Chapters II and IV.

CHAPTER VII. Chapter VI. Starting in §5: Chapter V. Starting in §8: complex manifolds (apart from complex Lie groups).

CHAPTER VIII. Chapter VII, differential forms, more Lebesgue integration.

APPENDIX B. Chapters I and V.

STANDARD NOTATION

Item	Meaning
$\#S$ or $\|S\|$	number of elements in S
$\emptyset$	empty set
E^c	complement of set, contragredient module
δ_{ij}	1 if $i = j$, 0 if $i \neq j$
n positive	$n > 0$
$\mathbb{Z}, \mathbb{Q}, \mathbb{R}, \mathbb{C}$	integers, rationals, reals, complex numbers
$\operatorname{Re} z, \operatorname{Im} z$	real and imaginary parts of z
$\bar{z}$	complex conjugate of z
1	multiplicative identity
1 or I	identity matrix
$\dim V$	dimension of vector space
V^*	dual of vector space
$\mathbb{R}^n, \mathbb{C}^n$	spaces of column vectors
$\operatorname{Tr} A$	trace of A
$\det A$	determinant of A
A^t	transpose of A
A^*	conjugate transpose of A
A diagonable	A has a basis of eigenvectors with eigenvalues in the given field
$\operatorname{diag}(a_1, \ldots, a_n)$	diagonal matrix
$\operatorname{End} V$	linear maps of V into itself
$GL(V)$	invertible linear maps of V into itself
$[A : B]$	index or multiplicity of B in A
$\bigoplus V_i$	direct sum of the V_i
$\operatorname{span}(S)$	linear span of S
G_0	identity component of group G
$Z_A(B)$	centralizer of B in A
$N_A(B)$	normalizer of B in A
C^∞	infinitely differentiable

Notation introduced in Appendix A and used throughout the book is generally defined at its first occurrence and appears in the Index of Notation at the end of the book.

CHAPTER I

Lie Algebras and Lie Groups

Abstract. The first part of this chapter treats Lie algebras, beginning with definitions and many examples. The notions of solvable, nilpotent, radical, semisimple, and simple are introduced, and these notions are followed by a discussion of the effect of a change of the underlying field.

The idea of a semidirect product begins the development of the main structural theorems for real Lie algebras—the iterated construction of all solvable Lie algebras from derivations and semidirect products, Lie's Theorem for solvable Lie algebras, Engel's Theorem in connection with nilpotent Lie algebras, and Cartan's criteria for solvability and semisimplicity in terms of the Killing form. From Cartan's Criterion for Semisimplicity, it follows that semisimple Lie algebras are direct sums of simple Lie algebras.

Cartan's Criterion for Semisimplicity is used also to provide a long list of classical examples of semisimple Lie algebras. Some of these examples are defined in terms of quaternion matrices. Quaternion matrices of size n-by-n may be related to complex matrices of size $2n$-by-$2n$.

The treatment of Lie algebras concludes with a study of the finite-dimensional complex-linear representations of $\mathfrak{sl}(2, \mathbb{C})$. There is a classification theorem for the irreducible representations of this kind, and the general representations are direct sums of irreducible ones.

Section 10 contains a review of the elementary theory of Lie groups and their Lie algebras. The abstract theory as in Chevalley [1946] is summarized, and the correspondence is made with the concrete theory of closed linear groups, where the Lie algebra is obtained as the space of derivatives at $t = 0$ of smooth curves in the group passing through the identity at $t = 0$. The section ends with a discussion of the adjoint representation.

The remainder of the chapter explores some aspects of the connection between Lie groups and Lie algebras. One aspect is the relationship between automorphisms and derivations. The derivations of a semisimple Lie algebra are inner, and consequently the identity component of the group of automorphisms of a semisimple Lie algebra consists of inner automorphisms. In addition, simply connected solvable Lie groups may be built one dimension at a time as semidirect products with $\mathbb{R}^1$, and consequently they are diffeomorphic to Euclidean space. For simply connected nilpotent groups the exponential map is itself a diffeomorphism. The earlier long list of classical semisimple Lie algebras corresponds to a list of the classical semisimple Lie groups. The issue that needs attention for these groups is their connectedness, and this is proved by using the polar decomposition of matrices.

1. Definitions and Examples

Let $\Bbbk$ be a field. An **algebra** $\mathfrak{g}$ (not necessarily associative) is a vector space over $\Bbbk$ with a product $[X, Y]$ that is linear in each variable. The algebra is a **Lie algebra** if the product satisfies also

(a) $[X, X] = 0$ for all $X \in \mathfrak{g}$ (and hence $[X, Y] = -[Y, X]$) and
(b) the **Jacobi identity**

$$[[X, Y], Z] + [[Y, Z], X] + [[Z, X], Y] = 0.$$

For any algebra $\mathfrak{g}$ we get a linear map $\operatorname{ad} : \mathfrak{g} \to \operatorname{End}_{\Bbbk} \mathfrak{g}$ given by

$$(\operatorname{ad} X)(Y) = [X, Y].$$

The fact that the image is in $\operatorname{End}_{\Bbbk} \mathfrak{g}$ follows from the linearity of the bracket in the second variable, and the fact that ad is linear follows from the linearity of bracket in the first variable. Whenever there is a possible ambiguity in what the underlying vector space is, we write $\operatorname{ad}_{\mathfrak{g}} X$ in place of $\operatorname{ad} X$.

Suppose (a) holds in the definition of Lie algebra. Then (b) holds if and only if

$$[Z, [X, Y]] = [X, [Z, Y]] + [[Z, X], Y],$$

which holds if and only if

$$(\operatorname{ad} Z)[X, Y] = [X, (\operatorname{ad} Z)Y] + [(\operatorname{ad} Z)X, Y]. \tag{1.1}$$

Any D in $\operatorname{End}_{\Bbbk} \mathfrak{g}$ for which

$$D[X, Y] = [X, DY] + [DX, Y] \tag{1.2}$$

is a **derivation**. We have just seen that in a Lie algebra, every $\operatorname{ad} X$ is a derivation. Conversely if (a) holds and if every $\operatorname{ad} X$ for $X \in \mathfrak{g}$ is a derivation, then $\mathfrak{g}$ is a Lie algebra.

Now let us make some definitions concerning a Lie algebra $\mathfrak{g}$. A **homomorphism** is a linear map $\varphi : \mathfrak{g} \to \mathfrak{h}$ such that

$$\varphi([X, Y]) = [\varphi(X), \varphi(Y)] \qquad \text{for all } X \text{ and } Y.$$

An isomorphism is a one-one homormorphism onto. If $\mathfrak{a}$ and $\mathfrak{b}$ are subsets of $\mathfrak{g}$, we write

$$[\mathfrak{a}, \mathfrak{b}] = \operatorname{span}\{[X, Y] \mid X \in \mathfrak{a},\ Y \in \mathfrak{b}\}.$$

A **subalgebra** or **Lie subalgebra** $\mathfrak{h}$ of $\mathfrak{g}$ is a subspace satisfying $[\mathfrak{h}, \mathfrak{h}] \subseteq \mathfrak{h}$; then $\mathfrak{h}$ is itself a Lie algebra. An **ideal** $\mathfrak{h}$ in $\mathfrak{g}$ is a subspace satisfying $[\mathfrak{h}, \mathfrak{g}] \subseteq \mathfrak{h}$; an ideal is automatically a subalgebra. The Lie algebra $\mathfrak{g}$ is said to be **abelian** if $[\mathfrak{g}, \mathfrak{g}] = 0$; a vector space with all brackets defined to be 0 is automatically an abelian Lie algebra.

EXAMPLES.

1) Let U be any open set in $\mathbb{R}^n$. A **smooth vector field** on U is any operator on smooth functions on U of the form $X = \sum_{i=1}^n a_i(x)\frac{\partial}{\partial x_i}$ with all $a_i(x)$ in $C^\infty(U)$. The real vector space $\mathfrak{g}$ of all smooth vector fields on U becomes a Lie algebra if the bracket is defined by $[X, Y] = XY - YX$. The skew-symmetry and the Jacobi identity follow from the next example applied to the associative algebra of all operators generated (under composition and linear combinations) by all smooth vector fields.

2) Let $\mathfrak{g}$ be an associative algebra. Then $\mathfrak{g}$ becomes a Lie algebra under $[X, Y] = XY - YX$. Certainly $[X, X] = 0$. For the Jacobi identity we have

$$\begin{aligned}
&[[X, Y], Z] + [[Y, Z], X] + [[Z, X], Y] \\
&\qquad = [X, Y]Z - Z[X, Y] + [Y, Z]X - X[Y, Z] + [Z, X]Y - Y[Z, X] \\
&\qquad = XYZ - YXZ - ZXY + ZYX + YZX - ZYX \\
&\qquad\quad - XYZ + XZY + ZXY - XZY - YZX + YXZ \\
&\qquad = 0.
\end{aligned}$$

3) Let $\mathfrak{g} = \mathfrak{gl}(n, \Bbbk)$ denote the associative algebra of all n-by-n matrices with entries in the field $\Bbbk$, and define a bracket product by $[X, Y] = XY - YX$. Then $\mathfrak{g}$ becomes a Lie algebra. This is a special case of Example 2. More generally, let $\mathfrak{g} = \operatorname{End}_\Bbbk V$ denote the associative algebra of all $\Bbbk$ linear maps from V to V, where V is a vector space over $\Bbbk$, and define a bracket product by $[X, Y] = XY - YX$. Then $\mathfrak{g}$ becomes a Lie algebra. The special case of $\mathfrak{gl}(n, \Bbbk)$ arises when V is the vector space $\Bbbk^n$ of all n-dimensional column vectors over $\Bbbk$.

4) Example 1 generalizes to any smooth manifold M. The vector space of all smooth vector fields on M becomes a real Lie algebra if the bracket is defined by $[X, Y] = XY - YX$.

5) (Review of the **Lie algebra of a Lie group**) Let G be a Lie group. If $f : G \to \mathbb{R}$ is a smooth function and if g is in G, let f_g be the left translate $f_g(x) = f(gx)$. A smooth vector field X on G is **left-invariant** if $(Xf)_g = X(f_g)$ for all f and g. The left-invariant smooth vector fields form a subalgebra $\mathfrak{g}$ of the Lie algebra of all smooth vector fields, and this is just the Lie algebra of G. We can regard a smooth vector field X as a (smoothly varying) family of tangent vectors X_g, one for every g in G. Then the map $X \to X_1$ is a vector-space isomorphism of $\mathfrak{g}$ onto the tangent space at the identity of G. Carrying the definition of bracket to the tangent space by this isomorphism, we may identify the tangent space at the identity of G with the Lie algebra of G. The elementary theory of Lie groups will be reviewed in more detail in §10.

6) (Review of the **Lie algebra of a Lie group of matrices**) Let G be a closed subgroup of nonsingular real or complex matrices. Consider smooth curves $c(t)$ of matrices with $c(0) = 1$ and $c(t) \in G$ for each t. Then $\mathfrak{g} = \{c'(0)\}$ is a real vector space of matrices closed under the bracket operation $[X, Y] = XY - YX$ in Example 3. Up to canonical isomorphism, $\mathfrak{g}$ is the Lie algebra of G. The isomorphism is given as follows: Let $e_{ij}(g)$ denote the $(i, j)^{\text{th}}$ entry of the matrix g. Then $\operatorname{Re} e_{ij}$ and $\operatorname{Im} e_{ij}$ are smooth functions on G to which we can apply smooth vector fields. If X is a left-invariant smooth vector field on G, then the associated matrix has $(i, j)^{\text{th}}$ entry $X_1(\operatorname{Re} e_{ij}) + iX_1(\operatorname{Im} e_{ij})$. Under this identification we may identify the Lie algebra of the general linear group $GL(n, \mathbb{R})$ with $\mathfrak{gl}(n, \mathbb{R})$ and the Lie algebra of $GL(n, \mathbb{C})$ with $\mathfrak{gl}(n, \mathbb{C})$. In a similar fashion if V is a finite-dimensional vector space over $\mathbb{R}$ or $\mathbb{C}$, we may identify the Lie algebra of the general linear group $GL(V)$ with $\operatorname{End} V$. The relationship between Examples 5 and 6 will be discussed in more detail in §10. See especially Proposition 1.76.

7) The space of n-by-n skew-symmetric matrices over the field $\mathbb{k}$, given by

$$\mathfrak{g} = \{X \in \mathfrak{gl}(n, \mathbb{k}) \mid X + X^t = 0\} = \mathfrak{so}(n, \mathbb{k}),$$

is a Lie subalgebra of the Lie algebra $\mathfrak{gl}(n, \mathbb{k})$ given in Example 3. To see closure under brackets, we compute that

$$[X, Y]^t = (XY - YX)^t = Y^tX^t - X^tY^t = YX - XY = -[X, Y].$$

When $\mathbb{k}$ is $\mathbb{R}$ or $\mathbb{C}$, this example arises as the Lie algebra in the sense of Example 6 of the orthogonal group over $\mathbb{R}$ or $\mathbb{C}$. The orthogonal group will be discussed in more detail in §14.

8) Fix an n-by-n matrix J over $\mathbb{k}$, and let

$$\mathfrak{g} = \{X \in \mathfrak{gl}(n, \mathbb{k}) \mid JX + X^tJ = 0\}.$$

This $\mathfrak{g}$ is a Lie subalgebra of $\mathfrak{gl}(n, \mathbb{k})$ that generalizes Example 7. To see closure under brackets, we compute that

$$[X, Y]^tJ = (XY - YX)^tJ = Y^tX^tJ - X^tY^tJ = JYX - JXY = -J[X, Y].$$

In the special case that $\mathbb{k}$ is $\mathbb{R}$ or $\mathbb{C}$ and n is even and J is of the block form $J = \begin{pmatrix} 0 & 1 \\ -1 & 0 \end{pmatrix}$, this example arises as the Lie algebra in the sense of Example 6 of the symplectic group over $\mathbb{R}$ or $\mathbb{C}$. The symplectic group will be discussed in more detail in §14.

9) Let

$$\mathfrak{g} = \{X \in \mathfrak{gl}(n, \Bbbk) \mid \operatorname{Tr}(X) = 0\} = \mathfrak{sl}(n, \Bbbk).$$

This $\mathfrak{g}$ is a Lie subalgebra of $\mathfrak{gl}(n, \Bbbk)$ because $\operatorname{Tr}[X, Y] = \operatorname{Tr} XY - \operatorname{Tr} YX = 0$ for any two matrices X and Y. This example arises as the Lie algebra in the sense of Example 6 of the special linear group (the group of matrices of determinant 1) over $\mathbb{R}$ or $\mathbb{C}$. The special linear group will be discussed in more detail in §14.

10) Examples in dimension 1. A 1-dimensional Lie algebra $\mathfrak{g}$ over $\Bbbk$ must have $[X, X] = 0$ if $\{X\}$ is a basis. Thus $\mathfrak{g}$ must be abelian and is unique up to isomorphism.

11) Examples in dimension 2. Let $\{U, V\}$ be a basis of the Lie algebra $\mathfrak{g}$. The expansion of $[U, V]$ in terms of U and V determines the Lie algebra up to isomorphism. If $[U, V] = 0$, then $\mathfrak{g}$ is abelian. Otherwise let $[U, V] = \alpha U + \beta V$. We shall produce a basis $\{X, Y\}$ with $[X, Y] = Y$. We have

$$\begin{aligned} X &= aU + bV \\ Y &= cU + dV \end{aligned} \qquad \text{with} \qquad \det \begin{pmatrix} a & b \\ c & d \end{pmatrix} \neq 0,$$

and we want $[X, Y] = Y$, i.e.,

$$[aU + bV, cU + dV] \stackrel{?}{=} cU + dV.$$

The left side is

$$= (ad - bc)[U, V] = (ad - bc)(\alpha U + \beta V).$$

Choose a and b so that $a\beta - b\alpha = 1$ and put $c = \alpha$ and $d = \beta$. Then $ad - bc = 1$, and so $(ad - bc)\alpha = c$ and $(ad - bc)\beta = d$. With these definitions, $[X, Y] = Y$. We conclude that the only possible 2-dimensional Lie algebras $\mathfrak{g}$ over the field $\Bbbk$, up to isomorphism, are

(a) $\mathfrak{g}$ abelian

(b) $\mathfrak{g}$ with a basis $\{X, Y\}$ such that $[X, Y] = Y$.

When $\Bbbk = \mathbb{R}$, the second example arises as the Lie algebra of the matrix group $G = \left\{ \begin{pmatrix} x & y \\ 0 & 1 \end{pmatrix} \right\}$, which is isomorphic to the group of affine transformations $t \mapsto xt + y$ of the line.

12) Some examples in dimension 3 with $\Bbbk = \mathbb{R}$. We give five examples; in each the variables are allowed to range arbitrarily through $\mathbb{R}$.

(a) The Lie algebra of all matrices

$$\begin{pmatrix} 0 & a & c \\ 0 & 0 & b \\ 0 & 0 & 0 \end{pmatrix}$$

is an example of what will be called a "nilpotent" Lie algebra. This Lie algebra is called the **Heisenberg Lie algebra**. This name is used even when the field is not $\mathbb{R}$.

(b) The Lie algebra of all matrices

$$\begin{pmatrix} t & 0 & x \\ 0 & t & y \\ 0 & 0 & 0 \end{pmatrix}$$

is an example of what will be called a "split solvable" Lie algebra. It is isomorphic with the Lie algebra of the group of translations and dilations of the plane.

(c) The Lie algebra of all matrices

$$\begin{pmatrix} 0 & \theta & x \\ -\theta & 0 & y \\ 0 & 0 & 0 \end{pmatrix}$$

is an example of a "solvable" Lie algebra that is not split solvable. It is isomorphic with the Lie algebra of the group of translations and rotations of the plane.

(d) The vector product Lie algebra has a basis $\mathbf{i}, \mathbf{j}, \mathbf{k}$ with bracket relations

$$[\mathbf{i}, \mathbf{j}] = \mathbf{k}, \quad [\mathbf{j}, \mathbf{k}] = \mathbf{i}, \quad [\mathbf{k}, \mathbf{i}] = \mathbf{j}. \tag{1.3}$$

It is an example of what will be called a "simple" Lie algebra, and it is isomorphic to the Lie algebra $\mathfrak{so}(3)$ of the (compact) group of rotations in $\mathbb{R}^3$, via the isomorphism

$$\mathbf{i} \mapsto \begin{pmatrix} 0 & 0 & 0 \\ 0 & 0 & 1 \\ 0 & -1 & 0 \end{pmatrix}, \quad \mathbf{j} \mapsto \begin{pmatrix} 0 & 0 & 1 \\ 0 & 0 & 0 \\ -1 & 0 & 0 \end{pmatrix}, \quad \mathbf{k} \mapsto \begin{pmatrix} 0 & 1 & 0 \\ -1 & 0 & 0 \\ 0 & 0 & 0 \end{pmatrix}. \tag{1.4}$$

(e) Finally

$$\mathfrak{sl}(2, \mathbb{R}) = \left\{ \begin{pmatrix} a & b \\ c & -a \end{pmatrix} \right\}$$

is another example of a simple Lie algebra. It is the Lie algebra of the group of 2-by-2 real matrices of determinant one, and it is not isomorphic to the Lie algebra of a compact group. In particular, it is not isomorphic to the previous example. We shall make use of its distinguished basis

$$h = \begin{pmatrix} 1 & 0 \\ 0 & -1 \end{pmatrix}, \quad e = \begin{pmatrix} 0 & 1 \\ 0 & 0 \end{pmatrix}, \quad f = \begin{pmatrix} 0 & 0 \\ 1 & 0 \end{pmatrix}. \tag{1.5}$$

In this basis the bracket relations are

$$[h, e] = 2e, \qquad [h, f] = -2f, \qquad [e, f] = h. \tag{1.6}$$

More generally if $\mathbb{k}$ is any field, then $\mathfrak{sl}(2, \mathbb{k})$ has (1.5) as basis, and the bracket relations are given by (1.6).

13) Centralizers. If $\mathfrak{g}$ is a Lie algebra and $\mathfrak{s}$ is a subset of $\mathfrak{g}$, then

$$Z_{\mathfrak{g}}(\mathfrak{s}) = \{X \in \mathfrak{g} \mid [X, Y] = 0 \text{ for all } Y \in \mathfrak{s}\}$$

is the **centralizer** of $\mathfrak{s}$ in $\mathfrak{g}$. This is a Lie subalgebra of $\mathfrak{g}$. If $\mathfrak{s}$ consists of one element S, we often write $Z_{\mathfrak{g}}(S)$ in place of $Z_{\mathfrak{g}}(\{S\})$.

14) Normalizers. If $\mathfrak{g}$ is a Lie algebra and $\mathfrak{s}$ is a Lie subalgebra of $\mathfrak{g}$, then

$$N_{\mathfrak{g}}(\mathfrak{s}) = \{X \in \mathfrak{g} \mid [X, Y] \in \mathfrak{s} \text{ for all } Y \in \mathfrak{s}\}$$

is the **normalizer** of $\mathfrak{s}$ in $\mathfrak{g}$. This is a Lie subalgebra of $\mathfrak{g}$.

2. Ideals

We shall now study ideals in a Lie algebra more closely. In the course of the study, we shall define the notions "nilpotent," "solvable," "simple," "semisimple," and "radical." The underlying field for our Lie algebras remains an arbitrary field $\mathbb{k}$. Our computations are made easier by using the following proposition.

Proposition 1.7. If $\mathfrak{a}$ and $\mathfrak{b}$ are ideals in a Lie algebra, then so are $\mathfrak{a} + \mathfrak{b}$, $\mathfrak{a} \cap \mathfrak{b}$, and $[\mathfrak{a}, \mathfrak{b}]$.

PROOF. The conclusions for $\mathfrak{a} + \mathfrak{b}$ and $\mathfrak{a} \cap \mathfrak{b}$ are obvious. In the case of $[\mathfrak{a}, \mathfrak{b}]$, we have

$$\begin{aligned} [\mathfrak{g}, [\mathfrak{a}, \mathfrak{b}]] &\subseteq [[\mathfrak{g}, \mathfrak{a}], \mathfrak{b}] + [\mathfrak{a}, [\mathfrak{g}, \mathfrak{b}]] \quad \text{by (1.1)} \\ &\subseteq [\mathfrak{a}, \mathfrak{b}] + [\mathfrak{a}, \mathfrak{b}] \\ &\subseteq [\mathfrak{a}, \mathfrak{b}]. \end{aligned}$$

EXAMPLES OF IDEALS.

1) $Z_{\mathfrak{g}} =$ **center** of $\mathfrak{g} = \{X \mid [X, Y] = 0 \text{ for all } Y \in \mathfrak{g}\}$. This is the centralizer of $\mathfrak{g}$ in $\mathfrak{g}$.

2) $[\mathfrak{g}, \mathfrak{g}] =$ **commutator ideal**. This is an ideal by Proposition 1.7.

3) $\ker \pi$ whenever $\pi : \mathfrak{g} \to \mathfrak{h}$ is a homomorphism of Lie algebras.

Let $\mathfrak{g}$ be a Lie algebra. Each $\operatorname{ad} X$ for $X \in \mathfrak{g}$ is a member of $\operatorname{End}_{\mathbb{k}} \mathfrak{g}$, and these members satisfy

$$\operatorname{ad}[X, Y] = \operatorname{ad} X \operatorname{ad} Y - \operatorname{ad} Y \operatorname{ad} X \tag{1.8}$$

as a consequence of the Jacobi identity. In view of the definition of bracket in $\operatorname{End}_{\mathbb{k}} V$ given in Example 3 of §1, we see from (1.8) that

$\operatorname{ad}: \mathfrak{g} \to \operatorname{End}_{\mathbb{k}} \mathfrak{g}$ is a homomorphism of Lie algebras. The kernel of this homomorphism is the center $Z_{\mathfrak{g}}$. The image can be written simply as $\operatorname{ad} \mathfrak{g}$ or $\operatorname{ad}_{\mathfrak{g}} \mathfrak{g}$, and Example 3 above notes that $\operatorname{ad} \mathfrak{g}$ is a Lie subalgebra of $\operatorname{End}_{\mathbb{k}} \mathfrak{g}$.

Let $\mathfrak{a}$ be an ideal in the Lie algebra $\mathfrak{g}$. Then $\mathfrak{g}/\mathfrak{a}$ as a vector space becomes a Lie algebra under the definition $[X + \mathfrak{a}, Y + \mathfrak{a}] = [X, Y] + \mathfrak{a}$ and is called the **quotient algebra** of $\mathfrak{g}$ and $\mathfrak{a}$. Checking that this bracket operation is independent of the choices uses that $\mathfrak{a}$ is an ideal, and then the defining properties of the bracket operation of a Lie algbera follow from the corresponding properties in $\mathfrak{g}$. The quotient map $\mathfrak{g} \to \mathfrak{g}/\mathfrak{a}$ is a homomorphism of Lie algebras, by definition, and hence every ideal is the kernel of a homomorphism.

Ideals and homomorphisms for Lie algebras have a number of properties in common with ideals and homomorphisms for rings. One such property is the construction of homomorphisms $\mathfrak{g}/\mathfrak{a} \to \mathfrak{h}$ when $\mathfrak{a}$ is an ideal in $\mathfrak{g}$: If a homomorphism $\pi : \mathfrak{g} \to \mathfrak{h}$ has the property that $\mathfrak{a} \subseteq \ker \pi$, then π factors through the quotient map $\mathfrak{g} \to \mathfrak{g}/\mathfrak{a}$, thus defining a homomorphism $\mathfrak{g}/\mathfrak{a} \to \mathfrak{h}$.

Another such property is the one-one correspondence of ideals in $\mathfrak{g}$ that contain $\mathfrak{a}$ with ideals in $\mathfrak{g}/\mathfrak{a}$, the correspondence being given by the quotient map.

Yet another such property is the **Second Isomorphism Theorem**. If $\mathfrak{g}$ is a Lie algebra and if $\mathfrak{a}$ and $\mathfrak{b}$ are ideals in $\mathfrak{g}$ such that $\mathfrak{a} + \mathfrak{b} = \mathfrak{g}$, then

$$\mathfrak{g}/\mathfrak{a} = (\mathfrak{a} + \mathfrak{b})/\mathfrak{a} \cong \mathfrak{b}/(\mathfrak{a} \cap \mathfrak{b}). \tag{1.9}$$

In fact, the map from left to right is $A + B + \mathfrak{a} \mapsto B + (\mathfrak{a} \cap \mathfrak{b})$. The map is known from linear algebra to be a vector space isomorphism, and we easily check that it respects brackets.

For the remainder of this section, let $\mathfrak{g}$ denote a finite-dimensional Lie algebra. We define recursively

$$\mathfrak{g}^0 = \mathfrak{g}, \qquad \mathfrak{g}^1 = [\mathfrak{g}, \mathfrak{g}], \qquad \mathfrak{g}^{j+1} = [\mathfrak{g}^j, \mathfrak{g}^j].$$

Then the decreasing sequence

$$\mathfrak{g} = \mathfrak{g}^0 \supseteq \mathfrak{g}^1 \supseteq \mathfrak{g}^2 \supseteq \cdots$$

is called the **commutator series** for $\mathfrak{g}$. Each $\mathfrak{g}^j$ is an ideal in $\mathfrak{g}$, by Proposition 1.7 and induction. We say that $\mathfrak{g}$ is **solvable** if $\mathfrak{g}^j = 0$ for some j. A nonzero solvable $\mathfrak{g}$ has a nonzero abelian ideal, namely the last nonzero $\mathfrak{g}^j$.

Next we define recursively

$$\mathfrak{g}_0 = \mathfrak{g}, \qquad \mathfrak{g}_1 = [\mathfrak{g}, \mathfrak{g}], \qquad \mathfrak{g}_{j+1} = [\mathfrak{g}, \mathfrak{g}_j].$$

Then the decreasing sequence

$$\mathfrak{g} = \mathfrak{g}_0 \supseteq \mathfrak{g}_1 \supseteq \mathfrak{g}_2 \supseteq \cdots$$

is called the **lower central series** for $\mathfrak{g}$. Each $\mathfrak{g}_j$ is an ideal in $\mathfrak{g}$, by Proposition 1.7 and induction. We say that $\mathfrak{g}$ is **nilpotent** if $\mathfrak{g}_j = 0$ for some j. A nonzero nilpotent $\mathfrak{g}$ has nonzero center, the last nonzero $\mathfrak{g}_j$ being in the center. Inductively we see that $\mathfrak{g}^j \subseteq \mathfrak{g}_j$, and it follows that nilpotent implies solvable.

Below are the standard examples of solvable and nilpotent Lie algebras. See also Example 12 in §1. Further examples appear in the exercises at the end of the chapter.

EXAMPLES.

1) The Lie algebra $\mathfrak{g} = \begin{pmatrix} a_1 & & * \\ & \ddots & \\ 0 & & a_n \end{pmatrix}$ is solvable.

2) The Lie algebra $\mathfrak{g} = \begin{pmatrix} 0 & & * \\ & \ddots & \\ 0 & & 0 \end{pmatrix}$ is nilpotent.

Proposition 1.10. Any subalgebra or quotient algebra of a solvable Lie algebra is solvable. Similarly any subalgebra or quotient algebra of a nilpotent Lie algebra is nilpotent.

PROOF. If $\mathfrak{h}$ is a subalgebra of $\mathfrak{g}$, then induction gives $\mathfrak{h}^k \subseteq \mathfrak{g}^k$. Hence $\mathfrak{g}$ solvable implies $\mathfrak{h}$ solvable. If $\pi : \mathfrak{g} \to \mathfrak{h}$ is a homomorphism of the Lie algebra $\mathfrak{g}$ onto the Lie algebra $\mathfrak{h}$, then $\pi(\mathfrak{g}^k) = \mathfrak{h}^k$. Hence $\mathfrak{g}$ solvable implies $\mathfrak{h}$ solvable. The arguments in the nilpotent case are similar.

Proposition 1.11. If $\mathfrak{a}$ is a solvable ideal in $\mathfrak{g}$ and if $\mathfrak{g}/\mathfrak{a}$ is solvable, then $\mathfrak{g}$ is solvable.

PROOF. Let $\pi : \mathfrak{g} \to \mathfrak{g}/\mathfrak{a}$ be the quotient homomorphism, and suppose that $(\mathfrak{g}/\mathfrak{a})^k = 0$. Since $\pi(\mathfrak{g}) = \mathfrak{g}/\mathfrak{a}$, $\pi(\mathfrak{g}^j) = (\mathfrak{g}/\mathfrak{a})^j$ for all j. Thus $\pi(\mathfrak{g}^k) = 0$, and we conclude that $\mathfrak{g}^k \subseteq \mathfrak{a}$. By assumption $\mathfrak{a}^l = 0$ for some l. Hence $\mathfrak{g}^{k+l} = (\mathfrak{g}^k)^l \subseteq \mathfrak{a}^l = 0$, and $\mathfrak{g}$ is solvable.

Proposition 1.12. If $\mathfrak{g}$ is a finite-dimensional Lie algebra, then there exists a unique solvable ideal $\mathfrak{r}$ of $\mathfrak{g}$ containing all solvable ideals in $\mathfrak{g}$.

PROOF. By finite-dimensionality it suffices to show that the sum of two solvable ideals, which is an ideal by Proposition 1.7, is solvable.

Thus let $\mathfrak{a}$ and $\mathfrak{b}$ be solvable ideals and let $\mathfrak{h} = \mathfrak{a}+\mathfrak{b}$. Then $\mathfrak{a}$ is a solvable ideal in $\mathfrak{h}$, and (1.9) gives

$$\mathfrak{h}/\mathfrak{a} = (\mathfrak{a}+\mathfrak{b})/\mathfrak{a} \cong \mathfrak{b}/(\mathfrak{a}\cap\mathfrak{b}).$$

This is solvable by Proposition 1.10 since $\mathfrak{b}$ is solvable. Hence $\mathfrak{h}$ is solvable by Proposition 1.11.

The ideal $\mathfrak{r}$ of Proposition 1.12 is called the **radical** of $\mathfrak{g}$ and is denoted $\operatorname{rad}\mathfrak{g}$.

A finite-dimensional Lie algebra $\mathfrak{g}$ is **simple** if $\mathfrak{g}$ is nonabelian and $\mathfrak{g}$ has no proper nonzero ideals. A finite-dimensional Lie algebra $\mathfrak{g}$ is **semisimple** if $\mathfrak{g}$ has no nonzero solvable ideals, i.e., if $\operatorname{rad}\mathfrak{g} = 0$.

Proposition 1.13. In a simple Lie algebra $[\mathfrak{g},\mathfrak{g}] = \mathfrak{g}$. Every simple Lie algebra is semisimple. Every semisimple Lie algebra has 0 center.

PROOF. Let $\mathfrak{g}$ be simple. The commutator $[\mathfrak{g},\mathfrak{g}]$ is an ideal and hence is 0 or $\mathfrak{g}$. It cannot be 0 since $\mathfrak{g}$ is nonabelian. So it is $\mathfrak{g}$. This proves the first statement. For the second statement, $\operatorname{rad}\mathfrak{g}$ is an ideal and so is 0 or $\mathfrak{g}$. If $\operatorname{rad}\mathfrak{g} = \mathfrak{g}$, then $\mathfrak{g}$ is solvable and $[\mathfrak{g},\mathfrak{g}] \subsetneqq \mathfrak{g}$, contradiction. So $\operatorname{rad}\mathfrak{g} = 0$, and $\mathfrak{g}$ is semisimple. For the third statement, $Z_{\mathfrak{g}}$ is an abelian ideal and must be 0, by definition of semisimplicity.

Proposition 1.14. If $\mathfrak{g}$ is a finite-dimensional Lie algebra, then $\mathfrak{g}/\operatorname{rad}\mathfrak{g}$ is semisimple.

PROOF. Let $\pi : \mathfrak{g} \to \mathfrak{g}/\operatorname{rad}\mathfrak{g}$ be the quotient homomorphism, and let $\mathfrak{h}$ be a solvable ideal in $\mathfrak{g}/\operatorname{rad}\mathfrak{g}$. Form the ideal $\mathfrak{a} = \pi^{-1}(\mathfrak{h}) \subseteq \mathfrak{g}$. Then $\pi(\mathfrak{a}) = \mathfrak{h}$ is solvable, and $\ker\pi|_{\mathfrak{a}}$ is solvable, being in $\operatorname{rad}\mathfrak{g}$. So $\mathfrak{a}$ is solvable by Proposition 1.11. Hence $\mathfrak{a} \subseteq \operatorname{rad}\mathfrak{g}$ and $\mathfrak{h} = 0$. Therefore $\mathfrak{g}/\operatorname{rad}\mathfrak{g}$ is semisimple.

EXAMPLES. Any 3-dimensional Lie algebra $\mathfrak{g}$ is either solvable or simple. In fact, Examples 10 and 11 in §1 show that any Lie algebra of dimension 1 or 2 is solvable. If $\mathfrak{g}$ is not simple, then it has a nontrivial ideal $\mathfrak{h}$. This $\mathfrak{h}$ is solvable, and so is $\mathfrak{g}/\mathfrak{h}$. Hence $\mathfrak{g}$ is solvable by Proposition 1.11.

To decide whether such a $\mathfrak{g}$ is solvable or simple, we have only to compute $[\mathfrak{g},\mathfrak{g}]$. If $[\mathfrak{g},\mathfrak{g}] = \mathfrak{g}$, then $\mathfrak{g}$ is simple (because the commutator series cannot end in 0), while if $[\mathfrak{g},\mathfrak{g}] \subsetneqq \mathfrak{g}$, then $\mathfrak{g}$ is solvable (because $[\mathfrak{g},\mathfrak{g}]$ has dimension at most 2 and is therefore solvable).

It follows from this analysis and from the bracket relations (1.6) that $\mathfrak{sl}(2,\mathbb{k})$ is simple, for any field $\mathbb{k}$ of characteristic $\neq 2$. For $\mathbb{k} = \mathbb{R}$, we see similarly that $\mathfrak{so}(3)$ is simple as a consequence of (1.3) and the isomorphism (1.4).

3. Field Extensions and the Killing Form

In this section we examine the effect of enlarging or shrinking the field of scalars for a Lie algebra. Let $\mathbb{k}$ be a field, and let $\mathbb{K}$ be an extension field.

If U and V are vector spaces over $\mathbb{k}$, then the **tensor product** $U \otimes_{\mathbb{k}} V$ is characterized up to canonical isomorphism by the **universal mapping property** that a $\mathbb{k}$ bilinear map L of $U \times V$ into a $\mathbb{k}$ vector space W extends uniquely to a $\mathbb{k}$ linear map $\tilde{L} : U \otimes_{\mathbb{k}} V \to W$. The sense in which $\tilde{L}$ is an extension of L is that $\tilde{L}(u \otimes v) = L(u, v)$ for all $u \in U$ and $v \in V$. Tensor products are described in more detail in Appendix A.

Let U' and V' be two further $\mathbb{k}$ vector spaces. If l is in $\operatorname{Hom}_{\mathbb{k}}(U, U')$ and m is in $\operatorname{Hom}_{\mathbb{k}}(V, V')$, then we can use this property to define the **tensor product** $l \otimes m : U \otimes_{\mathbb{k}} V \to U' \otimes_{\mathbb{k}} V'$ as the $\mathbb{k}$ linear extension of the bilinear map $(u, v) \mapsto l(u) \otimes m(v)$. See Appendix A for further discussion.

Of course, $l \otimes m$ can always be defined concretely in terms of bases. If $\{u_i\}$ is a basis of U and $\{v_j\}$ is a basis of V, then $\{u_i \otimes v_j\}$ is a basis of $U \otimes_{\mathbb{k}} V$. Hence if we let $(l \otimes m)(u_i \otimes v_j) = l(u_i) \otimes m(v_j)$, then $l \otimes m$ is defined on a basis, and we get a well defined linear transformation on all of $U \otimes_{\mathbb{k}} V$. But it is tedious to check that this way of defining $l \otimes m$ is independent of the choice of bases for U and V. The above approach using the universal mapping property avoids this problem.

Still with V as a vector space over $\mathbb{k}$, we are especially interested in the special case $V \otimes_{\mathbb{k}} \mathbb{K}$. If c is a member of $\mathbb{K}$, then multiplication by c, which we denote temporarily $m(c)$, is $\mathbb{k}$ linear from $\mathbb{K}$ to $\mathbb{K}$. Thus $1 \otimes m(c)$ defines a $\mathbb{k}$ linear map of $V \otimes_{\mathbb{k}} \mathbb{K}$ to itself, and we define this to be scalar multiplication by c in $V \otimes_{\mathbb{k}} \mathbb{K}$. With this definition we easily check that $V \otimes_{\mathbb{k}} \mathbb{K}$ becomes a vector space over $\mathbb{K}$. We write $V^{\mathbb{K}}$ for this vector space. The map $v \mapsto v \otimes 1$ allows us to identify V canonically with a subset of $V^{\mathbb{K}}$. If $\{v_i\}$ is a basis of V over $\mathbb{k}$, then $\{v_i \otimes 1\}$ (which we often write simply as $\{v_i\}$) is a basis of $V^{\mathbb{K}}$ over $\mathbb{K}$.

If W is a vector space over the extension field $\mathbb{K}$, we can restrict the definition of scalar multiplication to scalars in $\mathbb{k}$, thereby obtaining a vector space over $\mathbb{k}$. This vector space we denote by $W^{\mathbb{k}}$. There will be no possibility of confusing the notations $V^{\mathbb{K}}$ and $W^{\mathbb{k}}$ since the field is extended in one case and shrunk in the other.

In the special case that $\mathbb{k} = \mathbb{R}$ and $\mathbb{K} = \mathbb{C}$ and V is a real vector space, the complex vector space $V^{\mathbb{C}}$ is called the **complexification** of

V. If W is complex, then $W^{\mathbb{R}}$ is W regarded as a real vector space. The operations $(\cdot)^{\mathbb{C}}$ and $(\cdot)^{\mathbb{R}}$ are not inverse to each other: $(V^{\mathbb{C}})^{\mathbb{R}}$ has twice the real dimension of V, and $(W^{\mathbb{R}})^{\mathbb{C}}$ has twice the complex dimension of W. More precisely

$$(V^{\mathbb{C}})^{\mathbb{R}} = V \oplus iV \tag{1.15a}$$

as real vector spaces, where V means $V \otimes 1$ in $V \otimes_{\mathbb{R}} \mathbb{C}$ and the i refers to the real linear transformation multiplication-by-i. Often we abbreviate (1.15a) simply as

$$V^{\mathbb{C}} = V \oplus iV. \tag{1.15b}$$

When a complex vector space W and a real vector space V are related by

$$W^{\mathbb{R}} = V \oplus iV,$$

we say that V is a **real form** of the complex vector space W. Formula (1.15a) says that any real vector space is a real form of its complexification. In (1.15a) the $\mathbb{R}$ linear map that is 1 on V and -1 on iV is called the **conjugation** of the complex vector space $V^{\mathbb{C}}$ with respect to the real form V.

Now let us impose Lie algebra structures on these constructions. First suppose that $\mathfrak{g}_0$ is a Lie algebra over $\mathbb{k}$. We want to impose a Lie algebra structure on the $\mathbb{K}$ vector space $\mathfrak{g} = (\mathfrak{g}_0)^{\mathbb{K}}$. To do so canonically, we introduce the 4-linear map

$$\mathfrak{g}_0 \times \mathbb{K} \times \mathfrak{g}_0 \times \mathbb{K} \longrightarrow \mathfrak{g}_0 \otimes_{\mathbb{k}} \mathbb{K}$$

given by

$$(X, a, Y, b) \mapsto [X, Y] \otimes ab \in \mathfrak{g}_0 \otimes_{\mathbb{k}} \mathbb{K}.$$

This 4-linear map extends to a $\mathbb{k}$ linear map on $\mathfrak{g}_0 \otimes_{\mathbb{k}} \mathbb{K} \otimes_{\mathbb{k}} \mathfrak{g}_0 \otimes_{\mathbb{k}} \mathbb{K}$ that we can restrict to a $\mathbb{k}$ bilinear map

$$(\mathfrak{g}_0 \otimes_{\mathbb{k}} \mathbb{K}) \times (\mathfrak{g}_0 \otimes_{\mathbb{k}} \mathbb{K}) \longrightarrow \mathfrak{g}_0 \otimes_{\mathbb{k}} \mathbb{K}.$$

The result is our definition of the bracket product on $\mathfrak{g} = (\mathfrak{g}_0)^{\mathbb{K}} = \mathfrak{g}_0 \otimes_{\mathbb{k}} \mathbb{K}$. We readily check that it is $\mathbb{K}$ bilinear and extends the bracket product in $\mathfrak{g}_0$. Using bases, we see that it has the property $[X, X] = 0$ and satisfies the Jacobi identity. Hence $\mathfrak{g}$ is a Lie algebra over $\mathbb{K}$.

Starting from the Lie algebra $\mathfrak{g}$ over $\mathbb{K}$, we can forget about scalar multiplication other than by scalars in $\mathbb{k}$, and the result is a Lie algebra $\mathfrak{g}^{\mathbb{k}}$ over $\mathbb{k}$. The notation $\mathfrak{g}^{\mathbb{k}}$ is consistent with the notation for vector spaces, i.e., the underlying $\mathbb{k}$ vector space of $\mathfrak{g}^{\mathbb{k}}$ is the vector space constructed earlier by the operation $(\cdot)^{\mathbb{k}}$.

In the special case that $\mathbb{k} = \mathbb{R}$ and $\mathbb{K} = \mathbb{C}$ and $\mathfrak{g}_0$ is a real Lie algebra, the complex Lie algebra $(\mathfrak{g}_0)^{\mathbb{C}}$ is called the **complexification** of $\mathfrak{g}_0$. Similarly when a complex Lie algebra $\mathfrak{g}$ and a real Lie algebra $\mathfrak{g}_0$ are related as vector spaces over $\mathbb{R}$ by

$$\mathfrak{g}^{\mathbb{R}} = \mathfrak{g}_0 \oplus i\mathfrak{g}_0, \tag{1.16}$$

we say that $\mathfrak{g}_0$ is a **real form** of the complex Lie algebra $\mathfrak{g}$. Any real Lie algebra is a real form of its complexification. The conjugation of a complex Lie algebra $\mathfrak{g}$ with respect to a real form is a Lie algebra isomorphism of $\mathfrak{g}^{\mathbb{R}}$ with itself.

Proposition 1.17. Let $\mathfrak{g}$ be a Lie algebra over $\mathbb{k}$, and identify $\mathfrak{g}$ with the subset $\mathfrak{g} \otimes 1$ of $\mathfrak{g}^{\mathbb{K}}$. Then

$$[\mathfrak{g}, \mathfrak{g}]^{\mathbb{K}} = [\mathfrak{g}^{\mathbb{K}}, \mathfrak{g}^{\mathbb{K}}].$$

Consequently if $\mathfrak{g}$ is finite-dimensional, then $\mathfrak{g}$ is solvable if and only if $\mathfrak{g}^{\mathbb{K}}$ is solvable.

PROOF. From $\mathfrak{g} \subseteq \mathfrak{g}^{\mathbb{K}}$, we obtain $[\mathfrak{g}, \mathfrak{g}] \subseteq [\mathfrak{g}^{\mathbb{K}}, \mathfrak{g}^{\mathbb{K}}]$. The $\mathbb{K}$ subspace of $\mathfrak{g}^{\mathbb{K}}$ generated by $[\mathfrak{g}, \mathfrak{g}]$ is $[\mathfrak{g}, \mathfrak{g}]^{\mathbb{K}}$, and therefore $[\mathfrak{g}, \mathfrak{g}]^{\mathbb{K}} \subseteq [\mathfrak{g}^{\mathbb{K}}, \mathfrak{g}^{\mathbb{K}}]$. In the reverse direction let a and b be in $\mathbb{K}$, and let X and Y be in $\mathfrak{g}$. Then

$$[X \otimes a, Y \otimes b] = [X, Y] \otimes ab \in [\mathfrak{g}, \mathfrak{g}]^{\mathbb{K}}.$$

Passing to linear combinations in each factor of the bracket on the left, we obtain $[\mathfrak{g}^{\mathbb{K}}, \mathfrak{g}^{\mathbb{K}}] \subseteq [\mathfrak{g}, \mathfrak{g}]^{\mathbb{K}}$. Thus $[\mathfrak{g}, \mathfrak{g}]^{\mathbb{K}} = [\mathfrak{g}^{\mathbb{K}}, \mathfrak{g}^{\mathbb{K}}]$. It follows that the members of the commutator series satisfy $\mathfrak{g}^m = 0$ if and only if $(\mathfrak{g}^{\mathbb{K}})^m = 0$. Therefore $\mathfrak{g}$ is solvable if and only if $\mathfrak{g}^{\mathbb{K}}$ is solvable.

Now let $\mathfrak{g}$ be a finite-dimensional Lie algebra over $\mathbb{k}$. If X and Y are in $\mathfrak{g}$, then $\operatorname{ad} X \operatorname{ad} Y$ is a linear transformation from $\mathfrak{g}$ to itself, and it is meaningful to define

$$B(X, Y) = \operatorname{Tr}(\operatorname{ad} X \operatorname{ad} Y). \tag{1.18}$$

Then B is a symmetric bilinear form on $\mathfrak{g}$ known as the **Killing form** of $\mathfrak{g}$ after the person who introduced it. The Killing form is **invariant** in the sense that

$$B((\operatorname{ad} X)Y, Z) = -B(Y, (\operatorname{ad} X)Z) \tag{1.19a}$$

for all X, Y, and Z in $\mathfrak{g}$. An alternate way of writing (1.19a) is

$$B([X,Y],Z) = B(X,[Y,Z]). \tag{1.19b}$$

Equation (1.19) is straightforward to verify; we simply expand both sides and use the fact that $\mathrm{Tr}(LM) = \mathrm{Tr}(ML)$.

EXAMPLES.

1) Let $\mathfrak{g}$ be 2-dimensional nonabelian as in Example 11b of §1. Then $\mathfrak{g}$ has a basis $\{X, Y\}$ with $[X, Y] = Y$. To understand the Killing form B, it is enough to know what B is on every pair of basis vectors. Thus we have to compute the traces of $\operatorname{ad} X \operatorname{ad} X$, $\operatorname{ad} X \operatorname{ad} Y$, and $\operatorname{ad} Y \operatorname{ad} Y$. The matrix of $\operatorname{ad} X \operatorname{ad} X$ in the basis $\{X, Y\}$ is

$$\begin{matrix} & X \quad Y & \\ \operatorname{ad} X \operatorname{ad} X = & \begin{pmatrix} 0 & 0 \\ 0 & 1 \end{pmatrix} & \begin{matrix} X \\ Y \end{matrix} \end{matrix}$$

and hence $B(X, X) = 1$. Calculating $B(X, Y)$ and $B(Y, Y)$ similarly, we see that B is given by the matrix

$$\begin{matrix} & X \quad Y & \\ B = & \begin{pmatrix} 1 & 0 \\ 0 & 0 \end{pmatrix} & \begin{matrix} X \\ Y \end{matrix} \end{matrix} .$$

2) Let $\mathfrak{g} = \mathfrak{sl}(2, \mathbb{k})$ with basis $\{h, e, f\}$ as in (1.5) and bracket relations as in (1.6). Computing as in the previous example, we see that the matrix of B in this basis is

$$\begin{matrix} & h \quad e \quad f & \\ B = & \begin{pmatrix} 8 & 0 & 0 \\ 0 & 0 & 4 \\ 0 & 4 & 0 \end{pmatrix} & \begin{matrix} h \\ e \\ f \end{matrix} \end{matrix} .$$

Returning to a general finite-dimensional Lie algebra $\mathfrak{g}$ over $\mathbb{k}$, let us extend the scalars from $\mathbb{k}$ to $\mathbb{K}$, forming the Lie algebra $\mathfrak{g}^{\mathbb{K}}$. Let $B^{\mathbb{K}}$ be the Killing form of $\mathfrak{g}^{\mathbb{K}}$. If we fix a basis of $\mathfrak{g}$ over $\mathbb{k}$, then that same set is a basis of $\mathfrak{g}^{\mathbb{K}}$ over $\mathbb{K}$. Consequently if X and Y are in $\mathfrak{g}$, the matrix of $\operatorname{ad} X \operatorname{ad} Y$ is the same for $\mathfrak{g}$ as it is for $\mathfrak{g}^{\mathbb{K}}$, and it follows that

$$B^{\mathbb{K}}|_{\mathfrak{g}\times\mathfrak{g}} = B. \tag{1.20}$$

When we shrink the scalars from $\mathbb{K}$ to $\mathbb{k}$, passing from a Lie algebra $\mathfrak{h}$ over $\mathbb{K}$ to the Lie algebra $\mathfrak{h}^{\mathbb{k}}$, the dimension is not preserved. In fact, the $\mathbb{k}$ dimension of $\mathfrak{h}^{\mathbb{k}}$ is the product of the degree of $\mathbb{K}$ over $\mathbb{k}$ and the $\mathbb{K}$ dimension of $\mathfrak{h}$. Thus the Killing forms of $\mathfrak{h}$ and $\mathfrak{h}^{\mathbb{k}}$ are not related so simply. We shall be interested in this relationship only in the special case that $\mathbb{k} = \mathbb{R}$ and $\mathbb{K} = \mathbb{C}$, and we return to it in §8.

4. Semidirect Products of Lie Algebras

In this section the underlying field for our Lie algebras remains an arbitrary field $\mathbb{k}$.

Let $\mathfrak{a}$ and $\mathfrak{b}$ be Lie algebras, and let $\mathfrak{g}$ be the external direct sum of $\mathfrak{a}$ and $\mathfrak{b}$ as vector spaces, i.e., the set of ordered pairs with coordinate-wise addition and scalar multiplication. Then we can define a bracket operation on $\mathfrak{g}$ so that $\mathfrak{a}$ brackets with $\mathfrak{a}$ as before, $\mathfrak{b}$ brackets with $\mathfrak{b}$ as before, and $[\mathfrak{a}, \mathfrak{b}] = 0$. We say that $\mathfrak{g}$ is the **Lie algebra direct sum** of $\mathfrak{a}$ and $\mathfrak{b}$, and we write $\mathfrak{g} = \mathfrak{a} \oplus \mathfrak{b}$. Here $\mathfrak{a}$ and $\mathfrak{b}$ are ideals in $\mathfrak{g}$.

The above construction is what is sometimes called an "external direct sum," an object constructed out of the constituents $\mathfrak{a}$ and $\mathfrak{b}$. Now we consider an "internal direct sum," formed from $\mathfrak{g}$ by recognizing $\mathfrak{a}$ and $\mathfrak{b}$ within it. Let a Lie algebra $\mathfrak{g}$ be given, and let $\mathfrak{a}$ and $\mathfrak{b}$ be ideals in $\mathfrak{g}$ such that $\mathfrak{g} = \mathfrak{a} \oplus \mathfrak{b}$ as vector spaces. Then $\mathfrak{g}$ is isomorphic with the Lie algebra direct sum of $\mathfrak{a}$ and $\mathfrak{b}$ as defined above, and we shall simply say that $\mathfrak{g}$ is the **Lie algebra direct sum** of $\mathfrak{a}$ and $\mathfrak{b}$.

Generalizing these external and internal constructions, we shall now consider "semidirect products" of Lie algebras. We begin by examining derivations more closely. Recall that a derivation D of an algebra $\mathfrak{b}$ is a member of $\operatorname{End}_{\mathbb{k}} \mathfrak{b}$ satisfying the product rule (1.2). Let $\operatorname{Der}_{\mathbb{k}} \mathfrak{b}$ be the vector space of all derivations of $\mathfrak{b}$.

Proposition 1.21. If $\mathfrak{b}$ is any algebra over $\mathbb{k}$, then $\operatorname{Der}_{\mathbb{k}} \mathfrak{b}$ is a Lie algebra. If $\mathfrak{b}$ is a Lie algebra, then $\operatorname{ad} : \mathfrak{b} \to \operatorname{Der}_{\mathbb{k}} \mathfrak{b} \subseteq \operatorname{End}_{\mathbb{k}} \mathfrak{b}$ is a Lie algebra homomorphism.

PROOF. For D and E in $\operatorname{Der}_{\mathbb{k}} \mathfrak{b}$, we have

$$\begin{aligned}
[D, E][X, Y] &= (DE - ED)[X, Y] \\
&= D[EX, Y] + D[X, EY] - E[DX, Y] - E[X, DY] \\
&= [DEX, Y] + [EX, DY] + [DX, EY] + [X, DEY] \\
&\quad - [EDX, Y] - [DX, EY] - [EX, DY] - [X, EDY] \\
&= [[D, E]X, Y] + [X, [D, E]Y].
\end{aligned}$$

Thus $\operatorname{Der}_{\mathbb{k}} \mathfrak{b}$ is a Lie algebra. Finally we saw in (1.1) that $\operatorname{ad} \mathfrak{b} \subseteq \operatorname{Der}_{\mathbb{k}} \mathfrak{b}$, and in (1.8) that ad is a Lie algebra homomorphism.

We come to the notion of "internal semidirect product." Let $\mathfrak{g}$ be a Lie algebra, let $\mathfrak{g} = \mathfrak{a} \oplus \mathfrak{b}$ as vector spaces, and suppose that $\mathfrak{a}$ is a Lie subalgebra of $\mathfrak{g}$ and $\mathfrak{b}$ is an ideal. If A is in $\mathfrak{a}$, then $\operatorname{ad} A$ leaves $\mathfrak{b}$ stable (since $\mathfrak{b}$ is an ideal), and so $\operatorname{ad} A|_{\mathfrak{b}}$ is in $\operatorname{Der}_{\mathbb{k}} \mathfrak{b}$. By Proposition 1.21 we obtain a homomorphism π from $\mathfrak{a}$ to $\operatorname{Der}_{\mathbb{k}} \mathfrak{b}$. This homomorphism tells us the bracket of $\mathfrak{a}$ with $\mathfrak{b}$, namely $[A, B] = \pi(A)(B)$. Thus $\mathfrak{a}$, $\mathfrak{b}$, and π determine $\mathfrak{g}$. We say that $\mathfrak{g}$ is the **semidirect product** of $\mathfrak{a}$ and $\mathfrak{b}$, and we write $\mathfrak{g} = \mathfrak{a} \oplus_\pi \mathfrak{b}$.

The notion of "external semidirect product" is captured by the following proposition.

Proposition 1.22. Let Lie algebras $\mathfrak{a}$ and $\mathfrak{b}$ be given, and suppose π is a Lie algebra homomorphism from $\mathfrak{a}$ to $\operatorname{Der}_{\mathbb{k}} \mathfrak{b}$. Then there exists a unique Lie algebra structure on the vector space direct sum $\mathfrak{g} = \mathfrak{a} \oplus \mathfrak{b}$ retaining the old bracket in $\mathfrak{a}$ and $\mathfrak{b}$ and satisfying $[A, B] = \pi(A)(B)$ for $A \in \mathfrak{a}$ and $B \in \mathfrak{b}$. Within the Lie algebra $\mathfrak{g}$, $\mathfrak{a}$ is a subalgebra and $\mathfrak{b}$ is an ideal.

REMARK. The direct sum earlier in this section is $\mathfrak{g}$ in the special case that $\pi = 0$.

PROOF. Uniqueness is clear. Existence of an algebra structure with $[X, X] = 0$ is clear, and the problem is to prove the Jacobi identity. Thus let $X, Y, Z \in \mathfrak{g}$. If all three are in $\mathfrak{a}$ or all three are in $\mathfrak{b}$, we are done. By skew symmetry we reduce to two cases:

(i) X and Y are in $\mathfrak{a}$ and Z is in $\mathfrak{b}$. Then

$$\pi([X, Y]) = \pi(X)\pi(Y) - \pi(Y)\pi(X).$$

If we apply both sides to Z, we get

$$[[X, Y], Z] = [X, [Y, Z]] - [Y, [X, Z]],$$

which implies the Jacobi identity.

(ii) X is in $\mathfrak{a}$ and Y and Z are in $\mathfrak{b}$. Then $\pi(X)$ is a derivation of $\mathfrak{b}$, and so

$$\pi(X)[Y, Z] = [\pi(X)Y, Z] + [Y, \pi(X)Z]$$

or

$$[X, [Y, Z]] = [[X, Y], Z] + [Y, [X, Z]],$$

which implies the Jacobi identity.

Finally $\mathfrak{a}$ and $\mathfrak{b}$ both bracket $\mathfrak{b}$ into $\mathfrak{b}$, and consequently $\mathfrak{b}$ is an ideal.

EXAMPLES.

1) Let $\mathfrak{b}$ be any Lie algebra, and let D be in $\operatorname{Der}_{\mathbb{k}} \mathfrak{b}$. Then $(D, \mathfrak{b})$ defines a new Lie algebra $\mathfrak{g}$ unique up to isomorphism as follows. Let $\mathfrak{a} = \mathbb{k}X$ be a 1-dimensional algebra, and take $\mathfrak{g} = \mathfrak{a} \oplus_\pi \mathfrak{b}$, where $\pi(X) = D$. This condition on π means that $[X, Y] = D(Y)$ for $Y \in \mathfrak{b}$.

2) Let V be a vector space over $\mathbb{k}$, and let C be a bilinear form on $V \times V$. Let $\mathfrak{a}$ be the algebra of **derivations** of C, namely

$$\mathfrak{a} = \{D \in \operatorname{End}_{\mathbb{k}} V \mid C(DX, Y) + C(X, DY) = 0 \text{ for all } X, Y\}.$$

Let $\mathfrak{b}$ be V with abelian Lie algebra structure. Then $\operatorname{Der}_{\mathbb{k}} \mathfrak{b} = \operatorname{End}_{\mathbb{k}} V$, so that $\mathfrak{a} \subseteq \operatorname{Der}_{\mathbb{k}} \mathfrak{b}$. Thus we can form the semidirect product $\mathfrak{g} = \mathfrak{a} \oplus_\iota \mathfrak{b}$, where ι is inclusion. Here are two special cases:

(a) $V = \mathbb{R}^n$ with C as the usual dot product. In the standard basis of $\mathbb{R}^n$, $\mathfrak{a}$ gets identified with the Lie algebra of real skew-symmetric n-by-n matrices. The Lie algebra $\mathfrak{b}$ is just $\mathbb{R}^n$, and we can form the semidirect product $\mathfrak{g} = \mathfrak{a} \oplus_\iota \mathfrak{b}$. In this example, $\mathfrak{a}$ is the Lie algebra of the rotation group (about the origin) in $\mathbb{R}^n$, $\mathfrak{b}$ is the Lie algebra of the translation group in $\mathbb{R}^n$, and $\mathfrak{g}$ is the Lie algebra of the proper Euclidean motion group in $\mathbb{R}^n$ (the group containing all rotations and translations).

(b) Let V be $\mathbb{R}^4$, and define C by

$$C\left(\begin{pmatrix} x \\ y \\ z \\ t \end{pmatrix}, \begin{pmatrix} x' \\ y' \\ z' \\ t' \end{pmatrix}\right) = xx' + yy' + zz' - tt'.$$

In this case, $\mathfrak{a}$ is the Lie algebra of the homogeneous Lorentz group in space-time, $\mathfrak{b}$ is the Lie algebra of translations in $\mathbb{R}^4$, and $\mathfrak{g}$ is the Lie algebra of the inhomogeneous Lorentz group.

5. Solvable Lie Algebras and Lie's Theorem

In this section $\mathbb{k}$ and $\mathbb{K}$ are fields satisfying $\mathbb{k} \subseteq \mathbb{K} \subseteq \mathbb{C}$, and all Lie algebras have $\mathbb{k}$ as the underlying field and are finite-dimensional.

Proposition 1.23. An n-dimensional Lie algebra $\mathfrak{g}$ is solvable if and only if there exists a sequence of subalgebras

$$\mathfrak{g} = \mathfrak{a}_0 \supseteq \mathfrak{a}_1 \supseteq \mathfrak{a}_2 \supseteq \cdots \mathfrak{a}_n = 0$$

such that, for each i, $\mathfrak{a}_{i+1}$ is an ideal in $\mathfrak{a}_i$ and $\dim(\mathfrak{a}_i/\mathfrak{a}_{i+1}) = 1$.

PROOF. Let $\mathfrak{g}$ be solvable. Form the commutator series and interpolate subspaces $\mathfrak{a}_i$ in the sequence so that $\dim(\mathfrak{a}_i/\mathfrak{a}_{i+1}) = 1$ for all i. We have

$$\mathfrak{g} = \mathfrak{a}_0 \supseteq \mathfrak{a}_1 \supseteq \mathfrak{a}_2 \supseteq \cdots \mathfrak{a}_n = 0.$$

For any i, we can find j such that $\mathfrak{g}_j \supseteq \mathfrak{a}_i \supseteq \mathfrak{a}_{i+1} \supseteq \mathfrak{g}_{j+1}$. Then

$$[\mathfrak{a}_i, \mathfrak{a}_i] \subseteq [\mathfrak{g}_j, \mathfrak{g}_j] = \mathfrak{g}_{j+1} \subseteq \mathfrak{a}_{i+1}.$$

Hence $\mathfrak{a}_i$ is a subalgebra for each i, and $\mathfrak{a}_{i+1}$ is an ideal in $\mathfrak{a}_i$.

Conversely let the sequence exist. Choose x_i so that $\mathfrak{a}_i = \mathbb{k}x_i + \mathfrak{a}_{i+1}$. We show by induction that $\mathfrak{g}_i \subseteq \mathfrak{a}_i$, so that $\mathfrak{g}_n = 0$. In fact, $\mathfrak{g}_0 = \mathfrak{a}_0$. If $\mathfrak{g}_i \subseteq \mathfrak{a}_i$, then

$$\mathfrak{g}_{i+1} = [\mathfrak{g}_i, \mathfrak{g}_i] \subseteq [\mathbb{k}x_i + \mathfrak{a}_{i+1}, \mathbb{k}x_i + \mathfrak{a}_{i+1}] \subseteq [\mathbb{k}x_i, \mathfrak{a}_{i+1}] + [\mathfrak{a}_{i+1}, \mathfrak{a}_{i+1}] \subseteq \mathfrak{a}_{i+1},$$

and the induction is complete. Hence $\mathfrak{g}$ is solvable.

The kind of sequence in the theorem is called an **elementary sequence**. The existence of such a sequence has the following implication. Write $\mathfrak{a}_i = \mathbb{k}x_i \oplus \mathfrak{a}_{i+1}$. Then $\mathbb{k}x_i$ is a 1-dimensional subspace of $\mathfrak{a}_i$, hence a subalgebra. Also $\mathfrak{a}_{i+1}$ is an ideal in $\mathfrak{a}_i$. In view of Proposition 1.22, $\mathfrak{a}_i$ is exhibited as a semidirect product of a 1-dimensional Lie algebra and $\mathfrak{a}_{i+1}$. The proposition says that solvable Lie algebras are exactly those that can be obtained from semidirect products, starting from 0 and adding one dimension at a time.

Let V be a vector space over $\mathbb{K}$, and let $\mathfrak{g}$ be a Lie algebra. A **representation** of $\mathfrak{g}$ on V is a homomorphism of Lie algebras $\pi : \mathfrak{g} \to (\operatorname{End}_{\mathbb{K}} V)^{\mathbb{k}}$, which we often write simply as $\pi : \mathfrak{g} \to \operatorname{End}_{\mathbb{K}} V$. Because of the definition of bracket in $\operatorname{End}_{\mathbb{K}} V$, the conditions on π are that it be $\mathbb{k}$ linear and satisfy

$$\pi([X, Y]) = \pi(X)\pi(Y) - \pi(Y)\pi(X) \qquad \text{for all } X, Y \in \mathfrak{g}. \tag{1.24}$$

EXAMPLES.

1) ad is a representation of $\mathfrak{g}$ on $\mathfrak{g}$, by (1.8). Here $\mathbb{K} = \mathbb{k}$.

2) If $\mathfrak{g}$ is a Lie algebra of n-by-n matrices over $\mathbb{k}$, then the identity is a representation of $\mathfrak{g}$ on $\mathbb{K}^n$ whenever $\mathbb{K}$ contains $\mathbb{k}$.

3) A case often studied in later chapters is that $\mathbb{k} = \mathbb{R}$ and $\mathbb{K} = \mathbb{C}$. The vector space V is thus to be complex. A representation of a real Lie algebra $\mathfrak{g}$ on the complex vector space V is a homomorphism of $\mathfrak{g}$ into $\operatorname{End}_{\mathbb{C}} V$, which is to be regarded as the real Lie algebra $(\operatorname{End}_{\mathbb{C}} V)^{\mathbb{R}}$. [*Warning*: This space of endomorphisms is different from $\operatorname{End}_{\mathbb{R}}(V^{\mathbb{R}})$, whose real dimension is twice that of $(\operatorname{End}_{\mathbb{C}} V)^{\mathbb{R}}$.]

4) On other occasions in later chapters the Lie algebra $\mathfrak{g}$ under study will be complex. In such cases we shall need to say whether we are thinking of $\mathfrak{g}$ as a complex Lie algebra (so that representations are complex linear) or as the real Lie algebra $\mathfrak{g}^{\mathbb{R}}$ (so that representations are real linear).

Theorem 1.25 (Lie's Theorem). Let $\mathfrak{g}$ be solvable, let $V \neq 0$ be a finite-dimensional vector space over $\mathbb{K}$, and let $\pi : \mathfrak{g} \to \operatorname{End}_{\mathbb{K}} V$ be a representation. If $\mathbb{K}$ is algebraically closed, then there is a simultaneous eigenvector $v \neq 0$ for all the members of $\pi(\mathfrak{g})$. More generally (for $\mathbb{K}$), there is a simultaneous eigenvector if all the eigenvalues of all $\pi(X)$, $X \in \mathfrak{g}$, lie in $\mathbb{K}$.

REMARKS.

1) When $\mathfrak{g}$ is a solvable Lie algebra and π is a representation, $\pi(\mathfrak{g})$ is solvable. This follows immediately from Proposition 1.10. Consequently the theorem is really a result about solvable Lie algebras of matrices.

2) The theorem is the base step in an induction that will show that V has a basis in which all the matrices of $\pi(\mathfrak{g})$ are triangular. See Corollary 1.29 below. In particular, if $\mathfrak{g}$ is a solvable Lie algebra of matrices and π is the identity and one of the conditions on $\mathbb{K}$ is satisfied, then $\mathfrak{g}$ can be conjugated so as to be triangular.

PROOF. We induct on $\dim \mathfrak{g}$. If $\dim \mathfrak{g} = 1$, then $\pi(\mathfrak{g})$ consists of the multiples of a single transformation, and the result follows.

Assume the theorem for all solvable Lie algebras of dimension less than $\dim \mathfrak{g}$ satisfying the eigenvalue condition. Since $\mathfrak{g}$ is solvable, $[\mathfrak{g}, \mathfrak{g}] \subsetneq \mathfrak{g}$. Choose a subspace $\mathfrak{h}$ of codimension 1 in $\mathfrak{g}$ with $[\mathfrak{g}, \mathfrak{g}] \subseteq \mathfrak{h}$. Then $[\mathfrak{h}, \mathfrak{g}] \subseteq [\mathfrak{g}, \mathfrak{g}] \subseteq \mathfrak{h}$, and $\mathfrak{h}$ is an ideal. So $\mathfrak{h}$ is solvable. (Also the eigenvalue condition holds for $\mathfrak{h}$ if it holds for $\mathfrak{g}$.) By inductive hypothesis we can choose $e \in V$ with $\pi(H)e = \lambda(H)e$ for all $H \in \mathfrak{h}$, where $\lambda(H)$ is a scalar-valued function defined for $H \in \mathfrak{h}$.

Fix $X \in \mathfrak{g}$ with $\mathfrak{g} = \mathbb{k}X + \mathfrak{h}$. Define recursively

$$e_{-1} = 0, \qquad e_0 = e, \qquad e_p = \pi(X)e_{p-1},$$

and let $E = \operatorname{span}\{e_0, \ldots, e_p, \ldots\}$. Then $\pi(X)E \subseteq E$. Let v be an eigenvector for $\pi(X)$ in E. We show that v is an eigenvector for each $\pi(H)$, $H \in \mathfrak{h}$.

First we show that

$$\pi(H)e_p \equiv \lambda(H)e_p \mod \operatorname{span}\{e_0, \ldots, e_{p-1}\} \tag{1.26}$$

for all $H \in \mathfrak{h}$. We do so by induction on p. Formula (1.26) is valid for $p = 0$ by definition of e_0. Assume (1.26) for p. Then

$$\begin{aligned} \pi(H)e_{p+1} &= \pi(H)\pi(X)e_p \\ &= \pi([H, X])e_p + \pi(X)\pi(H)e_p \\ &\equiv \lambda([H, X])e_p + \pi(X)\pi(H)e_p \mod \operatorname{span}\{e_0, \ldots, e_{p-1}\} \\ &\qquad\qquad\qquad\qquad\qquad\qquad \text{by induction} \end{aligned}$$

$$\begin{aligned} &\equiv \lambda([H,X])e_p + \lambda(H)\pi(X)e_p \\ &\qquad \text{mod span}\{e_0,\dots,e_{p-1},\pi(X)e_0,\dots,\pi(X)e_{p-1}\} \\ &\qquad\qquad\qquad\qquad\qquad\qquad \text{by induction} \\ &\equiv \lambda(H)\pi(X)e_p \quad \text{mod span}\{e_0,\dots,e_p\} \\ &= \lambda(H)e_{p+1} \quad \text{mod span}\{e_0,\dots,e_p\}. \end{aligned}$$

This proves (1.26) for $p+1$ and completes the induction.

Next we show that

$$\lambda([H,X]) = 0 \qquad \text{for all } H \in \mathfrak{h}. \tag{1.27}$$

In fact, (1.26) says that $\pi(H)E \subseteq E$ and that, relative to the basis $e_0, e_1, \dots$, the linear transformation $\pi(H)$ has matrix

$$\pi(H) = \begin{pmatrix} \lambda(H) & & & * \\ & \lambda(H) & & \\ & & \ddots & \\ 0 & & & \lambda(H) \end{pmatrix}.$$

Thus $\operatorname{Tr}\pi(H) = \lambda(H)\dim E$, and we obtain

$$\lambda([H,X])\dim E = \operatorname{Tr}\pi([H,X]) = \operatorname{Tr}[\pi(H),\pi(X)] = 0.$$

Since our fields have characteristic 0, (1.27) follows.

Now we can sharpen (1.26) to

$$\pi(H)e_p = \lambda(H)e_p \qquad \text{for all } H \in \mathfrak{h}. \tag{1.28}$$

To prove (1.28), we induct on p. For $p = 0$, the formula is the definition of e_0. Assume (1.28) for p. Then

$$\begin{aligned} \pi(H)e_{p+1} &= \pi(H)\pi(X)e_p \\ &= \pi([H,X])e_p + \pi(X)\pi(H)e_p \\ &= \lambda([H,X])e_p + \pi(X)\lambda(H)e_p && \text{by induction} \\ &= 0 + \lambda(H)e_{p+1} && \text{by (1.27)} \\ &= \lambda(H)e_{p+1}. \end{aligned}$$

This completes the induction and proves (1.28). Because of (1.28), $\pi(H)x = \lambda(H)x$ for all $x \in E$ and in particular for $x = v$. Hence the eigenvector v of $\pi(X)$ is also an eigenvector of $\pi(\mathfrak{h})$. The theorem follows.

Before carrying out the induction indicated in Remark 2, we observe something about eigenvalues in connection with representations. Let π be a representation of $\mathfrak{g}$ on a finite-dimensional V, and let $U \subseteq V$ be an **invariant subspace**: $\pi(\mathfrak{g})U \subseteq U$. Then $\pi(X)(v+U) = \pi(X)v + U$ defines a **quotient representation** of $\mathfrak{g}$ on V/U. The characteristic polynomial of $\pi(X)$ on V is the product of the characteristic polynomial on U and that on V/U, and hence the eigenvalues for V/U are a subset of those for V.

Corollary 1.29 (Lie's Theorem). Under the assumptions on $\mathfrak{g}$, V, π, and $\mathbb{K}$ as in Theorem 1.25, there exists a sequence of subspaces

$$V = V_0 \supseteq V_1 \supseteq \cdots \supseteq V_m = 0$$

such that each V_i is stable under $\pi(\mathfrak{g})$ and $\dim V_i/V_{i+1} = 1$. Consequently V has a basis with respect to which all the matrices of $\pi(\mathfrak{g})$ are upper triangular.

REMARK. The sequence of subspaces in the corollary is called an **invariant flag** of V.

PROOF. We induct on $\dim V$, the case $\dim V = 1$ being trivial. If V is given, find by Theorem 1.13 an eigenvector $v \neq 0$ for $\pi(\mathfrak{g})$, and put $U = \mathbb{k}v$. Then U is an invariant subspace, and π provides a quotient representation on V/U, where $\dim(V/U) < \dim V$. Find by inductive hypothesis an invariant flag for V/U, say

$$V/U = W_0 \supseteq W_1 \supseteq \cdots \supseteq W_{m-1} = 0,$$

and put $V_i = \sigma^{-1}(W_i)$, where $\sigma : V \to V/U$ is the quotient map (which commutes with all $\pi(X)$ by definition). Taking $V_m = 0$, we have

$$V = V_0 \supseteq V_1 \supseteq \cdots \supseteq V_{m-1} \supseteq V_m = 0$$

as the required sequence.

A solvable Lie algebra $\mathfrak{g}$ is said to be **split-solvable** if there is an elementary sequence

$$\mathfrak{g} = \mathfrak{a}_0 \supseteq \mathfrak{a}_1 \supseteq \cdots \supseteq \mathfrak{a}_n = 0$$

in which each $\mathfrak{a}_i$ is an ideal in $\mathfrak{g}$ (rather than just in $\mathfrak{a}_{i-1}$). Notice that a subspace $\mathfrak{a} \subseteq \mathfrak{g}$ is an ideal if and only if $\mathfrak{a}$ is stable under $\operatorname{ad}\mathfrak{g}$. Thus in the terminology above, $\mathfrak{g}$ is split-solvable if and only if there is an invariant flag for the adjoint representation.

Corollary 1.30. If $\mathfrak{g}$ is solvable, then $\mathfrak{g}$ is split-solvable if and only if the eigenvalues of all $\operatorname{ad} X$, $X \in \mathfrak{g}$, are in $\mathbb{k}$.

PROOF. Sufficiency is by Corollary 1.29. For necessity let the sequence

$$\mathfrak{g} = \mathfrak{a}_0 \supseteq \mathfrak{a}_1 \supseteq \cdots \supseteq \mathfrak{a}_n = 0$$

exist. The eigenvalues of $\operatorname{ad} X$, $X \in \mathfrak{g}$, are those of $\operatorname{ad} X$ on the various $\mathfrak{a}_i/\mathfrak{a}_{i+1}$. Since $\dim(\mathfrak{a}_i/\mathfrak{a}_{i+1}) = 1$, the eigenvalue of any endomorphism of this space is in $\mathbb{k}$.

An example of a split-solvable Lie algebra is in Example 12b at the end of §1, and an example of a solvable Lie algebra that is not split-solvable is in Example 12c. The verifications of these properties use Corollary 1.30.

6. Nilpotent Lie Algebras and Engel's Theorem

In this section, $\mathbb{k}$ is any field, and all Lie algebras have $\mathbb{k}$ as underlying field and are finite-dimensional.

Let $\mathfrak{g}$ be a nilpotent Lie algebra. If $\mathfrak{g}_k = 0$, then

$$(\operatorname{ad} X)^k Y = [X, [X, [\cdots, [X, Y]\cdots]]] \in \mathfrak{g}_k = 0. \tag{1.31}$$

Hence $(\operatorname{ad} X)^k = 0$, and $\operatorname{ad} X$ is a nilpotent linear transformation on $\mathfrak{g}$.

Actually in (1.31) we can allow the k occurrences of X to be unequal, and then the conclusion is that $\operatorname{ad} \mathfrak{g}$ is a nilpotent Lie algebra. This result, along with a converse, are the subject of the following proposition.

Proposition 1.32. If $\mathfrak{g}$ is a Lie algebra, then $\mathfrak{g}$ is nilpotent if and only if the Lie algebra $\operatorname{ad} \mathfrak{g}$ is nilpotent.

PROOF. We have

$$[[\cdots[[X_{k+1}, X_k], X_{k-1}], \cdots], X_1] = \operatorname{ad}[\cdots[X_{k+1}, X_k], \cdots X_2](X_1) \tag{1.33}$$

and

$$\begin{aligned} \operatorname{ad}[\cdots[X_{k+1}, X_k], \cdots X_2] &= [\operatorname{ad}[\cdots[X_{k+1}, X_k], \cdots X_3], \operatorname{ad} X_2] \\ &= \cdots = [\cdots[\operatorname{ad} X_{k+1}, \operatorname{ad} X_k], \cdots \operatorname{ad} X_2]. \end{aligned} \tag{1.34}$$

If $\mathfrak{g}$ is nilpotent, then $\mathfrak{g}_k = 0$ for some k. Then the left side of (1.33) is always 0, and hence the right side of (1.34) is always 0. This says that $(\operatorname{ad} \mathfrak{g})_{k-1} = 0$, and hence $\operatorname{ad} \mathfrak{g}$ is nilpotent. Conversely if $\operatorname{ad} \mathfrak{g}$ is nilpotent, then by retracing the steps, we see that $\mathfrak{g}$ is nilpotent.

Engel's Theorem is a converse to (1.31), saying that if ad X is always a nilpotent transformation of $\mathfrak{g}$, then $\mathfrak{g}$ is a nilpotent Lie algebra. Actually we shall state Engel's Theorem more generally, in a way that lends itself better to an inductive proof, and then we shall derive this conclusion as a corollary.

Theorem 1.35 (Engel's Theorem). Let $V \neq 0$ be a finite-dimensional vector space over $\mathbb{k}$, and let $\mathfrak{g}$ be a Lie algebra of nilpotent endomorphisms of V. Then

(a) $\mathfrak{g}$ is a nilpotent Lie algebra
(b) there exists $v \neq 0$ in V with $X(v) = 0$ for all $X \in \mathfrak{g}$
(c) in a suitable basis of V, all X are upper triangular with 0's on the diagonal.

PROOF. The proof is by induction on $\dim \mathfrak{g}$. For $\dim \mathfrak{g} = 1$, (b) and (c) hold since X is nilpotent, and (a) is trivial. Suppose that (a), (b), and (c) hold for dimension $< \dim \mathfrak{g}$. We may assume that $\dim \mathfrak{g} > 1$.

We shall prove that (b) holds when the dimension equals $\dim \mathfrak{g}$. Then (c) follows by the argument of Corollary 1.29 (Lie's Theorem), and (a) follows from (c) since a subalgebra of a nilpotent Lie algebra is nilpotent (Proposition 1.10). Thus a proof of (b) will complete the induction.

The main step is to construct a nilpotent ideal $\mathfrak{h} \subseteq \mathfrak{g}$ of codimension 1 in $\mathfrak{g}$. To do so, let $\mathfrak{h}$ be a proper Lie subalgebra of $\mathfrak{g}$ of maximal dimension in $\mathfrak{g}$. By inductive hypothesis, $\mathfrak{h}$ is a nilpotent Lie algebra. We show that $\mathfrak{h}$ has codimension 1 and is an ideal. Since ad $\mathfrak{h}$ leaves $\mathfrak{h}$ stable, ad defines a representation ρ of $\mathfrak{h}$ on $\mathfrak{g}/\mathfrak{h}$ by

$$\rho(H)(X + \mathfrak{h}) = [H, X] + \mathfrak{h}.$$

We claim that each $\rho(H)$ is nilpotent. In fact, on $\mathfrak{g}$, ad H acts by

$$\begin{aligned}(\operatorname{ad} H)^{2m} X &= (\operatorname{ad} H)^{2m-1}(HX - XH) \\ &= (\operatorname{ad} H)^{2m-2}(H^2X - 2HXH + XH^2) \\ &= \cdots = \sum_{J=0}^{2m} c_j H^j X H^{2m-j}. \end{aligned} \tag{1.36}$$

If $H^m = 0$, we see that every term on the right side is 0 and hence that $(\operatorname{ad} H)^{2m} = 0$. Therefore ad H is nilpotent on $\mathfrak{g}$ and must be nilpotent on $\mathfrak{g}/\mathfrak{h}$.

Since $\dim \rho(\mathfrak{h}) < \dim \mathfrak{g}$, we can find by inductive hypothesis a coset $X_0 + \mathfrak{h} \neq \mathfrak{h}$ in $\mathfrak{g}/\mathfrak{h}$ with $\rho(H)(X_0 + \mathfrak{h}) = \mathfrak{h}$ for all $H \in \mathfrak{h}$. This condition says that

$$[H, X_0] \in \mathfrak{h} \qquad \text{for all } H \in \mathfrak{h}. \tag{1.37}$$

Let $\mathfrak{s} = \mathfrak{h} + \mathbb{k}X_0$. Then (1.37) shows that $\mathfrak{s}$ is a subalgebra of $\mathfrak{g}$ properly containing $\mathfrak{h}$, and hence $\mathfrak{s} = \mathfrak{g}$ by maximality of $\dim \mathfrak{h}$. Consequently $\mathfrak{h}$ has codimension 1 in $\mathfrak{g}$. Also (1.37) shows that $\mathfrak{h}$ is an ideal.

To complete the proof, let $V_0 = \{v \in V \mid Hv = 0 \text{ for all } H \in \mathfrak{h}\}$. Since $\mathfrak{h}$ acts as nilpotent endomorphisms, the inductive hypothesis shows that V_0 is not 0. If v is in V_0, then

$$HX_0v = [H, X_0]v + X_0Hv = 0 + 0 = 0$$

since $\mathfrak{h}$ is an ideal. Thus $X_0(V_0) \subseteq V_0$. By assumption X_0 is nilpotent, and thus 0 is its only eigenvalue. Hence X_0 has an eigenvector v_0 in V_0. Then $X_0(v_0) = 0$ and $\mathfrak{h}(v_0) = 0$, so that $\mathfrak{g}(v_0) = 0$. Consequently (b) holds for $\mathfrak{g}$, and the induction is complete.

Corollary 1.38. If $\mathfrak{g}$ is a Lie algebra such that each $\operatorname{ad} X$ for $X \in \mathfrak{g}$ is nilpotent, then $\mathfrak{g}$ is a nilpotent Lie algebra.

PROOF. Theorem 1.37 shows that $\operatorname{ad} \mathfrak{g}$ is nilpotent, and Proposition 1.32 allows us to conclude that $\mathfrak{g}$ is nilpotent.

7. Cartan's Criterion for Semisimplicity

In this section, $\mathbb{k}$ denotes a subfield of $\mathbb{C}$, and $\mathfrak{g}$ denotes a finite-dimensional Lie algebra over $\mathbb{k}$. We shall relate semisimplicity of $\mathfrak{g}$ to a nondegeneracy property of the Killing form of $\mathfrak{g}$, the Killing form having been defined in (1.18).

First we make some general remarks about bilinear forms. Let V be a finite-dimensional vector space, and let $C(\cdot\,,\,\cdot)$ be a bilinear form on $V \times V$. Define

$$\operatorname{rad} C = \{v \in V \mid C(v, u) = 0 \text{ for all } u \in V\}.$$

Writing $\langle \cdot\,,\,\cdot \rangle$ for the pairing of the dual V^* with V, define $\varphi : V \to V^*$ by $\langle \varphi(v), u\rangle = C(v, u)$. Then $\ker \varphi = \operatorname{rad} C$, and so φ is an isomorphism (onto) if and only if C is **nondegenerate** (i.e., $\operatorname{rad} C = 0$).

If U is a subspace of V, let

$$U^\perp = \{v \in V \mid C(v, u) = 0 \text{ for all } u \in U\}.$$

Then

$$U \cap U^\perp = \operatorname{rad}(C|_{U\times U}). \tag{1.39}$$

Even if C is nondegenerate, we may have $U \cap U^\perp \neq 0$. For example, take $\mathbb{k} = \mathbb{R}$, $V = \mathbb{R}^2$, $C(x, y) = x_1y_1 - x_2y_2$, and $U = \{(x_1, x_1)\}$; then C is nondegenerate, but $U = U^\perp \neq 0$. However, we can make the positive statement given in the following proposition.

Proposition 1.40. In the above notation, if C is nondegenerate, then

$$\dim U + \dim U^{\perp} = \dim V.$$

PROOF. Define $\psi : V \to U^*$ by

$$\langle \psi(v), u \rangle = C(v, u) \qquad \text{for } v \in V,\ u \in U.$$

Then $\ker \psi = U^{\perp}$. To see that image $\psi = U^*$, let U_1 be a linear complement for U in V. Let u^* be in U^*, and define $v^* \in V^*$ by

$$v^* = \begin{cases} u^* & \text{on } U \\ 0 & \text{on } U_1. \end{cases}$$

Since C is nondegenerate, φ is onto V^*. Thus choose $v \in V$ with $\varphi(v) = v^*$. Then

$$\langle \psi(v), u \rangle = C(v, u) = \langle v^*, u \rangle = \langle u^*, u \rangle,$$

and hence $\psi(v) = u^*$. Therefore image $\psi = U^*$, and

$$\begin{aligned} \dim V &= \dim(\ker \psi) + \dim(\text{image } \psi) \\ &= \dim U^{\perp} + \dim U^* \\ &= \dim U^{\perp} + \dim U. \end{aligned}$$

Corollary 1.41. In the above notation, if C is nondegenerate, then $V = U \oplus U^{\perp}$ if and only if $C|_{U \times U}$ is nondegenerate.

PROOF. This follows by combining (1.39) with Proposition 1.40.

Theorem 1.42 (Cartan's Criterion for Semisimplicity). The Lie algebra $\mathfrak{g}$ is semisimple if and only if the Killing form for $\mathfrak{g}$ is nondegenerate.

REMARKS. This theorem is a fairly easy consequence of Cartan's Criterion for Solvability, to be proved below. We shall state the criterion for solvability, state and prove a corollary of it, show how the corollary implies Theorem 1.42, and then prove the criterion for solvability. Let B denote the Killing form of $\mathfrak{g}$. We may assume that $\mathfrak{g} \neq 0$.

Proposition 1.43 (Cartan's Criterion for Solvability). The Lie algebra $\mathfrak{g}$ is solvable if and only if its Killing form B satisfies $B(X, Y) = 0$ for all $X \in \mathfrak{g}$ and $Y \in [\mathfrak{g}, \mathfrak{g}]$.

Corollary 1.44. For any Lie algebra $\mathfrak{g}$, $\operatorname{rad} B \subseteq \operatorname{rad} \mathfrak{g}$.

PROOF. We show that $\operatorname{rad} B$ is a solvable ideal, and then the corollary follows. To see that $\operatorname{rad} B$ is an ideal, let $H \in \operatorname{rad} B$ and let X_1 and X_2 be in $\mathfrak{g}$. Then

$$B([X_1, H], X_2) = -B(H, [X_1, X_2]) = 0,$$

and so $[X_1, H]$ is in $\operatorname{rad} B$. Thus $\operatorname{rad} B$ is an ideal. To see that $\operatorname{rad} B$ is solvable, let C be the Killing form of $\operatorname{rad} B$, and let $\mathfrak{s}$ be a vector subspace with $\mathfrak{g} = \operatorname{rad} B \oplus \mathfrak{s}$. If X is in $\operatorname{rad} B$, then the fact that $\operatorname{rad} B$ is an ideal forces $\operatorname{ad} X$ to have the matrix form

$$\begin{array}{c} \\ \operatorname{ad} X = \end{array} \begin{array}{c} \begin{array}{cc} \operatorname{rad} B & \mathfrak{s} \end{array} \\ \begin{pmatrix} * & * \\ 0 & 0 \end{pmatrix} \end{array} \begin{array}{c} \\ \operatorname{rad} B \\ \mathfrak{s} \end{array} .$$

Consequently if X and Y are in $\operatorname{rad} B$, then

$$C(X, Y) = \operatorname{Tr}((\operatorname{ad} X \operatorname{ad} Y)|_{\operatorname{rad} B}) = \operatorname{Tr}(\operatorname{ad} X \operatorname{ad} Y) = B(X, Y) = 0,$$

the last step holding since X is in $\operatorname{rad} B$. By Proposition 1.43, $\operatorname{rad} B$ is solvable.

PROOF OF THEOREM 1.42 GIVEN PROPOSITION 1.43. If B is degenerate, then $\operatorname{rad} B \neq 0$. By the corollary $\operatorname{rad} \mathfrak{g} \neq 0$. Hence $\mathfrak{g}$ is not semisimple.

Conversely let $\mathfrak{g}$ fail to be semisimple so that $\operatorname{rad} \mathfrak{g} \neq 0$. Since $\operatorname{rad} \mathfrak{g}$ is solvable, there is a least integer l such that the member $(\operatorname{rad} \mathfrak{g})^l$ of the commutator series is 0. Then $(\operatorname{rad} \mathfrak{g})^{l-1} = \mathfrak{a}$ is a nonzero abelian ideal in $\mathfrak{g}$. Let $\mathfrak{s}$ be a vector space complement to $\mathfrak{a}$, so that $\mathfrak{g} = \mathfrak{a} \oplus \mathfrak{s}$ as vector spaces. If X is in $\mathfrak{a}$ and Y is in $\mathfrak{g}$, then

$$\operatorname{ad} X = \overset{\begin{array}{cc}\mathfrak{a} & \mathfrak{s}\end{array}}{\begin{pmatrix} 0 & * \\ 0 & 0 \end{pmatrix}} \begin{array}{c} \mathfrak{a} \\ \mathfrak{s} \end{array} \qquad \text{and} \qquad \operatorname{ad} Y = \overset{\begin{array}{cc}\mathfrak{a} & \mathfrak{s}\end{array}}{\begin{pmatrix} * & * \\ 0 & * \end{pmatrix}} \begin{array}{c} \mathfrak{a} \\ \mathfrak{s} \end{array}$$

in matrix form. Then $\operatorname{ad} X \operatorname{ad} Y$ has 0's on the diagonal, and $B(X, Y) = \operatorname{Tr}(\operatorname{ad} X \operatorname{ad} Y) = 0$. Thus X is in $\operatorname{rad} B$. Hence $\mathfrak{a} \subseteq \operatorname{rad} B$, and B is degenerate.

We are left with proving Proposition 1.43. By way of preparation for the proof, let $\mathbb{K}$ be an extension field of $\mathbb{k}$ within $\mathbb{C}$, and form the Lie algebra $\mathfrak{g}^{\mathbb{K}} = \mathfrak{g} \otimes_{\mathbb{k}} \mathbb{K}$ (over $\mathbb{K}$) and its Killing form $B^{\mathbb{K}}$, as in §3.

We saw in §3 that $\mathfrak{g}$ is solvable if and only if $\mathfrak{g}^{\mathbb{K}}$ is solvable. Also $[\mathfrak{g},\mathfrak{g}]^{\mathbb{K}} = [\mathfrak{g}^{\mathbb{K}},\mathfrak{g}^{\mathbb{K}}]$. Next if Y is in $[\mathfrak{g},\mathfrak{g}]$ and Y is in $\operatorname{rad} B$, then Y is in $\operatorname{rad} B^{\mathbb{K}}$ since $B^{\mathbb{K}}$ is K bilinear and agrees with B on $\mathfrak{g}\times\mathfrak{g}$. So $[\mathfrak{g},\mathfrak{g}]\subseteq \operatorname{rad} B$ implies $[\mathfrak{g},\mathfrak{g}]\subseteq\operatorname{rad} B^{\mathbb{K}}$, and therefore $[\mathfrak{g}^{\mathbb{K}},\mathfrak{g}^{\mathbb{K}}]=[\mathfrak{g},\mathfrak{g}]^{\mathbb{K}}\subseteq \operatorname{rad} B^{\mathbb{K}}$. Conversely if $[\mathfrak{g}^{\mathbb{K}},\mathfrak{g}^{\mathbb{K}}]\subseteq\operatorname{rad} B^{\mathbb{K}}$, then $[\mathfrak{g},\mathfrak{g}]\subseteq\operatorname{rad} B^{\mathbb{K}}$ and so $[\mathfrak{g},\mathfrak{g}]\subseteq\operatorname{rad} B$. Consequently each direction of Proposition 1.43 holds for $\mathfrak{g}$ if and only if it holds for $\mathfrak{g}^{\mathbb{K}}$.

PROOF THAT $\mathfrak{g}$ SOLVABLE IMPLIES $[\mathfrak{g},\mathfrak{g}]\subseteq\operatorname{rad} B$. We may assume that $\mathbb{k}$ is algebraically closed in view of the previous paragraph. We apply Lie's Theorem (Theorem 1.25) to the representation ad. The theorem says that $\operatorname{ad}\mathfrak{g}$ is simultaneously triangular in a suitable basis of $\mathfrak{g}$. Then $\operatorname{ad}[\mathfrak{g},\mathfrak{g}]=[\operatorname{ad}\mathfrak{g},\operatorname{ad}\mathfrak{g}]$ has 0's on the diagonal, and so $X\in\mathfrak{g}$ and $Y\in[\mathfrak{g},\mathfrak{g}]$ imply that $\operatorname{ad}X\operatorname{ad}Y$ has 0's on the diagonal. Hence $B(X,Y)=0$.

For the converse we shall need to use the Jordan decomposition in the following sharp form.

Theorem 1.45 (Jordan decomposition). Let $\mathbb{k}$ be algebraically closed, and let V be a finite-dimensional vector space over $\mathbb{k}$. Each $l\in\operatorname{End}_{\mathbb{K}}V$ can be uniquely decomposed as $l=s+n$ with s diagonable, n nilpotent, and $sn=ns$. Moreover, $s=p(l)$ for some polynomial p without constant term.

PROOF. We omit the proof, which may be found with Theorem 8 in Hoffman-Kunze [1961], 217.

PROOF THAT $[\mathfrak{g},\mathfrak{g}]\subseteq\operatorname{rad} B$ IMPLIES $\mathfrak{g}$ SOLVABLE.
As we saw before the direct part of the proof, we may assume that $\mathbb{k}=\mathbb{C}$. We shall show that each $\operatorname{ad}Y$, for $Y\in[\mathfrak{g},\mathfrak{g}]$, is nilpotent. Then by Engel's Theorem, $[\mathfrak{g},\mathfrak{g}]$ is nilpotent, hence solvable. Consequently $\mathfrak{g}$ is solvable.

Arguing by contradiction, suppose that $\operatorname{ad}Y$ is not nilpotent. Let $\operatorname{ad}Y=s+n$ be its Jordan decomposition. Here $s\neq 0$. Let $\mu_1,\dots,\mu_m$ be the distinct eigenvalues of s, and let $V_1,\dots,V_m$ be the corresponding eigenspaces. Define $\bar{s}\in\operatorname{End}_{\mathbb{C}}\mathfrak{g}$ to be $\bar{\mu}_j$ on V_j. Then $\operatorname{Tr}(\bar{s}s)=\sum_{j=1}^m|\mu_j|^2>0$.

Since n and s commute, so do n and $\bar{s}$; thus $\bar{s}n$ is nilpotent. Consequently

$$\operatorname{Tr}(\bar{s}(\operatorname{ad}Y))=\operatorname{Tr}(\bar{s}s)+\operatorname{Tr}(\bar{s}n)=\operatorname{Tr}(\bar{s}s)>0, \tag{1.46}$$

the last equality following since $\bar{s}n$ is nilpotent.

Now we shall compute $\mathrm{Tr}(\bar{s}(\operatorname{ad} Y))$ another way. Since Y is in $[\mathfrak{g}, \mathfrak{g}]$, we can write $Y = \sum_{i=1}^{u} [X_i, Z_i]$ with X_i and Z_i in $\mathfrak{g}$. Then

$$\begin{aligned} \mathrm{Tr}(\bar{s} \operatorname{ad} Y) &= \sum_{i=1}^{u} \mathrm{Tr}(\bar{s} \operatorname{ad}[X_i, Z_i]) \\ &= \sum_{i=1}^{u} \mathrm{Tr}([\bar{s}, \operatorname{ad} X_i] \operatorname{ad} Z_i) \quad \text{as in the proof of (1.19)} \\ &= \sum_{i+1}^{u} \mathrm{Tr}(\{\operatorname{ad}_0 \bar{s}\, (\operatorname{ad} X_i)\} \operatorname{ad} Z_i), \end{aligned} \tag{1.47}$$

where ad_0 is ad for $\mathrm{End}_{\mathbb{C}}\, \mathfrak{g}$. Since $\operatorname{ad} Y = s + n$, we have

$$\operatorname{ad}_0(\operatorname{ad} Y) = \operatorname{ad}_0 s + \operatorname{ad}_0 n. \tag{1.48}$$

Since n is nilpotent, $\operatorname{ad}_0 n$ is nilpotent (by the same computation as for (1.36) in the proof of Engel's Theorem). Also

$$[\operatorname{ad}_0 s, \operatorname{ad}_0 n] = \operatorname{ad}_0[s, n] = 0.$$

So $\operatorname{ad}_0 s$ and $\operatorname{ad}_0 n$ commute. Choose a basis for $\mathfrak{g}$ compatible with the decomposition $\mathfrak{g} = \bigoplus_i V_i$, and let

$$E_{ij} = \begin{cases} 1 & \text{in the } (i, j)^{\text{th}} \text{ entry} \\ 0 & \text{elsewhere.} \end{cases}$$

Then

$$sE_{ij} = \mu_i E_{ij} \qquad \text{and} \qquad E_{ij} s = \mu_j E_{ij},$$

and so

$$(\operatorname{ad}_0 s) E_{ij} = (\mu_i - \mu_j) E_{ij}.$$

Since $\mathrm{End}_{\mathbb{C}}\, \mathfrak{g} = \bigoplus_{i,j} \mathbb{C} E_{ij}$, $\operatorname{ad}_0 s$ is diagonable. Thus (1.48) is the Jordan decomposition of $\operatorname{ad}_0(\operatorname{ad} Y)$, and it follows from Theorem 1.45 that

$$\operatorname{ad}_0 s = q(\operatorname{ad}_0(\operatorname{ad} Y)) \qquad \text{with } q(0) = 0.$$

If we choose a polynomial that maps all $\mu_i - \mu_j$ into $\bar{\mu}_i - \bar{\mu}_j$ (including 0 into 0) and if we compose it with q, the result is a polynomial r with

$$\operatorname{ad}_0 \bar{s} = r(\operatorname{ad}_0(\operatorname{ad} Y)) \qquad \text{and} \qquad r(0) = 0. \tag{1.49}$$

Now consider a term of the right side of (1.47), say the term $\mathrm{Tr}(\{\operatorname{ad}_0 \bar{s}\, (\operatorname{ad} X_i)\} \operatorname{ad} Z_i)$. Because of (1.49), it is a linear combination of terms

$$\mathrm{Tr}(\{(\operatorname{ad}_0(\operatorname{ad} Y))^k (\operatorname{ad} X_i)\} \operatorname{ad} Z_i) \qquad \text{with } k > 0.$$

For $k = 1$, we have

$$\begin{aligned}\{(\operatorname{ad}_0(\operatorname{ad} Y))^1(\operatorname{ad} X_i)\} &= (\operatorname{ad} Y)(\operatorname{ad} X_i) - (\operatorname{ad} X_i)(\operatorname{ad} Y)\\ &= \operatorname{ad}[Y, X_i].\end{aligned}$$

For $k = 2$, we have

$$\begin{aligned}\{(\operatorname{ad}_0(\operatorname{ad} Y))^2(\operatorname{ad} X_i)\} &= (\operatorname{ad} Y)(\operatorname{ad}[Y, X_i]) - \operatorname{ad}[Y, X_i] \operatorname{ad} Y\\ &= \operatorname{ad}[Y, [Y, X_i]].\end{aligned}$$

And so on. Iterating, we see that the right side of (1.47) is a linear combination of terms

$$\operatorname{Tr}(\operatorname{ad}([Y, [Y, [\cdots, [Y, X_i]\cdots]]])\operatorname{ad} Z_i),$$

and this is B of something in $[\mathfrak{g}, \mathfrak{g}]$ with something in $\mathfrak{g}$. By hypothesis, it is therefore 0. Thus the right side of (1.47) adds to 0, in contradiction with (1.46). We conclude that $\operatorname{ad} Y$ must indeed have been nilpotent, and the proof is complete.

Corollary 1.50. If $\mathfrak{g}_0$ is a real form of a complex Lie algebra $\mathfrak{g}$, then $\mathfrak{g}_0$ is semisimple if and only if $\mathfrak{g}$ is semisimple.

REMARK. By contrast the complexification of a simple real Lie algebra need not be simple. See §VI.9 for a full discussion.

PROOF. By (1.20) the Killing forms of $\mathfrak{g}_0$ and $\mathfrak{g}$ may be identified. Hence they are both nondegenerate or both degenerate, and the corollary follows from Theorem 1.42.

Theorem 1.51. The Lie algebra $\mathfrak{g}$ is semisimple if and only if $\mathfrak{g} = \mathfrak{g}_1 \oplus \cdots \oplus \mathfrak{g}_m$ with $\mathfrak{g}_j$ ideals that are each simple Lie algebras. In this case the decomposition is unique, and the only ideals of $\mathfrak{g}$ are the sums of various $\mathfrak{g}_j$.

PROOF IF $\mathfrak{g} = \mathfrak{g}_1 \oplus \cdots \oplus \mathfrak{g}_m$. Let P_i be the projection on $\mathfrak{g}_i$ along the other summands. Let $\mathfrak{a}$ be any ideal in $\mathfrak{g}$, and form $P_i\mathfrak{a} = \mathfrak{a}_i$. Then $\mathfrak{a}_i$ is an ideal in $\mathfrak{g}_i$ since

$$[P_i A, X_i] = P_i[A, X_i] \subseteq P_i\mathfrak{a} = \mathfrak{a}_i \qquad \text{for } A \in \mathfrak{a}.$$

Since $\mathfrak{g}_i$ is simple, either $\mathfrak{a}_i = 0$ or $\mathfrak{a}_i = \mathfrak{g}_i$. In the latter case, $\mathfrak{g}_i \subseteq \mathfrak{a}$ since Proposition 1.13 gives

$$\mathfrak{g}_i = [\mathfrak{g}_i, \mathfrak{g}_i] = [\mathfrak{g}_i, P_i\mathfrak{a}] = [\mathfrak{g}_i, \mathfrak{a}] \subseteq [\mathfrak{g}, \mathfrak{a}] \subseteq \mathfrak{a}.$$

Consequently $\mathfrak{g} = \bigoplus \mathfrak{g}_i$ implies

$$\mathfrak{a} = \mathfrak{a} \cap \mathfrak{g} = \bigoplus_{i=1}^{m} (\mathfrak{a} \cap \mathfrak{g}_i) = \bigoplus_{\mathfrak{g}_i \subseteq \mathfrak{a}} \mathfrak{g}_i.$$

This proves uniqueness and the structure of the ideals. Now

$$[\mathfrak{a}, \mathfrak{a}] = [\bigoplus_{\mathfrak{g}_i \subseteq \mathfrak{a}} \mathfrak{g}_i, \bigoplus_{\mathfrak{g}_j \subseteq \mathfrak{a}} \mathfrak{g}_j] = \bigoplus_{\mathfrak{g}_i \subseteq \mathfrak{a}} [\mathfrak{g}_i, \mathfrak{g}_i] = \bigoplus_{\mathfrak{g}_i \subseteq \mathfrak{a}} \mathfrak{g}_i = \mathfrak{a}.$$

So $\mathfrak{a}$ cannot be solvable unless $\mathfrak{a}$ is 0. Thus $\mathfrak{g}$ is semisimple.

PROOF IF $\mathfrak{g}$ IS SEMISIMPLE. Let $\mathfrak{a}$ be a minimal nonzero ideal. Form $\mathfrak{a}^\perp$ relative to the Killing form B. The subspace $\mathfrak{a}^\perp$ is an ideal because if H is in $\mathfrak{a}^\perp$, then

$$B([X, H], A) = B(H, -[X, A]) \subseteq B(H, \mathfrak{a}) = 0 \quad \text{for } A \in \mathfrak{a} \text{ and } X \in \mathfrak{g}.$$

Therefore $\operatorname{rad}(B|_{\mathfrak{a}\times\mathfrak{a}}) = \mathfrak{a} \cap \mathfrak{a}^\perp$ is 0 or $\mathfrak{a}$. Suppose $\operatorname{rad}(B|_{\mathfrak{a}\times\mathfrak{a}}) = \mathfrak{a}$. Then $B(A_1, A_2) = 0$ for all $A_1, A_2 \in \mathfrak{a}$. But for an ideal $\mathfrak{a}$, the Killing form $B_\mathfrak{a}$ for $\mathfrak{a}$ satisfies $B_\mathfrak{a} = B|_{\mathfrak{a}\times\mathfrak{a}}$, and the right side is 0. By Proposition 1.43, $\mathfrak{a}$ is solvable, in contradiction with the semisimplicity of $\mathfrak{g}$.

So $\mathfrak{a} \cap \mathfrak{a}^\perp = 0$. Then Corollary 1.41 and Theorem 1.42 show that $\mathfrak{g} = \mathfrak{a} \oplus \mathfrak{a}^\perp$ as vector spaces. Both $\mathfrak{a}$ and $\mathfrak{a}^\perp$ are ideals, as we have seen. Now $\mathfrak{a}$ is nonabelian; we show it is simple. If $\mathfrak{b} \subseteq \mathfrak{a}$ is an ideal of $\mathfrak{a}$, then $[\mathfrak{b}, \mathfrak{a}^\perp] = 0$ and so $\mathfrak{b}$ is an ideal of $\mathfrak{g}$. By minimality of $\mathfrak{a}$, either $\mathfrak{b} = 0$ or $\mathfrak{b} = \mathfrak{a}$. Thus $\mathfrak{a}$ is simple.

Similarly any ideal of $\mathfrak{a}^\perp$ is an ideal of $\mathfrak{g}$, and hence $\operatorname{rad}\mathfrak{a}^\perp = 0$. Thus $\mathfrak{a}^\perp$ is semisimple by Theorem 1.42. Therefore we can repeat the argument with $\mathfrak{a}^\perp$ and proceed by induction to complete the proof.

Corollary 1.52. If $\mathfrak{g}$ is semisimple, then $[\mathfrak{g}, \mathfrak{g}] = \mathfrak{g}$. If $\mathfrak{a}$ is any ideal in $\mathfrak{g}$, then $\mathfrak{a}^\perp$ is an ideal and $\mathfrak{g} = \mathfrak{a} \oplus \mathfrak{a}^\perp$.

This is immediate from Theorem 1.51.

We say that the Lie algebra $\mathfrak{g}$ is **reductive** if to each ideal $\mathfrak{a}$ in $\mathfrak{g}$ corresponds an ideal $\mathfrak{b}$ in $\mathfrak{g}$ with $\mathfrak{g} = \mathfrak{a} \oplus \mathfrak{b}$. Theorem 1.51 shows that the Lie algebra direct sum of a semisimple Lie algebra and an abelian Lie algebra is reductive. The next corollary shows that there are no other reductive Lie algebras.

Corollary 1.53. If $\mathfrak{g}$ is reductive, then $\mathfrak{g} = [\mathfrak{g}, \mathfrak{g}] \oplus Z_\mathfrak{g}$ with $[\mathfrak{g}, \mathfrak{g}]$ semisimple and $Z_\mathfrak{g}$ abelian.

PROOF. Among all direct sums of ideals in $\mathfrak{g}$ such that each contains no nonzero smaller ideals, let $\mathfrak{a}$ be one of the maximum possible dimension. By Proposition 1.7, $\mathfrak{a}$ is an ideal. Since $\mathfrak{g}$ is reductive, we can find an ideal $\mathfrak{b}$ with $\mathfrak{g} = \mathfrak{a} \oplus \mathfrak{b}$. If $\mathfrak{b} \neq 0$, then a nonzero ideal of $\mathfrak{g}$ in $\mathfrak{b}$ of smallest possible dimension can be adjoined to $\mathfrak{a}$ and exhibit a contradiction with the maximality of $\mathfrak{a}$. We conclude that $\mathfrak{b} = 0$ and that $\mathfrak{g}$ is the direct sum of ideals that contain no nonzero smaller ideals. Write

$$\mathfrak{g} = \mathfrak{a}_1 \oplus \cdots \oplus \mathfrak{a}_j \oplus \mathfrak{a}_{j+1} \oplus \cdots \oplus \mathfrak{a}_k,$$

where $\mathfrak{a}_1, \ldots, \mathfrak{a}_j$ are 1-dimensional and $\mathfrak{a}_{j+1}, \ldots, \mathfrak{a}_k$ are simple Lie algebras. By Proposition 1.13,

$$[\mathfrak{g}, \mathfrak{g}] = [\mathfrak{a}_{j+1}, \mathfrak{a}_{j+1}] \oplus \cdots \oplus [\mathfrak{a}_k, \mathfrak{a}_k] = \mathfrak{a}_{j+1} \oplus \cdots \oplus \mathfrak{a}_k,$$

and this is semisimple by Theorem 1.51. To complete the proof we show that $Z_\mathfrak{g} = \mathfrak{a}_1 \oplus \cdots \oplus \mathfrak{a}_j$. Certainly $Z_\mathfrak{g} \supseteq \mathfrak{a}_1 \oplus \cdots \oplus \mathfrak{a}_j$. In the reverse direction if $X = X_1 + \cdots + X_k$ is in $Z_\mathfrak{g}$ with $X_i \in \mathfrak{a}_i$, then X_i is in $Z_{\mathfrak{a}_i}$, which is 0 for $i > j$ by Proposition 1.13. Hence $X = X_1 + \cdots + X_j$, and we conclude that $Z_\mathfrak{g} \subseteq \mathfrak{a}_1 \oplus \cdots \oplus \mathfrak{a}_j$.

8. Examples of Semisimple Lie Algebras

Cartan's Criterion for Semisimplicity (Theorem 1.42) enables us to produce a long list of examples of semisimple matrix Lie algebras over $\mathbb{R}$ or $\mathbb{C}$. The examples that we shall produce are the Lie algebras of some groups of symmetries in geometry going under the name "classical groups." In this section we give only the Lie algebras, deferring discussion of the Lie groups to §14.

We shall work with Lie groups and real Lie algebras of matrices over the reals $\mathbb{R}$, the complex numbers $\mathbb{C}$, and the quaternions $\mathbb{H}$. Recall that $\mathbb{H}$ is a division algebra over $\mathbb{R}$ with $\mathbb{R}$ basis $1, i, j, k$ satisfying $i^2 = j^2 = k^2 = -1$, $ij = k$, $jk = i$, $ki = j$, $ji = -k$, $kj = -i$, and $ik = -j$. The real part of a quaternion is given by $\operatorname{Re}(a + bi + cj + dk) = a$. Despite the noncommutativity of $\mathbb{H}$, the real part satisfies

$$\operatorname{Re} xy = \operatorname{Re} yx \qquad \text{for all } x \text{ and } y \text{ in } \mathbb{H}. \tag{1.54}$$

Conjugation is given by

$$\overline{a + bi + cj + dk} = a - bi - cj - dk,$$

and it commutes with real part. Consequently (1.54) implies that

$$\operatorname{Re} x\bar{y} = \operatorname{Re} y\bar{x}. \tag{1.55}$$

If x is in $\mathbb{H}$, then $x\bar{x} = \bar{x}x = |x|^2$ is the square of the usual Euclidean norm from $\mathbb{R}^4$. The fact that the bracket operation $[X, Y] = XY - YX$ makes all square matrices of size n into a Lie algebra is valid over $\mathbb{H}$, as well as $\mathbb{R}$ and $\mathbb{C}$; this follows from Example 2 in §1.

Groups and Lie algebras of complex matrices can be realized as groups and Lie algebras of real matrices of twice the size. We shall write down an explicit isomorphism in (1.60) below. Similarly, groups and Lie algebras of quaternion matrices can be realized as groups of complex matrices of twice the size. We shall write down an explicit isomorphism in (1.62) below.

The technique for recognizing certain Lie algebras of matrices as semisimple begins with the following proposition.

Proposition 1.56. Let $\mathfrak{g}$ be a real Lie algebra of matrices over $\mathbb{R}$, $\mathbb{C}$, or $\mathbb{H}$. If $\mathfrak{g}$ is closed under the operation conjugate transpose, then $\mathfrak{g}$ is reductive.

REMARK. We write $(\cdot)^*$ for conjugate transpose.

PROOF. For matrices X and Y, define $\langle X, Y\rangle = \operatorname{Re}\operatorname{Tr}(XY^*)$. This is a real inner product on $\mathfrak{g}$, the symmetry following from (1.55). Let $\mathfrak{a}$ be an ideal in $\mathfrak{g}$, and let $\mathfrak{a}^\perp$ be the orthogonal complement of $\mathfrak{a}$ in $\mathfrak{g}$. Then $\mathfrak{g} = \mathfrak{a} \oplus \mathfrak{a}^\perp$ as vector spaces. To see that $\mathfrak{a}^\perp$ is an ideal in $\mathfrak{g}$, let X be in $\mathfrak{a}^\perp$, let Y be in $\mathfrak{g}$, and let Z be in $\mathfrak{a}$. Then

$$\begin{aligned}
\langle [X, Y], Z\rangle &= \operatorname{Re}\operatorname{Tr}(XYZ^* - YXZ^*) \\
&= -\operatorname{Re}\operatorname{Tr}(XZ^*Y - XYZ^*) \qquad \text{by (1.55)} \\
&= -\operatorname{Re}\operatorname{Tr}(X(Y^*Z)^* - X(ZY^*)^*) \\
&= -\langle X, [Y^*, Z]\rangle.
\end{aligned}$$

Since Y^* is in $\mathfrak{g}$, $[Y^*, Z]$ is in $\mathfrak{a}$. Thus the right side is 0 for all Z, and $[X, Y]$ is in $\mathfrak{a}^\perp$. Hence $\mathfrak{a}^\perp$ is an ideal, and $\mathfrak{g}$ is reductive.

All of our examples $\mathfrak{g}$ in this section will be closed under conjugate transpose. In view of Proposition 1.56 and Corollary 1.53, $\mathfrak{g}$ will be semisimple as a real Lie algebra if and only if its center is 0. Checking whether the center of $\mathfrak{g}$ is 0 is generally an easy matter.

Some of the examples will be complex Lie algebras. As we shall show in connection with the embedding of complex matrices into real matrices of twice the size, the Killing forms of a complex Lie algebra $\mathfrak{g}$ and its associated $\mathfrak{g}^{\mathbb{R}}$ are related by

$$B_{\mathfrak{g}^{\mathbb{R}}} = 2\operatorname{Re} B_{\mathfrak{g}}. \tag{1.57}$$

Consequently $B_{\mathfrak{g}^\mathbb{R}}$ and $B_\mathfrak{g}$ are both nondegenerate or both degenerate. By Cartan's Criterion for Semisimplicity (Theorem 1.42), we see that

(1.58) $\mathfrak{g}^\mathbb{R}$ is semisimple over $\mathbb{R}$ if and only if $\mathfrak{g}$ is semisimple over $\mathbb{C}$.

In Example 3 in §1, we defined $\mathfrak{gl}(n,\mathbb{R})$ and $\mathfrak{gl}(n,\mathbb{C})$ to be the Lie algebras of all n-by-n matrices over $\mathbb{R}$ and $\mathbb{C}$, respectively. These Lie algebras are reductive but not semisimple, the center in each case consisting of scalar matrices. Let $\mathfrak{gl}(n,\mathbb{H})$ be the real Lie algebra of all n-by-n matrices over $\mathbb{H}$. This Lie algebra is reductive but not semisimple, the center consisting of scalar matrices with real entries.

EXAMPLES.

1) These first examples will be seen in §14 to be Lie algebras of compact groups. They consist initially of all matrices that are skew Hermitian relative to $(\cdot)^*$. Specifically we define

$$\begin{aligned}\mathfrak{u}(n) &= \{X \in \mathfrak{gl}(n,\mathbb{C}) \mid X + X^* = 0\}\\ \mathfrak{so}(n) &= \{X \in \mathfrak{gl}(n,\mathbb{R}) \mid X + X^* = 0\}\\ \mathfrak{sp}(n) &= \{X \in \mathfrak{gl}(n,\mathbb{H}) \mid X + X^* = 0\}.\end{aligned}$$

To see that these examples are closed under bracket and hence are Lie algebras, we argue as in Example 7 in §1, replacing $(\cdot)^t$ by $(\cdot)^*$. All three of these examples are reductive. If $n \geq 3$, then $\mathfrak{so}(n)$ has 0 center, while if $n \geq 1$, $\mathfrak{sp}(n)$ has 0 center. For these values of n, $\mathfrak{so}(n)$ and $\mathfrak{sp}(n)$ are therefore semisimple. For $n \geq 1$, the Lie algebra $\mathfrak{u}(n)$ has imaginary scalar matrices as center; the commutator subalgebra is the semisimple Lie algebra

$$\mathfrak{su}(n) = \{X \in \mathfrak{gl}(n,\mathbb{C}) \mid X + X^* = 0 \text{ and } \operatorname{Tr} X = 0\}.$$

This is nonzero if $n \geq 2$.

2) These next examples are all complex Lie algebras with 0 center. Each such complex $\mathfrak{g}$ has the property that $\mathfrak{g}^\mathbb{R}$ is semisimple over $\mathbb{R}$, and it follows from (1.58) that $\mathfrak{g}$ is semisimple over $\mathbb{C}$. We define

$$\begin{aligned}\mathfrak{sl}(n,\mathbb{C}) &= \{X \in \mathfrak{gl}(n,\mathbb{C}) \mid \operatorname{Tr} X = 0\} && \text{for } n \geq 2\\ \mathfrak{so}(n,\mathbb{C}) &= \{X \in \mathfrak{gl}(n,\mathbb{C}) \mid X + X^t = 0\} && \text{for } n \geq 3\\ \mathfrak{sp}(n,\mathbb{C}) &= \{X \in \mathfrak{gl}(2n,\mathbb{C}) \mid X^tJ + JX = 0\} && \text{for } n \geq 1,\end{aligned}$$

where $J = J_{n,n}$ is the $2n$-by-$2n$ matrix $J = \begin{pmatrix} 0 & 1 \\ -1 & 0 \end{pmatrix}$. Each of these examples is a Lie algebra as a consequence of Examples 7, 8, and 9 in §1.

3) These final examples are real semisimple Lie algebras that are of neither of the types in Examples 1 and 2 (except for small values of the parameter). With $J = J_{n,n}$ as in Example 2, let

$$I_{m,n} = \begin{pmatrix} 1 & 0 \\ 0 & -1 \end{pmatrix} \begin{matrix} m \\ n \end{matrix} \quad \text{and} \quad I_{n,n}J_{n,n} = \begin{pmatrix} 0 & 1 \\ 1 & 0 \end{pmatrix} \begin{matrix} n \\ n \end{matrix} .$$

(column blocks: m, n and n, n)

The definitions are

$$\begin{aligned}
\mathfrak{sl}(n,\mathbb{R}) &= \{X \in \mathfrak{gl}(n,\mathbb{R}) \mid \operatorname{Tr} X = 0\} && \text{for } n \geq 2\\
\mathfrak{sl}(n,\mathbb{H}) &= \{X \in \mathfrak{gl}(n,\mathbb{H}) \mid \operatorname{Re}\operatorname{Tr} X = 0\} && \text{for } n \geq 1\\
\mathfrak{so}(m,n) &= \{X \in \mathfrak{gl}(m+n,\mathbb{R}) \mid X^*I_{m,n} + I_{m,n}X = 0\} && \text{for } m+n \geq 3\\
\mathfrak{su}(m,n) &= \{X \in \mathfrak{sl}(m+n,\mathbb{C}) \mid X^*I_{m,n} + I_{m,n}X = 0\} && \text{for } m+n \geq 2\\
\mathfrak{sp}(m,n) &= \{X \in \mathfrak{gl}(m+n,\mathbb{H}) \mid X^*I_{m,n} + I_{m,n}X = 0\} && \text{for } m+n \geq 1\\
\mathfrak{sp}(n,\mathbb{R}) &= \{X \in \mathfrak{gl}(2n,\mathbb{R}) \mid X^tJ_{n,n} + J_{n,n}X = 0\} && \text{for } n \geq 1\\
\mathfrak{so}^*(2n) &= \{X \in \mathfrak{su}(n,n) \mid X^tI_{n,n}J_{n,n} + I_{n,n}J_{n,n}X = 0\} && \text{for } n \geq 2.
\end{aligned}$$

It is necessary to check that each of these is closed under bracket, is closed under conjugate transpose, and has 0 center. For closure under bracket, we appeal to Example 9 of §1 for $\mathfrak{sl}(n,\mathbb{R})$, to (1.54) for $\mathfrak{sl}(n,\mathbb{H})$, and to Example 8 of §1 (possibly with $(\cdot)^t$ replaced by $(\cdot)^*$) for the remaining five classes of examples. Closure under conjugate transpose uses that $I^*_{m,n} = I_{m,n}$ and $J^*_{n,n} = -J_{n,n}$. Seeing that the center is 0 in each case is a matter of routine verification. Note that $\mathfrak{su}(m,n)$ is the commutator ideal of the reductive Lie algebra

$$\mathfrak{u}(m,n) = \{X \in \mathfrak{gl}(m+n,\mathbb{C}) \mid X^*I_{m,n} + I_{m,n}X = 0\}.$$

To complete our discussion of these examples, we give some of the details of how to write complex matrices as real matrices of twice the size, as well as quaternion matrices as complex matrices of twice the size.

We begin with the relationship between complex and real matrices. For v in $\mathbb{C}^n$, write $v = a+ib$ with $a \in \mathbb{R}^n$ and $b \in \mathbb{R}^n$, and define functions Re and Im from $\mathbb{C}^n$ to $\mathbb{R}^n$ by $\operatorname{Re} v = a$ and $\operatorname{Im} v = b$. Then set up the $\mathbb{R}$ isomorphism $\mathbb{C}^n \to \mathbb{R}^{2n}$ given in block form by

$$v \to \begin{pmatrix} \operatorname{Re} v \\ \operatorname{Im} v \end{pmatrix}. \tag{1.59}$$

Next let M be an n-by-n matrix over $\mathbb{C}$ and write $M = \operatorname{Re} M + i \operatorname{Im} M$. Under the isomorphism (1.59), left multiplication by M on $\mathbb{C}^n$ corresponds to left multiplication by

$$Z(M) = \begin{pmatrix} \operatorname{Re} M & -\operatorname{Im} M \\ \operatorname{Im} M & \operatorname{Re} M \end{pmatrix} \tag{1.60}$$

on $\mathbb{R}^{2n}$. This identification has the following properties:

(a) $Z(MN) = Z(M)Z(N)$
(b) $Z(M^*) = Z(M)^*$
(c) $\operatorname{Tr} Z(M) = 2\operatorname{Re} \operatorname{Tr} M$
(d) $\det Z(M) = |\det M|^2$.

For the proof, only (d) requires comment. Because of (a) it is enough to check (d) for elementary matrices. Matters come down to M of size 1-by-1, where the argument is that

$$(z) = (x + iy) \to \begin{pmatrix} x & -y \\ y & x \end{pmatrix} \qquad \text{with} \qquad \det \begin{pmatrix} x & -y \\ y & x \end{pmatrix} = |z|^2.$$

If $\mathfrak{g}$ is a complex Lie algebra, then $B_{\mathfrak{g}}(X, Y) = \operatorname{Tr}(\operatorname{ad} X \operatorname{ad} Y)$, while $B_{\mathfrak{g}^{\mathbb{R}}}(X, Y) = \operatorname{Tr}(Z(\operatorname{ad} X \operatorname{ad} Y))$. Hence (1.57) follows immediately from (c) above.

Next let us discuss the relationship between quaternion and complex matrices. Let $\mathbb{H}^n$ be the space of n-component column vectors with quaternion entries. Write v in $\mathbb{H}^n$ as $v = a + ib + jc + kd$ with a, b, c, d in $\mathbb{R}^n$, and define $z_1 : \mathbb{H}^n \to \mathbb{C}^n$ and $z_2 : \mathbb{H}^n \to \mathbb{C}^n$ by

$$z_1(v) = a + ib \qquad \text{and} \qquad z_2(v) = c - id, \tag{1.61a}$$

so that $v = z_1(v) + jz_2(v)$ if we allow i to be interpreted as in $\mathbb{H}$ or $\mathbb{C}$. Then the $\mathbb{R}$ isomorphism

$$v \to \begin{pmatrix} z_1(v) \\ z_2(v) \end{pmatrix} \tag{1.61b}$$

of $\mathbb{H}^n$ into $\mathbb{C}^{2n}$ is a $\mathbb{C}$ isomorphism if $\mathbb{H}$ is regarded as a right vector space over $\mathbb{C}$ (complex scalars multiplying as expected on the right). In fact, we have only to check that

$$z_1(vi) = z_1(v)i \qquad \text{and} \qquad z_2(vi) = z_2(v)i,$$

and then the $\mathbb{C}$ linearity of the isomorphism follows.

If M is an n-by-n matrix over $\mathbb{H}$, we define $z_1(M)$ and $z_2(M)$ similarly. Under the isomorphism (1.61), left multiplication by M on $\mathbb{H}^n$ corresponds to left multiplication by

$$Z(M) = \begin{pmatrix} z_1(M) & -\overline{z_2(M)} \\ z_2(M) & \overline{z_1(M)} \end{pmatrix} \tag{1.62}$$

on $\mathbb{C}^{2n}$. This identification has the following properties:

(a) $Z(MN) = Z(M)Z(N)$
(b) $Z(M^*) = Z(M)^*$
(c) $\operatorname{Tr} Z(M) = 2\operatorname{Re}\operatorname{Tr} M$.

From (a) it follows that the real Lie algebra $\mathfrak{gl}(n, \mathbb{H})$ is isomorphic to

$$\mathfrak{u}^*(2n) = \left\{ \begin{pmatrix} z_1 & -\overline{z_2} \\ z_2 & \overline{z_1} \end{pmatrix} \middle| \ z_1 \text{ and } z_2 \text{ are in } \mathfrak{gl}(n, \mathbb{C}) \right\}.$$

Taking also (c) into account, we see that $\mathfrak{sl}(n, \mathbb{H})$ is isomorphic to

$$\mathfrak{su}^*(2n) = \{X \in \mathfrak{u}^*(2n) \mid \operatorname{Tr} X = 0\}.$$

Similarly it follows from (a) and (b) that $\mathfrak{sp}(n)$ is isomorphic to

$$\left\{ \begin{pmatrix} z_1 & -\overline{z_2} \\ z_2 & \overline{z_1} \end{pmatrix} \middle| \ z_1 \text{ is skew Hermitian and } z_2 \text{ is symmetric} \right\}.$$

Then we obtain the important isomorphism

$$\mathfrak{sp}(n) \cong \mathfrak{sp}(n, \mathbb{C}) \cap \mathfrak{u}(2n)$$

by writing out the conditions on the n-by-n subblocks of members of the right side and comparing with the matrices above.

In addition, we can reinterpret $\mathfrak{sp}(m, n)$ in terms of complex matrices. We apply the function $Z(\cdot)$ to the identity defining $\mathfrak{sp}(m, n)$, noting that $Z(I_{m,n})$ is a diagonal matrix $I_{m,n,m,n}$ with m diagonal entries 1, followed by n diagonal entries -1, m diagonal entries 1, and n diagonal entries -1. Then (a) and (b) imply that $\mathfrak{sp}(m, n)$ is isomorphic to

$$\{X \in \mathfrak{u}^*(2m + 2n) \mid X^* I_{m,n,m,n} + I_{m,n,m,n} X = 0\}.$$

9. Representations of $\mathfrak{sl}(2, \mathbb{C})$

In Chapter II we shall see that complex semisimple Lie algebras $\mathfrak{g}$ are built out of many copies of $\mathfrak{sl}(2, \mathbb{C})$. The action of each $\operatorname{ad}(\mathfrak{sl}(2, \mathbb{C}))$ on $\mathfrak{g}$ will give us our first control over the bracket structure within $\mathfrak{g}$.

To prepare for this analysis, we shall study in this section complex-linear representations φ of the Lie algebra $\mathfrak{sl}(2, \mathbb{C})$ on finite-dimensional vector spaces V. An **invariant subspace** for such a representation is a complex vector subspace U such that $\varphi(X)U \subseteq U$ for all $X \in \mathfrak{sl}(2, \mathbb{C})$. We say that a representation on a nonzero space V is **irreducible** if the only invariant subspaces are 0 and V. Two representations φ and φ' are **equivalent** if there is an isomorphism E between the underlying vector spaces such that $E\varphi(X) = \varphi'(X)E$ for all X in the Lie algebra.

Let h, e, f be the basis of $\mathfrak{sl}(2, \mathbb{C})$ given in (1.5).

Theorem 1.63. For each integer $m \geq 1$, there exists up to equivalence a unique irreducible complex-linear representation π of $\mathfrak{sl}(2, \mathbb{C})$ on a complex vector space V of dimension m. In V there is a basis $\{v_0, \ldots, v_{m-1}\}$ such that (with $n = m - 1$)

(a) $\pi(h)v_i = (n - 2i)v_i$
(b) $\pi(e)v_0 = 0$
(c) $\pi(f)v_i = v_{i+1}$ with $v_{n+1} = 0$
(d) $\pi(e)v_i = i(n - i + 1)v_{i-1}$.

REMARK. Conclusion (a) gives the eigenvalues of $\pi(h)$. Note that the smallest eigenvalue is the negative of the largest.

PROOF OF UNIQUENESS. Let π be a complex-linear irreducible representation of $\mathfrak{sl}(2, \mathbb{C})$ on V with $\dim V = m$. Let $v \neq 0$ be an eigenvector for $\pi(h)$, say with $\pi(h)v = \lambda v$. Then $\pi(e)v, \pi(e)^2 v, \ldots$ are also eigenvectors because

$$\begin{aligned} \pi(h)\pi(e)v &= \pi(e)\pi(h)v + \pi([h, e])v && \text{by (1.24)} \\ &= \pi(e)\lambda v + 2\pi(e)v && \text{by (1.6)} \\ &= (\lambda + 2)\pi(e)v. \end{aligned}$$

Since $\lambda, \lambda+2, \lambda+4, \ldots$ are distinct, these eigenvectors are independent (or 0). By finite-dimensionality we can find v_0 in V with (λ redefined and)

(i) $v_0 \neq 0$
(ii) $\pi(h)v_0 = \lambda v_0$
(iii) $\pi(e)v_0 = 0$.

Define $v_i = \pi(f)^i v_0$. Then $\pi(h)v_i = (\lambda - 2i)v_i$, by the same argument as above, and so there is a minimum integer n with $\pi(f)^{n+1}v_0 = 0$. Then $v_0, \ldots, v_n$ are independent and

(a) $\pi(h)v_i = (\lambda - 2i)v_i$
(b) $\pi(e)v_0 = 0$
(c) $\pi(f)v_i = v_{i+1}$ with $v_{n+1} = 0$.

We claim $V = \operatorname{span}\{v_0, \ldots, v_n\}$. It is enough to show that $\operatorname{span}\{v_0, \ldots, v_n\}$ is stable under $\pi(e)$. In fact, we show that

(d) $\pi(e)v_i = i(\lambda - i + 1)v_{i-1}$ with $v_{-1} = 0$.

We proceed by induction for (d), the case $i = 0$ being (b). Assume (d) for case i. To prove case $i + 1$, we write

$$\begin{aligned}
\pi(e)v_{i+1} &= \pi(e)\pi(f)v_i \\
&= \pi([e, f])v_i + \pi(f)\pi(e)v_i \\
&= \pi(h)v_i + \pi(f)\pi(e)v_i \\
&= (\lambda - 2i)v_i + \pi(f)(i(\lambda - i + 1))v_{i-1} \\
&= (i + 1)(\lambda - i)v_i.
\end{aligned}$$

and the induction is complete.

To finish the proof of uniqueness, we show that $\lambda = n$. We have

$$\operatorname{Tr}\pi(h) = \operatorname{Tr}(\pi(e)\pi(f) - \pi(f)\pi(e)) = 0.$$

Thus $\sum_{i=0}^{n}(\lambda - 2i) = 0$, and we find that $\lambda = n$.

PROOF OF EXISTENCE. We define $\pi(h)$, $\pi(e)$, and $\pi(f)$ by (a) through (d) and extend linearly to obtain $\pi(\mathfrak{sl}(2, \mathbb{C})$. Easy computation verifies that

$$\begin{aligned}
\pi([h, e]) &= \pi(h)\pi(e) - \pi(e)\pi(h) \\
\pi([h, f]) &= \pi(h)\pi(f) - \pi(f)\pi(h) \\
\pi([e, f]) &= \pi(e)\pi(f) - \pi(f)\pi(e),
\end{aligned}$$

and consequently π is a representation. To see irreducibility, let U be a nonzero invariant subspace. Since U is invariant under $\pi(h)$, U is spanned by a subset of the basis vectors v_i. Taking one such v_i that is in U and applying $\pi(e)$ several times, we see that v_0 is in U. Repeated application of $\pi(f)$ then shows that $U = V$. Hence π is irreducible.

Theorem 1.64. Let φ be a complex-linear representation of $\mathfrak{sl}(2, \mathbb{C})$ on a finite-dimensional complex vector space V. Then V is **completely reducible** in the sense that there exist invariant subspaces $U_1, \ldots, U_r$ of V such that $V = U_1 \oplus \cdots \oplus U_r$ and such that the restriction of the representation to each U_i is irreducible.

At this time we shall give an algebraic proof. In Chapter VII we shall give another argument that is analytic in nature. The algebraic proof will be preceded by four lemmas. It is enough by induction to show that any invariant subspace U in V has an invariant complement U', i.e., an invariant subspace U' with $V = U \oplus U'$.

Lemma 1.65. If π is a representation of $\mathfrak{sl}(2, \mathbb{C})$, then

$$Z = \tfrac{1}{2}\pi(h)^2 + \pi(h) + 2\pi(f)\pi(e)$$

commutes with each $\pi(X)$ for X in $\mathfrak{sl}(2, \mathbb{C})$.

PROOF. For $X \in \mathfrak{sl}(2, \mathbb{C})$, we have

$$\begin{aligned} Z\pi(X) - \pi(X)Z &= \tfrac{1}{2}\pi(h)^2\pi(X) - \tfrac{1}{2}\pi(X)\pi(h)^2 + \pi[h, X] \\ &\quad + 2\pi(f)\pi(e)\pi(X) - 2\pi(X)\pi(f)\pi(e) \\ &= \tfrac{1}{2}\pi(h)\pi[h, X] - \tfrac{1}{2}\pi[X, h]\pi(h) + \pi[h, X] \\ &\quad + 2\pi(f)\pi[e, X] - 2\pi[X, f]\pi(e) \\ &= (*). \end{aligned}$$

Then the result follows from the following computations as we take X in succession to be h, e, and f:

$X = h$:

$$(*) = 0 - 0 + 0 - 4\pi(f)\pi(e) + 4\pi(f)\pi(e) = 0$$

$X = e$:

$$\begin{aligned} (*) &= \pi(h)\pi(e) + \pi(e)\pi(h) + 2\pi(e) + 0 - 2\pi(h)\pi(e) \\ &= 2\pi(h)\pi(e) + \pi[e, h] + 2\pi(e) - 2\pi(h)\pi(e) = 0 \end{aligned}$$

$X = f$:

$$\begin{aligned} (*) &= -\pi(h)\pi(f) - \pi(f)\pi(h) - 2\pi(f) + 2\pi(f)\pi(h) - 0 \\ &= -\pi[h, f] - 2\pi(f)\pi(h) - 2\pi(f) + 2\pi(f)\pi(h) = 0. \end{aligned}$$

Lemma 1.66 (Schur's Lemma). Let $\mathfrak{g} = \mathfrak{sl}(2, \mathbb{C})$. If $\pi : \mathfrak{g} \to \operatorname{End} V$ and $\pi' : \mathfrak{g} \to \operatorname{End} V'$ are irreducible finite-dimensional representations and if $L : V \to V'$ is a linear map such that $L\pi(X) = \pi'(X)L$ for all $X \in \mathfrak{g}$, then $L = 0$ or L is invertible. If $Z : V \to V$ is a linear map such that $Z\pi(X) = \pi(X)Z$ for all $X \in \mathfrak{g}$, then Z is scalar.

PROOF. The subspace $\ker L$ is $\pi(\mathfrak{g})$ invariant because $v \in \ker L$ implies

$$L(\pi(X)v) = \pi'(X)(Lv) = \pi'(X)(0) = 0.$$

The subspace image L is $\pi'(\mathfrak{g})$ invariant because $v = Lu$ implies

$$\pi'(X)v = \pi'(X)Lu = L(\pi(X)u).$$

By the assumed irreducibility, $L = 0$ or L is invertible.

Then Z, by the above, is 0 or invertible, and the same is true for $Z - \lambda 1$, for any complex constant λ. Choosing λ to be an eigenvalue of Z, we see that $Z - \lambda 1$ cannot be invertible. Therefore $Z - \lambda 1 = 0$ and $Z = \lambda 1$.

Lemma 1.67. If π is an irreducible representation of $\mathfrak{sl}(2, \mathbb{C})$ of dimension $n+1$, then the operator Z of Lemma 1.65 acts as the scalar $\frac{1}{2}n^2 + n$, which is not 0 unless π is trivial.

PROOF. The operator Z acts as a scalar, by Lemmas 1.65 and 1.66. To find the scalar, we identify π with the equivalent irreducible representation of dimension $n+1$ given in Theorem 1.63, and we compute Zv_0. We have

$$Zv_0 = \tfrac{1}{2}\pi(h)^2 v_0 + \pi(h)v_0 + 2\pi(f)\pi(e)v_0.$$

Since $\pi(h)v_0 = nv_0$ and $\pi(e)v_0 = 0$, the result follows.

Lemma 1.68. Let $\pi : \mathfrak{sl}(2, \mathbb{C}) \to \operatorname{End} V$ be a finite-dimensional representation, and let $U \subseteq V$ be an invariant subspace of codimension 1. Then there is a 1-dimensional invariant subspace W such that $V = U \oplus W$.

PROOF.

Case 1. Suppose $\dim U = 1$. Form the quotient representation π on V/U, with $\dim(V/U) = 1$. This quotient representation is irreducible of dimension 1, and Theorem 1.63 shows it is 0. Consequently

$$\pi(\mathfrak{sl}(2, \mathbb{C}))V \subseteq U \qquad \text{and} \qquad \pi(\mathfrak{sl}(2, \mathbb{C}))U = 0.$$

Hence if $Y = [X_1, X_2]$, we have

$$\begin{aligned}\pi(Y)V &\subseteq \pi(X_1)\pi(X_2)V + \pi(X_2)\pi(X_1)V \\ &\subseteq \pi(X_1)U + \pi(X_2)U = 0.\end{aligned}$$

Since $\mathfrak{sl}(2, \mathbb{C}) = [\mathfrak{sl}(2, \mathbb{C}), \mathfrak{sl}(2, \mathbb{C})]$, we conclude that $\pi(\mathfrak{sl}(2, \mathbb{C})) = 0$. Therefore any complementary subspace to U will serve as W.

Case 2. Suppose that $\pi(\cdot)|_U$ is irreducible and that $\dim U > 1$. Since $\dim V/U = 1$, the quotient representation is 0 and $\pi(\mathfrak{sl}(2, \mathbb{C}))V \subseteq U$. The formula for Z in Lemma 1.65 then shows that $Z(V) \subseteq U$, and Lemma 1.67 says that Z is a nonzero scalar on U. Therefore $\dim(\ker Z) = 1$ and $U \cap (\ker Z) = 0$. Since Z commutes with $\pi(\mathfrak{sl}(2, \mathbb{C}))$, $\ker Z$ is an invariant subspace. Taking $W = \ker Z$, we have $V = U \oplus W$ as required.

Case 3. Suppose that $\pi(\cdot)|_U$ is not necessarily irreducible and that $\dim U \geq 1$. We induct on $\dim V$. The base case is $\dim V = 2$ and is handled by Case 1. When $\dim V > 2$, let $U_1 \subseteq U$ be an irreducible invariant subspace, and form the quotient representations on

$$U/U_1 \subseteq V/U_1$$

with quotient V/U of dimension 1. By inductive hypothesis we can write

$$V/U_1 = U/U_1 \oplus Y/U_1,$$

where Y is an invariant subspace in V and $\dim Y/U_1 = 1$. Case 1 or Case 2 is applicable to the representation $\pi(\cdot)|_Y$ and the irreducible invariant subspace U_1. Then $Y = U_1 \oplus W$, where W is a 1-dimensional invariant subspace. Since $W \subseteq Y$ and $Y \cap U \subseteq U_1$, we find that

$$W \cap U = (W \cap Y) \cap U = W \cap (Y \cap U) \subseteq W \cap U_1 = 0.$$

Therefore $V = U \oplus W$ as required.

PROOF OF THEOREM 1.64. Let π be a representation of $\mathfrak{sl}(2, \mathbb{C})$ on M, and let $N \neq 0$ be an invariant subspace. Put

$$V = \{Y \in \operatorname{End} M \mid Y : M \to N \text{ and } Y|_N \text{ is scalar}\}.$$

Linear algebra shows that V is nonzero. Define a linear function $\sigma : \mathfrak{sl}(2, \mathbb{C}) \to \operatorname{End}(\operatorname{End} M)$ by

$$\sigma(X)\gamma = \pi(X)\gamma - \gamma\pi(X) \qquad \text{for } \gamma \in \operatorname{End} M \text{ and } X \in \mathfrak{sl}(2, \mathbb{C}).$$

Checking directly that $\sigma[X, Y]$ and $\sigma(X)\sigma(Y) - \sigma(Y)\sigma(X)$ are equal, we see that σ is a representation of $\mathfrak{sl}(2, \mathbb{C})$ on $\operatorname{End} M$.

We claim that the subspace $V \subseteq \operatorname{End} M$ is an invariant subspace under σ. In fact, let $\gamma(M) \subseteq N$ and $\gamma|_N = \lambda 1$. In the right side of the expression

$$\sigma(X)\gamma = \pi(X)\gamma - \gamma\pi(X),$$

the first term carries M to N since γ carries M to N and $\pi(X)$ carries N to N, and the second term carries M into N since $\pi(X)$ carries M to M and γ carries M to N. Thus $\sigma(X)\gamma$ carries M into N. On N, the action of $\sigma(X)\gamma$ is given by

$$\sigma(X)\gamma(n) = \pi(X)\gamma(n) - \gamma\pi(X)(n) = \lambda\pi(X)(n) - \lambda\pi(X)(n) = 0.$$

Thus V is an invariant subspace.

Actually the above argument shows also that the subspace U of V given by

$$U = \{\gamma \in V \mid \gamma = 0 \text{ on } N\}$$

is an invariant subspace. Clearly $\dim V/U = 1$. By Lemma 1.68, $V = U \oplus W$ for a 1-dimensional invariant subspace $W = \mathbb{C}\gamma$. Here γ is a nonzero scalar $\lambda 1$ on N. The invariance of W means that $\sigma(X)\gamma = 0$ since 1-dimensional representations are 0. Therefore γ commutes with $\pi(X)$ for all $X \in \mathfrak{sl}(2, \mathbb{C})$. But then $\ker \gamma$ is a nonzero invariant subspace of M. Since γ is nonsingular on N (being a nonzero scalar there), we must have $M = N \oplus \ker \gamma$. This completes the proof.

Corollary 1.69. Let π be a complex-linear representation of $\mathfrak{sl}(2, \mathbb{C})$ on a finite-dimensional complex vector space V. Then $\pi(h)$ is diagonable, all its eigenvalues are integers, and the multiplicity of an eigenvalue k equals the multiplicity of $-k$.

PROOF. This is immediate from Theorems 1.63 and 1.64.

To conclude the section, we sharpen the result about complete reducibility to include certain infinite-dimensional representations.

Corollary 1.70. Let φ be a complex-linear representation of $\mathfrak{sl}(2, \mathbb{C})$ on a complex vector space V, and suppose that each vector $v \in V$ lies in a finite-dimensional invariant subspace. Then V is the (possibly infinite) direct sum of finite-dimensional invariant subspaces on which $\mathfrak{sl}(2, \mathbb{C})$ acts irreducibly.

PROOF. By hypothesis and Theorem 1.64 each member of V lies in a finite direct sum of irreducible invariant subspaces. Thus $V = \sum_{s \in S} U_s$, where S is some (possibly infinite) index set and each U_s is an irreducible invariant subspace. Call a subset R of S independent if the sum $\sum_{r \in R} U_r$ is direct. This condition means that for every finite subset $\{r_1, \ldots, r_n\}$ of R and every set of elements $u_i \in U_{r_i}$, the equation

$$u_1 + \cdots + u_n = 0$$

implies that each u_i is 0. From this formulation it follows that the union of any increasing chain of independent subsets of S is itself independent. By Zorn's Lemma there is a maximal independent subset T of S. By definition the sum $V_0 = \sum_{t \in T} U_t$ is direct. We shall show that $V_0 = V$.

We do so by showing, for each $s \in S$, that $U_s \subseteq V_0$. If s is in T, this conclusion is obvious. If s is not in T, then the maximality of T implies that $T \cup \{s\}$ is not independent. Consequently the sum $U_s + V_0$ is not direct, and we must have $U_s \cap V_0 \neq 0$. But this intersection is an invariant subspace of U_s. Since U_s is irreducible and the intersection is not 0, the intersection must be U_s. Then it follows that $U_s \subseteq V_0$, as we wished to show.

10. Elementary Theory of Lie Groups

Now we turn to a discussion of Lie groups. This book assumes a familiarity with the elementary theory of Lie groups, as in Chapter IV of Chevalley [1946]. In this section we shall review most of that material briefly, discussing at length only certain aspects of the theory that are not treated fully in Chevalley.

The elementary theory of Lie groups uses manifolds and mappings, and these manifolds and maps may be assumed to be C^∞ or real analytic, depending on the version of the theory that one encounters. The two theories come to the same thing, because the C^∞ manifold structure of a Lie group is compatible with one and only one real analytic structure for the Lie group. Chevalley [1946] uses the real analytic theory, calling his manifolds and maps "analytic" as an abbreviation for "real analytic." We shall use the C^∞ theory for convenience, noting any aspects that need special attention in the real analytic theory. We use the terms "C^∞" and "smooth" interchangeably. A "manifold" for Chevalley is always connected, but for us it is not; we do, however, insist that a manifold have a countable base for its topology.

If M is a smooth manifold, smooth vector fields on M are sometimes defined as derivations of the algebra $C^\infty(M)$ of smooth real-valued functions on M, and then the tangent space is formed at each point of M out of the smooth vector fields. Alternatively the tangent space may be constructed first at each point, and a vector field may then be defined as a collection of tangent vectors, one for each point. In either case let us write $T_p(M)$ for the tangent space of M at p. If X is a vector field on M, let X_p be the value of X at p, i.e., the corresponding tangent vector in $T_p(M)$. If $F : M \to N$ is a smooth map between smooth manifolds, we write $d\Phi_p : T_p(M) \to T_{\Phi(p)}(N)$ for the differential of F at p. We may drop the subscript "p" on $d\Phi_p$ if p is understood.

A **Lie group** is a topological group with the structure of a smooth manifold such that multiplication and inversion are smooth. An **analytic group** is a connected Lie group.

Let G be a Lie group, and let $L_x : G \to G$ be left translation by x, i.e., the diffeomorphism from G to itself given by $L_x(y) = xy$. A vector field X on G is **left-invariant** if, for any x and y in G, $(dL_{yx^{-1}})(X_x) = X_y$. Equivalently X, as an operator on smooth real-valued functions, commutes with left translations.

If G is a Lie group, then the map $X \to X_1$ is an isomorphism of the real vector space of left-invariant vector fields on G onto $T_1(G)$, and the inverse map is $Xf(x) = X_1(L_{x^{-1}} f)$, where $L_{x^{-1}} f(y) = f(xy)$. Every left-invariant vector field on G is smooth, and the bracket of two left-invariant vector fields is left-invariant.

If G is a Lie group, set $\mathfrak{g} = T_1(G)$. Then $\mathfrak{g}$ becomes a Lie algebra over $\mathbb{R}$ with the bracket operation given in the previous paragraph, and $\mathfrak{g}$ is called the **Lie algebra** of G.

A closed subgroup G of nonsingular real or complex matrices will be called a **closed linear group**. As we recall in Proposition 1.75 below, such a group has canonically the structure of a Lie group. Example 6 in §1 mentioned how the Lie algebra of a closed linear group may be regarded as a Lie algebra of matrices. Let us carry out the details of this identification, referring to the Lie algebra of matrices eventually as the "linear Lie algebra" of G. For a closed linear group G, we define

$$\mathfrak{g} = \left\{ c'(0) \;\middle|\; \begin{array}{l} c : \mathbb{R} \to G \text{ is a curve with } c(0) = 1 \text{ that is} \\ \text{smooth as a function into matrices} \end{array} \right\}. \tag{1.71}$$

Use of $t \mapsto c(kt)$ shows that $\mathfrak{g}$ is closed under multiplication by the real number k, and use of $t \mapsto c(t)b(t)$ shows that $\mathfrak{g}$ is closed under addition. Use of the curve $t \mapsto gc(t)g^{-1}$, for $g \in G$, then shows that $\mathfrak{g}$ is closed under the operation $\text{Ad}(g) : \mathfrak{g} \to \mathfrak{g}$ given by

$$\text{Ad}(g)X = gXg^{-1}. \tag{1.72}$$

To see that $\mathfrak{g}$ is closed under the bracket operation on matrices, we combine three facts:

(i) $\mathfrak{g}$ contains $\text{Ad}(c(t))X$ if $c(t)$ is a curve as in (1.71) and X is in $\mathfrak{g}$
(ii) $\mathfrak{g}$ is topologically closed (being a vector subspace)
(iii) $\frac{d}{dt}\text{Ad}(c(t))X = c'(t)Xc(t)^{-1} - c(t)Xc(t)^{-1}c'(t)c(t)^{-1}$.

The first two facts are clear, and the third follows from (1.72) and the formula $\frac{d}{dt}c(t)^{-1} = -c(t)^{-1}c'(t)c(t)^{-1}$, which in turn follows by applying the product rule for differentiation to the identity $c(t)c(t)^{-1} = 1$. Now let us combine (i), (ii), and (iii). By (i) and (ii), $\frac{d}{dt}\text{Ad}(c(t))X$ is in $\mathfrak{g}$ for all t. Putting $t = 0$ in (iii), we see that $c'(0)X - Xc'(0)$ is in $\mathfrak{g}$. Consequently $\mathfrak{g}$ is closed under the bracket operation $[X, Y] = XY - YX$ and is a Lie algebra of matrices. We call $\mathfrak{g}$ the **linear Lie algebra** of G.

Still with G as a closed linear group, we work toward seeing that G is a Lie group and exhibiting an isomorphism of the linear Lie algebra of G and the Lie algebra of G. We use the exponential mapping for matrices, defined by

$$e^X = \sum_{n=0}^{\infty} \frac{1}{n!} X^n. \tag{1.73}$$

(See Chevalley [1946], pp. 5–9.) Part of the relevance of the matrix exponential mapping for matrices is that it provides specific curves in G of the kind in (1.71). According to the following proposition, e^{tX} is such a curve if X is in the linear Lie algebra.

Proposition 1.74. If G is a closed linear group and $\mathfrak{g}$ is its linear Lie algebra, then the matrix exponential function $e^{(\cdot)}$ carries $\mathfrak{g}$ into G. Consequently

$$\mathfrak{g} = \{X \in \mathfrak{gl}(n, \mathbb{C}) \mid e^{tX} \text{ is in } G \text{ for all real } t\}.$$

REFERENCE FOR PROOF: Knapp [1988], pp. 12–15.

Using this result and the Inverse Function Theorem, one constructs charts for the group G from the matrix exponential function. The result is as follows.

Proposition 1.75. If G is a closed linear group, then G (with its relative topology) becomes a Lie group in a unique way such that

(a) the restrictions from $GL(n, \mathbb{C})$ to G of the real and imaginary parts of each entry function are smooth and
(b) whenever $\Phi : M \to GL(n, \mathbb{C})$ is a smooth function on a smooth manifold M such that $\Phi(M) \subseteq G$, then $\Phi : M \to G$ is smooth.

Moreover, the dimension of the linear Lie algebra $\mathfrak{g}$ equals the dimension of the manifold G. And, in addition, there exists open neighborhoods U of 0 in $\mathfrak{g}$ and V of 1 in G such that $e^{(\cdot)} : U \to V$ is a homeomorphism onto and such that $(V, (e^{(\cdot)})^{-1})$ is a compatible chart.

REFERENCE FOR PROOF: Knapp [1988], pp. 20–25.

Finally we obtain the result that demonstrates the assertions in Example 6 of §1.

Proposition 1.76. Let G be a closed linear group of n-by-n matrices, regard the Lie algebra $\mathfrak{g}_1$ of the Lie group G as consisting of all left-invariant vector fields on G, and let $\mathfrak{g}_2$ be the linear Lie algebra of the matrix group G. Then the map $\mu : \mathfrak{g}_1 \to \mathfrak{gl}(n, \mathbb{C})$ given by

$$\mu(X)_{ij} = X_1(\operatorname{Re} e_{ij}) + iX_1(\operatorname{Im} e_{ij}) \qquad \text{with } e_{ij}(x) = x_{ij}$$

is a Lie algebra isomorphism of $\mathfrak{g}_1$ onto $\mathfrak{g}_2$.

REMARKS.

1) In this proof and later arguments it will be convenient to extend the definition of $X \in \mathfrak{g}_1$ from real-valued functions to complex-valued functions, using the rule $Xf = X(\operatorname{Re} f) + iX(\operatorname{Im} f)$. Then X still satisfies the product rule for differentiation.

2) The proposition makes a rigid distinction between the Lie algebra $\mathfrak{g}_1$ and the linear Lie algebra $\mathfrak{g}_2$, and we shall continue this distinction throughout this section. In practice, however, one makes the distinction only when clarity demands it, and we shall follow this more relaxed convention after the end of this section.

PROOF. To prove that μ is a Lie algebra homomorphism into matrices, we argue as follows. Let X be in $\mathfrak{g}_1$. We have

$$e_{ij} \circ L_x(y) = e_{ij}(xy) = \sum_k e_{ik}(x) e_{kj}(y).$$

Application of X gives

$$Xe_{ij}(x) = X_1(e_{ij} \circ L_x) = \sum_k e_{ik}(x) X_1 e_{kj} = \sum_k e_{ik}(x) \mu(X)_{kj}. \tag{1.77}$$

If also Y is in $\mathfrak{g}_1$, then

$$\begin{aligned} YXe_{ij}(x) &= Y_1((Xe_{ij}) \circ L_x) \\ &= Y_1\Big(\sum_{k,l} e_{il}(x) e_{lk}(y) \mu(X)_{kj}\Big) \quad \text{with } Y_1 \text{ acting in the } y \text{ variable} \\ &= \sum_{k,l} e_{il}(x) \mu(Y)_{lk} \mu(X)_{kj}. \end{aligned}$$

We reverse the roles of X and Y, evaluate at $x = 1$, and subtract. With δ_{ij} denoting the Kronecker delta, the result is that

$$\begin{aligned} \mu([X, Y])_{ij} &= ([X, Y]e_{ij})(1) \\ &= XYe_{ij}(1) - YXe_{ij}(1) \\ &= \sum_{k,l} \delta_{il}(\mu(X)_{lk}\mu(Y)_{kj} - \mu(Y)_{lk}\mu(X)_{kj}) \\ &= (\mu(X)\mu(Y) - \mu(Y)\mu(X))_{ij} \\ &= [\mu(X), \mu(Y)]_{ij}. \end{aligned}$$

Thus μ is a Lie algebra homomorphism into matrices.

Next we prove that

$$\operatorname{image} \mu \supseteq \mathfrak{g}_2. \tag{1.78}$$

Let A be in $\mathfrak{g}_2$, and choose a curve $c(t)$ as in (1.71) with $c'(0) = A$. Put

$$Xf(x) = \frac{d}{dt} f(xc(t))_{t=0}.$$

Then X is a left-invariant vector field on G, and

$$\begin{aligned}\mu(X)_{ij} = X_1 e_{ij} = X e_{ij}(1) &= \frac{d}{dt} e_{ij}(c(t))|_{t=0} \\ &= \frac{d}{dt} c(t)_{ij}|_{t=0} = c'(0)_{ij} = A_{ij}.\end{aligned}$$

This proves (1.78).

Finally we have $\dim G = \dim \mathfrak{g}_2$ by Proposition 1.75. Therefore (1.78) gives

$$\dim \mathfrak{g}_1 = \dim G = \dim \mathfrak{g}_2 \leq \dim(\text{image}\, \mu) \leq \dim(\text{domain}\, \mu) = \dim \mathfrak{g}_1,$$

and equality must hold throughout. Consequently μ is one-one, and its image is exactly $\mathfrak{g}_2$. This completes the proof.

Note that the proof shows what μ^{-1} is. If a matrix A is given, then $\mu^{-1}(A) = X$, where X is defined in terms of any curve (1.71) with $c'(0) = A$ by $Xf(x) = \frac{d}{dt} f(xc(t))|_{t=0}$. It is a consequence of the proof that the value of X does not depend on the particular choice of the curve c.

Let us return to general Lie groups.

An **analytic subgroup** H of a Lie group G is a subgroup with the structure of an analytic group such that the inclusion mapping is smooth and everywhere regular. If $\mathfrak{h}$ and $\mathfrak{g}$ denote the Lie algebras of H and G, then the differential of the inclusion at 1 carries $\mathfrak{h} = T_1(H)$ to a subspace $\tilde{T}_1(H)$ of $\mathfrak{g}$ and is a one-one Lie algebra homomorphism. Thus $\tilde{T}_1(H)$ is a Lie subalgebra of $\mathfrak{g}$, and $\mathfrak{h}$ can be identified with this subalgebra. This identification is normally made without specific comment.

The correspondence $H \to \tilde{T}_1(H) \subseteq \mathfrak{g}$ of analytic subgroups of G to Lie subalgebras of $\mathfrak{g}$, given as in the previous paragraph, is one-one onto. This fact is one of the cornerstones of the elementary theory of Lie groups. If M is a smooth manifold and $\Phi : M \to G$ is a smooth function such that $\Phi(M) \subseteq H$ for an analytic subgroup H, then $\Phi : M \to H$ is smooth.

Let $\Phi : G \to H$ be a smooth homomorphism between Lie groups, and let $d\Phi_x : \mathfrak{g} \to \mathfrak{h}$ be the differential at $x \in G$. Then $d\Phi$ has the following property: If X is a left-invariant vector field on G and if Y is the left-invariant vector field on H such that $(d\Phi)_1(X_1) = Y_1$, then

$$(d\Phi)_x(X_x) = Y_{\Phi(x)} \qquad \text{for all } x \in G. \tag{1.79}$$

It follows that $d\Phi_1$ is a Lie algebra homomorphism. Assume that G is connected. Then $d\Phi_1$ uniquely determines Φ, the image of Φ is an analytic subgroup H' of H, and $\Phi : G \to H'$ is smooth. When there is no possibility of confusion, we shall write $d\Phi$ in place of $d\Phi_1$.

In the case that G and H are closed linear groups, the Lie algebra homomorphism $d\Phi_1$, regarded as a homomorphism of the linear Lie algebras, may be computed in terms of derivatives of curves. The details are in the following lemma and proposition.

Lemma 1.80. Let G be a closed linear group, let $c(t)$ be a curve in G as in (1.71), and let μ be the isomorphism of Proposition 1.76. Then

$$\mu\big(dc_0(\tfrac{d}{dt})\big) = c'(0).$$

PROOF. The lemma follows from the computation

$$\mu\big(dc_0(\tfrac{d}{dt})\big)_{ij} = dc_0\big(\tfrac{d}{dt}\big)(e_{ij}) = \tfrac{d}{dt}\, e_{ij}(c(t))|_{t=0} = c'(0)_{ij}.$$

Proposition 1.81. Let $\Phi : G \to H$ be a smooth homomorphism between closed linear groups, and let μ_G and μ_H be the corresponding Lie algebra isomorphisms of Proposition 1.76. Let X be in the Lie algebra of G, and put $Y = (d\Phi_1)(X)$. If $c(t)$ is a curve in G as in (1.71) such that $\mu_G(X) = c'(0)$, then $\mu_H(Y) = \frac{d}{dt}\,\Phi(c(t))|_{t=0}$.

PROOF. By Lemma 1.80, $X = \mu_G^{-1}(c'(0))$ is given by $X = dc_0\big(\frac{d}{dt}\big)$. Then

$$\mu_H(Y) = \mu_H((d\Phi_1)(X)) = \mu_H\big((d\Phi_1)(dc_0)(\tfrac{d}{dt})\big) = \mu_H\big(d(\Phi\circ c)_0(\tfrac{d}{dt})\big),$$

and another application of Lemma 1.80 identifies the right side as $\frac{d}{dt}\,\Phi(c(t))|_{t=0}$.

Let G be an analytic group, let $\tilde{G}$ be the universal covering space with covering map $e : \tilde{G} \to G$, and let $\tilde{1}$ be in $e^{-1}(1)$. Then there exists a unique multiplication on $\tilde{G}$ that makes $\tilde{G}$ into a topological group in such a way that e is a group homomorphism and $\tilde{G}$ has $\tilde{1}$ as identity. Furthermore there exists a unique smooth manifold structure on $\tilde{G}$ of dimension equal to $\dim G$ in such a way that e is smooth and everywhere regular. The topological group structure and smooth manifold structure on $\tilde{G}$ are compatible, and $\tilde{G}$ becomes an analytic group. The group $\tilde{G}$ is called the **universal covering group** or **simply connected covering group** of G. The covering map e is a smooth homomorphism and exhibits the Lie algebras of $\tilde{G}$ and G as isomorphic.

If G and H are analytic groups with G simply connected and if $\varphi : \mathfrak{g} \to \mathfrak{h}$ is a homomorphism between their Lie algebras, then there exists a smooth homomorphism $\Phi : G \to H$ with $d\Phi_1 = \varphi$. Consequently any two simply connected analytic groups with isomorphic Lie algebras are isomorphic.

From the lifting of homomorphisms from Lie algebras to Lie groups, we can define the exponential mapping for a general analytic group. Let $\mathbb{R}$ denote the simply connected Lie group of additive reals with 1-dimensional abelian Lie algebra $\mathfrak{r}$ generated by $(\frac{d}{dt})_0$, and let G be an analytic group with Lie algebra $\mathfrak{g}$. If X is given in $\mathfrak{g}$, we can define a Lie algebra homomorphism of $\mathfrak{r}$ into $\mathfrak{g}$ by requiring that $(\frac{d}{dt})_0$ map to X. The corresponding smooth homomorphism $\mathbb{R} \to G$ is written $t \mapsto \exp tX$.

Write $c(t) = \exp tX$. Let $\frac{d}{dt}$ and $\tilde{X}$ be the left-invariant vector fields on $\mathbb{R}$ and G, respectively, that extend $(\frac{d}{dt})_0$ and X. According to (1.79), we have

$$(dc)_t\left(\tfrac{d}{dt}\right) = \tilde{X}_{c(t)}. \tag{1.82}$$

Also $c(0) = 1$. Thus (1.82) says that $c(t) = \exp tX$ is the **integral curve** for $\tilde{X}$ with $c(0) = 1$. On a function f, the left side of (1.82) is

$$= (dc)_t\left(\tfrac{d}{dt}\right)f = \tfrac{d}{dt}f(c(t)) = \tfrac{d}{dt}\, f(\exp tX).$$

Therefore we obtain the important formula

$$\tilde{X}f(\exp tX) = \frac{d}{dt}\, f(\exp tX). \tag{1.83}$$

From the system of differential equations satisfied by an integral curve, one sees that the map of the Lie algebra $\mathfrak{g}$ into G given by $X \mapsto \exp X$ is smooth. This is the **exponential map** for G. If $\Phi : G \to H$ is a smooth homomorphism, then Φ and the differential $d\Phi_1$ and the exponential map are connected by the formula

$$\exp_H \circ\, d\Phi_1 = \Phi \circ \exp_G\,. \tag{1.84}$$

The exponential map is locally invertible about $0 \mapsto 1$ in the sense that if $X_1, \dots, X_n$ is a basis of the Lie algebra $\mathfrak{g}$, then

$$(x_1, \dots, x_n) \mapsto g\, \exp(x_1X_1 + \cdots + x_nX_n) \tag{1.85}$$

carries a sufficiently small ball about 0 in $\mathbb{R}^n$ diffeomorphically onto an open neighborhood of g in G. The inverse is therefore a compatible chart about g and defines **canonical coordinates of the first kind** about the element g of G.

Let G be an analytic group with Lie algebra $\mathfrak{g}$, and suppose that $\mathfrak{g}$ is a direct sum of vector subspaces

$$\mathfrak{g} = \mathfrak{g}_1 \oplus \cdots \oplus \mathfrak{g}_k.$$

If U_j is a sufficiently small open neighborhood of 0 in $\mathfrak{g}_j$ for $1 \leq j \leq k$, then the map

$$(X_1, \ldots, X_k) \mapsto g(\exp X_1) \cdots (\exp X_k)$$

is a diffeomorphism of $U_1 \times \cdots \times U_k$ onto an open neighborhood of g in G. When the $\mathfrak{g}_j$'s are all 1-dimensional, the local coordinates given by the inverse map are called **canonical coordinates of the second kind** about g.

Proposition 1.86. Let G be a closed linear group, and let $\mathfrak{g}$ be its linear Lie algebra. If the exponential map is regarded as carrying $\mathfrak{g}$ to G, then it is given by the matrix exponential function (1.73).

PROOF. First consider $G = GL(n, \mathbb{C})$. Let X be in the Lie algebra, let $\mu(X)$ be the corresponding member of $\mathfrak{g}$, and let $\tilde{X}$ be the associated left-invariant vector field on G. We apply (1.83) to $f = e_{ij}$ and combine with (1.78) to obtain

$$\frac{d}{dt}(\exp tX)_{ij} = \tilde{X}e_{ij}(\exp tX) = \sum_k e_{ik}(\exp tX)\mu(X)_{kj}.$$

In other words, $c(t) = \exp tX$ satisfies

$$c'(t) = c(t)\mu(X) \qquad \text{with } c(0) = 1,$$

and it follows from the theory of linear systems of ordinary differential equations that $c(t) = e^{t\mu(X)}$. This completes the proof for $GL(n, \mathbb{C})$. We obtain the result for general G by applying (1.84) to the inclusion $\Phi : G \to GL(n, \mathbb{C})$ and the using the result for $GL(n, \mathbb{C})$.

Corollary 1.87. If G is an analytic group and $\Phi : G \to GL(n, \mathbb{C})$ is a smooth homomorphism, then $\Phi \circ \exp_G$ can be computed as $e^{d\Phi}$ if the Lie algebra of $GL(n, \mathbb{C})$ is identified with $\mathfrak{gl}(n, \mathbb{C})$.

PROOF. This follows by combining (1.84) and Proposition 1.86.

Let G be a Lie group, and let H be a closed subgroup. Then there exists a unique smooth manifold structure on H such that H, in its relative topology, is a Lie group and the identity component of H is an analytic subgroup of G. This result generalizes Proposition 1.75.

Let G be a Lie group of dimension n, let H be a closed subgroup of dimension s and let $\pi : G \to G/H$ be the quotient map. Then there exists a chart (U, φ) around 1 in G, say $\varphi = (x_1, \ldots, x_n)$, such that

(a) $\varphi(U) = \{(\xi_1, \ldots, \xi_n) \mid |\xi_j| < 2\varepsilon \text{ for all } j\}$ for some $\varepsilon > 0$
(b) each slice with $x_{s+1} = \xi_{s+1}, \ldots, x_n = \xi_n$ is a relatively open set in some coset gH and these cosets are all distinct
(c) the restriction of π to the slice $x_1 = 0, \ldots, x_s = 0$ is a homeomorphism onto an open set and therefore determines a chart about the identity coset in G/H.

If the translates in G/H of the chart in (c) are used as charts to cover G/H, then G/H becomes a smooth manifold such that π and the action of G are smooth. Moreover, any smooth map $\sigma : G \to M$ that factors continuously through G/H as $\sigma = \bar{\sigma} \circ \pi$ is such that $\bar{\sigma}$ is smooth.

Let G be a Lie group of dimension n, let H be a closed normal subgroup of dimension s, and let $\mathfrak{g}$ and $\mathfrak{h}$ be the respective Lie algebras. Then $\mathfrak{h}$ is an ideal in $\mathfrak{g}$, the manifold structure on G/H makes G/H a Lie group, the quotient map of G to G/H is a smooth homomorphism, and the differential of $G \to G/H$ at the identity may be regarded as the quotient map $\mathfrak{g} \to \mathfrak{g}/\mathfrak{h}$.

If G is any analytic group, then $G \cong \tilde{G}/H$, where $\tilde{G}$ is the universal covering group of G and H is a discrete subgroup of the center of $\tilde{G}$. Conversely if $\mathfrak{g}$ denotes the Lie algebra of G, then the most general analytic group with Lie algebra isomorphic to $\mathfrak{g}$ is isomorphic to $\tilde{G}/H'$ for some central discrete subgroup H' of $\tilde{G}$.

If G is an analytic group and H is an analytic subgroup and if $\mathfrak{g}$ and $\mathfrak{h}$ are their Lie algebras, then $H \subseteq \text{center}(G)$ if and only if $\mathfrak{h} \subseteq \text{center}(\mathfrak{g})$. An analytic group is abelian if and only if its Lie algebra is abelian. The most general abelian analytic group is of the form $\mathbb{R}^l \times T^k$, where T^k denotes the k-dimensional **torus** group (the product of k circle groups). In particular the most general compact abelian analytic group is a torus. If G is an analytic group and X and Y are in its Lie algebra, then $[X, Y] = 0$ if and only if $\exp sX$ and $\exp tY$ commute for all real s and t.

Any continuous homomorphism between Lie groups G and G' is automatically smooth. Once this result is known for the special case $G = \mathbb{R}$, the general case follows by using canonical coordinates of the second kind.

A corollary is that if two analytic groups have the same underlying topological group, then they coincide as analytic groups.

Shortly we shall develop the "adjoint representation" of a Lie group on its Lie algebra. In the development we shall need to use the following version of Taylor's Theorem.

Proposition 1.88 (Taylor's Theorem). Let G be a Lie group with Lie algebra $\mathfrak{g}$. If X is in $\mathfrak{g}$, if $\tilde{X}$ denotes the corresponding left-invariant vector field, and if f is a C^∞ function on G, then

$$(\tilde{X}^n f)(g \exp tX) = \frac{d^n}{dt^n}(f(g \exp tX)) \qquad \text{for } g \text{ in } G.$$

Moreover, if $|\cdot|$ denotes any norm on $\mathfrak{g}$ and if X is restricted to a bounded set in $\mathfrak{g}$, then

$$f(\exp X) = \sum_{k=0}^{n} \frac{1}{k!}(\tilde{X}^k f)(1) + R_n(X),$$

where $|R_n(X)| \le C_n |X|^{n+1}$.

REMARK. This proposition uses only the C^∞ manifold structure on G. In terms of the real analytic manifold structure on G, one can prove for real analytic functions f on G that

$$f(\exp X) = \sum_{k=0}^{\infty} \frac{1}{k!}(\tilde{X}^k f)(1)$$

for all X in a suitably small neighborhood of 0 in $\mathfrak{g}$. This variant of the result is needed in some applications to infinite-dimensional representation theory but will not play a role in this book.

PROOF. The first conclusion at $g = 1$ follows by iterating (1.83). Replacing $f(x)$ by $f_g(x) = f(gx)$ and using left invariance, we get the first conclusion for general g. For the second statement we expand $t \mapsto f(\exp tX)$ in Taylor series about $t = 0$ and evaluate at $t = 1$. Then

$$\begin{aligned} f(\exp X) &= \sum_{k=0}^{n} \frac{1}{k!} \left(\frac{d}{dt}\right)^k (f(\exp tX))|_{t=0} \\ &\quad + \frac{1}{n!} \int_0^1 (1-s)^n \left(\frac{d}{ds}\right)^{n+1} (f(\exp sX))\, ds \\ &= \sum_{k=0}^{n} \frac{1}{k!}(\tilde{X}^k f)(1) + \frac{1}{n!} \int_0^1 (1-s)^n (\tilde{X}^{n+1} f)(\exp sX)\, ds. \end{aligned}$$

In the second term on the right side, write $X = \sum_j \lambda_j X_j$ and expand $\tilde{X}^{n+1}$. Since X is restricted to lie in a compact set, so is $\exp sX$, and the integral is dominated by $|\lambda|^{n+1}$ times a harmless integral. This proves the estimate for the remainder.

Corollary 1.89. Let G be a Lie group with Lie algebra $\mathfrak{g}$. If X is in $\mathfrak{g}$, if $\tilde{X}$ denotes the corresponding left-invariant vector field, and if f is a C^∞ function on G, then

$$\tilde{X} f(g) = \frac{d}{dt} f(g \exp tX)|_{t=0}.$$

PROOF. Put $t = 0$ in the proposition.

The "adjoint representation" of a Lie group on its Lie algebra is defined as follows. Let G be a Lie group with Lie algebra $\mathfrak{g}$. Fix an element $g \in G$, and consider the smooth isomorphism $\Phi(x) = gxg^{-1}$ of G into itself. The corresponding isomorphism $d\Phi_1 : \mathfrak{g} \to \mathfrak{g}$ is denoted $\mathrm{Ad}(g)$. By (1.84), we have

$$\exp(\mathrm{Ad}(g)X) = g(\exp X)g^{-1}. \tag{1.90}$$

In the special case that G is a closed linear group, we can regard X and g in (1.90) as matrices, and we can use Proposition 1.86 to think of exp as the matrix exponential function. Let us replace X by tX, differentiate, and set $t = 0$. Then we see that $\mathrm{Ad}(g)X$, regarded as a member of the linear Lie algebra $\mathfrak{g}$, is given by gXg^{-1}.

Returning to the general case, let us combine (1.90) with the fact that exp has a smooth inverse in a neighborhood of the identity in G. Then we see that $\mathrm{Ad}(g)X$ is smooth as a function from a neighborhood of 1 in G to $\mathfrak{g}$ if X is small. That is, $g \mapsto \mathrm{Ad}(g)$ is smooth from a neighborhood of 1 in G into $GL(\mathfrak{g})$. But also it is clear that $\mathrm{Ad}(g_1g_2) = \mathrm{Ad}(g_1)\mathrm{Ad}(g_2)$. Thus the smoothness is valid everywhere on G, and we arrive at the following result.

Proposition 1.91. If G is a Lie group and $\mathfrak{g}$ is its Lie algebra, then Ad is a smooth homomorphism from G into $GL(\mathfrak{g})$.

We call Ad the **adjoint representation** of G on $\mathfrak{g}$. When we want to emphasize the space on which $\mathrm{Ad}(x)$ operates, we write $\mathrm{Ad}_{\mathfrak{g}}(x)$ for the linear transformation.

We shall now compute the differential of Ad.

Lemma 1.92. Let G be a Lie group with Lie algebra $\mathfrak{g}$. If X and Y are in $\mathfrak{g}$, then

(a) $\exp tX \exp tY = \exp\{t(X+Y) + \frac{1}{2}t^2[X,Y] + O(t^3)\}$
(b) $\exp tX \exp tY (\exp tX)^{-1} = \exp\{tY + t^2[X,Y] + O(t^3)\}$

as $t \to 0$. Here $O(t^3)$ denotes a smooth function from an interval of t's about $t = 0$ into $\mathfrak{g}$ such that the quotient by t^3 remains bounded as $t \to 0$.

PROOF. For (i) we use the local invertibility of exp near the identity to write $\exp tX \exp tY = \exp Z(t)$ for t near 0, where $Z(t)$ is smooth in t. Since $Z(0) = 0$, we have

$$Z(t) = tZ_1 + t^2 Z_2 + O(t^3),$$

and we are to identify Z_1 and Z_2. Let $\tilde{Z}_1$ and $\tilde{Z}_2$ be the corresponding left-invariant vector fields. If f is a smooth function near the identity of G, Taylor's Theorem (Proposition 1.88) gives

$$\begin{aligned} f(\exp Z(t)) &= \sum_{k=0}^{2} \frac{1}{k!}(t\tilde{Z}_1 + t^2\tilde{Z}_2 + O(t^3))^k f(1) + O(t^3) \\ &= f(1) + t(\tilde{Z}_1 f)(1) + t^2(\tfrac{1}{2}\tilde{Z}_1^2 + \tilde{Z}_2) f(1) + O(t^3). \end{aligned}$$

On the other hand, another application of Taylor's Theorem gives

$$\begin{aligned} f(\exp tX \exp sY) &= \sum_{k=0}^{2} \frac{1}{k!} s^k \tilde{Y}^k f(\exp tX) + O_t(s^3) \\ &= \sum_{k=0}^{2}\sum_{l=0}^{2} \frac{1}{k!}\frac{1}{l!} s^k t^l \tilde{X}^l \tilde{Y}^k f(1) + O_t(s^3) + O(t^3), \end{aligned}$$

where $O_t(s^3)$ denotes an expression $O(s^3)$ depending on t and having the property that the bound on $O(s^3)/s^3$ may be taken independent of t for t small. Setting $t = s$, we obtain

$$\begin{aligned} f(\exp Z(t)) &= f(\exp tX \exp tY) \\ &= f(1) + t(\tilde{X} + \tilde{Y}) f(1) + t^2(\tfrac{1}{2}\tilde{X}^2 + \tilde{X}\tilde{Y} + \tfrac{1}{2}\tilde{Y}^2) f(1) + O(t^3). \end{aligned}$$

Replacing f by the translate f_g with $f_g(x) = f(gx)$, we are led to the equalities of operators $\tilde{Z}_1 = \tilde{X} + \tilde{Y}$ and $\frac{1}{2}\tilde{Z}_1^2 + \tilde{Z}_2 = \frac{1}{2}\tilde{X}^2 + \tilde{X}\tilde{Y} + \frac{1}{2}\tilde{Y}^2$. Therefore $Z_1 = X + Y$ and $Z_2 = \frac{1}{2}[X, Y]$.

To prove (ii), we apply (i) twice, and the result follows.

Proposition 1.93. Let G be a Lie group with Lie algebra $\mathfrak{g}$. The differential of $\mathrm{Ad} : G \to GL(\mathfrak{g})$ is $\mathrm{ad} : \mathfrak{g} \to \mathrm{End}\,\mathfrak{g}$, where $\mathrm{ad}(X)Y = [X, Y]$ and where the Lie algebra of $GL(\mathfrak{g})$ has been identified with the linear Lie algebra $\mathrm{End}\,\mathfrak{g}$. Consequently

$$\mathrm{Ad}(\exp X) = e^{\mathrm{ad}\, X} \tag{1.94}$$

under this identification.

PROOF. Let $L : \mathfrak{g} \to \operatorname{End}(\mathfrak{g})$ be the differential of Ad. Fix X and Y in $\mathfrak{g}$. Applying Lemma 1.92b and using (1.90), we obtain

$$\operatorname{Ad}(\exp tX)tY = tY + t^2[X, Y] + O(t^3).$$

Division by t gives

$$\operatorname{Ad}(\exp tX)Y = Y + t[X, Y] + O(t^2).$$

Differentiating and putting $t = 0$, we see that

$$L(X)Y = [X, Y].$$

Therefore $L = \operatorname{ad}$ as asserted. Formula (1.94) then becomes a special case of Corollary 1.87.

We conclude this section with some remarks about complex Lie groups. A **complex Lie group** is a Lie group G possessing a complex analytic structure such that multiplication and inversion are holomorphic. For such a group the complex structure induces a multiplication-by-i mapping in the Lie algebra $\mathfrak{g} = T_1(G)$ such that $\mathfrak{g}$ becomes a Lie algebra over $\mathbb{C}$. Every left-invariant vector field has holomorphic coefficients, and exp is a holomorphic mapping. If Φ is a smooth homomorphism between complex Lie groups whose differential at 1 is complex linear, then Φ is holomorphic as a consequence of (1.84).

Within a complex Lie group G with Lie algebra $\mathfrak{g}$, suppose that H is an analytic subgroup whose Lie algebra is closed under the multiplication-by-i mapping for $\mathfrak{g}$. Then canonical coordinates of the first kind define charts on H that make H into a complex manifold, and multiplication and inversion are holomorphic. This complex structure for H is uniquely determined by the conditions that

(a) the inclusion $H \to G$ is holomorphic and
(b) whenever $\Phi : M \to G$ is a holomorphic function on a complex manifold M such that $\Phi(M) \subseteq H$, then $\Phi : M \to H$ is holomorphic.

Proofs of the existence and properties of the complex structure on H can be given in the style of the argument of Knapp [1988], 24.

11. Automorphisms and Derivations

In this section, $\mathfrak{g}$ denotes a finite-dimensional Lie algebra over $\mathbb{R}$ or $\mathbb{C}$.

First we define automorphisms. An **automorphism** of a Lie algebra is an invertible linear map L that preserves brackets: $[L(X), L(Y)] = [X, Y]$. For example if $\mathfrak{g}$ is the (real) Lie algebra of a Lie group G and if g is in G, then $\mathrm{Ad}(g)$ is an automorphism of $\mathfrak{g}$.

If $\mathfrak{g}$ is real, let $\mathrm{Aut}_{\mathbb{R}}\,\mathfrak{g} \subseteq GL_{\mathbb{R}}(\mathfrak{g})$ be the subgroup of $\mathbb{R}$ linear automorphisms of $\mathfrak{g}$. This is a closed subgroup of a general linear group, hence a Lie group. If $\mathfrak{g}$ is complex, we can regard

$$\mathrm{Aut}_{\mathbb{C}}\,\mathfrak{g} \subseteq GL_{\mathbb{C}}(\mathfrak{g}) \subseteq GL_{\mathbb{R}}(\mathfrak{g}^{\mathbb{R}}),$$

the subscript $\mathbb{C}$ referring to complex-linearity and $\mathfrak{g}^{\mathbb{R}}$ denoting the underlying real Lie algebra of $\mathfrak{g}$ as in §3. But also we have the option of regarding $\mathfrak{g}$ as the real Lie algebra $\mathfrak{g}^{\mathbb{R}}$ directly. Then we have

$$\mathrm{Aut}_{\mathbb{C}}\,\mathfrak{g} \subseteq \mathrm{Aut}_{\mathbb{R}}\,\mathfrak{g}^{\mathbb{R}} \subseteq GL_{\mathbb{R}}(\mathfrak{g}^{\mathbb{R}}).$$

Lemma 1.95. If a is an automorphism of $\mathfrak{g}$ and if X is in $\mathfrak{g}$, then $\mathrm{ad}(aX) = a(\mathrm{ad}\,X)a^{-1}$.

Proof. We have $\mathrm{ad}(aX)Y = [aX, Y] = a[X, a^{-1}Y] = (a(\mathrm{ad}\,X)a^{-1})Y$.

Proposition 1.96. If B is the Killing form of $\mathfrak{g}$ and if a is an automorphism of $\mathfrak{g}$, then $B(aX, aY) = B(X, Y)$ for all X and Y in $\mathfrak{g}$.

Proof. By Lemma 1.95 we have

$$\begin{aligned} B(aX, aY) &= \mathrm{Tr}(\mathrm{ad}(aX)\mathrm{ad}(aY)) \\ &= \mathrm{Tr}(a(\mathrm{ad}\,X)a^{-1}a(\mathrm{ad}\,Y)a^{-1}) \\ &= \mathrm{Tr}((\mathrm{ad}\,X)(\mathrm{ad}\,Y)) \\ &= B(X, Y), \end{aligned}$$

as required.

Next we recall that derivations of the Lie algebra $\mathfrak{g}$ were defined in (1.2). In §4 we introduced $\mathrm{Der}\,\mathfrak{g}$ as the Lie algebra of all derivations of $\mathfrak{g}$. If $\mathfrak{g}$ is real, then $\mathrm{Der}\,\mathfrak{g}$ has just one interpretation, namely the Lie subalgebra $\mathrm{Der}_{\mathbb{R}}\,\mathfrak{g} \subseteq \mathrm{End}_{\mathbb{R}}\,\mathfrak{g}$. If $\mathfrak{g}$ is complex, then two interpretations are possible, namely as $\mathrm{Der}_{\mathbb{R}}\,\mathfrak{g}^{\mathbb{R}} \subseteq \mathrm{End}_{\mathbb{R}}\,(\mathfrak{g}^{\mathbb{R}})$ or as $\mathrm{Der}_{\mathbb{C}}\,\mathfrak{g} \subseteq \mathrm{End}_{\mathbb{C}}(\mathfrak{g}) \subseteq \mathrm{End}_{\mathbb{R}}(\mathfrak{g}^{\mathbb{R}})$.

Proposition 1.97. If $\mathfrak{g}$ is real, the Lie algebra of $\mathrm{Aut}_{\mathbb{R}}\,\mathfrak{g}$ is $\mathrm{Der}_{\mathbb{R}}\,\mathfrak{g}$. If $\mathfrak{g}$ is complex, the Lie algebra of $\mathrm{Aut}_{\mathbb{C}}\,\mathfrak{g}$ is $\mathrm{Der}_{\mathbb{C}}\,\mathfrak{g}$. In either case the Lie algebra contains $\mathrm{ad}\,\mathfrak{g}$.

PROOF. First let $\mathfrak{g}$ be real. If $c(t)$ is a curve of automorphisms from 1 with $c'(0) = l$, then $c(t)[X, Y] = [c(t)X, c(t)Y]$ implies $l[X, Y] = [l(X), Y] + [X, l(Y)]$. Hence the Lie algebra in question is a Lie subalgebra of $\mathrm{Der}_{\mathbb{R}}(\mathfrak{g})$. For the reverse direction, we show that $l \in \mathrm{Der}_{\mathbb{R}}(\mathfrak{g})$ implies that e^{tl} is in $\mathrm{Aut}_{\mathbb{R}}\,\mathfrak{g}$, so that $\mathrm{Der}_{\mathbb{R}}\,\mathfrak{g}$ is a Lie subalgebra of the Lie algebra in question. Thus consider

$$y_1(t) = e^{tl}[X, Y] \qquad \text{and} \qquad y_2(t) = [e^{tl}X, e^{tl}Y]$$

as two curves in the real vector space $\mathfrak{g}$ with value $[X, Y]$ at $t = 0$. For any t we have

$$y_1'(t) = le^{tl}[X, Y] = ly_1(t)$$

and

$$\begin{aligned} y_2'(t) &= [le^{tl}X, e^{tl}Y] + [e^{tl}X, le^{tl}Y] \\ &= l[e^{tl}X, e^{tl}Y] \qquad\qquad \text{by the derivation property} \\ &= ly_2(t). \end{aligned}$$

Then $e^{tl}[X, Y] = [e^{tl}X, e^{tl}Y]$ by the uniqueness theorem for linear systems of ordinary differential equations.

If $\mathfrak{g}$ is complex, then the Lie algebra of $\mathrm{Aut}_{\mathbb{C}}\,\mathfrak{g}$ is contained in $\mathrm{Der}_{\mathbb{R}}\,\mathfrak{g}^{\mathbb{R}}$ by the above, and it is contained in $\mathrm{End}_{\mathbb{C}}\,\mathfrak{g}$, which is the Lie algebra of $GL_{\mathbb{C}}(\mathfrak{g})$. Hence the Lie algebra in question is contained in their intersection, which is $\mathrm{Der}_{\mathbb{C}}\,\mathfrak{g}$. In the reverse direction, if l is in $\mathrm{Der}_{\mathbb{C}}\,\mathfrak{g}$, then e^{tl} is contained in $\mathrm{Aut}_{\mathbb{R}}\,\mathfrak{g}^{\mathbb{R}}$ by the above, and it is contained in $GL_{\mathbb{C}}(\mathfrak{g})$ also. Hence it is contained in the intersection, which is $\mathrm{Aut}_{\mathbb{C}}\,\mathfrak{g}$.

Finally $\mathrm{ad}\,\mathfrak{g}$ is a Lie subalgebra of the Lie algebra of derivations, as a consequence of (1.8).

Define $\mathrm{Int}\,\mathfrak{g}$ to be the analytic subgroup of $\mathrm{Aut}_{\mathbb{R}}\,\mathfrak{g}$ with Lie algebra $\mathrm{ad}\,\mathfrak{g}$. If $\mathfrak{g}$ is complex, the definition is unaffected by using $\mathrm{Aut}_{\mathbb{C}}\,\mathfrak{g}$ instead of $\mathrm{Aut}_{\mathbb{R}}\,\mathfrak{g}^{\mathbb{R}}$ as the ambient group, since $\mathrm{ad}\,\mathfrak{g}$ is the same set of transformations as $\mathrm{ad}\,\mathfrak{g}^{\mathbb{R}}$.

The analytic group $\mathrm{Int}\,\mathfrak{g}$ is a universal version of the group of inner automorphisms. To be more precise, let us think of $\mathfrak{g}$ as real. Suppose $\mathfrak{g}$ is the Lie algebra of a Lie group G. As usual, we define $\mathrm{Ad}(g)$ to be the differential at the identity of the inner automorphism $x \mapsto gxg^{-1}$. Then Proposition 1.91 shows that $g \mapsto \mathrm{Ad}(g)$ is a smooth homomorphism of G into $\mathrm{Aut}_{\mathbb{R}}\,\mathfrak{g}$, and we may regard $\mathrm{Ad}(G)$ as a Lie subgroup of $\mathrm{Aut}_{\mathbb{R}}\,\mathfrak{g}$. As such, its Lie algebra is $\mathrm{ad}\,\mathfrak{g}$. By definition the analytic subgroup of $\mathrm{Aut}_{\mathbb{R}}(\mathfrak{g})$ with Lie algebra $\mathrm{ad}\,\mathfrak{g}$ is $\mathrm{Int}\,\mathfrak{g}$. Thus $\mathrm{Int}\,\mathfrak{g}$ is the identity component of $\mathrm{Ad}(G)$ and equals $\mathrm{Ad}(G)$ if G is connected. In this sense $\mathrm{Int}\,\mathfrak{g}$ is a universal version of $\mathrm{Ad}(G)$ that can be defined without reference to a particular group G.

EXAMPLE. If $\mathfrak{g} = \mathbb{R}^2$, then $\operatorname{Aut}_{\mathbb{R}} \mathfrak{g} = GL_{\mathbb{R}}(\mathfrak{g})$ and $\operatorname{Der}_{\mathbb{R}} \mathfrak{g} = \operatorname{End}_{\mathbb{R}} \mathfrak{g}$. Also $\operatorname{ad} \mathfrak{g} = 0$, and so $\operatorname{Int} \mathfrak{g} = \{1\}$. In particular $\operatorname{Int} \mathfrak{g}$ is strictly smaller than the identity component of $\operatorname{Aut}_{\mathbb{R}} \mathfrak{g}$ for this example.

Proposition 1.98. If $\mathfrak{g}$ is semisimple (real or complex), then $\operatorname{Der} \mathfrak{g} = \operatorname{ad} \mathfrak{g}$.

PROOF. Let D be a derivation of $\mathfrak{g}$. By Cartan's Criterion (Theorem 1.42) the Killing form B is nondegenerate. Thus we can find X in $\mathfrak{g}$ with $\operatorname{Tr}(D \operatorname{ad} Y) = B(X, Y)$ for all $Y \in \mathfrak{g}$. The derivation property

$$[DY, Z] = D[Y, Z] - [Y, DZ]$$

can be rewritten as

$$\operatorname{ad}(DY) = [D, \operatorname{ad} Y].$$

Therefore

$$\begin{aligned} B(DY, Z) &= \operatorname{Tr}(\operatorname{ad}(DY)\operatorname{ad} Z) \\ &= \operatorname{Tr}([D, \operatorname{ad} Y]\operatorname{ad} Z) \\ &= \operatorname{Tr}(D \operatorname{ad}[Y, Z]) && \text{by expanding both sides} \\ &= B(X, [Y, Z]) && \text{by definition of } X \\ &= B([X, Y], Z) && \text{by invariance of } B \text{ as in (1.19).} \end{aligned}$$

By a second application of nondegeneracy of B, $DY = [X, Y]$. Thus $D = \operatorname{ad} X$.

12. Semidirect Products of Lie Groups

In §4 we introduced semidirect products of Lie algebras. Now we shall introduce a parallel theory of semidirect products of Lie groups and make the correspondence with the theory for Lie algebras.

Proposition 1.99. If G is a Lie group with $G = H_1 \oplus H_2$ as Lie groups (i.e., simultaneously as groups and manifolds) and if $\mathfrak{g}$, $\mathfrak{h}_1$, and $\mathfrak{h}_2$ are the respective Lie algebras, then $\mathfrak{g} = \mathfrak{h}_1 \oplus \mathfrak{h}_2$ with $\mathfrak{h}_1$ and $\mathfrak{h}_2$ as ideals in $\mathfrak{g}$. Conversely if H_1 and H_2 are analytic subgroups of G whose Lie algebras satisfy $\mathfrak{g} = \mathfrak{h}_1 \oplus \mathfrak{h}_2$ and if G is connected and simply connected, then $G = H_1 \oplus H_2$ as Lie groups.

PROOF. For the direct part, H_1 and H_2 are closed and normal. Hence they are Lie subgroups, and their Lie algebras are ideals in $\mathfrak{g}$. The vector space direct sum relationship depends only on the product structure of the manifold G.

For the converse the inclusions of H_1 and H_2 into G give us a smooth homomorphism $H_1 \oplus H_2 \to G$. On the other hand, the isomorphism of $\mathfrak{g}$ with $\mathfrak{h}_1 \oplus \mathfrak{h}_2$, in combination with the fact that G is connected and simply connected, gives us a homomorphism $G \to H_1 \oplus H_2$. The composition of the two group homomorphisms in either order has differential the identity and is therefore the identity homomorphism.

As in §4 the next step is to expand the theory of direct sums to a theory of semidirect products. Let G and H be Lie groups. We say that G **acts on H by automorphisms** if a smooth map $\tau : G \times H \to H$ is specified such that $g \mapsto \tau(g, \cdot)$ is a homomorphism of G into the abstract group of automorphisms of H. In this case the **semidirect product** $G \times_\tau H$ is the Lie group with $G \times H$ as its underlying manifold and with multiplication and inversion given by

$$
\begin{aligned}
(g_1, h_1)(g_2, h_2) &= (g_1 g_2, \tau(g_2^{-1}, h_1)h_2) \\
(g, h)^{-1} &= (g^{-1}, \tau(g, h^{-1})).
\end{aligned}
\tag{1.100}
$$

(To understand the definition of multiplication, think of the formula as if it were written $g_1 h_1 g_2 h_2 = g_1 g_2 (g_2^{-1} h_1 g_2) h_2$.) A little checking shows that this multiplication is associative. Then $G \times_\tau H$ is a Lie group, G and H are closed subgroups, and H is normal.

EXAMPLE. Let $G = SO(n)$, $H = \mathbb{R}^n$, and $\tau(r, x) = r(x)$. Then $G \times_\tau H$ is the group of translations and rotations (with arbitrary center) in $\mathbb{R}^n$.

Let us compute the Lie algebra of a semidirect product $G \times_\tau H$. We consider the differential $\bar{\tau}(g)$ of $\tau(g, \cdot)$ at the identity of H. Then $\bar{\tau}(g)$ is a Lie algebra isomorphism of $\mathfrak{h}$. As with Ad in §8, we find that

$$
\begin{aligned}
&\bar{\tau} \text{ is smooth into } GL(\mathfrak{h}) \\
&\bar{\tau}(g_1 g_2) = \bar{\tau}(g_1)\bar{\tau}(g_2).
\end{aligned}
$$

Thus $\bar{\tau}$ is a smooth homomorphism of G into $\operatorname{Aut}_{\mathbb{R}}\mathfrak{h}$. Its differential $d\bar{\tau}$ is a homomorphism of $\mathfrak{g}$ into $\operatorname{Der}_{\mathbb{R}}\mathfrak{h}$, by Proposition 1.97, and Proposition 1.22 allows us to form the semidirect product of Lie algebras $\mathfrak{g} \oplus_{d\bar{\tau}} \mathfrak{h}$.

Proposition 1.101. The Lie algebra of $G \times_\tau H$ is $\mathfrak{g} \oplus_{d\bar\tau} \mathfrak{h}$.

PROOF. The tangent space at the identity of $G \times_\tau H$ is $\mathfrak{g} \oplus \mathfrak{h}$ as a vector space, and the inclusions of G and H into $G \times_\tau H$ exhibit the bracket structure on $\mathfrak{g}$ and $\mathfrak{h}$ as corresponding to the respective bracket structures on $(\mathfrak{g}, 0)$ and $(0, \mathfrak{h})$. We have to check the brackets of members of $(\mathfrak{g}, 0)$ with members of $(0, \mathfrak{h})$. Let X be in $\mathfrak{g}$, let Y be in $\mathfrak{h}$, and write $\tilde X = (X, 0)$ and $\tilde Y = (0, Y)$. Then

$$\begin{aligned}
\exp(\mathrm{Ad}(\exp t\tilde X)s\tilde Y) &= (\exp t\tilde X)(\exp s\tilde Y)(\exp t\tilde X)^{-1} && \text{by (1.84)}\\
&= (\exp tX, 1)(1, \exp sY)(\exp tX, 1)\\
&= (1, \tau(\exp tX, \exp sY)) && \text{by (1.100).}
\end{aligned}$$

For fixed t, both sides are one-parameter groups, and the corresponding identity on the Lie algebra level is

$$\mathrm{Ad}(\exp t\tilde X)\tilde Y = (0, \bar\tau(\exp tX)Y).$$

Differentiating with respect to t and putting $t = 0$, we obtain

$$[\tilde X, \tilde Y] = (\mathrm{ad}\,\tilde X)(\tilde Y) = (0, d\bar\tau(X)Y),$$

by Proposition 1.93. This completes the proof.

Theorem 1.102. Let G and H be simply connected analytic groups with Lie algebras $\mathfrak{g}$ and $\mathfrak{h}$, respectively, and let $\pi : \mathfrak{g} \to \mathrm{Der}\,\mathfrak{h}$ be a Lie algebra homomorphism. Then there exists a unique action τ of G on H by automorphisms such that $d\bar\tau = \pi$, and $G \times_\tau H$ is a simply connected analytic group with Lie algebra $\mathfrak{g} \oplus_\pi \mathfrak{h}$.

PROOF OF UNIQUENESS. If there exists an action τ with $d\bar\tau = \pi$, then $G \times_\tau H$ is a simply connected group and has Lie algebra $\mathfrak{g} \oplus_\pi \mathfrak{h}$, by Proposition 1.101. If τ' is an action different from τ, then $\bar\tau \neq \bar\tau'$ for some g, and consequently $d\bar\tau \neq d\bar\tau'$. Uniqueness follows.

PROOF OF EXISTENCE. Since G is simply connected, we can find a smooth $\bar\tau : G \to \mathrm{Aut}\,\mathfrak{h}$ such that $d\bar\tau = \pi$. Fix $g \in G$, and then $\bar\tau(g) : \mathfrak{h} \to \mathfrak{h}$ is an automorphism. Since H is simply connected, there exists an automorphism $\tau(g)$ of H such that $d(\tau(g)) = \bar\tau(g)$. Since $\tau(g_1g_2)$ and $\tau(g_1)\tau(g_2)$ both have $\bar\tau(g_1g_2)$ as differential, we see that $\tau(g_1g_2) = \tau(g_1)\tau(g_2)$. Thus τ is a homomorphism of G into $\mathrm{Aut}\,H$.

We are to prove that $\tau : G \times H \to H$ is smooth. First we observe that $\tau' : G \times \mathfrak{h} \to \mathfrak{h}$ given by $\tau'(g, Y) = \bar\tau(g)Y$ is smooth. In fact, we choose a basis Y_i of $\mathfrak{h}$ and write $\bar\tau(g)Y_j = \sum_i c_{ij}(g)Y_i$. If $Y = \sum_j a_jY_j$, then $\tau'(g, Y) = \sum_{i,j} c_{ij}(g)a_jY_i$, and this is smooth as a function of the pair $(g, \{a_j\})$.

Next we have $\tau(g, \exp Y) = \exp\bar{\tau}(g)Y = \exp\tau'(g, Y)$. Choose an open neighborhood W' of 0 in $\mathfrak{h}$ such that exp is a diffeomorphism of W' onto an open set W in H. Then τ is smooth on $G \times W$, being the composition

$$(g, \exp Y) \mapsto (g, Y) \mapsto \tau'(g, Y) \mapsto \exp\tau'(g, Y).$$

For $h \in H$, define $\tau^h : G \to H$ by $\tau^h = \tau(\cdot\,, h)$. To see that τ^h is smooth, write $h = h_1 \cdots h_k$ with $h_i \in W$. Since $\tau(g,\ \cdot)$ is an automorphism, $\tau^h(g) = \tau^{h_1}(g) \cdots \tau^{h_k}(g)$. Each $\tau^{h_i}(\cdot)$ is smooth, and thus τ^h is smooth. Finally $\tau|_{G\times Wh}$ is the composition

$$G \times Wh \xrightarrow{1\times\text{translation}} G \times W \xrightarrow{\tau\times\tau^h} H \times H \xrightarrow{\text{multiplication}} H$$

given by

$$(g, wh) \mapsto (g, w) \mapsto (\tau(g, w), \tau^h(g)) \mapsto \tau(g, w)\tau(g, h) = \tau(g, wh),$$

and so τ is smooth.

A Lie group is said to be **solvable**, **nilpotent**, or **semisimple** if it is connected and if its Lie algebra is solvable, nilpotent, or semisimple, respectively. (Occasionally an author will allow one or more of these terms to refer to a disconnected group, but we shall not do so. By contrast "reductive Lie groups," which will be defined in Chapter VII, will be allowed a certain amount of disconnectedness.) For the rest of this chapter, we shall consider special properties of solvable, nilpotent, and semisimple Lie groups.

Corollary 1.103. If $\mathfrak{g}$ is a finite-dimensional solvable Lie algebra over $\mathbb{R}$, then there exists a simply connected analytic group with Lie algebra $\mathfrak{g}$, and G is diffeomorphic to a Euclidean space via canonical coordinates of the second kind. Moreover, there exists a sequence of closed simply connected analytic subgroups

$$G = G_0 \supseteq G_1 \supseteq \cdots \supseteq G_{n-1} \supseteq G_n = \{1\}$$

such that G_i is a semidirect product $G_i = \mathbb{R}^1 \times_{\tau_i} G_{i+1}$ with G_{i+1} normal in G_i. If $\mathfrak{g}$ is split-solvable, then each G_i may be taken to be normal in G. Any nilpotent $\mathfrak{g}$ is split-solvable, and when G_{n-1} is chosen to be normal, it is contained in the center of G.

PROOF. By Proposition 1.23 we can find a sequence of subalgebras

$$\mathfrak{g} = \mathfrak{g}_0 \supseteq \mathfrak{g}_1 \supseteq \cdots \supseteq \mathfrak{g}_{n-1} \supseteq \mathfrak{g}_n = 0$$

such that $\dim(\mathfrak{g}_i/\mathfrak{g}_{i+1}) = 1$ and $\mathfrak{g}_{i+1}$ is an ideal in $\mathfrak{g}_i$. If we let X_i be a member of $\mathfrak{g}_i$ not in $\mathfrak{g}_{i+1}$, then Proposition 1.22 shows that $\mathfrak{g}_i$ is the semidirect product of $\mathbb{R}X_i$ and $\mathfrak{g}_{i+1}$. Using $\mathbb{R}^1$ as a simply connected Lie group with Lie algebra $\mathbb{R}X_i$, we can invoke Theorem 1.102 to define G_i inductively downward on i as a semidirect product of $\mathbb{R}^1$ with G_{i+1}. (Here the formula $G_n = \{1\}$ starts the induction.) The groups G_i are then diffeomorphic to Euclidean space and form the decreasing sequence in the statement of the corollary.

If $\mathfrak{g}$ is split-solvable in the sense of §5, then the $\mathfrak{g}_i$ may be taken as ideals in $\mathfrak{g}$, by definition, and in this case the G_i are normal subgroups of G.

If $\mathfrak{g}$ is nilpotent, then each $\operatorname{ad} X$ for $X \in \mathfrak{g}$ is nilpotent and has all eigenvalues 0. By Corollary 1.30, $\mathfrak{g}$ is split-solvable. Thus each $\mathfrak{g}_i$ may be assumed to be an ideal in $\mathfrak{g}$. Under the assumption that $\mathfrak{g}_{n-1}$ is an ideal, we must have $[\mathfrak{g}, \mathfrak{g}_{n-1}] = 0$ for $\mathfrak{g}$ nilpotent, since $[\mathfrak{g}, \mathfrak{h}]$ cannot equal all of $\mathfrak{h}$ for any nonzero ideal $\mathfrak{h}$. Therefore $\mathfrak{g}_{n-1}$ is contained in the center of $\mathfrak{g}$, and G_{n-1} is contained in the center of $\mathfrak{g}$.

13. Nilpotent Lie Groups

Since nilpotent Lie algebras are solvable, Corollary 1.103 shows that every simply connected nilpotent analytic group is diffeomorphic with a Euclidean space. In this section we shall prove for the nilpotent case that the exponential map itself gives the diffeomorphism. By contrast, for a simply connected solvable analytic group, the exponential map need not be onto, as the following example shows.

EXAMPLE. Let G_1 be the closed linear group of all 3-by-3 matrices

$$g_1(t, x, y) = \begin{pmatrix} \cos 2t & \sin 2t & x \\ -\sin 2t & \cos 2t & y \\ 0 & 0 & 1 \end{pmatrix}$$

with linear Lie algebra consisting of all 3-by-3 matrices

$$X_1(s, a, b) = \begin{pmatrix} 0 & 2s & a \\ -2s & 0 & b \\ 0 & 0 & 0 \end{pmatrix}.$$

This Lie algebra is solvable. For G_1, one can show that the exponential map is onto, but we shall show that it is not onto for the double cover G consisting of all 5-by-5 matrices

$$g(t,x,y)=\begin{pmatrix} g_1(t,x,y) & & \\ & \cos t & \sin t \\ & -\sin t & \cos t \end{pmatrix}.$$

By (1.84) the exponential map cannot be onto for the simply connected covering group of G.

The linear Lie algebra of G consists of all 5-by-5 matrices

$$X(s,a,b)=\begin{pmatrix} X_1(s,a,b) & & \\ & 0 & s \\ & -s & 0 \end{pmatrix}.$$

Suppose $\exp X(s,a,b) = g(\pi,1,0)$. Then $X_1(s,a,b)$ must commute with $g_1(\pi,1,0)$, and this condition forces $s=0$. But $\exp X(0,a,b) = g(0,a,b)$. Since $g(\pi,1,0)$ is not of the form $g(0,a,b)$ for any a and b, it follows that $g(\pi,1,0)$ is not in the image of the exponential map. Thus the exponential map is not onto for the solvable analytic group G, as asserted.

Theorem 1.104. If N is a simply connected nilpotent analytic group with Lie algebra $\mathfrak{n}$, then the exponential map is a diffeomorphism of $\mathfrak{n}$ onto N.

PROOF. The first step is to prove that the exponential map is one-one onto. We proceed by induction on the dimension of the group in question. The trivial case of the induction is dimension 1, where the group is $\mathbb{R}^1$ and the result is known.

For the inductive case let N be given. We begin to coordinatize the group N in question as in Corollary 1.103. Namely we form a decreasing sequence of subalgebras

$$\mathfrak{n}=\mathfrak{n}_0\supseteq\mathfrak{n}_1\supseteq\mathfrak{n}_2\supseteq\cdots\supseteq\mathfrak{n}_n=0 \tag{1.105}$$

with $\dim \mathfrak{n}_i/\mathfrak{n}_{i+1}=1$ and with each $\mathfrak{n}_i$ an ideal in $\mathfrak{n}$. The corresponding analytic subgroups are closed and simply connected, and we are interested in the analytic subgroup Z corresponding to $\mathfrak{z}=\mathfrak{n}_{n-1}$. Corollary 1.103 notes that Z is contained in the center of N, and therefore $\mathfrak{z}$ is contained in the center of $\mathfrak{n}$. Since Z is central, it is normal, and we can form the quotient homomorphism $\varphi : N\to N/Z$. The group N/Z is a connected nilpotent Lie group with Lie algebra $\mathfrak{n}/\mathfrak{z}$, and N/Z is simply connected since Z is connected and N is simply connected. The inductive hypothesis is thus applicable to N/Z.

We can now derive our conclusions inductively about N. First we prove "one-one." Let X and X' be in $\mathfrak{n}$ with $\exp_N X = \exp_N X'$. Application of φ gives $\exp_{N/Z}(X+\mathfrak{z}) = \exp_{N/Z}(X'+\mathfrak{z})$. By inductive hypothesis for N/Z, $X+\mathfrak{z} = X'+\mathfrak{z}$. Thus $X - X'$ is in the center and commutes with X'. Consequently

$$\exp_N X = \exp_N(X' + (X - X')) = (\exp_N X')(\exp_N(X - X')),$$

and we conclude that $\exp_N(X - X') = 1$. Since Z is simply connected, the result for dimension 1 implies that $X - X' = 0$. Hence $X = X'$, and the exponential map is one-one for N.

Next we prove "onto." Let $x \in N$ be given, and choose $X + \mathfrak{z}$ in $\mathfrak{n}/\mathfrak{z}$ with $\exp_{N/Z}(X+\mathfrak{z}) = \varphi(x)$. Put $x' = \exp_N X$. Then (1.84) gives

$$\varphi(x') = \varphi(\exp_N X) = \exp_{N/Z}(X+\mathfrak{z}) = \varphi(x),$$

so that $x = x'z$ with z in $\ker\varphi = Z$. Since Z is connected and abelian, we can find X'' in its Lie algebra $\mathfrak{z}$ with $\exp_N X'' = z$. Since X and X'' commute,

$$x = x'z = (\exp_N X)(\exp_N X'') = \exp_N(X + X'').$$

Thus the exponential map is onto N. This completes the inductive proof that exp is one-one onto.

To complete the proof of the theorem, we are to show that the exponential map is everywhere regular. We now more fully coordinatize the group N in question as in Corollary 1.103. With $\mathfrak{n}_i$ as in (1.105), let X_i be in $\mathfrak{n}_{i-1}$ but not $\mathfrak{n}_i$, $1 \le i \le n$. Corollary 1.103 says that the canonical coordinates of the second kind formed from the ordered basis $X_1, \dots, X_n$ exhibit N as diffeomorphic to $\mathbb{R}^n$. In other words we can write

(1.106)
$$\exp(x_1X_1+\cdots+x_nX_n) = \exp(y_1(x_1,\dots,x_n)X_1)\cdots\exp(y_n(x_1,\dots,x_n)X_n),$$

and what needs to be proved is that the matrix $(\partial y_i/\partial x_j)$ is everywhere nonsingular.

This nonsingularity will be an immediate consequence of the formula

$$y_i(x_1,\dots,x_n) = x_i + \tilde{y}_i(x_1,\dots,x_{i-1}) \qquad \text{for } i \le n. \tag{1.107}$$

To prove (1.107), we argue by induction on $n = \dim N$. The trivial case of the induction is the case $n = 1$, where we evidently have $y_1(x_1) = x_1$

as required. For the inductive case let N be given, and define $Z, \mathfrak{z}$, and φ as earlier. In terms of our basis $X_1, \ldots, X_n$, the Lie algebra $\mathfrak{z}$ is given by $\mathfrak{z} = \mathbb{R}X_n$. If we write $d\varphi$ for the differential at 1 of the homomorphism φ, then $d\varphi(X_1), \ldots, d\varphi(X_{n-1})$ is a basis of the Lie algebra of N/Z.

Let us apply φ to both sides of (1.106). Then (1.84) gives

$$\begin{aligned}&\exp(x_1 d\varphi(X_1) + \cdots + x_{n-1} d\varphi(X_{n-1}))\\&\quad = \exp(y_1(x_1, \ldots, x_n) d\varphi(X_1)) \cdots \exp(y_{n-1}(x_1, \ldots, x_n) d\varphi(X_{n-1})).\end{aligned}$$

The left side is independent of x_n, and therefore

$$y_1(x_1, \ldots, x_n), \ldots, y_{n-1}(x_1, \ldots, x_n)$$

are all independent of x_n. We can regard them as functions of $n-1$ variables, and our inductive hypothesis says that, as such, they are of the form

$$y_i(x_1, \ldots, x_{n-1}) = x_i + \tilde{y}_i(x_1, \ldots, x_{i-1}) \qquad \text{for } i \leq n-1.$$

In terms of the functions of n variables, the form is

$$y_i(x_1, \ldots, x_n) = x_i + \tilde{y}_i(x_1, \ldots, x_{i-1}) \qquad \text{for } i \leq n-1. \tag{1.108}$$

This proves (1.107) except for $i = n$.

Thus let us define $\tilde{y}_n$ by $y_n(x_1, \ldots, x_n) = x_n + \tilde{y}_n(x_1, \ldots, x_n)$. Then we have

$$\exp(y_n(x_1, \ldots, x_n)X_n) = \exp(\tilde{y}_n(x_1, \ldots, x_n)X_n)\exp(x_n X_n). \tag{1.109}$$

Since X_n is central, we have also

$$\exp(x_1 X_1 + \cdots + x_n X_n) = \exp(x_1 X_1 + \cdots + x_{n-1} X_{n-1})\exp(x_n X_n). \tag{1.110}$$

Substituting from (1.109) and (1.110) into (1.106), using (1.108), and canceling $\exp(x_n X_n)$ from both sides, we obtain

$$\begin{aligned}&\exp(x_1 X_1 + \cdots + x_{n-1} X_{n-1})\\&\quad = \exp((x_1 + \tilde{y}_1)X_1)\exp((x_2 + \tilde{y}_2(x_1))X_2)\\&\qquad \times \cdots \times \exp((x_{n-1} + \tilde{y}_{n-1}(x_1, \ldots, x_{n-2}))X_{n-1})\exp(\tilde{y}_n(x_1, \ldots, x_n)X_n).\end{aligned}$$

The left side is independent of x_n, and hence so is the right side. Therefore $\tilde{y}_n(x_1, \ldots, x_n)$ is independent of x_n, and the proof of (1.107) for $i = n$ is complete.

Corollary 1.111. If N is a simply connected nilpotent analytic group, then any analytic subgroup of N is simply connected and closed.

PROOF. Let $\mathfrak{n}$ be the Lie algebra of N. Let M be an analytic subgroup of N, let $\mathfrak{m} \subseteq \mathfrak{n}$ be its Lie algebra, let $\tilde{M}$ be the universal covering group of M, and let $\psi : \tilde{M} \to M$ be the covering homomorphism. Assuming that M is not simply connected, let $\tilde{m} \neq 1$ be in $\ker \psi$. Since exp is one-one onto for $\tilde{M}$ by Theorem 1.104, we can find $X \in \mathfrak{m}$ with $\exp_{\tilde{M}} X = \tilde{m}$. Evidently $X \neq 0$. By (1.84) applied to ψ, $\exp_M X = 1$. By (1.84) applied to the inclusion of M into N, $\exp_N X = 1$. But this identity contradicts the assertion in Theorem 1.104 that exp is one-one for N. We conclude that M is simply connected. Since $\exp_M$ and $\exp_N$ are consistent, the image of $\mathfrak{m}$ under the diffeomorphism $\exp_N : \mathfrak{n} \to N$ is M, and hence M is closed.

14. Classical Semisimple Lie Groups

The classical semisimple Lie groups are specific closed linear groups that are connected and have semisimple Lie algebras listed in §8. Technically we have insisted that closed linear groups be closed subgroups of $GL(n, \mathbb{R})$ or $GL(n, \mathbb{C})$ for some n, but it will be convenient to allow closed subgroups of the group $GL(n, \mathbb{H})$ of nonsingular quaternion matrices as well.

The groups will be topologically closed because they are in each case the sets of common zeros of some polynomial functions in the entries. Most of the verification that the groups have particular linear Lie algebras as in §8 will be routine. It is necessary to make a separate calculation for the **special linear group**

$$SL(n, \mathbb{C}) = \{g \in GL(n, \mathbb{C}) \mid \det g = 1\},$$

and we carry out this step in Proposition 1.113 below.

The issue that tends to be more complicated is the connectedness of the given group. If we neglect to prove connectedness, we do not end up with the conclusion that the given group is semisimple, only that its identity component is semisimple.

To handle connectedness, we proceed in two steps, first establishing connectedness for certain compact examples and then proving in general that the number of components of the given group is the same as for a particular compact subgroup. We return to this matter at the end of this section.

Lemma 1.112. If X is in $GL(n, \mathbb{C})$, then $\det e^X = e^{\operatorname{Tr} X}$.

PROOF. The identity is clear if X is upper triangular. For general X, Jordan form allows us to find $g \in GL(n, \mathbb{C})$ and $U \in \mathfrak{gl}(n, \mathbb{C})$ with U upper triangular and with $X = gUg^{-1}$. Applying the special case to U, we have

$$\det e^X = \det e^{gUg^{-1}} = \det(ge^U g^{-1}) = \det e^U = e^{\operatorname{Tr} U} = e^{\operatorname{Tr}(gUg^{-1})} = e^{\operatorname{Tr} X}.$$

Proposition 1.113. The linear Lie algebra of $SL(n, \mathbb{C})$ is $\mathfrak{sl}(n, \mathbb{C})$.

PROOF. The members of the linear Lie algebra are the matrices X such that e^{tX} is in $SL(n, \mathbb{C})$ for all real t. For any X, Lemma 1.112 shows that the determinant of e^{tX} is $e^{t \operatorname{Tr} X}$, and this is 1 for all t if and only if $\operatorname{Tr} X = 0$.

REMARK. In practice we use Proposition 1.113 by combining it with a result about intersections: If G_1 and G_2 are closed linear groups with respective linear Lie algebras $\mathfrak{g}_1$ and $\mathfrak{g}_2$, then the closed linear group $G_1 \cap G_2$ has linear Lie algebra $\mathfrak{g}_1 \cap \mathfrak{g}_2$. This fact follows immediately from the characterization of the linear Lie algebra as the set of all matrices X such that $\exp tX$ is in the corresponding group for all real t. Thus when "$\det g = 1$" appears as a defining condition for a closed linear group, the corresponding condition to impose for the linear Lie algebra is "$\operatorname{Tr} X = 0$."

We turn to a consideration of specific compact groups. Define

$$\begin{aligned} SO(n) &= \{g \in GL(n, \mathbb{R}) \mid g^*g = 1 \text{ and } \det g = 1\} \\ SU(n) &= \{g \in GL(n, \mathbb{C}) \mid g^*g = 1 \text{ and } \det g = 1\} \\ Sp(n) &= \{g \in GL(n, \mathbb{H}) \mid g^*g = 1\}. \end{aligned} \tag{1.114}$$

These are all closed linear groups, and they are compact by the Heine-Borel Theorem, their entries being bounded in absolute value by 1. The group $SO(n)$ is called the **rotation group**, and $SU(n)$ is called the **special unitary group**.

Notice that no determinant condition is imposed for $Sp(n)$. Artin [1957], 151–158, gives an exposition of Dieudonné's notion of determinant for square matrices with entries from $\mathbb{H}$. The determinant takes real values ≥ 0, is multiplicative, is 1 on the identity matrix, and is 0 exactly for singular matrices. For the members of $Sp(n)$, the determinant is automatically 1.

Proposition 1.115. The groups $SO(n)$, $SU(n)$, and $Sp(n)$ are all connected for $n \geq 1$.

PROOF. Consider $SO(n)$. For $n = 1$, this group is trivial and is therefore connected. For $n \geq 2$, $SO(n)$ acts transitively on the unit sphere in the space $\mathbb{R}^n$ of n-dimensional column vectors with entries from $\mathbb{R}$, and the isotropy subgroup at $v_0 = \begin{pmatrix} 0 \\ \vdots \\ 0 \\ 1 \end{pmatrix}$ is given in block form by

$$\begin{pmatrix} SO(n-1) & 0 \\ 0 & 1 \end{pmatrix}.$$

Thus the continuous map $g \mapsto gv_0$ of $SO(n)$ onto the unit sphere descends to a one-one continuous map of $SO(n)/SO(n-1)$ onto the unit sphere. Since $SO(n)/SO(n-1)$ is compact, this map is a homeomorphism. Consequently $SO(n)/SO(n-1)$ is connected. To complete the argument for $SO(n)$, we induct on n, using the fact about toplogical groups that if H and G/H are connected, then G is connected.

For $SU(n)$, we argue similarly, replacing $\mathbb{R}$ by $\mathbb{C}$. The group $SU(1)$ is trivial and connected, and the action of $SU(n)$ on the unit sphere in $\mathbb{C}^n$ is transitive for $n \geq 2$. For $Sp(n)$, we argue with $\mathbb{H}$ in place of $\mathbb{R}$. The group $Sp(1)$ is the unit quaternions and is connected, and the action of $Sp(n)$ on the unit sphere in $\mathbb{H}^n$ is transitive for $n \geq 2$.

It is clear from Proposition 1.115 and its remark that the linear Lie algebras of $SO(n)$ and $SU(n)$ are $\mathfrak{so}(n)$ and $\mathfrak{su}(n)$, respectively. In the case of matrices with quaternion entries, we did not develop a theory of closed linear groups, but we can use the correspondence in §8 of n-by-n matrices over $\mathbb{H}$ with certain $2n$-by-$2n$ matrices over $\mathbb{C}$ to pass from $Sp(n)$ to complex matrices of size $2n$, then to the linear Lie algebra, and then back to $\mathfrak{sp}(n)$. In this sense the linear Lie algebra of $Sp(n)$ is $\mathfrak{sp}(n)$.

Taking into account the values of n in §8 for which these Lie algebras are semisimple, we conclude that $SO(n)$ is compact semisimple for $n \geq 3$, $SU(n)$ is compact semisimple for $n \geq 2$, and $Sp(n)$ is compact semisimple for $n \geq 1$.

Two families of related compact groups are

$$\begin{aligned} O(n) &= \{g \in GL(n, \mathbb{R}) \mid g^*g = 1\} \\ U(n) &= \{g \in GL(n, \mathbb{C}) \mid g^*g = 1\}. \end{aligned} \tag{1.116}$$

These are the **orthogonal group** and the **unitary group**, respectively. The group $O(n)$ has two components; the Lie algebra is $\mathfrak{so}(n)$, and the identity component is $SO(n)$. The group $U(n)$ is connected by an argument like that in Proposition 1.115, and its Lie algebra is the reductive Lie algebra $\mathfrak{u}(n) \cong \mathfrak{su}(n) \oplus \mathbb{R}$.

Next we consider complex semisimple groups. According to §8, $\mathfrak{sl}(n, \mathbb{C})$ is semisimple for $n \geq 2$, $\mathfrak{so}(n, \mathbb{C})$ is semisimple for $n \geq 3$, and $\mathfrak{sp}(n, \mathbb{C})$ is semisimple for $n \geq 1$. Letting $J_{n,n}$ be as in §8, we define closed linear groups by

$$\begin{aligned} SL(n, \mathbb{C}) &= \{g \in GL(n, \mathbb{C}) \mid \det g = 1\} \\ SO(n, \mathbb{C}) &= \{g \in SL(n, \mathbb{C}) \mid g^t g = 1\} \\ Sp(n, \mathbb{C}) &= \{g \in SL(2n, \mathbb{C}) \mid g^t J_{n,n} g = J_{n,n}\}. \end{aligned} \tag{1.117}$$

We readily check that their linear Lie algebras are $\mathfrak{sl}(n, \mathbb{C})$, $\mathfrak{so}(n, \mathbb{C})$, and $\mathfrak{sp}(n, \mathbb{C})$, respectively. Since $GL(n, \mathbb{C})$ is a complex Lie group and each of these Lie subalgebras of $\mathfrak{gl}(n, \mathbb{C})$ is closed under multiplication by i, the remarks at the end of §10 say that each of these closed linear groups G has the natural structure of a complex manifold in such a way that multiplication and inversion are holomorphic.

Proposition 1.118. Under the identification $M \mapsto Z(M)$ in (1.62),

$$Sp(n) \cong Sp(n, \mathbb{C}) \cap U(2n).$$

PROOF. From (1.62) we see that a $2n$-by-$2n$ complex matrix W is of the form $Z(M)$ if and only if

$$JW = \bar{W} J. \tag{1.119}$$

Let g be in $Sp(n)$. From $g^* g = 1$, we obtain $Z(g)^* Z(g) = 1$. Thus $Z(g)$ is in $U(2n)$. Also (1.119) gives $Z(g)^t J Z(g) = Z(g)^t \overline{Z(g)} J = \overline{(Z(g)^* Z(g))} J = J$, and hence $Z(g)$ is in $Sp(n, \mathbb{C})$.

Conversely suppose that W is in $Sp(n, \mathbb{C}) \cap U(2n)$. From $W^* W = 1$ and $W^t J W = J$, we obtain $J = W^t \bar{W} \bar{W}^{-1} J W = \overline{(W^* W)} \bar{W}^{-1} J W = \bar{W}^{-1} J W$ and therefore $\bar{W} J = JW$. By (1.119), $W = Z(g)$ for some quaternion matrix g. From $W^* W = 1$, we obtain $Z(g^* g) = Z(g)^* Z(g) = 1$ and $g^* g = 1$. Therefore g is in $Sp(n)$.

We postpone to the end of this section a proof that the groups $SL(n, \mathbb{C})$, $SO(n, \mathbb{C})$, and $Sp(n, \mathbb{C})$ are connected for all n. We shall see that the proof of this connectivity reduces in the respective cases to the connectivity of $SU(n)$, $SO(n)$, and $Sp(n, \mathbb{C}) \cap U(2n)$, and this connectivity has been proved in Propositions 1.115 and 1.118. We conclude that $SL(n, \mathbb{C})$ is semisimple for $n \geq 2$, $SO(n, \mathbb{C})$ is semisimple for $n \geq 3$, and $Sp(n, \mathbb{C})$ is semisimple for $n \geq 1$.

The groups $SO(n, \mathbb{C})$ and $Sp(n, \mathbb{C})$ have interpretations in terms of bilinear forms. The group $SO(n, \mathbb{C})$ is the subgroup of matrices in $SL(n, \mathbb{C})$ preserving the symmetric bilinear form on $\mathbb{C}^n \times \mathbb{C}^n$ given by

$$\left\langle \begin{pmatrix} x_1 \\ \vdots \\ x_n \end{pmatrix}, \begin{pmatrix} y_1 \\ \vdots \\ y_n \end{pmatrix} \right\rangle = x_1 y_1 + \cdots + x_n y_n,$$

while the group $Sp(n, \mathbb{C})$ is the subgroup of matrices in $SL(2n, \mathbb{C})$ preserving the alternating bilinear form on $\mathbb{C}^{2n} \times \mathbb{C}^{2n}$ given by

$$\left\langle \begin{pmatrix} x_1 \\ \vdots \\ x_{2n} \end{pmatrix}, \begin{pmatrix} y_1 \\ \vdots \\ y_{2n} \end{pmatrix} \right\rangle = x_1 y_{n+1} + \cdots + x_n y_{2n} - x_{n+1} y_1 - \cdots - x_{2n} y_n.$$

Finally we consider noncompact noncomplex semisimple groups. With notation $I_{m,n}$ and $J_{n,n}$ as in §8, the definitions are

$$\begin{aligned} SL(n, \mathbb{R}) &= \{g \in GL(n, \mathbb{R}) \mid \det g = 1\} \\ SL(n, \mathbb{H}) &= \{g \in GL(n, \mathbb{H}) \mid \det g = 1\} \\ SO(m, n) &= \{g \in SL(m+n, \mathbb{R}) \mid g^* I_{m,n} g = I_{m,n}\} \\ SU(m, n) &= \{g \in SL(m+n, \mathbb{C}) \mid g^* I_{m,n} g = I_{m,n}\} \\ Sp(m, n) &= \{g \in GL(m+n, \mathbb{H}) \mid g^* I_{m,n} g = I_{m,n}\} \\ Sp(n, \mathbb{R}) &= \{g \in SL(2n, \mathbb{R}) \mid g^t J_{n,n} g = J_{n,n}\} \\ SO^*(2n) &= \{g \in SU(n, n) \mid g^t I_{n,n} J_{n,n} g = I_{n,n} J_{n,n}\}. \end{aligned} \tag{1.120}$$

Some remarks are in order about particular groups in this list. For $SL(n, \mathbb{H})$ and $Sp(m, n)$, the prescription at the end of §8 allows us to replace the realizations in terms of quaternion matrices by realizations in terms of complex matrices of twice the size. The realization of $SL(n, \mathbb{H})$ with complex matrices avoids the notion of determinant of a quaternion matrix that was mentioned before the statement of Proposition 1.115; the isomorphic group of complex matrices is

$$SU^*(2n) = \left\{ \begin{pmatrix} z_1 & -\bar{z}_2 \\ z_2 & \bar{z}_1 \end{pmatrix} \in SL(2n, \mathbb{C}) \right\}.$$

The groups $SO(m, n)$, $SU(m, n)$, and $Sp(m, n)$ are isometry groups of Hermitian forms. In more detail the group

$$O(m, n) = \{g \in GL(m+n, \mathbb{R}) \mid g^* I_{m,n} g = I_{m,n}\}$$

is the group of real matrices of size $m+n$ preserving the symmetric bilinear form on $\mathbb{R}^{m+n} \times \mathbb{R}^{m+n}$ given by

$$\left\langle \begin{pmatrix} x_1 \\ \vdots \\ x_n \end{pmatrix}, \begin{pmatrix} y_1 \\ \vdots \\ y_n \end{pmatrix} \right\rangle = x_1y_1 + \cdots + x_my_m - x_{m+1}y_{m+1} - \cdots - x_{m+n}y_{m+n},$$

and $SO(m,n)$ is the subgroup of members of $O(m,n)$ of determinant 1. The group

$$U(m,n) = \{g \in GL(m+n,\mathbb{C}) \mid g^* I_{m,n} g = I_{m,n}\}$$

is the group of complex matrices of size $m+n$ preserving the Hermitian form on $\mathbb{C}^{m+n} \times \mathbb{C}^{m+n}$ given by

$$\left\langle \begin{pmatrix} x_1 \\ \vdots \\ x_n \end{pmatrix}, \begin{pmatrix} y_1 \\ \vdots \\ y_n \end{pmatrix} \right\rangle = x_1\overline{y_1} + \cdots + x_m\overline{y_m} - x_{m+1}\overline{y_{m+1}} - \cdots - x_{m+n}\overline{y_{m+n}},$$

and $SU(m,n)$ is the subgroup of members of $U(m,n)$ of determinant 1. The group $Sp(m,n)$ is the group of quaternion matrices of size $m+n$ preserving the Hermitian form on $\mathbb{H}^{m+n} \times \mathbb{H}^{m+n}$ given by

$$\left\langle \begin{pmatrix} x_1 \\ \vdots \\ x_n \end{pmatrix}, \begin{pmatrix} y_1 \\ \vdots \\ y_n \end{pmatrix} \right\rangle = x_1\overline{y_1} + \cdots + x_m\overline{y_m} - x_{m+1}\overline{y_{m+1}} - \cdots - x_{m+n}\overline{y_{m+n}},$$

with no condition needed on the determinant.

The linear Lie algebras of the closed linear groups in (1.120) are given in a table in Example 3 of §8, and the table in §8 tells which values of m and n lead to semisimple Lie algebras. It will be a consequence of results below that all the closed linear groups in (1.120) are topologically connected except for $SO(m,n)$. In the case of $SO(m,n)$, one often works with the identity component $SO(m,n)_0$ in order to have access to the full set of results about semisimple groups in later chapters.

Let us now address the subject of connectedness in detail. We shall work with a closed linear group of complex matrices that is closed under adjoint and is defined by polynomial equations. We begin with a lemma.

Lemma 1.121. Let $P : \mathbb{R}^n \to \mathbb{R}$ be a polynomial, and suppose $(a_1, \ldots, a_n)$ has the property that $P(e^{ka_1}, \ldots, e^{ka_n}) = 0$ for all integers $k \geq 0$. Then $P(e^{ta_1}, \ldots, e^{ta_n}) = 0$ for all real t.

PROOF. A monomial $cx_1^{l_1}\cdots x_n^{l_n}$, when evaluated at $(e^{ta_1},\ldots,e^{ta_n})$, becomes $ce^{t\sum a_i l_i}$. Collecting terms with like exponentials, we may assume that we have an expression $\sum_{j=1}^N c_j e^{tb_j}$ that vanishes whenever t is an integer ≥ 0. We may further assume that all c_j are nonzero and that $b_1 < b_2 < \cdots < b_N$. We argue by contradiction and suppose $N > 0$. Multiplying by e^{-tb_N} and changing notation, we may assume that $b_N = 0$. We pass to the limit in the expression $\sum_{j=1}^N c_j e^{tb_j}$ as t tends to $+\infty$ through integer values, and we find that $c_N = 0$, contradiction.

Proposition 1.122. Let $G \subseteq GL(n,\mathbb{C})$ be a closed linear group that is the common zero locus of some set of real-valued polynomials in the real and imaginary parts of the matrix entries, and let $\mathfrak{g}$ be its linear Lie algebra. Suppose that G is closed under adjoints. Let K be the group $G \cap U(n)$, and let $\mathfrak{p}$ be the subspace of Hermitian matrices in $\mathfrak{g}$. Then the map $K \times \mathfrak{p} \to G$ given by $(k, X) \mapsto ke^X$ is a homeomorphism onto.

PROOF. For $GL(n,\mathbb{C})$, the map

$$U(n) \times \{\text{Hermitian matrices}\} \to GL(n,\mathbb{C})$$

given by $(k, X) \mapsto ke^X$ is known to be a homeomorphism; see Chevalley [1946], 14–15. The inverse map is the **polar decomposition** of $GL(n,\mathbb{C})$.

Let g be in G, and let $g = ke^X$ be the polar decomposition of g within $GL(n,\mathbb{C})$. To prove the proposition, we have only to show that k is in G and that X is in the linear Lie algebra $\mathfrak{g}$ of G.

Taking adjoints, we have $g^* = e^X k^{-1}$ and therefore $g^*g = e^{2X}$. Since G is closed under adjoints, e^{2X} is in G. By assumption, G is the zero locus of some set of real-valued polynomials in the real and imaginary parts of the matrix entries. Let us conjugate matters so that e^{2X} is diagonal, say $2X = \operatorname{diag}(a_1,\ldots,a_n)$ with each a_j real. Since e^{2X} and its integral powers are in G, the transformed polynomials vanish at

$$(e^{2X})^k = \operatorname{diag}(e^{ka_1},\ldots,e^{ka_n})$$

for every integer k. By Lemma 1.121 the transformed polynomials vanish at

$$\operatorname{diag}(e^{ta_1},\ldots,e^{ta_n})$$

for all real t. Therefore e^{tX} is in G for all real t. It follows from the definition of $\mathfrak{g}$ that X is in $\mathfrak{g}$. Since e^X and g are then in G, k is in G. This completes the proof.

Proposition 1.122 says that G is connected if and only if K is connected. To decide which of the groups in (1.117) and (1.120) are connected, we therefore compute K for each group. In the case of the groups of quaternion matrices, we compute K by converting to complex matrices, intersecting with the unitary group, and transforming back to

(1.123)

G	K up to isomorphism
$SL(n,\mathbb{C})$	$SU(n)$
$SO(n,\mathbb{C})$	$SO(n)$
$Sp(n,\mathbb{C})$	$Sp(n)$ or $Sp(n,\mathbb{C})\cap U(2n)$
$SL(n,\mathbb{R})$	$SO(n)$
$SL(n,\mathbb{H})$	$Sp(n)$
$SO(m,n)$	$S(O(m)\times O(n))$
$SU(m,n)$	$S(U(m)\times U(n))$
$Sp(m,n)$	$Sp(m)\times Sp(n)$
$Sp(n,\mathbb{R})$	$U(n)$
$SO^*(2n)$	$U(n)$

quaternion matrices. The results are in (1.123). In the K column of (1.123), the notation $S(\cdot)$ means the determinant-one subgroup of $(\cdot)$. By Propositions 1.115 and 1.118 and the connectedness of $U(n)$, we see that all the groups in the K column are connected except for $S(O(m)\times O(n))$. Using Proposition 1.122, we arrive at the following conclusion.

Proposition 1.124. All the classical groups $SL(n,\mathbb{C})$, $SO(n,\mathbb{C})$, $Sp(n,\mathbb{C})$, $SL(n,\mathbb{R})$, $SL(n,\mathbb{H})$, $SU(m,n)$, $Sp(m,n)$, $Sp(n,\mathbb{R})$, and $SO^*(2n)$ are connected. The group $SO(m,n)$ has two components if $m>0$ and $n>0$.

15. Problems

1. Verify that Example 12a in §1 is nilpotent and that Example 12b is split solvable.

2. For $\begin{pmatrix}\alpha & \beta\\ \gamma & \delta\end{pmatrix}$ any nonsingular matrix over $\mathbb{k}$, let $\mathfrak{g}_{\begin{pmatrix}\alpha & \beta\\ \gamma & \delta\end{pmatrix}}$ be the 3-dimensional algebra over $\mathbb{k}$ with basis X, Y, Z satisfying

$$[X,Y]=0$$
$$[X,Z]=\alpha X+\beta Y$$
$$[Y,Z]=\gamma X+\delta Y.$$

(a) Show that $\mathfrak{g}_{\begin{pmatrix}\alpha & \beta \\ \gamma & \delta\end{pmatrix}}$ is a *Lie* algebra by showing that $X \leftrightarrow \begin{pmatrix}0&0&1\\0&0&0\\0&0&0\end{pmatrix}$, $Y \leftrightarrow \begin{pmatrix}0&0&0\\0&0&1\\0&0&0\end{pmatrix}$, $Z \leftrightarrow -\begin{pmatrix}\alpha&\gamma&0\\\beta&\delta&0\\0&0&0\end{pmatrix}$ gives an isomorphism with a Lie algebra of matrices.

(b) Show that $\mathfrak{g}_{\begin{pmatrix}\alpha & \beta \\ \gamma & \delta\end{pmatrix}}$ is solvable but not nilpotent.

(c) Let $\mathbb{k} = \mathbb{R}$. Take $\delta = 1$ and $\beta = \gamma = 0$. Show that the various Lie algebras $\mathfrak{g}_{\begin{pmatrix}\alpha & 0 \\ 0 & 1\end{pmatrix}}$ for $\alpha > 1$ are mutually nonisomorphic. (Therefore for $k = \mathbb{R}$ that there are uncountably many nonisomorphic solvable real Lie algebras of dimension 3.)

3. Let

$$\mathfrak{s}(n, \mathbb{k}) = \left\{ X \in \mathfrak{gl}(n, \mathbb{k}) \;\middle|\; X = \begin{pmatrix} a_1 & & * \\ & \ddots & \\ 0 & & a_n \end{pmatrix} \right\}.$$

Define a bracket operation on $\mathfrak{g} = \bigoplus_{n=1}^{\infty} \mathfrak{s}(n, \mathbb{k})$ (a vector space in which each element has only finitely many nonzero coordinates) in such a way that each $\mathfrak{s}(n, \mathbb{k})$ is an ideal. Show that each member of $\mathfrak{g}$ lies in a finite-dimensional solvable ideal but that the commutator series of $\mathfrak{g}$ does not terminate in 0. (Hence there is no largest solvable ideal.)

4. Let $\mathfrak{g}$ be a real Lie algebra of complex matrices with the property that $X \in \mathfrak{g}$ and $X \neq 0$ imply $iX \notin \mathfrak{g}$. Make precise and verify the statement that $\mathfrak{g}^{\mathbb{C}}$ can be realized as a Lie algebra of matrices by complexifying the entries of $\mathfrak{g}$. Use this statement to prove directly that $\mathfrak{sl}(2, \mathbb{R})$ and $\mathfrak{su}(2)$ have isomorphic complexifications.

5. Under the isomorphism (1.4) of $\mathfrak{so}(3)$ with the vector product Lie algebra, show that the Killing form B for $\mathfrak{so}(3)$ gets identified with a multiple of the dot product in $\mathbb{R}^3$.

6. Let $\mathfrak{g}$ be a nonabelian 2-dimensional Lie algebra. Using the computation of the Killing form in Example 1 of §3, show that $\operatorname{rad} B \neq \operatorname{rad} \mathfrak{g}$.

7. Let $\mathfrak{g} = \mathfrak{sl}(2, \mathbb{k})$. Show that $B(X, X)$ is a multiple of $\det X$ independent of $X \in \mathfrak{g}$.

8. In $\mathfrak{sl}(n, \mathbb{R})$ the Killing form and the trace form $C(X, Y) = \operatorname{Tr}(XY)$ are multiples of one another. Identify the multiple.

9. Show that the solvable Lie algebra $\mathfrak{g} = \begin{pmatrix} 0 & \theta & x \\ -\theta & 0 & y \\ 0 & 0 & 0 \end{pmatrix}$ over $\mathbb{R}$ is not split solvable
   (a) by showing that $\mathfrak{g}$ has no 1-dimensional ideal.
   (b) by producing nonreal eigenvalues for some ad X with $X \in \mathfrak{g}$.

   Show also that $\mathfrak{g}^{\mathbb{C}}$ can be regarded as all complex matrices of the form $\mathfrak{g} = \begin{pmatrix} 0 & \theta & x \\ -\theta & 0 & y \\ 0 & 0 & 0 \end{pmatrix}$, and exhibit a 1-dimensional ideal in $\mathfrak{g}^{\mathbb{C}}$ (which exists since $\mathfrak{g}^{\mathbb{C}}$ has to be split solvable over $\mathbb{C}$).
10. Prove for that a finite-dimensional solvable Lie algebra over $\mathbb{R}$ that $[\mathfrak{g}, \mathfrak{g}]$ is nilpotent.
11. Prove that if $\mathfrak{g}$ is a finite-dimensional nilpotent Lie algebra over $\mathbb{R}$, then the Killing form of $\mathfrak{g}$ is identically 0.
12. Let $\mathfrak{g}$ be a complex Lie algebra of complex matrices, and suppose that $\mathfrak{g}$ is simple over $\mathbb{C}$. Let $C(X, Y) = \operatorname{Tr}(XY)$ for X and Y in $\mathfrak{g}$. Prove that C is a multiple of the Killing form.
13. For $k = \mathbb{R}$, prove that $\mathfrak{su}(2)$ and $\mathfrak{sl}(2, \mathbb{R})$ are not isomorphic.
14. (a) Show that $\mathfrak{so}(3)$ is isomorphic with $\mathfrak{su}(2)$.
    (b) Prove that $\mathfrak{su}(2)$ is simple.
    (c) Prove that there exists a covering homomorphism of $SU(2)$ onto $SO(3)$ with 2-element kernel.
15. Prove that $\mathfrak{so}(2, 1)$ is isomorphic with $\mathfrak{sl}(2, \mathbb{R})$.
16. For $\mathfrak{u}(n)$, we have an isomorphism $\mathfrak{u}(n) \cong \mathfrak{su}(n) \oplus \mathbb{R}$, where $\mathbb{R}$ is the center. Let Z be the analytic subgroup of $U(n)$ with Lie algebra the center. Is $U(n)$ isomorphic with the direct sum of $SU(n)$ and Z? Why or why not?
17. Let V_n be the complex vector space of all polynomials in two complex variables z_1 and z_2 homogeneous of degree n. Define a representation of $SL(2, \mathbb{C})$ by

$$\Phi_n \begin{pmatrix} a & b \\ c & d \end{pmatrix} P \begin{pmatrix} z_1 \\ z_2 \end{pmatrix} = P\left(\begin{pmatrix} a & b \\ c & d \end{pmatrix}^{-1} \begin{pmatrix} z_1 \\ z_2 \end{pmatrix} \right).$$

   Then $\dim V_n = n + 1$, Φ is a homomorphism, and Φ is holomorphic. Let φ be the differential of Φ at 1. Prove that φ is isomorphic with the irreducible complex-linear representation of $\mathfrak{sl}(2, \mathbb{C})$ of dimension $n + 1$ given in Theorem 1.63.

18. Let $\mathfrak{g}$ be the Heisenberg Lie algebra over $\mathbb{R}$ as in Example 12a of §1. Verify that $\mathfrak{g}$ is isomorphic with

$$\left\{ \begin{pmatrix} 0 & 0 & 0 \\ z & 0 & 0 \\ it & \bar{z} & 0 \end{pmatrix} \middle| z \in \mathbb{C},\, t \in \mathbb{R} \right\}.$$

19. The real Lie algebra

$$\mathfrak{g} = \left\{ \begin{pmatrix} i\theta & 0 & 0 \\ z & -2i\theta & 0 \\ it & \bar{z} & i\theta \end{pmatrix} \middle| z \in \mathbb{C},\, \theta \in \mathbb{R},\, t \in \mathbb{R} \right\}$$

is the Lie algebra of the "oscillator group." Show that $[\mathfrak{g}, \mathfrak{g}]$ is isomorphic with the Heisenberg Lie algebra over $\mathbb{R}$. (See Example 12a of §1.)

20. Let N be a simply connected nilpotent analytic group with Lie algebra $\mathfrak{n}$, and let $\mathfrak{n}_i$ be a sequence of ideals in $\mathfrak{n}$ such that

$$\mathfrak{n} = \mathfrak{n}_0 \supseteq \mathfrak{n}_1 \supseteq \cdots \supseteq \mathfrak{n}_{n-1} \supseteq \mathfrak{n}_n = 0$$

and $[\mathfrak{n}, \mathfrak{n}_i] \subseteq \mathfrak{n}_{i+1}$ for $0 \le i < n$. Suppose that $\mathfrak{s}$ and $\mathfrak{t}$ are vector subspaces of $\mathfrak{n}$ such that $\mathfrak{n} = \mathfrak{s} \oplus \mathfrak{t}$ and $\mathfrak{n}_i = (\mathfrak{s} \cap \mathfrak{n}_i) \oplus (\mathfrak{t} \cap \mathfrak{n}_i)$ for all i. Prove that the map $\mathfrak{s} \oplus \mathfrak{t} \to N$ given by $(X, Y) \mapsto \exp X \exp Y$ is a diffeomorphism onto.

21. Find the cardinality of the centers of $SU(n)$, $SO(n)$, $Sp(n)$, $SL(n, \mathbb{C})$, $SO(n, \mathbb{C})$, and $Sp(n, \mathbb{C})$.

22. Let $G = \{g \in SL(2n, \mathbb{C}) \mid g^t I_{n,n} g = I_{n,n}\}$. Prove that G is isomorphic to $SO(2n, \mathbb{C})$. (See §8 for the definition of $I_{n,n}$.)

23. Show that Proposition 1.122 can be applied to $GL(n, \mathbb{R})$ if $GL(n, \mathbb{R})$ is embedded in $SL(n+1, \mathbb{R})$ in block diagonal form as

$$g \mapsto \begin{pmatrix} g & 0 \\ 0 & (\det g)^{-1} \end{pmatrix}.$$

Deduce that $GL(n, \mathbb{R})$ has two connected components.

24. Give an example of a closed linear group $G \subseteq SL(n, \mathbb{C})$ such that G is closed under adjoints but G is not homeomorphic to the product of $G \cap U(n)$ and a Euclidean space.

Problems 25–27 concern the Heisenberg Lie algebra $\mathfrak{g}$ over $\mathbb{R}$ as in Example 12a of §1. Let V be the complex vector space of complex-valued functions on $\mathbb{R}$ of the form $e^{-\pi s^2}P(s)$, where P is a polynomial, and let $\hbar$ be a positive constant.

25. Show that the linear mappings $i\dfrac{d}{ds}$ and "multiplication by $-i\hbar s$" carry V into itself.

26. Define $\varphi\begin{pmatrix} 0 & 1 & 0 \\ 0 & 0 & 0 \\ 0 & 0 & 0 \end{pmatrix} = i\dfrac{d}{ds}$ and let $\varphi\begin{pmatrix} 0 & 0 & 0 \\ 0 & 0 & 1 \\ 0 & 0 & 0 \end{pmatrix}$ be multiplication by $-i\hbar s$. How should $\varphi\begin{pmatrix} 0 & 0 & 1 \\ 0 & 0 & 0 \\ 0 & 0 & 0 \end{pmatrix}$ be defined so that the linear extension of φ to $\mathfrak{g}$ is a representation of $\mathfrak{g}$ on V?

27. With φ defined as in Problem 26, prove that φ is irreducible.

Problems 28–30 classify the solvable Lie algebras $\mathfrak{g}$ of dimension 3 over $\mathbb{R}$.

28. Prove that if $\dim[\mathfrak{g}, \mathfrak{g}] = 1$, then $\mathfrak{g}$ is isomorphic with either the Heisenberg Lie algebra (Example 12a of §1) or the direct sum of a 1-dimensional (abelian) Lie algebra and a nonabelian 2-dimensional Lie algebra.

29. If $\dim[\mathfrak{g}, \mathfrak{g}] = 2$, use Problem 10 to show that $[\mathfrak{g}, \mathfrak{g}]$ is abelian. Let X, Y be a basis of $[\mathfrak{g}, \mathfrak{g}]$, and extend to a basis X, Y, Z of $\mathfrak{g}$. Define $\alpha, \beta, \gamma, \delta$ by
$$[X, Z] = \alpha X + \beta Y$$
$$[Y, Z] = \gamma X + \delta Y.$$
Show that $\begin{pmatrix} \alpha & \beta \\ \gamma & \delta \end{pmatrix}$ is nonsingular.

30. Conclude that the only nilpotent 3-dimensional Lie algebras over $\mathbb{R}$ are the abelian one and the Heisenberg Lie algebra, conclude that the only other solvable ones of dimension 3 are those given by Problem 2 and the one that is a direct sum of a 1-dimensional abelian Lie algebra with a nonabelian 2-dimensional algebra.

Problems 31–35 show that the only simple Lie algebras $\mathfrak{g}$ of dimension 3 over $\mathbb{R}$, up to isomorphism, are the ones in Examples 12d and 12e of §1. In view of the discussion at the end of §2, Problems 28–30 and Problems 31–35 together classify all the Lie algebras of dimension 3 over $\mathbb{R}$.

31. Show that $\operatorname{Tr}(\operatorname{ad} X) = 0$ for all X because $[\mathfrak{g}, \mathfrak{g}] = \mathfrak{g}$.

32. Using Engel's Theorem, choose X_0 such that $\operatorname{ad} X_0$ is not nilpotent. Show that the 1-dimensional space $\mathbb{R}X_0$ has a complementary subspace stable under $\operatorname{ad} X_0$.

33. Show by linear algebra that some real multiple X of X_0 is a member of a basis $\{X, Y, Z\}$ of $\mathfrak{g}$ in which $\operatorname{ad} X$ has matrix realization either

$$\begin{matrix} & \begin{matrix} X & Y & Z \end{matrix} & \\ \operatorname{ad} X = & \begin{pmatrix} 0 & 0 & 0 \\ 0 & 2 & 0 \\ 0 & 0 & -2 \end{pmatrix} & \begin{matrix} X \\ Y \\ Z \end{matrix} \end{matrix} \quad \text{or} \quad \begin{matrix} & \begin{matrix} X & Y & Z \end{matrix} & \\ \operatorname{ad} X = & \begin{pmatrix} 0 & 0 & 0 \\ 0 & 0 & -1 \\ 0 & 1 & 0 \end{pmatrix} & \begin{matrix} X \\ Y \\ Z \end{matrix} \end{matrix}.$$

34. Writing $[Y, Z]$ in terms of the basis and applying the Jacobi identity, show that Y can be multiplied by a constant so that the first case of Problem 33 leads to an isomorphism with $\mathfrak{sl}(2, \mathbb{R})$ and the second case of Problem 33 leads to an isomorphism with $\mathfrak{so}(3)$.

35. Using a simplified version of the argument in Problems 29–32, show that the only 3-dimensional simple Lie algebra over $\mathbb{C}$, up to isomorphism, is $\mathfrak{sl}(2, \mathbb{C})$.

CHAPTER II

Complex Semisimple Lie Algebras

Abstract. The theme of this chapter is an investigation of complex semisimple Lie algebras by a two-step process, first by passing from such a Lie algebra to a reduced abstract root system via a choice of Cartan subalgebra and then by passing from the root system to an abstract Cartan matrix and an abstract Dynkin diagram via a choice of an ordering.

The chapter begins by making explicit a certain amount of this structure for four infinite classes of classical complex semisimple Lie algebras. Then for a general finite-dimensional complex Lie algebra, it is proved that Cartan subalgebras exist and are unique up to conjugacy.

When the given Lie algebra is semisimple, the Cartan subalgebra is abelian. The adjoint action of the Cartan subalgebra on the given semisimple Lie algebra leads to a root-space decomposition of the given Lie algebra, and the set of roots forms a reduced abstract root system.

If a suitable ordering is imposed on the underlying vector space of an abstract root system, one can define simple roots as those positive roots that are not sums of positive roots. The simple roots form a particularly nice basis of the underlying vector space, and a Cartan matrix and Dynkin diagram may be defined in terms of them. The definitions of abstract Cartan matrix and abstract Dynkin diagram are arranged so as to include the matrix and diagram obtained from a root system.

Use of the Weyl group shows that the Cartan matrix and Dynkin diagram obtained from a root system by imposing a ordering are in fact independent of the ordering. Moreover, nonisomorphic reduced abstract root systems have distinct Cartan matrices. It is possible to classify the abstract Cartan matrices and then to see by a case-by-case argument that every abstract Cartan matrix arises from a reduced abstract root system. Consequently the correspondence between reduced abstract root systems and abstract Cartan matrices is one-one onto, up to isomorphism.

The correspondence between complex semisimple Lie algebras and reduced abstract root systems lies deeper. Apart from isomorphism, the correspondence does not depend upon the choice of Cartan subalgebra, as a consequence of the conjugacy of Cartan subalgebras proved earlier in the chapter. To examine the correspondence more closely, one first finds generators and relations for any complex semisimple Lie algebra. The Isomorphism Theorem then explains how much freedom there is in lifting an isomorphism between root systems to an isomorphism between complex semisimple Lie algebras. Finally the Existence Theorem says that every reduced abstract root system arises from some complex semisimple Lie algebra. Consequently the correspondence between complex semisimple Lie algebras and reduced abstract root systems is one-one onto, up to isomorphism.

1. Classical Root Space Decompositions

Recall from §I.8 that the complex Lie algebras $\mathfrak{sl}(n, \mathbb{C})$ for $n \geq 2$, $\mathfrak{so}(n, \mathbb{C})$ for $n \geq 3$, and $\mathfrak{sp}(n, \mathbb{C})$ for $n \geq 1$ are all semisimple. As we shall see in this section, each of these Lie algebras has an abelian subalgebra $\mathfrak{h}$ such that an analysis of $\operatorname{ad} \mathfrak{h}$ leads to a rather complete understanding of the bracket law in the full Lie algebra. We shall give the analysis of $\operatorname{ad} \mathfrak{h}$ in each example and then, to illustrate the power of the formulas we have, identify which of these Lie algebras are simple over $\mathbb{C}$.

EXAMPLE 1. The complex Lie algebra is $\mathfrak{g} = \mathfrak{sl}(n, \mathbb{C})$. Let

$$\begin{aligned} \mathfrak{h}_0 &= \text{real diagonal matrices in } \mathfrak{g} \\ \mathfrak{h} &= \text{all diagonal matrices in } \mathfrak{g}. \end{aligned}$$

Then $\mathfrak{h} = \mathfrak{h}_0 \oplus i\mathfrak{h}_0 = (\mathfrak{h}_0)^{\mathbb{C}}$. Define a matrix E_{ij} to be 1 in the $(i, j)^{\text{th}}$ place and 0 elsewhere, and define a member e_i of the dual space $\mathfrak{h}^*$ by

$$e_i \begin{pmatrix} h_1 & & \\ & \ddots & \\ & & h_n \end{pmatrix} = h_i.$$

For each $H \in \mathfrak{h}$, $\operatorname{ad} H$ is diagonalized by the basis of $\mathfrak{g}$ consisting of members of $\mathfrak{h}$ and the E_{ij} for $i \neq j$. We have

$$(\operatorname{ad} H)E_{ij} = [H, E_{ij}] = (e_i(H) - e_j(H))E_{ij}.$$

In other words, E_{ij} is a simultaneous eigenvector for all $\operatorname{ad} H$, with eigenvalue $e_i(H) - e_j(H)$. In its dependence on H, the eigenvalue is linear. Thus the eigenvalue is a linear functional on $\mathfrak{h}$, namely $e_i - e_j$. The $(e_i - e_j)$'s, for $i \neq j$, are called **roots**. The set of roots is denoted Δ. We have

$$\mathfrak{g} = \mathfrak{h} \oplus \bigoplus_{i \neq j} \mathbb{C}E_{ij},$$

which we can rewrite as

$$\mathfrak{g} = \mathfrak{h} \oplus \bigoplus_{i \neq j} \mathfrak{g}_{e_i - e_j}, \tag{2.1}$$

where

$$\mathfrak{g}_{e_i-e_j} = \{X \in \mathfrak{g} \mid (\operatorname{ad} H)X = (e_i - e_j)(H)X \text{ for all } H \in \mathfrak{h}\}.$$

The decomposition (2.1) is called a **root-space decomposition**. Notice that Δ spans $\mathfrak{h}^*$ over $\mathbb{C}$.

The bracket relations are easy, relative to (2.1). If α and β are roots, we can compute $[E_{ij}, E_{i'j'}]$ and see that

$$(2.2) \qquad [\mathfrak{g}_\alpha, \mathfrak{g}_\beta] \begin{cases} = \mathfrak{g}_{\alpha+\beta} & \text{if } \alpha+\beta \text{ is a root} \\ = 0 & \text{if } \alpha+\beta \text{ is not a root or } 0 \\ \subseteq \mathfrak{h} & \text{if } \alpha+\beta = 0. \end{cases}$$

In the last case the exact formula is

$$[E_{ij}, E_{ji}] = E_{ii} - E_{jj} \in \mathfrak{h}.$$

All the roots are real on $\mathfrak{h}_0$ and thus, by restriction, can be considered as members of $\mathfrak{h}_0^*$. The next step is to introduce a notion of positivity within $\mathfrak{h}_0^*$ such that

(i) for any nonzero $\varphi \in \mathfrak{h}_0^*$, exactly one of φ and $-\varphi$ is positive
(ii) the sum of positive elements is positive, and any positive multiple of a positive element is positive.

The way in which such a notion of positivity is introduced is not important, and we shall just choose one at this stage.

To do so, we observe a canonical form for members of $\mathfrak{h}_0^*$. The linear functionals $e_1, \ldots, e_n$ span $\mathfrak{h}_0$, and their sum is 0. Any member of $\mathfrak{h}_0^*$ can therefore be written nonuniquely as $\sum_j c_j e_j$, and $(\sum_i c_i)(e_1+\cdots+e_n) = 0$. Therefore our given linear functional equals

$$\sum_{j=1}^{n}\Big(c_j - \frac{1}{n}\sum_{i=1}^{n} c_i\Big)e_j.$$

In this latter representation the sum of the cofficients is 0. Thus any member of $\mathfrak{h}_0^*$ can be realized as $\sum_j a_j e_j$ with $\sum_j a_j = 0$. No such nonzero expression can vanish on $E_{ll} - E_{nn}$ for all l with $1 \le l < n$, and thus the realization as $\sum_j a_j e_j$ with $\sum_j a_j = 0$ is unique.

If $\varphi = \sum_j a_j e_j$ is given as a member of $\mathfrak{h}_0^*$ with $\sum_j a_j = 0$, we say that a nonzero φ is **positive** (written $\varphi > 0$) if the first nonzero coefficient a_j is > 0. It is clear that this notion of positivity satisfies properties (i) and (ii) above.

We say that $\varphi > \psi$ if $\varphi - \psi$ is positive. The result is a simple ordering on $\mathfrak{h}_0^*$ that is preserved under addition and under multiplication by positive scalars.

For the roots the effect is that

$$\begin{aligned} e_1 - e_n > e_1 - e_{n-1} > \cdots > e_1 - e_2 \\ > e_2 - e_n > e_2 - e_{n-1} > \cdots > e_2 - e_3 \\ > \cdots > e_{n-2} - e_n > e_{n-2} - e_{n-1} > e_{n-1} - e_n > 0, \end{aligned}$$

and afterward we have the negatives. The positive roots are the $e_i - e_j$ with $i < j$.

Now let us prove that $\mathfrak{g}$ is simple over $\mathbb{C}$ for $n \geq 2$. Let $\mathfrak{a} \subseteq \mathfrak{g}$ be an ideal, and first suppose $\mathfrak{a} \subseteq \mathfrak{h}$. Let $H \neq 0$ be in $\mathfrak{a}$. Since the roots span $\mathfrak{h}^*$, we can find a root α with $\alpha(H) \neq 0$. If X is in $\mathfrak{g}_\alpha$ and $X \neq 0$, then

$$\alpha(H)X = [H, X] \in [\mathfrak{a}, \mathfrak{g}] \subseteq \mathfrak{a} \subseteq \mathfrak{h},$$

and so X is in $\mathfrak{h}$, contradiction. Hence $\mathfrak{a} \subseteq \mathfrak{h}$ implies $\mathfrak{a} = 0$.

Next, suppose $\mathfrak{a}$ is not contained in $\mathfrak{h}$. Let $X = H + \sum X_\alpha$ be in $\mathfrak{a}$ with each X_α in $\mathfrak{g}_\alpha$ and with some $X_\alpha \neq 0$. For the moment assume that there is some root $\alpha < 0$ with $X_\alpha \neq 0$, and let β be the smallest such α. Say $X_\beta = cE_{ij}$ with $i > j$ and $c \neq 0$. Form

$$[E_{1i}, [X, E_{jn}]]. \tag{2.3}$$

The claim is that (2.3) is a nonzero multiple of E_{1n}. In fact, we cannot have $i = 1$ since $j < i$. If $i < n$, then $[E_{ij}, E_{jn}] = aE_{in}$ with $a \neq 0$, and also $[E_{1i}, E_{in}] = bE_{1n}$ with $b \neq 0$. Thus (2.3) has a nonzero component in $\mathfrak{g}_{e_1 - e_n}$ in the decomposition (2.1). The other components of (2.3) must correspond to larger roots than $e_1 - e_n$ if they are nonzero, but $e_1 - e_n$ is the largest root. Hence the claim follows if $i < n$. If $i = n$, then (2.3) is

$$= [E_{1n}, [cE_{nj} + \cdots, E_{jn}]] = c[E_{1n}, E_{nn} - E_{jj}] + \cdots = cE_{1n}.$$

Thus the claim follows if $i = n$.

In any case we conclude that E_{1n} is in $\mathfrak{a}$. For $i \neq j$, the formula

$$E_{kl} = c'[E_{k1}, [E_{1n}, E_{nl}]] \qquad \text{with } c' \neq 0$$

(with obvious changes if $k = 1$ or $l = n$) shows that E_{kl} is in $\mathfrak{a}$, and

$$[E_{kl}, E_{lk}] = E_{kk} - E_{ll}$$

shows that a spanning set of $\mathfrak{h}$ is in $\mathfrak{a}$. Hence $\mathfrak{a} = \mathfrak{g}$.

Thus an ideal $\mathfrak{a}$ that is not in $\mathfrak{h}$ has to be all of $\mathfrak{g}$ if there is some $\alpha < 0$ with $X_\alpha \neq 0$ above. Similarly if there is some $\alpha > 0$ with $X_\alpha \neq 0$, let β be the largest such α, say $\alpha = e_i - e_j$ with $i < j$. Form $[E_{ni}, [X, E_{j1}]]$ and argue with E_{n1} in the same way to get $\mathfrak{a} = \mathfrak{g}$. Thus $\mathfrak{g}$ is simple over $\mathbb{C}$. This completes the first example.

We can abstract these properties. The complex Lie algebra $\mathfrak{g}$ will be simple whenever we can arrange that

1) $\mathfrak{h}$ is an abelian subalgebra of $\mathfrak{g}$ such that $\mathfrak{g}$ has a simultaneous eigenspace decomposition relative to $\operatorname{ad}\mathfrak{h}$ and

(a) the 0 eigenspace is $\mathfrak{h}$
(b) the other eigenspaces are 1-dimensional
(c) with the set Δ of roots defined as before, (2.2) holds
(d) the roots are all real on some real form $\mathfrak{h}_0$ of $\mathfrak{h}$.

2) the roots span $\mathfrak{h}^*$. If α is a root, so is $-\alpha$.

3) $\sum_{\alpha\in\Delta}[\mathfrak{g}_\alpha, \mathfrak{g}_{-\alpha}] = \mathfrak{h}$.

4) each root $\beta < 0$ relative to an ordering of $\mathfrak{h}_0^*$ defined from a notion of positivity satisfying (i) and (ii) above has the following property: There exists a sequence of roots $\alpha_1, \dots, \alpha_k$ such that each partial sum from the left of $\beta + \alpha_1 + \cdots + \alpha_k$ is a root or 0 and the full sum is the largest root. If a partial sum $\beta + \cdots + \alpha_j$ is 0, then the member $[E_{\alpha_j}, E_{-\alpha_j}]$ of $\mathfrak{h}$ is such that $\alpha_{j+1}([E_{\alpha_j}, E_{-\alpha_j}]) \neq 0$.

We shall see that the other complex Lie algebras from §I.8, namely $\mathfrak{so}(n, \mathbb{C})$ and $\mathfrak{sp}(n, \mathbb{C})$, have the same kind of structure, provided n is restricted suitably.

EXAMPLE 2. The complex Lie algebra is $\mathfrak{g} = \mathfrak{so}(2n+1, \mathbb{C})$. Here a similar analysis by means of $\operatorname{ad}\mathfrak{h}$ for an abelian subalgebra $\mathfrak{h}$ is possible, and we shall say what the constructs are that lead to the conclusion that $\mathfrak{g}$ is simple for $n \geq 1$. We define

$$\mathfrak{h} = \{H \in \mathfrak{so}(2n+1, \mathbb{C}) \mid H = \text{matrix below}\}$$

$$H = \begin{pmatrix} \begin{pmatrix} 0 & ih_1 \\ -ih_1 & 0 \end{pmatrix} & & & & \\ & \begin{pmatrix} 0 & ih_2 \\ -ih_2 & 0 \end{pmatrix} & & & \\ & & \ddots & & \\ & & & \begin{pmatrix} 0 & ih_n \\ ih_n & 0 \end{pmatrix} & \\ & & & & 0 \end{pmatrix}$$

$$e_j(\text{above } H) = h_j, \qquad 1 \leq j \leq n$$
$$\mathfrak{h}_0 = \{H \in \mathfrak{h} \mid \text{entries are purely imaginary}\}$$
$$\Delta = \{\pm e_i \pm e_j \text{ with } i \neq j\} \cup \{\pm e_k\}.$$

The members of $\mathfrak{h}_0^*$ are the linear functionals $\sum_j a_j e_j$ with all a_j real, and every root is of this form. A member $\varphi = \sum_j a_j e_j$ of $\mathfrak{h}_0^*$ is defined to be positive if $\varphi \neq 0$ and if the first nonzero a_j is positive. In the resulting ordering the largest root is $e_1 + e_2$. The root space decomposition is

$$\mathfrak{g} = \mathfrak{h} \oplus \bigoplus_{\alpha \in \Delta} \mathfrak{g}_\alpha \qquad \text{with } \mathfrak{g}_\alpha = \mathbb{C} E_\alpha$$

and with E_α as defined below. To define E_α, first let $i < j$ and let $\alpha = \pm e_i \pm e_j$. Then E_α is 0 except in the sixteen entries corresponding to the ith and j^{th} pairs of indices, where it is

$$E_\alpha = \begin{array}{c} \begin{matrix} i & \; j \end{matrix} \\ \begin{pmatrix} 0 & X_\alpha \\ -X_\alpha^t & 0 \end{pmatrix} \end{array} \begin{matrix} \\ i \\ j \end{matrix}$$

with

$$X_{e_i - e_j} = \begin{pmatrix} 1 & i \\ -i & 1 \end{pmatrix}, \qquad X_{e_i + e_j} = \begin{pmatrix} 1 & -i \\ -i & -1 \end{pmatrix},$$

$$X_{-e_i + e_j} = \begin{pmatrix} 1 & -i \\ i & 1 \end{pmatrix}, \qquad X_{-e_i - e_j} = \begin{pmatrix} 1 & i \\ i & -1 \end{pmatrix}.$$

To define E_α for $\alpha = \pm e_k$, write

$$E_\alpha = \begin{array}{c} \begin{matrix} \text{pair} & \text{entry} \\ k & 2n+1 \end{matrix} \\ \begin{pmatrix} 0 & X_\alpha \\ -X_\alpha^t & 0 \end{pmatrix} \end{array}$$

with 0's elsewhere and with

$$X_{e_k} = \begin{pmatrix} 1 \\ -i \end{pmatrix} \qquad \text{and} \qquad X_{-e_k} = \begin{pmatrix} 1 \\ i \end{pmatrix}.$$

EXAMPLE 3. The complex Lie algebra is $\mathfrak{g} = \mathfrak{sp}(n, \mathbb{C})$. Again an analysis by means of $\operatorname{ad} \mathfrak{h}$ for an abelian subalgebra $\mathfrak{h}$ is possible, and we shall say what the constructs are that lead to the conclusion that $\mathfrak{g}$ is simple for $n \geq 1$. We define

$$\mathfrak{h} = \left\{ H = \begin{pmatrix} h_1 & & & & & \\ & \ddots & & & & \\ & & h_n & & & \\ & & & -h_1 & & \\ & & & & \ddots & \\ & & & & & -h_n \end{pmatrix} \right\}$$

$$e_j(\text{above } H) = h_j, \qquad 1 \leq j \leq n$$
$$\mathfrak{h}_0 = \{H \in \mathfrak{h} \mid \text{entries are real}\}$$
$$\Delta = \{\pm e_i \pm e_j \text{ with } i \neq j\} \cup \{\pm 2e_k\}$$

$$\begin{aligned} E_{e_i - e_j} &= E_{i,j} - E_{j+n,i+n}, & E_{2e_k} &= E_{k,k+n}, \\ E_{e_i + e_j} &= E_{i,j+n} + E_{j,i+n}, & E_{-2e_k} &= E_{k+n,k}, \\ E_{-e_i - e_j} &= E_{i+n,j} + E_{j+n,i}. & & \end{aligned}$$

EXAMPLE 4. The complex Lie algebra is $\mathfrak{g} = \mathfrak{so}(2n, \mathbb{C})$. The analysis is similar to that for $\mathfrak{so}(2n+1, \mathbb{C})$. The Lie algebra $\mathfrak{so}(2n, \mathbb{C})$ is simple over $\mathbb{C}$ for $n \geq 3$, the constructs for this example being

$\mathfrak{h}$ as with $\mathfrak{so}(2n+1, \mathbb{C})$ but with the last row and column deleted

$e_j(H) = h_j, \quad 1 \leq j \leq n, \quad$ as with $\mathfrak{so}(2n+1, \mathbb{C})$

$\mathfrak{h}_0 = \{H \in \mathfrak{h} \mid \text{entries are purely imaginary}\}$

$\Delta = \{\pm e_i \pm e_j \text{ with } i \neq j\}$

E_α as for $\mathfrak{so}(2n+1, \mathbb{C})$ when $\alpha = \pm e_i \pm e_j$.

When $n = 2$, condition (4) in the list of abstracted properties fails. In fact, take $\beta = -e_1 + e_2$. The only choice for α_1 is $e_1 - e_2$, and then $\beta + \alpha_1 = 0$. We have to choose $\alpha_2 = e_1 + e_2$, and $\alpha_2([E_{\alpha_1}, E_{-\alpha_1}]) = 0$. We shall see in §5 that $\mathfrak{so}(4, \mathbb{C})$ is actually not simple.

2. Existence of Cartan Subalgebras

The idea is to approach a general complex semisimple Lie algebra $\mathfrak{g}$ by imposing on it the same kind of structure as in §1. We try to construct an $\mathfrak{h}$, a set of roots, a real form $\mathfrak{h}_0$ on which the roots are real, and an ordering on $\mathfrak{h}_0^*$. Properties (1) through (3) in §1 turn out actually to be equivalent with $\mathfrak{g}$ semisimple. In the presence of the first three properties, property (4) will be equivalent with $\mathfrak{g}$ simple. But we shall obtain better formulations of property (4) later, and that property should be disregarded, at least for the time being.

The hypothesis of semisimplicity of $\mathfrak{g}$ enters the construction only by forcing special features of $\mathfrak{h}$ and the roots. Accordingly we work with a general finite-dimensional complex Lie algebra $\mathfrak{g}$ until near the end of this section.

Let $\mathfrak{h}$ be a finite-dimensional Lie algebra over $\mathbb{C}$. Recall from §I.5 that a representation π of $\mathfrak{h}$ on a complex vector space V is a complex-linear Lie algebra homomorphism of $\mathfrak{h}$ into $\mathrm{End}_{\mathbb{C}}(V)$. For such π and V, whenever α is in the dual $\mathfrak{h}^*$, we let V_α be defined as

$$\{v \in V \mid (\pi(H) - \alpha(H)1)^n v = 0 \text{ for all } H \in \mathfrak{h} \text{ and some } n = n(H, v)\}.$$

If $V_\alpha \neq 0$, V_α is called a **generalized weight space** and α is a **weight**. Members of V_α are called **generalized weight vectors.**

For now, we shall be interested only in the case that V is finite-dimensional. In this case $\pi(H) - \alpha(H)1$ has 0 as its only generalized eigenvalue on V_α and is nilpotent on this space, as a consequence of the theory of Jordan normal form. Therefore $n(H, v)$ can be taken to be $\dim V$.

Proposition 2.4. Suppose that $\mathfrak{h}$ is a nilpotent Lie algebra over $\mathbb{C}$ and that π is a representation of $\mathfrak{h}$ on a finite-dimensional complex vector space V. Then there are finitely many generalized weights, each generalized weight space is stable under $\pi(\mathfrak{h})$, and V is the direct sum of all the generalized weight spaces.

REMARKS.

1) The direct-sum decomposition of V as the sum of the generalized weight spaces is called a **weight-space decomposition** of V.

2) The weights need not be linearly independent. For example, they are dependent in our root-space decompositions in the previous section.

3) Since $\mathfrak{h}$ is nilpotent, it is solvable, and Lie's Theorem (Corollary 1.29) applies to it. In a suitable basis of V, $\pi(\mathfrak{h})$ is therefore simultaneously triangular. The generalized weights will be the distinct diagonal entries, as functions on $\mathfrak{h}$. To get the direct sum decomposition, however, is subtler; we need to make more serious use of the fact that $\mathfrak{h}$ is nilpotent.

PROOF. First we check that V_α is invariant under $\pi(\mathfrak{h})$. Fix $H \in \mathfrak{h}$ and let

$$V_{\alpha,H} = \{v \in V \mid (\pi(H) - \alpha(H)1)^n v = 0 \text{ for some } n = n(v)\},$$

so that $V_\alpha = \bigcap_{H\in\mathfrak{h}} V_{\alpha,H}$. It is enough to prove that $V_{\alpha,H}$ is invariant under $\pi(\mathfrak{h})$ if $H \neq 0$. Since $\mathfrak{h}$ is nilpotent, ad H is nilpotent. Let

$$\mathfrak{h}_{(m)} = \{Y \in \mathfrak{h} \mid (\operatorname{ad} H)^m Y = 0\},$$

so that $\mathfrak{h} = \bigcup_{m=0}^{d} \mathfrak{h}_{(m)}$ with $d = \dim \mathfrak{h}$. We prove that $\pi(Y)V_{\alpha,H} \subseteq V_{\alpha,H}$ for $Y \in \mathfrak{h}_{(m)}$ by induction on m.

For $m = 0$, we have $\mathfrak{h}_{(0)} = 0$ since $(\operatorname{ad} H)^0 = 1$. So $\pi(Y) = \pi(0) = 0$, and $\pi(Y)V_{\alpha,H} \subseteq V_{\alpha,H}$ trivially.

We now address general m under the assumption that our assertion is true for all $Z \in \mathfrak{h}_{(m-1)}$. Let Y be in $\mathfrak{h}_{(m)}$. Then $[H, Y]$ is in $\mathfrak{h}_{(m-1)}$, and we have

$$\begin{aligned}(\pi(H) - \alpha(H)1)\pi(Y) &= \pi([H, Y]) + \pi(Y)\pi(H) - \alpha(H)\pi(Y) \\ &= \pi(Y)(\pi(H) - \alpha(H)1) + \pi([H, Y])\end{aligned}$$

and

$$\begin{aligned}&(\pi(H) - \alpha(H)1)^2\pi(Y) \\ &= (\pi(H) - \alpha(H)1)\pi(Y)(\pi(H) - \alpha(H)1) + (\pi(H) - \alpha(H)1)\pi([H, Y]) \\ &= \pi(Y)(\pi(H) - \alpha(H)1)^2 + \pi([H, Y])(\pi(H) - \alpha(H)1) \\ &\quad + (\pi(H) - \alpha(H)1)\pi([H, Y]).\end{aligned}$$

Iterating, we obtain

$$\begin{aligned}&(\pi(H) - \alpha(H)1)^l\pi(Y) \\ &\qquad = \pi(Y)(\pi(H) - \alpha(H)1)^l \\ &\qquad\quad + \sum_{s=0}^{l-1}(\pi(H) - \alpha(H)1)^{l-1-s}\pi([H, Y])(\pi(H) - \alpha(H)1)^s.\end{aligned}$$

For $v \in V_{\alpha,H}$, we have $(\pi(H) - \alpha(H)1)^N v = 0$ if $N \geq \dim V$. Take $l = 2N$. When the above expression is applied to v, the only terms in the sum on the right side that can survive are those with $s < N$. For these we have $l - 1 - s \geq N$. Then $(\pi(H) - \alpha(H)1)^s v$ is in $V_{\alpha,H}$, $\pi([H, Y])$ leaves $V_{\alpha,H}$ stable since $[H, Y]$ is in $\mathfrak{h}_{(m-1)}$, and

$$(\pi(H) - \alpha(H)1)^{l-1-s}\pi([H, Y])(\pi(H) - \alpha(H)1)^s v = 0.$$

Hence $(\pi(H) - \alpha(H)1)^l\pi(Y)v = 0$, and $V_{\alpha,H}$ is stable under $\pi(Y)$. This completes the induction and the proof that V_α is invariant under $\pi(\mathfrak{h})$.

Now we can obtain the decomposition $V = \bigoplus_\alpha V_\alpha$. Let $H_1, \ldots, H_r$ be a basis for $\mathfrak{h}$. The Jordan decomposition of $\pi(H_1)$ gives us a generalized eigenspace decomposition that we can write as

$$V = \bigoplus_{\lambda} V_{\lambda, H_1}.$$

Here we can regard the complex number λ as running over all distinct values of $\alpha(H_1)$ for α arbitrary in $\mathfrak{h}^*$. Thus we can rewrite the Jordan decomposition as

$$V = \bigoplus_{\substack{\text{values of} \\ \alpha(H_1)}} V_{\alpha(H_1), H_1}.$$

For fixed $\alpha \in \mathfrak{h}^*$, $V_{\alpha(H_1),H_1}$ is nothing more than the space V_{α,H_1} defined at the start of the proof. From what we have already shown, the space $V_{\alpha(H_1),H_1} = V_{\alpha,H_1}$ is stable under $\pi(\mathfrak{h})$. Thus we can decompose it under $\pi(H_2)$ as

$$V = \bigoplus_{\alpha(H_1)} \bigoplus_{\alpha(H_2)} (V_{\alpha(H_1),H_1} \cap V_{\alpha(H_2),H_2}),$$

and we can iterate to obtain

$$V = \bigoplus_{\alpha(H_1),\ldots,\alpha(H_r)} \Big(\bigcap_{j=1}^{r} V_{\alpha(H_j),H_j} \Big)$$

with each of the spaces invariant under $\pi(\mathfrak{h})$. By Lie's Theorem (Corollary 1.29), we can regard all $\pi(H_i)$ as acting simultaneously by triangular matrices on $\bigcap_{j=1}^r V_{\alpha(H_j),H_j}$, evidently with all diagonal entries $\alpha(H_i)$. Then $\pi(\sum c_i H_i)$ must act as a triangular matrix with all diagonal entries $\sum c_i \alpha(H_i)$. Thus if we define a linear functional α by $\alpha(\sum c_i H_i) = \sum c_i \alpha(H_i)$, we see that $\bigcap_{j=1}^r V_{\alpha(H_j),H_j}$ is exactly V_α. Thus $V = \bigoplus_\alpha V_\alpha$, and in particular there are only finitely many weights.

Proposition 2.5. If $\mathfrak{g}$ is any finite-dimensional Lie algebra over $\mathbb{C}$ and if $\mathfrak{h}$ is a nilpotent Lie subalgebra, then the generalized weight spaces of $\mathfrak{g}$ relative to $\operatorname{ad}_{\mathfrak{g}} \mathfrak{h}$ satisfy

(a) $\mathfrak{g} = \bigoplus \mathfrak{g}_\alpha$, where $\mathfrak{g}_\alpha$ is defined as

$$\{X \in \mathfrak{g} \mid (\operatorname{ad} H - \alpha(H)1)^n X = 0 \text{ for all } H \in \mathfrak{h} \text{ and some } n = n(H, X)\}$$

(b) $\mathfrak{h} \subseteq \mathfrak{g}_0$

(c) $[\mathfrak{g}_\alpha, \mathfrak{g}_\beta] \subseteq \mathfrak{g}_{\alpha+\beta}$ (with $\mathfrak{g}_{\alpha+\beta}$ understood to be 0 if $\alpha + \beta$ is not a generalized weight)

PROOF.
(a) This is by Proposition 2.4.
(b) Since $\mathfrak{h}$ is nilpotent, $\operatorname{ad}\mathfrak{h}$ is nilpotent on $\mathfrak{h}$. Thus $\mathfrak{h} \subseteq \mathfrak{g}_0$.
(c) Let $X \in \mathfrak{g}_\alpha$, $Y \in \mathfrak{g}_\beta$, and $H \in \mathfrak{h}$. Then

$$\begin{aligned}(\operatorname{ad} H &- (\alpha(H) + \beta(H))1)[X, Y] \\ &= [H, [X, Y]] - \alpha(H)[X, Y] - \beta(H)[X, Y] \\ &= [(\operatorname{ad} H - \alpha(H)1)X, Y] + [X, (\operatorname{ad} H - \beta(H)1)Y],\end{aligned}$$

and we can readily set up an induction to see that

$$\begin{aligned}(\operatorname{ad} H &- (\alpha(H) + \beta(H))1)^n[X, Y] \\ &= \sum_{k=0}^{n} \binom{n}{k} [(\operatorname{ad} H - \alpha(H)1)^k X, (\operatorname{ad} H - \beta(H)1)^{n-k} Y].\end{aligned}$$

If $n \geq 2\dim\mathfrak{g}$, either k or $n-k$ is $\geq \dim\mathfrak{g}$, and hence every term on the right side is 0.

Corollary 2.6. $\mathfrak{g}_0$ is a subalgebra.

PROOF. This follows from Proposition 2.5c.

To match the behavior of our examples in the previous section, we make the following definition. A nilpotent Lie subalgebra $\mathfrak{h}$ of a finite-dimensional complex Lie algebra $\mathfrak{g}$ is a **Cartan subalgebra** if $\mathfrak{h} = \mathfrak{g}_0$. Note that $\mathfrak{h} \subseteq \mathfrak{g}_0$ is always guaranteed by Proposition 2.5b.

Proposition 2.7. A nilpotent Lie subalgebra $\mathfrak{h}$ of a finite-dimensional complex Lie algebra $\mathfrak{g}$ is a Cartan subalgebra if and only if $\mathfrak{h}$ equals the normalizer $N_\mathfrak{g}(\mathfrak{h}) = \{X \in \mathfrak{g} \mid [X, \mathfrak{h}] \subseteq \mathfrak{h}\}$.

PROOF. We always have

$$\mathfrak{h} \subseteq N_\mathfrak{g}(\mathfrak{h}) \subseteq \mathfrak{g}_0. \tag{2.8}$$

The first of these inclusions holds because $\mathfrak{h}$ is a Lie subalgebra. The second holds because $(\operatorname{ad} H)^n X = (\operatorname{ad} H)^{n-1}[H, X]$ and $\operatorname{ad} H$ is nilpotent on $\mathfrak{h}$.

Now assume that $\mathfrak{h}$ is a Cartan subalgebra. Then $\mathfrak{g}_0 = \mathfrak{h}$ by definition. By (2.8), $\mathfrak{h} = N_\mathfrak{g}(\mathfrak{h}) = \mathfrak{g}_0$. Conversely assume that $\mathfrak{h}$ is not a Cartan subalgebra, i.e., that $\mathfrak{g}_0 \neq \mathfrak{h}$. Form $\operatorname{ad}\mathfrak{h} : \mathfrak{g}_0/\mathfrak{h} \to \mathfrak{g}_0/\mathfrak{h}$ as a Lie algebra of transformations of the nonzero vector space $\mathfrak{g}_0/\mathfrak{h}$. Since $\mathfrak{h}$ is solvable, this Lie algebra of transformations is solvable. By Lie's Theorem (Theorem 1.25) there exists an $X + \mathfrak{h}$ in $\mathfrak{g}_0/\mathfrak{h}$ with $X \notin \mathfrak{h}$ that is a simultaneous eigenvector for $\operatorname{ad}\mathfrak{h}$, and we know that its simultaneous eigenvalue has to be 0. This means that $(\operatorname{ad} H)(X + \mathfrak{h}) \subseteq \mathfrak{h}$, i.e., $[H, X]$ is in $\mathfrak{h}$. Hence X is not in $\mathfrak{h}$ but X is in $N_\mathfrak{g}(\mathfrak{h})$. Thus $\mathfrak{h} \neq N_\mathfrak{g}(\mathfrak{h})$.

Theorem 2.9. Any finite-dimensional complex Lie algebra $\mathfrak{g}$ has a Cartan subalgebra.

Before coming to the proof, we introduce "regular" elements of $\mathfrak{g}$. In $\mathfrak{sl}(n, \mathbb{C})$ the regular elements will be the matrices with distinct eigenvalues. Let us consider matters more generally.

If π is a representation of $\mathfrak{g}$ on a finite-dimensional vector space V, we can regard each $X \in \mathfrak{g}$ as generating a 1-dimensional abelian subalgebra, and we can then form $V_{0,X}$, the generalized eigenspace for eigenvalue 0 under $\pi(X)$. Let

$$l_{\mathfrak{g}}(V) = \min_{X \in \mathfrak{g}} \dim V_{0,X}$$

$$R_{\mathfrak{g}}(V) = \{X \in \mathfrak{g} \mid \dim V_{0,X} = l_{\mathfrak{g}}(V)\}.$$

To understand $l_{\mathfrak{g}}(V)$ and $R_{\mathfrak{g}}(V)$ better, form the characteristic polynomial

$$\det(\lambda 1 - \pi(X)) = \lambda^n + \sum_{j=0}^{n-1} d_j(X)\lambda^j.$$

In any basis of $\mathfrak{g}$, the $d_j(X)$ are polynomial functions on $\mathfrak{g}$, as we see by expanding $\det(\lambda 1 - \sum \mu_i \pi(X_i))$. For given X, if j is the smallest value for which $d_j(X) \neq 0$, then $j = \dim V_{0,X}$, since the degree of the last term in the characteristic polynomial is the multiplicity of 0 as a generalized eigenvalue of $\pi(X)$. Thus $l_{\mathfrak{g}}(V)$ is the minimum j such that $d_j(X) \not\equiv 0$, and

$$R_{\mathfrak{g}}(V) = \{X \in \mathfrak{g} \mid d_{l_{\mathfrak{g}}(V)}(X) \neq 0\}.$$

Let us apply these considerations to the adjoint representation of $\mathfrak{g}$ on $\mathfrak{g}$. The elements of $R_{\mathfrak{g}}(\mathfrak{g})$, relative to the adjoint representation, are the **regular elements** of $\mathfrak{g}$. For any X in $\mathfrak{g}$, $\mathfrak{g}_{0,X}$ is a Lie subalgebra of $\mathfrak{g}$ by the corollary of Proposition 2.5, with $\mathfrak{h} = \mathbb{C}X$.

Theorem 2.9′. If X is a regular element of the finite-dimensional complex Lie algebra $\mathfrak{g}$, then the Lie algebra $\mathfrak{g}_{0,X}$ is a Cartan subalgebra of $\mathfrak{g}$.

PROOF. First we show that $\mathfrak{g}_{0,X}$ is nilpotent. Assuming the contrary, we construct two sets:

(i) the set of $Z \in \mathfrak{g}_{0,X}$ such that $(\operatorname{ad} Z)|_{\mathfrak{g}_{0,X}})^{\dim \mathfrak{g}_{0,X}} \neq 0$, which is nonempty by Engel's Theorem (Corollary 1.38) and is open

(ii) the set of $W \in \mathfrak{g}_{0,X}$ such that $\operatorname{ad} W|_{\mathfrak{g}/\mathfrak{g}_{0,X}}$ is nonsingular, which is nonempty since X is in it (regularity is not used here) and is the set where some polynomial is nonvanishing, hence is dense (because if a polynomial vanishes on a nonempty open set, it vanishes identically).

These two sets must have nonempty intersection, and so we can find $Z \in \mathfrak{g}_{0,X}$ such that

$$(\operatorname{ad} Z)|_{\mathfrak{g}_{0,X}})^{\dim \mathfrak{g}_{0,X}} \neq 0 \qquad \text{and} \qquad \operatorname{ad} Z|_{\mathfrak{g}/\mathfrak{g}_{0,X}} \text{ is nonsingular.}$$

Then the generalized multiplicity of the eigenvalue 0 for $\operatorname{ad} Z$ is less than $\dim \mathfrak{g}_{0,X}$, and hence $\dim \mathfrak{g}_{0,Z} < \dim \mathfrak{g}_{0,X}$, in contradiction with the regularity of X. We conclude that $\mathfrak{g}_{0,X}$ is nilpotent.

Since $\mathfrak{g}_{0,X}$ is nilpotent, we can use $\mathfrak{g}_{0,X}$ to decompose $\mathfrak{g}$ as in Proposition 2.4. Let $\mathfrak{g}_0$ be the 0 generalized weight space. Then we have

$$\mathfrak{g}_{0,X} \subseteq \mathfrak{g}_0 = \bigcap_{Y \in \mathfrak{g}_{0,X}} \mathfrak{g}_{0,Y} \subseteq \mathfrak{g}_{0,X}.$$

So $\mathfrak{g}_{0,X} = \mathfrak{g}_0$, and $\mathfrak{g}_{0,X}$ is a Cartan subalgebra.

In this book we shall be interested in Cartan subalgebras $\mathfrak{h}$ only when $\mathfrak{g}$ is semisimple. In this case $\mathfrak{h}$ has special properties, as follows.

Proposition 2.10. If $\mathfrak{g}$ is a complex semisimple Lie algebra and $\mathfrak{h}$ is a Cartan subalgebra, then $\mathfrak{h}$ is abelian.

PROOF. Since $\mathfrak{h}$ is nilpotent and therefore solvable, $\operatorname{ad} \mathfrak{h}$ is solvable as a Lie algebra of transformations of $\mathfrak{g}$. By Lie's Theorem (Corollary 1.29) it is simultaneously triangular in some basis. For any three triangular matrices A, B, C, we have $\operatorname{Tr}(ABC) = \operatorname{Tr}(BAC)$. Therefore

$$\operatorname{Tr}(\operatorname{ad}[H_1, H_2]\operatorname{ad} H) = 0 \qquad \text{for } H_1,\ H_2,\ H \in \mathfrak{h}. \tag{2.11}$$

Next let α be any nonzero generalized weight, let X be in $\mathfrak{g}_\alpha$, and let H be in $\mathfrak{h}$. By Proposition 2.5c, $\operatorname{ad} H \operatorname{ad} X$ carries $\mathfrak{g}_\beta$ to $\mathfrak{g}_{\alpha+\beta}$. Thus Proposition 2.5a shows that

$$\operatorname{Tr}(\operatorname{ad} H \operatorname{ad} X) = 0. \tag{2.12}$$

Specializing (2.12) to $H = [H_1, H_2]$ and using (2.11) and Proposition 2.5a, we see that the Killing form B of $\mathfrak{g}$ satisfies

$$B([H_1, H_2], X) = 0 \qquad \text{for all } X \in \mathfrak{g}.$$

By Cartan's Criterion for Semisimplicity (Theorem 1.42), B is nondegenerate. Therefore $[H_1, H_2] = 0$, and $\mathfrak{h}$ is abelian.

Corollary 2.13. In a complex semisimple Lie algebra $\mathfrak{g}$, a Lie subalgebra $\mathfrak{h}$ is a Cartan subalgebra if $\mathfrak{h}$ is maximal abelian and $\operatorname{ad}_{\mathfrak{g}} \mathfrak{h}$ is simultaneously diagonable.

REMARKS.
1) It is immediate from this corollary that the subalgebras $\mathfrak{h}$ in the examples of §1 are Cartan subalgebras.
2) In the direction converse to the corollary, Proposition 2.10 shows that a Cartan subalgebra $\mathfrak{h}$ is abelian, and it is maximal abelian since $\mathfrak{h} = \mathfrak{g}_0$. Corollary 2.22 will show for a Cartan subalgebra $\mathfrak{h}$ that $\operatorname{ad}_{\mathfrak{g}} \mathfrak{h}$ is simultaneously diagonable.

PROOF. Since $\mathfrak{h}$ is abelian and hence nilpotent, Proposition 2.4 shows that $\mathfrak{g}$ has a weight-space decomposition $\mathfrak{g} = \mathfrak{g}_0 \oplus \bigoplus_{\beta \neq 0} \mathfrak{g}_\beta$. Since $\operatorname{ad}_{\mathfrak{g}} \mathfrak{h}$ is simultaneously diagonable, $\mathfrak{g}_0 = \mathfrak{h} \oplus \mathfrak{r}$ with $[\mathfrak{h}, \mathfrak{r}] = 0$. In view of Proposition 2.7, we are to prove that $\mathfrak{h} = N_{\mathfrak{g}}(\mathfrak{h})$. Here $\mathfrak{h} \subseteq N_{\mathfrak{g}}(\mathfrak{h}) \subseteq \mathfrak{g}_0$ by (2.8), and it is enough to show that $\mathfrak{r} = 0$. If $X \neq 0$ is in $\mathfrak{r}$, then $\mathfrak{h} \oplus \mathbb{C}X$ is an abelian subalgebra properly containing $\mathfrak{h}$, in contradiction with $\mathfrak{h}$ maximal abelian. The result follows.

3. Uniqueness of Cartan Subalgebras

We turn to the question of uniqueness of Cartan subalgebras. We begin with a lemma about polynomial mappings.

Lemma 2.14. Let $P : \mathbb{C}^m \to \mathbb{C}^n$ be a holomorphic polynomial function not identically 0. Then the set of vectors z in $\mathbb{C}^m$ for which $P(z)$ is not the 0 vector is connected in $\mathbb{C}^m$.

PROOF. Suppose that z_0 and w_0 in $\mathbb{C}^m$ have $P(z_0) \neq 0$ and $P(w_0) \neq 0$. As a function of $z \in \mathbb{C}$, $P(z_0 + z(w_0 - z_0))$ is a vector-valued holomorphic polynomial nonvanishing at $z = 0$ and $z = 1$. The subset of $z \in \mathbb{C}$ where it vanishes is finite, and the complement in $\mathbb{C}$ is connected. Thus z_0 and w_0 lie in a connected set in $\mathbb{C}^m$ where P is nonvanishing. Taking the union of these connected sets with z_0 fixed and w_0 varying, we see that the set where $P(w_0) \neq 0$ is connected.

Theorem 2.15. If $\mathfrak{h}_1$ and $\mathfrak{h}_2$ are Cartan subalgebras of a finite-dimensional complex Lie algebra $\mathfrak{g}$, then there exists $a \in \operatorname{Int} \mathfrak{g}$ with $a(\mathfrak{h}_1) = \mathfrak{h}_2$.

REMARKS.
1) In particular any two Cartan subalgebras are conjugate by an automorphism of $\mathfrak{g}$. As was explained after the introduction of $\operatorname{Int} \mathfrak{g}$ in §I.11,

$\operatorname{Int}\mathfrak{g} = \operatorname{Int}\mathfrak{g}^{\mathbb{R}}$ is a universal version of $\operatorname{Ad}(G)$ for analytic groups G with Lie algebra $\mathfrak{g}^{\mathbb{R}}$. Thus if G is some analytic group with Lie algebra $\mathfrak{g}^{\mathbb{R}}$, the theorem asserts that the conjugacy can be achieved by some automorphism $\operatorname{Ad}(g)$ with $g \in G$.

2) By the theorem all Cartan subalgebras of $\mathfrak{g}$ have the same dimension. The common value of this dimension is called the **rank** of $\mathfrak{g}$.

PROOF. Let $\mathfrak{h}$ be a Cartan subalgebra of $\mathfrak{g}$. Under the definitions in §2,

$$R_{\mathfrak{h}}(\mathfrak{g}) = \{Y \in \mathfrak{h} \mid \dim \mathfrak{g}_{0,Y} \text{ is a minimum for elements of } \mathfrak{h}\}.$$

We shall show that

(a) two alternative formulas for $R_{\mathfrak{h}}(\mathfrak{g})$ are

$$\begin{aligned} R_{\mathfrak{h}}(\mathfrak{g}) &= \{Y \in \mathfrak{h} \mid \alpha(Y) \neq 0 \text{ for all generalized weights } \alpha \neq 0\} \\ &= \{Y \in \mathfrak{h} \mid \mathfrak{g}_{0,Y} = \mathfrak{h}\} \end{aligned}$$

(b) $Y \in R_{\mathfrak{h}}(\mathfrak{g})$ implies $\operatorname{ad} Y$ is nonsingular on $\bigoplus_{\alpha\neq 0} \mathfrak{g}_\alpha$

(c) the image of the map

$$\sigma : \operatorname{Int}\mathfrak{g} \times R_{\mathfrak{h}}(\mathfrak{g}) \to \mathfrak{g}$$

given by $\sigma(a, Y) = a(Y)$ is open in $\mathfrak{g}$ and is contained in $R_{\mathfrak{g}}(\mathfrak{g})$

(d) if $\mathfrak{h}_1$ and $\mathfrak{h}_2$ are Cartan subalgebras that are not conjugate by $\operatorname{Int}\mathfrak{g}$, then the corresponding images of the maps in (c) are disjoint

(e) every member of $R_{\mathfrak{g}}(\mathfrak{g})$ is in the image of the map in (c) for some Cartan subalgebra $\mathfrak{h}$

(f) $R_{\mathfrak{g}}(\mathfrak{g})$ is connected.

These six statements prove the theorem. In fact, (c) through (e) exhibit $R_{\mathfrak{g}}(\mathfrak{g})$ as a nontrivial disjoint union of open sets if we have nonconjugacy. But (f) says that such a nontrivial disjoint union is impossible. Thus let us prove the six statements.

(a) Since $\mathfrak{h}$ is a Cartan subalgebra, $\mathfrak{g} = \mathfrak{h} \oplus \bigoplus_{\alpha\neq 0} \mathfrak{g}_\alpha$. If Y is in $\mathfrak{h}$, then $\mathfrak{g}_{0,Y} = \{X \in \mathfrak{g} \mid (\operatorname{ad} Y)^n X = 0\}$, where $n = \dim \mathfrak{g}$. Thus elements X in $\mathfrak{g}_{0,Y}$ are characterized by being in the generalized eigenspace for $\operatorname{ad} Y$ with eigenvalue 0. So $\mathfrak{g}_{0,Y} = \mathfrak{h} \oplus \bigoplus_{\alpha\neq 0, \alpha(Y)=0} \mathfrak{g}_\alpha$. Since finitely many hyperplanes in $\mathfrak{h}$ cannot have union $\mathfrak{h}$ ($\mathbb{C}$ being an infinite field), we can find Y with $\alpha(Y) \neq 0$ for all $\alpha \neq 0$. Then we see that $\mathfrak{g}_{0,Y}$ is smallest when it is $\mathfrak{h}$, and (a) follows.

(b) The linear map $\operatorname{ad} Y$ acts on $\mathfrak{g}_\alpha$ with generalized eigenvalue $\alpha(Y) \neq 0$, by (a). Hence $\operatorname{ad} Y$ is nonsingular on each $\mathfrak{g}_\alpha$.

(c) Since $\operatorname{Int}\mathfrak{g}$ is a group, it is enough to show that $Y \in R_{\mathfrak{h}}(\mathfrak{g})$ implies that $(\operatorname{Int}\mathfrak{g})(R_{\mathfrak{h}}(\mathfrak{g}))$ contains a neighborhood of Y in $\mathfrak{g}$. Form the differential $d\sigma$ at the point $(1, Y)$. Since $R_{\mathfrak{h}}(\mathfrak{g})$ is open in $\mathfrak{h}$, the tangent space at Y may be regarded as $\mathfrak{h}$ (with $c_H(t) = Y + tH$ being a curve with derivative $H \in \mathfrak{h}$). Similarly the tangent space at the point $\sigma(1, Y)$ of $\mathfrak{g}$ may be identified with $\mathfrak{g}$. Finally the tangent space at the point 1 of $\operatorname{Int}\mathfrak{g}$ is the Lie algebra $\operatorname{ad}\mathfrak{g}$. Hence $d\sigma$ is a map

$$d\sigma : \operatorname{ad}\mathfrak{g} \times \mathfrak{h} \to \mathfrak{g}.$$

Now

$$\begin{aligned} d\sigma(\operatorname{ad}X, 0) &= \frac{d}{dt}\sigma(e^{t\operatorname{ad}X}, Y)|_{t=0} \\ &= \frac{d}{dt}(e^{t\operatorname{ad}X})Y|_{t=0} = (\operatorname{ad}X)Y = [X, Y] \end{aligned}$$

and

$$d\sigma(0, H) = \frac{d}{dt}\sigma(1, Y + tH)|_{t=0} = \frac{d}{dt}(Y + tH)|_{t=0} = H.$$

Thus $\operatorname{image}(d\sigma) = [Y, \mathfrak{g}] + \mathfrak{h}$. By (b), $d\sigma$ is onto $\mathfrak{g}$. Hence the image of σ includes a neighborhood of $\sigma(1, Y)$ in $\mathfrak{g}$. Therefore $\operatorname{image}(\sigma)$ is open. But $R_\mathfrak{g}(\mathfrak{g})$ is dense. So $\operatorname{image}(\sigma)$ contains a member X of $R_\mathfrak{g}(\mathfrak{g})$. Then $a(Y) = X$ for some $a \in \operatorname{Int}\mathfrak{g}$ and $Y \in \mathfrak{h}$. From $a(Y) = X$ we easily check that $a(\mathfrak{g}_{0,Y}) = \mathfrak{g}_{0,X}$. Hence $\dim \mathfrak{g}_{0,Y} = \dim \mathfrak{g}_{0,X}$. Since $\dim \mathfrak{g}_{0,Y} = l_\mathfrak{h}(\mathfrak{g})$ and $\dim \mathfrak{g}_{0,X} = l_\mathfrak{g}(\mathfrak{g})$, we obtain $l_\mathfrak{h}(\mathfrak{g}) = l_\mathfrak{g}(\mathfrak{g})$. Thus $R_\mathfrak{h}(\mathfrak{g}) \subseteq R_\mathfrak{g}(\mathfrak{g})$. Now $R_\mathfrak{g}(\mathfrak{g})$ is stable under $\operatorname{Aut}_\mathbb{C}\mathfrak{g}$, and so $\operatorname{image}(\sigma) \subseteq R_\mathfrak{g}(\mathfrak{g})$.

(d) Let $a_1(Y_1) = a_2(Y_2)$ with $Y_1 \in R_{\mathfrak{h}_1}(\mathfrak{g})$ and $Y_2 \in R_{\mathfrak{h}_2}(\mathfrak{g})$. Then $a = a_2^{-1}a_1$ has $a(Y_1) = Y_2$. As in the previous step, we obtain $a(\mathfrak{g}_{0,Y_1}) = \mathfrak{g}_{0,Y_2}$. By (a), $\mathfrak{g}_{0,Y_1} = \mathfrak{h}_1$ and $\mathfrak{g}_{0,Y_2} = \mathfrak{h}_2$. Hence $a(\mathfrak{h}_1) = \mathfrak{h}_2$.

(e) If X is in $R_\mathfrak{g}(\mathfrak{g})$, let $\mathfrak{h} = \mathfrak{g}_{0,X}$. This is a Cartan subalgebra, by Theorem 2.9′, and (a) says that X is in $R_\mathfrak{h}(\mathfrak{g})$ for this $\mathfrak{h}$. Then $\sigma(1, X) = X$ shows that X is in the image of the σ defined relative to this $\mathfrak{h}$.

(f) We have seen that $R_\mathfrak{g}(\mathfrak{g})$ is the complement of the set where a nonzero polynomial vanishes. By Lemma 2.14 this set is connected.

4. Roots

Throughout this section, $\mathfrak{g}$ denotes a complex semisimple Lie algebra, B is its Killing form, and $\mathfrak{h}$ is a Cartan subalgebra of $\mathfrak{g}$. We saw in Proposition 2.10 that $\mathfrak{h}$ is abelian. The nonzero generalized weights of $\operatorname{ad}\mathfrak{h}$ on $\mathfrak{g}$ are called the **roots** of $\mathfrak{g}$ with respect to $\mathfrak{h}$. We denote the set of roots by Δ or $\Delta(\mathfrak{g}, \mathfrak{h})$. Then we can rewrite the weight-space decomposition of Proposition 2.5a as

$$\mathfrak{g} = \mathfrak{h} \oplus \bigoplus_{\alpha \in \Delta} \mathfrak{g}_\alpha. \tag{2.16}$$

This decomposition is called the **root-space decomposition** of $\mathfrak{g}$ with respect to $\mathfrak{h}$. Members of $\mathfrak{g}_\alpha$ are called **root vectors** for the root α.

Proposition 2.17.

(a) If α and β are in $\Delta \cup \{0\}$ and $\alpha+\beta \neq 0$, then $B(\mathfrak{g}_\alpha, \mathfrak{g}_\beta) = 0$.

(b) If α is in $\Delta \cup \{0\}$, then B is nonsingular on $\mathfrak{g}_\alpha \times \mathfrak{g}_{-\alpha}$.

(c) If α is in Δ, then so is $-\alpha$.

(d) $B|_{\mathfrak{h}\times\mathfrak{h}}$ is nondegenerate; consequently to each root α corresponds H_α in $\mathfrak{h}$ with $\alpha(H) = B(H, H_\alpha)$ for all $H \in \mathfrak{h}$.

(e) Δ spans $\mathfrak{h}^*$.

PROOF.

(a) By Proposition 2.5c, $\operatorname{ad}\mathfrak{g}_\alpha \operatorname{ad}\mathfrak{g}_\beta$ carries $\mathfrak{g}_\lambda$ into $\mathfrak{g}_{\lambda+\alpha+\beta}$ and consequently, when written as a matrix in terms of a basis of $\mathfrak{g}$ compatible with (2.16), has zero in every diagonal entry. Therefore its trace is 0.

(b) Since B is nondegenerate (Theorem 1.42), $B(X, \mathfrak{g}) \neq 0$ for each $X \in \mathfrak{g}_\alpha$. Since (a) shows that $B(X, \mathfrak{g}_\beta) = 0$ for every β other than $-\alpha$, we must have $B(X, \mathfrak{g}_{-\alpha}) \neq 0$.

(c, d) These are immediate from (b).

(e) Suppose $H \in \mathfrak{h}$ has $\alpha(H) = 0$ for all $\alpha \in \Delta$. By (2.16), $\operatorname{ad} H$ is nilpotent. Since $\mathfrak{h}$ is abelian, $\operatorname{ad} H \operatorname{ad} H'$ is nilpotent for all $H' \in \mathfrak{h}$. Therefore $B(H, \mathfrak{h}) = 0$. By (d), $H = 0$. Consequently Δ spans $\mathfrak{h}^*$.

For each root α, choose and fix, by Lie's Theorem (Theorem 1.25) applied to the action of $\mathfrak{h}$ on $\mathfrak{g}_\alpha$, a vector $E_\alpha \neq 0$ in $\mathfrak{g}_\alpha$ with $[H, E_\alpha] = \alpha(H)E_\alpha$ for all $H \in \mathfrak{h}$.

Lemma 2.18.

(a) If α is a root and X is in $\mathfrak{g}_{-\alpha}$, then $[E_\alpha, X] = B(E_\alpha, X)H_\alpha$.

(b) If α and β are in Δ, then $\beta(H_\alpha)$ is a rational multiple of $\alpha(H_\alpha)$.

(c) If α is in Δ, then $\alpha(H_\alpha) \neq 0$.

PROOF.

(a) Since $[\mathfrak{g}_\alpha, \mathfrak{g}_{-\alpha}] \subseteq \mathfrak{g}_0$ by Proposition 2.5c, $[E_\alpha, X]$ is in $\mathfrak{h}$. For H in $\mathfrak{h}$, we have

$$\begin{aligned} B([E_\alpha, X], H) &= -B(X, [E_\alpha, H]) = B(X, [H, E_\alpha]) \\ &= \alpha(H)B(X, E_\alpha) = B(H_\alpha, H)B(E_\alpha, X) \\ &= B(B(E_\alpha, X)H_\alpha, H). \end{aligned}$$

Then the conclusion follows from Proposition 2.17d.

(b) By Proposition 2.17b, we can choose $X_{-\alpha}$ in $\mathfrak{g}_{-\alpha}$ so that $B(E_\alpha, X_{-\alpha}) = 1$. Then (a) shows that

$$[E_\alpha, X_{-\alpha}] = H_\alpha. \tag{2.19}$$

With β fixed in Δ, let $\mathfrak{g}' = \bigoplus_{n\in\mathbb{Z}} \mathfrak{g}_{\beta+n\alpha}$. This subspace is invariant under $\operatorname{ad} H_\alpha$, and we shall compute the trace of $\operatorname{ad} H_\alpha$ on this subspace in two ways. Noting that $\operatorname{ad} H_\alpha$ acts on $\mathfrak{g}_{\beta+n\alpha}$ with the single generalized eigenvalue $(\beta+n\alpha)(H_\alpha)$ and adding the contribution to the trace over all values of n, we obtain

$$\sum_{n\in\mathbb{Z}} (\beta(H_\alpha) + n\alpha(H_\alpha)) \dim \mathfrak{g}_{\beta+n\alpha} \tag{2.20}$$

as the trace. On the other hand, Proposition 2.5c shows that $\mathfrak{g}'$ is invariant under $\operatorname{ad} E_\alpha$ and $\operatorname{ad} X_{-\alpha}$. By (2.19) the trace is

$$= \operatorname{Tr} \operatorname{ad} H_\alpha = \operatorname{Tr}(\operatorname{ad} E_\alpha \operatorname{ad} X_{-\alpha} - \operatorname{ad} X_{-\alpha} \operatorname{ad} E_\alpha) = 0.$$

Thus (2.20) equals 0, and the conclusion follows.

(c) Suppose $\alpha(H_\alpha) = 0$. By (b), $\beta(H_\alpha) = 0$ for all $\beta \in \Delta$. By Proposition 2.17e every member of $\mathfrak{h}^*$ vanishes on H_α. Thus $H_\alpha = 0$. But this conclusion contradicts Proposition 2.17d, since α is assumed to be nonzero.

Proposition 2.21. If α is in Δ, then $\dim \mathfrak{g}_\alpha = 1$. Also $n\alpha$ is not in Δ for any integer $n \geq 2$.

REMARK. Thus we no longer need to use the cumbersome condition $(\operatorname{ad} H - \alpha(H)1)^k X = 0$ for $X \in \mathfrak{g}_\alpha$ but can work with $k = 1$. Briefly

$$\mathfrak{g}_\alpha = \{X \in \mathfrak{g} \mid (\operatorname{ad} H)X = \alpha(H)X\}. \tag{2.22}$$

PROOF. As in the proof of Lemma 2.18b, we can choose $X_{-\alpha}$ in $\mathfrak{g}_{-\alpha}$ with $B(E_\alpha, X_{-\alpha}) = 1$ and obtain the bracket relation (2.19). Put $\mathfrak{g}'' = \mathbb{C}E_\alpha \oplus \mathbb{C}H_\alpha \oplus \bigoplus_{n<0} \mathfrak{g}_{n\alpha}$. This subspace is invariant under $\operatorname{ad} H_\alpha$ and $\operatorname{ad} E_\alpha$, by Proposition 2.5c, and it is invariant under $\operatorname{ad} X_{-\alpha}$ by Proposition 2.5c and Lemma 2.18a. By (2.19), $\operatorname{ad} H_\alpha$ has trace 0 in its action on $\mathfrak{g}''$. But $\operatorname{ad} H_\alpha$ acts on each summand with a single generalized eigenvalue, and thus the trace is

$$= \alpha(H_\alpha) + 0 + \sum_{n<0} n\alpha(H_\alpha) \dim \mathfrak{g}_{n\alpha} = 0.$$

Using Lemma 2.18c, we see that

$$\sum_{n=1}^{\infty} n \dim \mathfrak{g}_{-n\alpha} = 1.$$

Consequently $\dim \mathfrak{g}_{-\alpha} = 1$ and $\dim \mathfrak{g}_{-n\alpha} = 0$ for $n \geq 2$. Proposition 2.17c shows that we may replace α by $-\alpha$ everywhere in the above argument, and then we obtain the conclusion of the proposition.

Corollary 2.23. The action of ad $\mathfrak{h}$ on $\mathfrak{g}$ is simultaneously diagonable.

REMARK. This corollary completes the promised converse to Corollary 2.13.

PROOF. This follows by combining (2.16), Proposition 2.10, and Proposition 2.21.

Corollary 2.24. On $\mathfrak{h} \times \mathfrak{h}$, the Killing form is given by

$$B(H, H') = \sum_{\alpha \in \Delta} \alpha(H)\alpha(H').$$

REMARK. This formula is a special property of the Killing form. By contrast the previous results of this section remain valid if B is replaced by any nondegenerate symmetric invariant bilinear form. We shall examine the role of special properties of B further when we come to Corollary 2.38.

PROOF. Let $\{H_i\}$ be a basis of $\mathfrak{h}$. By Proposition 2.21 and Corollary 2.23, $\{H_i\} \cup \{E_\alpha\}$ is a basis of $\mathfrak{g}$, and each ad H acts diagonally. Then ad H ad H' acts diagonally, and the respective eigenvalues are 0 and $\{\alpha(H)\alpha(H')\}$. Hence

$$B(H, H') = \operatorname{Tr}(\operatorname{ad} H \operatorname{ad} H') = \sum_{\alpha \in \Delta} \alpha(H)\alpha(H').$$

Corollary 2.25. The pair of vectors $\{E_\alpha, E_{-\alpha}\}$ selected before Lemma 2.18 may be normalized so that $B(E_\alpha, E_{-\alpha}) = 1$.

PROOF. By Proposition 2.17b, $\mathfrak{g}_\alpha$ and $\mathfrak{g}_{-\alpha}$ are nonsingularly paired. Since Proposition 2.21 shows each of these spaces to be 1-dimensional, the result follows.

The above results may be interpreted as saying that $\mathfrak{g}$ is built out of copies of $\mathfrak{sl}(2, \mathbb{C})$ in a certain way. To see this, let E_α and $E_{-\alpha}$ be normalized as in Corollary 2.25. Then Lemma 2.18a gives us the bracket relations

$$\begin{aligned}
[H_\alpha, E_\alpha] &= \alpha(H_\alpha)E_\alpha \\
[H_\alpha, E_{-\alpha}] &= -\alpha(H_\alpha)E_{-\alpha} \\
[E_\alpha, E_{-\alpha}] &= H_\alpha.
\end{aligned}$$

We normalize these vectors suitably, for instance by

$$H'_\alpha = \frac{2}{\alpha(H_\alpha)} H_\alpha, \quad E'_\alpha = \frac{2}{\alpha(H_\alpha)} E_\alpha, \quad E'_{-\alpha} = E_{-\alpha}. \tag{2.26}$$

Then

$$[H'_\alpha, E'_\alpha] = 2E'_\alpha$$
$$[H'_\alpha, E'_{-\alpha}] = -2E'_{-\alpha}$$
$$[E'_\alpha, E'_{-\alpha}] = H'_\alpha.$$

As in (1.5) let us define elements of $\mathfrak{sl}(2, \mathbb{C})$ by

$$h = \begin{pmatrix} 1 & 0 \\ 0 & -1 \end{pmatrix}, \quad e = \begin{pmatrix} 0 & 1 \\ 0 & 0 \end{pmatrix}, \quad f = \begin{pmatrix} 0 & 0 \\ 1 & 0 \end{pmatrix}.$$

These satisfy

$$[h, e] = 2e$$
$$[h, f] = -2f$$
$$[e, f] = h.$$

Consequently

$$H'_\alpha \mapsto h, \quad E'_\alpha \mapsto e, \quad E'_{-\alpha} \mapsto f \tag{2.27}$$

extends linearly to an isomorphism of $\operatorname{span}\{H_\alpha, E_\alpha, E_{-\alpha}\}$ onto $\mathfrak{sl}(2, \mathbb{C})$. Thus $\mathfrak{g}$ is spanned by embedded copies of $\mathfrak{sl}(2, \mathbb{C})$. The detailed structure of $\mathfrak{g}$ comes by understanding how these copies of $\mathfrak{sl}(2, \mathbb{C})$ fit together. To investigate this question, we study the action of such an $\mathfrak{sl}(2, \mathbb{C})$ subalgebra on all of $\mathfrak{g}$, i.e., we study a complex-linear representation of $\mathfrak{sl}(2, \mathbb{C})$ on $\mathfrak{g}$. We already know some invariant subspaces for this representation, and we study these one at a time.

Thus the representation to study is the one in the proof of Lemma 2.18b, with the version of $\mathfrak{sl}(2, \mathbb{C})$ built from a root α acting on the vector space $\mathfrak{g}' = \bigoplus_{n \in \mathbb{Z}} \mathfrak{g}_{\beta + n\alpha}$ by ad. Correspondingly we make the following definition of **root string**. Let α be in Δ, and let β be in $\Delta \cup \{0\}$. The α **string containing** β is the set of all members of $\Delta \cup \{0\}$ of the form $\beta + n\alpha$ for $n \in \mathbb{Z}$. Two examples of root strings appear in Figure 2.1.

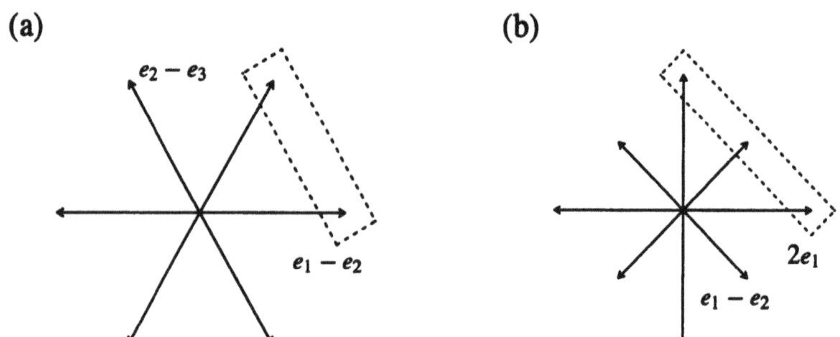

FIGURE 2.1. Root strings: (a) $e_2 - e_3$ string containing $e_1 - e_2$ for $\mathfrak{sl}(3, \mathbb{C})$, (b) $e_1 - e_2$ string through $2e_1$ for $\mathfrak{sp}(2, \mathbb{C})$

Also we transfer the restriction to $\mathfrak{h}$ of the Killing form to a bilinear form on the dual $\mathfrak{h}^*$ by the definition

$$\langle\varphi, \psi\rangle = B(H_\varphi, H_\psi) = \varphi(H_\psi) = \psi(H_\varphi) \tag{2.28}$$

for φ and ψ in $\mathfrak{h}^*$. Here H_φ and H_ψ are defined as in Proposition 2.17d.

Proposition 2.29. Let α be in Δ, and let β be in $\Delta \cup \{0\}$.

(a) The α string containing β has the form $\beta + n\alpha$ for $-p \leq n \leq q$ with $p \geq 0$ and $q \geq 0$. There are no gaps. Furthermore

$$p - q = \frac{2\langle\beta, \alpha\rangle}{\langle\alpha, \alpha\rangle},$$

and $\dfrac{2\langle\beta, \alpha\rangle}{\langle\alpha, \alpha\rangle}$ is in $\mathbb{Z}$.

(b) If $\beta + n\alpha$ is never 0, define $\mathfrak{sl}_\alpha$ to be the isomorphic copy of $\mathfrak{sl}(2, \mathbb{C})$ spanned by H'_α, E'_α, and $E'_{-\alpha}$ as in (2.26), and let $\mathfrak{g}' = \bigoplus_{n\in\mathbb{Z}} \mathfrak{g}_{\beta+n\alpha}$. Then the representation of $\mathfrak{sl}_\alpha$ on $\mathfrak{g}'$ by ad is irreducible.

PROOF. If $\beta + n\alpha = 0$ for some n, then conclusion (a) follows from Proposition 2.21, and there is nothing to prove for (b). Thus we may assume that $\beta + n\alpha$ is never 0, and we shall prove (a) and (b) together.

By Proposition 2.21 the transformation $\operatorname{ad} H'_\alpha$ is diagonable on $\mathfrak{g}'$ with distinct eigenvalues, and these eigenvalues are

$$\begin{aligned}(\beta + n\alpha)(H'_\alpha) &= \frac{2}{\langle\alpha, \alpha\rangle}(\beta + n\alpha)(H_\alpha)\\ &= \frac{2}{\langle\alpha, \alpha\rangle}(\langle\beta, \alpha\rangle + n\langle\alpha, \alpha\rangle)\\ &= \frac{2\langle\beta, \alpha\rangle}{\langle\alpha, \alpha\rangle} + 2n.\end{aligned} \tag{2.30}$$

Thus any $\operatorname{ad} H'_\alpha$ invariant subspace of $\mathfrak{g}'$ is a sum of certain $\mathfrak{g}_{\beta+n\alpha}$'s. Hence the same thing is true of any $\operatorname{ad}(\mathfrak{sl}_\alpha)$ invariant subspace.

Let V be an irreducible such subspace, and let $-p$ and q be the smallest and largest n's appearing for V. Theorem 1.63 shows that the eigenvalues of $\operatorname{ad} h = \operatorname{ad} H'_\alpha$ in V are $N - 2i$ with $0 \leq i \leq N$, where $N = \dim V - 1$. Since these eigenvalues jump by 2's, (2.30) shows that all n's between $-p$ and q are present. Also (2.30) gives

$$N = \frac{2\langle\beta, \alpha\rangle}{\langle\alpha, \alpha\rangle} + 2q$$

and

$$-N = \frac{2\langle\beta, \alpha\rangle}{\langle\alpha, \alpha\rangle} - 2p.$$

Adding, we obtain

$$p - q = \frac{2\langle\beta, \alpha\rangle}{\langle\alpha, \alpha\rangle}. \tag{2.31}$$

Theorem 1.64 shows that $\mathfrak{g}'$ is the direct sum of irreducible subspaces under $\mathfrak{sl}_\alpha$. If V' is another irreducible subspace, let $-p'$ and q' be the smallest and largest n's appearing for V'. Then (2.31), applied to V', gives

$$p' - q' = \frac{2\langle\beta, \alpha\rangle}{\langle\alpha, \alpha\rangle},$$

so that

$$p' - q' = p - q. \tag{2.32}$$

On the other hand, all the n's from $-p$ to q are accounted for by V, and we must therefore have either $-p' > q$ or $q' < -p$. By symmetry we may assume that $-p' > q$. This inequality implies that

$$p' < -q \tag{2.33}$$

and that $q' \geq -p' > q \geq -p$. From the latter inequality we obtain

$$-q' < p. \tag{2.34}$$

Adding (2.33) and (2.34), we obtain a contradiction with (2.32), and the proposition follows.

Corollary 2.35. If α and β are in $\Delta \cup \{0\}$ and $\alpha + \beta \neq 0$, then $[\mathfrak{g}_\alpha, \mathfrak{g}_\beta] = \mathfrak{g}_{\alpha+\beta}$.

PROOF. Without loss of generality, let $\alpha \neq 0$. Proposition 2.5c shows that

$$[\mathfrak{g}_\alpha, \mathfrak{g}_\beta] \subseteq \mathfrak{g}_{\alpha+\beta}. \tag{2.36}$$

We are to prove that equality holds in (2.36) We consider cases.

If β is an integral multiple of α and is not equal to $-\alpha$, then Proposition 2.21 shows that β must be α or 0. If $\beta = \alpha$, then $\mathfrak{g}_{\alpha+\beta} = 0$ by Proposition 2.21, and hence equality must hold in (2.36). If $\beta = 0$, then the equality $[\mathfrak{h}, \mathfrak{g}_\alpha] = \mathfrak{g}_\alpha$ says that equality holds in (2.36).

If β is not an integral multiple of α, then Proposition 2.29b is applicable and shows that $\mathfrak{sl}_\alpha$ acts irreducibly on $\mathfrak{g}' = \bigoplus_{n\in\mathbb{Z}} \mathfrak{g}_{\beta+n\alpha}$. Making the identification (2.27) and matching data with Theorem 1.63, we see that the root vectors $E_{\beta+n\alpha}$, except for constant factors, are the vectors v_i of Theorem 1.63. The only i for which $e = \begin{pmatrix} 0 & 1 \\ 0 & 0 \end{pmatrix}$ maps v_i to 0 is $i = 0$, and v_0 corresponds to $E_{\beta+q\alpha}$. Thus $[\mathfrak{g}_\alpha, \mathfrak{g}_\beta] = 0$ forces $q = 0$ and says that $\beta + \alpha$ is not a root. In this case, $\mathfrak{g}_{\alpha+\beta} = 0$, and equality must hold in (2.36).

Corollary 2.37. Let α and β be roots such that $\beta + n\alpha$ is never 0 for $n \in \mathbb{Z}$. Let E_α, $E_{-\alpha}$, and E_β be any root vectors for α, $-\alpha$, and β, respectively, and let p and q be the integers in Proposition 2.29a. Then

$$[E_{-\alpha}, [E_\alpha, E_\beta]] = \frac{q(1+p)}{2}\alpha(H_\alpha)B(E_\alpha, E_{-\alpha})E_\beta.$$

PROOF. Both sides are linear in E_α and $E_{-\alpha}$, and we may therefore normalize them as in Corollary 2.25 so that $B(E_\alpha, E_{-\alpha}) = 1$. If we then make the identification (2.27) of the span of $\{H_\alpha, E_\alpha, E_{-\alpha}\}$ with $\mathfrak{sl}(2, \mathbb{C})$, we can reinterpret the desired formula as

$$\frac{\langle\alpha, \alpha\rangle}{2}[f, [e, E_\beta]] \stackrel{?}{=} \frac{q(1+p)}{2}\alpha(H_\alpha)E_\beta,$$

i.e., as

$$[f, [e, E_\beta]] \stackrel{?}{=} q(1+p)E_\beta.$$

From Proposition 2.29b, the action of the span of $\{h, e, f\}$ on $\mathfrak{g}'$ is irreducible. The vector $E_{\beta+q\alpha}$ corresponds to a multiple of the vector v_0 in Theorem 1.63. Since E_β is a multiple of $(\operatorname{ad} f)^q E_{\beta+q\alpha}$, E_β corresponds to a multiple of v_q. By (d) and then (c) in Theorem 1.63, we obtain

$$(\operatorname{ad} f)(\operatorname{ad} e)E_\beta = q(N - q + 1)E_\beta,$$

where $N = \dim \mathfrak{g}' - 1 = (q + p + 1) - 1$. Then $q(N - q + 1) = q(1 + p)$, and the result follows.

Corollary 2.38. Let V be the $\mathbb{R}$ linear span of Δ in $\mathfrak{h}^*$. Then V is a real form of the vector space $\mathfrak{h}^*$, and the restriction of the bilinear form $\langle\cdot, \cdot\rangle$ to $V \times V$ is a positive definite inner product. Moreover, if $\mathfrak{h}_0$ denotes the $\mathbb{R}$ linear span of all H_α for $\alpha \in \Delta$, then $\mathfrak{h}_0$ is a real form of the vector space $\mathfrak{h}$, the members of V are exactly those linear functionals that are real on $\mathfrak{h}_0$, and restriction of the operation of those linear functionals from $\mathfrak{h}$ to $\mathfrak{h}_0$ is an $\mathbb{R}$ isomorphism of V onto $\mathfrak{h}_0^*$.

REMARK. The proof will make use of Corollary 2.24, which was the only result so far that used any properties of the Killing form other than that B is a nondegenerate symmetric invariant bilinear form. The present corollary will show that B is positive definite on $\mathfrak{h}_0$, and then Corollary 2.24 will no longer be needed. The remaining theory for complex semisimple Lie algebras in this chapter goes through if B is replaced by any nondegenerate symmetric invariant bilinear form that is positive definite on $\mathfrak{h}_0$. Because of Theorem 2.15, once such a form B is positive definite on the real form $\mathfrak{h}_0$ of the Cartan subalgebra $\mathfrak{h}$, it is positive definite on the corresponding real form of any other Cartan subalgebra.

PROOF. Combining Corollary 2.24 with the definition (2.28), we obtain

$$(2.39) \qquad \langle\varphi, \psi\rangle = B(H_\varphi, H_\psi) = \sum_{\beta\in\Delta} \beta(H_\varphi)\beta(H_\psi) = \sum_{\beta\in\Delta} \langle\beta, \varphi\rangle\langle\beta, \psi\rangle$$

for all φ and ψ in $\mathfrak{h}^*$. Let α be a root, and let p_β and q_β be the integers p and q associated to the α string containing β in Proposition 2.29a. Specializing (2.39) to $\varphi = \psi = \alpha$ gives

$$\langle\alpha, \alpha\rangle = \sum_{\beta\in\Delta} \langle\beta, \alpha\rangle^2 = \sum_{\beta\in\Delta} [(p_\beta - q_\beta)\tfrac{1}{2}\langle\alpha, \alpha\rangle]^2.$$

Since $\langle\alpha, \alpha\rangle \neq 0$ according to Lemma 2.18c, we obtain

$$\langle\alpha, \alpha\rangle = \frac{4}{\sum_{\beta\in\Delta}(p_\beta - q_\beta)^2},$$

and therefore $\langle\alpha, \alpha\rangle$ is rational. By Lemma 2.18b,

$$(2.40) \qquad \beta(H_\alpha) \text{ is rational for all } \alpha \text{ and } \beta \text{ in } \Delta.$$

Let $\dim_{\mathbb{C}} \mathfrak{h} = l$. By Proposition 2.17e we can choose l roots $\alpha_1, \ldots, \alpha_l$ such that $H_{\alpha_1}, \ldots, H_{\alpha_l}$ is a basis of $\mathfrak{h}$ over $\mathbb{C}$. Let $\omega_1, \ldots, \omega_l$ be the dual basis of $\mathfrak{h}^*$ satisfying $\omega_i(H_{\alpha_j}) = \delta_{ij}$, and let V be the real vector space of all members of $\mathfrak{h}^*$ that are real on all of $H_{\alpha_1}, \ldots, H_{\alpha_l}$. Then $V = \bigoplus_{j=1}^{l} \mathbb{R}\omega_j$, and it follows that V is a real form of the vector space $\mathfrak{h}^*$. By (2.40) all roots are in V. Since $\alpha_1, \ldots, \alpha_l$ are already linearly independent over $\mathbb{R}$, we conclude that V is the $\mathbb{R}$ linear span of the roots.

If φ is in V, then $\varphi(H_\beta)$ is real for each root β. Since (2.39) gives

$$\langle\varphi, \varphi\rangle = \sum_{\beta\in\Delta} \langle\beta, \varphi\rangle^2 = \sum_{\beta\in\Delta} \varphi(H_\beta)^2,$$

we see that the restriction of $\langle\cdot\,, \cdot\rangle$ to $V \times V$ is a positive definite inner product.

Now let $\mathfrak{h}_0$ denote the $\mathbb{R}$ linear span of all H_α for $\alpha \in \Delta$. Since $\varphi \mapsto H_\varphi$ is an isomorphism of $\mathfrak{h}^*$ with $\mathfrak{h}$ carrying V to $\mathfrak{h}_0$, it follows that $\mathfrak{h}_0$ is a real form of $\mathfrak{h}$. We know that the real linear span of the roots (namely V) has real dimension l, and consequently the real linear span of all H_α for $\alpha \in \Delta$ has real dimension l. Since $H_{\alpha_1}, \ldots, H_{\alpha_l}$ is linearly independent over $\mathbb{R}$, it is a basis of $\mathfrak{h}_0$ over $\mathbb{R}$. Hence V is the set of members of $\mathfrak{h}^*$ that are real on all of $\mathfrak{h}_0$. Therefore restriction from $\mathfrak{h}$ to $\mathfrak{h}_0$ is a vector-space isomorphism of V onto $\mathfrak{h}_0^*$.

Let $|\cdot|^2$ denote the norm squared associated to the inner product $\langle\cdot,\cdot\rangle$ on $\mathfrak{h}_0^* \times \mathfrak{h}_0^*$. Let α be a root. Relative to the inner product, we introduce the **root reflection**

$$s_\alpha(\varphi) = \varphi - \frac{2\langle\varphi,\alpha\rangle}{|\alpha|^2}\alpha \qquad \text{for } \varphi \in \mathfrak{h}_0^*.$$

This is an orthogonal transformation on $\mathfrak{h}_0^*$, is -1 on $\mathbb{R}\alpha$, and is $+1$ on the orthogonal complement of α.

Proposition 2.41. For any root α, the root reflection s_α carries Δ into itself.

PROOF. Let β be in Δ, and let p and q be as in Propositon 2.29a. Then

$$s_\alpha\beta = \beta - \frac{2\langle\beta,\alpha\rangle}{|\alpha|^2}\alpha = \beta - (p-q)\alpha = \beta + (q-p)\alpha.$$

Since $-p \le q - p \le q$, $\beta + (q-p)\alpha$ is in the α string containing β. Hence $s_\alpha\beta$ is a root or is 0. Since s_α is an orthogonal transformation on $\mathfrak{h}_0^*$, $s_\alpha\beta$ is not 0. Thus s_α carries Δ into Δ.

5. Abstract Root Systems

To examine roots further, it is convenient to abstract the results we have obtained so far. This approach will allow us to work more easily toward a classification of complex semisimple Lie algebras and also to apply the theory of roots in a different situation that will arise in Chapter VI.

An **abstract root system** in a finite-dimensional real inner product space V with inner product $\langle\cdot,\cdot\rangle$ and norm squared $|\cdot|^2$ is a finite set Δ of nonzero elements of V such that

(i) Δ spans V

(ii) the orthogonal transformations $s_\alpha(\varphi) = \varphi - \dfrac{2\langle\varphi,\alpha\rangle}{|\alpha|^2}\alpha$, for $\alpha \in \Delta$, carry Δ to itself

(iii) $\dfrac{2\langle\beta,\alpha\rangle}{|\alpha|^2}$ is an integer whenever α and β are in Δ.

An abstract root system is said to be **reduced** if $\alpha \in \Delta$ implies $2\alpha \notin \Delta$. Much of what we saw in §4 can be summarized in the following theorem.

Theorem 2.42. The root system of a complex semisimple Lie algebra $\mathfrak{g}$ with respect to a Cartan subalgebra $\mathfrak{h}$ forms a reduced abstract root system in $\mathfrak{h}_0^*$.

PROOF. With $V = \mathfrak{h}_0^*$, V is an inner product space spanned by Δ as a consequence of Corollary 2.38. Property (ii) follows from Proposition 2.41, and property (iii) follows from Proposition 2.29a. According to Proposition 2.21, the abstract root system Δ is reduced.

As a consequence of the theorem, the examples of §1 give us many examples of reduced abstract root systems. We recall them here and tell what names we shall use for them:

(2.43)

	Vector Space	Root System	$\mathfrak{g}$
A_n	$V = \left\{ \begin{matrix} \sum_{i=1}^{n+1} a_i e_i \text{ with} \\ \sum a_i e_i = 0 \end{matrix} \right\}$	$\Delta = \{e_i - e_j \mid i \neq j\}$	$\mathfrak{sl}(n+1, \mathbb{C})$
B_n	$V = \{\sum_{i=1}^n a_i e_i\}$	$\Delta = \{\pm e_i \pm e_j \mid i \neq j\} \cup \{\pm e_i\}$	$\mathfrak{so}(2n+1, \mathbb{C})$
C_n	$V = \{\sum_{i=1}^n a_i e_i\}$	$\Delta = \{\pm e_i \pm e_j \mid i \neq j\} \cup \{\pm 2e_i\}$	$\mathfrak{sp}(n, \mathbb{C})$
D_n	$V = \{\sum_{i=1}^n a_i e_i\}$	$\Delta = \{\pm e_i \pm e_j \mid i \neq j\}$	$\mathfrak{so}(2n, \mathbb{C})$

Some 2-dimensional examples of abstract root systems are given in Figure 2.2. All but $(BC)_2$ are reduced. The system $A_1 \oplus A_1$ arises as the root system for $\mathfrak{sl}(2, \mathbb{C}) \oplus \mathfrak{sl}(2, \mathbb{C})$.

We say that two abstract root systems Δ in V and Δ' in V' are **isomorphic** if there is a vector-space isomorphism of V onto V' carrying Δ onto Δ' and preserving the integers $2\langle\beta, \alpha\rangle/|\alpha|^2$ for α and β in Δ. The systems B_2 and C_2 in Figure 2.2 are isomorphic.

An abstract root system Δ is said to be **reducible** if Δ admits a nontrivial disjoint decomposition $\Delta = \Delta' \cup \Delta''$ with every member of Δ' orthogonal to every member of Δ''. We say that Δ is **irreducible** if it admits no such nontrivial decomposition. In Figure 2.2 all the abstract root systems are irreducible except $A_1 \oplus A_1$. The fact that this root system comes from a complex semisimple Lie algebra that is not simple generalizes as in Proposition 2.44 below.

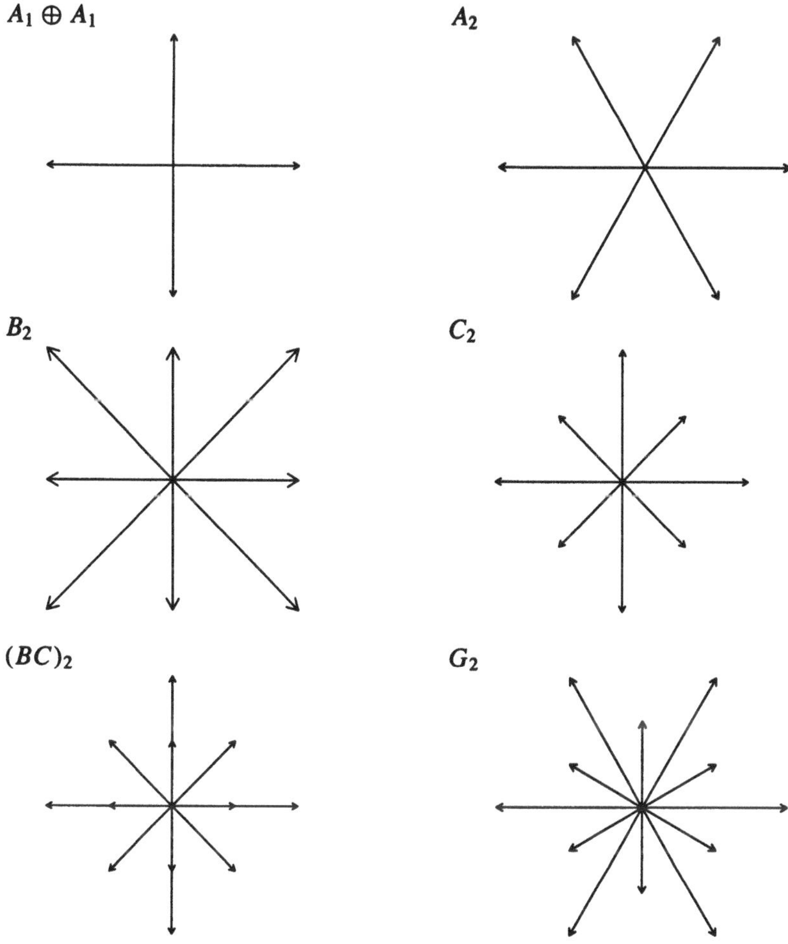

FIGURE 2.2. Abstract root systems with $V = \mathbb{R}^2$

Proposition 2.44. The root system Δ of a complex semisimple Lie algebra $\mathfrak{g}$ with respect to a Cartan subalgebra $\mathfrak{h}$ is irreducible as an abstract reduced root system if and only if $\mathfrak{g}$ is simple.

PROOF THAT Δ IRREDUCIBLE IMPLIES $\mathfrak{g}$ SIMPLE. Suppose that $\mathfrak{g}$ is a nontrivial direct sum of ideals

$$\mathfrak{g} = \mathfrak{g}' \oplus \mathfrak{g}''. \tag{2.45}$$

Let α be a root, and decompose the corresponding root vector E_α according to (2.45) as $E_\alpha = E'_\alpha + E''_\alpha$. For H in $\mathfrak{h}$, we have

$$0 = [H, E_\alpha] - \alpha(H)E_\alpha = ([H, E'_\alpha] - \alpha(H)E'_\alpha) + ([H, E''_\alpha] - \alpha(H)E''_\alpha).$$

Since $\mathfrak{g}'$ and $\mathfrak{g}''$ are ideals and have 0 intersection, the two terms on the right are separately 0. Thus E'_α and E''_α are both in the root space $\mathfrak{g}_\alpha$. Since $\dim \mathfrak{g}_\alpha = 1$, $E'_\alpha = 0$ or $E''_\alpha = 0$. Thus $\mathfrak{g}_\alpha \subseteq \mathfrak{g}'$ or $\mathfrak{g}_\alpha \subseteq \mathfrak{g}''$. Define

$$\Delta' = \{\alpha \in \Delta \mid \mathfrak{g}_\alpha \subseteq \mathfrak{g}'\}$$
$$\Delta'' = \{\alpha \in \Delta \mid \mathfrak{g}_\alpha \subseteq \mathfrak{g}''\}.$$

What we have just shown is that $\Delta = \Delta' \cup \Delta''$ disjointly. Now with obvious notation we have

$$\alpha'(H_{\alpha''})E_{\alpha'} = [H_{\alpha''}, E_{\alpha'}] \subseteq [H_{\alpha''}, \mathfrak{g}'] = [[E_{\alpha''}, E_{-\alpha''}], \mathfrak{g}'] \subseteq [\mathfrak{g}'', \mathfrak{g}'] = 0,$$

and thus $\alpha'(H_{\alpha''}) = 0$. Hence Δ' and Δ'' are mutually orthogonal.

PROOF THAT $\mathfrak{g}$ SIMPLE IMPLIES Δ IRREDUCIBLE. Suppose that $\Delta = \Delta' \cup \Delta''$ exhibits Δ as reducible. Define

$$\mathfrak{g}' = \sum_{\alpha \in \Delta'} \{\mathbb{C}H_\alpha + \mathfrak{g}_\alpha + \mathfrak{g}_{-\alpha}\}$$
$$\mathfrak{g}'' = \sum_{\alpha \in \Delta''} \{\mathbb{C}H_\alpha + \mathfrak{g}_\alpha + \mathfrak{g}_{-\alpha}\}.$$

Then $\mathfrak{g}'$ and $\mathfrak{g}''$ are vector subspaces of $\mathfrak{g}$, and $\mathfrak{g} = \mathfrak{g}' \oplus \mathfrak{g}''$ as vector spaces. To complete the proof, it is enough to show that $\mathfrak{g}'$ and $\mathfrak{g}''$ are ideals in $\mathfrak{g}$. It is clear that they are Lie subalgebras. For α' in Δ' and α'' in Δ'', we have

$$[H_{\alpha'}, E_{\alpha''}] = \alpha''(H_{\alpha'})E_{\alpha''} = 0 \tag{2.46}$$

by the assumed orthogonality. Also if $[\mathfrak{g}_{\alpha'}, \mathfrak{g}_{\alpha''}] \neq 0$, then $\alpha' + \alpha''$ is a root that is not orthogonal to every member of Δ' (α' for instance) and is not orthogonal to every member of Δ'' (α'' for instance), in contradiction with the given orthogonal decomposition of Δ. We conclude that

$$[\mathfrak{g}_{\alpha'}, \mathfrak{g}_{\alpha''}] = 0. \tag{2.47}$$

Combining (2.46) and (2.47), we see that $[\mathfrak{g}', \mathfrak{g}_{\alpha''}] = 0$. Since $[\mathfrak{g}', \mathfrak{h}] \subseteq \mathfrak{g}'$ and since $\mathfrak{g}'$ is a subalgebra, $\mathfrak{g}'$ is an ideal in $\mathfrak{g}$. Similarly $\mathfrak{g}''$ is an ideal. This completes the proof.

EXAMPLE. Let $\mathfrak{g} = \mathfrak{so}(4, \mathbb{C})$ with notation as in §1. Then $\Delta = \{\pm e_1 \pm e_2\}$. If we put $\Delta' = \{\pm(e_1 - e_2\}$ and $\Delta'' = \{\pm(e_1 + e_2)\}$, then $\Delta = \Delta' \cup \Delta''$ exhibits Δ as reducible. By Proposition 2.44, $\mathfrak{so}(4, \mathbb{C})$ is not simple. The root system is isomorphic to $A_1 \oplus A_1$.

We extend our earlier definition of **root string** to the context of an abstract root system Δ. For $\alpha \in \Delta$ and $\beta \in \Delta \cup \{0\}$, the **$\alpha$ string containing β** is the set of all members of $\Delta \cup \{0\}$ of the form $\beta + n\alpha$ with $n \in \mathbb{Z}$. Figure 2.1 in §4 showed examples of root strings. In the system G_2 as pictured in Figure 2.2, there are root strings containing four roots.

If α is a root and $\frac{1}{2}\alpha$ is not a root, we say that α is **reduced**.

Proposition 2.48. Let Δ be an abstract root system in the inner product space V.

(a) If α is in Δ, then $-\alpha$ is in Δ.

(b) If α is in Δ and is reduced, then the only members of $\Delta \cup \{0\}$ proportional to α are $\pm\alpha$, $\pm 2\alpha$, and 0, and $\pm 2\alpha$ cannot occur if Δ is reduced.

(c) If α is in Δ and β is in $\Delta \cup \{0\}$, then

$$\frac{2\langle \beta, \alpha\rangle}{|\alpha|^2} = 0,\ \pm 1,\ \pm 2,\ \pm 3,\ \text{or}\ \pm 4,$$

and ± 4 occurs only in a nonreduced system with $\beta = \pm 2\alpha$.

(d) If α and β are nonproportional members of Δ such that $|\alpha| \le |\beta|$, then $\dfrac{2\langle \beta, \alpha\rangle}{|\beta|^2}$ equals 0 or $+1$ or -1.

(e) If α and β are in Δ with $\langle \alpha, \beta\rangle > 0$, then $\alpha - \beta$ is a root or 0. If α and β are in Δ with $\langle \alpha, \beta\rangle < 0$, then $\alpha + \beta$ is a root or 0.

(f) If α and β are in Δ and neither $\alpha + \beta$ nor $\alpha - \beta$ is in $\Delta \cup \{0\}$, then $\langle \alpha, \beta\rangle = 0$.

(g) If α is in Δ and β is in $\Delta \cup \{0\}$, then the α string containing β has the form $\beta + n\alpha$ for $-p \le n \le q$ with $p \ge 0$ and $q \ge 0$. There are no gaps. Furthermore $p - q = \dfrac{2\langle \beta, \alpha\rangle}{|\alpha|^2}$. The α string containing β contains at most four roots.

PROOF.

(a) This follows since $s_\alpha(\alpha) = -\alpha$.

(b) Let α be in Δ, and let $c\alpha$ be in $\Delta \cup \{0\}$. We may assume that $c \ne 0$. Then $2\langle c\alpha, \alpha\rangle/|\alpha|^2$ and $2\langle \alpha, c\alpha\rangle/|c\alpha|^2$ are both integers, from which it follows that $2c$ and $2/c$ are integers. Since $c \ne \pm\frac{1}{2}$, the only possibilities are $c = \pm 1$ and $c = \pm 2$, as asserted. If Δ is reduced, $c = \pm 2$ cannot occur.

(c) We may assume that $\beta \ne 0$. From the Schwarz inequality we have

$$\left| \frac{2\langle \alpha, \beta\rangle}{|\alpha|^2} \frac{2\langle \alpha, \beta\rangle}{|\beta|^2} \right| \le 4$$

with equality only if $\beta = c\alpha$. The case of equality is handled by (b). If strict equality holds, then $\frac{2\langle\alpha,\beta\rangle}{|\alpha|^2}$ and $\frac{2\langle\alpha,\beta\rangle}{|\beta|^2}$ are two integers whose product is ≤ 3 in absolute value. The result follows in either case.

(d) We have an inequality of integers

$$\left|\frac{2\langle\alpha,\beta\rangle}{|\alpha|^2}\right| \geq \left|\frac{2\langle\alpha,\beta\rangle}{|\beta|^2}\right|,$$

and the proof of (c) shows that the product of the two sides is ≤ 3. Therefore the smaller side is 0 or 1.

(e) We may assume that α and β are not proportional. For the first statement, assume that $|\alpha| \leq |\beta|$. Then $s_\beta(\alpha) = \alpha - \frac{2\langle\alpha,\beta\rangle}{|\beta|^2}\beta$ must be $\alpha - \beta$, by (d). So $\alpha - \beta$ is in Δ. If $|\beta| \leq |\alpha|$ instead, we find that $s_\alpha(\beta) = \beta - \alpha$ is in Δ, and then $\alpha - \beta$ is in Δ as a consequence of (a). For the second statement we apply the first statement to $-\alpha$.

(f) This is immediate from (e).

(g) Let $-p$ and q be the smallest and largest values of n such that $\beta + n\alpha$ is in $\Delta \cup \{0\}$. If the string has a gap, we can find r and s with $r < s - 1$ such that $\beta + r\alpha$ is in $\Delta \cup \{0\}$, $\beta + (r+1)\alpha$ and $\beta + (s-1)\alpha$ are not in $\Delta \cup \{0\}$, and $\beta + s\alpha$ is in $\Delta \cup \{0\}$. By (e),

$$\langle\beta + r\alpha, \alpha\rangle \geq 0 \qquad \text{and} \qquad \langle\beta + s\alpha, \alpha\rangle \leq 0.$$

Subtracting these inequalities, we obtain $(r - s)|\alpha|^2 \geq 0$, and thus $r \geq s$, contradiction. We conclude that there are no gaps. Next

$$s_\alpha(\beta + n\alpha) = \beta + n\alpha - \frac{2\langle\beta + n\alpha, \alpha\rangle}{|\alpha|^2}\alpha = \beta - \left(n + \frac{2\langle\beta,\alpha\rangle}{|\alpha|^2}\right)\alpha,$$

and thus $-p \leq n \leq q$ implies $-q \leq n + \frac{2\langle\beta,\alpha\rangle}{|\alpha|^2} \leq p$. Taking $n = q$ and then $n = -p$, we obtain in turn

$$\frac{2\langle\beta,\alpha\rangle}{|\alpha|^2} \leq p - q \quad \text{and then} \quad p - q \leq \frac{2\langle\beta,\alpha\rangle}{|\alpha|^2}.$$

Thus $2\langle\beta,\alpha\rangle/|\alpha|^2 = p - q$. Finally, to investigate the length of the string, we may assume $q = 0$. The length of the string is then $p + 1$, with $p = 2\langle\beta,\alpha\rangle/|\alpha|^2$. The conclusion that the string has at most four roots then follows from (c) and (b).

We now introduce a notion of positivity in V that extends the notion in the examples in §1. The intention is to single out a subset of nonzero elements of V as **positive**, writing $\varphi > 0$ if φ is a positive element. The only properties of positivity that we need are that

(i) for any nonzero $\varphi \in V$, exactly one of φ and $-\varphi$ is positive
(ii) the sum of positive elements is positive, and any positive multiple of a positive element is positive.

The way in which such a notion of positivity is introduced is not important, and we shall give a sample construction shortly.

We say that $\varphi > \psi$ or $\psi < \varphi$ if $\varphi - \psi$ is positive. Then $>$ defines a simple ordering on V that is preserved under addition and under multiplication by positive scalars.

One way to define positivity is by means of a **lexicographic ordering**. Fix a spanning set $\varphi_1, \ldots, \varphi_m$ of V, and define positivity as follows: We say that $\varphi > 0$ if there exists an index k such that $\langle \varphi, \varphi_i \rangle = 0$ for $1 \leq i \leq k-1$ and $\langle \varphi, \varphi_k \rangle > 0$.

A lexicographic ordering sometimes arises disguised in a kind of dual setting. To use notation consistent with applications, think of V as the vector space dual of a space $\mathfrak{h}_0$, and fix a spanning set $H_1, \ldots, H_m$ for $\mathfrak{h}_0$. Then we say that $\varphi > 0$ if there exists an index k such that $\varphi(H_i) = 0$ for $1 \leq i \leq k-1$ and $\varphi(H_k) > 0$.

Anyway, we fix a notion of positivity and the resulting ordering for V. We say that a root α is **simple** if $\alpha > 0$ and if α does not decompose as $\alpha = \beta_1 + \beta_2$ with β_1 and β_2 both positive roots. A simple root is necessarily reduced.

Proposition 2.49. With $l = \dim V$, there are l simple roots $\alpha_1, \ldots, \alpha_l$, and they are linearly independent. If β is a root and is written as $\beta = x_1\alpha_1 + \cdots + x_l\alpha_l$, then all the x_j have the same sign (if 0 is allowed to be positive or negative), and all the x_j are integers.

REMARKS. Once this proposition has been proved, any positive root α can be written as $\alpha = \sum_{i=1}^{l} n_i\alpha_i$ with each n_i an integer ≥ 0. The integer $\sum_{i=1}^{l} n_i$ is called the **level** of α relative to $\{\alpha_1, \ldots, \alpha_l\}$ and is sometimes used in inductive proofs. The first example of such a proof will be with Proposition 2.54 below.

Before coming to the proof, let us review the examples in (2.43), which came from the complex semisimple Lie algebras in §1. In (2.50) we recall the choice of positive roots we made in §1 for each example and tell what the corresponding simple roots are:

(2.50)

	Positive Roots	Simple Roots
A_n	$e_i - e_j$, $i < j$	$e_1 - e_2,\ e_2 - e_3, \dots, e_n - e_{n+1}$
B_n	$e_i \pm e_j$ with $i < j$ e_i	$e_1 - e_2,\ e_2 - e_3, \dots, e_{n-1} - e_n,\ e_n$
C_n	$e_i \pm e_j$ with $i < j$ $2e_i$	$e_1 - e_2,\ e_2 - e_3, \dots, e_{n-1} - e_n,\ 2e_n$
D_n	$e_i \pm e_j$ with $i < j$	$e_1 - e_2, \dots, e_{n-2} - e_{n-1},\ e_{n-1} - e_n,\ e_{n-1} + e_n$

Lemma 2.51. If α and β are distinct simple roots, then $\alpha - \beta$ is not a root. Hence $\langle \alpha, \beta \rangle \le 0$.

PROOF. Assuming the contrary, suppose that $\alpha - \beta$ is a root. If $\alpha - \beta$ is positive, then $\alpha = (\alpha - \beta) + \beta$ exhibits α as a nontrivial sum of positive roots. If $\alpha - \beta$ is negative, then $\beta = (\beta - \alpha) + \alpha$ exhibits β as a nontrivial sum of positive roots. In either case we have a contradiction. Thus $\alpha - \beta$ is not a root, and Proposition 2.48e shows that $\langle \alpha, \beta \rangle \le 0$.

PROOF OF PROPOSITION 2.49. Let $\beta > 0$ be in Δ. If β is not simple, write $\beta = \beta_1 + \beta_2$ with β_1 and β_2 both positive in Δ. Then decompose β_1 and/or β_2, and then decompose each of their components if possible. Continue in this way. We can list the decompositions as tuples (β, β_1, component of β_1, etc.) with each entry a component of the previous entry. The claim is that no tuple has more entries than there are positive roots, and therefore the decomposition process must stop. In fact, otherwise some tuple would have the same $\gamma > 0$ in it at least twice, and we would have $\gamma = \gamma + \alpha$ with α a nonempty sum of positive roots, contradicting the properties of an ordering. Thus β is exhibited as $\beta = x_1\alpha_1 + \cdots + x_m\alpha_m$ with all x_j positive integers or 0 and with all α_j simple. Thus the simple roots span in the fashion asserted.

Finally we prove linear independence. Renumbering the α_j's, suppose that

$$x_1\alpha_1 + \cdots + x_s\alpha_s - x_{s+1}\alpha_{s+1} - \cdots - x_m\alpha_m = 0$$

with all $x_j \ge 0$ in $\mathbb{R}$. Put $\beta = x_1\alpha_1 + \cdots + x_s\alpha_s$. Then

$$0 \le \langle \beta, \beta \rangle = \left\langle \sum_{j=1}^{s} x_j\alpha_j,\ \sum_{k=s+1}^{m} x_k\alpha_k \right\rangle = \sum_{j,k} x_j x_k \langle \alpha_j, \alpha_k \rangle \le 0,$$

the last inequality holding by Lemma 2.51. We conclude that $\langle \beta, \beta \rangle = 0$, $\beta = 0$, and all the x_j's equal 0 since a positive combination of positive roots cannot be 0.

For the remainder of this section, we fix an abstract root system Δ, and we assume that Δ is reduced. Fix also an ordering coming from a notion of positivity as above, and let Π be the set of simple roots. We shall associate a "Cartan matrix" to the system Π and note some of the properties of this matrix. An "abstract Cartan matrix" will be any square matrix with this list of properties. Working with an abstract Cartan matrix is made easier by associating to the matrix a kind of graph known as an "abstract Dynkin diagram."

Enumerate Π as $\Pi = \{\alpha_1, \dots, \alpha_l\}$, where $l = \dim V$. The l-by-l matrix $A = (A_{ij})$ given by

$$A_{ij} = \frac{2\langle \alpha_i, \alpha_j \rangle}{|\alpha_i|^2}$$

is called the **Cartan matrix** of Δ and Π. The Cartan matrix depends on the enumeration of Π, and distinct enumerations evidently lead to Cartan matrices that are conjugate to one another by a permutation matrix.

For the examples in Figure 2.2 with $\dim V = 2$, the Cartan matrices are of course 2-by-2 matrices. For all the examples except G_2, an enumeration of the simple roots is given in (2.50). For G_2 let us agree to list the short simple root first. Then the Cartan matrices are as follows:

$$A_1 \oplus A_1 \qquad \begin{pmatrix} 2 & 0 \\ 0 & 2 \end{pmatrix}$$

$$A_2 \qquad \begin{pmatrix} 2 & -1 \\ -1 & 2 \end{pmatrix}$$

$$B_2 \qquad \begin{pmatrix} 2 & -1 \\ -2 & 2 \end{pmatrix}$$

$$C_2 \qquad \begin{pmatrix} 2 & -2 \\ -1 & 2 \end{pmatrix}$$

$$G_2 \qquad \begin{pmatrix} 2 & -3 \\ -1 & 2 \end{pmatrix}$$

Proposition 2.52. The Cartan matrix $A = (A_{ij})$ of Δ relative to the set Π of simple roots has the following properties:

(a) A_{ij} is in $\mathbb{Z}$ for all i and j
(b) $A_{ii} = 2$ for all i
(c) $A_{ij} \leq 0$ for $i \neq j$
(d) $A_{ij} = 0$ if and only if $A_{ji} = 0$
(e) there exists a diagonal matrix D with positive diagonal entries such that DAD^{-1} is symmetric positive definite.

PROOF. Properties (a), (b), and (d) are trivial, and (c) follows from Lemma 2.51. Let us prove (e). Put

$$D = \operatorname{diag}(|\alpha_1|, \ldots, |\alpha_l|), \tag{2.53}$$

so that $DAD^{-1} = \left(2\left\langle \frac{\alpha_i}{|\alpha_i|}, \frac{\alpha_j}{|\alpha_j|}\right\rangle\right)$. This is symmetric, and we can discard the 2 in checking positivity. But $(\langle\varphi_i, \varphi_j\rangle)$ is positive definite whenever $\{\varphi_i\}$ is a basis, since

$$(c_1 \quad \cdots \quad c_l)\left(\langle\varphi_i, \varphi_j\rangle\right)\begin{pmatrix} c_1 \\ \vdots \\ c_l \end{pmatrix} = \Big|\sum_i c_i\varphi_i\Big|^2.$$

The normalized simple roots may be taken as the basis φ_i of V, according to Proposition 2.49, and the result follows.

A square matrix A satifying properties (a) through (e) in Proposition 2.52 will be called an **abstract Cartan matrix**. Two abstract Cartan matrices are **isomorphic** if one is conjugate to the other by a permutation matrix.

Proposition 2.54. The abstract reduced root system Δ is reducible if and only if, for some enumeration of the indices, the Cartan matrix is block diagonal with more than one block.

PROOF. Suppose that $\Delta = \Delta' \cup \Delta''$ disjointly with every member of Δ' orthogonal to every member of Δ''. We enumerate the simple roots by listing all those in Δ' before all those in Δ'', and then the Cartan matrix is block diagonal.

Conversely suppose that the Cartan matrix is block diagonal, with the simple roots $\alpha_1, \ldots, \alpha_s$ leading to one block and the simple roots $\alpha_{s+1}, \ldots, \alpha_l$ leading to another block. Let Δ' be the set of all roots whose expansion in terms of the basis $\alpha_1, \ldots, \alpha_l$ involves only $\alpha_1, \ldots, \alpha_s$, and let Δ'' be the set of all roots whose expansion involves only $\alpha_{s+1}, \ldots, \alpha_l$. Then Δ' and Δ'' are nonempty and are orthogonal to each other, and it is enough to show that their union is Δ. Let $\alpha \in \Delta$ be given, and write $\alpha = \sum_{i=1}^{l} n_i\alpha_i$. We are to show that either $n_i = 0$ for $i > s$ or $n_i = 0$ for $i \leq s$. Proposition 2.49 says that all the n_i are integers and they have the same sign. Without loss of generality we may assume that α is positive, so that all n_i are ≥ 0.

We proceed by induction on the level $\sum_{i=1}^{l} n_i$. If the sum is 1, then $\alpha = \alpha_j$ for some j. Certainly either $n_i = 0$ for $i > s$ or $n_i = 0$ for $i \leq s$. Assume the result for level $n-1$, and let the level be $n > 1$ for α. We have

$$0 < |\alpha|^2 = \sum_{i=1}^{l} n_i \langle \alpha, \alpha_i \rangle,$$

and therefore $\langle \alpha, \alpha_j \rangle > 0$ for some j. To fix the notation, let us say that $1 \leq j \leq s$. By Proposition 2.48e, $\alpha - \alpha_j$ is a root, evidently of level $n-1$. By inductive hypothesis, $\alpha - \alpha_j$ is in Δ' or Δ''. If $\alpha - \alpha_j$ is in Δ', then α is in Δ', and the induction is complete. So we may assume that $\alpha - \alpha_j$ is in Δ''. Then $\langle \alpha - \alpha_j, \alpha_j \rangle = 0$. By Proposition 2.48g, the α_j string containing $\alpha - \alpha_j$ has $p = q$, and this number must be ≥ 1 since α is a root. Hence $\alpha - 2\alpha_j$ is in $\Delta \cup \{0\}$. We cannot have $\alpha - 2\alpha_j = 0$ since Δ is reduced, and we conclude that the coefficient of α_j in $\alpha - \alpha_j$ is > 0, in contradiction with the assumption that $\alpha - \alpha_j$ is in Δ''. Thus $\alpha - \alpha_j$ could not have been in Δ'', and the induction is complete.

Motivated by Proposition 2.54, we say that an abstract Cartan matrix is **reducible** if, for some enumeration of the indices, the matrix is block diagonal with more than one block. Otherwise the abstract Cartan matrix is said to be **irreducible**.

If we have several abstract Cartan matrices, we can arrange them as the blocks of a block-diagonal matrix, and the result is a new abstract Cartan matrix. The converse direction is addressed by the following proposition.

Proposition 2.55. After a suitable enumeration of the indices, any abstract Cartan matrix may be written in block-diagonal form with each block an irreducible abstract Cartan matrix.

PROOF. Call two indices i and j equivalent if there exists a sequence of integers $i = k_0, k_1, \ldots, k_{r-1}, k_r = j$ such that $A_{k_{s-1}k_s} \neq 0$ for $1 \leq s \leq r$. Enumerate the indices so that the members of each equivance class appear together, and then the abstract Cartan matrix will be in block-diagonal form with each block irreducible.

To our set Π of simple roots for the reduced abstract root system Δ, let us associate a kind of graph known as a "Dynkin diagram." We associate to each simple root α_i a vertex of a graph, and we attach to that vertex a weight proportional to $|\alpha_i|^2$. The vertices of the graph are connected by edges as follows. If two vertices are given, say corresponding to distinct simple roots α_i and α_j, we connect those vertices by $A_{ij}A_{ji}$ edges. The resulting graph is called the **Dynkin diagram** of Π. It follows from Proposition 2.54 that Δ is irreducible if and only if the Dynkin diagram is connected. Figure 2.3 gives the Dynkin diagrams for the root systems A_n, B_n, C_n, and D_n when the simple roots are chosen as in (2.50). Figure 2.3 shows also the Dynkin diagram for the root system G_2 of Figure 2.1 when the two simple roots are chosen so that $|\alpha_1| < |\alpha_2|$.

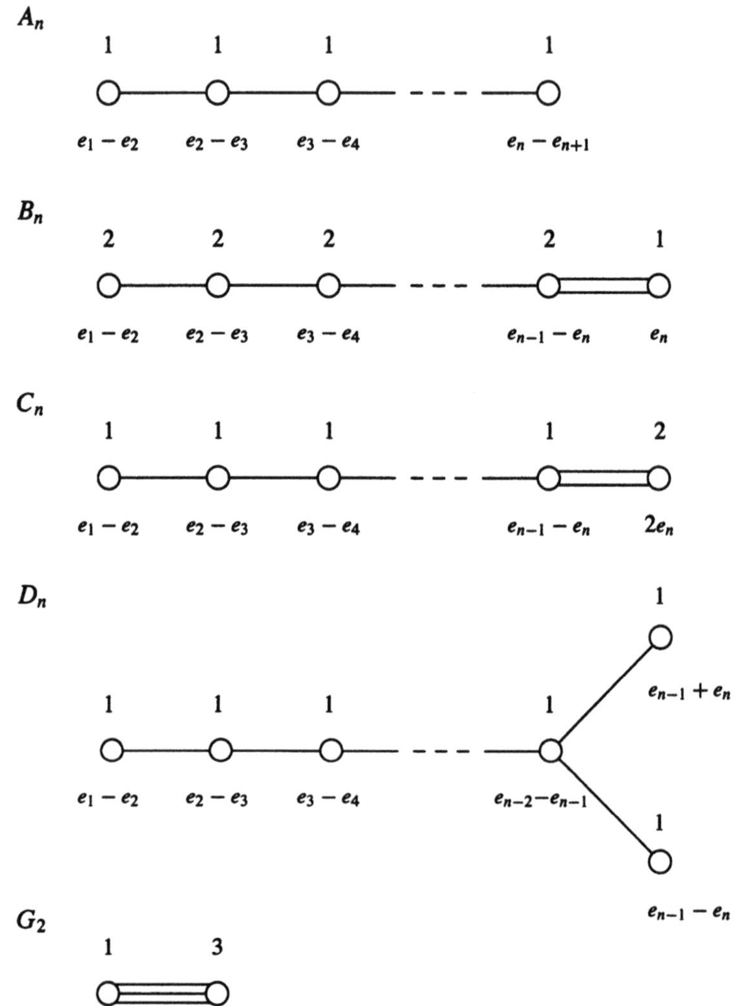

FIGURE 2.3. Dynkin diagrams for A_n, B_n, C_n, D_n, G_2

Let us indicate how we can determine the Dynkin diagram almost completely from the Cartan matrix. The key is the following lemma.

Lemma 2.56. Let A be an abstract Cartan matrix in block-diagonal form with each block an irreducible abstract Cartan matrix. Then the associated diagonal matrix D given in the defining property (e) of an abstract Cartan matrix is unique up to a multiplicative scalar on each block.

PROOF. Suppose that D and D' are two diagonal matrices with positive diagonal entries such that $P = DAD^{-1}$ and $P' = D'AD'^{-1}$ are symmetric positive definite. Then P and $P' = (D'D^{-1})P(D'D^{-1})^{-1}$ are both symmetric. Write $D'D^{-1} = \operatorname{diag}(b_1, \ldots, b_l)$. For any i and j, we have

$$b_i P_{ij} b_j^{-1} = P'_{ij} = P'_{ji} = b_j P_{ji} b_i^{-1} = b_j P_{ij} b_i^{-1}.$$

Thus either $P_{ij} = 0$ or $b_i = b_j$, i.e.,

$$A_{ij} = 0 \quad \text{or} \quad b_i = b_j. \tag{2.57}$$

If i and j are in the same block of A, then there exists a sequence of integers $i = k_0, k_1, \ldots, k_{r-1}, k_r = j$ such that $A_{k_{s-1}k_s} \neq 0$ for $1 \leq s \leq r$. From (2.57) we obtain

$$b_i = b_{k_0} = b_{k_1} = \cdots = b_{k_{r-1}} = b_{k_r} = b_j.$$

Thus the diagonal entries of D' are proportional to the diagonal entries of D within each block for A.

Returning to a Cartan matrix arising from the abstract reduced root system Δ and the set Π of simple roots, we note that the numbers $A_{ij}A_{ji}$ available from the Cartan matrix determine the numbers of edges between vertices in the Dynkin diagram. But the Cartan matrix also almost completely determines the weights in the Dynkin diagram. In fact, (2.53) says that the square roots of the weights are the diagonal entries of the matrix D of Proposition 2.52e. Lemma 2.56 says that D is determined by the properties of A up to a multiplicative scalar on each irreducible block, and irreducible blocks correspond to connected components of the Dynkin diagram. Thus by using A, we can determine the weights in the Dynkin diagram up to a proportionality constant on each connected component. These proportionality constants are the only ambiguity in obtaining the Dynkin diagram from the Cartan matrix.

The same considerations allow us to associate an "abstract Dynkin diagram" to an abstract Cartan matrix A. If A has size l-by-l, the **abstract Dynkin diagram** is a graph with l vertices, the ith and j^{th} vertices being connected by $A_{ij}A_{ji}$ edges. If D is the matrix given in defining property (e) of an abstract Cartan matrix (see Proposition 2.52), then we assign a weight to the vertex i equal to the square of the ith diagonal entry of D. Then A by itself determines the abstract Dynkin diagram up to a proportionality constant for the weights on each connected component.

Finally let us observe that we can recover an abstract Cartan matrix A from its abstract Dynkin diagram. Let the system of weights be $\{w_i\}$.

First suppose there are no edges from the ith vertex to the jth vertex. Then $A_{ij}A_{ji} = 0$. Since $A_{ij} = 0$ if and only if $A_{ji} = 0$, we obtain $A_{ij} = A_{ji} = 0$. Next suppose there exist edges between the ith vertex and the jth vertex. Then the number of edges tells us $A_{ij}A_{ji}$, while the symmetry of DAD^{-1} says that

$$w_i^{1/2}A_{ij}w_j^{-1/2} = w_j^{1/2}A_{ji}w_i^{-1/2},$$

i.e., that

$$\frac{A_{ij}}{A_{ji}} = \frac{w_j}{w_i}.$$

Since A_{ij} and A_{ji} are < 0, the number of edges and the ratio of weights together determine A_{ij} and A_{ji}.

6. Weyl Group

Schematically we can summarize our work so far in this chapter as constructing a two-step passage

(2.58)

$$\begin{matrix}\text{complex} \\ \text{semisimple} \\ \text{Lie algebra}\end{matrix} \xrightarrow[\text{Cartan subalgebra}]{\text{choice of}} \begin{matrix}\text{abstract reduced} \\ \text{root system}\end{matrix} \xrightarrow[\text{ordering}]{\text{choice of}} \begin{matrix}\text{abstract} \\ \text{Cartan} \\ \text{matrix.}\end{matrix}$$

Each step of the passage relies on a certain choice, and that choice is listed as part of the arrow. For this two-step passage to be especially useful, we should show that each step is independent of its choice, at least up to isomorphism. Then we will have a well defined way of passing from a complex semisimple Lie algebra first to an abstract reduced root system and then to an abstract Cartan matrix.

We can ask for even more. Once (2.58) is shown to be well defined independently of the choices, we can try to show that each step is one-one, up to isomorphism. In other words, two complex semisimple Lie algebras with isomorphic abstract reduced root systems are to be isomorphic, and two abstract reduced root systems leading to isomorphic abstract Cartan matrices are to be isomorphic. Then we can detect isomorphisms of complex semisimple Lie algebras by using Dynkin diagrams.

Finally we can ask that each step of the two-step passage be onto. In other words, every abstract reduced root system, up to isomorphism, is to come from a complex semisimple Lie algebra, and every abstract Cartan matrix is to come from an abstract reduced root system. Then a classification of abstract Cartan matrices will achieve a classification of complex semisimple Lie algebras.

We begin these steps in this section, starting by showing that each step in (2.58) is well defined, independently of the choices, up to isomorphism. For the first step, from the complex semisimple Lie algebra to the abstract reduced root system, the tool is Theorem 2.15, which says that any two Cartan subalgebras of our complex semisimple Lie algebra $\mathfrak{g}$ are conjugate via $\operatorname{Int}\mathfrak{g}$. It is clear that we can follow the effect of this conjugating automorphism through to its effect on roots and obtain an isomorphism of the associated root systems.

For the second step, from the abstract reduced root system to the abstract Cartan matrix up to isomorphism (or equivalently to the set Π of simple roots), the tool is the "Weyl group," which we study in this section.

Thus let Δ be an abstract root system in a finite-dimensional inner product space V. It will not be necessary to assume that Δ is reduced. We let $W = W(\Delta)$ be the subgroup of the orthogonal group on V generated by the reflections s_α for $\alpha \in \Delta$. This is the **Weyl group** of Δ. In the special case that Δ is the root system of a complex semisimple Lie algebra $\mathfrak{g}$ with respect to a Cartan subalgebra $\mathfrak{h}$, we sometimes write $W(\mathfrak{g}, \mathfrak{h})$ for the Weyl group.

We immediately see that W is a finite group of orthogonal transformations of V. In fact, any w in W maps the finite set Δ to itself. If w fixes each element of Δ, then w fixes a spanning set of V and hence fixes V. The assertion follows.

In addition, we have the formula

$$s_{r\alpha} = r s_\alpha r^{-1} \tag{2.59}$$

for any orthogonal transformation r of V. In fact,

$$s_{r\alpha}(r\varphi) = r\varphi - \frac{2\langle r\varphi, r\alpha\rangle}{|r\alpha|^2} r\alpha = r\varphi - \frac{2\langle \varphi, \alpha\rangle}{|\alpha|^2} r\alpha = r(s_\alpha\varphi).$$

As a consequence of (2.59), if r is in W and $r\alpha = \beta$, then

$$s_\beta = r s_\alpha r^{-1}. \tag{2.60}$$

EXAMPLES.

1) The root systems of types A_n, B_n, C_n, and D_n are described in (2.43). For A_n, $W(\Delta)$ consists of all permutations of $e_1, \dots, e_{n+1}$. For B_n and C_n, $W(\Delta)$ is generated by all permutations of $e_1, \dots, e_n$ and all sign changes (of the coefficients of $e_1, \dots, e_n$). For D_n, $W(\Delta)$ is generated by all permutations of $e_1, \dots, e_n$ and all even sign changes.

2) The nonreduced abstract root system $(BC)_2$ is pictured in Figure 2.2. For it, $W(\Delta)$ has order 8 and is the same group as for B_2 and C_2. The group contains the 4 rotations through multiples of angles $\pi/2$, together with the 4 reflections defined by sending a root to its negative and leaving the orthogonal complement fixed.

3) The reduced abstract root system G_2 is pictured in Figure 2.2. For it, $W(\Delta)$ has order 12 and consists of the 6 rotations through multiples of angles $\pi/3$, together with the 6 reflections defined by sending a root to its negative and leaving the orthogonal complement fixed.

Introduce a notion of positivity within V, such as from a lexicographic ordering, and let Δ^+ be the set of positive roots. The set Δ^+ determines a set $\Pi = \{\alpha_1, \dots, \alpha_l\}$ of simple roots, and in turn Π can be used to pick out the members of Δ^+ from Δ, since Proposition 2.49 says that the positive roots are those of the form $\alpha = \sum_i n_i\alpha_i$ with all $n_i \geq 0$.

Now suppose that $\Pi = \{\alpha_1, \dots, \alpha_l\}$ is any set of l independent reduced elements α_i such that every expression of a member α of Δ as $\sum_i c_i\alpha_i$ has all nonzero c_i of the same sign. We call Π a **simple system**. Given a simple system Π, we can define Δ^+ to be all roots of the form $\sum_i c_i\alpha_i$ with all $c_i \geq 0$. The claim is that Δ^+ is the set of positive roots in some lexicographic ordering. In fact, we can use the dual basis to $\{\alpha_i\}$ to get such an ordering. In more detail if $\langle\alpha_i, \omega_j\rangle = \delta_{ij}$ and if j is the first index with $\langle\alpha, \omega_j\rangle$ nonzero, then the fact that $\langle\alpha, \omega_j\rangle = c_j$ is positive implies that α is positive.

Thus we have an abstract characterization of the possible Π's that can arise as sets of simple roots: They are all possible simple systems.

Lemma 2.61. Let $\Pi = \{\alpha_1, \dots, \alpha_l\}$ be a simple system, and let $\alpha > 0$ be in Δ. Then

$$s_{\alpha_i}(\alpha) \text{ is } \begin{cases} = -\alpha & \text{if } \alpha = \alpha_i \text{ or } \alpha = 2\alpha_i \\ > 0 & \text{otherwise.} \end{cases}$$

PROOF. If $\alpha = \sum c_j\alpha_j$, then

$$s_{\alpha_i}(\alpha) = \sum_{j=1}^{l} c_j\alpha_j - \frac{2\langle\alpha, \alpha_i\rangle}{|\alpha_i|^2}\alpha_i.$$

If at least one c_j is > 0 for $j \neq i$, then $s_{\alpha_i}(\alpha)$ has the same coefficient for α_j that α does, and $s_{\alpha_i}(\alpha)$ must be positive. The only remaining case is that α is a multiple of α_i, and then α must be α_i or $2\alpha_i$, by Proposition 2.48b.

Proposition 2.62. Let $\Pi = \{\alpha_1, \ldots, \alpha_l\}$ be a simple system. Then $W(\Delta)$ is generated by the root reflections s_{α_i} for α_i in Π. If α is any reduced root, then there exist $\alpha_j \in \Pi$ and $s \in W(\Delta)$ such that $s\alpha_j = \alpha$.

PROOF. We begin by proving a seemingly sharper form of the second assertion. Let $W' \subseteq W$ be the group generated by the s_{α_i} for $\alpha_i \in \Pi$. We prove that any reduced root $\alpha > 0$ is of the form $s\alpha_j$ with $s \in W'$. Writing $\alpha = \sum n_j\alpha_j$, we proceed by induction on $\text{level}(\alpha) = \sum n_j$. The case of level one is the case of $\alpha = \alpha_i$ in Π, and we can take $s = 1$. Assume the assertion for level $<$ level(α), let level(α) be > 1, and write $\alpha = \sum n_j\alpha_j$. Since

$$0 < |\alpha|^2 = \sum n_j \langle \alpha, \alpha_j \rangle,$$

we must have $\langle \alpha, \alpha_i \rangle > 0$ for some $i = i_0$. By our assumptions, α is neither α_{i_0} nor $2\alpha_{i_0}$. Then $\beta = s_{\alpha_{i_0}}(\alpha)$ is > 0 by Lemma 2.61 and has

$$\beta = \sum_{j \neq i_0} n_j\alpha_j + \left(c_{i_0} - \frac{2\langle \alpha, \alpha_{i_0} \rangle}{|\alpha_{i_0}|^2} \right) \alpha_{i_0}.$$

Since $\langle \alpha, \alpha_{i_0} \rangle > 0$, level($\beta$) $<$ level(α). By inductive hypothesis, $\beta = s'\alpha_j$ for some $s' \in W'$ and some index j. Then $\alpha = s_{\alpha_{i_0}}\beta = s_{\alpha_{i_0}}s'\alpha_j$ with $s_{\alpha_{i_0}}s'$ in W'. This completes the induction.

If $\alpha < 0$, then we can write $-\alpha = s\alpha_j$, and it follows that $\alpha = ss_{\alpha_j}\alpha_j$. Thus each reduced member α of Δ is of the form $s'\alpha_j$ for some $s' \in W'$ and some $\alpha_j \in \Pi$.

To complete the proof, we show that each s_α, for $\alpha \in \Delta$, is in W'. There is no loss of generality in assuming that α is reduced. Write $\alpha = s\alpha_j$ with $s \in W'$. Then (2.60) shows that $s_\alpha = ss_{\alpha_j}s^{-1}$, which is in W'. Since W is generated by the reflections s_α for $\alpha \in \Delta$, $W \subseteq W'$ and $W = W'$.

Theorem 2.63. If Π and Π' are two simple systems for Δ, then there exists one and only one element $s \in W$ such that $s\Pi = \Pi'$.

PROOF OF EXISTENCE. Let Δ^+ and $\Delta^{+\prime}$ be the sets of positive roots in question. We have $|\Delta^+| = |\Delta^{+\prime}| = \frac{1}{2}|\Delta|$, which we write as q. Also $\Delta^+ = \Delta^{+\prime}$ if and only if $\Pi = \Pi'$, and $\Delta^+ \neq \Delta^{+\prime}$ implies $\Pi \not\subseteq \Delta^{+\prime}$ and $\Pi' \not\subseteq \Delta^+$. Let $r = |\Delta^+ \cap \Delta^{+\prime}|$. We induct downward on r, the case $r = q$ being handled by using $s = 1$. Let $r < q$. Choose $\alpha_i \in \Pi$ with $\alpha_i \notin \Delta^{+\prime}$, so that $-\alpha_i \in \Delta^{+\prime}$. If β is in $\Delta^+ \cap \Delta^{+\prime}$, then $s_{\alpha_i}\beta$ is in Δ^+ by Lemma 2.61. Thus $s_{\alpha_i}\beta$ is in $\Delta^+ \cap s_{\alpha_i}\Delta^{+\prime}$. Also $\alpha_i = s_{\alpha_i}(-\alpha_i)$ is in $\Delta^+ \cap s_{\alpha_i}\Delta^{+\prime}$. Hence $|\Delta^+ \cap s_{\alpha_i}\Delta^{+\prime}| \geq r + 1$. Now $s_{\alpha_i}\Delta^{+\prime}$ corresponds to the simple system $s_{\alpha_i}\Pi'$, and by inductive hypothesis we can find $t \in W$ with $t\Pi = s_{\alpha_i}\Pi'$. Then $s_{\alpha_i}t\Pi = \Pi'$, and the induction is complete.

PROOF OF UNIQUENESS. We may assume that $s\Pi = \Pi$, and we are to prove that $s = 1$. Write $\Pi = \{\alpha_1, \ldots, \alpha_l\}$, and abbreviate s_{α_j} as s_j. For $s = s_{i_m} \cdots s_{i_1}$, we prove by induction on m that $s\Pi = \Pi$ implies $s = 1$. If $m = 1$, then $s = s_{i_1}$ and $s\alpha_{i_1} < 0$. If $m = 2$, we obtain $s_{i_2}\Pi = s_{i_1}\Pi$, whence $-\alpha_{i_2}$ is in $s_{i_1}\Pi$ and so $-\alpha_{i_2} = -\alpha_{i_1}$, by Lemma 2.61; hence $s = 1$. Thus assume inductively that

$$t\Pi = \Pi \text{ with } t = s_{j_r} \cdots s_{j_1} \text{ and } r < m \quad \text{implies} \quad t = 1, \tag{2.64}$$

and let $s = s_{i_m} \cdots s_{i_1}$ satisfy $s\Pi = \Pi$ with $m > 2$.

Put $s' = s_{i_{m-1}} \cdots s_{i_1}$, so that $s = s_{i_m}s'$. Then $s' \neq 1$ by (2.64) for $t = s_{i_m}$. Also $s'\alpha_j < 0$ for some j by (2.64) applied to $t = s'$. The latter fact, together with

$$s_{i_m}s'\alpha_j = s\alpha_j > 0.$$

says that $-\alpha_{i_m} = s'\alpha_j$, by Lemma 2.61. Also if $\beta > 0$ and $s'\beta < 0$, then $s'\beta = -c\alpha_{i_m} = s'(c\alpha_j)$, so that $\beta = c\alpha_j$ with $c = 1$ or 2. Thus s' satisfies

(i) $s'\alpha_j = -\alpha_{i_m}$

(ii) $s'\beta > 0$ for every positive $\beta \in \Delta$ other than α_j and $2\alpha_j$.

Now $s_{i_{m-1}} \cdots s_{i_1}\alpha_j = -\alpha_{i_m} < 0$ by (i). Choose k so that $t = s_{i_{k-1}} \cdots s_{i_1}$ satisfies $t\alpha_j > 0$ and $s_{i_k}t\alpha_j < 0$. Then $t\alpha_j = \alpha_{i_k}$. By (2.60), $ts_jt^{-1} = s_{i_k}$. Hence $ts_j = s_{i_k}t$.

Put $t' = s_{i_{m-1}} \cdots s_{i_{k+1}}$, so that $s' = t's_{i_k}t = t'ts_j$. Then $t't = s's_j$. Now $\alpha > 0$ and $\alpha \neq c\alpha_j$ imply $s_j\alpha = \beta > 0$ with $\beta \neq c\alpha_j$. Thus

$$t't\alpha = s's_j\alpha = s'\beta > 0 \qquad \text{by (ii)}$$

and

$$t't\alpha_j = s'(-\alpha_j) = \alpha_{i_m} > 0 \qquad \text{by (i).}$$

Hence $t't\Pi = \Pi$. Now $t't$ is a product of $m - 2$ s_j's. By inductive hypothesis, $t't = 1$. Then $s's_j = 1$, $s' = s_j$, and $s = s_{i_m}s' = s_{i_m}s_j$. Since (2.64) has been proved for $r = 2$, we conclude that $s = 1$. This completes the proof.

Corollary 2.65. In the second step of the two-step passage (2.58), the resulting Cartan matrix is independent of the choice of positive system, up to permutation of indices.

PROOF. Let Π and Π' be the simple systems that result from two different positive systems. By Theorem 2.63, $\Pi' = s\Pi$ for some $s \in W(\Delta)$. Then we can choose enumerations $\Pi = \{\alpha_1, \ldots, \alpha_l\}$ and $\Pi' = \{\beta_1, \ldots, \beta_l\}$ so that $\beta_j = s\alpha_j$, and we have

$$\frac{2\langle\beta_i, \beta_j\rangle}{|\beta_i|^2} = \frac{2\langle s\alpha_i, s\alpha_j\rangle}{|s\alpha_i|^2} = \frac{2\langle\alpha_i, \alpha_j\rangle}{|\alpha_i|^2}$$

since s is orthogonal. Hence the resulting Cartan matrices match.

Consequently our use of the root-system names A_n, B_n, etc., with the Dynkin diagrams in Figure 2.3 was legitimate. The Dynkin diagram is not changed by changing the positive system (except that the names of roots attached to vertices change).

This completes our discussion of the fact that the steps in the passages (2.58) are well defined independently of the choices.

Let us take a first look at the uniqueness questions associated with (2.58). We want to see that each step in (2.58) is one-one, up to isomorphism. The following proposition handles the second step.

Proposition 2.66. The second step in the passage (2.58) is one-one, up to isomorphism. That is, the Cartan matrix determines the reduced root system up to isomorphism.

PROOF. First let us see that the Cartan matrix determines the set of simple roots, up to a linear transformation of V that is a scalar multiple of an orthogonal transformation on each irreducible component. In fact, we may assume that Δ is already irreducible, and we let $\alpha_1, \ldots, \alpha_l$ be the simple roots. Lemma 2.56 and (2.53) show that the Cartan matrix determines $|\alpha_1|, \ldots, |\alpha_l|$ up to a single proportionality constant. Suppose $\beta_1, \ldots, \beta_l$ is another simple system for the same Cartan matrix. Normalizing, we may assume that $|\alpha_j| = |\beta_j|$ for all j. From the Cartan matrix we obtain $\frac{2\langle\alpha_i,\alpha_j\rangle}{|\alpha_i|^2} = \frac{2\langle\beta_i,\beta_j\rangle}{|\beta_i|^2}$ for all i and j and hence $\langle\alpha_i, \alpha_j\rangle = \langle\beta_i, \beta_j\rangle$ for all i and j. In other words the linear transformation L defined by $L\alpha_i = \beta_i$ preserves inner products on a basis; it is therefore orthogonal.

To complete the proof, we want to see that the set $\{\alpha_1, \ldots, \alpha_l\}$ of simple roots determines the set of roots. Let W' be the group generated by the root reflections in the simple roots, and let $\Delta' = \bigcup_{j=1}^{l} W'\alpha_j$. Proposition 2.62 shows that $\Delta' = \Delta$ and that $W' = W(\Delta)$. The result follows.

Before leaving the subject of Weyl groups, we prove some further handy results. For the first result let us fix a system Δ^+ of positive roots and the corresponding simple system Π. We say that a member λ of V is **dominant** if $\langle\lambda, \alpha\rangle \geq 0$ for all $\alpha \in \Delta^+$. It is enough that $\langle\lambda, \alpha_i\rangle \geq 0$ for all $\alpha_i \in \Pi$.

Proposition 2.67. If λ is in V, then there exists a simple system Π for which λ is dominant.

PROOF. We may assume $\lambda \neq 0$. Put $\varphi_1 = \lambda$ and extend to an orthogonal basis $\varphi_1, \ldots, \varphi_l$ of V. Use this basis to define a lexicographic ordering and thereby to determine a simple system Π. Then λ is dominant relative to Π.

Corollary 2.68. If λ is in V and if a positive system Δ^+ is specified, then there is some element w of the Weyl group such that $w\lambda$ is dominant.

PROOF. This follows from Proposition 2.67 and Theorem 2.63.

For the remaining results we assume that Δ is reduced. Fix a positive system Δ^+, and let δ be half the sum of the members of Δ^+.

Proposition 2.69. Fix a positive system Δ^+ for the reduced abstract root system Δ. If α is a simple root, then $s_\alpha(\delta) = \delta - \alpha$ and $2\langle\delta, \alpha\rangle/|\alpha|^2 = 1$.

PROOF. By Lemma 2.61, s_α permutes the positive roots other than α and sends α to $-\alpha$. Therefore

$$s_\alpha(2\delta) = s_\alpha(2\delta - \alpha) + s_\alpha(\alpha) = (2\delta - \alpha) - \alpha = 2(\delta - \alpha),$$

and $s_\alpha(\delta) = \delta - \alpha$. Using the definition of s_α, we then see that

$$2\langle\delta, \alpha\rangle/|\alpha|^2 = 1.$$

For w in $W(\Delta)$, let $l(w)$ be the number of roots $\alpha > 0$ such that $w\alpha < 0$; $l(w)$ is called the **length** of the Weyl group element w relative to Π. In terms of a simple system $\Pi = \{\alpha_1, \dots, \alpha_l\}$ and its associated positive system Δ^+, let us abbreviate s_{α_j} as s_j.

Proposition 2.70. Fix a simple system $\Pi = \{\alpha_1, \dots, \alpha_l\}$ for the reduced abstract root system Δ. Then $l(w)$ is the smallest integer k such that w can be written as a product $w = s_{i_k} \cdots s_{i_1}$ of k reflections in simple roots.

REMARKS. Proposition 2.62 tells us that w has at least one expansion as a product of reflections in simple roots. Therefore the smallest integer k cited in the proposition exists. We prove Proposition 2.70 after first giving a lemma.

Lemma 2.71. Fix a simple system $\Pi = \{\alpha_1, \dots, \alpha_l\}$ for the reduced abstract root system Δ. If γ is a simple root and w is in $W(\Delta)$, then

$$l(ws_\gamma) = \begin{cases} l(w) - 1 & \text{if } w\gamma < 0 \\ l(w) + 1 & \text{if } w\gamma > 0. \end{cases}$$

PROOF. If α is a positive root other than γ, then Lemma 2.61 shows that $s_\gamma\alpha > 0$, and hence the correspondence $s_\gamma\alpha \leftrightarrow \alpha$ gives

$$\#\{\beta > 0 \mid \beta \neq \gamma \text{ and } ws_\gamma\beta < 0\} = \#\{\alpha > 0 \mid \alpha \neq \gamma \text{ and } w\alpha < 0\}.$$

To obtain $l(ws_\gamma)$, we add 1 to the left side if $w\gamma > 0$ and leave the left side alone if $w\gamma < 0$. To obtain $l(w)$, we add 1 to the right side if $w\gamma < 0$ and leave the right side alone if $w\gamma > 0$. The lemma follows.

PROOF OF PROPOSITION 2.70. Write $w = s_{i_k} \cdots s_{i_1}$ as a product of k reflections in simple roots. Then Lemma 2.71 implies that $l(w) \leq k$.

To get the equality asserted by the proposition, we need to show that if w sends exactly k positive roots into negative roots, then w can be expressed as a product of k factors $w = s_{i_k} \cdots s_{i_1}$. We do so by induction on k. For $k = 0$, this follows from the uniqueness in Theorem 2.63. Inductively assume the result for $k - 1$. If $k > 0$ and $l(w) = k$, then w must send some simple root α_j into a negative root. Set $w' = ws_j$. By Lemma 2.71, $l(w') = k - 1$. By inductive hypothesis, w' has an expansion $w' = s_{i_{k-1}} \cdots s_{i_1}$. Then $w = s_{i_{k-1}} \cdots s_{i_1} s_j$, and the induction is complete.

Proposition 2.72 (Chevalley's Lemma). Let the abstract root system Δ be reduced. Fix v in V, and let $W_0 = \{w \in W \mid wv = v\}$. Then W_0 is generated by the root reflections s_α such that $\langle v, \alpha \rangle = 0$.

PROOF. Choose an ordering with v first, so that $\langle \beta, v \rangle > 0$ implies $\beta > 0$. Arguing by contradiction, choose $w \in W_0$ with $l(w)$ as small as possible so that w is not a product of elements s_α with $\langle v, \alpha \rangle = 0$. Then $l(w) > 0$ by the uniqueness in Theorem 2.63. Let $\gamma > 0$ be a simple root such that $w\gamma < 0$. If $\langle v, \gamma \rangle > 0$, then

$$\langle v, w\gamma \rangle = \langle wv, w\gamma \rangle = \langle v, \gamma \rangle > 0,$$

in contradiction with the condition $w\gamma < 0$. Hence $\langle v, \gamma \rangle = 0$. That is, s_γ is in W_0. But then ws_γ is in W_0 with $l(ws_\gamma) < l(w)$, by Lemma 2.71. By assumption ws_γ is a product of the required root reflections, and therefore so is w.

Corollary 2.73. Let the abstract root system Δ be reduced. Fix v in V, and suppose that some element $w \neq 1$ of $W(\Delta)$ fixes v. Then some root is orthogonal to v.

PROOF. By Proposition 2.72, w is the product of root reflections s_α such that $\langle v, \alpha \rangle = 0$. Since $w \neq 1$, there must be such a root reflection.

7. Classification of Abstract Cartan Matrices

In this section we shall classify abstract Cartan matrices, and then we shall show that every abstract Cartan matrix arises from a reduced abstract root system. These results both contribute toward an understanding of the two-step passage (2.58), the second result showing that the second step of the passage is onto.

Recall that an abstract Cartan matrix is a square matrix satisfying properties (a) through (e) in Proposition 2.52. We continue to regard two such matrices as isomorphic if one can be obtained from the other by permuting the indices.

To each abstract Cartan matrix, we saw in §5 how to associate an abstract Dynkin diagram, the only ambiguity being a proportionality constant for the weights on each component of the diagram. We shall work simultaneously with a given abstract Cartan matrix and its associated abstract Dynkin diagram. Operations on the abstract Cartan matrix will correspond to operations on the abstract Dynkin diagram, and the diagram will thereby give us a way of visualizing what is happening. Our objective is to classify irreducible abstract Cartan matrices, since general abstract Cartan matrices can be obtaining by using irreducible such matrices as blocks. But we do not assume irreducibility yet.

We first introduce two operations on abstract Dynkin diagrams. Each operation will have a counterpart for abstract Cartan matrices, and we shall see that the counterpart carries abstract Cartan matrices to abstract Cartan matrices. Therefore each of our operations sends abstract Dynkin diagrams to abstract Dynkin diagrams:

1) Remove the ith vertex from the abstract Dynkin diagram, and remove all edges attached to that vertex.

2) Suppose that the ith and j^{th} vertices are connected by a single edge. Then the weights attached to the two vertices are equal. Collapse the two vertices to a single vertex and give it the common weight, remove the edge that joins the two vertices, and retain all other edges issuing from either vertex.

For Operation #1, the corresponding operation on a Cartan matrix A is to remove the ith row and column from A. It is clear that the new matrix satisfies the defining properties of an abstract Cartan matrix given in Proposition 2.52. This fact allows us to prove the following proposition.

Proposition 2.74. Let A be an abstract Cartan matrix. If $i \neq j$, then

(a) $A_{ij}A_{ji} < 4$.
(b) A_{ij} is 0 or -1 or -2 or -3.

PROOF.

(a) Let the diagonal matrix D of defining property (e) be given by $D = \operatorname{diag}(d_1, \ldots, d_l)$. Using Operation #1, remove all but the ith and j^{th} rows and columns from the abstract Cartan matrix A. Then

$$\begin{pmatrix} d_i & 0 \\ 0 & d_j \end{pmatrix} \begin{pmatrix} 2 & A_{ij} \\ A_{ji} & 2 \end{pmatrix} \begin{pmatrix} d_i^{-1} & 0 \\ 0 & d_j^{-1} \end{pmatrix}$$

is positive definite. So its determinant is > 0, and $A_{ij}A_{ji} < 4$.

(b) If $A_{ij} \neq 0$, then $A_{ji} \neq 0$, by defining property (d) in Proposition 2.52. Since A_{ij} and A_{ji} are integers ≤ 0, the result follows from (a).

We shall return presently to the verification that Operation #2 is a legitimate one on abstract Dynkin diagrams. First we derive some more subtle consequences of the use of Operation #1.

Let A be an l-by-l abstract Cartan matrix, and let $D = \operatorname{diag}(d_1, \ldots, d_l)$ be a diagonal matrix of the kind in defining condition (e) of Proposition 2.52. We shall define vectors $\alpha_i \in \mathbb{R}^l$ for $1 \leq i \leq l$ that will play the role of simple roots. Let us write $DAD^{-1} = 2Q$. Here $Q = (Q_{ij})$ is symmetric positive definite with 1's on the diagonal. Let $Q^{1/2}$ be its positive definite square root. Define vectors $\varphi \in \mathbb{R}^l$ for $1 \leq i \leq l$ by $\varphi_i = Q^{1/2}e_i$, where e_i is the ith standard basis vector of $\mathbb{R}^l$. Then

$$\langle \varphi_j, \varphi_i \rangle = \langle Q^{1/2}e_j, Q^{1/2}e_i \rangle = \langle Qe_j, e_i \rangle = Q_{ij},$$

and in particular φ_i is a unit vector. Put

$$\alpha_i = d_i \varphi_i, \tag{2.75}$$

so that

$$d_i = |\alpha_i|. \tag{2.76}$$

Then

$$\begin{aligned} A_{ij} &= 2(D^{-1}QD)_{ij} = 2d_i^{-1}Q_{ij}d_j \\ &= 2d_i^{-1}d_j\langle \varphi_j, \varphi_i \rangle = 2d_i^{-1}d_j\langle d_j^{-1}\alpha_j, d_i^{-1}\alpha_i \rangle \\ &= \frac{2\langle \alpha_i, \alpha_j \rangle}{|\alpha_i|^2}. \end{aligned} \tag{2.77}$$

Note that the vectors α_i are linearly independent since $\det A \neq 0$.

We shall find it convenient to refer to a vertex of the abstract Dynkin diagram either by its index i or by the associated vector α_i, depending on the context. We may write A_{ij} or $A_{\alpha_i,\alpha_{i+1}}$ for an entry of the abstract Cartan matrix.

Proposition 2.78. The abstract Dynkin diagram associated to the l-by-l abstract Cartan matrix A has the following properties:

(a) there are at most l pairs of vertices $i < j$ with at least one edge connecting them

(b) there are no loops

(c) at most three edges issue from any point of the diagram.

PROOF.

(a) With α_i as in (2.75), put $\alpha = \sum_{i=1}^{l} \frac{\alpha_i}{|\alpha_i|}$. Then

$$\begin{aligned} 0 < |\alpha|^2 &= \sum_{i,j} \left\langle \frac{\alpha_i}{|\alpha_i|}, \frac{\alpha_j}{|\alpha_j|} \right\rangle \\ &= \sum_{i} \left\langle \frac{\alpha_i}{|\alpha_i|}, \frac{\alpha_i}{|\alpha_i|} \right\rangle + 2\sum_{i<j} \left\langle \frac{\alpha_i}{|\alpha_i|}, \frac{\alpha_j}{|\alpha_j|} \right\rangle \\ &= l + \sum_{i<j} \frac{2\langle \alpha_i, \alpha_j \rangle}{|\alpha_i||\alpha_j|} \\ &= l - \sum_{i<j} \sqrt{A_{ij}A_{ji}}. \end{aligned} \tag{2.79}$$

By Proposition 2.74, $\sqrt{A_{ij}A_{ji}}$ is 0 or 1 or $\sqrt{2}$ or $\sqrt{3}$. When nonzero, it is therefore ≥ 1. Therefore the right side of (2.79) is

$$\leq l - \sum_{\substack{i<j, \\ \text{connected}}} 1.$$

Hence the number of connected pairs of vertices is $< l$.

(b) If there were a loop, we could use Operation #1 to remove all vertices except those in a loop. Then (a) would be violated for the loop.

(c) Fix $\alpha = \alpha_i$ as in (2.74). Consider the vertices that are connected by edges to the ith vertex. Write $\beta_1, \ldots, \beta_r$ for the α_j's associated to these vertices, and let there be $l_1, \ldots, l_r$ edges to the ith vertex. Let U be the $(r+1)$-dimensional vector subspace of $\mathbb{R}^l$ spanned by $\beta_1, \ldots, \beta_r, \alpha$. Then $\langle \beta_i, \beta_j \rangle = 0$ for $i \neq j$ by (b), and hence $\{\beta_k/|\beta_k|\}_{k=1}^{r}$ is an orthonormal set. Adjoin $\delta \in U$ to this set to make an orthonormal basis of U. Then $\langle \alpha, \delta \rangle \neq 0$ since $\{\beta_1, \ldots, \beta_r, \alpha\}$ is linearly independent. By Parseval's equality,

$$|\alpha|^2 = \sum_k \left\langle \alpha, \frac{\beta_k}{|\beta_k|} \right\rangle^2 + \langle \alpha, \delta \rangle^2 > \sum_k \left\langle \alpha, \frac{\beta_k}{|\beta_k|} \right\rangle^2$$

and hence

$$1 > \sum_k \frac{\langle \alpha, \beta_k \rangle^2}{|\alpha|^2 |\beta_k|^2} = \tfrac{1}{4} \sum_k l_k.$$

Thus $\sum_k l_k < 4$. This completes the proof.

We turn to Operation #2, which we have described in terms of abstract Dynkin diagrams. Let us describe the operation in terms of abstract Cartan matrices. We assume that $A_{ij} = A_{ji} = -1$, and we have asserted that the weights attached to the ith and j^{th} vertices, say w_i and w_j, are equal. The weights are given by $w_i = d_i^2$ and $w_j = d_j^2$. The symmetry of DAD^{-1} implies that

$$d_i A_{ij} d_j^{-1} = d_j A_{ji} d_i^{-1},$$

hence that $d_i^2 = d_j^2$ and $w_i = w_j$. Thus

$$A_{ij} = A_{ji} = -1 \qquad \text{implies} \qquad w_i = w_j. \tag{2.80}$$

Under the assumption that $A_{ij} = A_{ji} = -1$, Operation #2 replaces the abstract Cartan matrix A of size l by a square matrix of size $l - 1$, collapsing the ith and j^{th} indices. The replacement row is the sum of the ith and j^{th} rows of A in entries $k \notin \{i, j\}$, and similarly for the replacement column. The 2-by-2 matrix from the ith and j^{th} indices is $\begin{pmatrix} 2 & -1 \\ -1 & 2 \end{pmatrix}$ within A and gets replaced by the 1-by-1 matrix (2).

Proposition 2.81. Operation #2 replaces the abstract Cartan matrix A by another abstract Cartan matrix.

PROOF. Without loss of generality, let the indices i and j be $l - 1$ and l. Define E to be the $(l - 1)$-by-l matrix

$$E = \begin{pmatrix} 1 & 0 & \cdots & 0 & 0 & 0 \\ 0 & 1 & \cdots & 0 & 0 & 0 \\ \vdots & & \ddots & & \vdots & \\ 0 & 0 & \cdots & 1 & 0 & 0 \\ 0 & 0 & \cdots & 0 & 1 & 1 \end{pmatrix} = \begin{pmatrix} 1_{l-2} & 0 & 0 \\ 0 & 1 & 1 \end{pmatrix}.$$

The candidate for a new Cartan matrix is EAE^t, and we are to verify the five axioms in Proposition 2.52. The first four are clear, and we have to check (e). Let P be the positive definite matrix $P = DAD^{-1}$, and define

$$D' = EDE^t \operatorname{diag}(1, \ldots, 1, \tfrac{1}{2}),$$

which is square of size $l-1$. Remembering from (2.80) that the weights w_i satisfy $w_i = d_i^2$ and that $w_{l-1} = w_l$, we see that $d_{l-1} = d_l$. Write d for the common value of d_{l-1} and d_l. In block form, D is then of the form

$$D = \begin{pmatrix} D_0 & 0 & 0 \\ 0 & d & 0 \\ 0 & 0 & d \end{pmatrix}.$$

Therefore D' in block form is given by

$$\begin{aligned} D' &= \begin{pmatrix} 1_{l-2} & 0 & 0 \\ 0 & 1 & 1 \end{pmatrix} \begin{pmatrix} D_0 & 0 & 0 \\ 0 & d & 0 \\ 0 & 0 & d \end{pmatrix} \begin{pmatrix} 1_{l-2} & 0 \\ 0 & 1 \\ 0 & 1 \end{pmatrix} \begin{pmatrix} 1_{l-2} & 0 \\ 0 & \frac{1}{2} \end{pmatrix} \\ &= \begin{pmatrix} D_0 & 0 \\ 0 & d \end{pmatrix}. \end{aligned}$$

Meanwhile

$$\begin{aligned} E^t \operatorname{diag}(1, \dots, 1, \tfrac{1}{2}) E &= \begin{pmatrix} 1_{l-2} & 0 \\ 0 & 1 \\ 0 & 1 \end{pmatrix} \begin{pmatrix} 1_{l-2} & 0 \\ 0 & \frac{1}{2} \end{pmatrix} \begin{pmatrix} 1_{l-2} & 0 & 0 \\ 0 & 1 & 1 \end{pmatrix} \\ &= \begin{pmatrix} 1_{l-2} & 0 & 0 \\ 0 & \frac{1}{2} & \frac{1}{2} \\ 0 & \frac{1}{2} & \frac{1}{2} \end{pmatrix}, \end{aligned}$$

and it follows that $E^t \operatorname{diag}(1, \dots, 1, \frac{1}{2}) E$ commutes with D. Since $EE^t \operatorname{diag}(1, \dots, 1, \frac{1}{2}) = 1$, we therefore have

$$D'E = EDE^t \operatorname{diag}(1, \dots, 1, \tfrac{1}{2}) E = EE^t \operatorname{diag}(1, \dots, 1, \tfrac{1}{2}) ED = ED.$$

The same computation gives also $D'^{-1}E = ED^{-1}$, whose transpose is $E^t D'^{-1} = D^{-1}E^t$. Thus

$$D'(EAE^t)D'^{-1} = (D'E)A(E^t D'^{-1}) = EDAD^{-1}E^t = EPE^t,$$

and the right side is symmetric and positive semidefinite. To see that it is definite, let $\langle EPE^t v, v \rangle = 0$. Then $\langle PE^t v, E^t v \rangle = 0$. Since P is positive definite, $E^t v = 0$. But E^t is one-one, and therefore $v = 0$. We conclude that EPE^t is definite.

Now we specialize to irreducible abstract Cartan matrices, which correspond to connected abstract Dynkin diagrams. In five steps, we can obtain the desired classification.

1) No abstract Dynkin diagram contains a configuration

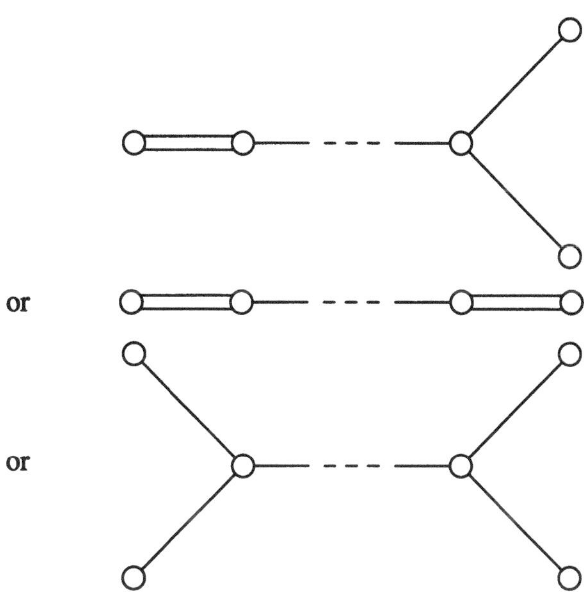

In fact, otherwise Operation #2 would allow us to collapse all the single-line part in the center to a single vertex, in violation of Proposition 2.78c.

2) The following are the only possibilities left for a connected abstract Dynkin diagram:

2a) There is a triple line. By Proposition 2.78c the only possibility is

(G_2)

2b) There is a double line, but there is no triple line. Then Step 1 shows that the diagram is

(B, C, F)

α_1 α_p β_q β_1

2c) There are only single lines. Call

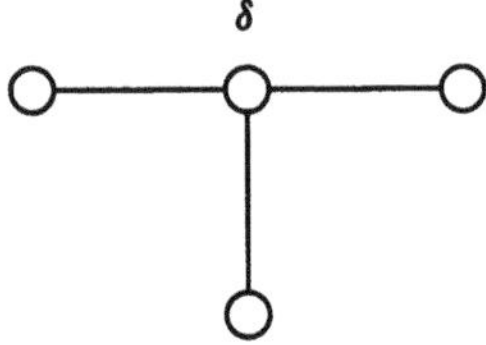

a **triple point**. If there is no triple point, then the absence of loops implies that the diagram is

(A) α_1 α_2 α_l

If there is a triple point, then there is only one, by Step 1, and the diagram is

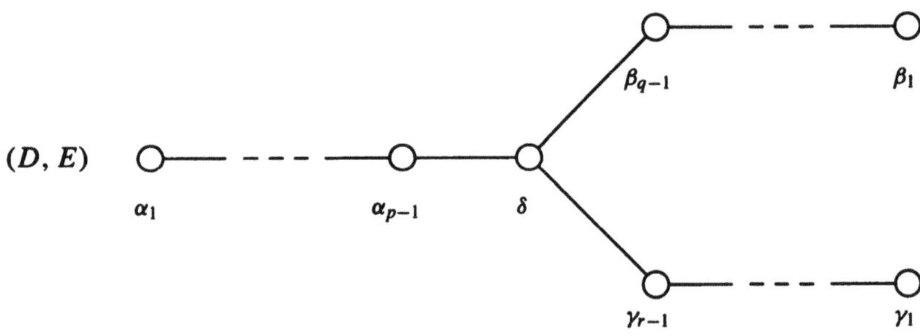

3) The following are the possibilities for weights:

3a) If the ith and j^{th} vertices are connected by a single line, then $A_{ij} = A_{ji} = -1$. By (2.80) the weights satisfy $w_i = w_j$. Thus in the cases (A) and (D, E) of Step 2, all the weights are equal, and we may take them to be 1. In this situation we shall omit the weights from the diagram.

3b) In the case (B, C, F) of Step 2, let $\alpha = \alpha_p$ and $\beta = \beta_q$. Also let us use α and β to denote the corresponding vertices. Possibly reversing the roles of α and β, we may assume that $A_{\alpha\beta} = -2$ and $A_{\beta\alpha} = -1$. Then

$$\begin{pmatrix} |\alpha| & 0 \\ 0 & |\beta| \end{pmatrix} \begin{pmatrix} 2 & -2 \\ -1 & 2 \end{pmatrix} \begin{pmatrix} |\alpha|^{-1} & 0 \\ 0 & |\beta|^{-1} \end{pmatrix}$$

is symmetric, so that $-2|\alpha||\beta|^{-1} = -1|\beta||\alpha|^{-1}$ and $|\beta|^2 = 2|\alpha|^2$. Apart from a proportionality constant, we obtain the diagram

$$\overset{1}{\underset{\alpha_1}{\circ}} — \cdots — \overset{1}{\underset{\alpha_p}{\circ}} = \overset{2}{\underset{\beta_q}{\circ}} — \cdots — \overset{2}{\underset{\beta_1}{\circ}}$$

3c) In the case (G_2) of Step 2, similar reasoning leads us to the diagram

$$\overset{1}{\circ} \equiv \overset{3}{\circ}$$

4) In case (B, C, F) of Step 2, the only possibilities are

$$(B) \qquad \overset{1}{\circ} = \overset{2}{\circ} — \cdots — \overset{2}{\circ}$$

$$(C) \qquad \overset{1}{\circ} — \cdots — \overset{1}{\circ} = \overset{2}{\circ}$$

$$(F_4) \qquad \overset{1}{\circ} — \overset{1}{\circ} = \overset{2}{\circ} — \overset{2}{\circ}$$

Let us prove this assertion. In the notation of Step 3b, it is enough to show that

$$(p-1)(q-1) < 2. \tag{2.82}$$

This inequality will follow by applying the Schwarz inequality to

$$\alpha = \sum_{i=1}^{p} i\alpha_i \qquad \text{and} \qquad \beta = \sum_{j=1}^{q} j\beta_j.$$

Since $|\alpha_1|^2 = \cdots = |\alpha_p|^2$, we have

$$-1 = A_{\alpha_i, \alpha_{i+1}} = \frac{2\langle\alpha_i, \alpha_{i+1}\rangle}{|\alpha_i|^2} = \frac{2\langle\alpha_i, \alpha_{i+1}\rangle}{|\alpha_p|^2}.$$

Thus

$$2\langle\alpha_i, \alpha_{i+1}\rangle = -|\alpha_p|^2.$$

Similarly

$$2\langle\beta_j, \beta_{j+1}\rangle = -|\beta_q|^2.$$

Also

$$2 = A_{\alpha_p,\beta_q} A_{\beta_q,\alpha_p} = \frac{4\langle \alpha_p, \beta_q \rangle^2}{|\alpha_p|^2 |\beta_q|^2}$$

and hence

$$\langle \alpha_p, \beta_q \rangle^2 = \tfrac{1}{2} |\alpha_p|^2 |\beta_q|^2.$$

Then

$$\langle \alpha, \beta \rangle = \sum_{i,j} \langle i\alpha_i, j\beta_j \rangle = pq \langle \alpha_p, \beta_q \rangle,$$

while

$$\begin{aligned} |\alpha|^2 &= \sum_{i,j} \langle i\alpha_i, j\alpha_j \rangle = \sum_{i=1}^{p} i^2 \langle \alpha_i, \alpha_i \rangle + 2 \sum_{i=1}^{p-1} i(i+1) \langle \alpha_i, \alpha_{i+1} \rangle \\ &= |\alpha_p|^2 \Big(\sum_{i=1}^{p} i^2 - \sum_{i=1}^{p-1} i(i+1) \Big) = |\alpha_p|^2 \Big(p^2 - \sum_{i=1}^{p-1} i \Big) \\ &= |\alpha_p|^2 (p^2 - \tfrac{1}{2}(p-1)p) = |\alpha_p|^2 (\tfrac{1}{2} p(p+1)). \end{aligned}$$

Similarly

$$|\beta|^2 = |\beta_q|^2 (\tfrac{1}{2} q(q+1)).$$

Since α and β are nonproportional, the Schwarz inequality gives $\langle \alpha, \beta \rangle^2 < |\alpha|^2 |\beta|^2$. Thus

$$\tfrac{1}{2} p^2 q^2 |\alpha_p|^2 |\beta_q|^2 = p^2 q^2 \langle \alpha_p, \beta_q \rangle^2 < |\alpha_p|^2 |\beta_q|^2 (\tfrac{1}{4} p(p+1) q(q+1)).$$

Hence $2pq < (p+1)(q+1)$ and $pq < p+q+1$, and (2.82) follows.

5) In case (D, E) of Step 2, we may take $p \geq q \geq r$, and then the only possibilities are

$$(D) \qquad r = 2, \ q = 2, \ p \text{ arbitrary} \geq 2$$

$$(E) \qquad r = 2, \ q = 3, \ p = 3 \text{ or } 4 \text{ or } 5.$$

Let us prove this assertion. In the notation of Step 2c, it is enough to show that

$$\frac{1}{p} + \frac{1}{q} + \frac{1}{r} > 1. \tag{2.83}$$

This inequality will follow by applying Parseval's equality to

$$\alpha = \sum_{i=1}^{p-1} i\alpha_i, \quad \beta = \sum_{j=1}^{q-1} j\beta_j, \quad \gamma = \sum_{k=1}^{r-1} k\gamma_k, \quad \text{and} \quad \delta.$$

As in Step 4 (but with p replaced by $p-1$), we have

$$2\langle\alpha_i,\alpha_{i+1}\rangle = -|\delta|^2 \qquad \text{and} \qquad |\alpha|^2 = |\delta|^2(\tfrac{1}{2}p(p-1)),$$

and similarly for β and γ. Also

$$\langle\alpha,\delta\rangle = \langle(p-1)\alpha_{p-1},\delta\rangle = (p-1)(-\tfrac{1}{2}|\delta|^2) = -\tfrac{1}{2}(p-1)|\delta|^2$$

and similarly for β and γ. The span U of $\{\alpha,\beta,\gamma,\delta\}$ is 4-dimensional since these four vectors are linear combinations of disjoint subsets of members of a basis. Within this span the set

$$\left\{\frac{\alpha}{|\alpha|},\frac{\beta}{|\beta|},\frac{\gamma}{|\gamma|}\right\}$$

is orthonormal. Adjoin ε to this set to obtain an orthonormal basis of U. Since δ is independent of $\{\alpha,\beta,\gamma\}$, we have $\langle\delta,\varepsilon\rangle \neq 0$. By the Bessel inequality

$$|\delta|^2 \geq \left\langle\delta,\frac{\alpha}{|\alpha|}\right\rangle^2 + \left\langle\delta,\frac{\beta}{|\beta|}\right\rangle^2 + \left\langle\delta,\frac{\gamma}{|\gamma|}\right\rangle^2 + \langle\delta,\varepsilon\rangle^2,$$

with the last term > 0. Thus

$$\begin{aligned}
1 &> \left(\frac{\langle\alpha,\delta\rangle}{|\alpha||\delta|}\right)^2 + \left(\frac{\langle\beta,\delta\rangle}{|\beta||\delta|}\right)^2 + \left(\frac{\langle\gamma,\delta\rangle}{|\gamma||\delta|}\right)^2 \\
&= \left(\frac{p-1}{2}\right)^2\frac{1}{\frac{1}{2}p(p-1)} + \left(\frac{q-1}{2}\right)^2\frac{1}{\frac{1}{2}q(q-1)} + \left(\frac{r-1}{2}\right)^2\frac{1}{\frac{1}{2}r(r-1)} \\
&= \frac{1}{2}\frac{p-1}{p} + \frac{1}{2}\frac{q-1}{q} + \frac{1}{2}\frac{r-1}{r}.
\end{aligned}$$

Thus $2 > 3 - (\frac{1}{p}+\frac{1}{q}+\frac{1}{r})$, and (2.83) follows.

Theorem 2.84. Up to isomorphism the connected abstract Dynkin diagrams are exactly those in Figure 2.4, specifically A_n for $n \geq 1$, B_n for $n \geq 2$, C_n for $n \geq 3$, D_n for $n \geq 4$, E_6, E_7, E_8, F_4, and G_2.

REMARKS.

1) The subscripts refer to the numbers of vertices in the various diagrams.

2) The names A_n, B_n, C_n, D_n, and G_2 are names of root systems, and Corollary 2.65 shows that the associated Dynkin diagrams are independent of the ordering. As yet, the names E_6, E_7, E_8, and F_4 are attached only to abstract Dynkin diagrams. At the end of this section, we show that these diagrams come from root systems, and then we may use these names unambiguously for the root systems.

PROOF. We have seen that any connected abstract Dynkin diagram has to be one of the ones in this list, up to isomorphism. Also we know that A_n, B_n, C_n, D_n, and G_2 come from abstract reduced root systems and are therefore legitimate Dynkin diagrams. To check that E_6 E_7, E_8, and F_4 are legitimate Dynkin diagrams, we write down the candidates for abstract Cartan matrices and observe the first four defining properties of an abstract Cartan matrix by inspection. For property (e) we exhibit vectors $\{\alpha_i\}$ for each case such that the matrix in question has entries $A_{ij} = \dfrac{2\langle\alpha_i, \alpha_j\rangle}{|\alpha_i|^2}$, and then property (e) follows.

For F_4, the matrix is

$$\begin{pmatrix} 2 & -1 & 0 & 0 \\ -1 & 2 & -2 & 0 \\ 0 & -1 & 2 & -1 \\ 0 & 0 & -1 & 2 \end{pmatrix}, \tag{2.85a}$$

and the vectors are the following members of $\mathbb{R}^4$:

$$\begin{aligned} \alpha_1 &= \tfrac{1}{2}(e_1 - e_2 - e_3 - e_4) \\ \alpha_2 &= e_4 \\ \alpha_3 &= e_3 - e_4 \\ \alpha_4 &= e_2 - e_3. \end{aligned} \tag{2.85b}$$

For reference we note that these vectors are attached to the vertices of the Dynkin diagram as follows:

(2.85c)

For E_8, the matrix is

$$\begin{pmatrix} 2 & 0 & -1 & 0 & 0 & 0 & 0 & 0 \\ 0 & 2 & 0 & -1 & 0 & 0 & 0 & 0 \\ -1 & 0 & 2 & -1 & 0 & 0 & 0 & 0 \\ 0 & -1 & -1 & 2 & -1 & 0 & 0 & 0 \\ 0 & 0 & 0 & -1 & 2 & -1 & 0 & 0 \\ 0 & 0 & 0 & 0 & -1 & 2 & -1 & 0 \\ 0 & 0 & 0 & 0 & 0 & -1 & 2 & -1 \\ 0 & 0 & 0 & 0 & 0 & 0 & -1 & 2 \end{pmatrix}, \tag{2.86a}$$

and the vectors are the following members of $\mathbb{R}^8$:

(2.86b)
$$\begin{aligned}
\alpha_1 &= \tfrac{1}{2}(e_8 - e_7 - e_6 - e_5 - e_4 - e_3 - e_2 + e_1)\\
\alpha_2 &= e_2 + e_1\\
\alpha_3 &= e_2 - e_1\\
\alpha_4 &= e_3 - e_2\\
\alpha_5 &= e_4 - e_3\\
\alpha_6 &= e_5 - e_4\\
\alpha_7 &= e_6 - e_5\\
\alpha_8 &= e_7 - e_6.
\end{aligned}$$

For reference we note that these vectors are attached to the vertices of the Dynkin diagram as follows:

(2.86c)
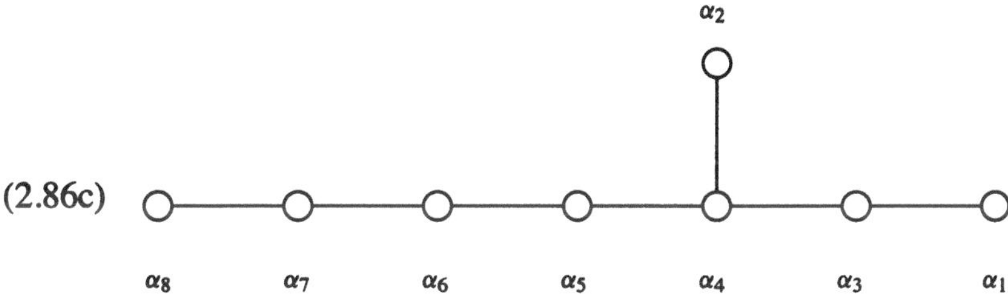

For E_7 or E_6, the matrix is the first 7 or 6 rows and columns of (2.86a), and the vectors are the first 7 or 6 of the vectors (2.86b).

This completes the classification of abstract Cartan matrices. The corresponding Dynkin diagrams are tabulated in Figure 2.4.

Actually we can see without difficulty that E_6, E_7, E_8, and F_4 are not just abstract Cartan matrices but actually come from abstract reduced root systems. As we remarked in connection with Theorem 2.84, we can then use the same names for the abstract root systems as for the Cartan matrices. The fact that E_6, E_7, E_8, and F_4 come from abstract reduced root systems enables us to complete our examination of the second step of the passage (2.58) from complex semisimple Lie algebras to abstract Cartan matrices.

Proposition 2.87. The second step in the passage (2.58) is onto. That is, every abstract Cartan matrix comes from a reduced root system.

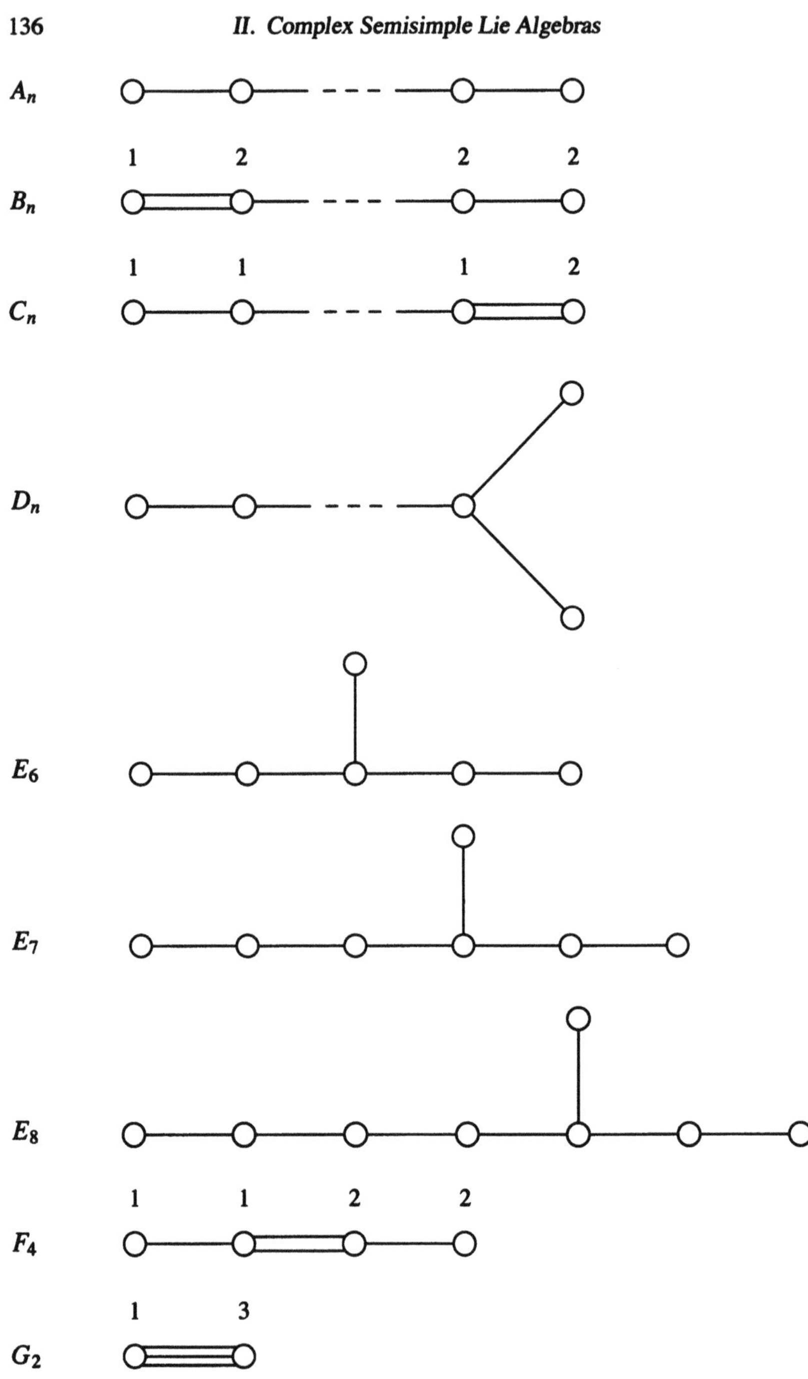

FIGURE 2.4. Classification of Dynkin diagrams

Proof. In the case of F_4, we take $V = \mathbb{R}^4$, and we let

$$\Delta = \begin{cases} \pm e_i \\ \pm e_i \pm e_j & \text{for } i \neq j \\ \frac{1}{2}(\pm e_1 \pm e_2 \pm e_3 \pm e_4) \end{cases} \tag{2.88}$$

with all possible signs allowed. We have to check the axioms for an abstract root system. Certainly the roots span $\mathbb{R}^4$, and it is a simple matter to check that $2\langle\beta, \alpha\rangle/|\alpha|^2$ is always an integer. The problem is to check that the root reflections carry roots to roots. The case that needs attention is $s_\alpha\beta$ with α of the third kind. If β is of the first kind, then $s_\alpha\beta = \pm s_\beta\alpha$, and there is no difficulty. If β is of the second kind, there is no loss of generality in assuming that $\beta = e_1 + e_2$. Then $s_\alpha\beta = \beta$ unless the coefficients of e_1 and e_2 in α are equal. In this case $s_\alpha\beta$ gives plus or minus the e_3, e_4 part of β, but without the factor of $\frac{1}{2}$.

Now suppose that α and β are both of the third kind. We need to consider $s_\alpha\beta$ when one or three of the signs in α and β match. In either case there is one exceptional sign, say as coefficient of e_i. Then $s_\alpha\beta = \pm e_i$, and hence the root reflections carry Δ to itself.

Therefore Δ is an abstract reduced root system. The vectors α_i in (2.85b) are the simple roots relative to the lexicographic ordering obtained from the ordered basis e_1, e_2, e_3, e_4, and then (2.85a) is the Cartan matrix.

In the case of E_8, we take $V = \mathbb{R}^8$, and we let

$$\Delta = \begin{cases} \pm e_i \pm e_j & \text{for } i \neq j \\ \frac{1}{2}\sum_{i=1}^{8}(-1)^{n(i)}e_i & \text{with } \sum_{i=1}^{8}(-1)^{n(i)} \text{ even.} \end{cases} \tag{2.89}$$

For the first kind of root, all possible signs are allowed. Again we have to check the axioms for an abstract root system, and again the problem is to check that the root reflections carry roots to roots. This time all roots have the same length. Thus when α and β are nonorthogonal and nonproportional, we have $s_\alpha\beta = \pm s_\beta\alpha$. Hence matters come down to checking the case that α and β are both of the second kind.

In this case we need to consider $s_\alpha\beta$ when two or six of the signs in α and β match. In either case there are two exceptional signs, say as coefficients of e_i and e_j. We readily check that $s_\alpha\beta = \pm e_i \pm e_j$ for a suitable choice of signs, and hence the root reflections carry Δ to itself.

Therefore Δ is an abstract reduced root system. The vectors α_i in (2.86b) are the simple roots relative to the lexicographic ordering obtained from the ordered basis $e_8, e_7, e_6, e_5, e_4, e_3, e_2, e_1$, and then (2.86a) is the Cartan matrix.

In the case of E_7, we take V to be the subspace of the space for E_8 orthogonal to $e_8 + e_7$, and we let Δ be the set of roots for E_8 that are in this space. Since E_8 is a root system, it follows that E_7 is a root system. All the α_i for E_8 except α_8 are roots for E_7, and they must remain simple. Since there are 7 such roots, we see that $\alpha_1, \dots, \alpha_7$ must be all of the simple roots. The associated Cartan matrix is then the part of (2.86a) that excludes α_8.

In the case of E_6, we take V to be the subspace of the space for E_8 orthogonal to $e_8 + e_7$ and $e_8 + e_6$, and we let Δ be the set of roots for E_8 that are in this space. Since E_8 is a root system, it follows that E_6 is a root system. All the α_i for E_8 except α_7 and α_8 are roots for E_6, and they must remain simple. Since there are 6 such roots, we see that $\alpha_1, \dots, \alpha_6$ must be all of the simple roots. The associated Cartan matrix is then the part of (2.86a) that excludes α_7 and α_8.

8. Classification of Nonreduced Abstract Root Systems

In this section we digress from considering the two-step passage (2.58) from complex semisimple Lie algebras to abstract Cartan matrices. Our topic will be nonreduced abstract root systems. Abstract root systems that are not necessarily reduced arise in the structure theory of real semisimple Lie algebras, as presented in Chapter VI; the root systems in question are the systems of "restricted roots" of the Lie algebra. In order not to attach special significance later to those real semisimple Lie algebras whose systems of restricted roots turn out to be reduced, we shall give a classification now of nonreduced abstract root systems. There is no loss of generality in assuming that such a system is irreducible.

An example arises by forming the union of the root systems B_n and C_n given in (2.43). The union is called $(BC)_n$ and is given as follows:

(2.90)

$$(BC)_n \quad V = \left\{ \sum_{i=1}^n a_i e_i \right\} \quad \Delta = \{\pm e_i \pm e_j \mid i \neq j\} \cup \{\pm e_i\} \cup \{\pm 2e_i\}.$$

A diagram of all of the roots of $(BC)_2$ appears in Figure 2.2.

In contrast with Proposition 2.66, the simple roots of an abstract root system that is not necessarily reduced do not determine the root system. For example, if B_n and $(BC)_n$ are taken to have the sets of positive roots as in (2.50), then they have the same sets of simple roots. Thus it is not helpful to associate an unadorned abstract Cartan matrix and Dynkin diagram to such a system. But we can associate the slightly more complicated diagram in Figure 2.5 to $(BC)_n$, and it conveys useful unambiguous information.

$(BC)_n$

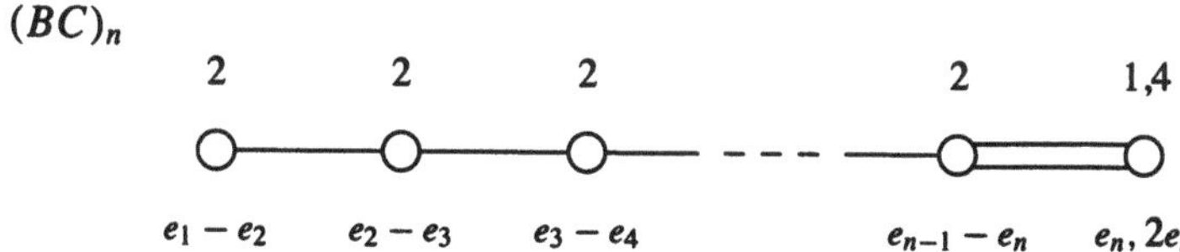

FIGURE 2.5. Substitute Dynkin diagram for $(BC)_n$

Now let Δ be any abstract root system in an inner product space V. Recall that if α is a root and $\frac{1}{2}\alpha$ is not a root, we say that α is **reduced**.

Lemma 2.91. The reduced roots $\alpha \in \Delta$ form a reduced abstract root system Δ_s in V. The roots $\alpha \in \Delta$ such that $2\alpha \notin \Delta$ form a reduced abstract root system Δ_l in V. The Weyl groups of Δ, Δ_s, and Δ_l coincide.

PROOF. It follows immediately from the definitions that Δ_s and Δ_l are abstract root systems. Also it is clear that Δ_s and Δ_l are reduced. The reflections for Δ, Δ_s, and Δ_l coincide, and hence the Weyl groups coincide.

Proposition 2.92. Up to isomorphism the only irreducible abstract root systems Δ that are not reduced are of the form $(BC)_n$ for $n \geq 1$.

PROOF. We impose a lexicographic ordering, thereby fixing a system of simple roots. Also we form Δ_s as in Lemma 2.91. Since Δ is not reduced, there exists a root α such that 2α is a root. By Proposition 2.62, α is conjugate via the Weyl group to a simple root. Thus there exists a simple root β such that 2β is a root. Evidently β is simple in Δ_s, and Δ_s is irreducible. Let $\gamma \neq \beta$ be any simple root of Δ_s such that $\langle \beta, \gamma \rangle \neq 0$. Then

$$\frac{2\langle \gamma, \beta \rangle}{|\beta|^2} \quad \text{and} \quad \frac{2\langle \gamma, 2\beta \rangle}{|2\beta|^2} = \frac{1}{2}\frac{2\langle \gamma, \beta \rangle}{|\beta|^2}$$

are negative integers, and it follows that $2\langle \gamma, \beta \rangle / |\beta|^2 = -2$. Referring to the classification in Theorem 2.84, we see that Δ_s is of type B_n, with β as the unique short simple root. Any Weyl group conjugate β' of β has $2\beta'$ in Δ, and the roots β' with $2\beta'$ in Δ are exactly those with $|\beta'| = |\beta|$. The result follows.

9. Serre Relations

We return to our investigation of the two-step passage (2.58), first from complex semisimple Lie algebras to reduced abstract root systems and then from reduced abstract root systems to abstract Cartan matrices.

We have completed our investigation of the second step, showing that that step is independent of the choice of ordering up to isomorphism, is one-one up to isomorphism, and is onto. Moreover, we have classified the abstract Cartan matrices.

For the remainder of this chapter we concentrate on the first step. Theorem 2.15 enabled us to see that the passage from complex semisimple Lie algebras to reduced abstract root systems is well defined up to isomorphism, and we now want to see that it is one-one and onto, up to isomorphism. First we show that it is one-one. Specifically we shall show that an isomorphism between the root systems of two complex semisimple Lie algebras lifts to an isomorphism between the Lie algebras themselves. More than one such isomorphism of Lie algebras exists, and we shall impose additional conditions so that the isomorphism exists and is unique. The result, known as the Isomorphism Theorem, will be the main result of the next section and will be the cornerstone of our development of structure theory for real semisimple Lie algebras and Lie groups in Chapter VI. The technique will be to use generators and relations, realizing any complex semisimple Lie algebra as the quotient of a "free Lie algebra" by an ideal generated by some "relations."

Thus let $\mathfrak{g}$ be a complex semisimple Lie algebra, fix a Cartan subalgebra $\mathfrak{h}$, let Δ be the set of roots, let B be a nondegenerate symmetric invariant bilinear form on $\mathfrak{g}$ that is positive definite on the real form of $\mathfrak{h}$ where the roots are real, let $\Pi = \{\alpha_1, \dots, \alpha_l\}$ be a simple system, and let $A = (A_{ij})_{i,j=1}^l$ be the Cartan matrix. For $1 \le i \le l$, let

$$(2.93)\qquad \begin{aligned} h_i &= \frac{2}{|\alpha_i|^2} H_{\alpha_i} \\ e_i &= \text{nonzero root vector for } \alpha_i \\ f_i &= \text{nonzero root vector for } -\alpha_i \text{ with } B(e_i, f_i) = 2/|\alpha_i|^2. \end{aligned}$$

Proposition 2.94. The set $X = \{h_i, e_i, f_i\}_{i=1}^l$ generates $\mathfrak{g}$ as a Lie algebra.

REMARK. We call X a set of **standard generators** of $\mathfrak{g}$ relative to $\mathfrak{h}$, Δ, B, Π, and $A = (A_{ij})_{i,j=1}^l$.

PROOF. The linear span of the h_i's is all of $\mathfrak{h}$ since the α_i form a basis of $\mathfrak{h}^*$. Let α be a positive root, and let e_α be a nonzero root vector. If $\alpha = \sum_i n_i \alpha_i$, we show by induction on the level $\sum_i n_i$ that e_α is a multiple of an iterated bracket of the e_i's. If the level is 1, then $\alpha = \alpha_j$ for some j, and e_α is a multiple of e_j. Assume the result for level $< n$ and let the level of α be $n > 1$. Since

$$0 < |\alpha|^2 = \sum_i n_i \langle \alpha, \alpha_i \rangle,$$

we must have $\langle \alpha, \alpha_j \rangle > 0$ for some j. By Proposition 2.48e, $\beta = \alpha - \alpha_j$ is a root, and Proposition 2.49 shows that β is positive. If e_β is a nonzero root vector for β, then the induction hypothesis shows that e_β is a multiple of an iterated bracket of the e_i's. Corollary 2.35 shows that e_α is a multiple of $[e_\beta, e_j]$, and the induction is complete.

Thus all the root spaces for positive roots are in the Lie subalgebra of $\mathfrak{g}$ generated by X. A similar argument with negative roots, using the f_i's, shows that the root spaces for the negative roots are in this Lie subalgebra, too. Therefore X generates all of $\mathfrak{g}$.

Proposition 2.95. The set $X = \{h_i, e_i, f_i\}_{i=1}^{l}$ satisfies the following properties within $\mathfrak{g}$.

(a) $[h_i, h_j] = 0$
(b) $[e_i, f_j] = \delta_{ij} h_i$
(c) $[h_i, e_j] = A_{ij} e_j$
(d) $[h_i, f_j] = -A_{ij} f_j$
(e) $(\operatorname{ad} e_i)^{-A_{ij}+1} e_j = 0$ when $i \neq j$
(f) $(\operatorname{ad} f_i)^{-A_{ij}+1} f_j = 0$ when $i \neq j$.

REMARK. Relations (a) through (f) are called the **Serre relations** for $\mathfrak{g}$. We shall refer to them by letter.

PROOF.
(a) The subalgebra $\mathfrak{h}$ is abelian.
(b) For $i = j$, we use Lemma 2.18a. When $i \neq j$, $\alpha_i - \alpha_j$ cannot be a root, by Proposition 2.49.
(c,d) We observe that $[h_i, e_j] = \alpha_j(h_i) e_j = \frac{2}{|\alpha_i|^2} \alpha_j(H_{\alpha_i}) e_j = A_{ij} e_j$, and we argue similarly for $[h_i, f_j]$.
(e,f) When $i \neq j$, the α_i string containing α_j is

$$\alpha_j, \alpha_j + \alpha_i, \ldots, \alpha_j + q\alpha_i \qquad \text{since } \alpha_j - \alpha_i \notin \Delta.$$

Thus $p = 0$ for the root string, and

$$-q = p - q = \frac{2\langle \alpha_j, \alpha_i \rangle}{|\alpha_i|^2} = A_{ij}.$$

Hence $1 - A_{ij} = q + 1$, and $\alpha_j + (1 - A_{ij})\alpha_i$ is not a root. Then (e) follows, and (f) is proved similarly.

Now we look at (infinite-dimensional) complex Lie algebras with no relations. A **free Lie algebra** on a set X is a pair $(\mathfrak{F}, \iota)$ consisting of a Lie algebra $\mathfrak{F}$ and a function $\iota : X \to \mathfrak{F}$ with the following universal mapping property: Whenever $\mathfrak{l}$ is a complex Lie algebra and $\varphi : X \to \mathfrak{l}$

is a function, then there exists a unique Lie algebra homomorphism $\tilde{\varphi}$ such that the diagram

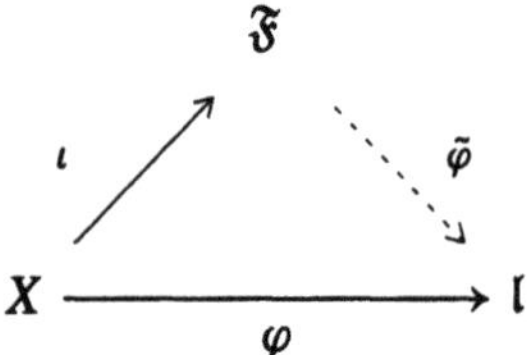

commutes.

Proposition 2.96. If X is a nonempty set, then there exists a free Lie algebra $\mathfrak{F}$ on X, and the image of X in $\mathfrak{F}$ generates $\mathfrak{F}$. Any two free Lie algebras on X are canonically isomorphic.

REMARK. The proof is elementary but uses the Poincaré-Birkhoff-Witt Theorem, which will be not be proved until Chapter III. We therefore postpone the proof of Proposition 2.96 until that time.

Now we can express our Lie algebra in terms of generators and relations. With $\mathfrak{g}$, $\mathfrak{h}$, Δ, B, Π, and $A = (A_{ij})_{i,j=1}^{l}$ as before, let $\mathfrak{F}$ be the free Lie algebra on the set $X = \{h_i, e_i, f_i\}_{i=1}^{l}$, and let $\mathfrak{R}$ be the ideal in $\mathfrak{F}$ generated by the Serre relations (a) through (f), i.e., generated by the differences of the left sides and right sides of all equalities (a) through (f) in Proposition 2.95. We set up the diagram

(2.97)

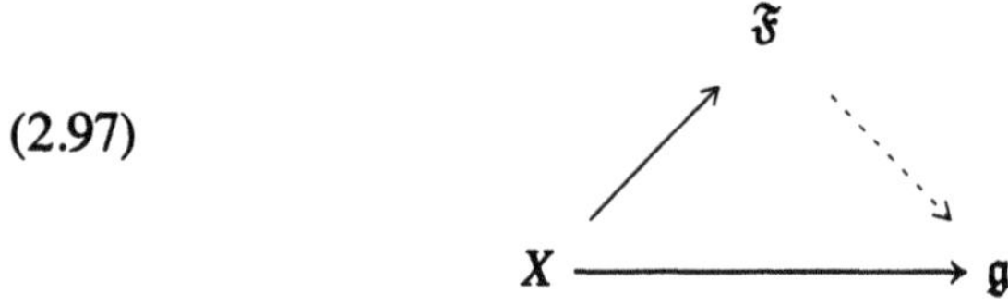

and obtain a Lie algebra homomorphism of $\mathfrak{F}$ into $\mathfrak{g}$. This homomorphism carries $\mathfrak{R}$ to 0 as a consequence of Proposition 2.95, and therefore it descends to a Lie algebra homomorphism

$$\mathfrak{F}/\mathfrak{R} \longrightarrow \mathfrak{g}$$

that is onto $\mathfrak{g}$ by Proposition 2.94 and is one-one on the linear span of $X = \{h_i, e_i, f_i\}_{i=1}^{l}$. We call this map the **canonical homomorphism** of $\mathfrak{F}/\mathfrak{R}$ onto $\mathfrak{g}$ relative to $\{h_i, e_i, f_i\}_{i=1}^{l}$.

Theorem 2.98 (Serre). Let $\mathfrak{g}$ be a complex semisimple Lie algebra, and let $X = \{h_i, e_i, f_i\}_{i=1}^l$ be a set of standard generators. Let $\mathfrak{F}$ be the free Lie algebra on $3l$ generators h_i, e_i, f_i with $1 \le i \le l$, and let $\mathfrak{R}$ be the ideal generated in $\mathfrak{F}$ by the Serre relations (a) through (f). Then the canonical homomorphism of $\mathfrak{F}/\mathfrak{R}$ onto $\mathfrak{g}$ is an isomorphism.

REMARK. The proof will be preceded by two lemmas that will play a role both here and in §11.

Lemma 2.99. Let $A = A = (A_{ij})_{i,j=1}^l$ be an abstract Cartan matrix, let $\mathfrak{F}$ be the free Lie algebra on $3l$ generators h_i, e_i, f_i with $1 \le i \le l$, and let $\tilde{\mathfrak{R}}$ be the ideal generated in $\mathfrak{F}$ by the Serre relations (a) through (d). Define $\tilde{\mathfrak{g}} = \mathfrak{F}/\tilde{\mathfrak{R}}$, and write h_i, e_i, f_i also for the images of the generators in $\tilde{\mathfrak{g}}$. In $\tilde{\mathfrak{g}}$, put

$$\begin{aligned} \tilde{\mathfrak{h}} &= \operatorname{span}\{h_i\}, \quad \text{an abelian Lie subalgebra} \\ \tilde{\mathfrak{e}} &= \text{Lie subalgebra generated by all } e_i \\ \tilde{\mathfrak{f}} &= \text{Lie subalgebra generated by all } f_i. \end{aligned}$$

Then

$$\tilde{\mathfrak{g}} = \tilde{\mathfrak{h}} \oplus \tilde{\mathfrak{e}} \oplus \tilde{\mathfrak{f}}.$$

PROOF. Proposition 2.96 shows that X generates $\mathfrak{F}$, and consequently the image of X in $\tilde{\mathfrak{g}}$ generates $\tilde{\mathfrak{g}}$. Therefore $\tilde{\mathfrak{g}}$ is spanned by iterated brackets of elements from X. In $\tilde{\mathfrak{g}}$, each generator from X is an eigenvector under $\operatorname{ad} h_i$, by Serre relations (a), (c), and (d). Hence so is any iterated bracket, the eigenvalue for an iterated bracket being the sum of the eigenvalues from the factors.

To see that

$$\tilde{\mathfrak{g}} = \tilde{\mathfrak{h}} + \tilde{\mathfrak{e}} + \tilde{\mathfrak{f}}, \tag{2.100}$$

we note that X is contained in the right side of (2.100). Thus it is enough to see that the right side is invariant under the operation $\operatorname{ad} x$ for each $x \in X$. Each of $\tilde{\mathfrak{h}}, \tilde{\mathfrak{e}}, \tilde{\mathfrak{f}}$ is invariant under $\operatorname{ad} h_i$, from the previous paragraph. Also $\tilde{\mathfrak{h}} + \tilde{\mathfrak{e}}$ is invariant under $\operatorname{ad} e_i$. We prove that $(\operatorname{ad} f_i)\tilde{\mathfrak{e}} \subseteq \tilde{\mathfrak{h}} + \tilde{\mathfrak{e}}$. We do so by treating the iterated brackets that span $\tilde{\mathfrak{e}}$, proceeding inductively on the number of factors. When we have one factor, Serre relation (b) gives us

$$(\operatorname{ad} f_i)e_j = -\delta_{ij}h_i \in \tilde{\mathfrak{h}} + \tilde{\mathfrak{e}}.$$

When we have more than one factor, let the iterated bracket from $\tilde{\mathfrak{e}}$ be $[x, y]$ with n factors, where x and y have $< n$ factors. Then $(\operatorname{ad} f_i)x$ and $(\operatorname{ad} f_i)y$ are in $\tilde{\mathfrak{h}} + \tilde{\mathfrak{e}}$ by inductive hypothesis, and hence

$$(\operatorname{ad} f_i)[x, y] = [(\operatorname{ad} f_i)x, y] + [x, (\operatorname{ad} f_i)y] \in [\tilde{\mathfrak{h}} + \tilde{\mathfrak{e}}, \tilde{\mathfrak{e}}] + [\tilde{\mathfrak{e}}, \tilde{\mathfrak{h}} + \tilde{\mathfrak{e}}] \subseteq \tilde{\mathfrak{e}}.$$

Therefore $(\operatorname{ad} x)\tilde{\mathfrak{e}} \subseteq \tilde{\mathfrak{h}}+\tilde{\mathfrak{e}}+\tilde{\mathfrak{f}}$ for each $x \in X$. Similarly $(\operatorname{ad} x)\tilde{\mathfrak{f}} \subseteq \tilde{\mathfrak{h}}+\tilde{\mathfrak{e}}+\tilde{\mathfrak{f}}$ for each $x \in X$, and we obtain (2.100).

Now let us prove that the sum (2.100) is direct. As we have seen, each term on the right side of (2.100) is spanned by simultaneous eigenvectors for $\operatorname{ad}\tilde{\mathfrak{h}}$. Let us be more specific. As a result of Serre relation (c), an iterated bracket in $\tilde{\mathfrak{e}}$ involving $e_{j_1}, \dots, e_{j_k}$ has eigenvalue under $\operatorname{ad} h_i$ given by

$$A_{ij_1} + \cdots + A_{ij_k} = \sum_{j=1}^{l} m_j A_{ij} \qquad \text{with } m_j \geq 0 \text{ an integer.}$$

If an eigenvalue for $\tilde{\mathfrak{e}}$ coincides for all i with an eigenvalue for $\tilde{\mathfrak{h}} + \tilde{\mathfrak{f}}$, we obtain an equation $\sum_{j=1}^{l} m_j A_{ij} = -\sum_{j=1}^{l} n_j A_{ij}$ for all i with $m_j \geq 0$, $n_j \geq 0$, and not all m_j equal to 0. Consequently $\sum_{i=1}^{l}(m_j+n_j)A_{ij} = 0$ for all i. Since (A_{ij}) is nonsingular, $m_j + n_j = 0$ for all j. Then $m_j = n_j = 0$ for all j, contradiction. Therefore the sum (2.100) is direct.

Lemma 2.101. Let $A = (A_{ij})_{i,j=1}^{l}$ be an abstract Cartan matrix, let $\mathfrak{F}$ be the free Lie algebra on $3l$ generators h_i, e_i, f_i with $1 \leq i \leq l$, and let $\mathfrak{R}$ be the ideal generated in $\mathfrak{F}$ by the Serre relations (a) through (f). Define $\mathfrak{g}' = \mathfrak{F}/\mathfrak{R}$, and suppose that $\operatorname{span}\{h_i\}_{i=1}^{l}$ maps one-one from $\mathfrak{F}$ into $\mathfrak{g}'$. Write h_i also for the images of the generators h_i in $\mathfrak{g}'$. Then $\mathfrak{g}'$ is a (finite-dimensional) complex semisimple Lie algebra, the subspace $\mathfrak{h}' = \operatorname{span}\{h_i\}$ is a Cartan subalgebra, the linear functionals $\alpha_j \in \mathfrak{h}'^*$ given by $\alpha_j(h_i) = A_{ij}$ form a simple system within the root system, and the Cartan matrix relative to this simple system is exactly A.

PROOF. Use the notation e_i and f_i also for the images of the generators e_i and f_i in $\mathfrak{g}'$. Let us observe that under the quotient map from $\mathfrak{F}$ to $\mathfrak{g}'$, all the e_i's and f_i's map to nonzero elements in $\mathfrak{g}'$. In fact, $\{h_i\}$ maps to a linearly independent set by hypothesis, and hence the images of the h_i's are nonzero. Then Serre relation (b) shows that $[e_i, f_i] = h_i \neq 0$ in $\mathfrak{g}'$, and hence e_i and f_i are nonzero in $\mathfrak{g}'$, as asserted..

Because the h_i are linearly independent in $\mathfrak{g}$, we can define $\alpha_j \in \mathfrak{h}'^*$ by $\alpha_j(h_i) = A_{ij}$. These linear functionals are a basis of $\mathfrak{h}'^*$. For $\varphi \in \mathfrak{h}'^*$, put

$$\mathfrak{g}'_\varphi = \{x \in \mathfrak{g}' \mid (\operatorname{ad} h)x = \varphi(h)x \text{ for all } h \in \mathfrak{h}'\}.$$

We call φ a **root** if $\varphi \neq 0$ and $\mathfrak{g}'_\varphi \neq 0$, and we call $\mathfrak{g}'_\varphi$ the corresponding **root space**. The Lie algebra $\mathfrak{g}'$ is a quotient of the Lie algebra $\tilde{\mathfrak{g}}$ of Lemma 2.99, and it follows from Lemma 2.99 that

$$\mathfrak{g}' = \mathfrak{h}' \oplus \bigoplus_{\varphi \doteq \text{root}} \mathfrak{g}'_\varphi$$

and that all roots are of the form $\varphi = \sum n_j\alpha_j$ with all nonzero n_j given as integers of the same sign. Let Δ' be the set of all roots, Δ'^+ the set of all roots with all $n_j \geq 0$, and Δ'^- the set of all roots with all $n_j \leq 0$. We have just established that

$$\Delta' = \Delta'^+ \cup \Delta'^-. \tag{2.102}$$

Let us show that $\mathfrak{g}'_\varphi$ is finite-dimensional for each root φ. First consider $\varphi = \sum n_j\alpha_j$ in Δ'^+. Lemma 2.99 shows that $\mathfrak{g}'_\varphi$ is spanned by the images of all iterated brackets of e_i's in $\tilde{\mathfrak{g}}$ involving n_j instances of e_j, and there are only finitely many such iterated brackets. Therefore $\mathfrak{g}'_\varphi$ is finite-dimensional when φ is in Δ'^+. Similarly $\mathfrak{g}'_\varphi$ is finite-dimensional when φ is in Δ'^-, and it follows from (2.102) that $\mathfrak{g}'_\varphi$ is finite-dimensional for each root φ.

The vectors e_i and f_i, which we have seen are nonzero, are in the respective spaces $\mathfrak{g}'_{\alpha_i}$ and $\mathfrak{g}'_{-\alpha_i}$, and hence each α_i and $-\alpha_i$ is a root. For these roots the root spaces have dimension 1.

Next let us show for each $\varphi \in \mathfrak{h}'^*$ that

$$\dim \mathfrak{g}'_\varphi = \dim \mathfrak{g}'_{-\varphi} \quad \text{and hence} \quad \Delta'^- = -\Delta'^+. \tag{2.103}$$

In fact, we set up the diagram

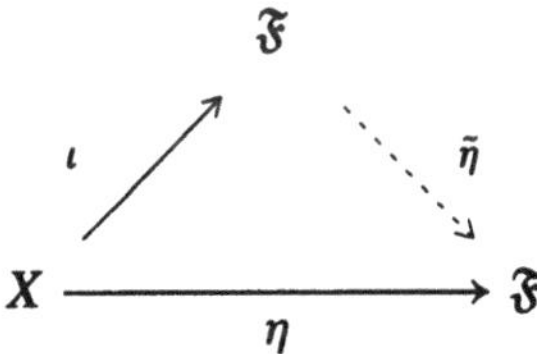

where η is the function $\eta(e_i) = f_i$, $\eta(f_i) = e_i$, and $\eta(h_i) = -h_i$. By the universal mapping property of $\mathfrak{F}$, η extends to a Lie algebra homomorphism $\tilde{\eta}$ of $\mathfrak{F}$ into itself. If we next observe that $\tilde{\eta}^2$ is an extension of the inclusion ι of X into $\mathfrak{F}$ in the diagram

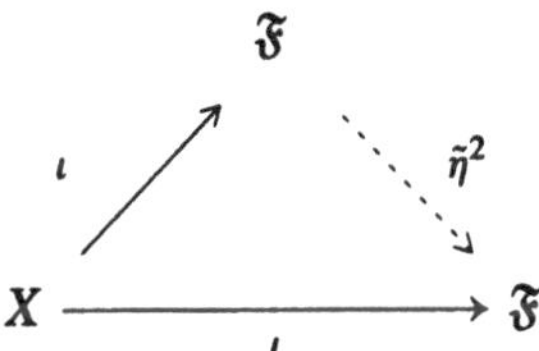

then we conclude from the uniqueness of the extension that $\tilde{\eta}^2 = 1$. We readily check that $\tilde{\eta}(\mathfrak{R}) \subseteq \mathfrak{R}$, and hence $\tilde{\eta}$ descends to a homomorphism $\tilde{\eta} : \mathfrak{g}' \to \mathfrak{g}'$ that is -1 on $\mathfrak{h}'$ and interchanges e_i with f_i for all i. Moreover $\tilde{\eta}^2 = 1$. Since $\tilde{\eta}$ is -1 on $\mathfrak{h}'$ and is invertible, we see that $\tilde{\eta}(\mathfrak{g}'_\varphi) = \mathfrak{g}'_{-\varphi}$ for all $\varphi \in \mathfrak{h}'^*$, and then (2.103) follows.

We shall introduce an inner product on the real form of $\mathfrak{h}'^*$ given by $\mathfrak{h}_0'^* = \sum \mathbb{R}\alpha_i$. We saw in (2.75) and (2.77) how to construct vectors $\beta_i \in \mathbb{R}^l$ for $1 \le i \le l$ such that

$$A_{ij} = 2\langle \beta_i, \beta_j \rangle / |\beta_i|^2. \tag{2.104}$$

We define a linear map $\mathbb{R}^l \to \mathfrak{h}_0'^*$ by $\beta_i \mapsto \alpha_i$, and we carry the inner product from $\mathbb{R}^l$ to $\mathfrak{h}_0'^*$. Then we have

$$\alpha_j(h_i) = A_{ij} = \frac{2\langle \beta_i, \beta_j \rangle}{|\beta_i|^2} = \frac{2\langle \alpha_i, \alpha_j \rangle}{|\alpha_i|^2} = \alpha_j\left(\frac{2H_{\alpha_i}}{|\alpha_i|^2}\right)$$

for all j, and it follows that

$$h_i = \frac{2H_{\alpha_i}}{|\alpha_i|^2} \tag{2.105}$$

in this inner product.

Next we define a Weyl group. For $1 \le i \le l$, let $s_{\alpha_i} : \mathfrak{h}_0'^* \to \mathfrak{h}_0'^*$ be the linear transformation given by

$$s_{\alpha_i}(\varphi) = \varphi - \varphi(h_i)\alpha_i = \varphi - \frac{2\langle \varphi, \alpha_i \rangle}{|\alpha_i|^2}\alpha_i.$$

This is an orthogonal transformation on $\mathfrak{h}_0'^*$. Let W' be the group of orthogonal transformations generated by the s_{α_i}, $1 \le i \le l$.

Let us prove that W' is a finite group. From the correspondence of reduced abstract root systems to abstract Cartan matrices established in §7, we know that the members $\beta_i \in \mathbb{R}^l$ in (2.104) have reflections generating a finite group W such that $\Delta = \bigcup_{i=1}^l W\beta_i$ is the reduced abstract root system associated to the abstract Cartan matrix A. Under the isomorphism $\beta_i \mapsto \alpha_i$, W is identified with W', and Δ is identified with the subset $\bigcup_{i=1}^l W\alpha_i$ of $\mathfrak{h}_0'^*$. Since $W \cong W'$, W' is finite.

We now work toward the conclusion that $\mathfrak{g}'$ is finite-dimensional. Fix i, and let $\mathfrak{sl}_i$ be the span of $\{h_i, e_i, f_i\}$ within $\mathfrak{g}'$. This is a Lie subalgebra of $\mathfrak{g}'$ isomorphic to $\mathfrak{sl}(2, \mathbb{C})$. We shall first show that every element of $\mathfrak{g}'$ lies in a finite-dimensional subspace invariant under $\mathfrak{sl}_i$.

If $j \ne i$, consider the subspace of $\mathfrak{g}'$ spanned by

$$f_j, (\operatorname{ad} f_i) f_j, \ldots, (\operatorname{ad} f_i)^{-A_{ij}} f_j.$$

These vectors are eigenvectors for $\operatorname{ad} h_i$ with respective eigenvalues

$$\alpha_j(h_i),\ \alpha_j(h_i)-2,\ \ldots,\alpha_j(h_i)+2A_{ij},$$

and hence the subspace is invariant under $\operatorname{ad} h_i$. It is invariant under $\operatorname{ad} f_i$ since $(\operatorname{ad} f_i)^{-A_{ij}+1} f_j = 0$ by Serre relation (f). Finally it is invariant under $\operatorname{ad} e_i$ by induction, starting from the fact that $(\operatorname{ad} e_i) f_j = 0$ (Serre relation (b)). Thus the subspace is invariant under $\mathfrak{sl}_i$.

Similarly for $j \neq i$, the subspace of $\mathfrak{g}'$ spanned by

$$e_j,\ (\operatorname{ad} e_i)e_j, \ldots, (\operatorname{ad} e_i)^{-A_{ij}} e_j$$

is invariant under $\mathfrak{sl}_i$, by Serre relations (e) and (b). And also $\operatorname{span}\{h_i, e_i, f_i\}$ is invariant under $\mathfrak{sl}_i$. Therefore a generating subset of $\mathfrak{g}'$ lies in a finite-dimensional subspace invariant under $\mathfrak{sl}_i$.

Now consider the set of all elements in $\mathfrak{g}'$ that lie in some finite-dimensional space invariant under $\mathfrak{sl}_i$. Say r and s are two such elements, lying in spaces R and S. Form the finite-dimensional subspace $[R, S]$ generated by all brackets from R and S. If x is in $\mathfrak{sl}_i$, then

$$(\operatorname{ad} x)[R,S] \subseteq [(\operatorname{ad} x)R, S] + [R, (\operatorname{ad} x)S] \subseteq [R,S],$$

and hence $[r, s]$ is such an element of $\mathfrak{g}'$. We conclude that every element of $\mathfrak{g}'$ lies in a finite-dimensional subspace invariant under $\mathfrak{sl}_i$.

Continuing toward the conclusion that $\mathfrak{g}'$ is finite-dimensional, let us introduce an analog of the root string analysis done in §4. Fix i, let φ be in $\Delta' \cup \{0\}$, and consider the subspace $\bigoplus_{n\in\mathbb{Z}} \mathfrak{g}'_{\varphi+n\alpha_i}$ of $\mathfrak{g}'$. This is invariant under $\mathfrak{sl}_i$, and what we have just shown implies that every member of it lies in a finite-dimensional subspace invariant under $\mathfrak{sl}_i$. By Corollary 1.70 it is the direct sum of irreducible invariant subspaces. Let U be one of the irreducible summands. Since U is invariant under $\operatorname{ad} h_i$, we have

$$U = \bigoplus_{n=-p}^{q} (U \cap \mathfrak{g}'_{\varphi+n\alpha_i})$$

with $U \cap \mathfrak{g}'_{\varphi-p\alpha_i} \neq 0$ and $U \cap \mathfrak{g}'_{\varphi+q\alpha_i} \neq 0$. By Corollary 1.69,

$$(\varphi + q\alpha_i)(h_i) = -(\varphi - p\alpha_i)(h_i)$$

and hence

$$p - q = \varphi(h_i). \tag{2.106}$$

Moreover Theorem 1.63 shows that $U \cap \mathfrak{g}'_{\varphi+n\alpha_i}$ has dimension 1 for $-p \leq n \leq q$ and has dimension 0 otherwise.

In our direct sum decomposition of $\bigoplus_{n\in\mathbb{Z}}\mathfrak{g}'_{\varphi+n\alpha_i}$ into irreducible subspaces U, suppose that the root space $\mathfrak{g}'_{\varphi+n\alpha_i}$ has dimension m. Then it meets a collection of exactly m such U's, say $U_1,\dots,U_m$. The root space

$$\mathfrak{g}'_{s_{\alpha_i}(\varphi+n\alpha_i)} = \mathfrak{g}'_{\varphi-(n+\varphi(h_i))\alpha_i}$$

must meet the same $U_1,\dots,U_m$ since (2.106) shows that

$$-p\le n\le q \quad \text{implies} -p\le -n-\varphi(h_i) = -n+q-p\le q.$$

We conclude that

$$\dim\mathfrak{g}'_\varphi = \dim\mathfrak{g}'_{s_{\alpha_i}\varphi}. \tag{2.107}$$

From (2.107), we see that $W'\Delta'\subseteq\Delta'$. Since W' mirrors for $\mathfrak{h}_0'^*$ the action of W on $\mathbb{R}^l$, the linear extension of the map $\beta_i\mapsto\alpha_i$ carries Δ into Δ'. Since $\dim\mathfrak{g}'_{\alpha_i}=1$ for all i, we see that $\dim\mathfrak{g}'_\varphi=1$ for every root φ in the finite set $\bigcup_{i=1}^l W'\alpha_i$.

To complete the proof of finite-dimensionality of $\mathfrak{g}'$, we show that every root lies in $\bigcup_{i=1}^l W'\alpha_i$. Certainly $\bigcup_{i=1}^l W'\alpha_i$ is closed under negatives, since it is generated by the α_i's and contains the $-\alpha_i$'s. Arguing by contradiction, assume that $\bigcup_{i=1}^l W'\alpha_i$ does not exhaust Δ'. By (2.103) there is some $\alpha=\sum_{j=1}^l n_j\alpha_j$ in Δ'^+ not in $\bigcup_{i=1}^l W'\alpha_i$, and we may assume that $\sum_{j=1}^l n_j$ is as small as possible. From

$$0<|\alpha|^2=\sum_{j=1}^l n_j\langle\alpha,\alpha_j\rangle,$$

we see that there is some k such that $n_k>0$ and $\langle\alpha,\alpha_k\rangle>0$. Then

$$s_{\alpha_k}(\alpha)=\alpha-\frac{2\langle\alpha,\alpha_k\rangle}{|\alpha_k|^2}\alpha_k=\sum_{j\ne k}n_j\alpha_j+\Big(n_k-\frac{2\langle\alpha,\alpha_k\rangle}{|\alpha_k|^2}\Big)\alpha_k.$$

We must have $n_j>0$ for some $j\ne k$ since otherwise $\alpha=n_k\alpha_k$, from which we obtain $n_k=1$ since $[e_k,e_k]=0$. Thus $s_{\alpha_k}(\alpha)$ is in Δ'^+. Since the sum of coefficients for $s_{\alpha_k}(\alpha)$ is less than $\sum_{j=1}^l n_j$, we conclude by minimality that $s_{\alpha_k}(\alpha)$ is in $\bigcup_{i=1}^l W'\alpha_i$. But then so is α, contradiction. We conclude that $\Delta'=\bigcup_{i=1}^l W'\alpha_i$ and hence that Δ' is finite and $\mathfrak{g}'$ is finite-dimensional.

Now that $\mathfrak{g}'$ is finite-dimensional, we prove that it is semisimple and has the required structure. In fact, $\operatorname{rad}\mathfrak{g}'$ is $\operatorname{ad}\mathfrak{h}'$ invariant and therefore satisfies

$$\operatorname{rad}\mathfrak{g}'=(\mathfrak{h}'\cap\operatorname{rad}\mathfrak{g}')\oplus\bigoplus_{\varphi\in\Delta'}(\mathfrak{g}'_\varphi\cap\operatorname{rad}\mathfrak{g}').$$

Suppose $h \neq 0$ is in $\mathfrak{h} \cap \operatorname{rad} \mathfrak{g}'$. Choose j with $\alpha_j(h) \neq 0$. Since $\operatorname{rad} \mathfrak{g}'$ is an ideal, $e_j = \alpha_j(h)^{-1}[h, e_j]$ and $f_j = -\alpha_j(h)^{-1}[h, f_j]$ are in $\operatorname{rad} \mathfrak{g}'$, and so is $h_j = [e_j, f_j]$. Thus $\operatorname{rad} \mathfrak{g}'$ contains the semisimple subalgebra $\mathfrak{sl}_j$, contradiction. We conclude that $\mathfrak{h}' \cap \operatorname{rad} \mathfrak{g}' = 0$.

Since the root spaces are 1-dimensional, we obtain

$$\operatorname{rad} \mathfrak{g}' = \bigoplus_{\varphi \in \Delta_0'} \mathfrak{g}_\varphi'$$

for some subset Δ_0' of Δ'. The Lie algebra $\mathfrak{g}'/\operatorname{rad} \mathfrak{g}'$ is semisimple, according to Proposition 1.14, and we can write it as

$$\mathfrak{g}'/\operatorname{rad} \mathfrak{g}' = \mathfrak{h} \oplus \bigoplus_{\varphi \in \Delta' - \Delta_0'} \mathfrak{g}_\varphi' \mod (\operatorname{rad} \mathfrak{g}').$$

From this decomposition we see that $\mathfrak{h}$ is a Cartan subalgebra of $\mathfrak{g}'/\operatorname{rad} \mathfrak{g}'$ and that the root system is $\Delta' - \Delta_0'$. On the other hand, no α_j is in Δ_0' since $\mathfrak{sl}_j$ is semisimple. Thus $\Delta' - \Delta_0'$ contains each α_j, and these are the simple roots. We have seen that the simple roots determine Δ' as the corresponding abstract root system. Thus Δ_0' is empty. It follows that $\mathfrak{g}'$ is semisimple, and then the structural conclusions about $\mathfrak{g}'$ are obvious. This completes the proof of Lemma 2.101.

PROOF OF THEOREM 2.98. In the diagram (2.97), X maps to a linearly independent subset of $\mathfrak{g}$, and hence the embedded subset X of $\mathfrak{F}$ maps to a linearly independent subset of $\mathfrak{g}$. Since the map $\mathfrak{F} \to \mathfrak{g}$ factors through $\mathfrak{g}' = \mathfrak{F}/\mathfrak{R}$, $\operatorname{span}\{h_i\}_{i=1}^l$ maps one-one from $\mathfrak{F}$ to $\mathfrak{g}'$ and one-one from $\mathfrak{g}'$ to $\mathfrak{g}$. Since $\operatorname{span}\{h_i\}_{i=1}^l$ maps one-one from $\mathfrak{F}$ to $\mathfrak{g}'$, Lemma 2.101 is applicable and shows that $\mathfrak{g}'$ is finite-dimensional semisimple and that $\mathfrak{h}' = \operatorname{span}\{h_i\}_{i=1}^l$ is a Cartan subalgebra.

The map $\mathfrak{F} \to \mathfrak{g}$ is onto by Proposition 2.94, and hence the map $\mathfrak{g}' \to \mathfrak{g}$ is onto. Thus $\mathfrak{g}$ is isomorphic with a quotient of $\mathfrak{g}'$. If $\mathfrak{a}$ is a simple ideal in $\mathfrak{g}'$, it follows from Corollary 2.13 that $\mathfrak{h}' \cap \mathfrak{a}$ is a Cartan subalgebra of $\mathfrak{g}'$. Since $\mathfrak{h}'$ maps one-one under the quotient map from $\mathfrak{g}'$ to $\mathfrak{g}$, $\mathfrak{h}' \cap \mathfrak{a}$ does not map to 0. Thus $\mathfrak{a}$ does not map to 0. Hence the map of $\mathfrak{g}'$ onto $\mathfrak{g}$ has 0 kernel and is an isomorphism.

10. Isomorphism Theorem

Theorem 2.98 enables us to lift isomorphisms of reduced root systems to isomorphisms of complex semisimple Lie algebras with little effort. The result is as follows.

Theorem 2.108 (Isomorphism Theorem). Let $\mathfrak{g}$ and $\mathfrak{g}'$ be complex semisimple Lie algebras with respective Cartan subalgebras $\mathfrak{h}$ and $\mathfrak{h}'$ and respective root systems Δ and Δ'. Suppose that a vector space isomorphism $\varphi : \mathfrak{h} \to \mathfrak{h}'$ is given with the property that its transpose $\varphi^t : \mathfrak{h}'^* \to \mathfrak{h}^*$ has $\varphi^t(\Delta') = \Delta$. For α in Δ, write α' for the member $(\varphi^t)^{-1}(\alpha)$ of Δ'. Fix a simple system Π for Δ. For each α in Π, select nonzero root vectors $E_\alpha \in \mathfrak{g}$ for α and $E_{\alpha'} \in \mathfrak{g}'$ for α'. Then there exists one and only one Lie algebra isomorphism $\tilde{\varphi} : \mathfrak{g} \to \mathfrak{g}'$ such that $\tilde{\varphi}|_{\mathfrak{h}} = \varphi$ and $\tilde{\varphi}(E_\alpha) = E_{\alpha'}$ for all $\alpha \in \Pi$.

PROOF OF UNIQUENESS. If $\tilde{\varphi}_1$ and $\tilde{\varphi}_2$ are two such isomorphisms, then $\tilde{\varphi}_0 = \tilde{\varphi}_2^{-1}\tilde{\varphi}_1$ is an automorphism of $\mathfrak{g}$ fixing $\mathfrak{h}$ and the root vectors for the simple roots. If $\{h_i, e_i, f_i\}$ is a triple associated to the simple root α_i by (2.93), then $\tilde{\varphi}_0(f_i)$ must be a root vector for $-\alpha_i$ and hence must be a multiple of f_i, say $c_i f_i$. Applying $\tilde{\varphi}_0$ to the relation $[e_i, f_i] = h_i$, we see that $c_i = 1$. Therefore $\tilde{\varphi}_0$ fixes all h_i, e_i, and f_i. By Proposition 2.94, $\tilde{\varphi}_0$ is the identity on $\mathfrak{g}$.

PROOF OF EXISTENCE. The linear map $(\varphi^t)^{-1}$ is given by $(\varphi^t)^{-1}(\alpha) = \alpha' = \alpha \circ \varphi^{-1}$. By assumption this map carries Δ to Δ', hence root strings to root strings. Proposition 2.29a therefore gives

$$\frac{2\langle\beta,\alpha\rangle}{|\alpha|^2} = \frac{2\langle\beta',\alpha'\rangle}{|\alpha'|^2} \qquad \text{for all } \alpha, \beta \in \Delta. \tag{2.109}$$

Write $\Pi = \{\alpha_1, \dots, \alpha_l\}$, and let $\Pi' = (\varphi^t)^{-1}(\Pi) = \{\alpha'_1, \dots, \alpha'_l\}$. Define h_i and h'_i to be the respective members of $\mathfrak{h}$ and $\mathfrak{h}'$ with $\alpha_j(h_i) = 2\langle\beta,\alpha\rangle/|\alpha|^2$ and $\alpha'_j(h'_i) = 2\langle\beta',\alpha'\rangle/|\alpha'|^2$. These are the elements of the Cartan subalgebras appearing in (2.93). By (2.109), $\alpha'_j(h'_i) = \alpha_j(h_i)$ and hence $(\varphi^t)^{-1}(\alpha_j)(h'_i) = \alpha_j(h_i)$ and $\alpha_j(\varphi^{-1}(h'_i)) = \alpha_j(h_i)$. Therefore

$$\varphi(h_i) = h'_i \qquad \text{for all } i. \tag{2.110}$$

Take e_i in (2.93) to be E_{α_i}, and let $e'_i = E_{\alpha'_i}$. Define $f_i \in \mathfrak{g}$ to be a root vector for $-\alpha_i$ with $[e_i, f_i] = h_i$, and define $f'_i \in \mathfrak{g}'$ to be a root vector for $-\alpha'_i$ with $[e'_i, f'_i] = h'_i$. Then $X = \{h_i, e_i, f_i\}_{i=1}^l$ and $X' = \{h'_i, e'_i, f'_i\}_{i=1}^l$ are standard sets of generators for $\mathfrak{g}$ and $\mathfrak{g}'$ as in (2.93) and Proposition 2.94.

Let $\mathfrak{F}$ and $\mathfrak{F}'$ be the free Lie algebras on X and X', and let $\mathfrak{R}$ and $\mathfrak{R}'$ be the ideals in $\mathfrak{F}$ and $\mathfrak{F}'$ generated by the Serre relations (a) through (f) in Theorem 2.98. Let us define $\psi : X \to \mathfrak{F}'$ by $\psi(h_i) = h'_i$, $\psi(e_i) = e'_i$, and $\psi(f_i) = f'_i$. Setting up the diagram

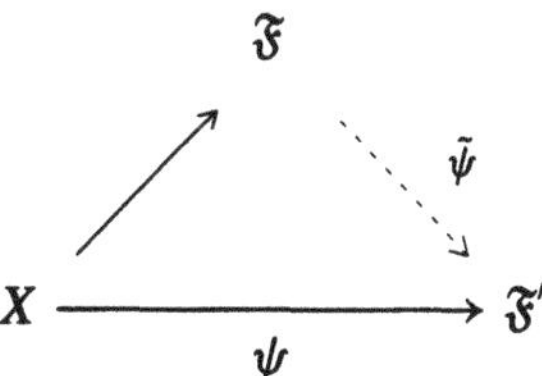

we see from the universal mapping property of $\mathfrak{F}$ that ψ extends to a Lie algebra homomorphism $\tilde{\psi} : \mathfrak{F} \to \mathfrak{F}'$. By (2.109), $\tilde{\psi}(\mathfrak{R}) \subseteq \mathfrak{R}'$. Therefore $\tilde{\psi}$ descends to Lie algebra homomorphism $\mathfrak{F}/\mathfrak{R} \to \mathfrak{F}'/\mathfrak{R}'$, and we denote this homomorphism by $\tilde{\psi}$ as well.

Meanwhile the canonical maps $\tilde{\varphi}_1 : \mathfrak{F}/\mathfrak{R} \to \mathfrak{g}$ and $\tilde{\varphi}_2 : \mathfrak{F}'/\mathfrak{R}' \to \mathfrak{g}'$, which are isomorphisms by Theorem 2.98, satisfy

$$\begin{aligned} \tilde{\varphi}_1^{-1}(h_i) = h_i \mod \mathfrak{R} \qquad &\text{and} \qquad \tilde{\varphi}_1^{-1}(E_{\alpha_i}) = e_i \mod \mathfrak{R}, \\ \tilde{\varphi}_2(h_i' \mod \mathfrak{R}') = h_i' \qquad &\text{and} \qquad \tilde{\varphi}_2(e_i' \mod \mathfrak{R}') = E_{\alpha_i'}. \end{aligned}$$

Therefore $\tilde{\varphi} = \tilde{\varphi}_2 \circ \tilde{\psi} \circ \tilde{\varphi}_1^{-1}$ is a Lie algebra homomorphism from $\mathfrak{g}$ to $\mathfrak{g}'$ with $\tilde{\varphi}(h_i) = h_i'$ and $\tilde{\varphi}(E_{\alpha_i}) = E_{\alpha_i'}$ for all i. By (2.110), $\tilde{\varphi}|_{\mathfrak{h}} = \varphi$.

To see that $\tilde{\varphi}$ is an isomorphism, we observe that $\tilde{\varphi} : \mathfrak{h} \to \mathfrak{h}'$ is an isomorphism. By the same argument as in the last paragraph of §9, it follows that $\tilde{\varphi} : \mathfrak{g} \to \mathfrak{g}'$ is one-one. Finally

$$\dim \mathfrak{g} = \dim \mathfrak{h} + |\Delta| = \dim \mathfrak{h}' + |\Delta'| = \dim \mathfrak{g}',$$

and we conclude that $\tilde{\varphi}$ is an isomorphism.

EXAMPLES.

1) One-oneness of first step in (2.58). We are to show that if $\mathfrak{g}$ and $\mathfrak{g}'$ are two complex semisimple Lie algebras with isomorphic root systems, then $\mathfrak{g}$ and $\mathfrak{g}'$ are isomorphic. To do so, we apply Theorem 2.108, mapping the root vector E_α for each simple root α to any nonzero root vector for the corresponding simple root for $\mathfrak{g}'$. We conclude that the first step of the two-step passage (2.58) is one-one, up to isomorphism.

2) Automorphisms of Dynkin diagram. Let $\mathfrak{g}$, $\mathfrak{h}$, Δ, and $\Pi = \{\alpha_1, \dots, \alpha_l\}$ be arbitrary. Suppose that σ is an automorphism of the Dynkin diagram, i.e., a permutation of the indices $1, \dots, l$ such that the Cartan matrix satisfies $A_{ij} = A_{\sigma(i)\sigma(j)}$. Define $\varphi : \mathfrak{h} \to \mathfrak{h}$ to be the linear extension of the map $h_i \to h_{\sigma(i)}$, and apply Theorem 2.108. The result is an automorphism $\tilde{\varphi}$ of $\mathfrak{g}$ that normalizes $\mathfrak{h}$, maps the set of positive roots to itself, and has the effect σ on the Dynkin diagram.

3) An automorphism constructed earlier. With $\mathfrak{g}$, $\mathfrak{h}$, and Δ given, define $\varphi = -1$ on $\mathfrak{h}$. Then Δ gets carried to Δ, and hence φ extends to an automorphism $\tilde{\varphi}$ of $\mathfrak{g}$. This automorphism has already been constructed directly (as $\tilde{\eta}$ in the course of the proof of Lemma 2.101).

11. Existence Theorem

We have now shown that the first step in the passage (2.58), i.e., the step from complex semisimple Lie algebras to abstract reduced root systems, is well defined independently of the choice of Cartan subalgebra and is one-one up to isomorphism. To complete our discussion of (2.58), we show that this step is onto, i.e., that any reduced abstract root system is the root system of a complex semisimple Lie algebra.

The Existence Theorem accomplishes this step, actually showing that any abstract Cartan matrix comes via the two steps of (2.58) from a complex semisimple Lie algebra. However, the theorem does not substitute for our case-by-case argument in §7 that the second step of (2.58) is onto. The fact that the second step is onto was used critically in the proof of Lemma 2.101 to show that W' is a finite group.

The consequence of the Existence Theorem is that there exist complex simple Lie algebras with root systems of the five exceptional types E_6, E_7, E_8, F_4, and G_2. We shall have occasion to use these complex Lie algebras in Chapter VI and then shall refer to them as complex simple Lie algebras of types E_6, etc.

Theorem 2.111 (Existence Theorem). If $A = (A_{ij})_{i,j=1}^{l}$ is an abstract Cartan matrix, then there exists a complex semisimple Lie algebra $\mathfrak{g}$ whose root system has A as Cartan matrix.

PROOF. Let $\mathfrak{F}$ be the free Lie algebra on the set $X = \{h_i, e_i, f_i\}_{i=1}^{l}$, and let $\mathfrak{R}$ be the ideal in $\mathfrak{F}$ generated by the Serre relations (a) through (f) given in Proposition 2.95. Put $\mathfrak{g} = \mathfrak{F}/\mathfrak{R}$. According to Lemma 2.101, $\mathfrak{g}$ will be the required Lie algebra if it is shown that $\operatorname{span}\{h_i\}_{i=1}^{l}$ maps one-one from $\mathfrak{F}$ to its image in $\mathfrak{F}/\mathfrak{R}$.

We shall establish this one-one behavior by factoring the quotient map into two separate maps and showing that $\operatorname{span}\{h_i\}_{i=1}^{l}$ maps one-one in each case. The first map is from $\mathfrak{F}$ to $\mathfrak{F}/\tilde{\mathfrak{R}}$, where $\tilde{\mathfrak{R}}$ is the ideal in $\mathfrak{F}$ generated by the Serre relations (a) through (d). Write h_i, e_i, f_i also for the images of the generators in $\mathfrak{F}/\tilde{\mathfrak{R}}$. Define $\tilde{\mathfrak{h}}$, $\tilde{\mathfrak{e}}$, and $\tilde{\mathfrak{f}}$ as in the statement of Lemma 2.99. The lemma says that

$$\mathfrak{F}/\tilde{\mathfrak{R}} = \tilde{\mathfrak{h}} \oplus \tilde{\mathfrak{e}} \oplus \tilde{\mathfrak{f}}, \tag{2.112}$$

but it does not tell us how large $\tilde{\mathfrak{h}}$ is.

To get at the properties of the first map, we introduce an l-dimensional complex vector space V with basis $\{v_1, \ldots, v_l\}$, and we let $T(V)$ be the tensor algebra over V. (See Appendix A for the definition and elementary properties of $T(V)$.) We drop tensor signs in writing products within $T(V)$ in order to simplify the notation. In view of the diagram

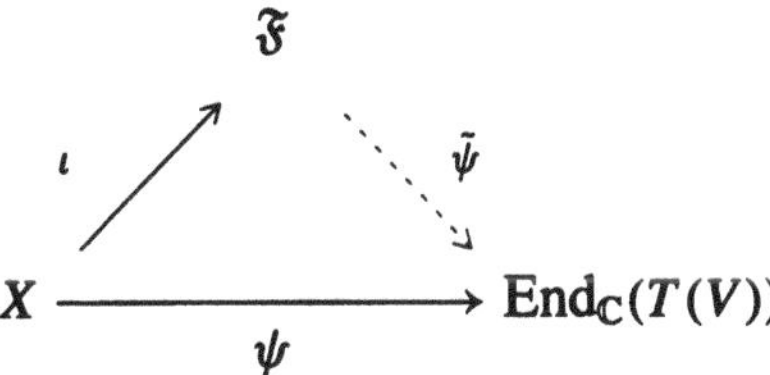

we can construct a homomorphism $\tilde{\psi} : \mathfrak{F} \to \text{End}_{\mathbb{C}}(T(V))$ by telling how x acts in $T(V)$ for each x in X. Dropping the notation ψ from the action, we define

$$\begin{aligned}
h_i(1) &= 0 \\
h_i(v_{j_1} \cdots v_{j_k}) &= -(A_{ij_1} + \cdots + A_{ij_k}) v_{j_1} \cdots v_{j_k} \\
f_i(1) &= v_i \\
f_i(v_{j_1} \cdots v_{j_k}) &= v_i v_{j_1} \cdots v_{j_k} \\
e_i(1) &= 0 \\
e_i(v_j) &= 0 \\
e_i(v_{j_1} \cdots v_{j_k}) &= v_{j_1} \cdot e_i(v_{j_2} \cdots v_{j_k}) - \delta_{ij_1}(A_{ij_2} + \cdots + A_{ij_k}) v_{j_2} \cdots v_{j_k}.
\end{aligned}$$

(The last three lines, defining the action of e_i, are made recursively on the order of the tensor.)

We show that this homomorphism defined on $\mathfrak{F}$ descends to a homomorphism $\mathfrak{F}/\mathfrak{R} \to \text{End}_{\mathbb{C}}(T(V))$ by showing that the generators of $\mathfrak{R}$ act by 0. We check the generators of types (a), (d), (b), and (c) in turn.

For (a) the generator is $[h_i, h_j]$. The span of the h_i's acts diagonally, and thus

$$\tilde{\psi}[h_i, h_j] = [\tilde{\psi}(h_i), \tilde{\psi}(h_j)] = \tilde{\psi}(h_i)\tilde{\psi}(h_j) - \tilde{\psi}(h_j)\tilde{\psi}(h_i) = 0.$$

For (d) the generator is $[h_i, f_j] + A_{ij} f_j$, and we have

$$\tilde{\psi}([h_i, f_j] + A_{ij} f_j) = \tilde{\psi}(h_i)\tilde{\psi}(f_j) - \tilde{\psi}(f_j)\tilde{\psi}(h_i) + A_{ij}\tilde{\psi}(f_j).$$

On 1, the right side gives

$$\tilde{\psi}(h_i) v_j - 0 + A_{ij} v_j = 0.$$

On $v_{j_1} \cdots v_{j_k}$, the right side gives

$$-(A_{ij} + A_{ij_1} + \cdots + A_{ij_k}) v_j v_{j_1} \cdots v_{j_k} \\ + (A_{ij_1} + \cdots + A_{ij_k}) v_j v_{j_1} \cdots v_{j_k} + A_{ij} v_j v_{j_1} \cdots v_{j_k} = 0.$$

For (b) the generator is $[e_i, f_j] - \delta_{ij} h_i$, and we have

$$\tilde{\psi}([e_i, f_j] - \delta_{ij} h_i) = \tilde{\psi}(e_i)\tilde{\psi}(f_j) - \tilde{\psi}(f_j)\tilde{\psi}(e_i) - \delta_{ij}\tilde{\psi}(h_i).$$

On 1, each term on the right side acts as 0. On a monomial $v_{j_2} \cdots v_{j_k}$, the right side gives

$$e_i(v_j v_{j_2} \cdots v_{j_k}) - v_j \cdot e_i(v_{j_2} \cdots v_{j_k}) + \delta_{ij}(A_{ij_2} + \cdots + A_{ij_k}) v_{j_2} \cdots v_{j_k},$$

and this is 0 by the recursive definition of the action of e_i.

For (c) the generator is $[h_i, e_j] - A_{ij} e_j$. Let us observe by induction on k that

$$h_i e_j(v_{j_1} \cdots v_{j_k}) = -(A_{ij_1} + \cdots + A_{ij_k} - A_{ij}) e_j(v_{j_1} \cdots v_{j_k}). \tag{2.113}$$

Formula (2.113) is valid for $k = 0$ and $k = 1$ since e_j acts as 0 on monomials of degrees 0 and 1. For general k, the recursive definition of the action of e_i and the inductive hypothesis combine to show that the left side of (2.113) is

$$\begin{aligned} h_i e_j(v_{j_1} \cdots v_{j_k}) &= h_i(v_{j_1} \cdot e_j(v_{j_2} \cdots v_{j_k})) - \delta_{jj_1}(A_{jj_2} + \cdots + A_{jj_k}) h_i(v_{j_2} \cdots v_{j_k}) \\ &= -(A_{ij_1} + \cdots + A_{ij_k} - A_{ij}) v_{j_1} \cdot e_j(v_{j_2} \cdots v_{j_k}) \\ &\quad + \delta_{jj_1}(A_{jj_2} + \cdots + A_{jj_k})(A_{ij_2} + \cdots + A_{ij_k}) v_{j_2} \cdots v_{j_k}, \end{aligned}$$

and that the right side of (2.113) is

$$\begin{aligned} &-(A_{ij_1} + \cdots + A_{ij_k} - A_{ij}) e_j(v_{j_1} \cdots v_{j_k}) \\ &\quad = -(A_{ij_1} + \cdots + A_{ij_k} - A_{ij}) v_{j_1} \cdot e_j(v_{j_2} \cdots v_{j_k}) \\ &\quad\quad + (A_{ij_1} + \cdots + A_{ij_k} - A_{ij}) \delta_{jj_1}(A_{jj_2} + \cdots + A_{jj_k}) v_{j_2} \cdots v_{j_k}. \end{aligned}$$

Subtraction shows that the difference of the left side and the right side of (2.113) is

$$= -\delta_{jj_1}(A_{ij_1} - A_{ij})(A_{jj_2} + \cdots + A_{jj_k}) v_{j_2} \cdots v_{j_k} = 0.$$

The induction is complete, and (2.113) is established. Returning to our generator, we have

$$\tilde{\psi}([h_i, e_j] - A_{ij} e_j) = \tilde{\psi}(h_i)\tilde{\psi}(e_j) - \tilde{\psi}(e_j)\tilde{\psi}(h_i) - A_{ij}\tilde{\psi}(e_j).$$

On 1, each term on the right side acts as 0. On $v_{j_1}\cdots v_{j_k}$, (2.113) shows that the effect of the right side is

$$\begin{aligned}&= -(A_{ij_1}+\cdots+A_{ij_k}-A_{ij})e_j(v_{j_1}\cdots v_{j_k})\\&\quad+(A_{ij_1}+\cdots+A_{ij_k})e_j(v_{j_1}\cdots v_{j_k})-A_{ij}e_j(v_{j_1}\cdots v_{j_k})=0.\end{aligned}$$

Thus $\tilde{\psi}$ descends to $\mathfrak{F}/\tilde{\mathfrak{R}}$.

Now we can prove that $\operatorname{span}\{h_i\}_{i=1}^l$ maps one-one from $\mathfrak{F}$ to $\mathfrak{F}/\tilde{\mathfrak{R}}$. If a nontrivial $\sum c_i h_i$ maps to 0, then we have

$$0=\Big(\sum_i c_i h_i\Big)(v_j)=-\Big(\sum_i c_i A_{ij}\Big)v_j$$

for all j. Hence $\sum_i c_i A_{ij}=0$ for all j, in contradiction with the linear independence of the rows of (A_{ij}). We conclude that $\operatorname{span}\{h_i\}_{i=1}^l$ maps one-one from $\mathfrak{F}$ to $\mathfrak{F}/\tilde{\mathfrak{R}}$.

Now we bring in Serre relations (e) and (f), effectively imposing them directly on $\mathfrak{F}/\tilde{\mathfrak{R}}$ to obtain $\mathfrak{g}$ as quotient. Define $\tilde{\mathfrak{g}}=\mathfrak{F}/\tilde{\mathfrak{R}}$. Let $\mathfrak{R}'$ be the ideal in $\tilde{\mathfrak{g}}$ generated by all

$$(\operatorname{ad} e_i)^{-A_{ij}+1}e_j \quad\text{and all}\quad (\operatorname{ad} f_i)^{-A_{ij}+1}f_j \qquad \text{for } i\neq j.$$

Then indeed $\mathfrak{g}\cong\tilde{\mathfrak{g}}/\mathfrak{R}'$.

We define subalgebras $\tilde{\mathfrak{h}}$, $\tilde{\mathfrak{e}}$, and $\tilde{\mathfrak{f}}$ of $\tilde{\mathfrak{g}}$ as in the statement of Lemma 2.99. Let $\tilde{\mathfrak{e}}'$ be the ideal in $\tilde{\mathfrak{e}}$ generated by all $(\operatorname{ad} e_i)^{-A_{ij}+1}e_j$, and let $\tilde{\mathfrak{f}}'$ be the ideal in $\tilde{\mathfrak{f}}$ generated by all $(\operatorname{ad} f_i)^{-A_{ij}+1}f_j$. Then

$$(2.114)\qquad (\text{generators of } \mathfrak{R}')\subseteq\tilde{\mathfrak{e}}'+\tilde{\mathfrak{f}}'\subseteq\tilde{\mathfrak{e}}+\tilde{\mathfrak{f}}.$$

We shall prove that $\tilde{\mathfrak{e}}'$ is actually an ideal in $\tilde{\mathfrak{g}}$. We observe that $\tilde{\mathfrak{e}}'$ is invariant under all $\operatorname{ad} h_k$ (since the generators of $\tilde{\mathfrak{e}}'$ are eigenvectors) and all e_k (since $\tilde{\mathfrak{e}}'\subseteq\tilde{\mathfrak{e}}$). Thus we are to show that

$$(\operatorname{ad} f_k)(\operatorname{ad} e_i)^{-A_{ij}+1}e_j$$

is in $\tilde{\mathfrak{e}}'$ if $i\neq j$. In fact, we show it is 0.

If $k\neq i$, then $[f_k,e_i]=0$ shows that $\operatorname{ad} f_k$ commutes with $\operatorname{ad} e_i$. Thus we are led to

$$(\operatorname{ad} e_i)^{-A_{ij}+1}[f_k,e_j].$$

If $k\neq j$, this is 0 by Serre relation (b). If $k=j$, it is

$$(2.115)\qquad =-(\operatorname{ad} e_i)^{-A_{ij}+1}h_j=A_{ji}(\operatorname{ad} e_i)^{-A_{ij}}e_i.$$

If $A_{ij} < 0$, then the right side of (2.115) is 0 since $[e_i, e_i] = 0$; if $A_{ij} = 0$, then the right side of (2.115) is 0 because the coefficient A_{ji} is 0.

If $k = i$, we are to consider

$$(\operatorname{ad} f_i)(\operatorname{ad} e_i)^{-A_{ij}+1} e_j.$$

Now

$$(\operatorname{ad} f_i)(\operatorname{ad} e_i)^n e_j = -(\operatorname{ad} h_i)(\operatorname{ad} e_i)^{n-1} e_j + (\operatorname{ad} e_i)(\operatorname{ad} f_i)(\operatorname{ad} e_i)^{n-1} e_j.$$

Since $(\operatorname{ad} f_i)e_j = 0$, an easy induction with this equation shows that

$$(\operatorname{ad} f_i)(\operatorname{ad} e_i)^n e_j = -n(A_{ij} + n - 1)(\operatorname{ad} e_i)^{n-1} e_j.$$

For $n = -A_{ij} + 1$, the right side is 0, as asserted. This completes the proof that $\tilde{\mathfrak{e}}'$ is an ideal in $\tilde{\mathfrak{g}}$.

Similarly $\tilde{\mathfrak{f}}'$ is an ideal in $\tilde{\mathfrak{g}}$, and so is the sum $\tilde{\mathfrak{e}}' + \tilde{\mathfrak{f}}'$. From (2.114) we therefore obtain

$$\mathfrak{R}' \subseteq \tilde{\mathfrak{e}}' + \tilde{\mathfrak{f}}' \subseteq \tilde{\mathfrak{e}} + \tilde{\mathfrak{f}}.$$

In view of the direct sum decomposition (2.112), $\mathfrak{R}' \cap \tilde{\mathfrak{h}} = 0$. Therefore $\operatorname{span}\{h_i\}_{i=1}^{l}$ maps one-one from $\tilde{\mathfrak{g}}$ to $\tilde{\mathfrak{g}}/\mathfrak{R}' \cong \mathfrak{g}$, and the proof of the theorem is complete.

12. Problems

1. According to Problem 12 in Chapter I, the trace form is a multiple of the Killing form for $\mathfrak{sl}(n+1, \mathbb{C})$ if $n \geq 1$, for $\mathfrak{so}(2n+1, \mathbb{C})$ if $n \geq 2$, $\mathfrak{sp}(n, \mathbb{C})$ if $n \geq 3$, and $\mathfrak{so}(2n, \mathbb{C})$ if $n \geq 4$. Find the multiple in each case.

2. Since the Dynkin diagrams of $A_1 \oplus A_1$ and D_2 are isomorphic, the Isomorphism Theorem predicts that $\mathfrak{sl}(2, \mathbb{C}) \oplus \mathfrak{sl}(2, \mathbb{C})$ is isomorphic with $\mathfrak{so}(4, \mathbb{C})$. Using the explicit root-space decomposition for $\mathfrak{so}(4, \mathbb{C})$ found in §1, exhibit two 3-dimensional ideals in $\mathfrak{so}(4, \mathbb{C})$, proving that they are indeed ideals.

3. Let $\mathfrak{g}$ be the 2-dimensional complex Lie algebra with a basis $\{X, Y\}$ such that $[X, Y] = Y$.
   (a) Identify the regular elements.
   (b) Prove that $\mathbb{C}X$ is a Cartan subalgebra but that $\mathbb{C}Y$ is not.
   (c) Find the weight space decomposition of $\mathfrak{g}$ relative to the Cartan subalgebra $\mathbb{C}X$.

4. Let $\mathfrak{g} = \mathfrak{h} \oplus \bigoplus_{\alpha\in\Delta} \mathfrak{g}_\alpha$ be a root-space decomposition for a complex semisimple Lie algebra, and let Δ' be a subset of Δ that forms a root system in $\mathfrak{h}_0^*$.
   (a) Show by example that $\mathfrak{s} = \mathfrak{h} \oplus \bigoplus_{\alpha\in\Delta'} \mathfrak{g}_\alpha$ need not be a subalgebra of $\mathfrak{g}$.
   (b) Suppose that $\Delta' \subseteq \Delta$ is a root subsystem with the following property. Whenever α and β are in Δ' and $\alpha + \beta$ is in Δ, then $\alpha + \beta$ is in Δ'. Prove that $\mathfrak{s} = \mathfrak{h} \oplus \bigoplus_{\alpha\in\Delta'} \mathfrak{g}_\alpha$ is a subalgebra of $\mathfrak{g}$ and that it is semisimple.

5. Exhibit complex semisimple Lie algebras of dimensions 8, 9, and 10. Deduce that there are complex semisimple Lie algebras of every dimension ≥ 8.

6. Using results from §§4–5 but not the classification, show that there are no complex semisimple Lie algebras of dimensions 4, 5, or 7.

7. Let Δ be a root system, and fix a simple system Π. Show that any positive root can be written in the form

$$\alpha = \alpha_{i_1} + \alpha_{i_2} + \cdots + \alpha_{i_k}$$

with each α_{i_j} in Π and with each partial summand from the left equal to a positive root.

8. Let Δ be a root system, and fix a lexicographic ordering. Show that the largest root α_0 has $\langle\alpha_0, \alpha\rangle \geq 0$ for all positive roots α. If Δ is of type B_n with $n \geq 2$, find a positive root β_0 other than α_0 with $\langle\beta_0, \alpha\rangle \geq 0$ for all positive roots α.

9. Write down the Cartan matrices for A_n, B_n, C_n, and D_n.

10. The root system G_2 is pictured in Figure 2.2. According to Theorem 2.63, there are exactly 12 simple systems for this root system.
    (a) Identify them in Figure 2.2.
    (b) Fix one of them, letting the short simple root be α and the long simple root be β. Identify the positive roots, and express each of them as a linear combination of α and β.

11. (a) Prove that two simple roots in a Dynkin diagram that are connected by a single edge are in the same orbit under the Weyl group.
    (b) For an irreducible root system, prove that all roots of a particular length form a single orbit under the Weyl group.

12. In a reduced root system with a positive system imposed, let α and β be distinct simple roots connected by n edges ($0 \le n \le 3$) in the Dynkin diagram, and let s_α and s_β be the corresponding reflections in the Weyl group. Show that

$$(s_\alpha s_\beta)^k = 1, \quad \text{where } k = \begin{cases} 2 & \text{if } n = 0 \\ 3 & \text{if } n = 1 \\ 4 & \text{if } n = 2 \\ 6 & \text{if } n = 3. \end{cases}$$

13. (a) Prove that any element of order 2 in a Weyl group is the product of commuting root reflections.
    (b) Prove that the only reflections in a Weyl group are the root reflections.

14. Let Δ be an abstract root system in V, and fix an ordering. Suppose that λ is in V and w is in the Weyl group. Prove that if λ and $w\lambda$ are both dominant, then $w\lambda = \lambda$.

15. Verify the following table of values for the number of roots, the dimension of $\mathfrak{g}$, and the order of the Weyl group for the classical irreducible reduced root systems:

Type of Δ	$\lvert\Delta\rvert$	$\dim \mathfrak{g}$	$\lvert W\rvert$
A_n	$n(n+1)$	$n(n+2)$	$(n+1)!$
B_n	$2n^2$	$n(2n+1)$	$n!2^n$
C_n	$2n^2$	$n(2n+1)$	$n!2^n$
D_n	$2n(n-1)$	$n(2n-1)$	$n!2^{n-1}$

16. Verify the following table of values for the number of roots and the dimension of $\mathfrak{g}$ for the exceptional irreducible reduced root systems. These systems are described explicitly in Figure 2.2 and Proposition 2.87:

Type of Δ	$\lvert\Delta\rvert$	$\dim \mathfrak{g}$
E_6	72	78
E_7	126	133
E_8	240	248
F_4	48	52
G_2	12	14

17. If Δ is an abstract root system and α is in Δ, let $\alpha^\vee = 2|\alpha|^{-2}\alpha$. Define $\Delta^\vee = \{\alpha^\vee \mid \alpha \in \Delta\}$.
    (a) Prove that $\Delta^\vee$ is an abstract root system with the same Weyl group as Δ.
    (b) If Π is a simple system for Δ, prove that $\Pi^\vee = \{\alpha^\vee \mid \alpha \in \Pi\}$ is a simple system for $\Delta^\vee$.
    (c) For any reduced irreducible root system Δ other than B_n and C_n, show from the classification that $\Delta^\vee \cong \Delta$. For B_n and C_n, show that $(B_n)^\vee \cong C_n$ and $(C_n)^\vee \cong B_n$.

18. Let Π be a simple system in a root system Δ, and let Δ^+ be the corresponding set of positive roots.
    (a) Prove that the negatives of the members of Π form another simple system, and deduce that there is a unique member w_0 of the Weyl group sending Δ^+ to $-\Delta^+$.
    (b) Prove that $-w_0$ gives an automorphism of the Dynkin diagram, and conclude that -1 is in the Weyl group for B_n, C_n, E_7, E_8, F_4, and G_2.
    (c) Prove that -1 is not in the Weyl group of A_n for $n \geq 2$.
    (d) Prove that -1 is in the Weyl group of D_n if $n \geq 2$ is even but not if $n \geq 3$ is odd.

19. Using the classification theorems, show that Figure 2.2 exhibits all but two of the root systems in 2-dimensional spaces, up to isomorphism. What are the two that are missing?

20. Let $\mathfrak{g}$ be a complex semisimple Lie algebra, let $\mathfrak{h}$ be a Cartan subalgebra, let Δ be the roots, let W be the Weyl group, and let w be in W. Using the Isomorphism Theorem, prove that there is a member of $\text{Aut}_{\mathbb{C}}\,\mathfrak{g}$ whose restriction to $\mathfrak{h}$ is w.

Problems 21–24 concern the length function $l(w)$ on the Weyl group W. Fix a reduced root system Δ and an ordering, and let $l(w)$ be defined as in §6 before Proposition 2.70.

21. Prove that $l(w) = l(w^{-1})$.

22. (a) Define $\operatorname{sgn} w = (-1)^{l(w)}$. Prove that the function sgn carrying W to $\{\pm 1\}$ is a homomorphism.
    (b) Prove that $\operatorname{sgn} w = \det w$ for all $w \in W$.
    (c) Prove that $l(s_\alpha)$ is odd for any root reflection s_α.

23. For w_1 and w_2 in W, prove that

$$l(w_1 w_2) = l(w_1) + l(w_2) - 2\#\{\beta \in \Delta \mid \beta > 0,\ w_1\beta < 0,\ w_2^{-1}\beta < 0\}.$$

24. If α is a root, prove that $l(ws_\alpha) < l(w)$ if $w\alpha < 0$ and that $l(ws_\alpha) > l(w)$ if $w\alpha > 0$.

Problems 25–30 compute the determinants of all irreducible Cartan matrices.

25. Let M_l be an l-by-l Cartan matrix whose first two rows and columns look like
$$\begin{pmatrix} 2 & -1 & 0 \\ -1 & 2 & -1 \\ 0 & -1 & * \end{pmatrix},$$
the other entries in those rows and columns being 0. Let M_{l-1} be the Cartan matrix obtained by deleting the first row and column from M_l, and let M_{l-2} be the Cartan matrix obtained by deleting the first row and column from M_{l-1}. Prove that
$$\det M_l = 2 \det M_{l-1} - \det M_{l-2}.$$

26. Reinterpret the condition on the Cartan matrix M_l in Problem 25 as a condition on the corresponding Dynkin diagram.

27. Calculate explicitly the determinants of the irreducible Cartan matrices of types A_1, A_2, B_2, B_3, C_3, and D_4, showing that they are $2, 3, 2, 2, 2$, and 4, respectively.

28. Using the inductive formula in Problem 25 and the initial data in Problem 27, show that the determinants of the irreducible Cartan matrices of types A_n for $n \geq 1$, B_n for $n \geq 2$, C_n for $n \geq 3$, and D_n for $n \geq 4$ are $n+1, 2$, 2, and 4, respectively.

29. Using the inductive formula in Problem 25 and the initial data for A_4 and D_5 computed in Problem 28, show that the determinants of the irreducible Cartan matrices of types E_6, E_7, and E_8 are $3, 2$, and 1, respectively.

30. Calculate explicitly the determinants of the Cartan matrices for F_4 and G_2, showing that they are both 1.

Problems 31–34 compute the order of the Weyl group for the root systems F_4, E_6, E_7, and E_8. In each case the idea is to identify a transitive group action by the Weyl group, compute the number of elements in an orbit, compute the order of the subgroup fixing an element, and multiply.

31. The root system F_4 is given explicitly in (2.88).
    (a) Show that the long roots form a root system of type D_4.

(b) By (a) the Weyl group W_D of D_4 is a subgroup of the Weyl group W_F of F_4. Show that every element of W_F leaves the system D_4 stable and therefore carries an ordered system of simple roots for D_4 to another ordered simple system. Conclude that $|W_F/W_D|$ equals the number of symmetries of the Dynkin diagram of D_4 that can be implemented by W_F.

(c) Show that reflection in e_4 and reflection in $\frac{1}{2}(e_1 - e_2 - e_3 - e_4)$ are members of W_F that permute the standard simple roots of D_4 as given in (2.50), and deduce that $|W_F/W_D| = 6$.

(d) Conclude that $|W_F| = 2^7 \cdot 3^2$.

32. The root system $\Delta = E_6$ is given explicitly in the proof of Proposition 2.87. Let W be the Weyl group.
   (a) Why is the orbit of $\frac{1}{2}(e_8 - e_7 - e_6 + e_5 + e_4 + e_3 + e_2 + e_1)$ under W equal exactly to Δ?
   (b) Show that the subset of Δ orthogonal to the root in (a) is a root system of type A_5.
   (c) The element -1 is not in the Weyl group of A_5. Why does it follow from this fact and (b) that -1 is not in the Weyl group of E_6?
   (d) Deduce from (b) that the subgroup of W fixing the root in (a) is isomorphic to the Weyl group of A_5.
   (e) Conclude that $|W| = 2^7 \cdot 3^4 \cdot 5$.

33. The root system $\Delta = E_7$ is given explicitly in the proof of Proposition 2.87. Let W be the Weyl group.
   (a) Why is the orbit of $e_8 - e_7$ under W equal exactly to Δ?
   (b) Show that the subset of Δ orthogonal to $e_8 - e_7$ is a root system of type D_6.
   (c) Deduce from (b) that the subgroup of W fixing $e_8 - e_7$ is isomorphic to the Weyl group of D_6.
   (d) Conclude that $|W| = 2^{10} \cdot 3^4 \cdot 5 \cdot 7$.

34. The root system $\Delta = E_8$ is given explicitly in (2.89). Let W be the Weyl group.
   (a) Why is the orbit of $e_8 + e_7$ under W equal exactly to Δ?
   (b) Show that the subset of Δ orthogonal to $e_8 + e_7$ is a root system of type E_7.
   (c) Deduce from (b) that the subgroup of W fixing $e_8 + e_7$ is isomorphic to the Weyl group of E_7.
   (d) Conclude that $|W| = 2^{14} \cdot 3^5 \cdot 5^2 \cdot 7$.

Problems 35–37 exhibit an explicit isomorphism of $\mathfrak{sl}(4, \mathbb{C})$ with $\mathfrak{so}(6, \mathbb{C})$. Such an isomorphism is predicted by the Isomorphism Theorem since the Dynkin diagrams of A_3 and D_3 are isomorphic.

35. Let $I_{3,3}$ be the 6-by-6 diagonal matrix defined in Example 3 in §I.8, and define $\mathfrak{g} = \{X \in \mathfrak{gl}(6, \mathbb{C}) \mid X^t I_{3,3} + I_{3,3} X = 0\}$. Let $S = \operatorname{diag}(i, i, i, 1, 1, 1)$. For $X \in \mathfrak{g}$, let $Y = SXS^{-1}$. Prove that the map $X \mapsto Y$ is an isomorphism of $\mathfrak{g}$ onto $\mathfrak{so}(6, \mathbb{C})$.

36. Any member of $\mathfrak{sl}(4, \mathbb{C})$ acts on the 6-dimensional complex vector space of alternating tensors of rank 2 by

$$M(e_i \wedge e_j) = Me_i \wedge e_j + e_i \wedge Me_j,$$

where $\{e_i\}_{i=1}^4$ is the standard basis of $\mathbb{C}^4$. Using

$$(e_1 \wedge e_2) \pm (e_3 \wedge e_4), \quad (e_1 \wedge e_3) \pm (e_2 \wedge e_4), \quad (e_1 \wedge e_4) \pm (e_2 \wedge e_3)$$

in some particular order as an ordered basis for the alternating tensors, show that the action of M is given by an element of the Lie algebra of $\mathfrak{g}$ in Problem 35.

37. The previous two problems combine to give a Lie algebra homomorphism of $\mathfrak{sl}(4, \mathbb{C})$ into $\mathfrak{so}(6, \mathbb{C})$. Show that no nonzero element of $\mathfrak{sl}(4, \mathbb{C})$ acts as the 0 operator on alternating tensors, and deduce from the simplicity of $\mathfrak{sl}(4, \mathbb{C})$ that the homomorphism is an isomorphism.

Problems 38–39 exhibit an explicit isomorphism of $\mathfrak{sp}(2, \mathbb{C})$ with $\mathfrak{so}(5, \mathbb{C})$. Such an isomorphism is predicted by the Isomorphism Theorem since the Dynkin diagrams of C_2 and B_2 are isomorphic.

38. The composition of the inclusion $\mathfrak{sp}(2, \mathbb{C}) \hookrightarrow \mathfrak{sl}(4, \mathbb{C})$ followed by the mapping of Problem 36 gives a homomorphism of $\mathfrak{sp}(2, \mathbb{C})$ into the Lie algebra $\mathfrak{g}$ of Problem 35. Show that there is some index i, $1 \leq i \leq 6$, such that the ith row and column of the image in $\mathfrak{g}$ are always 0.

39. Deduce that the composition of the homomorphism of Problem 38 followed by the isomorphism $\mathfrak{g} \cong \mathfrak{so}(6, \mathbb{C})$ of Problem 35 may be regarded as an isomorphism of $\mathfrak{sp}(2, \mathbb{C})$ with $\mathfrak{so}(5, \mathbb{C})$.

Problems 40–42 give an explicit construction of a simple complex Lie algebra of type G_2.

40. Let Δ be the root system of type B_3 given in a space V as in (2.43). Prove that the orthogonal projection of Δ on the subspace of V orthogonal to $e_1 + e_2 + e_3$ is a root system of type G_2.

41. Let $\mathfrak{g}$ be a simple complex Lie algebra of type B_3. Let $\mathfrak{h}$ be a Cartan subalgebra, let the root system be as in Problem 40, and let B be the Killing form. Prove that the centralizer of $H_{e_1+e_2+e_3}$ is the direct sum of $\mathbb{C}H_{e_1+e_2+e_3}$ and a simple complex Lie algebra of type A_2 and dimension 8.

42. In Problem 41 normalize root vectors X_α so that $B(X_\alpha, X_{-\alpha}) = 1$. From the two vectors $[X_{e_1}, X_{e_2}] + 2X_{-e_3}$ and $[X_{-e_1}, X_{-e_2}] - 2X_{e_3}$, obtain four more vectors by permuting the indices cyclically. Let $\mathfrak{g}'$ be the 14-dimensional linear span of these six vectors and the A_2 Lie subalgebra of Problem 41. Prove that $\mathfrak{g}'$ is a Lie subalgebra of $\mathfrak{g}$ of type G_2.

CHAPTER III

Universal Enveloping Algebra

Abstract. For a complex Lie algebra $\mathfrak{g}$, the universal enveloping algebra $U(\mathfrak{g})$ is an explicit complex associative algebra with identity having the property that any Lie algebra homomorphism of $\mathfrak{g}$ into an associative algebra A with identity "extends" to an associative algebra homomorphism of $U(\mathfrak{g})$ into A and carrying 1 to 1. The algebra $U(\mathfrak{g})$ is a quotient of the tensor algebra $T(\mathfrak{g})$ and is a filtered algebra as a consequence of this property. The Poincaré-Birkhoff-Witt Theorem gives a vector-space basis of $U(\mathfrak{g})$ in terms of an ordered basis of $\mathfrak{g}$.

One consequence of this theorem is to identify the associated graded algebra for $U(\mathfrak{g})$ as canonically isomorphic to the symmetric algebra $S(\mathfrak{g})$. This identification allows the construction of a vector-space isomorphism called "symmetrization" from $S(\mathfrak{g})$ onto $U(\mathfrak{g})$. When $\mathfrak{g}$ is a direct sum of subspaces, the symmetrization mapping exhibits $U(\mathfrak{g})$ canonically as a tensor product.

Another consequence of the Poincaré-Birkhoff-Witt Theorem is the existence of a free Lie algebra on any set X. This is a Lie algebra $\mathfrak{F}$ with the property that any function from X into a Lie algebra extends uniquely to a Lie algebra homomorphism of $\mathfrak{F}$ into the Lie algebra.

1. Universal Mapping Property

Throughout this chapter we suppose that $\mathfrak{g}$ is a complex Lie algebra. We shall be interested only in Lie algebras whose dimension is at most countable, but our discussion will apply in general. Usually, but not always, $\mathfrak{g}$ will be finite-dimensional. When we are studying a Lie group G with Lie algebra $\mathfrak{g}_0$, $\mathfrak{g}$ will be the complexification of $\mathfrak{g}_0$.

If we have a (complex-linear) representation π of $\mathfrak{g}$ on a complex vector space V, then the investigation of invariant subspaces in principle involves writing down all iterates $\pi(X_1)\pi(X_2)\cdots\pi(X_n)$ for members of $\mathfrak{g}$, applying them to members of V, and seeing what elements of V result. In the course of computing the resulting elements of V, one might be able to simplify an expression by using the identity $\pi(X)\pi(Y) = \pi(Y)\pi(X) + \pi[X, Y]$. This identity really has little to do with π, and our objective in this section will be to introduce a setting in which we can make such calculations without reference to π; to obtain an identity for the representation π, one simply applies π to both sides of a universal identity.

For a first approximation of what we want, we can use the tensor algebra $T(\mathfrak{g}) = \bigoplus_{k=0}^{\infty} T^k(\mathfrak{g})$. (See Appendix A for the definition and elementary properties of $T(\mathfrak{g})$.) The representation π is a linear map of $\mathfrak{g}$ into the associative algebra $\mathrm{End}_{\mathbb{C}} V$ and extends to an algebra homomorphism $\tilde{\pi} : T(\mathfrak{g}) \to \mathrm{End}_{\mathbb{C}} V$ with $\tilde{\pi}(1) = 1$. Then $\pi(X_1)\pi(X_2)\cdots\pi(X_n)$ can be replaced by $\tilde{\pi}(X_1 \otimes X_2 \otimes \cdots \otimes X_n)$. The difficulty with using $T(\mathfrak{g})$ is that it does not take advantage of the Lie algebra structure of $\mathfrak{g}$ and does not force the identity $\pi(X)\pi(Y) = \pi(Y)\pi(X) + \pi[X, Y]$ for all X and Y in $\mathfrak{g}$ and all π. Thus instead of the tensor algebra, we use the following quotient of $T(\mathfrak{g})$:

$$U(\mathfrak{g}) = T(\mathfrak{g})/J, \tag{3.1a}$$

where

$$J = \begin{pmatrix} \text{two-sided ideal generated by all} \\ X \otimes Y - Y \otimes X - [X, Y] \text{ with} \\ X \text{ and } Y \text{ in } T^1(\mathfrak{g}) \end{pmatrix}. \tag{3.1b}$$

The quotient $U(\mathfrak{g})$ is an associative algebra with identity and is known as the **universal enveloping algebra** of $\mathfrak{g}$. Products in $U(\mathfrak{g})$ are written without multiplication signs.

The canonical map $\mathfrak{g} \to U(\mathfrak{g})$ given by embedding $\mathfrak{g}$ into $T^1(\mathfrak{g})$ and then passing to $U(\mathfrak{g})$ is denoted ι. Because of (3.1), ι satisfies

$$\iota[X, Y] = \iota(X)\iota(Y) - \iota(Y)\iota(X) \qquad \text{for } X \text{ and } Y \text{ in } \mathfrak{g}. \tag{3.2}$$

The algebra $U(\mathfrak{g})$ is harder to work with than the exterior algebra $\bigwedge(\mathfrak{g})$ or the symmetric algebra $S(\mathfrak{g})$, which are both quotients of $T(\mathfrak{g})$ and are discussed in Appendix A. The reason is that the ideal in (3.1b) is not generated by homogeneous elements. Thus, for example, it is not evident that the canonical map $\iota : \mathfrak{g} \to U(\mathfrak{g})$ is one-one. However, when $\mathfrak{g}$ is *abelian*, $U(\mathfrak{g})$ reduces to $S(\mathfrak{g})$, and we have a clear notion of what to expect of $U(\mathfrak{g})$. Even when $\mathfrak{g}$ is nonabelian, $U(\mathfrak{g})$ and $S(\mathfrak{g})$ are still related, and we shall make the relationship precise later in this chapter.

Let $U_n(\mathfrak{g})$ be the image of $T_n(\mathfrak{g}) = \bigoplus_{k=0}^{n} T^k(\mathfrak{g})$ under the passage to the quotient in (3.1). Then $U(\mathfrak{g}) = \bigcup_{n=0}^{\infty} U_n(\mathfrak{g})$. Since $\{T_n(\mathfrak{g})\}$ exhibits $T(\mathfrak{g})$ as a filtered algebra, $\{U_n(\mathfrak{g})\}$ exhibits $U(\mathfrak{g})$ as a filtered algebra. If $\mathfrak{g}$ is finite-dimensional, each $U_n(\mathfrak{g})$ is finite-dimensional.

Proposition 3.3. The algebra $U(\mathfrak{g})$ and the canonical map $\iota : \mathfrak{g} \to U(\mathfrak{g})$ have the following universal mapping property: Whenever

A is a complex associative algebra with identity and $\pi : \mathfrak{g} \to A$ is a linear mapping such that

$$(3.4) \qquad \pi(X)\pi(Y) - \pi(Y)\pi(X) = \pi[X, Y] \qquad \text{for all } X \text{ and } Y \text{ in } \mathfrak{g},$$

then there exists a unique algebra homomorphism $\tilde{\pi} : U(\mathfrak{g}) \to A$ such that $\tilde{\pi}(1) = 1$ and the diagram

(3.5)

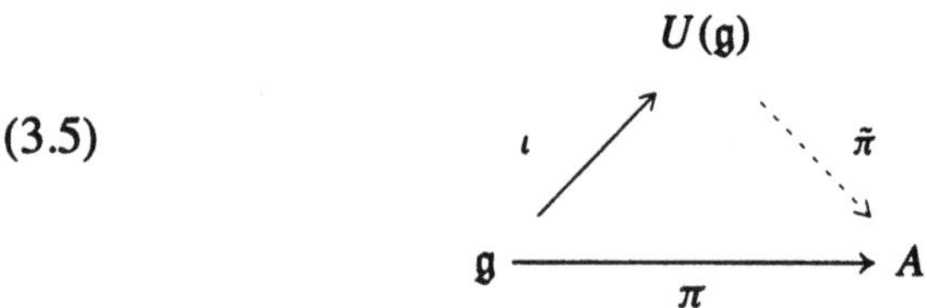

commutes.

REMARKS.

1) We regard $\tilde{\pi}$ as an "extension" of π. This notion will be more appropriate after we prove that ι is one-one.

2) This proposition allows us to make an alternative definition of **universal enveloping algebra** for $\mathfrak{g}$. It is a pair $(U(\mathfrak{g}), \iota)$ such that $U(\mathfrak{g})$ is an associative algebra with identity, $\iota : \mathfrak{g} \to U(\mathfrak{g})$ is a linear mapping satisfying (3.2), and whenever $\pi : \mathfrak{g} \to A$ is a linear mapping satisfying (3.4), then there exists a unique algebra homomorphism $\tilde{\pi} : U(\mathfrak{g}) \to A$ such that $\tilde{\pi}(1) = 1$ and the diagram (3.5) commutes. The proposition says that the constructed $U(\mathfrak{g})$ has this property, and we can use this property to see that any other candidate, say $(U'(\mathfrak{g}), \iota')$, has $U'(\mathfrak{g})$ canonically isomorphic with the constructed $U(\mathfrak{g})$. In fact, if we use (3.5) with $A = U'(\mathfrak{g})$ and $\pi = \iota'$, we obtain an algebra map $\tilde{\iota}' : U(\mathfrak{g}) \to U'(\mathfrak{g})$. Reversing the roles of $U(\mathfrak{g})$ and $U'(\mathfrak{g})$ yields $\tilde{\iota} : U'(\mathfrak{g}) \to U(\mathfrak{g})$. To see that $\tilde{\iota} \circ \tilde{\iota}' = 1_{U(\mathfrak{g})}$, we use the uniqueness of the extension $\tilde{\pi}$ in (3.5) when $A = U(\mathfrak{g})$ and $\pi = 1$. Similarly $\tilde{\iota}' \circ \tilde{\iota} = 1_{U'(\mathfrak{g})}$.

PROOF. Uniqueness follows from the fact that 1 and $\iota(\mathfrak{g})$ generate $U(\mathfrak{g})$. For existence let $\pi_1 : T(\mathfrak{g}) \to A$ be the extension given by the universal mapping property of $T(\mathfrak{g})$ in Proposition A.14. To obtain $\tilde{\pi}$, we are to show that π_1 annihilates the ideal J in (3.1b). It is enough to consider π_1 on a typical generator of J, where we have

$$\begin{aligned}
\pi_1(\iota X \otimes \iota Y - \iota Y \otimes \iota X - \iota[X, Y]) \\
&= \pi_1(\iota X)\pi_1(\iota Y) - \pi_1(\iota Y)\pi_1(\iota X) - \pi_1(\iota[X, Y]) \\
&= \pi(X)\pi(Y) - \pi(Y)\pi(X) - \pi[X, Y] \\
&= 0.
\end{aligned}$$

Corollary 3.6. Representations of $\mathfrak{g}$ on complex vector spaces stand in one-one correspondence with unital left $U(\mathfrak{g})$ modules (under the correspondence $\pi \to \tilde{\pi}$ of Proposition 3.3).

REMARK. **Unital** means that 1 operates as 1.

PROOF. If π is a representation of $\mathfrak{g}$ on V, we apply Proposition 3.3 to $\pi : \mathfrak{g} \to \operatorname{End}_{\mathbb{C}} V$, and then V becomes a unital left $U(\mathfrak{g})$ module under $uv = \tilde{\pi}(u)v$ for $u \in U(\mathfrak{g})$ and $v \in V$. Conversely if V is a unital left $U(\mathfrak{g})$ module, then V is already a complex vector space with scalar multiplication given by the action of the scalar multiples of 1 in $U(\mathfrak{g})$. If we define $\pi(X)v = (\iota X)v$, then (3.2) implies that π is a representation of $\mathfrak{g}$. The two constructions are inverse to each other since $\tilde{\pi} \circ \iota = \pi$ in Proposition 3.3.

Proposition 3.7. There exists a unique antiautomorphism $u \mapsto u^t$ of $U(\mathfrak{g})$ such that $\iota(X)^t = -\iota(X)$ for all $X \in \mathfrak{g}$.

REMARK. The map $(\cdot)^t$ is called **transpose**.

PROOF. It is unique since $\iota(\mathfrak{g})$ and 1 generate $U(\mathfrak{g})$. Let us prove existence. For each $n \geq 1$, the map

$$(X_1, \dots, X_n) \mapsto (-1)^n X_n \otimes \cdots \otimes X_1$$

is n-multilinear from $\mathfrak{g} \times \cdots \times \mathfrak{g}$ into $T^n(\mathfrak{g})$ and hence extends to a linear map of $T^n(\mathfrak{g})$ into itself with

$$X_1 \otimes \cdots \otimes X_n \mapsto (-1)^n X_n \otimes \cdots \otimes X_1.$$

Taking the direct sum of these maps as n varies, we obtain a linear map $x \mapsto x^t$ of $T(\mathfrak{g})$ into itself sending 1 into 1. It is clear that this map is an antiautomorphism and extends $X \mapsto -X$ in $T^1(\mathfrak{g})$. Composing with passage to the quotient by J, we obtain an antihomomorphism of $T(\mathfrak{g})$ into $U(\mathfrak{g})$. Its kernel is an ideal. To show that the map descends to $U(\mathfrak{g})$, it is enough to show that each generator

$$X \otimes Y - Y \otimes X - [X, Y]$$

maps to 0. But this element maps in $T(\mathfrak{g})$ to itself and then maps to 0 in $U(\mathfrak{g})$. Hence the transpose map descends to $U(\mathfrak{g})$. It is clearly of order two and thus is one-one onto.

The transpose map $u \mapsto u^t$ allows us to regard left $U(\mathfrak{g})$ modules V also as right $U(\mathfrak{g})$ modules, and vice versa: To convert a left $U(\mathfrak{g})$ module into a right $U(\mathfrak{g})$ module, we just define $vu = u^t v$ for $u \in U(\mathfrak{g})$ and $v \in V$. Conversion in the opposite direction is accomplished by $uv = vu^t$.

2. Poincaré-Birkhoff-Witt Theorem

The main theorem about $U(\mathfrak{g})$ gives a basis for $U(\mathfrak{g})$ as a vector space. Let $\{X_i\}_{i\in A}$ be a basis of $\mathfrak{g}$. A set such as A always admits a **simple ordering**, i.e., a partial ordering in which every pair of elements is comparable. In cases of interest, the dimension of $\mathfrak{g}$ is at most countable, and we can think of this ordering as quite elementary. For example, it might be the ordering of the positive integers, or it might be something quite different but still reasonable.

Theorem 3.8 (Poincaré-Birkhoff-Witt). Let $\{X_i\}_{i\in A}$ be a basis of $\mathfrak{g}$, and suppose a simple ordering has been imposed on the index set A. Then the set of all monomials

$$(\iota X_{i_1})^{j_1}\cdots(\iota X_{i_n})^{j_n}$$

with $i_1 < \cdots < i_n$ and with all $j_k \geq 0$, is a basis of $U(\mathfrak{g})$. In particular the canonical map $\iota : \mathfrak{g} \to U(\mathfrak{g})$ is one-one.

REMARKS.

1) If A is finite, say $A = \{1, \ldots, N\}$, the basis consists of all monomials $(\iota X_1)^{j_1}\cdots(\iota X_N)^{j_N}$ with all $j_k \geq 0$.

2) The proof will be preceded by two lemmas, which will essentially establish the spanning. The main step will be to prove the linear independence. For this we have to prove that $U(\mathfrak{g})$ is suitably large. The motivation for carrying out this step comes from assuming the theorem to be true. Then we might as well drop ι from the notation, and monomials $X_{i_1}^{j_1}\cdots X_{i_n}^{j_n}$ with $i_1 < \cdots < i_n$ will form a basis. These same monomials, differently interpreted, are a basis of $S(\mathfrak{g})$. Thus the theorem is asserting a particular vector-space isomorphism $U(\mathfrak{g}) \to S(\mathfrak{g})$. Since $U(\mathfrak{g})$ is naturally a unital left $U(\mathfrak{g})$ module, this isomorphism suggests that $S(\mathfrak{g})$ should be a left unital $U(\mathfrak{g})$ module. By Corollary 3.6 we should look for a natural representation of $\mathfrak{g}$ on $S(\mathfrak{g})$ consistent with left multiplication of $\mathfrak{g}$ on $U(\mathfrak{g})$ and consistent with the particular isomorphism $U(\mathfrak{g}) \to S(\mathfrak{g})$. The proof consists of constructing this representation, and then the linear independence follows easily. Actually the proof will make use of a polynomial algebra, but the polynomial algebra is canonically isomorphic to $S(\mathfrak{g})$ once a basis of $\mathfrak{g}$ has been specified.

Lemma 3.9. Let $Z_1, \ldots, Z_p$ be in $\mathfrak{g}$, and let σ be a permutation of $\{1, \ldots, p\}$. Then

$$(\iota Z_1)\cdots(\iota Z_p) - (\iota Z_{\sigma(1)})\cdots(\iota Z_{\sigma(p)})$$

is in $U_{p-1}(\mathfrak{g})$.

PROOF. Without loss of generality, let σ be the transposition of j with $j+1$. Then the lemma follows from

$$(\iota Z_j)(\iota Z_{j+1}) - (\iota Z_{j+1})(\iota Z_j) = \iota[Z_j, Z_{j+1}]$$

by multiplying through on the left by $(\iota Z_1)\cdots(\iota Z_{j-1})$ and on the right by $(\iota Z_{j+2})\cdots(\iota Z_p)$.

For the remainder of the proof of Theorem 3.8, we shall use the following notation: For $i \in A$, let $Y_i = \iota X_i$. For any tuple $I = (i_1, \ldots, i_p)$ of members of A, we say that I is **increasing** if $i_1 \leq \cdots \leq i_p$. Whether or not I is increasing, we write $Y_I = Y_{i_1} \cdots Y_{i_p}$. Also $i \leq I$ means $i \leq \min\{i_1, \ldots, i_p\}$.

Lemma 3.10. The Y_I, for all increasing tuples from A of length $\leq p$, span $U_p(\mathfrak{g})$.

PROOF. If we use *all* tuples of length $\leq p$, we certainly have a spanning set, since the obvious preimages in $T(\mathfrak{g})$ span $T_p(\mathfrak{g})$. Lemma 3.9 then implies inductively that the increasing tuples suffice.

PROOF OF THEOREM 3.8. Let P be the polynomial algebra over $\mathbb{C}$ in the variables z_i, $i \in A$, and let P_p be the subspace of members of total degree $\leq p$. For a tuple $I = (i_1, \ldots, i_p)$, define $z_I = z_{i_1} \cdots z_{i_p}$ as a member of P_p. We shall construct a representation π of $\mathfrak{g}$ on P such that

$$\pi(X_i)z_I = z_i z_I \qquad \text{if } i \leq I. \tag{3.11}$$

Let us see that the theorem follows from the existence of such a representation. In fact, let us use Corollary 3.6 to regard P as a unital left $U(\mathfrak{g})$ module. Then (3.11) and the identity $\pi(X)v = (\iota X)v$ imply that

$$Y_i z_I = z_i z_I \qquad \text{if } i \leq I.$$

If $i_1 \leq \cdots \leq i_p$, then as a consequence we obtain

$$\begin{aligned}(Y_{i_1} \cdots Y_{i_p})1 &= (Y_{i_1} \cdots Y_{i_{p-1}})Y_{i_p}1 \\ &= (Y_{i_1} \cdots Y_{i_{p-1}})z_{i_p} \\ &= (Y_{i_1} \cdots Y_{i_{p-2}})Y_{i_{p-1}}z_{i_p} \\ &= (Y_{i_1} \cdots Y_{i_{p-2}})(z_{i_{p-1}}z_{i_p}) \\ &= \cdots = z_{i_1} \cdots z_{i_p}.\end{aligned}$$

Thus $\{Y_I 1 \mid I \text{ increasing}\}$ is a linearly independent set in P, and $\{Y_I \mid I \text{ increasing}\}$ must be independent in $U(\mathfrak{g})$. The independence in Theorem 3.8 follows, and the spanning is given in Lemma 3.10.

Thus we have to construct π satisfying (3.11). We shall define linear maps $\pi(X) : P_p \to P_{p+1}$ for X in $\mathfrak{g}$, by induction on p so that they are compatible and satisfy

(A_p) $\pi(X_i)z_I = z_i z_I$ for $i \leq I$ and z_I in P_p

(B_p) $\pi(X_i)z_I - z_i z_I$ is in P_p for all I with z_I in P_p

(C_p) $\pi(X_i)(\pi(X_j)z_J) = \pi(X_j)(\pi(X_i)z_J) + \pi[X_i, X_j]z_J$ for all J with z_J in P_{p-1}.

With $\pi(X)$ defined on P as the union of its definitions on the P_p's, π will be a representation by (C_p) and will satisfy (3.11) by (A_p). Hence we will be done.

For $p = 0$, we define $\pi(X_i)1 = z_i$. Then (A_0) holds, (B_0) is valid, and (C_0) is vacuous.

Inductively assume that $\pi(X)$ has been defined on P_{p-1} for all $X \in \mathfrak{g}$ in such a way that (A_{p-1}), (B_{p-1}), and (C_{p-1}) hold. We are to define $\pi(X_i)z_I$ for each increasing sequence I of p indices in such a way that (A_p), (B_p), and (C_p) hold. If $i \leq I$, we make the definition according to (A_p). Otherwise we can write in obvious notation $I = (j, J)$ with $j < i$, $j \leq J$, $|J| = p - 1$. We are forced to define

$$
\begin{aligned}
\pi(X_i)z_I &= \pi(X_i)(z_j z_J) \\
&= \pi(X_i)\pi(X_j)z_J && \text{since } \pi(X_j)z_J \text{ is already defined by } (A_{p-1}) \\
&= \pi(X_j)\pi(X_i)z_J + \pi[X_i, X_j]z_J && \text{by } (C_p) \\
&= \pi(X_j)(z_i z_J + w) + \pi[X_i, X_j]z_J && \text{with } w \text{ in } P_{p-1} \text{ by } (B_{p-1}) \\
&= z_j z_i z_J + \pi(X_j)w + \pi[X_i, X_j]z_J && \text{by } (A_p) \\
&= z_i z_I + \pi(X_j)w + \pi[X_i, X_j]z_J.
\end{aligned}
$$

We make this definition, and then (B_p) holds. Therefore $\pi(X_i)z_I$ has now been defined in all cases on P_p, and we have to show that (C_p) holds.

Our construction above was made so that (C_p) holds if $j < i$, $j \leq J$, $|J| = p - 1$. Since $[X_j, X_i] = -[X_i, X_j]$, it holds also if $i < j$, $i \leq J$, $|J| = p-1$. Also (C_p) is trivial if $i = j$. Thus it holds whenever $i \leq J$ or $j \leq J$. So we may assume that $J = (k, K)$, where $k \leq K$, $k < i$, $k < j$, $|K| = p - 2$. We know that

$$
\begin{aligned}
\pi(X_j)z_J &= \pi(X_j)z_k z_K \\
&= \pi(X_j)\pi(X_k)z_K \\
&= \pi(X_k)\pi(X_j)z_K + \pi[X_j, X_k]z_K && \text{by } (C_{p-1}) \\
&= \pi(X_k)(z_j z_K + w) + \pi[X_j, X_k]z_K
\end{aligned}
\tag{3.12}
$$

for a certain element w in P_{p-2} given by (B_{p-2}), which is assumed valid since $(B_{p-2}) \subseteq (B_{p-1})$. We apply $\pi(X_i)$ to both sides of this equation, calling the three terms on the right T_1, T_2, and T_3. We can use what we already know for (C_p) to handle $\pi(X_i)$ of T_1 because $k \leq (j, K)$, and we can use (C_{p-1}) with $\pi(X_i)$ of T_2 and T_3. Reassembling T_1 and T_2 as in line (3.12), we conclude that we can use known cases of (C_p) with the sum $\pi(X_i)\pi(X_k)\pi(X_j)z_K$, and we can use (C_{p-1}) with $\pi(X_i)$ of T_3. Thus we have

$$\begin{aligned}
\pi(X_i)\pi(X_j)z_J &= \pi(X_i)\pi(X_k)\pi(X_j)z_K + \pi(X_i)\pi[X_j, X_k]z_K \qquad \text{from (3.12)}\\
&= \pi(X_k)\pi(X_i)\pi(X_j)z_K + \pi[X_i, X_k]\pi(X_j)z_K\\
&\quad + \pi[X_j, X_k]\pi(X_i)z_K + \pi[X_i, [X_j, X_k]]z_K\\
&\qquad\qquad \text{by known cases of } (C_p)\\
&= T_1' + T_2' + T_3' + T_4'.
\end{aligned}$$

Interchanging i and j and subtracting, we see that the terms of type T_2' and T_3' cancel and that we get

$$\begin{aligned}
&\pi(X_i)\pi(X_j)z_J - \pi(X_j)\pi(X_i)z_J\\
&\quad = \pi(X_k)\{\pi(X_i)\pi(X_j)z_K - \pi(X_j)\pi(X_i)z_K\}\\
&\qquad + \{\pi[X_i, [X_j, X_k]] - \pi[X_j, [X_i, X_k]]\}z_K\\
&\quad = \pi(X_k)\pi[X_i, X_j]z_K + \pi[[X_i, X_j], X_k]z_K \qquad \text{by } (C_{p-1}) \text{ and Jacobi}\\
&\quad = \pi[X_i, X_j]\pi(X_k)z_K \qquad \text{by } (C_{p-1})\\
&\quad = \pi[X_i, X_j]z_k z_K\\
&\quad = \pi[X_i, X_j]z_J.
\end{aligned}$$

We have obtained (C_p) in the remaining case, and the proof of Theorem 3.8 is complete.

Now that ι is known to be one-one, there is no danger in dropping it from the notation. We shall freely use Corollary 3.6, identifying representations of $\mathfrak{g}$ with unital left $U(\mathfrak{g})$ modules. Moreover we shall feel free either to drop the name of a representation from the notation (to emphasize the module structure) or to include it even when the argument is in $U(\mathfrak{g})$ (to emphasize the representation structure).

The Poincaré-Birkhoff-Witt Theorem appears in a number of guises. Here is one such.

Corollary 3.13. If $\mathfrak{h}$ is a Lie subalgebra of $\mathfrak{g}$, then the associative subalgebra of $U(\mathfrak{g})$ generated by 1 and $\mathfrak{h}$ is canonically isomorphic to $U(\mathfrak{h})$.

PROOF. If $\rho : \mathfrak{h} \to \mathfrak{g}$ denotes inclusion, then ρ yields an inclusion (also denoted ρ) of $\mathfrak{h}$ into $U(\mathfrak{g})$ such that $\rho(X)\rho(Y) - \rho(Y)\rho(X) = \rho[X, Y]$ for X and Y in $\mathfrak{h}$. By the universal mapping property of $U(\mathfrak{h})$, we obtain a corresponding algebra map $\tilde{\rho} : U(\mathfrak{h}) \to U(\mathfrak{g})$ with $\tilde{\rho}(1) = 1$. The image of $\tilde{\rho}$ is certainly the subalgebra of $U(\mathfrak{g})$ generated by 1 and $\rho(\mathfrak{h})$. Theorem 3.8 says that monomials in an ordered basis of $\mathfrak{h}$ span $U(\mathfrak{h})$, and a second application of the theorem says that these monomials in $U(\mathfrak{g})$ are linearly independent. Thus $\tilde{\rho}$ is one-one and the corollary follows.

If $\mathfrak{g}$ happens to be the vector-space direct sum of two Lie subalgebras $\mathfrak{a}$ and $\mathfrak{b}$, then it follows that we have a vector-space isomorphism

$$U(\mathfrak{g}) \cong U(\mathfrak{a}) \otimes_{\mathbb{C}} U(\mathfrak{b}). \tag{3.14}$$

Namely we obtain a linear map from right to left from the inclusions in Corollary 3.13. To see that the map is an isomorphism, we apply Theorem 3.8 to a basis of $\mathfrak{a}$ followed by a basis of $\mathfrak{b}$. The monomials in the separate bases are identified within $U(\mathfrak{g})$ as bases for $U(\mathfrak{a})$ and $U(\mathfrak{b})$, respectively, by Corollary 3.13, while the joined-together bases give both a basis of the tensor product and a basis of $U(\mathfrak{g})$, again by Theorem 3.8. Thus our map sends basis to basis and is an isomorphism.

3. Associated Graded Algebra

If A is a complex associative algebra with identity and if A is filtered in the sense of Appendix A, say as $A = \cup_{n=0}^{\infty} A_n$, then Appendix A shows how to define the associated graded algebra $\operatorname{gr} A = \bigoplus_{n=0}^{\infty}(A_n/A_{n-1})$, where $A_{-1} = 0$. In this section we shall compute $\operatorname{gr} U(\mathfrak{g})$, showing that it is canonically isomorphic with the symmetric algebra $S(\mathfrak{g})$. Then we shall derive some consequences of this isomorphism.

The idea is to use the Poincaré-Birkhoff-Witt Theorem. The theorem implies that a basis of $U_n(\mathfrak{g})/U_{n-1}(\mathfrak{g})$ is all monomial cosets

$$X_{i_1}^{j_1} \cdots X_{i_k}^{j_k} + U_{n-1}(\mathfrak{g})$$

for which the indices have $i_1 < \cdots < i_k$ and the sum of the exponents is exactly n. The monomials $X_{i_1}^{j_1} \cdots X_{i_k}^{j_k}$, interpreted as in $S(\mathfrak{g})$, are a basis of $S^n(\mathfrak{g})$, and the linear map that carries basis to basis ought to be the desired isomorphism. In fact, this statement is true, but this approach does not conveniently show that the isomorphism is independent of basis. We shall therefore proceed somewhat differently.

We shall construct the map in the opposite direction without using the Poincaré-Birkhoff-Witt Theorem, appeal to the theorem to show that we have an isomorphism, and then compute what the map is in terms of a basis. Let $T_n(\mathfrak{g}) = \bigoplus_{k=0}^n T^n(\mathfrak{g})$ be the nth member of the usual filtration of $T(\mathfrak{g})$. We have defined $U_n(\mathfrak{g})$ to be the image in $U(\mathfrak{g})$ of $T_n(\mathfrak{g})$ under the passage $T(\mathfrak{g}) \to T(\mathfrak{g})/J$. Thus we can form the composition

$$T_n(\mathfrak{g}) \to (T_n(\mathfrak{g}) + J)/J = U_n(\mathfrak{g}) \to U_n(\mathfrak{g})/U_{n-1}(\mathfrak{g}).$$

This composition is onto and carries $T_{n-1}(\mathfrak{g})$ to 0. Since $T^n(\mathfrak{g})$ is a vector-space complement to $T_{n-1}(\mathfrak{g})$ in $T_n(\mathfrak{g})$, we obtain an onto linear map

$$T^n(\mathfrak{g}) \to U_n(\mathfrak{g})/U_{n-1}(\mathfrak{g}).$$

Taking the direct sum over n gives an onto linear map

$$\tilde{\psi} : T(\mathfrak{g}) \to \operatorname{gr} U(\mathfrak{g})$$

that respects the grading.

Appendix A uses the notation I for the two-sided ideal in $T(\mathfrak{g})$ such that $S(\mathfrak{g}) = T(\mathfrak{g})/I$:

$$I = \begin{pmatrix} \text{two-sided ideal generated by all} \\ X \otimes Y - Y \otimes X \text{ with } X \text{ and } Y \\ \text{in } T^1(\mathfrak{g}) \end{pmatrix}. \tag{3.15}$$

Proposition 3.16. The linear map $\tilde{\psi} : T(\mathfrak{g}) \to \operatorname{gr} U(\mathfrak{g})$ respects multiplication and annihilates the defining ideal I for $S(\mathfrak{g})$. Therefore $\tilde{\psi}$ descends to an algebra homomorphism

$$\psi : S(\mathfrak{g}) \to \operatorname{gr} U(\mathfrak{g}) \tag{3.17}$$

that respects the grading. This homomorphism is an isomorphism.

PROOF. Let x be in $T^r(\mathfrak{g})$ and let y be in $T^s(\mathfrak{g})$. Then $x + J$ is in $U_r(\mathfrak{g})$, and we may regard $\tilde{\psi}(x)$ as the coset $x + T_{r-1}(\mathfrak{g}) + J$ in $U_r(\mathfrak{g})/U_{r-1}(\mathfrak{g})$, with 0 in all other coordinates of $\operatorname{gr} U(\mathfrak{g})$ since x is homogeneous. Arguing in a similar fashion with y and xy, we obtain

$$\tilde{\psi}(x) = x + T_{r-1}(\mathfrak{g}) + J, \qquad \tilde{\psi}(y) = y + T_{s-1}(\mathfrak{g}) + J,$$
$$\text{and} \qquad \tilde{\psi}(xy) = xy + T_{r+s-1}(\mathfrak{g}) + J.$$

Since J is an ideal, $\tilde{\psi}(x)\tilde{\psi}(y) = \tilde{\psi}(xy)$. General members x and y of $T(\mathfrak{g})$ are sums of homogeneous elements, and hence $\tilde{\psi}$ respects multiplication.

Consequently $\ker \tilde{\psi}$ is a two-sided ideal. To show that $\ker \tilde{\psi} \supseteq I$, it is enough to show that $\ker \tilde{\psi}$ contains all generators $X \otimes Y - Y \otimes X$. We have

$$\begin{aligned}\tilde{\psi}(X \otimes Y - Y \otimes X) &= X \otimes Y - Y \otimes X + T_1(\mathfrak{g}) + J \\ &= [X, Y] + T_1(\mathfrak{g}) + J \\ &= T_1(\mathfrak{g}) + J,\end{aligned}$$

and thus $\tilde{\psi}$ maps the generator to 0. Hence $\tilde{\psi}$ descends to a homomorphism ψ as in (3.17).

Now let $\{X_i\}$ be an ordered basis of $\mathfrak{g}$. The monomials $X_{i_1}^{j_1} \cdots X_{i_k}^{j_k}$ in $S(\mathfrak{g})$ with $i_1 < \cdots < i_k$ and with $\sum_m j_m = n$ form a basis of $S^n(\mathfrak{g})$. Let us follow the effect of (3.17) on such a monomial. A preimage of this monomial in $T^n(\mathfrak{g})$ is the element

$$X_{i_1} \otimes \cdots \otimes X_{i_1} \otimes \cdots \otimes X_{i_k} \otimes \cdots \otimes X_{i_k},$$

in which there are j_m factors of X_{i_m} for $1 \le m \le k$. This element maps to the monomial in $U_n(\mathfrak{g})$ that we have denoted $X_{i_1}^{j_1} \cdots X_{i_k}^{j_k}$, and then we pass to the quotient $U_n(\mathfrak{g})/U_{n-1}(\mathfrak{g})$. Theorem 3.8 shows that such monomials modulo $U_{n-1}(\mathfrak{g})$ form a basis of $U_n(\mathfrak{g})/U_{n-1}(\mathfrak{g})$. Consequently (3.17) is an isomorphism.

Inspecting the proof of Proposition 3.16, we see that if $i_1 < \cdots < i_k$ and $\sum_m j_m = n$, then

$$\psi(X_{i_1}^{j_1} \cdots X_{i_k}^{j_k}) = X_{i_1}^{j_1} \cdots X_{i_k}^{j_k} + U_{n-1}(\mathfrak{g}). \tag{3.18a}$$

Hence

$$\psi^{-1}(X_{i_1}^{j_1} \cdots X_{i_k}^{j_k} + U_{n-1}(\mathfrak{g})) = X_{i_1}^{j_1} \cdots X_{i_k}^{j_k}, \tag{3.18b}$$

as asserted in the second paragraph of this section. Note that the restriction $i_1 < \cdots < i_k$ can be dropped in (3.18) as a consequence of Lemma 3.9.

Corollary 3.19. Let W be a subspace of $T^n(\mathfrak{g})$, and suppose that the quotient map $T^n(\mathfrak{g}) \to S^n(\mathfrak{g})$ sends W isomorphically onto $S^n(\mathfrak{g})$. Then the image of W in $U_n(\mathfrak{g})$ is a vector-space complement to $U_{n-1}(\mathfrak{g})$.

PROOF. Consider the diagram

$$\begin{array}{ccc} T^n(\mathfrak{g}) & \longrightarrow & U_n(\mathfrak{g}) \\ \downarrow & & \downarrow \\ S^n(\mathfrak{g}) & \xrightarrow{\psi} & U_n(\mathfrak{g})/U_{n-1}(\mathfrak{g}) \end{array}$$

The fact that this diagram is commutative is equivalent with the conclusion in Proposition 3.16 that $\tilde{\psi} : T^n(\mathfrak{g}) \to U_n(\mathfrak{g})/U_{n-1}(\mathfrak{g})$ descends to a map $\psi : S^n(\mathfrak{g}) \to U_n(\mathfrak{g})/U_{n-1}(\mathfrak{g})$. The proposition says that ψ on the bottom of the diagram is an isomorphism, and the hypothesis is that the map on the left, when restricted to W, is an isomorphism onto $S^n(\mathfrak{g})$. Therefore the composition of the map on the top followed by the map on the right is an isomorphism when restricted to W, and the corollary follows.

We apply Corollary 3.19 to the space $\tilde{S}^n(\mathfrak{g})$ of symmetrized tensors within $T^n(\mathfrak{g})$. As in §A.2, $\tilde{S}^n(\mathfrak{g})$ is the linear span, for all n-tuples $X_1, \ldots, X_n$ from $\mathfrak{g}$, of the elements

$$\frac{1}{n!} \sum_{\tau \in \mathfrak{S}_n} X_{\tau(1)} \cdots X_{\tau(n)},$$

where $\mathfrak{S}_n$ is the symmetric group on n letters. According to Proposition A.23, we have a direct sum decomposition

$$T^n(\mathfrak{g}) = \tilde{S}^n(\mathfrak{g}) \oplus (T^n(\mathfrak{g}) \cap I). \tag{3.20}$$

We shall use this decomposition to investigate a map known as "symmetrization."

For $n \geq 1$, define a symmetric n-multilinear map

$$\sigma_n : \mathfrak{g} \times \cdots \times \mathfrak{g} \to U(\mathfrak{g})$$

by

$$\sigma_n(X_1, \ldots, X_n) = \frac{1}{n!} \sum_{\tau \in \mathfrak{S}_n} X_{\tau(1)} \cdots X_{\tau(n)}.$$

By Proposition A.18a we obtain a corresponding linear map, also denoted σ_n, from $S^n(\mathfrak{g})$ into $U(\mathfrak{g})$. The image of $S^n(\mathfrak{g})$ in $U(\mathfrak{g})$ is clearly the same as the image of the subspace $\tilde{S}^n(\mathfrak{g})$ of $T^n(\mathfrak{g})$ in $U_n(\mathfrak{g})$. By (3.20) and Corollary 3.19, σ_n is one-one from $S^n(\mathfrak{g})$ onto a vector-space complement to $U_{n-1}(\mathfrak{g})$ in $U_n(\mathfrak{g})$, i.e.,

$$U_n(\mathfrak{g}) = \sigma_n(S^n(\mathfrak{g})) \oplus U_{n-1}(\mathfrak{g}). \tag{3.21}$$

The direct sum of the maps σ_n for $n \geq 0$ (with $\sigma_0(1) = 1$) is a linear map $\sigma : S(\mathfrak{g}) \to U(\mathfrak{g})$ such that

$$\sigma(X_1 \cdots X_n) = \frac{1}{n!} \sum_{\tau \in \mathfrak{S}_n} X_{\tau(1)} \cdots X_{\tau(n)}.$$

The map σ is called **symmetrization**.

Lemma 3.22. The symmetrization map $\sigma : S(\mathfrak{g}) \to U(\mathfrak{g})$ has associated graded map $\psi : S(\mathfrak{g}) \to \operatorname{gr} U(\mathfrak{g})$, with ψ as in (3.17).

REMARK. See §A.4 for "associated graded map."

PROOF. Let $\{X_i\}$ be a basis of $\mathfrak{g}$, and let $X_{i_1}^{j_1} \cdots X_{i_k}^{j_k}$, with $\sum_m j_m = n$, be a basis vector of $S^n(\mathfrak{g})$. Under σ, this vector is sent to a symmetrized sum, but each term of the sum is congruent mod$U_{n-1}(\mathfrak{g})$ to $(n!)^{-1} X_{i_1}^{j_1} \cdots X_{i_k}^{j_k}$, by Lemma 3.9. Hence the image of $X_{i_1}^{j_1} \cdots X_{i_k}^{j_k}$ under the associated graded map is

$$= X_{i_1}^{j_1} \cdots X_{i_k}^{j_k} + U_{n-1}(\mathfrak{g}) = \psi(X_{i_1}^{j_1} \cdots X_{i_k}^{j_k}),$$

as asserted.

Proposition 3.23. Symmetrization σ is a vector-space isomorphism of $S(\mathfrak{g})$ onto $U(\mathfrak{g})$ satisfying

$$U_n(\mathfrak{g}) = \sigma(S^n(\mathfrak{g})) \oplus U_{n-1}(\mathfrak{g}). \tag{3.24}$$

PROOF. Formula (3.24) is a restatement of (3.21), and the other conclusion follows by combining Lemma 3.22 and Proposition A.37.

The canonical decomposition of $U(\mathfrak{g})$ from $\mathfrak{g} = \mathfrak{a} \oplus \mathfrak{b}$ when $\mathfrak{a}$ and $\mathfrak{b}$ are merely vector spaces is given in the following proposition.

Proposition 3.25. Suppose $\mathfrak{g} = \mathfrak{a} \oplus \mathfrak{b}$ and suppose $\mathfrak{a}$ and $\mathfrak{b}$ are subspaces of $\mathfrak{g}$. Then the mapping $a \otimes b \mapsto \sigma(a)\sigma(b)$ of $S(\mathfrak{a}) \otimes_{\mathbb{C}} S(\mathfrak{b})$ into $U(\mathfrak{g})$ is a vector-space isomorphism onto.

PROOF. The vector space $S(\mathfrak{a}) \otimes_{\mathbb{C}} S(\mathfrak{b})$ is graded consistently for the given mapping, the nth space of the grading being $\bigoplus_{p=0}^n S^p(\mathfrak{a}) \otimes_{\mathbb{C}} S^{n-p}(\mathfrak{b})$. The given mapping operates on an element of this space by

$$\sum_{p=0}^n a_p \otimes b_{n-p} \mapsto \sum_{p=0}^n \sigma(a_p)\sigma(b_{n-p}),$$

and the image of this under the associated graded map is

$$= \sum_{p=0}^{n} \sigma(a_p)\sigma(b_{n-p}) + U_{n-1}(\mathfrak{g}).$$

In turn this is

$$= \sigma\Big(\sum_{p=0}^{n} a_p \otimes b_{n-p}\Big) + U_{n-1}(\mathfrak{g})$$

by Lemma 3.9. In other words the associated graded map is just the same as for σ. Hence the result follows by combining Propositions 3.23 and A.37.

Corollary 3.26. Suppose that $\mathfrak{g} = \mathfrak{k} \oplus \mathfrak{p}$ and that $\mathfrak{k}$ is a Lie subalgebra of $\mathfrak{g}$. Then the mapping $(u, p) \mapsto u\sigma(p)$ of $U(\mathfrak{k}) \otimes_{\mathbb{C}} S(\mathfrak{p})$ into $U(\mathfrak{g})$ is a vector-space isomorphism onto.

PROOF. The composition

$$(k, p) \mapsto (\sigma(k), p) \mapsto \sigma(k)\sigma(p),$$

sending

$$S(\mathfrak{k}) \otimes_{\mathbb{C}} S(\mathfrak{p}) \to U(\mathfrak{k}) \otimes_{\mathbb{C}} S(\mathfrak{p}) \to U(\mathfrak{g}),$$

is an isomorphism by Proposition 3.25, and the first map is an isomorphism by Proposition 3.23. Therefore the second map is an isomorphism, and the notation corresponds to the statement of the corollary when we write $u = \sigma(k)$.

4. Free Lie Algebras

Using the Poincaré-Birkhoff-Witt Theorem, we can establish the existence of "free Lie algebras." A **free Lie algebra** on a set X is a pair $(\mathfrak{F}, \iota)$ consisting of a Lie algebra $\mathfrak{F}$ and a function $\iota : X \to \mathfrak{F}$ with the following universal mapping property: Whenever $\mathfrak{l}$ is a complex Lie algebra and $\varphi : X \to \mathfrak{l}$ is a function, there exists a unique Lie algebra homomorphism $\tilde{\varphi}$ such that the diagram

(3.27)

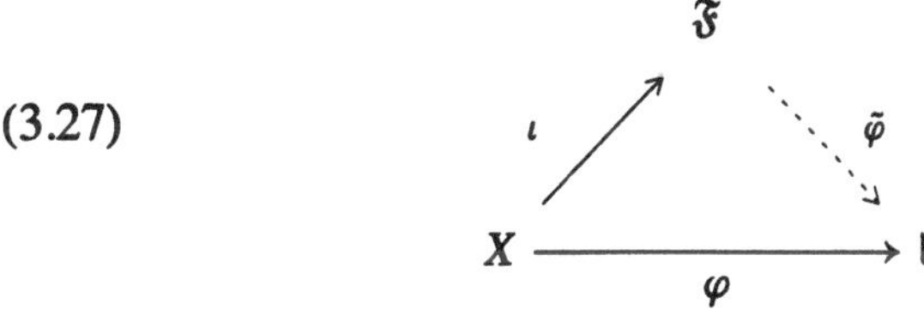

commutes. We regard $\tilde{\varphi}$ as an extension of φ.

Let us construct such a Lie algebra. Let V consist of all formal complex linear combinations of the members of X, so that V can be regarded as a complex vector space with X as basis. We embed V in its tensor algebra via $\iota_V : V \to T(V)$, obtaining $T^1(V) = \iota_V(V)$ as usual. Since $T(V)$ is an associative algebra, we can regard it as a Lie algebra in the manner of Example 2 in §I.1. Let $\mathfrak{F}$ be the Lie subalgebra of $T(V)$ generated by $T^1(V)$.

In the setting of (3.27), we are to construct a Lie algebra homomorphism $\tilde{\varphi}$ so that (3.27) commutes, and we are to show that $\tilde{\varphi}$ is unique. Extend $\varphi : X \to \mathfrak{l}$ to a linear map $\varphi : V \to \mathfrak{l}$, and let $\iota_{\mathfrak{l}} : \mathfrak{l} \to U(\mathfrak{l})$ be the canonical map. The universal mapping property of $T(V)$ allows us in the diagram

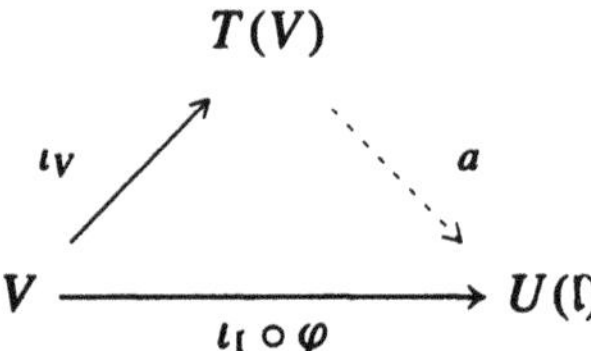

to extend $\iota_{\mathfrak{l}} \circ \varphi$ to an associative algebra homomorphism a with $a(1) = 1$. For $x \in X$, the commutativity of this diagram implies that

$$a(\iota_V(x)) = \iota_{\mathfrak{l}}(\varphi(x)). \tag{3.28}$$

Let us think of a as a Lie algebra homomorphism in (3.28). The right side of (3.28) is in image $\iota_{\mathfrak{l}}$, and it follows that $a(\mathfrak{F}) \subseteq$ image $\iota_{\mathfrak{l}}$.

Now we use the Poincaré-Birkhoff-Witt Theorem, which implies that $\iota_{\mathfrak{l}} : \mathfrak{l} \to$ image $\iota_{\mathfrak{l}}$ is one-one. We write $\iota_{\mathfrak{l}}^{-1}$ for the inverse of this Lie algebra isomorphism, and we put $\tilde{\varphi} = \iota_{\mathfrak{l}}^{-1} \circ a$. Then $\tilde{\varphi}$ is the required Lie algebra homomorphism making (3.27) commute.

To see that $\tilde{\varphi}$ is unique when $\mathfrak{F}$ is defined this way, we observe that (3.27) forces $\tilde{\varphi}(\iota_V(x)) = \varphi(x)$ for all $x \in X$. Since the elements $\iota_V(x)$ generate $\mathfrak{F}$ and since $\tilde{\varphi}$ is a Lie algebra homomorphism, $\tilde{\varphi}$ is completely determined on all of $\mathfrak{F}$. This proves the first statement in the following proposition.

Proposition 3.29. If X is a nonempty set, then there exists a free Lie algebra $\mathfrak{F}$ on X, and the image of X in $\mathfrak{F}$ generates $\mathfrak{F}$. Any two free Lie algebras on X are canonically isomorphic.

REMARK. This result was stated in Chapter II as Proposition 2.96, and the proof was deferred until now.

PROOF. Existence of $\mathfrak{F}$ was proved before the statement of the proposition. We still have to prove that $\mathfrak{F}$ is unique up to canonical isomorphism. Let $(\mathfrak{F}, \iota)$ and $(\mathfrak{F}, \iota')$ be two free Lie algebras on X. We set up the diagram (3.27) with $\mathfrak{l} = \mathfrak{F}'$ and $\varphi = \iota'$ and invoke existence to obtain a Lie algebra homomorphism $\tilde{\iota}' : \mathfrak{F} \to \mathfrak{F}'$. Reversing the roles of $\mathfrak{F}$ and $\mathfrak{F}'$, we obtain a Lie algebra homomorphism $\tilde{\iota} : \mathfrak{F}' \to \mathfrak{F}$. To see that $\tilde{\iota} \circ \tilde{\iota}' = 1_{\mathfrak{F}}$, we set up the diagram (3.27) with $\mathfrak{l} = \mathfrak{F}$ and $\varphi = \iota_X$ to see that $\tilde{\iota} \circ \tilde{\iota}'$ is an extension of ι. By uniqueness of the extension, $\tilde{\iota} \circ \tilde{\iota}' = 1_{\mathfrak{F}}$. Similarly $\tilde{\iota}' \circ \tilde{\iota} = 1_{\mathfrak{F}'}$.

5. Problems

1. For $\mathfrak{g} = \mathfrak{sl}(2, \mathbb{C})$, let Ω be the member of $U(\mathfrak{g})$ given by $\Omega = \frac{1}{2}h^2 + ef + fe$, where h, e, and f are as in (1.5).
   (a) Prove that Ω is in the center of $U(\mathfrak{g})$.
   (b) Let π be a representation of $\mathfrak{sl}(2, \mathbb{C})$ on a complex vector space V, and regard V as a $U(\mathfrak{g})$ module. Show that Ω acts in V by the operator Z of Lemma 1.65.

2. Let $\mathfrak{g}$ be a finite-dimensional complex Lie algebra, and define ad X on $U(\mathfrak{g})$ for $X \in \mathfrak{g}$ by $(\operatorname{ad} X)u = Xu - uX$. Prove that ad is a representation of $\mathfrak{g}$ and that each element of $U(\mathfrak{g})$ lies in a finite-dimensional space invariant under ad $\mathfrak{g}$.

3. Let $U(\mathfrak{g})$ be the universal enveloping algebra of a complex Lie algebra $\mathfrak{g}$. Prove that $U(\mathfrak{g})$ has no zero divisors.

4. (a) Identify a free Lie algebra on a set consisting of one element.
   (b) Prove that a free Lie algebra on a set consisting of two elements is infinite-dimensional.

5. Let $\mathfrak{F}$ be a free Lie algebra on the set $\{X_1, X_2, X_3\}$, and let $\mathfrak{g}$ be the quotient obtained by setting to 0 all brackets involving three or more members of $\mathfrak{F}$.
   (a) Prove that $\dim \mathfrak{g} = 6$ and that $\mathfrak{g}$ is nilpotent but not abelian.
   (b) Define $B(X_i, X_j) = 0$, $B([X_i, X_j], [X_{i'}, X_{j'}] = 0$, and

$$B(X_3, [X_1, X_2]) = B(X_2, [X_3, X_1]) = B(X_1, [X_2, X_3]) = 1.$$

   Prove that B extends to a nondegenerate symmetric invariant bilinear form on $\mathfrak{g}$.

6. Say that a complex Lie algebra $\mathfrak{h}$ is two-step nilpotent if $[\mathfrak{h}, [\mathfrak{h}, \mathfrak{h}]] = 0$. Prove for each integer $n \geq 1$ that that there is a finite-dimensional two-step nilpotent Lie algebra $\mathfrak{g}$ such that every two-step nilpotent Lie algebra of dimension $\leq n$ is isomorphic to a homomorphic image of $\mathfrak{g}$.

Problems 7–9 concern the diagonal mapping for a universal enveloping algebra. Fix a complex Lie algebra $\mathfrak{g}$ and its universal enveloping algebra $U(\mathfrak{g})$.

7. Use the 4-multilinear map $(u_1, u_2, u_3, u_4) \mapsto u_1u_2 \otimes u_3u_4$ of $U(\mathfrak{g}) \times U(\mathfrak{g}) \times U(\mathfrak{g}) \times U(\mathfrak{g})$ into $U(\mathfrak{g}) \otimes_{\mathbb{C}} U(\mathfrak{g})$ to define a multiplication in $U(\mathfrak{g}) \otimes_{\mathbb{C}} U(\mathfrak{g})$. Prove that $U(\mathfrak{g}) \otimes_{\mathbb{C}} U(\mathfrak{g})$ becomes an associative algebra with identity.
8. Prove that there exists a unique associative algebra homomorphism $\Delta : U(\mathfrak{g}) \to U(\mathfrak{g}) \otimes_{\mathbb{C}} U(\mathfrak{g})$ such that $\Delta(X) = X \otimes 1 + 1 \otimes X$ for all $X \in \mathfrak{g}$ and such that $\Delta(1) = 1$.
9. If φ_1 and φ_2 are in the dual space $U(\mathfrak{g})^*$, then $\varphi_1 \otimes \varphi_2$ is well defined as a linear functional on $U(\mathfrak{g}) \otimes_{\mathbb{C}} U(\mathfrak{g})$ since $\mathbb{C} \otimes_{\mathbb{C}} \mathbb{C} \cong \mathbb{C}$ canonically. Define a product $\varphi_1\varphi_2$ in $U(\mathfrak{g})^*$ by

$$(\varphi_1\varphi_2)(u) = (\varphi_1 \otimes \varphi_2)(\Delta(u)),$$

where Δ is as in Problem 8. Prove that this product makes $U(\mathfrak{g})^*$ into a commutative associative algebra (without necessarily an identity).

Problems 10–12 identify $U(\mathfrak{g})$ with an algebra of differential operators. Let G be a Lie group, let $\mathfrak{g}_0$ be the Lie algebra, and let $\mathfrak{g}$ be the complexification of $\mathfrak{g}_0$. For $X \in \mathfrak{g}_0$, let $\tilde{X}$ be the left-invariant vector field on G corresponding to X, regarded as acting in the space $C^\infty(G)$ of all *complex*-valued functions on G. The vector field $\tilde{X}$ is a **left-invariant differential operator** in the sense that it is a member D of $\operatorname{End}_{\mathbb{C}}(C^\infty(G))$ commuting with left translations such that, for each $g \in G$, there is a chart (φ, V) about g, say $\varphi = (x_1, \dots, x_n)$, and there are functions $a_{k_1 \cdots k_n}$ in $C^\infty(V)$ with the property that

$$Df(x) = \sum_{\text{bounded}} a_{k_1 \cdots k_n}(x) \frac{\partial^{k_1 + \cdots + k_n} f}{\partial x_1^{k_1} \cdots \partial x_n^{k_n}}(x)$$

for all $x \in V$ and $f \in C^\infty(G)$. Such operators form a complex subalgebra $D(G)$ of $\operatorname{End}_{\mathbb{C}}(C^\infty(G))$ containing the identity. Moreover, any D of this kind has such an expansion in any chart about x.

10. Prove that the map $X \mapsto \tilde{X}$ extends to an algebra homomorphism of $U(\mathfrak{g})$ into $D(G)$ sending 1 to 1.
11. Prove that the map in Problem 10 is onto.
12. Let $X_1, \dots, X_n$ be a basis of $\mathfrak{g}_0$.
    (a) For each tuple $(i_1, \dots, i_n)$ of integers ≥ 0, prove that there is a function $f \in C^\infty(G)$ with the property that $(\tilde{X}_1)^{j_1} \cdots (\tilde{X}_n)^{j_n} f(1) = 1$ if $j_1 = i_1, \dots, j_n = i_n$, and $= 0$ if not.
    (b) Deduce that the map in Problem 10 is one-one.

CHAPTER IV

Compact Lie Groups

Abstract. This chapter is about structure theory for compact Lie groups, and a certain amount of representation theory is needed for the development. The first section gives examples of group representations and shows how to construct new representations from old ones by using tensor products and the symmetric and exterior algebras.

In the abstract representation theory for compact groups, the basic result is Schur's Lemma, from which the Schur orthogonality relations follow. A deeper result is the Peter-Weyl Theorem, which guarantees a rich supply of irreducible representations. From the Peter-Weyl Theorem it follows that any compact Lie group can be realized as a group of real or complex matrices.

The Lie algebra of a compact Lie group admits an invariant inner product, and consequently such a Lie algebra is reductive. From Chapter I it is known that a reductive Lie algebra is always the direct sum of its center and its commutator subalgebra. In the case of the Lie algebra of a compact connected Lie group, the analytic subgroups corresponding to the center and the commutator subalgebra are closed. Consequently the structure theory of compact connected Lie groups in many respects reduces to the semisimple case.

If T is a maximal torus of a compact connected Lie group G, then each element of G is conjugate to a member of T. It follows that the exponential map for G is onto and that the centralizer of a torus is always connected. The analytically defined Weyl group $W(G, T)$ is the quotient of the normalizer of T by the centralizer of T, and it coincides with the Weyl group of the underlying root system.

Weyl's Theorem says that the fundamental group of a compact semsimple Lie group G is finite. Hence the universal covering group of G is compact.

1. Examples of Representations

The subject of this chapter is structure theory for compact Lie groups, but the structure theory is closely tied with representation theory. In fact, one of several equivalent formulations of the first main structure theorem says that the exponential map for a compact connected Lie group is onto. In our treatment this theorem makes critical use of the fact that any compact Lie group is a matrix group, and this is a theorem of representation theory.

We shall begin with the representation theory, providing some examples and constructions of representations in this section. We are interested in representations of both a Lie group and its Lie algebra; all representations for us will be finite-dimensional.

A **finite-dimensional representation** of a compact group G on a finite-dimensional complex vector space V is a continuous homomorphism Φ of G into $GL_{\mathbb{C}}(V)$. If G is a Lie group, as it always will be in this section, then Φ is automatically smooth (§I.10). The differential at the identity provides us with a representation of the (real) Lie algebra $\mathfrak{g}_0$ of G on the space V.

For any G the **trivial representation** of G on V is the representation Φ of G for which $\Phi(g) = 1$ for all $g \in G$. Sometimes when the term "trivial representation" is used, it is understood that $V = \mathbb{C}$; sometimes the case $V = \mathbb{C}$ is indicated by referring to the "trivial 1-dimensional representation."

Let us now consider specific examples.

EXAMPLES FOR $G = U(n)$ OR $SU(n)$.

1) Let $V = \mathbb{C}^n$, and let G act on $\mathbb{C}^n$ by matrix multiplication, i.e.,

$$\Phi(g)\begin{pmatrix} z_1 \\ \vdots \\ z_n \end{pmatrix} = g\begin{pmatrix} z_1 \\ \vdots \\ z_n \end{pmatrix}.$$

The result is what is called the **standard representation** of G. If on the right side of this equation g is replaced by $(g^t)^{-1} = \bar{g}$, then the result is the **contragredient** or **conjugate** of the standard representation.

2) Let V consist of all polynomials in $z_1, \ldots, z_n, \bar{z}_1, \ldots, \bar{z}_n$ homogeneous of degree N, and let

$$\Phi(g)P(\begin{pmatrix} z_1 \\ \vdots \\ z_n \end{pmatrix}, \begin{pmatrix} \bar{z}_1 \\ \vdots \\ \bar{z}_n \end{pmatrix}) = P(g^{-1}\begin{pmatrix} z_1 \\ \vdots \\ z_n \end{pmatrix}, \bar{g}^{-1}\begin{pmatrix} \bar{z}_1 \\ \vdots \\ \bar{z}_n \end{pmatrix}).$$

The subspace V' of holomorphic polynomials (those with no $\bar{z}$'s) is carried to itself by all $\Phi(g)$, and therefore we call V' an **invariant subspace**. The restriction of the $\Phi(g)$'s to V' is thus itself a representation. When $N = 1$, this representation is essentially the contragredient of the standard representation. When antiholomorphic polynomials are used (those with no z's) and N is taken to be 1, the result is essentially the standard representation itself.

3) Let $V = \bigwedge^k \mathbb{C}^n$. This vector space is discussed in §A.3. A basis over $\mathbb{C}$ of $\bigwedge^k \mathbb{C}^n$ consists of all alternating tensors $\varepsilon_{i_1} \wedge \cdots \wedge \varepsilon_{i_k}$ with $i_1 < \cdots < i_k$, where $\{\varepsilon_j\}_{j=1}^n$ is the standard basis of $\mathbb{C}^n$. If we define

$$\Phi(g)(\varepsilon_{i_1} \wedge \cdots \wedge \varepsilon_{i_k}) = g\varepsilon_{i_1} \wedge \cdots \wedge g\varepsilon_{i_k},$$

then we can see that $\Phi(g)$ extends to a linear map of $\bigwedge^k \mathbb{C}^n$ into itself, and Φ is a representation. What we should do to get a construction that does not use a basis is first to define $\tilde{\Phi}(g)$ on $T^k(\mathbb{C}^n)$ by

$$\tilde{\Phi}(g) = g \otimes \cdots \otimes g$$

as in (A.7). The result is multiplicative in g by (A.8), and the continuity follows by examining the effect on a basis. Hence we have a representation of G on $T^k(\mathbb{C}^n)$. Next we easily check that each $\tilde{\Phi}(g)$ carries $T^k(\mathbb{C}^n) \cap I'$ to itself, where I' is the defining ideal (A.24b) for the exterior algebra. Consequently $\tilde{\Phi}(g)$ descends to a linear transformation $\Phi(g)$ from $\bigwedge^k \mathbb{C}^n$ to itself, and Φ is a representation on $\bigwedge^k \mathbb{C}^n$.

4) For $G = SU(2)$, let V be the space of homogeneous holomorphic polynomials of degree N in z_1 and z_2, let Φ be the representation as in Example 2, and let V' be the space of all holomorphic polynomials in z of degree N with

$$\Phi'\begin{pmatrix} \alpha & \beta \\ -\bar{\beta} & \bar{\alpha} \end{pmatrix} Q(z) = (\bar{\beta}z + \alpha)^N Q\left(\frac{\bar{\alpha}z - \beta}{\bar{\beta}z + \alpha}\right).$$

Define $E : V \to V'$ by $(EP)(z) = P\begin{pmatrix} z \\ 1 \end{pmatrix}$. Then E is an invertible linear mapping and satisfies $E\Phi(g) = \Phi'(g)E$ for all g, and we say that E exhibits Φ and Φ' as **equivalent** (i.e., isomorphic).

EXAMPLES FOR $G = O(n)$ OR $SO(n)$.

1) Let $V = \mathbb{C}^n$, and let G act on $\mathbb{C}^n$ by matrix multiplication, i.e.,

$$\Phi(g)\begin{pmatrix} z_1 \\ \vdots \\ z_n \end{pmatrix} = g\begin{pmatrix} z_1 \\ \vdots \\ z_n \end{pmatrix}.$$

The result is what is called the **standard representation** of G.

2) Let V consist of all polynomials in $x_1, \ldots, x_n$ homogeneous of degree N, and let

$$\Phi(g)P(\begin{pmatrix} x_1 \\ \vdots \\ x_n \end{pmatrix}) = P(g^{-1}\begin{pmatrix} x_1 \\ \vdots \\ x_n \end{pmatrix}).$$

Then Φ is a representation. When we want to emphasize the degree, let us write Φ_N and V_N. Define the Laplacian operator by

$$\Delta = \frac{\partial^2}{\partial x_1^2} + \cdots + \frac{\partial^2}{\partial x_n^2}.$$

This carries V_N to V_{N-2} and satisfies $\Delta \Phi_N(g) = \Phi_{N-2}(g)\Delta$. This commutativity property implies that the kernel of Δ is an invariant subspace of V_N, the space of homogeneous harmonic polynomials of degree N.

3) Let $V = \bigwedge^k \mathbb{C}^n$. For $g \in G$, if we define

$$\Phi(g)(\varepsilon_{i_1} \wedge \cdots \wedge \varepsilon_{i_k}) = g\varepsilon_{i_1} \wedge \cdots \wedge g\varepsilon_{i_k},$$

then we can see that $\Phi(g)$ extends to a linear map of $\bigwedge^k \mathbb{C}^n$ into itself, and Φ is a representation. Unlike the case with $G = SU(n)$, the representations in $\bigwedge^k \mathbb{C}^n$ and $\bigwedge^{n-k} \mathbb{C}^n$ are equivalent when $G = SO(n)$.

Now let us consider matters more generally. Fix a compact group G. If Φ is a finite-dimensional representation of G on V, then the **contragredient** Φ^c takes place on the dual space V^* and is given by

$$(4.1) \qquad \langle \Phi^c(g)v^*, v\rangle = \langle v^*, \Phi(g)^{-1}v\rangle \qquad \text{for } v^* \in V^* \text{ and } v \in V.$$

Here $\langle \cdot, \cdot \rangle$ is the natural pairing of V^* and V.

If Φ_1 and Φ_2 are finite-dimensional representations on V_1 and V_2, then their tensor product is the representation $\Phi_1 \otimes \Phi_2$ of G given on $V_1 \otimes_{\mathbb{C}} V_2$ by

$$(4.2) \qquad (\Phi_1 \otimes \Phi_2)(g) = \Phi_1(g) \otimes \Phi_2(g).$$

Then $(\Phi_1 \otimes \Phi_2)(g)$ is multiplicative in g by (A.8), and the continuity follows by examining the effect on a basis. Hence $\Phi_1 \otimes \Phi_2$ is a representation.

If Φ is a finite-dimensional representation on V, we can define representations on the spaces $S^k(V)$ and $\bigwedge^k(V)$ of symmetric and alternating tensors for every $k \geq 0$. The argument is just as in Example 3 for $U(n)$ and $SU(n)$ above. In the case of $S^k(V)$, we start with the representation on the k-fold tensor product $T^k(V)$. If I is the defining ideal (A.16b) for $S(V)$, the representation on $T^k(V)$ will descend to $S^k(V)$ if it is shown that $T^k(V) \cap I$ is an invariant subspace. The space $T^k(V) \cap I$ is spanned by vectors

$$x \otimes u \otimes v \otimes y - x \otimes v \otimes u \otimes y$$

with $x \in T^r(V)$, u and v in $T^1(V)$, and $y \in T^s(V)$, where $r+2+s=k$. When we apply g to this element, we get the element

$$gx \otimes gu \otimes gv \otimes gy - gx \otimes gv \otimes gu \otimes gy,$$

which another element of the spanning set for $T^k(V) \cap I$. Hence the representation on $T^k(V)$ descends to $S^k(V)$. To get a representation on $\bigwedge^k(V)$, we argue similarly. The descent from $T^k(V)$ to $\bigwedge^k(V)$ is possible since $T^k(V) \cap I'$, with I' as in (A.24b), is spanned by elements

$$x \otimes v \otimes v \otimes y$$

with $x \in T^r(V)$, $v \in T^1(V)$, and $y \in T^s(V)$, and since g of this element is another element of this form.

The motivation for the definitions of Lie algebra representations comes from thinking of G as a closed linear group and differentiating the Lie group formulas. For example, if Φ_1 and Φ_2 are finite-dimensional representations on V_1 and V_2, then we have

$$(\Phi_1 \otimes \Phi_2)(g)(v_1 \otimes v_2) = \Phi_1(g)v_1 \otimes \Phi_2(g)v_2.$$

If $c(t)$ is a curve in G with $c(0) = 1$ and $c'(0) = X$, then the product rule for differentiation gives

$$d(\Phi_1 \otimes \Phi_2)(X)(v_1 \otimes v_2) = d\Phi_1(X)v_1 \otimes v_2 + v_1 \otimes d\Phi_2(X)v_2$$

for X in the real Lie algebra $\mathfrak{g}_0$ of G. It will be convenient to pass to the complexification of $\mathfrak{g}_0$, thereby obtaining a (complex-linear) representation of $(\mathfrak{g}_0)^{\mathbb{C}}$ on $V_1 \otimes_{\mathbb{C}} V_2$. Once we have formulas for representations of this particular kind of complex Lie algebra, we may as well make our definitions for all complex Lie algebras.

Thus let $\mathfrak{g}$ be a finite-dimensional complex Lie algebra. If φ_1 and φ_2 are representations of $\mathfrak{g}$ on vector spaces V_1 and V_2, then we define a representation $\varphi_1 \otimes \varphi_2$ on $V_1 \otimes_{\mathbb{C}} V_2$ by

$$(\varphi_1 \otimes \varphi_2)X = \varphi_1(X) \otimes 1 + 1 \otimes \varphi_2(X). \tag{4.3}$$

A little calculation using (A.8) shows that indeed $\varphi_1 \otimes \varphi_2$ is a representation.

In a similar fashion we can define the tensor product of n representations of $\mathfrak{g}$. Also in a similar fashion if φ is a representation of $\mathfrak{g}$ on V, the contragredient is the representation φ^c of $\mathfrak{g}$ on V^* given by

$$\langle \varphi^c(X)v^*, v\rangle = -\langle v^*, \varphi(X)v\rangle. \tag{4.4}$$

If φ is a representation of $\mathfrak{g}$ on V, we can construct corresponding representations on $S^k(V)$ and $\bigwedge^k(V)$. We start with the representations of $\mathfrak{g}$ on $T^k(E)$ and show that X of a member of $T^k(V) \cap I$ or $T^k(V) \cap I'$ is a member of the same space. Then the representation of $\mathfrak{g}$ on $T^k(V)$ descends to $S^k(V)$ or $\bigwedge^k(V)$.

2. Abstract Representation Theory

Let us be more systematic about some of the definitions in §1. A **finite-dimensional representation** of a topological group G is a continuous homomorphism Φ of G into the group $GL_{\mathbb{C}}(V)$ of invertible linear transformations on a finite-dimensional complex vector space V. The continuity condition means that in any basis of V the matrix entries of $\Phi(g)$ are continuous for $g \in G$. It is equivalent to say that $g \mapsto \Phi(g)v$ is a continuous function from G into V for each v in V.

An equivalent definition of finite-dimensional representation is that Φ is a continuous group action of G on a finite-dimensional complex vector space V by linear transformations. In this case the assertion about continuity is that the map $G \times V \to V$ is continuous jointly, rather than continuous only as a function of the first variable. To deduce the joint continuity from continuity in the first variable, it is enough to verify continuity of $G \times V \to V$ at $g = 1$ and $v = 0$. Let $\dim_{\mathbb{C}} V = n$. We fix a basis $v_1, \ldots, v_n$ and recall that the map $\{c_i\}_{i=1}^n \mapsto \sum_{i=1}^n c_i v_i$ is a homeomorphism of $\mathbb{C}^n$ onto V. Put $\| \sum_{i=1}^n c_i v_i \| = \left(\sum_{i=1}^n c_i^2\right)^{1/2}$. Given $\epsilon > 0$, choose for each i between 1 and n a neighborhood U_i of 1 in G such that $\|\Phi(g)v_i - v_i\| < 1$ for $g \in U_i$. If g is in $\bigcap_{i=1}^n U_i$ and if $v = \sum_i c_i v_i$ has $\|v\| < \varepsilon$, then

$$\begin{aligned}
\|\Phi(g)v\| &\le \|\Phi(g)\left(\sum c_i v_i\right) - \left(\sum c_i v_i\right)\| + \|v\| \\
&\le \sum |c_i| \|\Phi(g)v_i - v_i\| + \|v\| \\
&\le \left(\sum |c_i|^2\right)^{1/2} n^{1/2} + \|v\| \qquad \text{by the Schwarz inequality} \\
&\le (n^{1/2} + 1)\varepsilon.
\end{aligned}$$

An **invariant subspace** for such a Φ is a vector subspace U such that $\Phi(g)(U) \subseteq U$ for all $g \in G$. A representation on a nonzero vector space V is **irreducible** if it has no invariant subspaces other than 0 and V.

A representation Φ on the finite-dimensional complex vector space V is called **unitary** if a Hermitian inner product has been specified for V and if each $\Phi(g)$ is unitary relative to that inner product (i.e., has $\Phi(g)^*\Phi(g) =$

1 for all $g \in G$). For a unitary representation the orthogonal complement $U^\perp$ of an invariant subspace is an invariant subspace because

(4.5)
$$\langle \Phi(g)u^\perp, u\rangle = \langle u^\perp, \Phi(g)^{-1}u\rangle \in \langle u^\perp, U\rangle = 0 \qquad \text{for } u^\perp \in U^\perp,\ u \in U.$$

Two representations of G, Φ on V and Φ' on V' are **equivalent** if there is a linear invertible $E : V \to V'$ such that $\Phi'(g)E = E\Phi(g)$ for all $g \in G$.

Now let us suppose that the topological group G is compact. One of the critical properties of such a group for representation theory is that it has a left **Haar measure**, i.e., a nonzero regular Borel measure that is invariant under left translation. We shall take for granted this existence; it may be proved by techniques of functional analysis or, in the case of compact Lie groups, by an argument using differential forms. Let μ_l be a left Haar measure. Then G possesses also a right Haar measure in the obvious sense, for example $\mu_r(E) = \mu_l(E^{-1})$, where E^{-1} denotes the set of inverses of elements of the set E. Let A be a set in the σ-algebra generated by the compact subsets of G that are countable intersections of open sets, and let I_A be the characteristic function of A. Fubini's Theorem is applicable to the function $(x, y) \mapsto I_A(xy)$, and we have

$$\begin{aligned}
\mu_l(G)\mu_r(A) &= \int_G \Big[\int_G I_A(x)\, d\mu_r(x)\Big] d\mu_l(y) \\
&= \int_G \Big[\int_G I_A(xy)\, d\mu_r(x)\Big] d\mu_l(y) \\
&= \int_G \Big[\int_G I_A(xy)\, d\mu_l(y)\Big] d\mu_r(x) \\
&= \int_G \Big[\int_G I_A(y)\, d\mu_l(y)\Big] d\mu_r(x) \\
&= \mu_r(G)\mu_l(A).
\end{aligned}$$

Since μ_l and μ_r are regular as Borel measures, this equality extends to be valid for all Borel sets A. In other words any left Haar measure is proportional to any right Haar measure. Consequently there is only one left Haar measure up to a proportionality constant, and it is a right Haar measure. We may thus speak of a **Haar measure**, understanding that it is two-sided invariant. Let us normalize it so that it has total measure 1. Since the normalized measure is unambiguous, we write integrals with respect to normalized Haar measure by expressions like $\int_G f(x)\, dx$, dropping μ from the notation.

Proposition 4.6. If Φ is a representation of G on a finite-dimensional V, then V admits a Hermitian inner product such that Φ is unitary.

PROOF. Let $\langle \cdot, \cdot \rangle$ be any Hermitian inner product on V, and define

$$(u, v) = \int_G \langle \Phi(x)u, \Phi(x)v \rangle \, dx.$$

It is straightforward to see that $(\cdot, \cdot)$ has the required properties.

Corollary 4.7. If Φ is a representation of G on a finite-dimensional V, then Φ is the direct sum of irreducible representations. In other words, $V = V_1 \oplus \cdots \oplus V_k$, with each V_j an invariant subspace on which Φ acts irreducibly.

PROOF. Form $(\cdot, \cdot)$ as in Proposition 4.6. Find an invariant subspace $U \neq 0$ of minimal dimension and take its orthogonal complement $U^\perp$. Then (4.5) shows that $U^\perp$ is invariant. Repeating the argument with $U^\perp$ and iterating, we obtain the required decomposition.

Proposition 4.8 (Schur's Lemma). Suppose Φ and Φ' are irreducible representations of G on finite-dimensional vector spaces V and V', respectively. If $L : V \to V'$ is a linear map such that $\Phi'(g)L = L\Phi(g)$ for all $g \in G$, then L is one-one onto or $L = 0$.

PROOF. We see easily that $\ker L$ and $\operatorname{image} L$ are invariant subspaces of V and V', respectively, and then the only possibilities are the ones listed.

Corollary 4.9. Suppose Φ is an irreducible representation of G on a finite-dimensional V. If $L : V \to V$ is a linear map such that $\Phi(g)L = L\Phi(g)$ for all $g \in G$, then L is scalar.

PROOF. Let λ be an eigenvalue of L. Then $L - \lambda I$ is not one-one onto, but it does commute with $\Phi(g)$ for all $g \in G$. By Proposition 4.8, $L - \lambda I = 0$.

EXAMPLE. If G is abelian, then it follows from Corollary 4.9 (applied to $L = \Phi(g_0)$) that every irreducible finite-dimensional representation of G is 1-dimensional. For the circle group $S^1 = \{e^{i\theta}\}$, the 1-dimensional representations are parametrized by $n \in \mathbb{Z}$, the nth representation being

$$e^{i\theta} \mapsto \text{multiplication by } e^{in\theta}.$$

Corollary 4.10 (Schur orthogonality relations).

(a) Let Φ and Φ' be inequivalent irreducible unitary representations of G on finite-dimensional spaces V and V', respectively, and let the understood Hermitian inner products be denoted $(\cdot, \cdot)$. Then

$$\int_G (\Phi(x)u, v)\overline{(\Phi'(x)u', v')}\, dx = 0 \qquad \text{for all } u, v \in V \text{ and } u', v' \in V.$$

(b) Let Φ be an irreducible unitary representation on a finite-dimensional V, and let the understood Hermitian inner product be denoted $(\cdot, \cdot)$. Then

$$\int_G (\Phi(x)u_1, v_1)\overline{(\Phi(x)u_2, v_2)}\, dx = \frac{(u_1, u_2)\overline{(v_1, v_2)}}{\dim V} \quad \text{for } u_1, v_1, u_2, v_2 \in V.$$

PROOF.
(a) Let $l : V' \to V$ be linear and form

$$L = \int_G \Phi(x) l \Phi'(x^{-1})\, dx.$$

(This integration can be regarded as occurring for matrix-valued functions and is to be handled entry-by-entry.) Then it follows that $\Phi(y)L\Phi'(y^{-1}) = L$, so that $\Phi(y)L = L\Phi'(y)$ for all $y \in G$. By Proposition 4.8, $L = 0$. Thus $(Lv', v) = 0$. Choose $l(w') = (w', u')u$, and then we have

$$\begin{aligned} 0 &= (Lv', v) \\ &= \int_G (\Phi(x) l \Phi'(x^{-1})v', v)\, dx \\ &= \int_G (\Phi(x)(\Phi'(x^{-1})v', u')u, v)\, dx \\ &= \int_G (\Phi(x)u, v)(\Phi'(x^{-1})v', u')\, dx, \end{aligned}$$

and (a) results.

(b) We proceed in the same way, starting from $l : V \to V$ and obtain $L = \lambda I$ from Corollary 4.9. Taking the trace of both sides, we find

$$\lambda \dim V = \operatorname{Tr} L = \operatorname{Tr} l,$$

so that $\lambda = (\operatorname{Tr} l)/\dim V$. Thus

$$(Lv_2, v_1) = \frac{\operatorname{Tr} l}{\dim V}\overline{(v_1, v_2)}.$$

Choosing $l(w) = (w, u_2)u_1$, we have

$$\begin{aligned} \frac{(u_1, u_2)\overline{(v_1, v_2)}}{\dim V} &= \frac{\operatorname{Tr} l}{\dim V}\overline{(v_1, v_2)} \\ &= (Lv_2, v_1) \\ &= \int_G (\Phi(x) l \Phi(x^{-1}) v_2, v_1)\, dx \\ &= \int_G (\Phi(x) u_1, v_1)(\Phi(x^{-1}) v_2, u_2)\, dx, \end{aligned}$$

and (b) results.

We can interpret Corollary 4.10 as follows. Let $\{\Phi^{(\alpha)}\}$ be a maximal set of mutually inequivalent finite-dimensional irreducible unitary representations of G. For each $\Phi^{(\alpha)}$, choose an orthonormal basis for the underlying vector space, and let $\Phi_{ij}^{(\alpha)}(x)$ be the matrix of $\Phi^{(\alpha)}(x)$ in this basis. Then the functions $\{\Phi_{ij}^{(\alpha)}(x)\}_{i,j,\alpha}$ form an orthogonal set in $L^2(G)$. In fact, if $d^{(\alpha)}$ denotes the **degree** of $\Phi^{(\alpha)}$ (i.e., the dimension of the underlying vector space), then $\{(d^{(\alpha)})^{1/2}\Phi_{ij}^{(\alpha)}(x)\}_{i,j,\alpha}$ is an orthonormal set in $L^2(G)$. The Peter-Weyl Theorem in the next section will show that this orthonormal set is an orthonormal basis.

We can use Schur orthogonality to get a qualitative idea of the decomposition into irreducibles in Corollary 4.7 when Φ is a given finite-dimensional representation of G. By Proposition 4.6 there is no loss of generality in assuming that Φ is unitary. If Φ is a unitary finite-dimensional representation of G, a **matrix coefficient** of Φ is any function on G of the form $(\Phi(x)u, v)$. The **character** of Φ is the function

$$\chi_\Phi(x) = \operatorname{Tr}\Phi(x) = \sum_i (\Phi(x)u_i, u_i), \tag{4.11}$$

where $\{u_i\}$ is an orthonormal basis. This function depends only on the equivalence class of Φ and satisfies

$$\chi_\Phi(gxg^{-1}) = \chi_\Phi(x) \qquad \text{for all } g, x \in G. \tag{4.12}$$

If a finite-dimensional Φ is the direct sum of representations $\Phi_1, \dots, \Phi_n$, then

$$\chi_\Phi = \chi_{\Phi_1} + \cdots + \chi_{\Phi_n}. \tag{4.13}$$

The corresponding formulas for characters of contragredients and tensor products are

$$\chi_{\Phi^c} = \overline{\chi_\Phi} \tag{4.14}$$

$$\chi_{\Phi\otimes\Phi'} = \chi_\Phi \chi_{\Phi'}. \tag{4.15}$$

Corollary 4.16. If G is a compact group, then the character χ of an irreducible finite-dimensional representation has $\|\chi\|_2 = 1$. If χ and χ' are characters of inequivalent irreducible finite-dimensional representations, then $\int_G \chi(x)\overline{\chi'(x)}\,dx = 0$.

PROOF. These formulas are immediate from Corollary 4.10 since characters are sums of matrix coefficients.

Now let Φ be a given finite-dimensional representation of G, and write Φ as the direct sum of irreducible representations $\Phi_1, \ldots, \Phi_n$. If τ is an irreducible finite-dimensional representation of G, then (4.13) and Corollary 4.16 show that $\int_G \chi_\Phi(x)\overline{\chi_\tau(x)}\,dx$ is the number of summands Φ_i equivalent with τ. Evidently this integer is independent of the decomposition of Φ into irreducible representations. We call it the **multiplicity** of τ in Φ.

To make concrete use of characters in determining reducibility, it is helpful to have explicit formulas for characters. Formula (4.12) says that characters are constant on conjugacy classes and therefore need to be determined only on one representative of each conjugacy class. The Weyl Character Formula in Chapter V will provide the required character formulas when G is a compact connected Lie group.

3. Peter-Weyl Theorem

The goal of this section is to establish the more analytic parts of the abstract representation theory of compact groups. At the end of this section we deduce the important consequence that any compact Lie group can be realized as a group of complex matrices.

Lemma 4.17. If G is a compact group and h is in $L^2(G)$, then the function $y \mapsto h(y^{-1}x)$ of G into $L^2(G)$ is continuous.

PROOF. Given $\epsilon > 0$, we shall produce an open neighborhood U of 1 in G such that $\|h(y_1^{-1}x) - h(y_2^{-1}x)\|_{2,x} < \epsilon$ whenever $y_1^{-1}y_2$ is in U. Let $h \in L^2(G)$ be given, and find a continuous function c such that $\|h - c\|_2 < \epsilon/3$. (This density property is valid for all regular Borel measures, not just the invariant ones.) The function c, being continuous on G, is uniformly continuous. Thus we can find an open neighborhood U of 1 in G such that

$$|c(y_1^{-1}x) - c(y_2^{-1}x)| < \epsilon/3$$

for all $x \in G$ whenever $y_1^{-1}y_2$ is in U. Then

$$\begin{aligned}\|h(y_1^{-1}x) - h(y_2^{-1}x)\|_{2,x} &\leq \|h(y_1^{-1}x) - c(y_1^{-1}x)\|_{2,x} \\ &\quad + \|c(y_1^{-1}x) - c(y_2^{-1}x)\|_{2,x} \\ &\quad + \|c(y_2^{-1}x) - h(y_2^{-1}x)\|_{2,x} \\ &= 2\|h - c\|_2 + \|c(y_1^{-1}x) - c(y_2^{-1}x)\|_{2,x} \\ &\leq 2\|h - c\|_2 + \sup_{x \in G} |c(y_1^{-1}x) - c(y_2^{-1}x)| \\ &< 2\epsilon/3 + \epsilon/3 = \epsilon.\end{aligned}$$

Lemma 4.18. Let G be a compact group, and let h be in $L^2(G)$. For any $\epsilon > 0$, there exist finitely many $y_i \in G$ and Borel sets $E_i \subseteq G$ such that the E_i disjointly cover G and

$$\|h(y^{-1}x) - h(y_i^{-1}x)\|_{2,x} < \epsilon \qquad \text{for all } i \text{ and for all } y \in E_i.$$

PROOF. By Lemma 4.17 choose an open neighborhood U of 1 so that $\|h(gx) - h(x)\|_{2,x} < \epsilon$ whenever g is in U. For each $z_0 \in G$, $\|h(gz_0x) - h(z_0x)\|_{2,x} < \epsilon$ whenever g is in U. The set Uz_0 is an open neighborhood of z_0, and such sets cover G as z_0 varies. Find a finite subcover, say $Uz_1, \ldots, Uz_n$, and let $U_i = Uz_i$. Define $F_j = U_j - \bigcup_{i=1}^{j-1} U_i$. Then the lemma follows with $y_i = z_i^{-1}$ and $E_i = F_i^{-1}$.

Lemma 4.19. Let G be a compact group, let f be in $L^1(G)$, and let h be in $L^2(G)$. Put $F(x) = \int_G f(y)h(y^{-1}x)\,dy$. Then F is the limit in $L^2(G)$ of a sequence of functions, each of which is a finite linear combination of left translates of h.

PROOF. Given $\epsilon > 0$, choose y_i and E_i as in Lemma 4.18, and put $c_i = \int_{E_i} f(y)\,dy$. Then

$$\begin{aligned}&\Big\| \int_G f(y)h(y^{-1}x)\,dy - \sum_i c_i h(y_i^{-1}x) \Big\|_{2,x} \\ &\qquad \leq \Big\| \sum_i \int_{E_i} |f(y)||h(y^{-1}x) - h(y_i^{-1}x)|\,dy \Big\|_{2,x} \\ &\qquad \leq \sum_i \int_{E_i} |f(y)|\|h(y^{-1}x) - h(y_i^{-1}x)\|_{2,x}\,dy \\ &\qquad \leq \sum_i \int_{E_i} |f(y)|\epsilon\,dy = \epsilon\|f\|_1.\end{aligned}$$

Theorem 4.20 (Peter-Weyl Theorem). If G is a compact group, then the linear span of all matrix coefficients for all finite-dimensional irreducible unitary representations of G is dense in $L^2(G)$.

PROOF. If $h(x) = (\Phi(x)u, v)$ is a matrix coefficient, then the following functions of x are also matrix coefficients for the same representation:

$$\overline{h(x^{-1})} = (\Phi(x)v, u)$$
$$h(gx) = (\Phi(x)u, \Phi(g^{-1})v)$$
$$h(xg) = (\Phi(x)\Phi(g)u, v).$$

Then the closure U in $L^2(G)$ of the linear span of all matrix coefficients of all finite-dimensional irreducible unitary representations is stable under $h(x) \mapsto \overline{h(x^{-1})}$ and under left and right translation. Arguing by contradiction, suppose $U \neq L^2(G)$. Then $U^\perp \neq 0$ and $U^\perp$ is closed under $h(x) \mapsto \overline{h(x^{-1})}$ and under left and right translation.

We first prove that there is a nonzero continuous function in $U^\perp$. Thus let $H \neq 0$ be in $U^\perp$. For each open neighborhood N of 1, we define

$$F_N(x) = \frac{1}{|N|}\int_G I_N(y)H(y^{-1}x)\,dy,$$

where I_N is the characteristic function of N and $|N|$ is the Haar measure of N. Use of the Schwarz inequality and the fact that I_N and H are in $L^2(G)$ shows that F_N is continuous. As N shrinks to $\{1\}$, the functions F_N tend to H in L^2; hence some F_N is not 0. Finally each linear combination of left translates of H is in $U^\perp$, and hence F_N is in $U^\perp$ by Lemma 4.19.

Thus $U^\perp$ contains a nonzero continuous function. Using translations and scalar multiplications, we can adjust this function so that it becomes a continuous F_1 in $U^\perp$ with $F_1(1)$ real and nonzero. Set

$$F_2(x) = \int_G F_1(yxy^{-1})\,dy.$$

Then F_2 is continuous and is in $U^\perp$, $F_2(gxg^{-1}) = F_2(x)$ for all $g \in G$, and $F_2(1) = F_1(1)$ is real and nonzero. Finally put

$$F(x) = F_2(x) + \overline{F_2(x^{-1})}.$$

Then F is continuous and is in $U^\perp$, $F(gxg^{-1}) = F(x)$ for all $g \in G$, $F(1) = 2F_2(1)$ is real and nonzero, and $\overline{F(x)} = F(x^{-1})$. In particular, F is not the 0 function in $L^2(G)$.

Form the function $k(x, y) = F(x^{-1}y)$ and the integral operator

$$Tf(x) = \int_G k(x, y)f(y)\,dy = \int_G F(x^{-1}y)f(y)\,dy \qquad \text{for } f \in L^2(G).$$

Then $k(x,y) = \overline{k(y,x)}$ and $\int_{G\times G} |k(x,y)|^2\,dx\,dy < \infty$, and hence T is a Hilbert-Schmidt operator from $L^2(G)$ into itself. Also T is not 0 since $F \neq 0$. According to the Hilbert-Schmidt Theorem (Riesz-Nagy [1955], 242), such an operator has a real nonzero eigenvalue λ and the corresponding eigenspace $V_\lambda \subseteq L^2(G)$ is finite-dimensional.

Let us see that the subspace V_λ is invariant under left translation by g, which we write as $(L(g)f)(x) = f(g^{-1}x)$. In fact, f in V_λ implies

$$\begin{aligned} TL(g)f(x) &= \int_G F(x^{-1}y)f(g^{-1}y)\,dy = \int_G F(x^{-1}gy)f(y)\,dy \\ &= Tf(g^{-1}x) = \lambda f(g^{-1}x) = \lambda L(g)f(x). \end{aligned}$$

By Lemma 4.17, $g \mapsto L(g)f$ is continuous, and therefore L is a representation of G in the finite-dimensional space V_λ. By Corollary 4.7, V_λ contains an irreducible invariant subspace $W_\lambda \neq 0$.

Let $f_1, \ldots, f_n$ be an orthonormal basis of W_λ. The matrix coefficients for W_λ are

$$h_{ij}(x) = (L(x)f_j, f_i) = \int_G f_j(x^{-1}y)\overline{f_i(y)}\,dy$$

and by definition are in U. Since F is in $U^\perp$, we have

$$\begin{aligned} 0 &= \int_G F(x)\overline{h_{ii}(x)}\,dx \\ &= \int_G\int_G F(x)\overline{f_i(x^{-1}y)}f_i(y)\,dy\,dx \\ &= \int_G\int_G F(x)\overline{f_i(x^{-1}y)}f_i(y)\,dx\,dy \\ &= \int_G\int_G F(yx^{-1})\overline{f_i(x)}f_i(y)\,dx\,dy \\ &= \int_G\left[\int_G F(x^{-1}y)f_i(y)\,dy\right]\overline{f_i(x)}\,dx \qquad \text{since } F(gxg^{-1}) = F(x) \\ &= \int_G [Tf_i(x)]\overline{f_i(x)}\,dx \\ &= \lambda\int_G |f_i(x)|^2\,dx \end{aligned}$$

for all i, in contradiction with the fact that $W_\lambda \neq 0$. We conclude that $U^\perp = 0$ and therefore that $U = L^2(G)$.

EXAMPLE. For $S^1 = \{e^{i\theta}\}$, we observed after Corollary 4.9 that the irreducible finite-dimensional representations are 1-dimensional. The matrix coefficients are just the functions $e^{in\theta}$. For this group the Peter-Weyl Theorem says that the finite linear combinations of these functions are dense in $L^2(S^1)$. An equivalent formulation of this result is that $\{e^{in\theta}\}_{n=-\infty}^{\infty}$ is an orthonormal basis of $L^2(S^1)$. This equivalent formulation is generalized in Corollary 4.21 below.

Corollary 4.21. If $\{\Phi^{(\alpha)}\}$ is a maximal set of mutually inequivalent finite-dimensional irreducible unitary representations of a compact group G and if $\{(d^{(\alpha)})^{1/2}\Phi_{ij}^{(\alpha)}(x)\}_{i,j,\alpha}$ is a corresponding orthonormal set of matrix coefficients, then $\{(d^{(\alpha)})^{1/2}\Phi_{ij}^{(\alpha)}(x)\}_{i,j,\alpha}$ is an orthonormal basis of $L^2(G)$.

PROOF. The linear span of the functions in question is the linear span considered in the theorem. Then the theorem and general Hilbert space theory imply the corollary.

Now we specialize to Lie groups. Recall from §I.10 that any continuous homomorphism between Lie groups is automatically smooth. Therefore the Lie algebra of a Lie group is an important tool in working with arbitrary finite-dimensional representations of the group. This idea will be used implicitly in the proof of the next corollary and more explicitly in the proofs in the next section.

Corollary 4.22. Any compact Lie group G has a one-one finite-dimensional representation and hence is isomorphic to a closed linear group.

PROOF. It follows from Theorem 4.20 that for each $x \neq 1$ in G, there is a finite-dimensional representation Φ_x of G such that $\Phi_x(x) \neq 1$. If the identity component G_0 is not $\{1\}$, pick $x_1 \neq 1$ in the identity component G_0. Then $G_1 = \ker \Phi_{x_1}$ is a closed subgroup of G, and its identity component is a proper subgroup of G_0. If $(G_1)_0 \neq \{1\}$, pick $x_2 \neq 1$ in $(G_1)_0$. Then $G_2 = \ker(\Phi_{x_1} \oplus \Phi_{x_2})$ is a closed subgroup of G_1, and its identity component is a proper subgroup of $(G_1)_0$. Continuing in this way and using the finite-dimensionality of G, we see that we can find a finite-dimensional representation Φ_0 of G such that $\ker \Phi_0$ is 0-dimensional. Then $\ker \Phi_0$ is finite, being a compact 0-dimensional Lie group. Say $\ker \Phi_0 = \{y_1, \ldots, y_n\}$. Then

$$\Phi = \Phi_0 \oplus \bigoplus_{j=1}^{n} \Phi_{y_j}$$

is a one-one finite-dimensional representation of G.

4. Compact Lie Algebras

Let $\mathfrak{g}$ be a real Lie algebra. We say that $\mathfrak{g}$ is a **compact Lie algebra** if the group $\operatorname{Int}\mathfrak{g}$ is compact. More generally let $\mathfrak{k}$ be a Lie subalgebra of $\mathfrak{g}$, and let $\operatorname{Int}_{\mathfrak{g}}(\mathfrak{k})$ be the analytic subgroup of $GL(\mathfrak{g})$ with Lie algebra $\operatorname{ad}_{\mathfrak{g}}(\mathfrak{k})$. We say that $\mathfrak{k}$ is **compactly embedded** in $\mathfrak{g}$ if $\operatorname{Int}_{\mathfrak{g}}(\mathfrak{k})$ is compact.

Proposition 4.23. If G is a Lie group with Lie algebra $\mathfrak{g}$ and if K is a compact subgroup with corresponding Lie subalgebra $\mathfrak{k}$, then $\mathfrak{k}$ is a compactly embedded subalgebra of $\mathfrak{g}$. In particular, the Lie algebra of a compact Lie group is always a compact Lie algebra.

PROOF. Since K is compact, so is the identity component K_0. Then $\operatorname{Ad}_{\mathfrak{g}}(K_0)$ must be compact, being the continuous image of a compact group. The groups $\operatorname{Ad}_{\mathfrak{g}}(K_0)$ and $\operatorname{Int}_{\mathfrak{g}}(\mathfrak{k})$ are both analytic subgroups of $GL(\mathfrak{g})$ with Lie algebra $\operatorname{ad}_{\mathfrak{g}}(\mathfrak{k})$ and hence are isomorphic as Lie groups. Therefore $\operatorname{Int}_{\mathfrak{g}}(\mathfrak{k})$ is compact.

The next proposition and its two corollaries give properties of compact Lie algebras.

Proposition 4.24. Let G be a compact Lie group, and let $\mathfrak{g}$ be its Lie algebra. Then the real vector space $\mathfrak{g}$ admits an inner product $(\cdot\,,\,\cdot)$ that is invariant under $\operatorname{Ad}(G)$: $(\operatorname{Ad}(g)u, \operatorname{Ad}(g)v) = (u, v)$. Relative to this inner product the members of $\operatorname{Ad}(G)$ act by orthogonal transformations, and the members of $\operatorname{ad}\mathfrak{g}$ act by skew-symmetric transformations.

PROOF. Proposition 4.6 applies to complex vector spaces, but the same argument can be used here to obtain the first conclusion. Then $\operatorname{Ad}(G)$ acts by orthogonal transformations. Differentiating the identity $(\operatorname{Ad}(\exp tX)u, \operatorname{Ad}(\exp tX)v) = (u, v)$ at $t = 0$, we see that $((\operatorname{ad} X)u, v) = -(u, (\operatorname{ad} X)v)$ for all $X \in \mathfrak{g}$. In other words, $\operatorname{ad} X$ is skew symmetric.

Corollary 4.25. Let G be a compact Lie group, and let $\mathfrak{g}$ be its Lie algebra. Then $\mathfrak{g}$ is reductive, and hence $\mathfrak{g} = \mathfrak{z}_{\mathfrak{g}} \oplus [\mathfrak{g}, \mathfrak{g}]$, where $\mathfrak{z}_{\mathfrak{g}}$ is the center and where $[\mathfrak{g}, \mathfrak{g}]$ is semisimple.

PROOF. Define $(\cdot\,,\,\cdot)$ as in Proposition 4.24. The invariant subspaces of $\mathfrak{g}$ under $\operatorname{ad}\mathfrak{g}$ are the ideals of $\mathfrak{g}$. If $\mathfrak{a}$ is an ideal, then $\mathfrak{a}$ is an invariant subspace. By (4.5), $\mathfrak{a}^\perp$ relative to this inner product is an invariant subspace, and thus $\mathfrak{a}^\perp$ is an ideal. Since $(\cdot\,,\,\cdot)$ is definite, $\mathfrak{g} = \mathfrak{a} \oplus \mathfrak{a}^\perp$. Hence $\mathfrak{a}$ has $\mathfrak{a}^\perp$ as a complementary ideal, and $\mathfrak{g}$ is reductive. The rest follows from Corollary 1.53.

Corollary 4.26. If G is a compact Lie group with Lie algebra $\mathfrak{g}$, then the Killing form of $\mathfrak{g}$ is negative semidefinite.

REMARKS. Starting in the next section, we shall bring roots into the analysis of $\mathfrak{g}$, and we shall want the theory in the semisimple case to be consistent with the theory in Chapter II. Recall from the remarks with Corollary 2.38 that the Killing form can be replaced in the theory of Chapter II by any nondegenerate invariant symmetric bilinear form that yields a positive definite form on the real subspace of the Cartan subalgebra where the roots are real-valued. Once we see from (4.32) below that this space is contained in $i\mathfrak{g}_0$, Corollary 4.26 will imply that the negative of the invariant inner product, rather than the inner product itself, is a valid substitute for the Killing form in the semisimple case.

PROOF. Define $(\cdot\,,\,\cdot)$ as in Proposition 4.24. By the proposition, $\operatorname{ad} X$ is skew symmetric for $X \in \mathfrak{g}$. The eigenvalues of $\operatorname{ad} X$ are therefore purely imaginary, and the eigenvalues of $(\operatorname{ad} X)^2$ must be ≤ 0. If B is the Killing form, then it follows that $B(X, X) = \operatorname{Tr}((\operatorname{ad} X)^2)$ is ≤ 0.

The next proposition provides a kind of converse to Corollary 4.26.

Proposition 4.27. If the Killing form of a real Lie algebra $\mathfrak{g}$ is negative definite, then $\mathfrak{g}$ is a compact Lie algebra.

PROOF. By Cartan's Criterion for Semisimplicity (Theorem 1.42), $\mathfrak{g}$ is semisimple. By Propositions 1.97 and 1.98, $\operatorname{Int}\mathfrak{g} = (\operatorname{Aut}_{\mathbb{R}}\mathfrak{g})_0$. Consequently $\operatorname{Int}\mathfrak{g}$ is a closed subgroup of $GL(\mathfrak{g})$. On the other hand, the negative of the Killing form is an inner product on $\mathfrak{g}$ in which every member of $\operatorname{ad}\mathfrak{g}$ is skew symmetric. Therefore the corresponding analytic group $\operatorname{Int}\mathfrak{g}$ acts by orthogonal transformations. Since $\operatorname{Int}\mathfrak{g}$ is then exhibited as a closed subgroup of the orthogonal group, $\operatorname{Int}\mathfrak{g}$ is compact.

Lemma 4.28. Any 1-dimensional representation of a semisimple Lie algebra is 0. Consequently any 1-dimensional representation of a semisimple Lie group is trivial.

REMARK. Recall that semisimple Lie groups are connected by definition.

PROOF. Let $\mathfrak{g}$ be the Lie algebra. A 1-dimensional representation of $\mathfrak{g}$ is a Lie algebra homomorphism of $\mathfrak{g}$ into the abelian real Lie algebra $\mathbb{C}$. Commutators must map to commutators, which are 0 in $\mathbb{C}$. Thus $[\mathfrak{g}, \mathfrak{g}]$ maps to 0. But $[\mathfrak{g}, \mathfrak{g}] = \mathfrak{g}$ by Corollary 1.52, and thus $\mathfrak{g}$ maps to 0. The conclusion about groups follows from the conclusion about Lie algebras, since any continuous homomorphism between Lie groups is smooth (§I.10).

Theorem 4.29. Let G be a compact connected Lie group with center Z_G, let $\mathfrak{g}$ be its Lie algebra, and let G_{ss} be the analytic subgroup of G with Lie algebra $[\mathfrak{g}, \mathfrak{g}]$. Then G_{ss} has finite center, $(Z_G)_0$ and G_{ss} are closed subgroups, and G is the commuting product $G = (Z_G)_0 G_{ss}$.

PROOF. By Corollary 4.25 we have $\mathfrak{g} = \mathfrak{z}_\mathfrak{g} \oplus [\mathfrak{g}, \mathfrak{g}]$. If $(Z_G)_0^\sim$ and $G_{ss}^\sim$ denote simply connected covers of $(Z_G)_0$ and G_{ss}, then $(Z_G)_0^\sim \times G_{ss}^\sim$ is a simply connected group with Lie algebra $\mathfrak{g}$ and hence is a covering group $\tilde{G}$ of G. The covering homomorphism carries $\tilde{G}$ onto G, $(Z_G)_0^\sim$ onto $(Z_G)_0$, and $G_{ss}^\sim$ onto G_{ss}. Since $\tilde{G} = (Z_G)_0^\sim G_{ss}^\sim$, it follows that $G = (Z_G)_0 G_{ss}$.

Since $(Z_G)_0$ is the identity component of the center and since the center is closed, $(Z_G)_0$ is closed.

Let us show that the center Z_{ss} of G_{ss} is finite. By Corollary 4.22, G has a one-one representation Φ on some finite-dimensional complex vector space V. By Corollary 4.7 we can write $V = V_1 \oplus \cdots \oplus V_n$ with G acting irreducibly on each V_j. Let $d_j = \dim V_j$, and put $\Phi_j(g) = \Phi(g)|_{V_j}$. Since $G = (Z_G)_0 G_{ss}$, the members of Z_{ss} are central in G, and Proposition 4.8 shows that $\Phi_j(x)$ is a scalar operator for each $x \in Z_{ss}$. On the other hand, Lemma 4.28 shows that $\det \Phi_j(x) = 1$ for all $x \in G_{ss}$. Thus $\Phi_j(x)$ for $x \in Z_{ss}$ acts on V_j by a scalar that is a power of $\exp 2\pi i/d_j$. Therefore there are at most $\prod_{j=1}^n d_j$ possibilities for the operator $\Phi(x)$ when x is in Z_{ss}. Since Φ is one-one, Z_{ss} has at most $\prod_{j=1}^n d_j$ elements.

Finally we prove that G_{ss} is closed. By Corollary 4.26 the Killing form of $\mathfrak{g}$ is negative semidefinite. The Killing form of an ideal is obtained by restriction, and thus the Killing form of $[\mathfrak{g}, \mathfrak{g}]$ is negative semidefinite. But $[\mathfrak{g}, \mathfrak{g}]$ is semisimple, and Cartan's Criterion for Semisimplicity (Theorem 1.42) says that the Killing form of $[\mathfrak{g}, \mathfrak{g}]$ is nondegenerate. Consequently the Killing form of $[\mathfrak{g}, \mathfrak{g}]$ is negative definite. By Proposition 4.27, $[\mathfrak{g}, \mathfrak{g}]$ is a compact Lie algebra. That is, $\operatorname{Int}([\mathfrak{g}, \mathfrak{g}])$ is compact. But $\operatorname{Int}([\mathfrak{g}, \mathfrak{g}]) \cong \operatorname{Ad}(G_{ss})$. Since, as we have seen, G_{ss} has finite center, the covering $G_{ss} \to \operatorname{Ad}(G_{ss})$ is a finite covering. Therefore G_{ss} is a compact group. Consequently G_{ss} is closed as a subgroup of G.

5. Centralizers of Tori

Throughout this section, G will denote a compact connected Lie group, $\mathfrak{g}_0$ will be its Lie algebra, and $\mathfrak{g}$ will be $(\mathfrak{g}_0)^\mathbb{C}$. Fix an invariant inner product $\mathfrak{g}_0$ as in Proposition 4.24, and, in accordance with the remarks with Corollary 4.26, write B for its negative.

A **torus** is a product of circle groups. From §I.10 we know that every compact connected abelian Lie group is a torus.

Within G we can look for tori as subgroups. These are ordered by inclusion, and any torus is contained in a maximal torus just by dimensional considerations. A key role in the structure theory for G is played by maximal tori, and we begin by explaining their significance. In the discussion we shall make use of the following proposition without specific reference.

Proposition 4.30. The maximal tori in G are exactly the analytic subgroups corresponding to the maximal abelian subalgebras of $\mathfrak{g}_0$.

PROOF. If T is a maximal torus and $\mathfrak{t}_0$ is its Lie algebra, we show that $\mathfrak{t}_0$ is maximal abelian. Otherwise let $\mathfrak{h}_0$ be a strictly larger abelian subalgebra. The corresponding analytic subgroup H will be abelian and will strictly contain T. Hence $\bar{H}$ will be a torus strictly containing T.

Conversely if $\mathfrak{t}_0$ is maximal abelian in $\mathfrak{g}_0$, then the corresponding analytic subgroup T is abelian. If T were not closed, then $\bar{T}$ would have a strictly larger abelian Lie algebra than $\mathfrak{t}_0$, in contradiction with maximality. Hence T is closed and is a torus, clearly maximal.

EXAMPLES.

1) For $G = SU(n)$, $\mathfrak{g}_0$ is $\mathfrak{su}(n)$ and $\mathfrak{g}$ is $\mathfrak{sl}(n, \mathbb{C})$. Let

$$T = \operatorname{diag}(e^{i\theta_1}, \ldots, e^{i\theta_n}).$$

The Lie algebra of T is

$$\mathfrak{t}_0 = \operatorname{diag}(i\theta_1, \ldots, i\theta_n),$$

and the complexification is the Cartan subalgebra of $\mathfrak{g}$ that was denoted $\mathfrak{h}$ in Example 1 of §II.1. Then $\mathfrak{h}$ is maximal abelian in $\mathfrak{g}$, and hence $\mathfrak{t}_0$ is maximal abelian in $\mathfrak{g}_0$. By Proposition 4.30, T is a maximal torus of G.

2) For $G = SO(2n+1)$, $\mathfrak{g}_0$ is $\mathfrak{so}(2n+1)$ and $\mathfrak{g}$ is $\mathfrak{so}(2n+1, \mathbb{C})$. Referring to Example 2 of §II.1 and using Proposition 4.30, we see that

$$T = \left\{ \begin{pmatrix} \begin{pmatrix} \cos\theta_1 & \sin\theta_1 \\ -\sin\theta_1 & \cos\theta_1 \end{pmatrix} & & & \\ & \ddots & & \\ & & \begin{pmatrix} \cos\theta_n & \sin\theta_n \\ -\sin\theta_n & \cos\theta_n \end{pmatrix} & \\ & & & 1 \end{pmatrix} \right\}$$

is a maximal torus of G.

3) For $G = Sp(n, \mathbb{C}) \cap U(2n)$, $\mathfrak{g}_0$ is $\mathfrak{sp}(n, \mathbb{C}) \cap \mathfrak{u}(2n)$ and $\mathfrak{g}$ is $\mathfrak{sp}(n, \mathbb{C})$. Referring to Example 3 of §II.1 and using Proposition 4.30, we see that

$$T = \operatorname{diag}(e^{i\theta_1}, \ldots, e^{i\theta_n}, e^{-i\theta_1}, \ldots, e^{-i\theta_n})$$

is a maximal torus of G. From Proposition 1.118 we know that $G \cong Sp(n)$.

4) For $G = SO(2n)$, $\mathfrak{g}_0$ is $\mathfrak{so}(2n)$ and $\mathfrak{g}$ is $\mathfrak{so}(2n, \mathbb{C})$. Referring to Example 4 of §II.1 and using Proposition 4.30, we see that

$$T = \left\{ \begin{pmatrix} \begin{pmatrix} \cos\theta_1 & \sin\theta_1 \\ -\sin\theta_1 & \cos\theta_1 \end{pmatrix} & & \\ & \ddots & \\ & & \begin{pmatrix} \cos\theta_n & \sin\theta_n \\ -\sin\theta_n & \cos\theta_n \end{pmatrix} \end{pmatrix} \right\}$$

is a maximal torus of G.

Let T be a maximal torus in G, and let $\mathfrak{t}_0$ be its Lie algebra. By Corollary 4.25 we know that $\mathfrak{g}_0 = \mathfrak{z}_{\mathfrak{g}_0} \oplus [\mathfrak{g}_0, \mathfrak{g}_0]$ with $[\mathfrak{g}_0, \mathfrak{g}_0]$ semisimple. Since $\mathfrak{t}_0$ is maximal abelian in $\mathfrak{g}_0$, $\mathfrak{t}_0$ is of the form $\mathfrak{t}_0 = \mathfrak{z}_{\mathfrak{g}_0} \oplus \mathfrak{t}_0'$, where $\mathfrak{t}_0'$ is maximal abelian in $[\mathfrak{g}_0, \mathfrak{g}_0]$. Dropping subscripts 0 to indicate complexifications, we have $\mathfrak{g} = \mathfrak{z}_\mathfrak{g} \oplus [\mathfrak{g}, \mathfrak{g}]$ with $[\mathfrak{g}, \mathfrak{g}]$ semisimple. Also $\mathfrak{t} = \mathfrak{z}_\mathfrak{g} \oplus \mathfrak{t}'$ with $\mathfrak{t}'$ maximal abelian in $[\mathfrak{g}, \mathfrak{g}]$. The members of $\operatorname{ad}_{\mathfrak{g}_0}(\mathfrak{t}_0)$ are diagonable over $\mathbb{C}$ by Proposition 4.24, and hence the members of $\operatorname{ad}_\mathfrak{g}(\mathfrak{t})$ are diagonable. Consequently the members of $\operatorname{ad}_{[\mathfrak{g},\mathfrak{g}]}(\mathfrak{t}')$ are diagonable. By Corollary 2.13, $\mathfrak{t}'$ is a Cartan subalgebra of the complex semisimple Lie algebra $[\mathfrak{g}, \mathfrak{g}]$.

Using the root-space decomposition of $[\mathfrak{g}, \mathfrak{g}]$, we can write

$$\mathfrak{g} = \mathfrak{z}_\mathfrak{g} \oplus \mathfrak{t}' \oplus \bigoplus_{\alpha \in \Delta([\mathfrak{g},\mathfrak{g}],\mathfrak{t}')} [\mathfrak{g}, \mathfrak{g}]_\alpha.$$

From this decomposition and the definition of Cartan subalgebra, it is clear that $\mathfrak{t}$ is a Cartan subalgebra of $\mathfrak{g}$. If we extend the members α of $\Delta([\mathfrak{g}, \mathfrak{g}], \mathfrak{t}')$ to $\mathfrak{t}$ by defining them to be 0 on $\mathfrak{z}_\mathfrak{g}$, then the weight-space decomposition of $\mathfrak{g}$ relative to $\operatorname{ad}\mathfrak{t}$ may be written

$$\mathfrak{g} = \mathfrak{t} \oplus \bigoplus_{\alpha \in \Delta(\mathfrak{g},\mathfrak{t})} \mathfrak{g}_\alpha. \tag{4.31}$$

Here, as in the semisimple case, $\mathfrak{g}_\alpha$ is given by

$$\mathfrak{g}_\alpha = \{X \in \mathfrak{g} \mid [H, X] = \alpha(H)X \text{ for all } H \in \mathfrak{t}\},$$

and we say that α is a **root** if $\mathfrak{g}_\alpha \neq 0$ and $\alpha \neq 0$. The members of $\mathfrak{g}_\alpha$ are the **root vectors** for the root α, and we refer to (4.31) as the **root-space decomposition** of $\mathfrak{g}$ relative to $\mathfrak{t}$. The set $\Delta(\mathfrak{g}, \mathfrak{t})$ has all the usual properties of roots except that the roots do not span the dual of $\mathfrak{t}$.

In general for a real reductive Lie algebra $\mathfrak{g}_0'$, we call a Lie subalgebra of $\mathfrak{g}_0'$ a **Cartan subalgebra** if its complexification is a Cartan subalgebra of $\mathfrak{g}'$. The dimension of the Cartan subalgebra is called the **rank** of $\mathfrak{g}_0'$ and of any corresponding analytic group. In this terminology the Lie algebra $\mathfrak{t}_0$ of the maximal torus T of the compact connected Lie group G is a Cartan subalgebra of $\mathfrak{g}_0$, and the rank of $\mathfrak{g}_0$ and of G is the dimension of $\mathfrak{t}_0$.

Because of the presence of the groups G and T, we get extra information about the root-space decomposition (4.31). In fact, $\operatorname{Ad}(T)$ acts by orthogonal transformations on $\mathfrak{g}_0$ relative to our given inner product. If we extend this inner product on $\mathfrak{g}_0$ to a Hermitian inner product on $\mathfrak{g}$, then $\operatorname{Ad}(T)$ acts on $\mathfrak{g}$ by a commuting family of unitary transformations. Such a family must have a simultaneous eigenspace decomposition, and it is clear that (4.31) is that eigenspace decomposition. The action of $\operatorname{Ad}(T)$ on the 1-dimensional space $\mathfrak{g}_\alpha$ is a 1-dimensional representation of T, necesarily of the form

$$\operatorname{Ad}(t)X = \xi_\alpha(t)X \qquad \text{for } t \in T, \tag{4.32}$$

where $\xi_\alpha : T \to S^1$ is a continuous homomorphism of T into the group of complex numbers of modulus 1. We call ξ_α a **multiplicative character**. From (4.32) the differential of ξ_α is $\alpha|_{\mathfrak{t}_0}$. In particular, $\alpha|_{\mathfrak{t}_0}$ is imaginary-valued, and the roots are real-valued on $i\mathfrak{t}_0$.

Although we can easily deduce from Theorem 2.15 that the Cartan subalgebras $\mathfrak{t}$ of $\mathfrak{g}$ obtained from all maximal tori are conjugate via $\operatorname{Int}\mathfrak{g}$, we need a separate argument to get the conjugacy of the $\mathfrak{t}_0$'s to take place within $\mathfrak{g}_0$.

Lemma 4.33. Let T be a maximal torus in G, let $\mathfrak{t}$ be its complexified Lie algebra, and let Δ be the set of roots. If $H \in \mathfrak{t}$ has $\alpha(H) \neq 0$ for all $\alpha \in \Delta$, then the centralizer $Z_{\mathfrak{g}}(H)$ is just $\mathfrak{t}$.

PROOF. Let X be in $Z_{\mathfrak{g}}(H)$, and let $X = H' + \sum_{\alpha\in\Delta} X_\alpha$ be the decomposition of X according to (4.31). Then

$$0 = [H, X] = 0 + \sum_{\alpha\in\Delta}[H, X_\alpha] = \sum_{\alpha\in\Delta} \alpha(H)X_\alpha,$$

and it follows that $\alpha(H)X_\alpha = 0$ for all $\alpha \in \Delta$. Since $\alpha(H) \neq 0$, $X_\alpha = 0$. Thus $X = H'$ is in $\mathfrak{t}$.

Theorem 4.34. For a compact connected Lie group G, any two maximal abelian subalgebras of $\mathfrak{g}_0$ are conjugate via $\mathrm{Ad}(G)$.

Proof. Let $\mathfrak{t}_0$ and $\mathfrak{t}'_0$ be maximal abelian subalgebras with T and T' the corresponding maximal tori. There are only finitely many roots relative to $\mathfrak{t}'$, and the union of their kernels therefore cannot exhaust $\mathfrak{t}'$. By Lemma 4.33 we can find $X \in \mathfrak{t}'$ such that $Z_{\mathfrak{g}}(X) = \mathfrak{t}'$. Similarly we can find $Y \in \mathfrak{t}$ such that $Z_{\mathfrak{g}}(Y) = \mathfrak{t}$. Remembering that B is defined to be the negative of the invariant inner product on $\mathfrak{g}_0$, choose g_0 in G such that $B(\mathrm{Ad}(g)X, Y)$ is minimized for $g = g_0$. For any Z in $\mathfrak{g}_0$, $r \mapsto B(\mathrm{Ad}(\exp rZ)\mathrm{Ad}(g_0)X, Y)$ is then a smooth function of r that is minimized for $r = 0$. Differentiating and setting $r = 0$, we obtain

$$0 = B((\mathrm{ad}\, Z)\mathrm{Ad}(g_0)X, Y) = B([Z, \mathrm{Ad}(g_0)X], Y) = B(Z, [\mathrm{Ad}(g_0)X, Y]).$$

Since Z is arbitrary, $[\mathrm{Ad}(g_0)X, Y] = 0$. Thus $\mathrm{Ad}(g_0)X$ is in $Z_{\mathfrak{g}_0}(Y) = \mathfrak{t}_0$. Since $\mathfrak{t}_0$ is abelian, this means

$$\mathfrak{t}_0 \subseteq Z_{\mathfrak{g}_0}(\mathrm{Ad}(g_0)X) = \mathrm{Ad}(g_0)Z_{\mathfrak{g}_0}(X) = \mathrm{Ad}(g_0)\mathfrak{t}'_0.$$

Equality must hold since $\mathfrak{t}_0$ is maximal abelian. Thus $\mathfrak{t}_0 = \mathrm{Ad}(g_0)\mathfrak{t}'_0$.

Corollary 4.35. For a compact connected Lie group, any two maximal tori are conjugate.

Proof. This follows by combining Proposition 4.30 and Theorem 4.34.

We come to the main result of this section.

Theorem 4.36. If G is a compact connected Lie group and T is a maximal torus, then each element of G is conjugate to a member of T.

Remark. If $G = U(n)$ and T is the diagonal subgroup, this theorem says that each unitary matrix is conjugate by a unitary matrix to a diagonal matrix. This is just the finite-dimensional Spectral Theorem and is an easy result. Theorem 4.36 lies deeper, and thus $U(n)$ is not completely typical. Even for $SO(n)$ and $Sp(n)$, the theorem is a little harder, and the full power of the theorem comes in handling those compact connected Lie groups that have not yet arisen as examples.

Proof. We use the following notation for conjugates:

$$y^x = xyx^{-1}$$

$$T^g = \{t^g \mid t \in T\}$$

$$T^G = \bigcup_{g \in G} T^g.$$

The statement of the theorem is that $T^G = G$.

We prove the theorem by induction on $\dim G$, the trivial case for the induction being $\dim G = 0$. Suppose that the theorem is known when the dimension is $< n$, and suppose that $\dim G = n > 0$. By Theorem 4.29 and the inductive hypothesis, there is no loss of generality in assuming that G is semisimple.

Let Z_G be the center of G, and write $T^\times = T - (T \cap Z_G)$ and $G^\times = G - Z_G$. Since G is compact semisimple, Theorem 4.29 notes that $|Z_G| < \infty$. From the examples in §I.1, we know that no Lie algebra of dimension 1 or 2 is semisimple, and hence $\dim G \geq 3$. Therefore $G^\times$ is open connected dense in G. And, of course, $(T^\times)^G$ is nonempty. We shall prove that

$$(T^\times)^G \text{ is open and closed in } G^\times, \tag{4.37}$$

and then it follows that

$$G^\times = (T^\times)^G. \tag{4.38}$$

To obtain the conclusion of the theorem from this result, we observe that T^G is the image of the compact set $G \times T$ under the map $(x, t) \mapsto xtx^{-1}$. Hence T^G is a closed set, and (4.38) shows that it contains $G^\times$. Since $G^\times$ is dense, we obtain $T^G = G$. This conclusion will complete the induction and the proof of the theorem.

Thus we are to prove (4.37). To prove that $(T^\times)^G$ is closed in $G^\times$, let $\{t_n\}$ and $\{x_n\}$ be sequences in T and G with $\lim (t_n)^{x_n} = g \in G^\times$. Passing to a subsequence (by compactness) if necessary, we may assume that $\lim t_n = t \in T$ and $\lim x_n = x \in G$. Then continuity gives $g = t^x$. To see that t is in $T^\times$, suppose on the contrary that t is in $T \cap Z_G$. Then $g = t$ is in Z_G and is not in $G^\times$, contradiction. We conclude that t is in $T^\times$ and that g is in $(T^\times)^G$. Hence $(T^\times)^G$ is closed in $G^\times$.

To prove that $(T^\times)^G$ is open in $G^\times$, it is enough to prove that any $t \in T^\times$ is an interior point of $(T^\times)^G$. Fix t in $T^\times$. Let $Z = (Z_G(t))_0$. This is a compact connected group with $T \subseteq Z \subseteq G$. Let $\mathfrak{z}_0$ be its Lie algebra. Since by assumption t is not in Z_G, we see that $\mathfrak{z}_0 \neq \mathfrak{g}_0$ and hence $\dim Z < \dim G$. By inductive hypothesis,

$$T^Z = Z. \tag{4.39}$$

Let $Z^\times = Z - (Z \cap Z_G)$. Then (4.39) gives

$$\begin{aligned} Z^\times &= \bigcup_{y \in Z} T^y - (Z \cap Z_G) = \bigcup_{y \in Z} T^y - ((\bigcup_{y \in Z} T^y) \cap Z_G) \\ &= \bigcup_{y \in Z} T^y - \bigcup_{y \in Z} (T^y \cap Z_G) = \bigcup_{y \in Z} T^y - \bigcup_{y \in Z} (T \cap Z_G)^y \\ &\subseteq \bigcup_{y \in Z} (T - (T \cap Z_G))^y = \bigcup_{y \in Z} (T^\times)^y. \end{aligned} \tag{4.40}$$

The right side of (4.40) is contained in Z. Also it does not contain any member of $Z \cap Z_G$. In fact, if $(t')^y = z$ is in $Z \cap Z_G$ with $y \in Z$, then $t' = z^{y^{-1}} = z$ shows that t' is in Z_G, contradiction. Consequently the right side of (4.40) is contained in $Z^\times$. Then equality must hold throughout (4.40), and we find that $Z^\times = (T^\times)^Z$. Hence

$$(T^\times)^G = (Z^\times)^G. \tag{4.41}$$

We shall introduce a certain open subset Z_1 of $Z^\times$ containing t. Let $\mathfrak{q}_0$ be the orthogonal complement to $\mathfrak{z}_0$ in $\mathfrak{g}_0$ relative to our given inner product $-B$. We have

$$\mathfrak{z}_0 = \{X \in \mathfrak{g}_0 \mid \operatorname{Ad}(t)X = X\} = \ker(\operatorname{Ad}(t) - 1).$$

Since $\operatorname{Ad}(t)$ is an orthogonal transformation, we have

$$\mathfrak{q}_0 = \operatorname{image}(\operatorname{Ad}(t) - 1).$$

For any $z \in Z$, the orthogonal transformation $\operatorname{Ad}(z)$ leaves $\mathfrak{z}_0$ stable, and therefore it leaves $\mathfrak{q}_0$ stable also. Put

$$Z_1 = \{z \in Z \mid \det(\operatorname{Ad}(z) - 1)|_{\mathfrak{q}_0} \neq 0\}.$$

This set is open in Z, and no member of Z_G is in it (since $\mathfrak{q}_0 \neq 0$). Since $\operatorname{Ad}(t)$ does not have the eigenvalue 1 on $\mathfrak{q}_0$, t is in Z_1. Thus Z_1 is an open subset of $Z^\times$ containing t.

By (4.41), we obtain

$$t \in Z_1^G \subseteq (Z^\times)^G = (T^\times)^G.$$

Thus it is enough to prove that Z_1^G is open in G. To do so, we shall prove that the map $\psi : G \times Z \to G$ given by $\psi(y, x) = x^y$ has differential mapping onto at every point of $G \times Z_1$. Thus fix $y \in G$ and $x \in Z_1$. We identify the tangent spaces at y, x, and x^y with $\mathfrak{g}_0$, $\mathfrak{z}_0$, and $\mathfrak{g}_0$ by left translation. First let Y be in $\mathfrak{g}_0$. To compute $(d\psi)_{(y,x)}(Y, 0)$, we observe from (1.90) that

$$x^{y \exp rY} = x^y \exp(r\operatorname{Ad}(yx^{-1})Y) \exp(-r\operatorname{Ad}(y)Y). \tag{4.42}$$

We know from Lemma 1.92a that

$$\exp rX' \exp rY' = \exp\{r(X' + Y') + O(r^2)\} \qquad \text{as } r \to 0.$$

Hence the right side of (4.42) is

$$= x^y \exp(r\operatorname{Ad}(y)(\operatorname{Ad}(x^{-1}) - 1)Y + O(r^2)),$$

and

$$d\psi(Y, 0) = \mathrm{Ad}(y)(\mathrm{Ad}(x^{-1}) - 1)Y. \tag{4.43}$$

Next let X be in $\mathfrak{z}_0$. Then (1.90) gives

$$(x \exp rX)^y = x^y \exp(r\mathrm{Ad}(y)X),$$

and hence

$$d\psi(0, X) = \mathrm{Ad}(y)X. \tag{4.44}$$

Combining (4.43) and (4.44), we obtain

$$d\psi(Y, X) = \mathrm{Ad}(y)((\mathrm{Ad}(x^{-1}) - 1)Y + X). \tag{4.45}$$

Since x is in Z_1, $\mathrm{Ad}(x^{-1}) - 1$ is invertible on $\mathfrak{q}_0$, and thus the set of all $(\mathrm{Ad}(x^{-1}) - 1)Y$ contains $\mathfrak{q}_0$. Since X is arbitrary in $\mathfrak{z}_0$, the set of all $(\mathrm{Ad}(x^{-1}) - 1)Y + X$ is all of $\mathfrak{g}_0$. But $\mathrm{Ad}(y)$ is invertible, and thus (4.45) shows that $d\psi$ is onto $\mathfrak{g}_0$. This completes the proof that $(T^\times)^G$ is open in $G^\times$, and the theorem follows.

Corollary 4.46. Every element of a compact connected Lie group G lies in some maximal torus.

PROOF. Let T be a maximal torus. If y is given, then Theorem 4.36 gives $y = xtx^{-1}$ for some $x \in G$ and $t \in T$. Then y is in T^x, and T^x is the required maximal torus.

Corollary 4.47. The center Z_G of a compact connected Lie group lies in every maximal torus.

PROOF. Let T be a maximal torus. If $z \in Z_G$ is given, then Theorem 4.36 gives $z = xtx^{-1}$ for some $x \in G$ and $t \in T$. Multiplying on the left by x^{-1} and on the right by x and using that z is central, we see that $z = t$. Hence z is in T.

Corollary 4.48. For any compact connected Lie group G, the exponential map is onto G.

PROOF. The exponential map is onto for each maximal torus, and hence this corollary follows from Corollary 4.46.

Lemma 4.49. Let A be a compact abelian Lie group such that A/A_0 is cyclic, where A_0 denotes the identity component of A. Then A has an element whose powers are dense in A.

PROOF. Since A_0 is a torus, we can choose a_0 in A_0 such that the powers of a_0 are dense in A_0. Let $N = |A/A_0|$, and let b be a representative of A of a generating coset of A/A_0. Since b^N is in A_0, we can find c in A_0 with $b^N c^N = a_0$. Then the closure of the powers of bc is a subgroup containing A_0 and a representative of each coset of A/A_0, hence is all of A.

Theorem 4.50. Let G be a compact connected Lie group, and let S be a torus of G. If g in G centralizes S, then there is a torus S' in G containing both S and g.

PROOF. Let A be the closure of $\bigcup_{n=-\infty}^{\infty} g^n S$. Then the identity component A_0 is a torus. Since A_0 is open in A, $\bigcup_{n=-\infty}^{\infty} g^n A_0$ is an open subgroup of A containing $\bigcup_{n=-\infty}^{\infty} g^n S$. Hence $\bigcup_{n=-\infty}^{\infty} g^n A_0 = A$. By compactness of A_0, some nonzero power of g is in A_0. If N denotes the smallest positive such power, then A/A_0 is cyclic of order N. Applying Lemma 4.49, we can find a in A whose powers are dense in A. By Corollary 4.48 we can write $a = \exp X$ for some $X \in \mathfrak{g}_0$. Then the closure of $\{\exp rX, -\infty < r < \infty\}$ is a torus S' containing A, hence containing both S and g.

Corollary 4.51. In a compact connected Lie group, the centralizer of a torus is connected.

PROOF. Theorem 4.50 shows that the centralizer is the union of the tori containing the given torus.

Corollary 4.52. A maximal torus in a compact connected Lie group is equal to its own centralizer.

PROOF. Apply Theorem 4.50.

6. Analytic Weyl Group

We continue with the notation of §5: G is a compact connected Lie group, $\mathfrak{g}_0$ is its Lie algebra, and B is the negative of an invariant inner product on $\mathfrak{g}_0$. Let T be a maximal torus, and let $\mathfrak{t}_0$ be its Lie algebra. We indicate complexifications of Lie algebras by dropping subscripts 0. Let $\Delta(\mathfrak{g}, \mathfrak{t})$ be the set of roots of $\mathfrak{g}$ with respect to $\mathfrak{t}$. The center $Z_{\mathfrak{g}_0}$ of $\mathfrak{g}_0$ is contained in $\mathfrak{t}_0$, and all roots vanish on $Z_{\mathfrak{g}}$.

The roots are purely imaginary on $\mathfrak{t}_0$, as a consequence of (4.32) and passage to differentials. We define $\mathfrak{t}_{\mathbb{R}} = i\mathfrak{t}_0$; this is a real form of $\mathfrak{t}$ on which all the roots are real. We may then regard all roots as members of $\mathfrak{t}_{\mathbb{R}}^*$. The set $\Delta(\mathfrak{g}, \mathfrak{t})$ is an abstract reduced root system in the subspace of $\mathfrak{t}_{\mathbb{R}}^*$ coming from the semisimple Lie algebra $[\mathfrak{g}, \mathfrak{g}]$.

The negative definite form B on $\mathfrak{t}_0$ leads by complexification to a positive definite form on $\mathfrak{t}_{\mathbb{R}}$. Thus, for $\lambda \in \mathfrak{t}_{\mathbb{R}}^*$, let H_λ be the member of $\mathfrak{t}_{\mathbb{R}}$ such that

$$\lambda(H) = B(H, H_\lambda) \qquad \text{for } H \in \mathfrak{t}_{\mathbb{R}}.$$

The resulting linear map $\lambda \mapsto H_\lambda$ is a vector-space isomorphism of $\mathfrak{t}_{\mathbb{R}}^*$ with $\mathfrak{t}_{\mathbb{R}}$. Under this isomorphism let $(iZ_{\mathfrak{g}_0})^*$ be the subspace of $\mathfrak{t}_{\mathbb{R}}^*$ corresponding to $iZ_{\mathfrak{g}_0}$. The inner product on $\mathfrak{t}_{\mathbb{R}}$ induces an inner product on $\mathfrak{t}_{\mathbb{R}}^*$ denoted by $\langle \cdot, \cdot \rangle$. Relative to this inner product, the members of $\Delta(\mathfrak{g}, \mathfrak{t})$ span the orthogonal complement of $(iZ_{\mathfrak{g}_0})^*$, and $\Delta(\mathfrak{g}, \mathfrak{t})$ is an abstract reduced root system in this orthogonal complement. Also we have

$$\langle \lambda, \mu \rangle = \lambda(H_\mu) = \mu(H_\lambda) = B(H_\lambda, H_\mu).$$

For $\alpha \in \Delta(\mathfrak{g}, \mathfrak{t})$, the **root reflection** s_α is given as in the semisimple case by

$$s_\alpha(\lambda) = \lambda - \frac{2\langle \lambda, \alpha \rangle}{|\alpha|^2}\alpha.$$

The linear transformation s_α is the identity on $(iZ_{\mathfrak{g}_0})^*$ and is the usual root reflection in the orthogonal complement. The **Weyl group** $W(\Delta(\mathfrak{g}, \mathfrak{t}))$ is the group generated by the s_α's for $\alpha \in \Delta(\mathfrak{g}, \mathfrak{t})$. This consists of the members of the usual Weyl group of the abstract root system, with each member extended to be the identity on $(iZ_{\mathfrak{g}_0})^*$.

We might think of $W(\Delta(\mathfrak{g}, \mathfrak{t}))$ as an algebraically defined Weyl group. There is also an analytically defined **Weyl group** $W(G, T)$, defined as the quotient of normalizer by centralizer

$$W(G, T) = N_G(T)/Z_G(T).$$

The group $W(G, T)$ acts by automorphisms of T, hence by invertible linear transformations on $\mathfrak{t}_0$ and the associated spaces $\mathfrak{t}_{\mathbb{R}} = i\mathfrak{t}_0$, $\mathfrak{t}$, $\mathfrak{t}_{\mathbb{R}}^*$, and $\mathfrak{t}^*$. Only 1 acts as the identity. In the definition of $W(G, T)$, we can replace $Z_G(T)$ by T, according to Corollary 4.52. The group $W(G, T)$ plays the following role in the theory.

Proposition 4.53. For a compact connected Lie group G with maximal torus T, every element of G is conjugate to a member of T, and two elements of T are conjugate within G if and only if they are conjugate via $W(G, T)$. Thus the conjugacy classes in G are parametrized by $T/W(G,T)$. This correspondence respects the topologies in that a continuous complex-valued function f on T extends to a continuous function F on G invariant under group conjugation if and only if f is invariant under $W(G,T)$.

PROOF. Theorem 4.36 says that every conjugacy class meets T. Suppose that s and t are in T and g is in G and $gtg^{-1} = s$. We show that there is an element g_0 of $N_G(T)$ with $g_0 t g_0^{-1} = s$. In fact, consider the centralizer $Z_G(s)$. This is a closed subgroup of G with Lie algebra

$$Z_{\mathfrak{g}}(s) = \{X \in \mathfrak{g} \mid \mathrm{Ad}(s)X = X\}.$$

The identity component $(Z_G(s))_0$ is a group to which we can apply Theorem 4.34. Both $\mathfrak{t}$ and $\mathrm{Ad}(g)\mathfrak{t}$ are in $Z_{\mathfrak{g}}(s)$, and they are maximal abelian; hence there exists $z \in (Z_G(s))_0$ with

$$\mathfrak{t} = \mathrm{Ad}(zg)\mathfrak{t}.$$

Then $g_0 = zg$ is in $N_G(T)$ and $(zg)t(zg)^{-1} = s$.

Thus the conjugacy classes in G are given by $T/W(G,T)$. Let us check that continuous functions correspond. If F is continuous on G, then certainly its restriction f to T is continuous. Conversely suppose f is continuous on T and invariant under $W(G,T)$. Define F on G by $F(xtx^{-1}) = f(t)$; we have just shown that F is well defined. Let $\{g_n\}$ be a sequence in G with limit g, and write $g_n = x_n t_n x_n^{-1}$. Using the compactness of G and T, we can pass to a subsequence so that $\{x_n\}$ and $\{t_n\}$ have limits, say $\lim x_n = x$ and $\lim t_n = t$. Then $g = xtx^{-1}$, and the continuity of f gives

$$\lim F(g_n) = \lim f(t_n) = f(t) = F(g).$$

Hence F is continuous.

The discussion of characters of finite-dimensional representations in §2 showed the importance of understanding the conjugacy classes in G, and Proposition 4.53 has now reduced that question to an understanding of $W(G,T)$. There is a simple description of $W(G,T)$, which is the subject of the following theorem.

Theorem 4.54. For a compact connected Lie group G with maximal torus T, the analytically defined Weyl group $W(G,T)$, when considered as acting on $\mathfrak{t}_{\mathbb{R}}^*$, coincides with the algebraically defined Weyl group $W(\Delta(\mathfrak{g},\mathfrak{t}))$.

REMARK. Most of the argument consists in exhibiting the root reflections as occurring in $W(G,T)$. The calculation for this part is motivated by what happens in $SU(2)$. For this group, T is diagonal and s_α is given by $\begin{pmatrix} 0 & 1 \\ -1 & 0 \end{pmatrix}$. Let bar denote conjugation of $\mathfrak{sl}(2,\mathbb{C})$ with respect to $\mathfrak{su}(2)$. With $E_\alpha = \begin{pmatrix} 0 & 1 \\ 0 & 0 \end{pmatrix}$, we have $\overline{E_\alpha} = \begin{pmatrix} 0 & 0 \\ -1 & 0 \end{pmatrix}$ and $\begin{pmatrix} 0 & 1 \\ -1 & 0 \end{pmatrix} = \exp \frac{\pi}{2}(E_\alpha + \overline{E_\alpha})$. The general case amounts to embedding this argument in G.

PROOF. In view of Theorem 4.29, we may assume that G is semisimple. To show that $W(\Delta(\mathfrak{g},\mathfrak{t})) \subseteq W(G,T)$, it is enough to show that α in $\Delta(\mathfrak{g},\mathfrak{t})$ implies s_α in $W(G,T)$. Thus let bar denote conjugation of $\mathfrak{g}$ with respect to $\mathfrak{g}_0$, and extend B to a complex bilinear form on $\mathfrak{g}$. Let E_α be in $\mathfrak{g}_\alpha$, and write $E_\alpha = X_\alpha + iY_\alpha$ with X_α and Y_α in $\mathfrak{g}_0$. Then $\overline{E_\alpha} = X_\alpha - iY_\alpha$ is in $\mathfrak{g}_{-\alpha}$. For H in $\mathfrak{t}$, we have

$$[X_\alpha, H] = -\tfrac{1}{2}[H, E_\alpha + \overline{E_\alpha}] = -\tfrac{1}{2}\alpha(H)(E_\alpha - \overline{E_\alpha}) = -i\alpha(H)Y_\alpha. \tag{4.55}$$

Also Lemma 2.18a, reinterpreted with B in place of the Killing form, gives

$$\begin{aligned} [X_\alpha, Y_\alpha] &= \frac{1}{4i}[E_\alpha + \overline{E_\alpha}, E_\alpha - \overline{E_\alpha}] \\ &= -\frac{1}{2i}[E_\alpha, \overline{E_\alpha}] = -\frac{1}{2i}B(E_\alpha, \overline{E_\alpha})H_\alpha. \end{aligned} \tag{4.56}$$

Since $B(E_\alpha, \overline{E_\alpha}) < 0$, we can define a real number r by

$$r = \frac{\sqrt{2}\pi}{|\alpha|\sqrt{-B(E_\alpha, \overline{E_\alpha})}}.$$

Since X_α is in $\mathfrak{g}_0$, $g = \exp rX_\alpha$ is in G. We compute $\mathrm{Ad}(g)H$ for $H \in \mathfrak{t}_{\mathbb{R}}$. We have

$$\mathrm{Ad}(g)H = e^{\mathrm{ad}\, rX_\alpha} H = \sum_{k=0}^{\infty} \frac{r^k}{k!}(\mathrm{ad}\, X_\alpha)^k H. \tag{4.57}$$

If $\alpha(H) = 0$, then (4.55) shows that the series (4.57) collapses to H. If $H = H_\alpha$, then we obtain

$$r^2(\mathrm{ad}\, X_\alpha)^2 H_\alpha = \tfrac{1}{2}|\alpha|^2 B(E_\alpha, \overline{E_\alpha}) r^2 H_\alpha = -\pi^2 H_\alpha$$

from (4.55) and (4.56). Therefore (4.57) shows that

$$\begin{aligned} \mathrm{Ad}(g)H_\alpha &= \sum_{m=0}^{\infty} \frac{r^{2m}}{(2m)!}(\mathrm{ad}\, X_\alpha)^{2m} H_\alpha + \sum_{m=0}^{\infty} \frac{r^{2m+1}}{(2m+1)!}(\mathrm{ad}\, X_\alpha)^{2m+1} H_\alpha \\ &= \sum_{m=0}^{\infty} \frac{(-1)^m \pi^{2m}}{(2m!)} H_\alpha + r \sum_{m=0}^{\infty} \frac{(-1)^m \pi^{2m}}{(2m+1)!}[X_\alpha, H_\alpha] \\ &= (\cos \pi) H_\alpha + r\pi^{-1}(\sin \pi)[X_\alpha, H_\alpha] \\ &= -H_\alpha. \end{aligned}$$

Thus every $H \in \mathfrak{t}_{\mathbb{R}}$ satisfies

$$\mathrm{Ad}(g)H = H - \frac{2\alpha(H)}{|\alpha|^2} H_\alpha,$$

and $\mathrm{Ad}(g)$ normalizes $\mathfrak{t}_{\mathbb{R}}$, operating as s_α on $\mathfrak{t}_{\mathbb{R}}^*$.

It follows that $W(\Delta(\mathfrak{g}, \mathfrak{t})) \subseteq W(G, T)$. Next let us observe that $W(G, T)$ permutes the roots. In fact, let g be in $N_G(T) = N_G(\mathfrak{t})$, let α be in Δ, and let E_α be in $\mathfrak{g}_\alpha$. Then

$$\begin{aligned}[H, \mathrm{Ad}(g)E_\alpha] &= \mathrm{Ad}(g)[\mathrm{Ad}(g)^{-1}H, E_\alpha] = \mathrm{Ad}(g)(\alpha(\mathrm{Ad}(g)^{-1}H)E_\alpha) \\ &= \alpha(\mathrm{Ad}(g)^{-1}H)\mathrm{Ad}(g)E_\alpha = (g\alpha)(H)\mathrm{Ad}(g)E_\alpha\end{aligned}$$

shows that $g\alpha$ is in Δ and that $\mathrm{Ad}(g)E_\alpha$ is a root vector for $g\alpha$. Thus $W(G, T)$ permutes the roots.

Fix a simple system Π for Δ, let g be given in $W(G, T)$, and let $\tilde{g}$ be a representative of g in $N_G(T)$. It follows from the previous paragraph that $g\Pi$ is another simple system for Δ. By Theorem 2.63 choose w in $W(\Delta(\mathfrak{g}, \mathfrak{t}))$ with $wg\Pi = \Pi$. We show that wg fixes $\mathfrak{t}_{\mathbb{R}}^*$. If so, then wg is the identity in $W(G, T)$, and $g = w^{-1}$. So g is exhibited as in $W(\Delta(\mathfrak{g}, \mathfrak{t}))$, and $W(\Delta(\mathfrak{g}, \mathfrak{t})) = W(G, T)$.

Thus let $wg\Pi = \Pi$. Since $W(\Delta(\mathfrak{g}, \mathfrak{t})) \subseteq W(G, T)$, w has a representative $\tilde{w}$ in $N_G(T)$. Let Δ^+ be the positive system corresponding to Π, and define $\delta = \frac{1}{2}\sum_{\alpha \in \Delta^+} \alpha$. Then $wg\delta = \delta$, and so $\mathrm{Ad}(\tilde{w}\tilde{g})H_\delta = H_\delta$. If S denotes the closure of $\{\exp irH_\delta \mid r \in \mathbb{R}\}$, then S is a torus, and $\tilde{w}\tilde{g}$ is in its centralizer $Z_G(S)$. Let $\mathfrak{s}_0$ be the Lie algebra of S. We claim that $Z_{\mathfrak{g}_0}(\mathfrak{s}_0) = \mathfrak{t}_0$. If so, then $Z_G(S) = T$ by Corollary 4.51. Hence $\tilde{w}\tilde{g}$ is in T, $wg = \mathrm{Ad}(\tilde{w}\tilde{g})$ fixes $\mathfrak{t}_{\mathbb{R}}^*$, and the proof is complete.

To see that $Z_{\mathfrak{g}_0}(\mathfrak{s}_0) = \mathfrak{t}_0$, let α_i be a simple root. Proposition 2.69 shows that $2\langle\delta, \alpha_i\rangle/|\alpha_i|^2 = 1$. For a general positive root α, we therefore have $\langle\delta, \alpha\rangle > 0$. Thus $\alpha(H_\delta) \neq 0$ for all $\alpha \in \Delta(\mathfrak{g}, \mathfrak{t})$. By Lemma 4.33, $Z_{\mathfrak{g}}(H_\delta) = \mathfrak{t}$. Hence

$$Z_{\mathfrak{g}_0}(\mathfrak{s}_0) = \mathfrak{g}_0 \cap Z_{\mathfrak{g}}(\mathfrak{s}_0) = \mathfrak{g}_0 \cap Z_{\mathfrak{g}}(H_\delta) = \mathfrak{g}_0 \cap \mathfrak{t} = \mathfrak{t}_0,$$

as required.

7. Integral Forms

We continue with the notation of §§5–6 for a compact connected Lie group G with maximal torus T. We saw in (4.32) that roots $\alpha \in \Delta(\mathfrak{g}, \mathfrak{t})$ have the property that they lift from imaginary-valued linear functionals on $\mathfrak{t}_0$ to multiplicative characters of T. Let us examine this phenomenon more systematically.

Proposition 4.58. If λ is in $\mathfrak{t}^*$, then the following conditions on λ are equivalent:

(i) Whenever $H \in \mathfrak{t}_0$ satisfies $\exp H = 1$, $\lambda(H)$ is in $2\pi i\mathbb{Z}$.
(ii) There is a multiplicative character ξ_λ of T with $\xi_\lambda(\exp H) = e^{\lambda(H)}$ for all $H \in \mathfrak{t}_0$.

All roots have these properties. If λ has these properties, then λ is real-valued on $\mathfrak{t}_{\mathbb{R}}$.

REMARK. A linear functional λ satisfying (i) and (ii) is said to be **analytically integral**.

PROOF. If $\mathbb{R}^n$ denotes the universal covering group of T, then $\exp : \mathfrak{t}_0 \to \mathbb{R}^n$ is an isomorphism and $\tilde{\xi}_\lambda(\exp H) = e^{\lambda(H)}$ is a well-defined homomorphism of $\mathbb{R}^n$ into $\mathbb{C}^\times$. Then $\tilde{\xi}_\lambda$ descends to T if and only if (i) holds, and so (i) and (ii) are equivalent. If $\tilde{\xi}_\lambda$ descends to ξ_λ on T, then ξ_λ has compact image in $\mathbb{C}^\times$, hence image in the unit circle, and it follows that λ is real-valued on $\mathfrak{t}_{\mathbb{R}}$. We saw in (4.32) that roots satisfy (ii).

Proposition 4.59. If λ in $\mathfrak{t}^*$ is analytically integral, then λ satisfies the condition

$$\frac{2\langle\lambda, \alpha\rangle}{|\alpha|^2} \quad \text{is in } \mathbb{Z} \text{ for each } \alpha \in \Delta. \tag{4.60}$$

REMARK. A linear functional λ satisfying (4.60) is said to be **algebraically integral**.

PROOF. Let bar denote conjugation of $\mathfrak{g}$ with respect to $\mathfrak{g}_0$, and extend B to a complex bilinear form on $\mathfrak{g}$. Fix $\alpha \in \Delta(\mathfrak{g}, \mathfrak{t})$, and let E_α be a nonzero root vector. Since $B(E_\alpha, \overline{E_\alpha}) < 0$, we can normalize E_α so that $B(E_\alpha, \overline{E_\alpha}) = -2/|\alpha|^2$. Write $E_\alpha = X_\alpha + iY_\alpha$ with X_α and Y_α in $\mathfrak{g}_0$. Put $Z_\alpha = -i|\alpha|^{-2}H_\alpha \in \mathfrak{g}_0$. Then X_α, Y_α, and Z_α are in $\mathfrak{g}_0$, and (4.55) and (4.56) respectively give

$$[Z_\alpha, X_\alpha] = \frac{i}{|\alpha|^2}[X_\alpha, H_\alpha] = \frac{i}{|\alpha|^2}(-i\alpha(H_\alpha)Y_\alpha) = Y_\alpha$$

and

$$[X_\alpha, Y_\alpha] = \frac{1}{2i}\frac{2}{|\alpha|^2}H_\alpha = Z_\alpha.$$

Similarly

$$[Y_\alpha, Z_\alpha] = \frac{i}{|\alpha|^2}[H_\alpha, \tfrac{1}{2i}(E_\alpha - \overline{E_\alpha})] = \frac{1}{2|\alpha|^2}\alpha(H_\alpha)(E_\alpha + \overline{E_\alpha}) = X_\alpha.$$

Hence the correspondence

$$\tfrac{1}{2}\begin{pmatrix} 0 & i \\ i & 0 \end{pmatrix} \leftrightarrow X_\alpha, \qquad \tfrac{1}{2}\begin{pmatrix} 0 & 1 \\ -1 & 0 \end{pmatrix} \leftrightarrow Y_\alpha, \qquad \tfrac{1}{2}\begin{pmatrix} -i & 0 \\ 0 & i \end{pmatrix} \leftrightarrow Z_\alpha$$

gives us an isomorphism

$$\mathfrak{su}(2) \cong \mathbb{R}X_\alpha + \mathbb{R}Y_\alpha + \mathbb{R}Z_\alpha. \tag{4.61}$$

Since $SU(2)$ is simply connected, there exists a homomorphism $\Phi : SU(2) \to G$ whose differential implements (4.61). Under the complexification of (4.61), $h = \begin{pmatrix} 1 & 0 \\ 0 & -1 \end{pmatrix}$ maps to $2iZ_\alpha = 2|\alpha|^{-2}H_\alpha$. Thus $d\Phi(ih) = -2Z_\alpha = 2i|\alpha|^{-2}H_\alpha$. By (1.84),

$$1 = \Phi(1) = \Phi(\exp 2\pi i h) = \exp(d\Phi(2\pi i h)) = \exp(2\pi i(2|\alpha|^{-2}H_\alpha)).$$

Since λ is analytically integral, (i) of Proposition 4.58 shows that $\lambda(2\pi i(2|\alpha|^{-2}H_\alpha))$ is in $2\pi i\mathbb{Z}$. This means that $2\langle\lambda, \alpha\rangle/|\alpha|^2$ is in $\mathbb{Z}$.

Proposition 4.62. Fix a simple system of roots $\{\alpha_1, \ldots, \alpha_l\}$. Then $\lambda \in \mathfrak{t}^*$ is algebraically integral if and only if $2\langle\lambda, \alpha_i\rangle/|\alpha_i|^2$ is in $\mathbb{Z}$ for each simple root α_i.

PROOF. If λ is algebraically integral, then $2\langle\lambda, \alpha_i\rangle/|\alpha_i|^2$ is in $\mathbb{Z}$ for each α_i by definition. Conversely if $2\langle\lambda, \alpha_i\rangle/|\alpha_i|^2$ is an integer for each α_i, let $\alpha = \sum c_i\alpha_i$ be a positive root. We prove by induction on the level $\sum c_i$ that $2\langle\lambda, \alpha\rangle/|\alpha|^2$ is an integer. Level 1 is the given case. Assume the assertion when the level is $< n$, and let the level be $n > 1$ for α. Choose α_i with $\langle\alpha, \alpha_i\rangle > 0$. By Lemma 2.61, $\beta = s_{\alpha_i}\alpha$ is positive, and it certainly has level $< n$. Then

$$\frac{2\langle\lambda, \alpha\rangle}{|\alpha|^2} = \frac{2\langle s_{\alpha_i}\lambda, \beta\rangle}{|\beta|^2} = \frac{2\langle\lambda, \beta\rangle}{|\beta|^2} - \frac{2\langle\lambda, \alpha_i\rangle}{|\alpha_i|^2}\frac{2\langle\alpha_i, \beta\rangle}{|\beta|^2},$$

and the right side is an integer by inductive hypothesis. The proposition follows.

Propositions 4.58 and 4.59 tell us that we have inclusions

$$\begin{aligned} \mathbb{Z} \text{ combinations of roots} &\subseteq \text{analytically integral forms} \\ &\subseteq \text{algebraically integral forms.} \end{aligned} \tag{4.63}$$

Each of these three sets may be regarded as an additive group in $\mathfrak{t}^*_\mathbb{R}$.

Let us specialize to the case that G is semisimple. Propositions 2.49 and 4.62 show that the right member of (4.63) is a **lattice** in $\mathfrak{t}^*_\mathbb{R}$, i.e., a discrete subgroup with compact quotient. Proposition 2.49 shows that the left member of (4.63) spans $\mathfrak{t}^*_\mathbb{R}$ over $\mathbb{R}$ and hence is a sublattice. Thus (4.63) provides us with an inclusion relation for three lattices. Matters are controlled somewhat by the following result.

Proposition 4.64. If G is semisimple, then the index of the lattice of $\mathbb{Z}$ combinations of roots in the lattice of algebraically integral forms is exactly the determinant of the Cartan matrix.

Lemma 4.65. Let F be a free abelian group of rank l, and let R be a subgroup of rank l. Then it is possible to choose $\mathbb{Z}$ bases $\{t_i\}$ of F and $\{u_i\}$ of R such that $u_i = \delta_i t_i$ (with $\delta_i \in \mathbb{Z}$) for all i and such that δ_i divides δ_j if $i < j$.

REMARK. This result tells what happens in a standard proof of the Fundamental Theorem for Finitely Generated Abelian Groups. We omit the argument. See Artin [1991], 458.

Lemma 4.66. Let F be a free abelian group of rank l, and let R be a subgroup of rank l. Let $\{t_i\}$ and $\{u_i\}$ be $\mathbb{Z}$ bases of F and R, respectively, and suppose that $u_j = \sum_{i=1}^{l} d_{ij} t_i$. Then F/R has order $|\det(d_{ij})|$.

PROOF. A change of basis in the t's or in the u's corresponds to multiplying (d_{ij}) by an integer matrix of determinant ± 1. In view of Lemma 4.65, we may therefore assume (d_{ij}) is diagonal. Then the result is obvious.

PROOF OF PROPOSITION 4.64. Fix a simple system $\{\alpha_1, \ldots, \alpha_l\}$ and define $\{\lambda_1, \ldots, \lambda_l\}$ by $2\langle \lambda_i, \alpha_j \rangle / |\alpha_j|^2 = \delta_{ij}$. The $\{\lambda_i\}$ form a $\mathbb{Z}$ basis of the lattice of algebraically integral forms by Proposition 4.62, and the $\{\alpha_i\}$ form a $\mathbb{Z}$ basis of the lattice generated by the roots by Proposition 2.49. Write

$$\alpha_j = \sum_{k=1}^{l} d_{kj} \lambda_k,$$

and apply $2\langle \alpha_i, \cdot \rangle / |\alpha_i|^2$ to both sides. Then we see that $d_{ij} = 2\langle \alpha_i, \alpha_j \rangle / |\alpha_i|^2$. Proposition 2.52e shows that the determinant of the Cartan matrix is positive; thus the result follows from Lemma 4.66.

Proposition 4.67. If G is a compact connected Lie group and $\tilde{G}$ is a finite covering group, then the index of the group of analytically integral forms for G in the group of analytically integral forms for $\tilde{G}$ equals the order of the kernel of the covering homomorphism $\tilde{G} \to G$.

PROOF. This follows by combining Corollary 4.47 and Proposition 4.58.

Proposition 4.68. If G is a compact semisimple Lie group with trivial center, then every analytically integral form is a $\mathbb{Z}$ combination of roots.

PROOF. Let λ be analytically integral. Let $\alpha_1, \ldots, \alpha_l$ be a simple system, and define $H_j \in \mathfrak{t}_0$ by $\alpha_k(H_j) = 2\pi i \delta_{kj}$. If $\alpha = \sum_m n_m \alpha_m$ is a root and if X is in $\mathfrak{g}_\alpha$, then

$$\operatorname{Ad}(\exp H_j)X = e^{\operatorname{ad} H_j} X = e^{\alpha(H_j)} X = e^{\sum n_m \alpha_m(H_j)} X = e^{2\pi i n_j} X = X.$$

So $\exp H_j$ is in Z_G and therefore is 1. Since λ is analytically integral, Proposition 4.58 shows that $\lambda(H_j)$ is in $2\pi i \mathbb{Z}$. Write $\lambda = \sum_m c_m \alpha_m$. Evaluating both sides on H_j, we see that $c_j 2\pi i$ is in $2\pi i \mathbb{Z}$. Hence c_j is in $\mathbb{Z}$ for each j, and λ is a $\mathbb{Z}$ combination of roots.

8. Weyl's Theorem

We now combine a number of results from this chapter to prove the following theorem.

Theorem 4.69 (Weyl's Theorem). If G is a compact semisimple Lie group, then the fundamental group of G is finite. Consequently the universal covering group of G is compact.

Lemma 4.70. If G is a compact connected Lie group, then its fundamental group is finitely generated.

PROOF. We can write $G = \tilde{G}/Z$ where $\tilde{G}$ is the univeral covering group of G and Z is a discrete subgroup of the center of $\tilde{G}$. Here Z is isomorphic to the fundamental group of G. Let $e : \tilde{G} \to G$ be the covering homomorphism. About each point $x \in G$, choose a connected simply connected open neighborhood N_x and a connected simply connected neighborhood N'_x with closure in N_x. Extract a finite subcover of G from among the N'_x, say $N'_{x_1}, \ldots, N'_{x_n}$. Since N'_{x_j} is connected and simply connected, the components of $e^{-1}(N'_{x_j})$ in $\tilde{G}$ are homeomorphic to N'_{x_j}. Let M_{x_j} be one of them. Since N_{x_j} is connected and simply connected, the homeomorphism of M_{x_j} with N'_{x_j} extends to a homeomorphism of the closures. Therefore $U = \bigcup_{j=1}^n M_{x_j}$ is an open set in $\tilde{G}$ such that $\bar{U}$ is compact and $\tilde{G} = ZU$. By enlarging U, we may suppose also that 1 is in U and $U = U^{-1}$.

The set $\bar{U}\bar{U}^{-1}$ is compact in $\tilde{G}$ and is covered by the open sets zU, $z \in Z$, since $\tilde{G} = ZU$. Thus we can find $z_1, \ldots, z_k$ in Z such that

$$\bar{U}\bar{U}^{-1} \subseteq \bigcup_{j=1}^{k} z_j U. \tag{4.71}$$

Let Z_1 be the subgroup of Z generated by $z_1, \ldots, z_k$, and let E be the image of $\bar{U}$ in $\tilde{G}/Z_1$. Then E contains the identity and $E = E^{-1}$, and (4.71) shows that $EE^{-1} \subseteq E$. Thus E is a subgroup of $\tilde{G}/Z_1$. Since E contains the image of U, E is open, and thus $E = \tilde{G}/Z_1$ by connectedness. Since $\bar{U}$ is compact, E is compact. Consequently E is a finite-sheeted covering group of G. That is, G has a finite-sheeted covering group whose fundamental group Z_1 is finitely generated. The lemma follows.

PROOF OF THEOREM 4.69. Let $G = \tilde{G}/Z$, where $\tilde{G}$ is the universal covering group of G and Z is a discrete subgroup of the center of $\tilde{G}$. Here Z is a finitely generated abelian group by Lemma 4.70. If Z is finite, we are done. Otherwise Z has an infinite cyclic direct summand, and we can find a subgroup Z_1 of Z such that Z_1 has finite index in Z greater than the determinant of the Cartan matrix. Then $\tilde{G}/Z_1$ is a compact covering group of G with a number of sheets exceeding the determinant of the Cartan matrix. By Proposition 4.67 the index of the lattice of analytically integral forms for G in the corresponding lattice for $\tilde{G}/Z_1$ exceeds the determinant of the Cartan matrix. Comparing this conclusion with (4.63) and Proposition 4.64, we arrive at a contradiction. Theorem 4.69 follows.

More is true. As we noted in the proof of Weyl's Theorem, Propositions 4.64 and 4.67 show for a compact semisimple Lie group that the order of the center is $\leq$ the determinant of the Cartan matrix. Actually equality holds when the group is simply connected. This result may be regarded as a kind of existence theorem. For example, the group $SO(2n)$ has the 2-element center $\{\pm 1\}$, while the determinant of the Cartan matrix (type D_n) is 4. It follows that $SO(2n)$ is not simply connected and has a double cover. The relevant existence theorem will be proved in Chapter V as the Theorem of the Highest Weight, and the consequence about the order of the center of a compact semisimple Lie group will be proved at the end of that chapter.

9. Problems

1. Example 2 for $SU(n)$ in §1 gives a representation of $SU(2)$ in the space of holomorphic polynomials in z_1, z_2 homogeneous of degree N. Call this representation Φ_N, and let χ_N be its character. Let T be the maximal torus $T = \{t_\theta\}$ with $t_\theta = \begin{pmatrix} e^{i\theta} & 0 \\ 0 & e^{-i\theta} \end{pmatrix}$.
   (a) Compute $\Phi_N(t_\theta)$ on each monomial $z_1^k z_2^{N-k}$.

(b) Compute $d\Phi_N\begin{pmatrix}1 & 0\\ 0 & -1\end{pmatrix}$, and use Theorem 1.63 to deduce that Φ_N is irreducible.

(c) Give an explicit $(N+1)$-term geometric series for $\chi_N(t_\theta)$, and write the sum as a quotient.

(d) Decompose $\chi_M\chi_N$ as a sum of $\chi_{N'}$'s, and give a formula for the multiplicity of $\Phi_{N'}$ in $\Phi_M \otimes \Phi_N$.

2. Deduce Theorem 4.36 from Corollary 4.48 and Theorem 4.34.
3. Give direct proofs of Theorem 4.36 for $SO(n)$ and $Sp(n)$ in the spirit of the remarks with that theorem.
4. In $SO(3)$, show that there is an element whose centralizer in $SO(3)$ is not connected.
5. For $G = U(n)$ with T equal to the diagonal subgroup, what elements are in the normalizer $N_G(T)$?
6. Let G be a compact semisimple Lie group, and suppose that every algebraically integral form is analytically integral. Prove that G is simply connected.
7. Let Φ be an irreducible unitary finite-dimensional representation of the compact group G on the space V. The linear span of the matrix coefficients $(\Phi(x)u, v)$ is a vector space $\tilde{V}$, and it was noted in the proof of Theorem 4.20 that this space is invariant under the representation $r(g)f(x) = f(xg)$. Find the multiplicity of Φ in the space $\tilde{V}$ when $\tilde{V}$ is acted upon by r.

Problems 8–13 concern Example 2 for $G = SO(n)$ in §1. Let V_N be the space of complex-valued polynomials in $x_1, \dots, x_n$ homogeneous of degree N. For any homogeneous polynomial p, we define a differential operator $\partial(p)$ with constant coefficients by requiring that $\partial(\cdot)$ is linear in $(\cdot)$ and that

$$\partial(x_1^{k_1}\cdots x_n^{k_n}) = \frac{\partial^{k_1+\cdots+k_n}}{\partial x_1^{k_1}\cdots\partial x_n^{k_n}}.$$

For example, if we write $|x|^2 = x_1^2 + \cdots + x_n^2$, then $\partial(|x|^2) = \Delta$. If p and q are in the same V_N, then $\partial(\bar{q})p$ is a constant polynomial, and we define $\langle p, q\rangle$ to be that constant.

8. Prove that $\langle\cdot,\cdot\rangle$ is G invariant on V_N.
9. Prove that distinct monomials in V_N are orthogonal relative to $\langle\cdot,\cdot\rangle$ and that $\langle p, p\rangle$ is > 0 for such a monomial. Deduce that $\langle\cdot,\cdot\rangle$ is a Hermitian inner product.
10. Call $p \in V_N$ **harmonic** if $\partial(|x|^2)p = 0$, and let H_N be the subspace of harmonic polynomials. Prove that the orthogonal complement of $|x|^2 V_{N-2}$ in V_N relative to $\langle\cdot,\cdot\rangle$ is H_N.

11. Deduce from Problem 10 that Δ carries V_N onto V_{N-2}.
12. Deduce from Problem 10 that each $p \in V_N$ decomposes uniquely as

$$p = h_N + |x|^2 h_{N-2} + |x|^2 h_{N-4} + \cdots$$

with $h_N, h_{N-2}, h_{N-4}, \ldots$ homogeneous harmonic of the indicated degrees.
13. Compute the dimension of H_N.

Problems 14–16 concern Example 2 for $SU(n)$ in §1. Let V_N be the space of polynomials in $z_1, \ldots, z_n, \bar{z}_1, \ldots, \bar{z}_n$ that are homogeneous of degree N.

14. Show for each pair (p, q) with $p + q = N$ that the subspace $V_{p,q}$ of polynomials with p z-type factors and q $\bar{z}$-type factors is an invariant subspace under $SU(n)$.
15. The Laplacian in these coordinates is a multiple of $\sum_j \dfrac{\partial^2}{\partial z_j \partial \bar{z}_j}$. Using the result of Problem 11, prove that the Laplacian carries $V_{p,q}$ onto $V_{p-1,q-1}$.
16. Compute the dimension of the subspace of harmonic polynomials in $V_{p,q}$.

Problems 17–20 deal with integral forms. In each case the maximal torus T is understood to be as in the corresponding example of §5, and the notation for members of $\mathfrak{t}^*$ is to be as in the corresponding example of §II.1 (with $\mathfrak{h} = \mathfrak{t}$).

17. For $SU(n)$, a general member of $\mathfrak{t}^*$ may be written uniquely as $\sum_{j=1}^n c_j e_j$ with $\sum_{j=1}^n c_j = 0$.
    (a) Prove that the $\mathbb{Z}$ combinations of roots are those forms with all c_j in $\mathbb{Z}$.
    (b) Prove that the algebraically integral forms are those for which all c_j are in $\mathbb{Z} + \frac{k}{n}$ for some k.
    (c) Prove that every algebraically integral form is analytically integral.
    (d) Prove that the quotient of the lattice of algebraically integral forms by the lattice of $\mathbb{Z}$ combinations of roots is a cyclic group of order n.
18. For $SO(2n+1)$, a general member of $\mathfrak{t}^*$ is $\sum_{j=1}^n c_j e_j$.
    (a) Prove that the $\mathbb{Z}$ combinations of roots are those forms with all c_j in $\mathbb{Z}$.
    (b) Prove that the algebraically integral forms are those forms with all c_j in $\mathbb{Z}$ or all c_j in $\mathbb{Z} + \frac{1}{2}$.
    (c) Prove that every analytically integral form is a $\mathbb{Z}$ combination of roots.
19. For $Sp(n, \mathbb{C}) \cap U(2n)$, a general member of $\mathfrak{t}^*$ is $\sum_{j=1}^n c_j e_j$.
    (a) Prove that the $\mathbb{Z}$ combinations of roots are those forms with all c_j in $\mathbb{Z}$ and with $\sum_{j=1}^n c_j$ even.

(b) Prove that the algebraically integral forms are those forms with all c_j in $\mathbb{Z}$.

(c) Prove that every algebraically integral form is analytically integral.

20. For $SO(2n)$, a general member of $\mathfrak{t}^*$ is $\sum_{j=1}^n c_j e_j$.

(a) Prove that the $\mathbb{Z}$ combinations of roots are those forms with all c_j in $\mathbb{Z}$ and with $\sum_{j=1}^n c_j$ even.

(b) Prove that the algebraically integral forms are those forms with all c_j in $\mathbb{Z}$ or all c_j in $\mathbb{Z} + \frac{1}{2}$.

(c) Prove that the analytically integral forms are those forms with all c_j in $\mathbb{Z}$.

(d) The quotient of the lattice of algebraically integral forms by the lattice of $\mathbb{Z}$ combinations of roots is a group of order 4. Identify the group.

CHAPTER V

Finite-Dimensional Representations

Abstract. In any finite-dimensional representation of a complex semisimple Lie algebra $\mathfrak{g}$, a Cartan subalgebra $\mathfrak{h}$ acts completely reducibly, the simultaneous eigenvalues being called "weights." Once a positive system for the roots $\Delta^+(\mathfrak{g}, \mathfrak{h})$ has been fixed, one can speak of highest weights. The Theorem of the Highest Weight says that irreducible finite-dimensional representations are characterized by their highest weights and that the highest weight can be any dominant algebraically integral linear functional on $\mathfrak{h}$. The hard step in the proof is the construction of an irreducible representation corresponding to a given dominant algebraically integral form. This step is carried out by using "Verma modules," which are universal highest weight modules.

All finite-dimensional representations of $\mathfrak{g}$ are completely reducible. Consequently the nature of such a representation can be determined from the representation of $\mathfrak{h}$ in the space of "$\mathfrak{n}$ invariants." The Harish-Chandra Isomorphism identifies the center of the universal enveloping algebra $U(\mathfrak{g})$ with the Weyl-group invariant members of $U(\mathfrak{h})$. The proof uses the complete reducibility of finite-dimensional representations of $\mathfrak{g}$.

The center of $U(\mathfrak{g})$ acts by scalars in any irreducible representation of $\mathfrak{g}$, whether finite-dimensional or infinite-dimensional. The result is a homormorphism of the center into $\mathbb{C}$ and is known as the "infinitesimal character" of the representation. The Harish-Chandra Isomorphism makes it possible to parametrize all possible homomorphisms of the center into $\mathbb{C}$, thus to parametrize all possible infinitesimal characters. The parametrization is by the quotient of $\mathfrak{h}^*$ by the Weyl group.

The Weyl Character Formula attaches to each irreducible finite-dimensional representation a formal exponential sum corresponding to the character of the representation. The proof uses infinitesimal characters. The formula encodes the multiplicity of each weight, and this multiplicity is made explicit by the Kostant Multiplicity Formula. The formula encodes also the dimension of the representation, which is made explicit by the Weyl Dimension Formula.

Parabolic subalgebras provide a framework for generalizing the Theorem of the Highest Weight so that the Cartan subalgebra is replaced by a larger subalgebra called the "Levi factor" of the parabolic subalgebra.

The theory of finite-dimensional representations of complex semisimple Lie algebras has consequences for compact connected Lie groups. One of these is a formula for the order of the fundamental group. Another is a version of the Theorem of the Highest Weight that takes global properties of the group into account. The Weyl Character Formula becomes more explicit, giving an expression for the character of any irreducible representation when restricted to a maximal torus.

1. Weights

For most of this chapter we study finite-dimensional representations of complex semisimple Lie algebras. As introduced in Example 4 of §I.5, these are complex-linear homomorphisms of a complex semisimple Lie algebra into $\operatorname{End}_{\mathbb{C}} V$, where V is a finite-dimensional complex vector space. Historically the motivation for studying such representations comes from two sources—representations of $\mathfrak{sl}(2,\mathbb{C})$ and representations of compact Lie groups. Representations of $\mathfrak{sl}(2,\mathbb{C})$ were studied in §I.9, and the theory of the present chapter may be regarded as generalizing the results of that section to all complex semisimple Lie algebras.

Representations of compact connected Lie groups were studied in Chapter IV. If G is a compact connected Lie group, then a representation of G on a finite-dimensional complex vector space V yields a representation of the Lie algebra $\mathfrak{g}_0$ on V and then a representation of the complexification $\mathfrak{g}$ of $\mathfrak{g}_0$ on V. The Lie algebra $\mathfrak{g}_0$ is the direct sum of an abelian Lie algebra and a semisimple Lie algebra, and the same thing is true of $\mathfrak{g}$. Through studying the representations of the semisimple part of $\mathfrak{g}$, we shall be able, with only little extra effort, to complete the study of the representations of G at the end of this chapter.

The examples of representations in Chapter IV give us examples for the present chapter, as well as clues for how to proceed. The easy examples, apart from the trivial representation with $\mathfrak{g}$ acting as 0, are the standard representations of $\mathfrak{su}(n)^{\mathbb{C}}$ and $\mathfrak{so}(n)^{\mathbb{C}}$. These are obtained by differentiation of the standard representations of $SU(n)$ and $SO(n)$ and just amount to multiplication of a matrix by a column vector, namely

$$\varphi(X)\begin{pmatrix} z_1 \\ \vdots \\ z_n \end{pmatrix} = X\begin{pmatrix} z_1 \\ \vdots \\ z_n \end{pmatrix}.$$

The differentiated versions of the other examples in §IV.1 are more complicated because they involve tensor products. Although tensor products on the group level (4.2) are fairly simple, they become more complicated on the Lie algebra level (4.3) because of the product rule for differentiation. This complication persists for representations in spaces of symmetric or alternating tensors, since such spaces are subspaces of tensor products. Thus the usual representation of $SU(n)$ on $\bigwedge^l\mathbb{C}^n$ is given simply by

$$\Phi(g)(\varepsilon_{j_1}\wedge\cdots\wedge\varepsilon_{j_l}) = g\varepsilon_{j_1}\wedge\cdots\wedge g\varepsilon_{j_l},$$

while the corresponding representation of $\mathfrak{su}(n)^{\mathbb{C}}$ on $\bigwedge^l\mathbb{C}^n$ is given by

$$\varphi(X)(\varepsilon_{j_1}\wedge\cdots\wedge\varepsilon_{j_l}) = \sum_{k=1}^{l}\varepsilon_{j_1}\wedge\cdots\wedge\varepsilon_{j_{k-1}}\wedge X\varepsilon_{j_k}\wedge\varepsilon_{j_{k+1}}\wedge\cdots\wedge\varepsilon_{j_l}.$$

The second construction that enters the examples of §IV.1 is contragredient, given on the Lie group level by (4.1) and on the Lie algebra level by (4.4). Corollary A.22b, with $E = \mathbb{C}^n$, shows that the representation in a space $S^n(E^*)$ of polynomials may be regarded as the contragredient of the representation in the space $S^n(E)$ of symmetric tensors.

The clue for how to proceed comes from the representation theory of compact connected Lie groups G in Chapter IV. Let $\mathfrak{g}_0$ be the Lie algebra of G, and let $\mathfrak{g}$ be the complexification. If T is a maximal torus in G, then the complexified Lie algebra of T is a Cartan subalgebra $\mathfrak{t}$ of $\mathfrak{g}$. Insight into $\mathfrak{g}$ comes from roots relative to $\mathfrak{t}$, which correspond to simultaneous eigenspaces for the action of T, according to (4.32). If Φ is any finite-dimensional representation of G on a complex vector space V, then Φ may be regarded as unitary by Proposition 4.6. Hence $\Phi|_T$ is unitary, and Corollary 4.7 shows that $\Phi|_T$ splits as the direct sum of irreducible representations of T. By Corollary 4.9 each of these irreducible representations of T is 1-dimensional. Thus V is the direct sum of simultaneous eigenspaces for the action of T, hence also for the action of $\mathfrak{t}$.

At first this kind of decomposition seems unlikely to persist when the compact groups are dropped and we have only a representation of a complex semisimple Lie algebra, since Proposition 2.4 predicts only a generalized weight-space decomposition. But a decomposition into simultaneous eigenspaces is nonetheless valid and is the starting point for our investigation. Before coming to this, let us note that the proofs of Schur's Lemma and its corollary in §IV.2 are valid for representations of Lie algebras.

Proposition 5.1 (Schur's Lemma). Suppose φ and φ' are irreducible representations of a Lie algebra $\mathfrak{g}$ on finite-dimensional vector spaces V and V', respectively. If $L : V \to V'$ is a linear map such that $\varphi'(X)L = L\varphi(X)$ for all $X \in \mathfrak{g}$, then L is one-one onto or $L = 0$.

PROOF. We see easily that $\ker L$ and $\operatorname{image} L$ are invariant subspaces of V and V', respectively, and then the only possibilities are the ones listed.

Corollary 5.2. Suppose φ is an irreducible representation of a Lie algebra $\mathfrak{g}$ on a finite-dimensional complex vector space V. If $L : V \to V$ is a linear map such that $\varphi(X)L = L\varphi(X)$ for all $X \in \mathfrak{g}$, then L is scalar.

PROOF. Let λ be an eigenvalue of L. Then $L - \lambda I$ is not one-one onto, but it does commute with $\varphi(X)$ for all $X \in \mathfrak{g}$. By Proposition 5.1, $L - \lambda I = 0$.

Let $\mathfrak{g}$ be a complex semisimple Lie algebra. Fix a Cartan subalgebra $\mathfrak{h}$, and let $\Delta = \Delta(\mathfrak{g}, \mathfrak{h})$ be the set of roots. Following the notation first introduced in Corollary 2.38, let $\mathfrak{h}_0$ be the real form of $\mathfrak{h}$ on which all roots are real-valued. Let B be any nondegenerate symmetric invariant bilinear form on $\mathfrak{g}$ that is positive definite on $\mathfrak{h}_0$. Relative to B, we can define members H_α of $\mathfrak{h}$ for each $\alpha \in \Delta$. Then $\mathfrak{h}_0 = \sum_{\alpha\in\Delta} \mathbb{R}H_\alpha$.

Let φ be a representation on the complex vector space V. Recall from §II.2 that if λ is in $\mathfrak{h}^*$, we define V_λ to be the subspace

$$\{v \in V \mid (\varphi(H) - \lambda(H)1)^n v = 0 \text{ for all } H \in \mathfrak{h} \text{ and some } n = n(H, V)\}.$$

If $V_\lambda \neq 0$, then V_λ is called a **generalized weight space** and λ is a **weight**. Members of V_λ are called **generalized weight vectors**. When V is finite-dimensional, V is the direct sum of its generalized weight spaces by Proposition 2.4.

The **weight space** corresponding to λ is

$$\{v \in V \mid \varphi(H)v = \lambda(H)v \text{ for all } H \in \mathfrak{h}\},$$

i.e., the subspace of V_λ for which n can be taken to be 1. Members of the weight space are called **weight vectors**. The examples of weight vectors below continue the discussion of examples in §IV.1.

EXAMPLES FOR $G = SU(n)$. Here $\mathfrak{g} = \mathfrak{su}(n)^{\mathbb{C}} = \mathfrak{sl}(n, \mathbb{C})$. As in Example 1 of §II.1, we define $\mathfrak{h}$ to be the diagonal subalgebra. The roots are all $e_i - e_j$ with $i \neq j$.

1) Let V consist of all polynomials in $z_1, \dots, z_n, \bar{z}_1, \dots, \bar{z}_n$ homogeneous of degree N. Let $H = \operatorname{diag}(it_1, \dots, it_n)$ with $\sum t_j = 0$. Then the Lie algebra representation φ has

$$\begin{aligned}
\varphi(H)P(z, \bar{z}) &= \frac{d}{dr} P\left(e^{-rH}\begin{pmatrix} z_1 \\ \vdots \\ z_n \end{pmatrix}, e^{rH}\begin{pmatrix} \bar{z}_1 \\ \vdots \\ \bar{z}_n \end{pmatrix}\right)\Bigg|_{r=0} \\
&= \frac{d}{dr} P\left(\begin{pmatrix} e^{-irt_1} z_1 \\ \vdots \\ e^{-irt_n} z_n \end{pmatrix}, \begin{pmatrix} e^{irt_1} \bar{z}_1 \\ \vdots \\ e^{irt_n} \bar{z}_n \end{pmatrix}\right)\Bigg|_{r=0} \\
&= \sum_{j=1}^n (-it_j z_j) \frac{\partial P}{\partial z_j}(z, \bar{z}) + \sum_{j=1}^n (it_j \bar{z}_j) \frac{\partial P}{\partial \bar{z}_j}(z, \bar{z}).
\end{aligned}$$

If P is a monomial of the form

$$P(z, \bar{z}) = z_1^{k_1} \cdots z_n^{k_n} \bar{z}_1^{l_1} \cdots \bar{z}_n^{l_n} \qquad \text{with } \sum_{j=1}^n (k_j + l_j) = N,$$

then the above expression simplifies to

$$\varphi(H)P = \Big(\sum_{j=0}^{n} (l_j - k_j)(it_j)\Big)P.$$

Thus the monomial P is a weight vector of weight $\sum_{j=0}^{n} (l_j - k_j)e_j$.

2) Let $V = \bigwedge^l \mathbb{C}^n$. Again let $H = \operatorname{diag}(it_1, \dots, it_n)$ with $\sum t_j = 0$. Then the Lie algebra representation φ has

$$\begin{aligned}\varphi(H)(\varepsilon_{j_1} \wedge \cdots \wedge \varepsilon_{j_l}) &= \sum_{k=1}^{l} \varepsilon_{j_1} \wedge \cdots \wedge H\varepsilon_{j_k} \wedge \cdots \wedge \varepsilon_{j_l} \\ &= \sum_{k=1}^{l} (it_{j_k})(\varepsilon_{j_1} \wedge \cdots \wedge \varepsilon_{j_l}).\end{aligned}$$

Thus $\varepsilon_{j_1} \wedge \cdots \wedge \varepsilon_{j_l}$ is a weight vector of weight $\sum_{k=1}^{l} e_{j_k}$.

EXAMPLES FOR $G = SO(2n+1)$. Here $\mathfrak{g} = \mathfrak{so}(2n+1)^{\mathbb{C}} = \mathfrak{so}(2n+1, \mathbb{C})$. As in Example 2 of §II.1, we define $\mathfrak{h}$ to be built from the first n diagonal blocks of size 2. The roots are $\pm e_j$ and $\pm e_i \pm e_j$ with $i \neq j$.

1) Let $m = 2n+1$, and let V consist of all complex-valued polynomials on $\mathbb{R}^m$ of degree $\leq N$. Let H_1 be the member of $\mathfrak{h}$ equal to $\begin{pmatrix} 0 & 1 \\ -1 & 0 \end{pmatrix}$ in the first 2-by-2 block and 0 elsewhere. Then the Lie algebra representation φ has

(5.3)

$$\varphi(H_1)P\begin{pmatrix} x_1 \\ \vdots \\ x_m \end{pmatrix} = \frac{d}{dr} P \begin{pmatrix} x_1 \cos r - x_2 \sin r \\ x_1 \sin r + x_2 \cos r \\ x_3 \\ \vdots \\ x_m \end{pmatrix}\Bigg|_{r=0} = -x_2 \frac{\partial P}{\partial x_1}(x) + x_1 \frac{\partial P}{\partial x_2}(x).$$

For $P(x) = (x_1 + ix_2)^k$, $\varphi(H_1)$ thus acts as the scalar ik. The other 2-by-2 blocks of $\mathfrak{h}$ annihilate this P, and it follows that $(x_1 + ix_2)^k$ is a weight vector of weight $-ke_1$. Similarly $(x_1 - ix_2)^k$ is a weight vector of weight $+ke_1$.

Replacing P in (5.3) by $(x_{2j-1} \pm x_{2j})Q$ and making the obvious adjustments in the computation, we obtain

$$\varphi(H)((x_{2j-1} \pm ix_{2j})Q) = (x_{2j-1} \pm ix_{2j})(\varphi(H) \mp e_j(H))Q \qquad \text{for } H \in \mathfrak{h}.$$

Since $x_{2j-1} + ix_{2j}$ and $x_{2j-1} - ix_{2j}$ together generate x_{2j-1} and x_{2j} and since $\varphi(H)$ acts as 0 on x_{2n+1}^k, this equation tells us how to compute $\varphi(H)$ on any monomial, hence on any polynomial.

It is clear that the subspace of polynomials homogeneous of degree N is an invariant subspace under the representation. This invariant subspace is spanned by the weight vectors

$$(x_1 + ix_2)^{k_1}(x_1 - ix_2)^{l_1}(x_3 + ix_4)^{k_2} \cdots (x_{2n-1} - ix_{2n})^{l_n} x_{2n+1}^{k_0},$$

where $\sum_{j=0}^n k_j + \sum_{j=1}^n l_j = N$. Hence the weights of the subspace are all expressions $\sum_{j=1}^n (l_j - k_j)e_j$ with $\sum_{j=0}^n k_j + \sum_{j=1}^n l_j = N$.

2) Let $V = \bigwedge^l \mathbb{C}^{2n+1}$. The element H_1 of $\mathfrak{h}$ in the above example acts on $\varepsilon_1 + i\varepsilon_2$ by the scalar $-i$ and on $\varepsilon_1 - i\varepsilon_2$ by the scalar $+i$. Thus $\varepsilon_1 + i\varepsilon_2$ and $\varepsilon_1 - i\varepsilon_2$ are weight vectors in $\mathbb{C}^{2n+1}$ of respective weights $+e_1$ and $-e_1$. Also ε_{2n+1} has weight 0. Then the product rule for differentiation allows us to compute the weights in $\bigwedge^l \mathbb{C}^{2n+1}$ and find that they are all expressions

$$\pm e_{j_1} \pm \cdots \pm e_{j_r}$$

with

$$j_1 < \cdots < j_r \qquad \text{and} \qquad \begin{cases} r \le l & \text{if } l \le n \\ r \le 2n+1-l & \text{if } l > n. \end{cases}$$

Motivated by Proposition 4.59 for compact Lie groups, we say that a member λ of $\mathfrak{h}^*$ is **algebraically integral** if $2\langle \lambda, \alpha \rangle / |\alpha|^2$ is in $\mathbb{Z}$ for each $\alpha \in \Delta$.

Proposition 5.4. Let $\mathfrak{g}$ be a complex semisimple Lie algebra, let $\mathfrak{h}$ be a Cartan subalgebra, let $\Delta = \Delta(\mathfrak{g}, \mathfrak{h})$ be the roots, and let $\mathfrak{h}_0 = \sum_{\alpha \in \Delta} \mathbb{R} H_\alpha$. If φ is a representation of $\mathfrak{g}$ on the finite-dimensional complex vector space V, then

(a) $\varphi(\mathfrak{h})$ acts diagonably on V, so that every generalized weight vector is a weight vector and V is the direct sum of all the weight spaces

(b) every weight is real-valued on $\mathfrak{h}_0$ and is algebraically integral

(c) roots and weights are related by $\varphi(\mathfrak{g}_\alpha) V_\lambda \subseteq V_{\lambda+\alpha}$.

PROOF.

(a, b) If α is a root and E_α and $E_{-\alpha}$ are nonzero root vectors for α and $-\alpha$, then $\{H_\alpha, E_\alpha, E_{-\alpha}\}$ spans a subalgebra $\mathfrak{sl}_\alpha$ of $\mathfrak{g}$ isomorphic to $\mathfrak{sl}(2, \mathbb{C})$, with $2|\alpha|^{-2} H_\alpha$ corresponding to $h = \begin{pmatrix} 1 & 0 \\ 0 & -1 \end{pmatrix}$. Then the restriction of φ to $\mathfrak{sl}_\alpha$ is a finite-dimensional representation of $\mathfrak{sl}_\alpha$, and Corollary 1.69

shows that $\varphi(2|\alpha|^{-2}H_\alpha)$ is diagonable with integer eigenvalues. This proves (a) and the first half of (b). If λ is a weight and $v \in V_\lambda$ is nonzero, then we have just seen that $\varphi(2|\alpha|^{-2}H_\alpha)v = 2|\alpha|^{-2}\langle\lambda, \alpha\rangle v$ is an integral multiple of v. Hence $2\langle\lambda, \alpha\rangle/|\alpha|^2$ is an integer, and λ is algebraically integral.

(c) Let E_α be in $\mathfrak{g}_\alpha$, let v be in V_λ, and let H be in $\mathfrak{h}$. Then

$$\begin{aligned}\varphi(H)\varphi(E_\alpha)v &= \varphi(E_\alpha)\varphi(H)v + \varphi([H, E_\alpha])v \\ &= \lambda(H)\varphi(E_\alpha)v + \alpha(H)\varphi(E_\alpha)v \\ &= (\lambda + \alpha)(H)\varphi(E_\alpha)v.\end{aligned}$$

Hence $\varphi(E_\alpha)v$ is in $V_{\lambda+\alpha}$.

2. Theorem of the Highest Weight

In this section let $\mathfrak{g}$ be a complex semisimple Lie algebra, let $\mathfrak{h}$ be a Cartan subalgebra, let $\Delta = \Delta(\mathfrak{g}, \mathfrak{h})$ be the set of roots, and let $W(\Delta)$ be the Weyl group. Let $\mathfrak{h}_0$ be the real form of $\mathfrak{h}$ on which all roots are real-valued, and let B be any nondegenerate symmetric invariant bilinear form on $\mathfrak{g}$ that is positive definite on $\mathfrak{h}_0$. Introduce an ordering in $\mathfrak{h}_0^*$ in the usual way, and let Π be the resulting simple system.

If φ is a representation of $\mathfrak{g}$ on a finite-dimensional complex vector space V, then the weights of V are in $\mathfrak{h}_0^*$ by Proposition 5.4b. The largest weight in the ordering is called the **highest weight** of φ.

Theorem 5.5 (Theorem of the Highest Weight). Apart from equivalence the irreducible finite-dimensional representations φ of $\mathfrak{g}$ stand in one-one correspondence with the dominant algebraically integral linear functionals λ on $\mathfrak{h}$, the correspondence being that λ is the highest weight of φ_λ. The highest weight λ of φ_λ has these additional properties:

(a) λ depends only on the simple system Π and not on the ordering used to define Π

(b) the weight space V_λ for λ is 1-dimensional

(c) each root vector E_α for arbitrary $\alpha \in \Delta^+$ annihilates the members of V_λ, and the members of V_λ are the only vectors with this property

(d) every weight of φ_λ is of the form $\lambda - \sum_{i=1}^{l} n_i\alpha_i$ with the integers ≥ 0 and the α_i in Π

(e) each weight space V_μ for φ_λ has $\dim V_{w\mu} = \dim V_\mu$ for all w in the Weyl group $W(\Delta)$, and each weight μ has $|\mu| \leq |\lambda|$ with equality only if μ is in the orbit $W(\Delta)\lambda$.

REMARKS.

1) Because of (e) the weights in the orbit $W(\Delta)\lambda$ are said to be **extreme**. The set of extreme weights does not depend on the choice of Π.

2) Much of the proof of Theorem 5.5 will be given in this section after some examples. The proof will be completed in §3. The examples continue the notation of the examples in §1.

EXAMPLES.

1) With $\mathfrak{g} = \mathfrak{sl}(n, \mathbb{C})$, let V consist of all polynomials in $z_1, \dots, z_n$, $\bar{z}_1, \dots, \bar{z}_n$ homogeneous of degree N. The weights are all expressions $\sum_{j=1}^n (l_j - k_j)e_j$ with $\sum_{j=1}^n (k_j + l_j) = N$. The highest weight relative to the usual positive system is Ne_1. The subspace of holomorphic polynomials is an invariant subspace, and it has highest weight $-Ne_n$. The subspace of antiholomorphic polynomials is another invariant subspace, and it has highest weight Ne_1.

2) With $\mathfrak{g} = \mathfrak{sl}(n, \mathbb{C})$, let $V = \bigwedge^l \mathbb{C}^n$. The weights are all expressions $\sum_{k=1}^l e_{j_k}$. The highest weight relative to the usual positive system is $\sum_{k=1}^l e_k$.

3) With $\mathfrak{g} = \mathfrak{so}(2n+1, \mathbb{C})$, let the representation space consist of all complex-valued polynomials in $x_1, \dots, x_{2n+1}$ homogeneous of degree N. The weights are all expressions $\sum_{j=1}^n (l_j - k_j)e_j$ with $k_0 + \sum_{j=1}^n (k_j + l_j) = N$. The highest weight relative to the usual positive system is Ne_1.

4) With $\mathfrak{g} = \mathfrak{so}(2n+1, \mathbb{C})$, let $V = \bigwedge^l \mathbb{C}^{2n+1}$. If $l \le n$, the weights are all expressions $\pm e_{j_1} \pm \cdots \pm e_{j_r}$ with $j_1 < \cdots < j_r$ and $r \le l$, and the highest weight relative to the usual positive system is $\sum_{k=1}^l e_k$.

PROOF OF EXISTENCE OF THE CORRESPONDENCE. Let φ be an irreducible finite-dimensional representation of $\mathfrak{g}$ on a space V. The representation φ has weights by Proposition 2.4, and we let λ be the highest. Then λ is algebraically integral by Proposition 5.4b.

If α is in Δ^+, then $\lambda + \alpha$ exceeds λ and cannot be a weight. Thus $E_\alpha \in \mathfrak{g}_\alpha$ and $v \in V_\lambda$ imply $\varphi(E_\alpha)v = 0$ by Proposition 5.4c. This proves the first part of (c).

Extend φ multiplicatively to be defined on all of $U(\mathfrak{g})$ with $\varphi(1) = 1$ by Corollary 3.6. Since φ is irreducible, $\varphi(U(\mathfrak{g}))v = V$ for each $v \ne 0$ in V. Let $\beta_1, \dots, \beta_k$ be an enumeration of Δ^+, and let $H_1, \dots, H_l$ be a basis of $\mathfrak{h}$. By the Poincaré-Birkhoff-Witt Theorem (Theorem 3.8) the monomials

$$E_{-\beta_1}^{q_1} \cdots E_{-\beta_k}^{q_k} H_1^{m_1} \cdots H_l^{m_l} E_{\beta_1}^{p_1} \cdots E_{\beta_k}^{p_k} \tag{5.6}$$

form a basis of $U(\mathfrak{g})$. Let us apply φ of each of these monomials to v in V_λ. The E_β's give 0, the H's multiply by constants (by Proposition 5.4a), and the $E_{-\beta}$'s push the weight down (by Proposition 5.4c). Consequently the only members of V_λ that can be obtained by applying φ of (5.6) to v are the vectors of $\mathbb{C}v$. Thus V_λ is 1-dimensional, and (b) is proved.

The effect of φ of (5.6) applied to v in V_λ is to give a weight vector with weight

$$\lambda - \sum_{j=1}^{k} q_j \beta_j, \tag{5.7}$$

and these weight vectors span V. Thus the weights (5.7) are the only weights of φ, and (d) follows from Proposition 2.49. Also (d) implies (a).

To prove the second half of (c), let $v \notin V_\lambda$ satisfy $\varphi(E_\alpha)v = 0$ for all $\alpha \in \Delta^+$. Subtracting the component in V_λ, we may assume that v has 0 component in V_λ. Let λ_0 be the largest weight such that v has a nonzero component in V_{λ_0}, and let v' be the component. Then $\varphi(E_\alpha)v' = 0$ for all $\alpha \in \Delta^+$, and $\varphi(\mathfrak{h})v' \subseteq \mathbb{C}v'$. Applying φ of (5.6), we see that

$$V = \sum \mathbb{C}\varphi(E_{-\beta_1})^{q_1} \cdots \varphi(E_{-\beta_k})^{q_k} v'.$$

Every weight of vectors on the right side is strictly lower than λ, and we have a contradiction with the fact that λ occurs as a weight.

Next we prove that λ is dominant. Let α be in Δ^+, and form H'_α, E'_α, and $E'_{-\alpha}$ as in (2.26). These vectors span a Lie subalgebra $\mathfrak{sl}_\alpha$ of $\mathfrak{g}$ isomorphic to $\mathfrak{sl}(2, \mathbb{C})$, and the isomorphism carries H'_α to $h = \begin{pmatrix} 1 & 0 \\ 0 & -1 \end{pmatrix}$. For $v \neq 0$ in V_λ, the subspace of V spanned by all

$$\varphi(E'_{-\alpha})^p \varphi(H'_\alpha)^q \varphi(E'_\alpha)^r v$$

is stable under $\mathfrak{sl}_\alpha$, and (c) shows that it is the same as the span of all $\varphi(E'_{-\alpha})^p v$. On these vectors $\varphi(H'_\alpha)$ acts with eigenvalue

$$(\lambda - p\alpha)(H'_\alpha) = \frac{2\langle \lambda, \alpha \rangle}{|\alpha|^2} - 2p,$$

and the largest eigenvalue of $\varphi(H'_\alpha)$ is therefore $2\langle \lambda, \alpha \rangle / |\alpha|^2$. By Corollary 1.69 the largest eigenvalue for h in any finite-dimensional representation of $\mathfrak{sl}(2, \mathbb{C})$ is ≥ 0, and λ is therefore dominant.

Finally we prove (e). Fix $\alpha \in \Delta$, and form $\mathfrak{sl}_\alpha$ as above. Proposition 5.4a shows that V is the direct sum of its simultaneous eigenspaces under $\mathfrak{h}$ and hence also under the subspace $\ker\alpha$ of $\mathfrak{h}$. In turn, since $\ker\alpha$ commutes with $\mathfrak{sl}_\alpha$, each of these simultaneous eigenspaces under $\ker\alpha$ is invariant under $\mathfrak{sl}_\alpha$ and is completely reducible by Theorem 1.64.

Thus V is the direct sum of subspaces invariant and irreducible under $\mathfrak{sl}_\alpha \oplus \ker\alpha$. Let V' be one of these irreducible subspaces. Since $\mathfrak{h} \subseteq \mathfrak{sl}_\alpha \oplus \ker\alpha$, V' is the direct sum of its weight spaces: $V' = \bigoplus_\nu (V' \cap V_\nu)$. If ν and ν' are two weights occurring in V', then the irreducibility under $\mathfrak{sl}_\alpha \oplus \ker\alpha$ forces $\nu' - \nu = n\alpha$ for some integer n.

Fix a weight μ, and consider such a space V'. The weights of V' are $\mu + n\alpha$, and these are distinguished from one another by their values on H'_α. By Corollary 1.69, $\dim(V' \cap V_\mu) = \dim(V' \cap V_{s_\alpha\mu})$. Summing over V', we obtain $\dim V_\mu = \dim V_{s_\alpha\mu}$. Since the root reflections generate $W(\Delta)$, it follows that $\dim V_\mu = \dim V_{w\mu}$ for all $w \in W(\Delta)$. This proves the first half of (e).

For the second half of (e), Corollary 2.68 and the result just proved show that there is no loss of generality in assuming that μ is dominant. Under this restriction on μ, let us use (d) to write $\lambda = \mu + \sum_{i=1}^l n_i\alpha_i$ with all $n_i \geq 0$. Then

$$|\lambda|^2 = |\mu|^2 + \sum_{i=1}^l n_i\langle\mu, \alpha_i\rangle + \Big|\sum_{i=1}^l n_i\alpha_i\Big|^2$$

$$\geq |\mu|^2 + \Big|\sum_{i=1}^l n_i\alpha_i\Big|^2 \qquad \text{by dominance of } \mu.$$

The right side is $\geq |\mu|^2$ with equality only if $\sum_{i=1}^l n_i\alpha_i = 0$. In this case $\mu = \lambda$.

PROOF THAT THE CORRESPONDENCE IS ONE-ONE. Let φ and φ' be irreducible finite-dimensional on V and V', respectively, both with highest weight λ, and regard φ and φ' as representations of $U(\mathfrak{g})$. Let v_0 and v'_0 be nonzero highest weight vectors. Form $\varphi \oplus \varphi'$ on $V \oplus V'$. We claim that

$$S = (\varphi \oplus \varphi')(U(\mathfrak{g}))(v_0 \oplus v'_0)$$

is an irreducible invariant subspace of $V \oplus V'$.

Certainly S is invariant. Let $T \subseteq S$ be an irreducible invariant subspace, and let $v \oplus v'$ be a nonzero highest weight vector. For $\alpha \in \Delta^+$, we have

$$0 = (\varphi \oplus \varphi')(E_\alpha)(v \oplus v') = \varphi(E_\alpha)v \oplus \varphi'(E_\alpha)v',$$

and thus $\varphi(E_\alpha)v = 0$ and $\varphi'(E_\alpha)v' = 0$. By (c), $v = cv_0$ and $v' = c'v'_0$. Hence $v \oplus v' = cv_0 \oplus c'v'_0$. This vector by assumption is in $\varphi(U(\mathfrak{g}))(v_0 \oplus v'_0)$.

When we apply φ of (5.6) to $v_0 \oplus v_0'$, the E_β's give 0, while the H's multiply by constants, namely

$$(\varphi \oplus \varphi')(H)(v_0 \oplus v_0') = \varphi(H)v_0 \oplus \varphi'(H)v_0' = \lambda(H)(v_0 \oplus v_0').$$

Also the $E_{-\beta}$'s push weights down by Proposition 5.4c. We conclude that $c' = c$. Hence $T = S$, and S is irreducible.

The projection of S to V commutes with the representations and is not identically 0. By Schur's Lemma (Proposition 5.1), $\varphi \oplus \varphi'|_S$ is equivalent with φ. Similarly it is equivalent with φ'. Hence φ and φ' are equivalent.

To complete the proof of Theorem 5.5, we need to prove an existence result. The existence result says that for any dominant algebraically integral λ, there exists an irreducible finite-dimensional representation φ_λ of $\mathfrak{g}$ with highest weight λ. We carry out this step in the next section.

3. Verma Modules

In this section we complete the proof of the Theorem of the Highest Weight (Theorem 5.5): Under the assumption that λ is algebraically integral, we give an algebraic construction of an irreducible finite-dimensional representation of $\mathfrak{g}$ with highest weight λ.

By means of Corollary 3.6, we can identify representations of $\mathfrak{g}$ with unital left $U(\mathfrak{g})$ modules, and henceforth we shall often drop the name of the representation when working in this fashion. The idea is to consider all $U(\mathfrak{g})$ modules, finite-dimensional or infinite-dimensional, that possess a vector that behaves like a highest weight vector with weight λ. Among these we shall see that there is one (called a "Verma module") with a universal mapping property. A suitable quotient of the Verma module will give us our irreducible representation, and the main step will be to prove that it is finite-dimensional.

We retain the notation of §2, and we write $\Pi = \{\alpha_1, \ldots, \alpha_l\}$. In addition we let

$$\begin{aligned} \mathfrak{n} &= \bigoplus_{\alpha \in \Delta^+} \mathfrak{g}_\alpha \\ \mathfrak{n}^- &= \bigoplus_{\alpha \in \Delta^+} \mathfrak{g}_{-\alpha} \\ \mathfrak{b} &= \mathfrak{h} \oplus \mathfrak{n} \\ \delta &= \tfrac{1}{2} \sum_{\alpha \in \Delta^+} \alpha. \end{aligned} \tag{5.8}$$

Then $\mathfrak{n}$, $\mathfrak{n}^-$, and $\mathfrak{b}$ are Lie subalgebras of $\mathfrak{g}$, and $\mathfrak{g} = \mathfrak{b} \oplus \mathfrak{n}^-$ as a direct sum of vector spaces.

Let the complex vector space V be a unital left $U(\mathfrak{g})$ module. We allow V to be infinite-dimensional. Because of Corollary 3.6 we have already defined in §1 the notions "weight," "weight space," and "weight vector" for V. Departing slightly from the notation of that section, let V_μ be the weight space for the weight μ. The sum $\sum V_\mu$ is necessarily a direct sum. As in Proposition 5.4c, we have

$$\mathfrak{g}_\alpha(V_\mu) \subseteq V_{\mu+\alpha} \tag{5.9}$$

if α is in Δ and μ is in $\mathfrak{h}^*$. Moreover, (5.9) and the root-space decomposition of $\mathfrak{g}$ show that

$$\mathfrak{g}\Big(\bigoplus_{\mu\in\mathfrak{h}^*} V_\mu\Big) \subseteq \Big(\bigoplus_{\mu\in\mathfrak{h}^*} V_\mu\Big). \tag{5.10}$$

A **highest weight vector** for V is by definition a weight vector $v \neq 0$ with $\mathfrak{n}(v) = 0$. Notice that $\mathfrak{n}(v)$ will be 0 as soon as $E_\alpha v = 0$ for the root vectors E_α of simple roots α. In fact, we easily see this assertion by expanding any positive α in terms of simple roots as $\sum_i n_i\alpha_i$ and proceeding by induction on the level $\sum_i n_i$.

A **highest weight module** is a $U(\mathfrak{g})$ module generated by a highest weight vector. "Verma modules," to be defined below, will be universal highest weight modules.

Proposition 5.11. Let M be a highest weight module for $U(\mathfrak{g})$, and let v be a highest weight vector generating M. Suppose v is of weight λ. Then

(a) $M = U(\mathfrak{n}^-)v$

(b) $M = \bigoplus_{\mu\in\mathfrak{h}^*} M_\mu$ with each M_μ finite-dimensional and with $\dim M_\lambda = 1$

(c) every weight of M is of the form $\lambda - \sum_{i=1}^{l} n_i\alpha_i$ with the α_i's in Π and with each n_i an integer ≥ 0.

PROOF.

(a) We have $\mathfrak{g} = \mathfrak{n}^- \oplus \mathfrak{h} \oplus \mathfrak{n}$. By the Poincaré-Birkhoff-Witt Theorem (Theorem 3.8 and (3.14)), $U(\mathfrak{g}) = U(\mathfrak{n}^-)U(\mathfrak{h})U(\mathfrak{n})$. On the vector v, $U(\mathfrak{n})$ and $U(\mathfrak{h})$ act to give multiples of v. Thus $U(\mathfrak{g})v = U(\mathfrak{n}^-)v$. Since v generates M, $M = U(\mathfrak{g})v = U(\mathfrak{n}^-)v$.

(b, c) By (5.10), $\bigoplus M_\mu$ is $U(\mathfrak{g})$ stable, and it contains v. Since $M = U(\mathfrak{g})v$, $M = \bigoplus M_\mu$. By (a), $M = U(\mathfrak{n}^-)v$, and (5.9) shows that any expression

$$E_{-\beta_1}^{q_1}\cdots E_{-\beta_k}^{q_k}v \qquad \text{with all } \beta_j \in \Delta^+ \tag{5.12}$$

is a weight vector with weight $\mu - q_1\beta_1 - \cdots - q_k\beta_k$, from which (c) follows. The number of expressions (5.12) leading to this μ is finite, and so $\dim M_\mu < \infty$. The number of expressions (5.12) leading to λ is 1, from v itself, and so $\dim M_\lambda = 1$.

Before defining Verma modules, we recall some facts about tensor products of associative algebras. (A special case has already been treated in §I.3.) Let M_1 and M_2 be complex vector spaces, and let A and B be complex associative algebras with identity. Suppose that M_1 is a right B module and M_2 is a left B module, and suppose that M_1 is also a left A module in such a way that $(am_1)b = a(m_1b)$. We define

$$M_1 \otimes_B M_2 = \frac{M_1 \otimes_{\mathbb{C}} M_2}{\text{subspace generated by all } m_1b \otimes m_2 - m_1 \otimes bm_2},$$

and we let A act on the quotient by $a(m_1 \otimes m_2) = (am_1) \otimes m_2$. Then $M_1 \otimes_B M_2$ is a left A module, and it has the following universal mapping property: Whenever $\psi : M_1 \times M_2 \to E$ is a bilinear map into a complex vector space E such that $\psi(m_1b, m_2) = \psi(m_1, bm_2)$, then there exists a unique linear map $\tilde{\psi} : M_1 \otimes_B M_2 \to E$ such that $\psi(m_1, m_2) = \tilde{\psi}(m_1 \otimes m_2)$.

Now let λ be in $\mathfrak{h}^*$, and make $\mathbb{C}$ into a left $U(\mathfrak{b})$ module $\mathbb{C}_{\lambda-\delta}$ by defining

$$(5.13) \qquad \begin{aligned} Hz &= (\lambda - \delta)(H)z \qquad && \text{for } H \in \mathfrak{h},\ z \in \mathbb{C} \\ Xz &= 0 && \text{for } X \in \mathfrak{n}. \end{aligned}$$

(Equation (5.13) defines a 1-dimensional representation of $\mathfrak{b}$, and thus $\mathbb{C}_{\lambda-\delta}$ becomes a left $U(\mathfrak{b})$ module.) The algebra $U(\mathfrak{g})$ itself is a right $U(\mathfrak{b})$ module and a left $U(\mathfrak{g})$ module under multiplication, and we define the **Verma module** $V(\lambda)$ to be the left $U(\mathfrak{g})$ module

$$V(\lambda) = U(\mathfrak{g}) \otimes_{U(\mathfrak{b})} \mathbb{C}_{\lambda-\delta}.$$

Proposition 5.14. Let λ be in $\mathfrak{h}^*$.

(a) $V(\lambda)$ is a highest weight module under $U(\mathfrak{g})$ and is generated by $1 \otimes 1$ (the **canonical generator**), which is of weight $\lambda - \delta$.

(b) The map of $U(\mathfrak{n}^-)$ into $V(\lambda)$ given by $u \mapsto u(1 \otimes 1)$ is one-one onto.

(c) If M is any highest weight module under $U(\mathfrak{g})$ generated by a highest weight vector $v \neq 0$ of weight $\lambda - \delta$, then there exists one and only one $U(\mathfrak{g})$ homomorphism $\tilde{\psi}$ of $V(\lambda)$ into M such that $\tilde{\psi}(1 \otimes 1) = v$. The map $\tilde{\psi}$ is onto. Also $\tilde{\psi}$ is one-one if and only if $u \neq 0$ in $U(\mathfrak{n}^-)$ implies $u(v) \neq 0$ in M.

PROOF.

(a) Clearly $V(\lambda) = U(\mathfrak{g})(1 \otimes 1)$. Also

$$\begin{aligned} H(1 \otimes 1) &= H \otimes 1 = 1 \otimes H(1) = (\lambda - \delta)(H)(1 \otimes 1) && \text{for } H \in \mathfrak{h} \\ X(1 \otimes 1) &= X \otimes 1 = 1 \otimes X(1) = 0 && \text{for } X \in \mathfrak{n}, \end{aligned}$$

and so $1 \otimes 1$ is a highest weight vector of weight $\lambda - \delta$.

(b) By the Poincaré-Birkhoff-Witt Theorem (Theorem 3.8 and (3.14)), we have $U(\mathfrak{g}) \cong U(\mathfrak{n}^-) \otimes_{\mathbb{C}} U(\mathfrak{b})$, and this isomorphism is clearly an isomorphism of left $U(\mathfrak{n}^-)$ modules. Thus we obtain a chain of canonical left $U(\mathfrak{n}^-)$ module isomorphisms

$$\begin{aligned} V(\lambda) = U(\mathfrak{g}) \otimes_{U(\mathfrak{b})} \mathbb{C} &\cong (U(\mathfrak{n}^-) \otimes_{\mathbb{C}} U(\mathfrak{b})) \otimes_{U(\mathfrak{b})} \mathbb{C} \\ &\cong U(\mathfrak{n}^-) \otimes_{\mathbb{C}} (U(\mathfrak{b}) \otimes_{U(\mathfrak{b})} \mathbb{C}) \cong U(\mathfrak{n}^-) \otimes_{\mathbb{C}} \mathbb{C} \cong U(\mathfrak{n}^-), \end{aligned}$$

and (b) follows.

(c) We consider the bilinear map of $U(\mathfrak{g}) \times \mathbb{C}_{\lambda-\delta}$ into M given by $(u, z) \mapsto u(zv)$. In terms of the action of $U(\mathfrak{b})$ on $\mathbb{C}_{\lambda-\delta}$, we check for b in $\mathfrak{h}$ and then for b in $\mathfrak{n}$ that

$$(u, b(z)) \mapsto u(b(z)v) = zu((b(1))v)$$

and

$$(ub, z) \mapsto ub(zv) = zub(v) = zu((b(1))v).$$

By the universal mapping property, there exists one and only one linear map

$$\tilde{\psi} : U(\mathfrak{g}) \otimes_{U(\mathfrak{b})} \mathbb{C}_{\lambda-\delta} \to M$$

such that $u(zv) = \tilde{\psi}(u \otimes z)$ for all $u \in U(\mathfrak{g})$ and $z \in \mathbb{C}$, i.e., such that $u(v) = \tilde{\psi}(u(1 \otimes 1))$. This condition says that $\tilde{\psi}$ is a $U(\mathfrak{g})$ homomorphism and that $1 \otimes 1$ maps to v. Hence existence and uniqueness follow. Clearly $\tilde{\psi}$ is onto.

Let u be in $U(\mathfrak{n}^-)$. If $u(v) = 0$ with $u \neq 0$, then $\tilde{\psi}(u(1 \otimes 1)) = 0$ while $u(1 \otimes 1) \neq 0$, by (b). Hence $\tilde{\psi}$ is not one-one. Conversely if $\tilde{\psi}$ is not one-one, then Proposition 5.11a implies that there exists $u \in U(\mathfrak{n}^-)$ with $u \neq 0$ and $\tilde{\psi}(u \otimes 1) = 0$. Then

$$u(v) = u(\tilde{\psi}(1 \otimes 1)) = \tilde{\psi}(u(1 \otimes 1)) = \tilde{\psi}(u \otimes 1) = 0.$$

This completes the proof.

Proposition 5.15. Let λ be in $\mathfrak{h}$, and let $V(\lambda)_+ = \bigoplus_{\mu \neq \lambda - \delta} V(\lambda)_\mu$. Then every proper $U(\mathfrak{g})$ submodule of $V(\lambda)$ is contained in $V(\lambda)_+$. Consequently the sum S of all proper $U(\mathfrak{g})$ submodules is a proper $U(\mathfrak{g})$ submodule, and $L(\lambda) = V(\lambda)/S$ is an irreducible $U(\mathfrak{g})$ submodule. Moreover, $L(\lambda)$ is a highest weight module with highest weight $\lambda - \delta$.

PROOF. If N is a $U(\mathfrak{h})$ submodule, then $N = \bigoplus_\mu (N \cap V(\lambda)_\mu)$. Since $V(\lambda)_{\lambda-\delta}$ is 1-dimensional and generates $V(\lambda)$ (by Proposition 5.14a), the $\lambda - \delta$ term must be 0 in the sum for N if N is proper. Thus $N \subseteq V(\lambda)_+$. Hence S is proper, and $L(\lambda) = V(\lambda)/S$ is irreducible. The image of $1 \otimes 1$ in $L(\lambda)$ is not 0, is annihilated by $\mathfrak{n}$, and is acted upon by $\mathfrak{h}$ according to $\lambda - \delta$. Thus $L(\lambda)$ has all the required properties.

Theorem 5.16. Suppose that $\lambda \in \mathfrak{h}^*$ is real-valued on $\mathfrak{h}_0$ and is dominant and algebraically integral. Then the irreducible highest weight module $L(\lambda + \delta)$ is an irreducible finite-dimensional representation of $\mathfrak{g}$ with highest weight λ.

REMARKS. Theorem 5.16 will complete the proof of the Theorem of the Highest Weight (Theorem 5.5). The proof of Theorem 5.16 will be preceded by two lemmas.

Lemma 5.17. In $U(\mathfrak{sl}(2, \mathbb{C}))$, $[e, f^n] = nf^{n-1}(h - (n-1))$.

PROOF. Let

$$\begin{aligned} Lf &= \text{left by } f \text{ in } U(\mathfrak{sl}(2,\mathbb{C})) \\ Rf &= \text{right by } f \\ \operatorname{ad} f &= Lf - Rf. \end{aligned}$$

Then $Rf = Lf - \operatorname{ad} f$, and the terms on the right commute. By the binomial theorem,

$$\begin{aligned} (Rf)^n e &= \sum_{j=0}^{n} \binom{n}{j} (Lf)^{n-j}(-\operatorname{ad} f)^j e \\ &= (Lf)^n e + n(Lf)^{n-1}(-\operatorname{ad} f)e + \frac{n(n-1)}{2}(Lf)^{n-2}(-\operatorname{ad} f)^2 e \end{aligned}$$

since $(\operatorname{ad} f)^3 e = 0$, and this expression is

$$\begin{aligned} &= (Lf)^n e + nf^{n-1}h + \frac{n(n-1)}{2} f^{n-2}(-2f) \\ &= (Lf)^n e + nf^{n-1}(h - (n-1)). \end{aligned}$$

Thus

$$[e, f^n] = (Rf)^n e - (Lf)^n e = nf^{n-1}(h - (n-1)).$$

Lemma 5.18. For general complex semisimple $\mathfrak{g}$, let λ be in $\mathfrak{h}^*$, let α be a simple root, and suppose that $m = 2\langle\lambda, \alpha\rangle/|\alpha|^2$ is a positive integer. Let $v_{\lambda-\delta}$ be the canonical generator of $V(\lambda)$, and let M be the $U(\mathfrak{g})$ submodule generated by $(E_{-\alpha})^m v_{\lambda-\delta}$, where $E_{-\alpha}$ is a nonzero root vector for the root $-\alpha$. Then M is isomorphic to $V(s_\alpha\lambda)$.

PROOF. The vector $v = (E_{-\alpha})^m v_{\lambda-\delta}$ is not 0 by Proposition 5.14b. Since $s_\alpha\lambda = \lambda - m\alpha$, v is in $V(\lambda)_{\lambda-\delta-m\alpha} = V(\lambda)_{s_\alpha\lambda-\delta}$. Thus the result will follow from Proposition 5.14c if we show that $E_\beta v = 0$ whenever E_β is a root vector for a simple root β. For $\beta \neq \alpha$, $[E_\beta, E_{-\alpha}] = 0$ since $\beta - \alpha$ is not a root (Lemma 2.51). Thus

$$E_\beta v = E_\beta(E_{-\alpha})^m v_{\lambda-\delta} = (E_{-\alpha})^m E_\beta v_{\lambda-\delta} = 0.$$

For $\beta = \alpha$, let us introduce a root vector E_α for α so that $[E_\alpha, E_{-\alpha}] = 2|\alpha|^{-2}H_\alpha$. The isomorphism (2.27) identifies $\operatorname{span}\{H_\alpha, E_\alpha, E_{-\alpha}\}$ with $\mathfrak{sl}(2,\mathbb{C})$, and then Lemma 5.17 gives

$$\begin{aligned} E_\alpha(E_{-\alpha})^m v_{\lambda-\delta} &= [E_\alpha, E_{-\alpha}^m] v_{\lambda-\delta} \\ &= m(E_{-\alpha})^{m-1}(2|\alpha|^{-2}H_\alpha - (m-1)) v_{\lambda-\delta} \\ &= m\left(\frac{2\langle\lambda-\delta, \alpha\rangle}{|\alpha|^2} - (m-1)\right) E_{-\alpha}^{m-1} v_{\lambda-\delta} \\ &= 0, \end{aligned}$$

the last step following from Proposition 2.69.

PROOF OF THEOREM 5.16. Let $v_\lambda \neq 0$ be a highest weight vector in $L(\lambda+\delta)$, with weight λ. We proceed in three steps.

First we show: For every simple root α, $E_{-\alpha}^n v_\lambda = 0$ for all n sufficiently large. Here $E_{-\alpha}$ is a nonzero root vector for $-\alpha$. In fact, for $n = \dfrac{2\langle\lambda+\delta, \alpha\rangle}{|\alpha|^2}$ (which is positive by Proposition 2.69), the member $E_{-\alpha}^n(1 \otimes 1)$ of $V(\lambda+\delta)$ lies in a proper $U(\mathfrak{g})$ submodule, according to Lemma 5.18, and hence is in the submodule S in Proposition 5.15. Thus $E_{-\alpha}^n v_\lambda = 0$ in $L(\lambda+\delta)$.

Second we show: The set of weights is stable under the Weyl group $W = W(\Delta)$. In fact, let α be a simple root, let $\mathfrak{sl}_\alpha$ be the copy of $\mathfrak{sl}(2,\mathbb{C})$ given by $\mathfrak{sl}_\alpha = \operatorname{span}\{H_\alpha, E_\alpha, E_{-\alpha}\}$, set $v^{(i)} = E_{-\alpha}^i v_\lambda$, and let n be the largest integer such that $v^{(n)} \neq 0$ (existence by the first step above). Then $\mathbb{C}v^{(0)} + \cdots + \mathbb{C}v^{(n)}$ is stable under $\mathfrak{sl}_\alpha$. The sum of all finite-dimensional $U(\mathfrak{sl}_\alpha)$ submodules thus contains $v^{(0)} = v_\lambda$, and we claim it is $\mathfrak{g}$ stable.

In fact, if W is a finite-dimensional $U(\mathfrak{sl}_\alpha)$ submodule, then

$$\mathfrak{g}W = \left\{\sum Xw \mid X \in \mathfrak{g} \text{ and } w \in W\right\}$$

is finite-dimensional and for $Y \in \mathfrak{sl}_\alpha$ and $X \in \mathfrak{g}$ we have

$$YXw = XYw + [Y, X]w = Xw' + [Y, X]w \in \mathfrak{g}W.$$

So $\mathfrak{g}W$ is $\mathfrak{sl}_\alpha$ stable, and the claim follows.

Since the sum of all finite-dimensional $U(\mathfrak{sl}_\alpha)$ submodules of $L(\lambda+\delta)$ is $\mathfrak{g}$ stable, the irreducibility of $L(\lambda + \delta)$ implies that this sum is all of $L(\lambda + \delta)$. By Corollary 1.70, $L(\lambda + \delta)$ is the direct sum of finite-dimensional irreducible $U(\mathfrak{sl}_\alpha)$ submodules.

Let μ be a weight, and let $t \neq 0$ be in V_μ. We have just shown that t lies in a finite direct sum of finite-dimensional irreducible $U(\mathfrak{sl}_\alpha)$ submodules. Let us write $t = \sum_{i\in I} t_i$ with t_i in a $U(\mathfrak{sl}_\alpha)$ submodule T_i and $t_i \neq 0$. Then

$$\sum H_\alpha t_i = H_\alpha t = \mu(H_\alpha)t = \sum \mu(H_\alpha)t_i,$$

and so
$$\frac{2H_\alpha}{|\alpha|^2}t_i = \frac{2\langle\mu,\alpha\rangle}{|\alpha|^2}t_i \qquad \text{for each } i \in I.$$

If $\langle\mu,\alpha\rangle > 0$, we know that $(E_{-\alpha})^{2\langle\mu,\alpha\rangle/|\alpha|^2}t_i \neq 0$ from Theorem 1.63. Hence $(E_{-\alpha})^{2\langle\mu,\alpha\rangle/|\alpha|^2}t \neq 0$, and we see that

$$\mu - \frac{2\langle\mu,\alpha\rangle}{|\alpha|^2}\alpha = s_\alpha\mu$$

is a weight. If $\langle\mu,\alpha\rangle < 0$ instead, we know that $(E_\alpha)^{-2\langle\mu,\alpha\rangle/|\alpha|^2}t_i \neq 0$ from Theorem 1.63. Hence $(E_\alpha)^{-2\langle\mu,\alpha\rangle/|\alpha|^2}t \neq 0$, and so

$$\mu - \frac{2\langle\mu,\alpha\rangle}{|\alpha|^2}\alpha = s_\alpha\mu$$

is a weight. If $\langle\mu,\alpha\rangle = 0$, then $s_\alpha\mu = \mu$. In any case $s_\alpha\mu$ is a weight. So the set of weights is stable under each reflection s_α for α simple, and Proposition 2.62 shows that the set of weights is stable under W.

Third we show: The set of weights of $L(\lambda + \delta)$ is finite, and $L(\lambda + \delta)$ is finite-dimensional. In fact, Corollary 2.68 shows that any linear functional on $\mathfrak{h}_0$ is W conjugate to a dominant one. Since the second step above says that the set of weights is stable under W, the number of weights is at most $|W|$ times the number of dominant weights, which are of the form $\lambda - \sum_{i=1}^{l} n_i\alpha_i$ by Proposition 5.11c. Each such dominant form must satisfy

$$\langle \lambda, \delta \rangle \geq \sum_{i=1}^{l} n_i \langle \alpha_i, \delta \rangle,$$

and Proposition 2.69 shows that $\sum n_i$ is bounded; thus the number of dominant weights is finite. Then $L(\lambda + \delta)$ is finite-dimensional by Proposition 5.11b.

4. Complete Reducibility

Let $\mathfrak{g}$ be a finite-dimensional complex Lie algebra, and let $U(\mathfrak{g})$ be its universal enveloping algebra. As a consequence of the generalization of Schur's Lemma given in Proposition 5.19 below, the center $Z(\mathfrak{g})$ of $U(\mathfrak{g})$ acts by scalars in any irreducible unital left $U(\mathfrak{g})$ module, even an infinite-dimensional one. The resulting homomorphism $\chi : Z(\mathfrak{g}) \to \mathbb{C}$ is the first serious algebraic invariant of an irreducible representation of $\mathfrak{g}$ and is called the **infinitesimal character**. This invariant is most useful in situations where $Z(\mathfrak{g})$ can be shown to be large, which will be the case when $\mathfrak{g}$ is semisimple.

Proposition 5.19 (Dixmier). Let $\mathfrak{g}$ be a complex Lie algebra, and let V be an irreducible unital left $U(\mathfrak{g})$ module. Then the only $U(\mathfrak{g})$ linear maps $L : V \to V$ are the scalars.

PROOF. The space $E = \operatorname{End}_{U(\mathfrak{g})}(V, V)$ is an associative algebra over $\mathbb{C}$, and Schur's Lemma (Proposition 5.1) shows that every nonzero element of E has a two-sided inverse, i.e., E is a division algebra.

If $v \neq 0$ is in V, then the irreducibility implies that $V = U(\mathfrak{g})v$. Hence the dimension of V is at most countable. Since every nonzero element of E is invertible, the $\mathbb{C}$ linear map $L \mapsto L(v)$ of E into V is one-one. Therefore the dimension of E over $\mathbb{C}$ is at most countable.

Let L be in E. Arguing by contradiction, suppose that L is not a scalar multiple of the identity. Form the field extension $\mathbb{C}(L) \subseteq E$. Since $\mathbb{C}$ is algebraically closed, L is not algebraic over $\mathbb{C}$. Thus L is transcendental over $\mathbb{C}$. In the transcendental extension $\mathbb{C}(X)$, the set $\{(X - c)^{-1} \mid c \in \mathbb{C}\}$ is linearly independent, and consequently the dimension of $\mathbb{C}(X)$ is uncountable. Therefore $\mathbb{C}(L)$ has uncountable dimension, and so does E, contradiction.

Let us introduce **adjoint representations** on the universal enveloping algebra $U(\mathfrak{g})$ when $\mathfrak{g}$ is a finite-dimensional complex Lie algebra. We define a representation ad of $\mathfrak{g}$ on $U(\mathfrak{g})$ by

$$(\operatorname{ad} X)u = Xu - uX \qquad \text{for } X \in \mathfrak{g} \text{ and } u \in U(\mathfrak{g}).$$

(The representation property follows from the fact that $XY - YX = [X, Y]$ in $U(\mathfrak{g})$.) Lemma 3.9 shows that $\operatorname{ad} X$ carries $U_n(\mathfrak{g})$ to itself. Therefore ad provides for all n a consistently defined family of representations of $\mathfrak{g}$ on $U_n(\mathfrak{g})$.

Each $g \in \operatorname{Int}\mathfrak{g}$ gives an automorphism of $\mathfrak{g}$. Composing with the inclusion of $\mathfrak{g}$ into $U(\mathfrak{g})$, we obtain a complex-linear map of $\mathfrak{g}$ into $U(\mathfrak{g})$, and it will be convenient to call this map $\operatorname{Ad}(g)$. This composition has the property that

$$\begin{aligned}\operatorname{Ad}(g)[X, Y] &= [\operatorname{Ad}(g)X, \operatorname{Ad}(g)Y] \\ &= (\operatorname{Ad}(g)X)(\operatorname{Ad}(g)Y) - (\operatorname{Ad}(g)Y)(\operatorname{Ad}(g)X).\end{aligned}$$

By Proposition 3.3 (with $A = U(\mathfrak{g})$), $\operatorname{Ad}(g)$ extends to a homomorphism of $U(\mathfrak{g})$ into itself carrying 1 to 1. Moreover

$$\operatorname{Ad}(g_1)\operatorname{Ad}(g_2) = \operatorname{Ad}(g_1 g_2) \tag{5.20}$$

because of the uniqueness of the extension and the validity of this formula on $U_1(\mathfrak{g})$. Therefore each $\operatorname{Ad}(g)$ is an automorphism of $U(\mathfrak{g})$. Because $\operatorname{Ad}(g)$ leaves $U_1(\mathfrak{g})$ stable, it leaves each $U_n(\mathfrak{g})$ stable. Its smoothness in g on $U_1(\mathfrak{g})$ implies its smoothness in g on $U_n(\mathfrak{g})$. Thus we obtain for all n a consistently defined family Ad of smooth representations of G on $U_n(\mathfrak{g})$.

Proposition 5.21. Let $\mathfrak{g}$ be a finite-dimensional complex Lie algebra. Then

(a) the differential at 1 of Ad on $U_n(\mathfrak{g})$ is ad
(b) on each $U_n(\mathfrak{g})$, $\operatorname{Ad}(\exp X) = e^{\operatorname{ad} X}$ for all $X \in \mathfrak{g}$.

PROOF. For (a) let $u = X_1^{k_1} \cdots X_n^{k_n}$ be a monomial in $U_n(\mathfrak{g})$. For X in $\mathfrak{g}$, we have

$$\operatorname{Ad}(\exp rX)u = (\operatorname{Ad}(\exp rX)X_1)^{k_1} \cdots (\operatorname{Ad}(\exp rX)X_n)^{k_n}$$

since each $\operatorname{Ad}(g)$ for $g \in \operatorname{Int}\mathfrak{g}$ is an automorphism of $U(\mathfrak{g})$. Differentiating both sides with respect to r and applying the product rule for differentiation, we obtain at $r = 0$

$$\begin{aligned}
&\frac{d}{dr}\operatorname{Ad}(\exp rX)u\Big|_{r=0} \\
&\quad = \sum_{i=1}^{n}\sum_{j=1}^{k_i} X_1^{k_1}\cdots X_{i-1}^{k_{i-1}}X_i^{j-1}\Big(\frac{d}{dr}\operatorname{Ad}(\exp rX)X_i\Big)_{r=0}X_i^{k_i-j}X_{i+1}^{k_{i+1}}\cdots X_n^{k_n} \\
&\quad = (\operatorname{ad}X)u.
\end{aligned}$$

Then (a) follows from Proposition 1.93, and (b) follows from Corollary 1.87.

Proposition 5.22. If $\mathfrak{g}$ is a finite-dimensional complex Lie algebra, then the following conditions on an element u of $U(\mathfrak{g})$ are equivalent:

(a) u is in the center $Z(\mathfrak{g})$
(b) $uX = Xu$ for all $X \in \mathfrak{g}$
(c) $e^{\operatorname{ad}X}u = u$ for all $X \in \mathfrak{g}$
(d) $\operatorname{Ad}(g)u = u$ for all $g \in \operatorname{Int}\mathfrak{g}$.

PROOF. Conclusion (a) implies (b) trivially, and (b) implies (a) since $\mathfrak{g}$ generates $U(\mathfrak{g})$. If (b) holds, then $(\operatorname{ad}X)u = 0$, and (c) follows by summing the series for the exponential. Conversely if (c) holds, then we can replace X by rX in (c) and differentiate to obtain (b). Finally (c) follows from (d) by taking $g = \exp X$ and applying Proposition 5.21b, while (d) follows from (c) by (5.20) and Proposition 5.21b.

In the case that $\mathfrak{g}$ is semisimple, we shall construct some explicit elements of $Z(\mathfrak{g})$ and use them to extend to all semisimple $\mathfrak{g}$ the theorem of complete reducibility proved for $\mathfrak{sl}(2,\mathbb{C})$ in Theorem 1.64. To begin with, here is an explicit element of $Z(\mathfrak{g})$ when $\mathfrak{g} = \mathfrak{sl}(2,\mathbb{C})$.

EXAMPLE. $\mathfrak{g} = \mathfrak{sl}(2,\mathbb{C})$. Let $Z = \frac{1}{2}h^2 + ef + fe$ with h, e, f as in (1.5). The action of Z in a representation already appeared in Lemma 1.65. We readily check that Z is in $Z(\mathfrak{g})$ by seeing that Z commutes with h, e, and f. The element Z is a multiple of the Casimir element Ω defined below.

For a general semisimple $\mathfrak{g}$, let B be the Killing form. (To fix the definitions in this section, we shall not allow more general invariant forms in place of B.) Let X_i be any basis of $\mathfrak{g}$ over $\mathbb{C}$, and let $\tilde{X}_i$ be the dual basis relative to B, i.e., the basis with

$$B(\tilde{X}_i, X_j) = \delta_{ij}.$$

The **Casimir element** Ω is defined by

$$\Omega = \sum_{i,j} B(X_i, X_j)\tilde{X}_i\tilde{X}_j. \tag{5.23}$$

Proposition 5.24. In a complex semisimple Lie algebra $\mathfrak{g}$, the Casimir element Ω is defined independently of the basis X_i and is a member of the center $Z(\mathfrak{g})$ of $U(\mathfrak{g})$.

PROOF. Let a second basis X_i' be given by means of a nonsingular complex matrix (a_{ij}) as

$$X_j' = \sum_m a_{mj} X_m. \tag{5.25a}$$

Let (b_{ij}) be the inverse of the matrix (a_{ij}), and define

$$\tilde{X}_i' = \sum_l b_{il}\tilde{X}_l. \tag{5.25b}$$

Then

$$B(\tilde{X}_i', X_j') = \sum_{l,m} b_{il}a_{mj}B(\tilde{X}_l, X_m) = \sum_l b_{il}a_{lj} = \delta_{ij}.$$

Thus $\tilde{X}_i'$ is the dual basis of X_j'. The element to consider is

$$\begin{aligned}\Omega' &= \sum_{i,j} B(X_i', X_j')\tilde{X}_i'\tilde{X}_j' \\ &= \sum_{m,m'}\sum_{l,l'}\sum_{i,j} a_{mi}a_{m'j}b_{il}b_{jl'}B(X_m, X_{m'})\tilde{X}_l\tilde{X}_{l'} \\ &= \sum_{m,m'}\sum_{l,l'} \delta_{ml}\delta_{m'l'}B(X_m, X_{m'})\tilde{X}_l\tilde{X}_{l'} \\ &= \sum_{l,l'} B(X_l, X_{l'})\tilde{X}_l\tilde{X}_{l'} \\ &= \Omega.\end{aligned}$$

This proves that Ω is independent of the basis.

Let g be in $\operatorname{Int}\mathfrak{g}$, and take the second basis to be $X_i' = gX_i = \operatorname{Ad}(g)X_i$. Because of Proposition 1.96 the invariance of the Killing form gives

$$B(\operatorname{Ad}(g)\tilde{X}_i, X_j') = B(\tilde{X}_i, \operatorname{Ad}(g)^{-1}X_j') = B(\tilde{X}_i, X_j) = \delta_{ij}, \tag{5.26}$$

and we conclude that $\tilde{X}_i' = \operatorname{Ad}(g)\tilde{X}_i$. Therefore

$$\begin{aligned}
\operatorname{Ad}(g)\Omega &= \sum_{i,j} B(X_i, X_j)\operatorname{Ad}(g)(\tilde{X}_i\tilde{X}_j) \\
&= \sum_{i,j} B(\operatorname{Ad}(g)X_i, \operatorname{Ad}(g)X_j)\tilde{X}'_i\tilde{X}'_j \qquad \text{by Proposition 1.96} \\
&= \sum_{i,j} B(X'_i, X'_j)\tilde{X}'_i\tilde{X}'_j \\
&= \sum_{i,j} B(X_i, X_j)\tilde{X}_i\tilde{X}_j \qquad \text{by change of basis} \\
&= \Omega.
\end{aligned}$$

By Proposition 5.22, Ω is in $Z(\mathfrak{g})$.

EXAMPLE. $\mathfrak{g} = \mathfrak{sl}(2, \mathbb{C})$. We take as basis the elements h, e, f as in (1.5). The Killing form has already been computed in Example 2 of §I.3, and we find that $\tilde{h} = \frac{1}{8}h$, $\tilde{e} = \frac{1}{4}f$, $\tilde{f} = \frac{1}{4}e$. Then

$$\begin{aligned}
\Omega &= B(h,h)\tilde{h}^2 + B(e,f)\tilde{e}\tilde{f} + B(f,e)\tilde{f}\tilde{e} \\
&= 8\tilde{h}^2 + 4\tilde{e}\tilde{f} + 4\tilde{f}\tilde{e} \\
&= \tfrac{1}{8}h^2 + \tfrac{1}{4}ef + \tfrac{1}{4}fe, \qquad (5.27)
\end{aligned}$$

which is $\frac{1}{4}$ of the element $Z = \frac{1}{2}h^2 + ef + fe$ whose action in a representation appeared in Lemma 1.65.

Let φ be an irreducible finite-dimensional representation of $\mathfrak{g}$ on a space V. Schur's Lemma (Proposition 5.1) and Proposition 5.24 imply that Ω acts as a scalar in V. We shall compute this scalar, making use of the Theorem of the Highest Weight (Theorem 5.5). Thus let us introduce a Cartan subalgebra $\mathfrak{h}$, the set $\Delta = \Delta(\mathfrak{g}, \mathfrak{h})$ of roots, and a positive system $\Delta^+ = \Delta^+(\mathfrak{g}, \mathfrak{h})$.

Proposition 5.28. In the complex semisimple Lie algebra $\mathfrak{g}$, let $\mathfrak{h}_0$ be the real form of $\mathfrak{h}$ on which all roots are real-valued, and let $\{H_i\}_{i=1}^l$ be an orthonormal basis of $\mathfrak{h}_0$ relative to the Killing form B of $\mathfrak{g}$. Choose root vectors E_α so that $B(E_\alpha, E_{-\alpha}) = 1$ for all $\alpha \in \Delta$. Then

(a) $\Omega = \sum_{i=1}^l H_i^2 + \sum_{\alpha\in\Delta} E_\alpha E_{-\alpha}$

(b) Ω operates by the scalar $|\lambda|^2 + 2\langle\lambda, \delta\rangle = |\lambda + \delta|^2 - |\delta|^2$ in an irreducible finite-dimensional representation of $\mathfrak{g}$ of highest weight λ, where δ is half the sum of the positive roots

(c) the scalar value by which Ω operates in an irreducible finite-dimensional representation of $\mathfrak{g}$ is nonzero if the representation is not trivial.

PROOF.

(a) Since $B(\mathfrak{h}, E_\alpha) = 0$ for all $\alpha \in \Delta$, $\tilde{H}_i = H_i$. Also the normalization $B(E_\alpha, E_{-\alpha}) = 1$ makes $\tilde{E}_\alpha = E_{-\alpha}$. Then (a) follows immediately from (5.23).

(b) Let φ be an irreducible finite-dimensional representation of $\mathfrak{g}$ with highest weight λ, and let v_λ be a nonzero vector of weight λ. Using the relation $[E_\alpha, E_{-\alpha}] = H_\alpha$ from Lemma 2.18a, we rewrite Ω from (a) as

$$\begin{aligned}\Omega &= \sum_{i=1}^{l} H_i^2 + \sum_{\alpha\in\Delta^+} E_\alpha E_{-\alpha} + \sum_{\alpha\in\Delta^+} E_{-\alpha} E_\alpha \\ &= \sum_{i=1}^{l} H_i^2 + \sum_{\alpha\in\Delta^+} H_\alpha + 2\sum_{\alpha\in\Delta^+} E_{-\alpha} E_\alpha \\ &= \sum_{i=1}^{l} H_i^2 + 2H_\delta + 2\sum_{\alpha\in\Delta^+} E_{-\alpha} E_\alpha .\end{aligned}$$

When we apply Ω to v_λ and use Theorem 5.5c, the last term gives 0. Thus

$$\Omega v_\lambda = \sum_{i=1}^{l} \lambda(H_i)^2 v_\lambda + 2\lambda(H_\delta) v_\lambda = (|\lambda|^2 + 2\langle\lambda, \delta\rangle) v_\lambda .$$

Schur's Lemma (Proposition 5.1) shows that Ω acts by a scalar, and hence that scalar must be $|\lambda|^2 + 2\langle\lambda, \delta\rangle$.

(c) We have $\langle\lambda, \delta\rangle = \frac{1}{2}\sum_{\alpha\in\Delta^+}\langle\lambda, \alpha\rangle$. Since λ is dominant, this is ≥ 0 with equality only if $\langle\lambda, \alpha\rangle = 0$ for all α, i.e., only if $\lambda = 0$. Thus the scalar in (b) is $\geq |\lambda|^2$ and can be 0 only if λ is 0.

Theorem 5.29. Let φ be a complex-linear representation of the complex semisimple Lie algebra $\mathfrak{g}$ on a finite-dimensional complex vector space V. Then V is completely reducible in the sense that there exist invariant subspaces $U_1, \ldots, U_r$ of V such that $V = U_1 \oplus \cdots \oplus U_r$ and the restriction of the representation to each U_i is irreducible.

REMARKS. The proof is very similar to the proof of Theorem 1.64. It is enough by induction to show that any invariant subspace U in V has an invariant complement U'. For the case that U has codimension 1, we shall prove this result as a lemma. Then we return to the proof of Theorem 5.29.

Lemma 5.30. Let $\varphi : \mathfrak{g} \to \operatorname{End} V$ be a finite-dimensional representation, and let $U \subseteq V$ be an invariant subspace of codimension 1. Then there is a 1-dimensional invariant subspace W such that $V = U \oplus W$.

PROOF.

Case 1. Suppose $\dim U = 1$. Form the quotient representation φ on V/U, with $\dim(V/U) = 1$. This quotient representation is irreducible of dimension 1, and Lemma 4.28 shows that it is 0. Consequently

$$\varphi(\mathfrak{g})V \subseteq U \qquad \text{and} \qquad \varphi(\mathfrak{g})U = 0.$$

Hence if $Y = [X_1, X_2]$, we have

$$\begin{aligned}\varphi(Y)V &\subseteq \varphi(X_1)\varphi(X_2)V + \varphi(X_2)\varphi(X_1)V \\ &\subseteq \varphi(X_1)U + \varphi(X_2)U = 0.\end{aligned}$$

Since Corollary 1.52 gives $\mathfrak{g} = [\mathfrak{g}, \mathfrak{g}]$, we conclude that $\varphi(\mathfrak{g}) = 0$. Therefore any complementary subspace to U will serve as W.

Case 2. Suppose that $\varphi(\cdot)|_U$ is irreducible and $\dim U > 1$. Since $\dim V/U = 1$, the quotient representation is 0 and $\varphi(\mathfrak{g})V \subseteq U$. The formula for Ω in (5.23) then shows that $\Omega(V) \subseteq U$, and Proposition 5.28c says that Ω is a nonzero scalar on U. Therefore $\dim(\ker \Omega) = 1$ and $U \cap (\ker \Omega) = 0$. Since Ω commutes with $\varphi(\mathfrak{g})$, $\ker \Omega$ is an invariant subspace. Taking $W = \ker \Omega$, we have $V = U \oplus W$ as required.

Case 3. Suppose that $\varphi(\cdot)|_U$ is not necessarily irreducible and that $\dim U \geq 1$. We induct on $\dim V$. The base case is $\dim V = 2$ and is handled by Case 1. When $\dim V > 2$, let $U_1 \subseteq U$ be an irreducible invariant subspace, and form the quotient representations on

$$U/U_1 \subseteq V/U_1$$

with quotient V/U of dimension 1. By inductive hypothesis we can write

$$V/U_1 = U/U_1 \oplus Y/U_1,$$

where Y is an invariant subspace in V and $\dim Y/U_1 = 1$. Case 1 or Case 2 is applicable to the representation $\varphi(\cdot)|_Y$ and the irreducible invariant subspace U_1. Then $Y = U_1 \oplus W$, where W is a 1-dimensional invariant subspace. Since $W \subseteq Y$ and $Y \cap U \subseteq U_1$, we find that

$$W \cap U = (W \cap Y) \cap U = W \cap (Y \cap U) \subseteq W \cap U_1 = 0.$$

Therefore $V = U \oplus W$ as required.

PROOF OF THEOREM 5.29. Let φ be a representation of $\mathfrak{g}$ on M, and let $N \neq 0$ be an invariant subspace. Put

$$V = \{\gamma \in \operatorname{End} M \mid \gamma(M) \subseteq N \text{ and } \gamma|_N \text{ is scalar}\}.$$

Linear algebra shows that V is nonzero. Define a linear function $\sigma : \mathfrak{g} \to \operatorname{End}(\operatorname{End} M)$ by

$$\sigma(X)\gamma = \varphi(X)\gamma - \gamma\varphi(X) \qquad \text{for } \gamma \in \operatorname{End} M \text{ and } X \in \mathfrak{g}.$$

Checking directly that $\sigma[X, Y]$ and $\sigma(X)\sigma(Y) - \sigma(Y)\sigma(X)$ are equal, we see that σ is a representation of $\mathfrak{g}$ on $\operatorname{End} M$.

We claim that the subspace $V \subseteq \operatorname{End} M$ is an invariant subspace under σ. In fact, let $\gamma(M) \subseteq N$ and $\gamma|_N = \lambda 1$. In the right side of the expression

$$\sigma(X)\gamma = \varphi(X)\gamma - \gamma\varphi(X),$$

the first term carries M to N since γ carries M to N and $\varphi(X)$ carries N to N. The second term carries M into N since $\varphi(X)$ carries M to M and γ carries M to N. Thus $\sigma(X)\gamma$ carries M into N. On N, the action of $\sigma(X)\gamma$ is given by

$$\sigma(X)\gamma(n) = \varphi(X)\gamma(n) - \gamma\varphi(X)(n) = \lambda\varphi(X)(n) - \lambda\varphi(X)(n) = 0.$$

Thus V is an invariant subspace.

Actually the above argument shows also that the subspace U of V given by

$$U = \{\gamma \in V \mid \gamma = 0 \text{ on } N\}$$

is an invariant subspace. Clearly $\dim V/U = 1$. By Lemma 5.30, $V = U \oplus W$ for a 1-dimensional invariant subspace $W = \mathbb{C}\gamma$. Here γ is a nonzero scalar $\lambda 1$ on N. The invariance of W means that $\sigma(X)\gamma = 0$ since 1-dimensional representations are 0 by Lemma 4.28. Therefore γ commutes with $\varphi(X)$ for all $X \in \mathfrak{g}$. But then $\ker\gamma$ is a nonzero invariant subspace of M. Since γ is nonsingular on N (being a nonzero scalar there), we must have $M = N \oplus \ker\gamma$. This completes the proof.

Let us return to the notation introduced before Proposition 5.28, taking $\mathfrak{h}$ to be a Cartan subalgebra, $\Delta = \Delta(\mathfrak{g}, \mathfrak{h})$ to be the set of roots, and $\Delta^+ = \Delta^+(\mathfrak{g}, \mathfrak{h})$ to be a positive system. Define $\mathfrak{n}$ and $\mathfrak{n}^-$ as in (5.8).

Corollary 5.31. Let a finite-dimensional representation of $\mathfrak{g}$ be given on a space V, and let $V^{\mathfrak{n}}$ be the subspace of $\mathfrak{n}$ **invariants** given by

$$V^{\mathfrak{n}} = \{v \in V \mid Xv = 0 \text{ for all } X \in \mathfrak{n}\}.$$

Then the subspace $V^{\mathfrak{n}}$ is a $U(\mathfrak{h})$ module, and

(a) $V = V^{\mathfrak{n}} \oplus \mathfrak{n}^- V$ as $U(\mathfrak{h})$ modules
(b) the natural map $V^{\mathfrak{n}} \to V/(\mathfrak{n}^- V)$ is an isomorphism of $U(\mathfrak{h})$ modules
(c) the $U(\mathfrak{h})$ module $V^{\mathfrak{n}}$ determines the $U(\mathfrak{g})$ module V up to equivalence; the dimension of $V^{\mathfrak{n}}$ equals the number of irreducible constituents of V, and the multiplicity of a weight in $V^{\mathfrak{n}}$ equals the multiplicity in V of the irreducible representation of $\mathfrak{g}$ with that highest weight.

PROOF. To see that $V^{\mathfrak{n}}$ is a $U(\mathfrak{h})$ module, let H be in $\mathfrak{h}$ and v be in $V^{\mathfrak{n}}$. If X is in $\mathfrak{n}$, then $X(Hv) = H(Xv) + [X, H]v = 0 + X'v$ with X' in $\mathfrak{n}$, and it follows that Hv is in $V^{\mathfrak{n}}$. Thus $V^{\mathfrak{n}}$ is a $U(\mathfrak{h})$ module. Similarly $\mathfrak{n}^- V$ is a $U(\mathfrak{h})$ module. Conclusion (b) is immediate from (a), and conclusion (c) is immediate from Theorems 5.29 and 5.5. Thus we are left with proving (a).

By Theorem 5.29, V is a direct sum of irreducible representations, and hence there is no loss of generality for the proof of (a) in assuming that V is irreducible, say of highest weight λ. With V irreducible, choose nonzero root vectors E_α for every root α, and let $H_1, \ldots, H_l$ be a basis of $\mathfrak{h}$. By the Poincaré-Birkhoff-Witt Theorem (Theorem 3.8), $U(\mathfrak{g})$ is spanned by all elements

$$E_{-\beta_1} \cdots E_{-\beta_p} H_{i_1} \cdots H_{i_q} E_{\alpha_1} \cdots E_{\alpha_r},$$

where the α_i and β_j are positive roots, not necessarily distinct. Since V is irreducible, V is spanned by all elements

$$E_{-\beta_1} \cdots E_{-\beta_p} H_{i_1} \cdots H_{i_q} E_{\alpha_1} \cdots E_{\alpha_r} v$$

with v in V_λ. Since V_λ is annihilated by $\mathfrak{n}$, such an element is 0 unless $r = 0$. The space V_λ is mapped into itself by $\mathfrak{h}$, and we conclude that V is spanned by all elements

$$E_{-\beta_1} \cdots E_{-\beta_p} v$$

with v in V_λ. If $p > 0$, such an element is in $\mathfrak{n}^- V$ and has weight less than λ, while if $p = 0$, it is in V_λ. Consequently

$$V = V_\lambda \oplus \mathfrak{n}^- V.$$

Theorem 5.5c shows that $V^{\mathfrak{n}}$ is just the λ weight space of V, and (a) follows. This completes the proof of the corollary.

We conclude this section by giving a generalization of Proposition 5.24 that yields many elements in $Z(\mathfrak{g})$ when $\mathfrak{g}$ is semisimple. We shall use this result in the next section.

Proposition 5.32. Let φ be a finite-dimensional representation of a complex semisimple Lie algebra $\mathfrak{g}$, and let B be the Killing form of $\mathfrak{g}$. If X_i is a basis of $\mathfrak{g}$ over $\mathbb{C}$, let $\tilde{X}_i$ be the dual basis relative to B. Fix an integer $n \geq 1$ and define

$$z = \sum_{i_1, \ldots, i_n} \operatorname{Tr} \varphi(X_{i_1} \cdots X_{i_n}) \tilde{X}_{i_1} \cdots \tilde{X}_{i_n}$$

as a member of $U(\mathfrak{g})$. Then z is independent of the choice of basis X_i and is a member of the center $Z(\mathfrak{g})$ of $U(\mathfrak{g})$.

Proof. The proof is modeled on the argument for Proposition 5.24. Let a second basis X_i' be given by (5.25a), with dual basis $\tilde{X}_i'$ given by (5.25b). The element to consider is

$$\begin{aligned}
z' &= \sum_{i_1,\dots,i_n} \operatorname{Tr}\varphi(X_{i_1}'\cdots X_{i_n}')\tilde{X}_{i_1}'\cdots\tilde{X}_{i_n}' \\
&= \sum_{m_1,\dots,m_n}\sum_{l_1,\dots,l_n}\sum_{i_1,\dots,i_n} a_{m_1i_1}\cdots a_{m_ni_n}\operatorname{Tr}\varphi(X_{m_1}\cdots X_{m_n}) \\
&\qquad\qquad \times b_{i_1l_1}\cdots b_{i_nl_n}\tilde{X}_{l_1}\cdots\tilde{X}_{l_n} \\
&= \sum_{m_1,\dots,m_n}\sum_{l_1,\dots,l_n} \delta_{m_1l_1}\cdots\delta_{m_nl_n}\operatorname{Tr}\varphi(X_{m_1}\cdots X_{m_n})\tilde{X}_{l_1}\cdots\tilde{X}_{l_n} \\
&= \sum_{l_1,\dots,l_n} \operatorname{Tr}\varphi(X_{l_1}\cdots X_{l_n})\tilde{X}_{l_1}\cdots\tilde{X}_{l_n} \\
&= z.
\end{aligned}$$

This proves that z is independent of the basis.

The group $G = \operatorname{Int}\mathfrak{g}$ has Lie algebra $(\operatorname{ad}\mathfrak{g})^{\mathbb{R}}$, and its simply connected cover $\tilde{G}$ is a simply connected analytic group with Lie algebra $\mathfrak{g}^{\mathbb{R}}$. Regarding the representation φ of $\mathfrak{g}$ as a representation of $\mathfrak{g}^{\mathbb{R}}$, we can lift it to a representation Φ of $\tilde{G}$ since $\tilde{G}$ is simply connected. Fix $g \in \tilde{G}$. In the earlier part of the proof let the new basis be $X_i' = \operatorname{Ad}(g)X_i$. Then (5.26) shows that $\tilde{X}_i' = \operatorname{Ad}(g)\tilde{X}_i$. Consequently

$$\begin{aligned}
\operatorname{Ad}(g)z &= \sum_{i_1,\dots,i_n} \operatorname{Tr}\varphi(X_{i_1}\cdots X_{i_n})\operatorname{Ad}(g)(\tilde{X}_{i_1}\cdots\tilde{X}_{i_n}) \\
&= \sum_{i_1,\dots,i_n} \operatorname{Tr}(\Phi(g)\varphi(X_{i_1}\cdots X_{i_n})\Phi(g)^{-1})\tilde{X}_{i_1}'\cdots\tilde{X}_{i_n}' \\
&= \sum_{i_1,\dots,i_n} \operatorname{Tr}((\Phi(g)\varphi(X_{i_1})\Phi(g)^{-1})\cdots(\Phi(g)\varphi(X_{i_n})\Phi(g)^{-1}))\tilde{X}_{i_1}'\cdots\tilde{X}_{i_n}' \\
&= \sum_{i_1,\dots,i_n} \operatorname{Tr}(\varphi(\operatorname{Ad}(g)X_{i_1})\cdots\varphi(\operatorname{Ad}(g)X_{i_n}))\tilde{X}_{i_1}'\cdots\tilde{X}_{i_n}' \\
&= \sum_{l_1,\dots,l_n} \operatorname{Tr}(\varphi((\operatorname{Ad}(g)X_{i_1})\cdots(\operatorname{Ad}(g)X_{i_n})))\tilde{X}_{i_1}'\cdots\tilde{X}_{i_n}' \\
&= \sum_{i_1,\dots,i_n} \operatorname{Tr}(\varphi(X_{i_1}'\cdots X_{i_n}'))\tilde{X}_{i_1}'\cdots\tilde{X}_{i_n}',
\end{aligned}$$

and this equals z, by the result of the earlier part of the proof. By Proposition 5.22, z is in $Z(\mathfrak{g})$.

5. Harish-Chandra Isomorphism

Let $\mathfrak{g}$ be a complex semisimple Lie algebra, and let $\mathfrak{h}$, $\Delta = \Delta(\mathfrak{g}, \mathfrak{h})$, $W = W(\Delta)$, and B be as in §2. Define $\mathcal{H} = U(\mathfrak{h})$. Since $\mathfrak{h}$ is abelian, the algebra $\mathcal{H}$ coincides with the symmetric algebra $S(\mathfrak{h})$. By Proposition A.18b every linear transformation of $\mathfrak{h}$ into an associative commutative algebra A with identity extends uniquely to a homomorphism of $\mathcal{H}$ into A sending 1 into 1. Consequently

(i) W acts on $\mathcal{H}$ (since it maps $\mathfrak{h}$ into $\mathfrak{h} \subseteq \mathcal{H}$, with $\lambda^w(H) = \lambda(H^{w^{-1}})$),

(ii) $\mathcal{H}$ may be regarded as the space of polynomial functions on $\mathfrak{h}^*$ (because if λ is in $\mathfrak{h}^*$, λ is linear from $\mathfrak{h}$ into $\mathbb{C}$ and so extends to a homomorphism of $\mathcal{H}$ into $\mathbb{C}$; we can think of λ on a member of $\mathcal{H}$ as the value of the member of $\mathcal{H}$ at the point λ).

Let $\mathcal{H}^W = U(\mathfrak{h})^W = S(\mathfrak{h})^W$ be the subalgebra of Weyl-group invariants of $\mathcal{H}$. In this section we shall establish the "Harish-Chandra isomorphism" $\gamma : Z(\mathfrak{g}) \to \mathcal{H}^W$, and we shall see an indication of how this isomorphism allows us to work with infinitesimal characters when $\mathfrak{g}$ is semisimple.

The Harish-Chandra mapping is motivated by considering how an element $z \in Z(\mathfrak{g})$ acts in an irreducible finite-dimensional representation with highest weight λ. The action is by scalars, by Proposition 5.19, and we compute those scalars by testing the action on a nonzero highest-weight vector.

First we use the Poincaré-Birkhoff-Witt Theorem (Theorem 3.8) to introduce a suitable basis of $U(\mathfrak{g})$ for making the computation. Introduce a positive system $\Delta^+ = \Delta^+(\mathfrak{g}, \mathfrak{h})$, and define $\mathfrak{n}$, $\mathfrak{n}^-$, $\mathfrak{b}$, and δ as in (5.8). As in (5.6), enumerate the positive roots as $\beta_1, \ldots, \beta_k$, and let $H_1, \ldots, H_l$ be a basis of $\mathfrak{h}$ over $\mathbb{C}$. For each root $\alpha \in \Delta$, let E_α be a nonzero root vector. Then the monomials

$$E_{-\beta_1}^{q_1} \cdots E_{-\beta_k}^{q_k} H_1^{m_1} \cdots H_l^{m_l} E_{\beta_1}^{p_1} \cdots E_{\beta_k}^{p_k} \tag{5.33}$$

are a basis of $U(\mathfrak{g})$ over $\mathbb{C}$.

If we expand the central element z in terms of the above basis of $U(\mathfrak{g})$ and consider the effect of the term (5.33), there are two possibilities. One is that some p_j is > 0, and then the term acts as 0. The other is that all p_j are 0. In this case, as we shall see in Proposition 5.34b below, all q_j are 0. The $U(\mathfrak{h})$ part acts on a highest weight vector v_λ by the scalar

$$\lambda(H_1)^{m_1} \cdots \lambda(H_l)^{m_l},$$

and that is the total effect of the term. Hence we can compute the effect of z if we can extract those terms in the expansion relative to the basis

(5.33) such that only the $U(\mathfrak{h})$ part is present. This idea was already used in the proof of Proposition 5.28b.

Thus define

$$\mathcal{P} = \sum_{\alpha\in\Delta^+} U(\mathfrak{g})E_\alpha \qquad \text{and} \qquad \mathcal{N} = \sum_{\alpha\in\Delta^+} E_{-\alpha}U(\mathfrak{g}).$$

Proposition 5.34.
(a) $U(\mathfrak{g}) = \mathcal{H} \oplus (\mathcal{P} + \mathcal{N})$
(b) Any member of $Z(\mathfrak{g})$ has its $\mathcal{P} + \mathcal{N}$ component in $\mathcal{P}$.

PROOF.
(a) The fact that $U(\mathfrak{g}) = \mathcal{H}+(\mathcal{P}+\mathcal{N})$ follows by the Poincaré-Birkhoff-Witt Theorem (Theorem 3.8) from the fact that the elements (5.33) span $U(\mathfrak{g})$. Fix the basis of elements (5.33). For any nonzero element of $U(\mathfrak{g})E_\alpha$ with $\alpha \in \Delta^+$, write out the $U(\mathfrak{g})$ factor in terms of the basis (5.33), and consider a single term of the product, say

$$cE_{-\beta_1}^{q_1}\cdots E_{-\beta_k}^{q_k}H_1^{m_1}\cdots H_l^{m_l}E_{\beta_1}^{p_1}\cdots E_{\beta_k}^{p_k}E_\alpha. \tag{5.35}$$

The factor $E_{\beta_1}^{p_1}\cdots E_{\beta_k}^{p_k}E_\alpha$ is in $U(\mathfrak{n})$ and has no constant term. By the Poincaré-Birkhoff-Witt Theorem, we can rewrite it as a linear combination of terms $E_{\beta_1}^{r_1}\cdots E_{\beta_k}^{r_k}$ with $r_1 + \cdots + r_k > 0$. Putting

$$cE_{-\beta_1}^{q_1}\cdots E_{-\beta_k}^{q_k}H_1^{m_1}\cdots H_l^{m_l}$$

in place on the left of each term, we see that (5.35) is a linear combination of terms (5.33) with $p_1 + \cdots + p_k > 0$. Similarly any member of $\mathcal{N}$ is a linear combination of terms (5.33) with $q_1 + \cdots + q_k > 0$. Thus any member of $\mathcal{P} + \mathcal{N}$ is a linear combination of terms (5.33) with $p_1+\cdots+p_k > 0$ or $q_1+\cdots+q_k > 0$. Any member of $\mathcal{H}$ has $p_1+\cdots+p_k = 0$ and $q_1+\cdots+q_k = 0$ in every term of its expansion, and thus (a) follows.

(b) In terms of the representation ad on $U(\mathfrak{g})$ given in Proposition 5.21, the monomials (5.33) are a basis of $U(\mathfrak{g})$ of weight vectors for ad $\mathfrak{h}$, the weight of (5.33) being

$$-q_1\beta_1 - \cdots - q_k\beta_k + p_1\beta_1 + \cdots + p_k\beta_k. \tag{5.36}$$

Any member z of $Z(\mathfrak{g})$ satisfies $(\operatorname{ad} H)z = Hz - zH = 0$ for $H \in \mathfrak{h}$ and thus is of weight 0. Hence its expansion in terms of the basis (5.33) involves only terms of weight 0. In the proof of (a) we saw that any member of $\mathcal{P}+\mathcal{N}$ has each term with $p_1+\cdots+p_k > 0$ or $q_1+\cdots+q_k > 0$. Since the p's and q's are constrained by the condition that (5.36) equal 0, each term must have both $p_1 + \cdots + p_k > 0$ and $q_1 + \cdots + q_k > 0$. Hence each term is in $\mathcal{P}$.

Let γ_n' be the projection of $Z(\mathfrak{g})$ into the $\mathcal{H}$ term in Proposition 5.34a. Applying the basis elements (5.33) to a highest weight vector of a finite-dimensional representation, we see that

(5.37) $\lambda(\gamma_n'(z))$ is the scalar by which z acts in an irreducible finite-dimensional representation of highest weight λ.

Despite the tidiness of this result, Harish-Chandra found that a slight adjustment of γ_n' leads to an even more symmetric formula. Define a linear map $\tau_n : \mathfrak{h} \to \mathcal{H}$ by

$$\tau_n(H) = H - \delta(H)1, \tag{5.38}$$

and extend τ_n to an algebra automorphism of $\mathcal{H}$ by the universal mapping property for symmetric algebras. The **Harish-Chandra map** γ is defined by

$$\gamma = \tau_n \circ \gamma_n' \tag{5.39}$$

as a mapping of $Z(\mathfrak{g})$ into $\mathcal{H}$.

Any element $\lambda \in \mathfrak{h}^*$ defines an algebra homomorphism $\lambda : \mathcal{H} \to \mathbb{C}$ with $\lambda(1) = 1$, because the universal mapping property of symmetric algebras allows us to extend $\lambda : \mathfrak{h} \to \mathbb{C}$ to $\mathcal{H}$. In terms of this extension, the maps γ and γ_n' are related by

$$\lambda(\gamma(z)) = (\lambda - \delta)(\gamma_n'(z)) \qquad \text{for } z \in Z(\mathfrak{g}),\ \lambda \in \mathfrak{h}^*. \tag{5.40a}$$

If instead we think of $\mathcal{H}$ as the space of polynomial functions on $\mathfrak{h}^*$, this formula may be rewritten as

$$\gamma(z)(\lambda) = \gamma_n'(z)(\lambda - \delta) \qquad \text{for } z \in Z(\mathfrak{g}),\ \lambda \in \mathfrak{h}^*. \tag{5.40b}$$

We define

$$\chi_\lambda(z) = \lambda(\gamma(z)) \qquad \text{for } z \in Z(\mathfrak{g}), \tag{5.41}$$

so that χ_λ is a map of $Z(\mathfrak{g})$ into $\mathbb{C}$. This map has the following interpretation.

Proposition 5.42. For $\lambda \in \mathfrak{h}^*$ and $z \in Z(\mathfrak{g})$, $\chi_\lambda(z)$ is the scalar by which z operates on the Verma module $V(\lambda)$.

REMARK. In this notation we can restate (5.37) as follows:

(5.43) $\chi_{\lambda+\delta}(z)$ is the scalar by which z acts in an irreducible finite-dimensional representation of highest weight λ.

PROOF. Write $z = \gamma'_{\mathfrak{n}}(z) + p$ with $p \in \mathcal{P}$. If $v_{\lambda-\delta}$ denotes the canonical generator of $V(\lambda)$, then

$$\begin{aligned} zv_{\lambda-\delta} &= \gamma'_{\mathfrak{n}}(z)v_{\lambda-\delta} + pv_{\lambda-\delta} \\ &= (\lambda-\delta)(\gamma'_{\mathfrak{n}}(z))v_{\lambda-\delta} \\ &= \lambda(\gamma(z))v_{\lambda-\delta} && \text{by (5.40)} \\ &= \chi_\lambda(z)v_{\lambda-\delta} && \text{by (5.41).} \end{aligned}$$

For $u \in U(\mathfrak{g})$, we therefore have $zuv_{\lambda-\delta} = uzv_{\lambda-\delta} = \chi_\lambda(z)uv_{\lambda-\delta}$. Since $V(\lambda) = U(\mathfrak{g})v_{\lambda-\delta}$, the result follows.

Theorem 5.44 (Harish-Chandra). The mapping γ in (5.40) is an algebra isomorphism of $Z(\mathfrak{g})$ onto the algebra $\mathcal{H}^W$ of Weyl-group invariants in $\mathcal{H}$, and it does not depend on the choice of the positive system Δ^+.

EXAMPLE. $\mathfrak{g} = \mathfrak{sl}(2, \mathbb{C})$. Let $Z = \frac{1}{2}h^2 + ef + fe$ with h, e, f as in (1.5). We noted in the first example in §4 that Z is in $Z(\mathfrak{sl}(2, \mathbb{C}))$. Let us agree that e corresponds to the positive root α. Then $ef = fe + [e, f] = fe + h$ implies

$$Z = \tfrac{1}{2}h^2 + ef + fe = (\tfrac{1}{2}h^2 + h) + 2fe \in \mathcal{H} \oplus \mathcal{P}.$$

Hence

$$\gamma'_{\mathfrak{n}}(Z) = \tfrac{1}{2}h^2 + h.$$

Now $\delta(h) = \frac{1}{2}\alpha\begin{pmatrix} 1 & 0 \\ 0 & -1 \end{pmatrix} = 1$, and so

$$\tau_{\mathfrak{n}}(h) = h - 1.$$

Thus

$$\gamma(Z) = \tfrac{1}{2}(h-1)^2 + (h-1) = \tfrac{1}{2}h^2 - \tfrac{1}{2}.$$

The nontrivial element of the 2-element Weyl group acts on $\mathcal{H}$ by sending h to $-h$, and thus we have a verification that $\gamma(Z)$ is invariant under the Weyl group. Moreover it is now clear that $\mathcal{H}^W = \mathbb{C}[h^2]$ and that $\gamma(\mathbb{C}[Z]) = \mathbb{C}[h^2]$. Theorem 5.44 therefore implies that $Z(\mathfrak{sl}(2, \mathbb{C})) = \mathbb{C}[Z]$.

The proof of Theorem 5.44 will occupy the remainder of this section and will take five steps.

PROOF THAT image$(\gamma) \subseteq \mathcal{H}^W$.

Since members of $\mathcal{H}$ are determined by the effect of all $\lambda \in \mathfrak{h}^*$ on them, we need to prove that

$$\lambda(w(\gamma(z))) = \lambda(\gamma(z))$$

for all $\lambda \in \mathfrak{h}^*$ and $w \in W$. In other words, we need to see that every $w \in W$ has

$$(w^{-1}\lambda)(\gamma(z)) = \lambda(\gamma(z)), \tag{5.45}$$

and it is enough to handle w equal to a reflection in a simple root by Proposition 2.62. Moreover each side for fixed z is a polynomial in λ, and thus it is enough to prove (5.45) for λ dominant integral.

Form the Verma module $V(\lambda)$. We know from Proposition 5.42 that z acts in $V(\lambda)$ by the scalar $\lambda(\gamma(z))$. Also z acts in $V(s_\alpha\lambda)$ by the scalar $(s_\alpha\lambda)(\gamma(z))$. Since $2\langle\lambda,\alpha\rangle/|\alpha|^2$ is an integer ≥ 0, Lemma 5.18 says that $V(s_\alpha\lambda)$ is isomorphic to a (clearly nonzero) $U(\mathfrak{g})$ submodule of $V(\lambda)$. Thus the two scalars must match, and (5.45) is proved.

PROOF THAT γ DOES NOT DEPEND ON THE CHOICE OF Δ^+.

Let λ be algebraically integral and dominant for Δ^+, let V be a finite-dimensional irreducible representation of $\mathfrak{g}$ with highest weight λ (Theorem 5.5), and let χ be the infinitesimal character of V. Temporarily, let us drop the subscript $\mathfrak{n}$ from γ'. By Theorem 2.63 any other positive system of roots is related in Δ^+ by a member of $W(\Delta)$. Thus let w be in $W(\Delta)$, and let $\tilde{\gamma}'$ and $\tilde{\gamma}$ be defined relative to $\Delta^{+\sim} = w\Delta^+$. We are to prove that $\gamma = \tilde{\gamma}$. The highest weight of V relative to $w\Delta^+$ is $w\lambda$. If z is in $Z(\mathfrak{g})$, then (5.37) gives

$$\lambda(\gamma'(z)) = \chi(z) = w\lambda(\tilde{\gamma}'(z)). \tag{5.46}$$

Since $\gamma(z)$ is invariant under $W(\Delta)$,

$$\begin{aligned}(w\lambda + w\delta)(\gamma(z)) &= (\lambda+\delta)(\gamma(z)) = \lambda(\gamma'(z)) \\ &= w\lambda(\tilde{\gamma}'(z)) = (w\lambda + w\delta)(\tilde{\gamma}(z)),\end{aligned}$$

the next-to-last step following from (5.46). Since $\gamma(z)$ and $\tilde{\gamma}(z)$ are polynomial functions equal at the lattice points of an octant, they are equal everywhere.

PROOF THAT γ IS MULTIPLICATIVE.

Since τ_n is an algebra isomorphism, we need to show that

$$\gamma'_n(z_1z_2) = \gamma'_n(z_1)\gamma'_n(z_2). \tag{5.47}$$

We have

$$z_1z_2 - \gamma'_n(z_1)\gamma'_n(z_2) = z_1(z_2 - \gamma'_n(z_2)) + \gamma'_n(z_2)(z_1 - \gamma'_n(z_1)),$$

which is in $\mathcal{P}$, and therefore (5.47) follows.

PROOF THAT γ IS ONE-ONE.

If $\gamma(z) = 0$, then $\gamma'_n(z) = 0$, and (5.37) shows that z acts as 0 in every irreducible finite-dimensional representation of $\mathfrak{g}$. By Theorem 5.29, z acts as 0 in every finite-dimensional representation of $\mathfrak{g}$.

In the representation ad of $\mathfrak{g}$ on $U_n(\mathfrak{g})$, $U_{n-1}(\mathfrak{g})$ is an invariant subspace. Thus we obtain a representation ad of $\mathfrak{g}$ on $U_n(\mathfrak{g})/U_{n-1}(\mathfrak{g})$ for each n. It is enough to show that if $u \in U(\mathfrak{g})$ acts as 0 in each of these representations, then $u = 0$. Specifically let us expand u in terms of the basis

$$E_{-\beta_1}^{q_1} \cdots E_{-\beta_k}^{q_k} H_1^{m_1} \cdots H_l^{m_l} E_{\beta_1}^{p_1} \cdots E_{\beta_k}^{p_k} \tag{5.48}$$

of $U(\mathfrak{g})$. We show that if ad u is 0 on all elements

$$H_\delta^m E_{\beta_1}^{r_1} \cdots E_{\beta^k}^{r_k} \mod U^{m+\sum r_j - 1}(\mathfrak{g}), \tag{5.49}$$

then $u = 0$. (Here as usual, δ is half the sum of the positive roots.)

In (5.48) let $m' = \sum_{j=1}^k (p_j + q_j)$. The effect of a monomial term of u on (5.49) will be to produce a sum of monomials, all of whose $\mathcal{H}$ factors have total degree $> m - m'$. There will be one monomial whose $\mathcal{H}$ factors have total degree $= m - m'$, and we shall be able to identify that monomial and its coefficient exactly.

Let us verify this assertion. If X is in $\mathfrak{g}$, the action of ad X on a monomial $X_1 \cdots X_n$ is

(5.50)
$$\begin{aligned}(\operatorname{ad} X)(X_1 \cdots X_n) &= XX_1 \cdots X_n - X_1 \cdots X_n X \\ &= [X, X_1]X_2 \cdots X_n + X_1[X, X_2]X_3 \cdots X_n + \cdots + X_1 \cdots X_{n-1}[X, X_n].\end{aligned}$$

If $X_1, \ldots, X_n$ are root vectors or members of $\mathfrak{h}$ and if X has the same property, then so does each $[X, X_j]$. Moreover, Lemma 3.9 allows us to commute a bracket into its correct position in (5.49), modulo lower-order terms.

Consider the effect of ad $E_{\pm\alpha}$ when applied to an expression (5.49). The result is a sum of terms as in (5.50). When ad $E_{\pm\alpha}$ acts on the $\mathcal{H}$ part, the degree of the $\mathcal{H}$ part of the resulting term goes down by 1, whereas if ad $E_{\pm\alpha}$ acts on a root vector, the degree of the $\mathcal{H}$ part of the resulting term goes up by 1 or stays the same. When some ad H_j acts on an expression (5.49), the degree of the $\mathcal{H}$ part of each term stays the same.

Thus when ad of (5.48) acts on (5.49), every term of the result has $\mathcal{H}$ part of degree $\geq m - m'$, and degree $= m - m'$ arises only when all ad $E_{\pm\alpha}$'s act on one of the factors H_δ. To compute exactly the term at the end with $\mathcal{H}$ part of degree $= m - m'$, let us follow this process step by step. When we apply ad E_{β_k} to (5.49), we get a contribution of $\langle -\beta_k, \delta\rangle$ from each factor of H_δ in (5.49), plus irrelevant terms. Thus ad E_{β_k} of (5.49) gives

$$m\langle -\beta_n, \delta\rangle H_\delta^{m-1} E_{\beta_1}^{r_1} \cdots E_{\beta_k}^{r_k+1} + \text{irrelevant terms.}$$

By the time we have applied all of $\operatorname{ad}(E_{\beta_1}^{p_1} \cdots E_{\beta_k}^{p_k})$ to (5.49), the result is

(5.51)

$$\frac{m!}{(m - \sum p_j)!}\Big(\prod_{j=1}^{k} \langle -\beta_j, \delta\rangle^{p_j}\Big) H_\delta^{m-\sum p_j} E_{\beta_1}^{p_1+r_1} \cdots E_{\beta_k}^{p_k+r_k} + \text{irrelevant terms.}$$

Next we apply ad H_l to (5.51). The main term is multiplied by the constant $\sum_{j=1}^{k}(p_j + r_j)\beta_j(H_l)$. Repeating this kind of computation for the other factors from $\operatorname{ad}(\mathcal{H})$, we see that $\operatorname{ad}(H_1^{m_1} \cdots H_l^{m_l})$ of (5.51) is

$$\text{(5.52)}\qquad \frac{m!}{m - \sum p_j}\Big(\prod_{j=1}^{k} \langle -\beta_j, \delta\rangle^{p_j}\Big) \prod_{i=1}^{l}\Big(\sum_{j=1}^{k}(p_j + r_j)\beta_j(H_i)\Big)^{m_i} \times H_\delta^{m-\sum p_j} E_{\beta_1}^{p_1+r_1} \cdots E_{\beta_k}^{p_k+r_k} + \text{irrelevant terms.}$$

Finally we apply ad $E_{-\beta_k}$ to (5.52). The main term gets multiplied by $(m - \sum p_j)\langle \beta_k, \delta\rangle$, another factor of H_δ gets dropped, and a factor of $E_{-\beta_k}$ appears. Repeating this kind of computation for the other factors ad $E_{-\beta_j}$, we see that $\operatorname{ad}(E_{-\beta_1}^{q_1} \cdots E_{-\beta_k}^{q_k})$ of (5.52) is

$$\frac{m!}{(m - m')!}\Big(\prod_{j=1}^{k} (-1)^{p_j}\langle \beta_j, \delta\rangle^{p_j+q_j}\Big) \prod_{i=1}^{l}\Big(\sum_{j=1}^{k}(p_j + r_j)\beta_j(H_i)\Big)^{m_i}$$

$$\text{(5.53)}\qquad \times E_{-\beta_1}^{q_1} \cdots E_{-\beta_k}^{q_k} H_\delta^{m-m'} E_{\beta_1}^{p_1+r_1} \cdots E_{\beta_k}^{p_k+r_k} + \text{irrelevant terms.}$$

This completes our exact computation of the main term of ad of (5.48) on (5.49).

We regard m and the r_j's fixed for the present. Among the terms of u, we consider the effect of ad of only those with m' as large as possible. From these, the powers of the root vectors in (5.53) allow us to reconstruct the p_j's and q_j's. The question is whether the different terms of u for which m' is maximal and the p_j's and q_j's take on given values can have their main contributions to (5.53) add to 0. Thus we ask whether a finite sum

$$\sum_{m_1,\dots,m_l} c_{m_1,\dots,m_l} \prod_{i=1}^{l} \Big(\sum_{j=1}^{k} (p_j + r_j)\beta_j(H_i) \Big)^{m_i}$$

can be 0 for all choices of integers $r_j \geq 0$.

Assume it is 0 for all such choices. Then

$$\sum_{m_1,\dots,m_l} c_{m_1,\dots,m_l} \prod_{i=1}^{l} \Big(\sum_{j=1}^{k} z_j\beta_j(H_i) \Big)^{m_i} = 0$$

for all complex $z_1, \dots, z_k$. Hence

$$\sum_{m_1,\dots,m_l} c_{m_1,\dots,m_l} \prod_{i=1}^{l} (\mu(H_i))^{m_i} = 0$$

for all $\mu \in \mathfrak{h}^*$, and we obtain

$$\mu\Big(\sum_{m_1,\dots,m_l} c_{m_1,\dots,m_l} H_1^{m_1} \cdots H_l^{m_l} \Big) = 0$$

for all $\mu \in \mathfrak{h}^*$. Therefore

$$\sum_{m_1,\dots,m_l} c_{m_1,\dots,m_l} H_1^{m_1} \cdots H_l^{m_l} = 0,$$

and it follows that all the terms under consideration in u were 0. Thus γ is one-one.

PROOF THAT γ IS ONTO.

To prove that γ is onto $\mathcal{H}^W$, we need a supply of members of $Z(\mathfrak{g})$. Proposition 5.32 will fulfill this need. Let $\mathcal{H}_n$ and $\mathcal{H}_n^W$ be the subspaces of $\mathcal{H}$ and $\mathcal{H}^W$ of elements homogeneous of degree n. It is clear from the Poincaré-Birkhoff-Witt Theorem that

$$\gamma(Z(\mathfrak{g}) \cap U_n(\mathfrak{g})) \subseteq \bigoplus_{d=0}^{n} \mathcal{H}_d^W. \tag{5.54}$$

Let λ be any dominant algebraically integral member of $\mathfrak{h}^*$, and let φ_λ be the irreducible finite-dimensional representation of $\mathfrak{g}$ with highest weight λ. Let $\Lambda(\lambda)$ be the weights of φ_λ, repeated as often as their multiplicities. In Proposition 5.32 let X_i be the ordered basis dual to one consisting of a basis $H_1, \ldots, H_l$ of $\mathfrak{h}$ followed by the root vectors E_α. The proposition says that the following element z is in $Z(\mathfrak{g})$:

$$z = \sum_{i_1,\ldots,i_n} \operatorname{Tr} \varphi_\lambda(\tilde{X}_{i_1} \cdots \tilde{X}_{i_n}) X_{i_1} \cdots X_{i_n}$$

$$= \sum_{\substack{i_1,\ldots,i_n,\\ \text{all } \le l}} \operatorname{Tr} \varphi_\lambda(\tilde{H}_{i_1} \cdots \tilde{H}_{i_n}) H_{i_1} \cdots H_{i_n} + \sum_{\substack{j_1,\ldots,j_n,\\ \text{at least one } > l}} \operatorname{Tr} \varphi_\lambda(\tilde{X}_{j_1} \cdots \tilde{X}_{j_n}) X_{j_1} \cdots X_{j_n}.$$

In the second sum on the right side of the equality, some factor of $X_{j_1} \cdots X_{j_n}$ is a root vector. Commuting the factors into their positions to match terms with the basis vectors (5.33) of $U(\mathfrak{g})$, we see that

$$X_{j_1} \cdots X_{j_n} \equiv u \mod U_{n-1}(\mathfrak{g}) \qquad \text{with } u \in \mathcal{P} + \mathcal{N},$$

i.e.,

$$X_{j_1} \cdots X_{j_n} \equiv 0 \mod \Big(\bigoplus_{d=0}^{n-1} \mathcal{H}_d \oplus (\mathcal{P} + \mathcal{N}) \Big).$$

Application of γ_n' to z therefore gives

$$\gamma_n'(z) \equiv \sum_{\substack{i_1,\ldots,i_n,\\ \text{all } \le l}} \operatorname{Tr} \varphi_\lambda(\tilde{H}_{i_1} \cdots \tilde{H}_{i_n}) H_{i_1} \cdots H_{i_n} \mod \Big(\bigoplus_{d=0}^{n-1} \mathcal{H}_d \Big).$$

The automorphism τ_n of $\mathcal{H}$ affects elements only modulo lower-order terms, and thus

$$\gamma(z) \equiv \sum_{\substack{i_1,\ldots,i_n,\\ \text{all } \le l}} \operatorname{Tr} \varphi_\lambda(\tilde{H}_{i_1} \cdots \tilde{H}_{i_n}) H_{i_1} \cdots H_{i_n} \mod \Big(\bigoplus_{d=0}^{n-1} \mathcal{H}_d \Big)$$

$$= \sum_{\mu \in \Lambda(\lambda)} \sum_{\substack{i_1,\ldots,i_n,\\ \text{all } \le l}} \mu(\tilde{H}_{i_1}) \cdots \mu(\tilde{H}_{i_n}) H_{i_1} \cdots H_{i_n} \mod \Big(\bigoplus_{d=0}^{n-1} \mathcal{H}_d \Big).$$

Now

$$\sum_i \mu(\tilde{H}_i) H_i = H_\mu \tag{5.55}$$

since

$$\left\langle \sum_i \mu(\tilde{H}_i)H_i,\ \tilde{H}_j \right\rangle = \mu(\tilde{H}_j) = \langle H_\mu, \tilde{H}_j \rangle \qquad \text{for all } j.$$

Thus

$$\gamma(z) \equiv \sum_{\mu\in\Lambda(\lambda)} (H_\mu)^n \mod \Big(\bigoplus_{d=0}^{n-1} \mathcal{H}_d\Big).$$

The set of weights of φ_λ, together with their multiplicities, is invariant under W by Theorem 5.5e. Hence $\sum_{\mu\in\Lambda(\lambda)}(H_\mu)^n$ is in $\mathcal{H}^W$, and we can write

$$\gamma(z) \equiv \sum_{\mu\in\Lambda(\lambda)} (H_\mu)^n \mod \Big(\bigoplus_{d=0}^{n-1} \mathcal{H}_d^W\Big). \tag{5.56}$$

To prove that γ is onto $\mathcal{H}^W$, we show that the image of γ contains $\bigoplus_{d=0}^m \mathcal{H}_d^W$ for every m. For $m = 0$, we have $\gamma(1) = 1$, and there is nothing further to prove. Assuming the result for $m = n-1$, we see from (5.56) that we can choose $z_1 \in Z(\mathfrak{g})$ with

$$\gamma(z - z_1) = \sum_{\mu\in\Lambda(\lambda)} (H_\mu)^n. \tag{5.57}$$

To complete the induction, we shall show that

$$\text{the elements } \sum_{\mu\in\Lambda(\lambda)} (H_\mu)^n \text{ span } \mathcal{H}_n^W. \tag{5.58}$$

Let $\Lambda_D(\lambda)$ be the set of dominant weights of φ_λ, repeated according to their multiplicities. Since again the set of weights, together with their multiplicities, is invariant under W, we can rewrite the right side of (5.58) as

$$= \sum_{\mu\in\Lambda_D(\lambda)} c_\mu \sum_{w\in W} (H_{w\mu})^n, \tag{5.59}$$

where c_μ^{-1} is the order of the stablizer of μ in W. We know that φ_λ contains the weight λ with multiplicity 1. Equation (5.57) shows that the elements (5.59) are in the image of γ in $\mathcal{H}_n^W$. To complete the induction, it is thus enough to show that

$$\text{the elements (5.59) span } \mathcal{H}_n^W. \tag{5.60}$$

We do so by showing that

(5.61a) the span of all elements (5.59) includes all elements $\sum_{w\in W}(H_{w\nu})^n$ for ν dominant and algebraically integral,

(5.61b) $\mathcal{H}_n^W$ is spanned by all elements $\sum_{w\in W}(H_{w\nu})^n$ for ν dominant and algebraically integral.

To prove (5.61a), note that the set of dominant algebraically integral ν in a compact set is finite because the set of integral points forms a lattice in the real linear span of the roots. Hence it is permissible to induct on $|\nu|$. The trivial case for the induction is $|\nu| = 0$. Suppose inductively that (5.61a) has been proved for all dominant algebraically integral ν with $|\nu| < |\lambda|$. If μ is any dominant weight of φ_λ other than λ, then $|\mu| < |\lambda|$ by Theorem 5.5e. Thus the expression (5.59) involving λ is the sum of $c_\lambda \sum_{w\in W}(H_{w\lambda})^n$ and a linear combination of terms for which (5.61a) is assumed by induction already to be proved. Since $c_\lambda \neq 0$, (5.61a) holds for $\sum_{w\in W}(H_{w\lambda})^n$. This completes the induction and the proof of (5.61a).

To prove (5.61b), it is enough (by summing over $w \in W$) to prove that

(5.61c) $\mathcal{H}_n$ is spanned by all elements $(H_\nu)^n$ for ν dominant and algebraically integral,

and we do so by induction on n. The trivial case of the induction is $n = 0$.

For $1 \leq i \leq \dim \mathfrak{h}$, we can choose dominant algebraically integral forms λ_i such that $\{\lambda_i\}$ is a $\mathbb{C}$ basis for $\mathfrak{h}^*$. Since the λ_i's span $\mathfrak{h}^*$, the H_{λ_i} span $\mathfrak{h}$. Consequently the nth degree monomials in the H_{λ_i} span $\mathcal{H}_n$.

Assuming (5.61c) inductively for $n - 1$, we now prove it for n. Let $\nu_1, \ldots, \nu_n$ be dominant and algebraically integral. It is enough to show that the monomial $H_{\nu_1} \cdots H_{\nu_n}$ is a linear combination of elements $(H_\nu)^n$ with ν dominant and algebraically integral. By the induction hypothesis,

$$(H_{\nu_1} \cdots H_{\nu_{n-1}})H_{\nu_n} = \sum_\nu c_\nu H_\nu^{n-1} H_{\nu_n},$$

and it is enough to show that $H_\nu^{n-1} H_{\nu'}$ is a linear combination of terms $(H_{\nu+r\nu'})^n$ with $r \geq 0$ in $\mathbb{Z}$. By the invertibility of a Vandermonde matrix, choose constants $c_1, \ldots, c_n$ with

$$\begin{pmatrix} 1 & 1 & 1 & \cdots & 1 \\ 1 & 2 & 3 & \cdots & n \\ 1 & 2^2 & 3^2 & \cdots & n^2 \\ & & \vdots & & \\ 1 & 2^{n-1} & 3^{n-1} & \cdots & n^{n-1} \end{pmatrix} \begin{pmatrix} c_1 \\ c_2 \\ c_3 \\ \vdots \\ c_n \end{pmatrix} = \begin{pmatrix} 0 \\ 1 \\ 0 \\ \vdots \\ 0 \end{pmatrix}.$$

Then

$$\begin{aligned}\sum_{j=1}^{n} c_j (H_{\nu+j\nu'})^n &= \sum_{j=1}^{n} c_j (H_\nu + jH_{\nu'})^n \\ &= \sum_{k=0}^{n} \binom{n}{k} H_\nu^{n-k} H_{\nu'}^k \sum_{j=1}^{n} c_j j^k \\ &= nH_\nu^{n-1} H_{\nu'}.\end{aligned}$$

Thus $H_\nu^{n-1} H_{\nu'}$ has the required expansion, and the induction is complete. This proves (5.61c), and consequently γ is onto $\mathcal{H}^W$. This completes the proof of Theorem 5.44.

For $\mathfrak{g}$ complex semisimple we say that a unital left $U(\mathfrak{g})$ module V "has an infinitesimal character" if $Z(\mathfrak{g})$ acts by scalars in V. In this case the **infinitesimal character** of V is the homomorphism $\chi : Z(\mathfrak{g}) \to \mathbb{C}$ with $\chi(z)$ equal to the scalar by which z acts. Proposition 5.19 says that every irreducible unital left $U(\mathfrak{g})$ module has an infinitesimal character.

The Harish-Chandra isomorphism allows us to determine explicitly all possible infinitesimal characters. Let $\mathfrak{h}$ be a Cartan subalgebra of $\mathfrak{g}$. If λ is in $\mathfrak{h}^*$, then λ is meaningful on the element $\gamma(z)$ of $\mathcal{H}$. Earlier we defined in (5.41) a homomorphism $\chi_\lambda : Z(\mathfrak{g}) \to \mathbb{C}$ by $\chi_\lambda(z) = \lambda(\gamma(z))$.

Theorem 5.62. If $\mathfrak{g}$ is a reductive Lie algebra and $\mathfrak{h}$ is a Cartan subalgebra, then every homomorphism of $Z(\mathfrak{g})$ into $\mathbb{C}$ sending 1 into 1 is of the form χ_λ for some $\lambda \in \mathfrak{h}^*$. If λ' and λ are in $\mathfrak{h}^*$, then $\chi_{\lambda'} = \chi_\lambda$ if and only if λ' and λ are in the same orbit under the Weyl group $W = W(\mathfrak{g}, \mathfrak{h})$.

PROOF. Let $\chi : Z(\mathfrak{g}) \to \mathbb{C}$ be a homomorphism with $\chi(1) = 1$. By Theorem 5.44, γ carries $Z(\mathfrak{g})$ onto $\mathcal{H}^W$, and therefore $\gamma(\ker\chi)$ is an ideal in $\mathcal{H}^W$. Let us check that the corresponding ideal $I = \mathcal{H}\gamma(\ker\chi)$ in $\mathcal{H}$ is proper. Assuming the contrary, suppose $u_1, \ldots, u_n$ in $\mathcal{H}$ and $H_1, \ldots, H_n$ in $\gamma(\ker\chi)$ are such that $\sum_i u_i H_i = 1$. Application of $w \in W$ gives $\sum_i (wu_i)H_i = 1$. Summing on w, we obtain

$$\sum_i \Big(\sum_{w\in W} wu_i\Big) H_i = |W|.$$

Since $\sum_{w\in W} wu_i$ is in $\mathcal{H}^W$, we can apply $\chi \circ \gamma^{-1}$ to both sides. Since $\chi(1) = 1$, the result is

$$\sum_i \chi\Big(\gamma^{-1}\Big(\sum_{w\in W} wu_i\Big)\Big)\chi(\gamma^{-1}(H_i)) = |W|.$$

But the left side is 0 since $\chi(\gamma^{-1}(H_i)) = 0$ for all i, and we have a contradiction. We conclude that the ideal I is proper.

By Zorn's Lemma, extend I to a maximal ideal $\tilde{I}$ of $\mathcal{H}$. The Hilbert Nullstellensatz tells us that there is some $\lambda \in \mathfrak{h}^*$ with

$$\tilde{I} = \{H \in \mathcal{H} \mid \lambda(H) = 0\}.$$

Since $\gamma(\ker\chi) \subseteq I \subseteq \tilde{I}$, we have $\chi_\lambda(z) = \lambda(\gamma(z)) = 0$ for all $z \in \ker\chi$. In other words, $\chi(z) = \chi_\lambda(z)$ for $z \in \ker\chi$ and for $z = 1$. These z's span $\mathcal{H}^W$, and hence $\chi = \chi_\lambda$.

If λ' and λ are in the same orbit under W, say $\lambda' = w\lambda$, then the identity $w(\gamma(z)) = \gamma(z)$ for $w \in W$ forces

$$\chi_{\lambda'}(z) = \lambda'(\gamma(z)) = \lambda'(w(\gamma(z))) = w^{-1}\lambda'(\gamma(z)) = \lambda(\gamma(z)) = \chi_\lambda(z).$$

Finally suppose λ' and λ are not in the same orbit under W. Choose a polynomial p on $\mathfrak{h}^*$ that is 1 on $W\lambda$ and 0 on $W\lambda'$. The polynomial p on $\mathfrak{h}^*$ is nothing more than an element H of $\mathcal{H}$ with

$$w\lambda(H) = 1 \quad \text{and} \quad w\lambda'(H) = 0 \quad \text{for all } w \in W. \tag{5.63}$$

The element $\tilde{H}$ of $\mathcal{H}$ with $\tilde{H} = |W|^{-1}\sum_{w\in W} wH$ is in $\mathcal{H}^W$ and satisfies the same properties (5.63) as H. By Theorem 5.44 we can choose $z \in Z(\mathfrak{g})$ with $\gamma(z) = \tilde{H}$. Then $\chi_\lambda(z) = \lambda(\gamma(z)) = \lambda(\tilde{H}) = 1$ while $\chi_{\lambda'}(z) = 0$. Hence $\chi_{\lambda'} \neq \chi_\lambda$.

Now suppose that V is a $U(\mathfrak{g})$ module with infinitesimal character χ. By Theorem 5.62, $\chi = \chi_\lambda$ for some $\lambda \in \mathfrak{h}^*$. We often abuse notation and say that V has **infinitesimal character** λ. The element λ is determined up to the operation of the Weyl group, again by Theorem 5.62.

EXAMPLES.

1) Let V be a finite-dimensional irreducible $U(\mathfrak{g})$ module with highest weight λ. By (5.43), V has infinitesimal character $\lambda + \delta$.

2) If λ is in $\mathfrak{h}^*$, then the Verma module $V(\lambda)$ has infinitesimal character λ by Proposition 5.42.

3) When B is the Killing form and Ω is the Casimir element, Proposition 5.28b shows that $\lambda(\gamma'_{\mathfrak{n}}(\Omega)) = |\lambda - \delta|^2 - |\delta|^2$ if λ is dominant and algebraically integral. The same proof shows that this formula remains valid as long as λ is in the real linear span of the roots. Combining this result with the definition (5.41), we obtain

$$\chi_\lambda(\Omega) = |\lambda|^2 - |\delta|^2 \tag{5.64}$$

for λ in the real linear span of the roots.

6. Weyl Character Formula

We saw in §IV.2 that the character of a finite-dimensional representation of a compact group determines the representation up to equivalence. Thus characters provide an effective tool for working with representations in a canonical fashion. In this section we shall deal with characters in a formal way, working in the context of complex semisimple Lie algebras, deferring until §8 the interpretation in terms of compact connected Lie groups.

To understand where the formalism comes from, it is helpful to think of the group $SL(2, \mathbb{C})$ and its compact subgroup $SU(2)$. The group $SU(2)$ is simply connected, being homeomorphic to the 3-sphere, and it follows from Proposition 1.122 that $SL(2, \mathbb{C})$ is simply connected also. A finite-dimensional representation of $SU(2)$ is automatically smooth. Thus it leads via differentiation to a representation of $\mathfrak{su}(2)$, then via complexification to a representation of $\mathfrak{sl}(2, \mathbb{C})$, and then via passage to the simply connected group to a holomorphic representation of $SL(2, \mathbb{C})$. We can recover the original representation of $SU(2)$ by restriction, and we can begin this cycle at any stage, continuing all the way around. This construction is an instance of "Weyl's unitary trick," which we shall study later.

Let us see the effect of this construction as we follow the character of an irreducible representation Φ with differential φ. Let $h = \begin{pmatrix} 1 & 0 \\ 0 & -1 \end{pmatrix}$. The diagonal subalgebra $\mathfrak{h} = \{zh \mid z \in \mathbb{C}\}$ is a Cartan subalgebra of $\mathfrak{sl}(2, \mathbb{C})$, and the roots are 2 and -2 on h. We take the root that is 2 on h (and has $e = \begin{pmatrix} 0 & 1 \\ 0 & 0 \end{pmatrix}$ as root vector) to be positive, and we call it α. The weights of φ are determined by the eigenvalues of $\varphi(h)$. According to Theorem 1.63, the eigenvalues are of the form $n, n-2, \ldots, -n$. Hence if we define $\lambda \in \mathfrak{h}^*$ by $\lambda(zh) = zn$, then the weights are

$$\lambda, \lambda - \alpha, \lambda - 2\alpha, \ldots, -\lambda.$$

Thus the matrix of $\varphi(zh)$ relative to a basis of weight vectors is

$$\varphi(zh) = \operatorname{diag}(\lambda(zh), (\lambda - \alpha)(zh), (\lambda - 2\alpha)(zh), \ldots, -\lambda(zh)).$$

Exponentiating this formula in order to pass to the group $SL(2, \mathbb{C})$, we obtain

$$\Phi(\exp zh) = \operatorname{diag}(e^{\lambda(zh)}, e^{(\lambda-\alpha)(zh)}, e^{(\lambda-2\alpha)(zh)}, \ldots, e^{-\lambda(zh)}).$$

This formula makes sense within $SU(2)$ if z is purely imaginary. In any event if χ_Φ denotes the character of Φ (i.e., the trace of Φ of a group element), then we obtain

$$\chi_\Phi(\exp zh) = e^{\lambda(zh)} + e^{(\lambda-\alpha)(zh)} + e^{(\lambda-2\alpha)(zh)} + \cdots + e^{-\lambda(zh)}$$
$$= \frac{e^{(\lambda+\delta)(zh)} - e^{-(\lambda+\delta)(zh)}}{e^{\delta(zh)} - e^{-\delta(zh)}},$$

where $\delta = \frac{1}{2}\alpha$ takes the value 1 on h. We can drop the group element from the notation if we introduce formal exponentials. Then we can write

$$\chi_\Phi = e^\lambda + e^{\lambda-\alpha} + e^{\lambda-2\alpha} + \cdots + e^{-\lambda} = \frac{e^{\lambda+\delta} - e^{-(\lambda+\delta)}}{e^\delta - e^{-\delta}}.$$

In this section we shall derive a similar expression involving formal exponentials for the character of an irreducible representation of a complex semisimple Lie algebra with a given highest weight. This result is the "Weyl Character Formula." We shall interpret the result in terms of compact connected Lie groups in §8.

The first step is to develop the formalism of exponentials. We fix a complex semisimple Lie algebra $\mathfrak{g}$, a Cartan subalgebra $\mathfrak{h}$, the set Δ of roots, the Weyl group W, and a simple system $\Pi = \{\alpha_1, \ldots, \alpha_l\}$. Let Δ^+ be the set of positive roots, and let δ be half the sum of the positive roots.

Following customary set-theory notation, let $\mathbb{Z}^{\mathfrak{h}^*}$ be the additive group of all functions from $\mathfrak{h}^*$ to $\mathbb{Z}$. If f is in $\mathbb{Z}^{\mathfrak{h}^*}$, then the **support** of f is the set of $\lambda \in \mathfrak{h}^*$ where $f(\lambda) \neq 0$. For $\lambda \in \mathfrak{h}^*$, define e^λ to be the member of $\mathbb{Z}^{\mathfrak{h}^*}$ that is 1 at λ and 0 elsewhere.

Within $\mathbb{Z}^{\mathfrak{h}^*}$, let $\mathbb{Z}[\mathfrak{h}^*]$ be the subgroup of elements of finite support. For such elements we can write $f = \sum_{\lambda \in \mathfrak{h}^*} f(\lambda) e^\lambda$ since the sum is really a finite sum. However, it will be convenient to allow this notation also for f in the larger group $\mathbb{Z}^{\mathfrak{h}^*}$, since the notation is unambiguous in this larger context.

Let Q^+ be the set of all members of $\mathfrak{h}^*$ given as $\sum_{i=1}^l n_i \alpha_i$ with all the n_i equal to integers ≥ 0. The **Kostant partition function** $\mathcal{P}$ is the function from Q^+ to the nonnegative integers that tells the number of ways, apart from order, that a member of Q^+ can be written as the sum of positive roots. By convention, $\mathcal{P}(0) = 1$.

Let $\mathbb{Z}\langle\mathfrak{h}^*\rangle$ be the set of all $f \in \mathbb{Z}^{\mathfrak{h}^*}$ whose support is contained in the union of a finite number of sets $\nu_i - Q^+$ with each ν_i in $\mathfrak{h}^*$. This is an abelian group, and we have

$$\mathbb{Z}[\mathfrak{h}^*] \subseteq \mathbb{Z}\langle\mathfrak{h}^*\rangle \subseteq \mathbb{Z}^{\mathfrak{h}^*}.$$

Within $\mathbb{Z}\langle\mathfrak{h}^*\rangle$, we introduce the multiplication

$$(5.65) \qquad \Big(\sum_{\lambda \in \mathfrak{h}^*} c_\lambda e^\lambda\Big)\Big(\sum_{\mu \in \mathfrak{h}^*} \tilde{c}_\mu e^\mu\Big) = \sum_{\nu \in \mathfrak{h}^*}\Big(\sum_{\lambda+\mu=\nu} c_\lambda \tilde{c}_\mu\Big) e^\nu.$$

To see that (5.65) makes sense, we have to check that the interior sum on the right side is finite. Because we are working within $\mathbb{Z}\langle\mathfrak{h}^*\rangle$, we can write $\lambda = \lambda_0 - q_\lambda^+$ with $q_\lambda^+ \in Q^+$ and with only finitely many possibilities for λ_0, and we can similarly write $\mu = \mu_0 - q_\mu^+$. Then

$$(\lambda_0 - q_\lambda^+) + (\mu_0 - q_\mu^+) = \nu$$

and hence
$$q_\lambda^+ + q_\mu^+ = \nu - \lambda_0 - \mu_0.$$

Finiteness follows since there are only finitely many possibilities for λ_0 and μ_0 and since $\mathcal{P}(\nu - \lambda_0 - \mu_0) < \infty$ for each.

Under the definition of multiplication in (5.65), $\mathbb{Z}\langle\mathfrak{h}^*\rangle$ is a commutative ring with identity e^0. Since $e^\lambda e^\mu = e^{\lambda+\mu}$, the natural multiplication in $\mathbb{Z}[\mathfrak{h}^*]$ is consistent with the multiplication in $\mathbb{Z}\langle\mathfrak{h}^*\rangle$.

The Weyl group W acts on $\mathbb{Z}^{\mathfrak{h}^*}$. The definition is $wf(\mu) = f(w^{-1}\mu)$ for $f \in \mathbb{Z}^{\mathfrak{h}^*}$, $\mu \in \mathfrak{h}^*$, and $w \in W$. Then $w(e^\lambda) = e^{w\lambda}$. Each $w \in W$ leaves $\mathbb{Z}[\mathfrak{h}^*]$ stable, but in general w does not leave $\mathbb{Z}\langle\mathfrak{h}^*\rangle$ stable.

We shall make use of the sign function on W. Let $\varepsilon(w) = \det w$ for $w \in W$. This is always ± 1. Any root reflection s_α has $\varepsilon(s_\alpha) = -1$. Thus if w is written as the product of k root reflections, then $\varepsilon(w) = (-1)^k$. By Proposition 2.70,

$$\varepsilon(w) = (-1)^{l(w)}, \tag{5.66}$$

where $l(w)$ is the length of w as defined in §II.6.

When φ is a representation of $\mathfrak{g}$ on V, we shall sometimes abuse notation and refer to V as the representation. If V is a representation, we say that V **has a character** if V is the direct sum of its weight spaces under $\mathfrak{h}$, i.e., $V = \bigoplus_{\mu\in\mathfrak{h}^*} V_\mu$, and if $\dim V_\mu < \infty$ for $\mu \in \mathfrak{h}^*$. In this case the **character** is

$$\operatorname{char}(V) = \sum_{\mu\in\mathfrak{h}^*} (\dim V_\mu)e^\mu.$$

EXAMPLE. Let $V(\lambda)$ be a Verma module, and let $v_{\lambda-\delta}$ be the canonical generator. Let $\mathfrak{n}^-$ be the sum of the root spaces in $\mathfrak{g}$ for the negative roots. By Proposition 5.14b the map of $U(\mathfrak{n}^-)$ into $V(\lambda)$ given by $u \mapsto uv_{\lambda-\delta}$ is one-one onto. Also the action of $U(\mathfrak{h})$ on $V(\lambda)$ matches the action of $U(\mathfrak{h})$ on $U(\mathfrak{n}^-) \otimes \mathbb{C}v_{\lambda-\delta}$. Thus

$$\dim V(\lambda)_\mu = \dim U(\mathfrak{n}^-)_{\mu-\lambda+\delta}.$$

Let $E_{-\beta_1}, \ldots, E_{-\beta_k}$ be a basis of $\mathfrak{n}^-$ consisting of root vectors. The Poincaré-Birkhoff-Witt Theorem (Theorem 3.8) shows that monomials in this basis form a basis of $U(\mathfrak{n}^-)$, and it follows that $\dim U(\mathfrak{n}^-)_{-\nu} = \mathcal{P}(\nu)$. Therefore

$$\dim V(\lambda)_\mu = \mathcal{P}(\lambda - \delta - \mu),$$

and $V(\lambda)$ has a character. The character is given by

$$\text{char}(V(\lambda)) = \sum_{\mu \in \mathfrak{h}^*} \mathcal{P}(\lambda - \delta - \mu) e^\mu = e^{\lambda - \delta} \sum_{\gamma \in Q^+} \mathcal{P}(\gamma) e^{-\gamma}. \tag{5.67}$$

Let us establish some properties of characters. Let V be a representation of $\mathfrak{g}$ with a character, and suppose that V' is a subrepresentation. Then the representations V' and V/V' have characters, and

$$\text{char}(V) = \text{char}(V') + \text{char}(V/V'). \tag{5.68}$$

In fact, we just extend a basis of weight vectors for V' to a basis of weight vectors of V. Then it is apparent that

$$\dim V_\mu = \dim V'_\mu + \dim(V/V')_\mu,$$

and (5.68) follows.

The relationship among V, V', and V/V' is summarized by saying that

$$0 \longrightarrow V' \longrightarrow V \longrightarrow V/V' \longrightarrow 0$$

is an **exact sequence**. This means that the kernel of each map going out equals the image of each map going in.

In these terms, we can generalize (5.68) as follows. Whenever

$$0 \longrightarrow V_1 \xrightarrow{\varphi_1} V_2 \xrightarrow{\varphi_2} V_3 \xrightarrow{\varphi_3} \cdots \xrightarrow{\varphi_{n-1}} V_n \longrightarrow 0$$

is an exact sequence of representations of $\mathfrak{g}$ with characters, then

$$\sum_{j=1}^{n} (-1)^j \text{char}(V_j) = 0. \tag{5.69}$$

To prove (5.69), we note that the following are exact sequences; in each case "inc" denotes an inclusion:

$$0 \longrightarrow \text{image}(\varphi_1) \xrightarrow{\text{inc}} V_2 \xrightarrow{\varphi_2} \text{image}(\varphi_2) \longrightarrow 0,$$

$$0 \longrightarrow \text{image}(\varphi_2) \xrightarrow{\text{inc}} V_3 \xrightarrow{\varphi_3} \text{image}(\varphi_3) \longrightarrow 0,$$

$$\vdots$$

$$0 \longrightarrow \text{image}(\varphi_{n-2}) \xrightarrow{\text{inc}} V_{n-1} \xrightarrow{\varphi_{n-1}} \text{image}(\varphi_{n-1}) \longrightarrow 0.$$

For $2 \le j \le n-1$, (5.68) gives

$$-\text{char}(\text{image}(\varphi_{j-1})) + \text{char}(V_j) - \text{char}(\text{image}(\varphi_j)) = 0.$$

Multiplying by $(-1)^j$ and summing, we obtain

$$\begin{aligned} 0 = -\text{char}(\text{image}(\varphi_1)) &+ \text{char}(V_2) - \text{char}(V_3) \\ &+ \cdots + (-1)^{n-1}\text{char}(V_{n-1}) + (-1)^n \text{char}(\text{image}(\varphi_{n-1})). \end{aligned}$$

Since $V_1 \cong \text{image}(\varphi_1)$ and $V_n \cong \text{image}(\varphi_{n-1})$, (5.69) follows.

Suppose that V_1 and V_2 are representations of $\mathfrak{g}$ having characters that are in $\mathbb{Z}\langle\mathfrak{h}^*\rangle$. Then $V_1 \otimes V_2$, which is a representation under the definition (4.3), has a character, and

$$(V_1 \otimes V_2) = (\text{char}(V_1))(\text{char}(V_2)). \tag{5.70}$$

In fact, the tensor product of weight vectors is a weight vector, and we can form a basis of $V_1 \otimes V_2$ from such tensor-product vectors. Hence (5.70) is an immediate consequence of (5.65).

The **Weyl denominator** is the member of $\mathbb{Z}[\mathfrak{h}^*]$ given by

$$d = e^{\delta} \prod_{\alpha \in \Delta^+} (1 - e^{-\alpha}). \tag{5.71}$$

Define

$$K = \sum_{\gamma \in Q^+} \mathcal{P}(\gamma) e^{-\gamma}.$$

This is a member of $\mathbb{Z}\langle\mathfrak{h}^*\rangle$.

Lemma 5.72. In the ring $\mathbb{Z}\langle\mathfrak{h}^*\rangle$, $Ke^{-\delta}d = 1$. Hence d^{-1} exists in $\mathbb{Z}\langle\mathfrak{h}^*\rangle$.

PROOF. From the definition in (5.71), we have

$$e^{-\delta} d = \prod_{\alpha \in \Delta^+} (1 - e^{-\alpha}). \tag{5.73}$$

Meanwhile

$$\prod_{\alpha \in \Delta^+} (1 + e^{-\alpha} + e^{-2\alpha} + \cdots) = \sum_{\gamma \in Q^+} \mathcal{P}(\gamma) e^{-\gamma} = K. \tag{5.74}$$

Since $(1 - e^{-\alpha})(1 + e^{-\alpha} + e^{-2\alpha} + \cdots) = 1$ for α positive, the lemma follows by multiplying (5.74) by (5.73).

Theorem 5.75 (Weyl Character Formula). Let V be an irreducible finite-dimensional representation of the complex semisimple Lie algebra $\mathfrak{g}$ with highest weight λ. Then

$$\operatorname{char}(V) = d^{-1} \sum_{w \in W} \varepsilon(w) e^{w(\lambda+\delta)}.$$

REMARKS. We shall prove this theorem below after giving three lemmas. But first we deduce an alternate formulation of the theorem.

Corollary 5.76 (Weyl Denominator Formula).

$$e^{\delta} \prod_{\alpha \in \Delta^+} (1 - e^{-\alpha}) = \sum_{w \in W} \varepsilon(w) e^{w\delta}.$$

PROOF. Take $\lambda = 0$ in Theorem 5.75. Then V is the 1-dimensional trivial representation, and $\operatorname{char}(V) = e^0 = 1$.

Theorem 5.77 (Weyl Character Formula, alternate formulation). Let V be an irreducible finite-dimensional representation of the complex semisimple Lie algebra $\mathfrak{g}$ with highest weight λ. Then

$$\Big(\sum_{w \in W} \varepsilon(w) e^{w\delta}\Big) \operatorname{char}(V) = \sum_{w \in W} \varepsilon(w) e^{w(\lambda+\delta)}.$$

PROOF. This follows by substituting the result of Corollary 5.76 into the formula of Theorem 5.75.

Lemma 5.78. If λ in $\mathfrak{h}^*$ is dominant, then no $w \neq 1$ in W fixes $\lambda + \delta$.

PROOF. If $w \neq 1$ fixes $\lambda + \delta$, then Chevalley's Lemma in the form of Corollary 2.73 shows that some root α has $\langle \lambda + \delta, \alpha \rangle = 0$. We may assume that α is positive. But then $\langle \lambda, \alpha \rangle \geq 0$ by dominance and $\langle \delta, \alpha \rangle > 0$ by Proposition 2.69, and we have a contradiction.

Lemma 5.79. The Verma module $V(\lambda)$ has a character belonging to $\mathbb{Z}\langle \mathfrak{h}^* \rangle$, and $\operatorname{char}(V(\lambda)) = d^{-1} e^{\lambda}$.

PROOF. Formula (5.67) shows that

$$\operatorname{char}(V(\lambda)) = e^{\lambda-\delta} \sum_{\gamma \in Q^+} \mathcal{P}(\gamma) e^{-\gamma} = K e^{-\delta} e^{\lambda},$$

and thus the result follows by substituting from Lemma 5.72.

Lemma 5.80. Let λ_0 be in $\mathfrak{h}^*$, and suppose that M is a representation of $\mathfrak{g}$ such that

(i) M has infinitesimal character λ_0
(ii) M has a character belonging to $\mathbb{Z}\langle\mathfrak{h}^*\rangle$.

Let

$$D_M = \{\lambda \in W\lambda_0 \mid (\lambda - \delta + Q^+) \cap \text{support}(\text{char}(M)) \neq \emptyset\}.$$

Then $\text{char}(M)$ is a finite $\mathbb{Z}$ linear combination of $\text{char}(V(\lambda))$ for λ in D_M.

REMARK. D_M is a finite set, being a subset of an orbit of the finite group W.

PROOF. We may assume that $M \neq 0$, and we proceed by induction on $|D_M|$. First assume that $|D_M| = 0$. Since M has a character belonging to $\mathbb{Z}\langle\mathfrak{h}^*\rangle$, we can find μ in $\mathfrak{h}^*$ such that $\mu - \delta$ is a weight of M but $\mu - \delta + q^+$ is not a weight of M for any $q^+ \neq 0$ in Q^+. Set $m = \dim M_{\mu-\delta}$. Since the root vectors for positive roots evidently annihilate $M_{\mu-\delta}$, the universal mapping property for Verma modules (Proposition 5.14c) shows that we can find a $U(\mathfrak{g})$ homomorphism $\varphi : V(\mu)^m \to M$ such that $(V(\mu)^m)_{\mu-\delta}$ maps one-one onto $M_{\mu-\delta}$. The infinitesimal character λ_0 of M must match the infinitesimal character of $V(\mu)$, which is μ by Proposition 5.42. By Theorem 5.62, μ is in $W\lambda_0$. Then μ is in D_M, and $|D_M| = 0$ is impossible. This completes the base case of the induction.

Now assume the result of the lemma for modules N satisfying (i) and (ii) such that D_N has fewer than $|D_M|$ members. Construct μ, m, and φ as above. Let L be the kernel of φ, and put $N = M/\text{image}\,\varphi$. Then

$$0 \longrightarrow L \longrightarrow V(\mu)^m \xrightarrow{\varphi} M \xrightarrow{\psi} N \longrightarrow 0$$

is an exact sequence of representations. By (5.68), $\text{char}(L)$ and $\text{char}(N)$ exist. Thus (5.69) gives

$$\text{char}(M) = -\text{char}(L) + m\,\text{char}(V(\mu)) + \text{char}(N).$$

Moreover L and N satisfy (i) and (ii). The induction will be complete if we show that $|D_L| < |D_M|$ and $|D_N| < |D_M|$.

In the case of N, we clearly have $D_N \subseteq D_M$. Since ψ is onto, the equality $M_{\mu-\delta} = \text{image}\,\varphi$ implies that $N_{\mu-\delta} = 0$. Thus μ is not in D_N, and $|D_N| < |D_M|$.

In the case of L, if λ is in D_L, then $\lambda - \delta + Q^+$ has nonempty intersection with $\text{support}(\text{char}(L))$ and hence with $\text{support}(\text{char}(V(\mu)))$. Then $\mu - \delta$ is in $\lambda - \delta + Q^+$, and hence $\mu - \delta$ is a member of the intersection $(\lambda - \delta + Q^+) \cap \text{support}(\text{char}(M))$. That is, λ is in D_M. Therefore $D_L \subseteq D_M$. But μ is not in D_L, and hence $|D_L| < |D_M|$. This completes the proof.

PROOF OF THEOREM 5.75. By (5.43), V has infinitesimal character $\lambda+\delta$. Lemma 5.80 applies to V with λ_0 replaced by $\lambda+\delta$, and Lemma 5.79 allows us to conclude that

$$\operatorname{char}(V) = d^{-1} \sum_{w \in W} c_w e^{w(\lambda+\delta)}$$

for some unknown integers c_w. We rewrite this formula as

$$d \operatorname{char}(V) = \sum_{w \in W} c_w e^{w(\lambda+\delta)}. \tag{5.81}$$

Let us say that a member f of $\mathbb{Z}[\mathfrak{h}^*]$ is **even** (under W) if $wf = f$ for all w in W. It is **odd** if $wf = \varepsilon(w) f$ for all w in W. Theorem 5.5e shows that $\operatorname{char}(V)$ is even. Let us see that d is odd. In fact, we can write d as

$$d = \prod_{\alpha \in \Delta^+} (e^{\alpha/2} - e^{-\alpha/2}). \tag{5.82}$$

If we replace each α by $w\alpha$, we get the same factors on the right side of (5.82) except for minus signs, and the number of minus signs is the number of positive roots α such that $w\alpha$ is negative. By (5.66) this product of minus signs is just $\varepsilon(w)$.

Consequently the left side of (5.81) is odd under W, and application of w_0 to both sides of (5.81) gives

$$\begin{aligned}\sum_{w \in W} c_w \varepsilon(w_0) e^{w(\lambda+\delta)} &= \varepsilon(w_0) d \operatorname{char}(V) = w_0(d \operatorname{char}(V)) \\ &= \sum_{w \in W} c_w e^{w_0 w(\lambda+\delta)} = \sum_{w \in W} c_{w_0^{-1} w} e^{w(\lambda+\delta)}.\end{aligned}$$

By Lemma 5.78 the two sides of this formula are equal term by term. Thus we have $c_{w_0^{-1}w} = c_w \varepsilon(w_0)$ for w in W. Taking $w = 1$ gives $c_{w_0^{-1}} = c_1 \varepsilon(w_0) = c_1 \varepsilon(w_0^{-1})$, and hence $c_{w_0} = c_1 \varepsilon(w_0)$. Therefore

$$d \operatorname{char}(V) = c_1 \sum_{w \in W} \varepsilon(w) e^{w(\lambda+\delta)}.$$

Expanding the left side and taking Theorem 5.5b into account, we see that the coefficient of $e^{\lambda+\delta}$ on the left side is 1. Thus another application of Lemma 5.78 gives $c_1 = 1$.

Corollary 5.83 (Kostant Multiplicity Formula). Let V be an irreducible finite-dimensional representation of the complex semisimple Lie algebra $\mathfrak{g}$ with highest weight λ. If μ is in $\mathfrak{h}^*$, then the multiplicity of μ as a weight of V is

$$\sum_{w\in W} \varepsilon(w)\mathcal{P}(w(\lambda+\delta)-(\mu+\delta)).$$

REMARK. By convention in this formula, $\mathcal{P}(\nu)=0$ if ν is not in Q^+.

PROOF. Lemma 5.72 and Theorem 5.75 combine to give

$$\begin{aligned}\operatorname{char}(V) &= d^{-1}(d\operatorname{char}(V))\\ &= (Ke^{-\delta})(d\operatorname{char}(V))\\ &= \Big(\sum_{\gamma\in Q^+}\mathcal{P}(\gamma)e^{-\delta-\gamma}\Big)\Big(\sum_{w\in W}\varepsilon(w)e^{w(\lambda+\delta)}\Big).\end{aligned}$$

Hence the required multiplicity is

$$\sum_{\substack{\gamma\in Q^+,\, w\in W\\ -\delta-\gamma+w(\lambda+\delta)=\mu}} \mathcal{P}(\gamma)\varepsilon(w) = \sum_{w\in W}\varepsilon(w)\mathcal{P}(w(\lambda+\delta)-\mu-\delta).$$

Theorem 5.84 (Weyl Dimension Formula). Let V be an irreducible finite-dimensional representation of the complex semisimple Lie algebra $\mathfrak{g}$ with highest weight λ. Then

$$\dim V = \frac{\prod_{\alpha\in\Delta^+}\langle\lambda+\delta,\alpha\rangle}{\prod_{\alpha\in\Delta^+}\langle\delta,\alpha\rangle}.$$

PROOF. For $H\in\mathfrak{h}^*$, we introduce the ring homomorphism called "evaluation at H," which is written $\epsilon_H:\mathbb{Z}[\mathfrak{h}^*]\to\mathbb{C}$ and is given by

$$f=\sum f(\lambda)e^{\lambda}\mapsto \sum f(\lambda)e^{\lambda(H)}.$$

Then $\dim V=\epsilon_0(\operatorname{char}(V))$. The idea is thus to apply ϵ_0 to the Weyl Character Formula as given in Theorem 5.75 or Theorem 5.77. But a direct application will give $0/0$ for the value of $\epsilon_0(\operatorname{char}(V))$, and we have to proceed more carefully. In effect, we shall use a version of l'Hôpital's Rule.

For $f\in\mathbb{Z}[\mathfrak{h}^*]$ and $\varphi\in\mathfrak{h}^*$, we define

$$\partial_\varphi f(H)=\frac{d}{dr}f(H+rH_\varphi)|_{r=0}.$$

Then

$$\partial_\varphi e^{\lambda(H)} = \frac{d}{dr} e^{\lambda(H+rH_\varphi)}|_{r=0} = \langle \lambda, \varphi \rangle e^{\lambda(H)}. \tag{5.85}$$

Consider any derivative $\partial_{\varphi_1} \cdots \partial_{\varphi_n}$ of order less than the number of positive roots, and apply it to the Weyl denominator (5.71), evaluating at H. We are then considering

$$\partial_{\varphi_1} \cdots \partial_{\varphi_n} \Big(e^{-\delta(H)} \prod_{\alpha \in \Delta^+} (e^{\alpha(H)} - 1) \Big).$$

Each ∂_{φ_j} operates by the product rule and differentiates one factor, leaving the others alone. Thus each term in the derivative has an undifferentiated $e^{\alpha(H)} - 1$ and will give 0 when evaluated at $H = 0$.

We apply $\prod_{\alpha \in \Delta^+} \partial_\alpha$ to both sides of the identity given by the Weyl Character Formula

$$d \operatorname{char}(V) = \sum_{w \in W} \varepsilon(w) e^{w(\lambda+\delta)}.$$

Then we evaluate at $H = 0$. The result on the left side comes from the Leibniz rule and involves many terms, but all of them give 0 (according to the previous paragraph) except the one that comes from applying all the derivatives to d and evaluating the other factor at $H = 0$. Thus we obtain

$$\Big(\Big(\prod_{\alpha \in \Delta^+} \partial_\alpha \Big) d(H) \Big)(0) \dim V = \Big(\Big(\prod_{\alpha \in \Delta^+} \partial_\alpha \Big) \sum_{w \in W} \varepsilon(w) e^{w(\lambda+\delta)(H)} \Big)(0).$$

By Corollary 5.76 we can rewrite this formula as

$$\begin{aligned} \Big(\Big(\prod_{\alpha \in \Delta^+} \partial_\alpha \Big) \sum_{w \in W} \varepsilon(w) e^{(w\delta)(H)} \Big)(0) \dim V & \\ = \Big(\Big(\prod_{\alpha \in \Delta^+} \partial_\alpha \Big) \sum_{w \in W} \varepsilon(w) e^{w(\lambda+\delta)(H)} \Big)(0). & \end{aligned} \tag{5.86}$$

We calculate

$$\begin{aligned} \Big(\prod_{\alpha \in \Delta^+} \partial_\alpha \Big) \Big(\sum_{w \in W} \varepsilon(w) e^{w(\lambda+\delta)(H)} \Big) & \\ &= \sum_{w \in W} \varepsilon(w) \prod_{\alpha \in \Delta^+} \langle w(\lambda+\delta), \alpha \rangle e^{w(\lambda+\delta)(H)} \qquad \text{by (5.85)} \\ &= \sum_{w \in W} \varepsilon(w^{-1}) \prod_{\alpha \in \Delta^+} \langle \lambda+\delta, w^{-1}\alpha \rangle e^{w(\lambda+\delta)(H)} \\ &= \sum_{w \in W} \prod_{\alpha \in \Delta^+} \langle \lambda+\delta, \alpha \rangle e^{w(\lambda+\delta)(H)} \qquad \text{by (5.66)} \\ &= \Big(\prod_{\alpha \in \Delta^+} \langle \lambda+\delta, \alpha \rangle \Big) \sum_{w \in W} e^{w(\lambda+\delta)(H)}. \qquad (5.87) \end{aligned}$$

When $\lambda = 0$, (5.87) has a nonzero limit as H tends to 0 by Proposition 2.69. Therefore we can evaluate $\dim V$ from (5.86) by taking the quotient with H in place and then letting H tend to 0. By (5.87) the result is the formula of the theorem.

The Weyl Dimension Formula provides a convenient tool for deciding irreducibility. Let φ be a finite-dimensional representation of $\mathfrak{g}$, and suppose that λ is the highest weight of φ. Theorem 5.29 shows that φ is completely reducible, and one of the irreducible summands must have λ as highest weight. Call this summand φ_λ. Theorem 5.84 allows us to compute $\dim \varphi_\lambda$. Then it follows that φ is irreducible if and only if $\dim \varphi$ matches the value of $\dim \varphi_\lambda$ given by Theorem 5.84.

EXAMPLE. With $\mathfrak{g} = \mathfrak{sl}(n, \mathbb{C})$, let φ be the representation on the space consisting of all holomorphic polynomials in $z_1, \ldots, z_n$ homogeneous of degree N. We shall prove that this representation is irreducible. From the first example in §2, we know that this representation has highest weight $-Ne_n$. Its dimension is $\binom{N+n-1}{N}$, the number of ways of labeling $n-1$ of $N+n-1$ objects as dividers and the others as monomials z_j. To check that φ is irreducible, it is enough to see from the Weyl Dimension Formula that the irreducible representation φ_{-Ne_n} with highest weight $\lambda = -Ne_n$ has dimension $\binom{N+n-1}{N}$. Easy calculation gives

$$\delta - \tfrac{1}{2}(n-1)e_1 + \tfrac{1}{2}(n-3)e_2 + \cdots + \tfrac{1}{2}(1-n)e_n.$$

A quotient $\dfrac{\langle \lambda + \delta, \alpha \rangle}{\langle \delta, \alpha \rangle}$ will be 1 unless $\langle \lambda, \alpha \rangle \neq 0$. Therefore

$$\dim \varphi_{-Ne_n} = \prod_{j=1}^{n-1} \frac{\langle -Ne_n + \delta, e_j - e_n \rangle}{\langle \delta, e_j - e_n \rangle} = \prod_{j=1}^{n-1} \frac{N+n-j}{n-j} = \binom{N+n-1}{N},$$

as required.

7. Parabolic Subalgebras

Let $\mathfrak{g}$ be a complex semisimple Lie algebra, and let $\mathfrak{h}$, $\Delta = \Delta(\mathfrak{g}, \mathfrak{h})$, and B be as in §2. A **Borel subalgebra** of $\mathfrak{g}$ is a subalgebra $\mathfrak{b} = \mathfrak{h} \oplus \mathfrak{n}$, where $\mathfrak{n} = \bigoplus_{\alpha \in \Delta^+} \mathfrak{g}_\alpha$ for some positive system Δ^+ within Δ. Any subalgebra $\mathfrak{q}$ of $\mathfrak{g}$ containing a Borel subalgebra is called a **parabolic subalgebra** of $\mathfrak{g}$. Our goal in this section is to classify parabolic subalgebras and to relate them to finite-dimensional representations of $\mathfrak{g}$.

We regard $\mathfrak{h}$ and $\mathfrak{n}$ as fixed in our discussion, and we study only parabolic subalgebras $\mathfrak{q}$ that contain $\mathfrak{b} = \mathfrak{h} \oplus \mathfrak{n}$. Let Π be the simple system determining Δ^+ and $\mathfrak{n}$, and define $\mathfrak{n}^-$ as in (5.8). Since $\mathfrak{q} \supseteq \mathfrak{h}$ and since the root spaces are 1-dimensional, $\mathfrak{q}$ is necessarily of the form

$$\mathfrak{q} = \mathfrak{h} \oplus \bigoplus_{\alpha \in \Gamma} \mathfrak{g}_\alpha, \tag{5.88}$$

where Γ is a subset of $\Delta(\mathfrak{g}, \mathfrak{h})$ containing $\Delta^+(\mathfrak{g}, \mathfrak{h})$. The extreme cases are $\mathfrak{q} = \mathfrak{b}$ (with $\Gamma = \Delta^+(\mathfrak{g}, \mathfrak{h})$) and $\mathfrak{q} = \mathfrak{g}$ (with $\Gamma = \Delta(\mathfrak{g}, \mathfrak{h})$).

To obtain further examples of parabolic subalgebras, we fix a subset Π' of the set Π of simple roots and let

$$\Gamma = \Delta^+(\mathfrak{g}, \mathfrak{h}) \cup \{\alpha \in \Delta(\mathfrak{g}, \mathfrak{h}) \mid \alpha \in \operatorname{span}(\Pi')\}. \tag{5.89}$$

Then (5.88) is a parabolic subalgebra containing the given Borel subalgebra $\mathfrak{b}$. (Closure under brackets follows from the fact that if α and β are in Γ and if $\alpha + \beta$ is a root, then $\alpha + \beta$ is in Γ; this fact is an immediate consequence of Proposition 2.49.) All examples are of this form, according to Proposition 5.90 below. With Γ as in (5.88), define $-\Gamma$ to be the set of negatives of the members of Γ.

Proposition 5.90. The parabolic subalgebras $\mathfrak{q}$ containing $\mathfrak{b}$ are parametrized by the set of subsets of simple roots; the one corresponding to a subset Π' is of the form (5.88) with Γ as in (5.89).

PROOF. If $\mathfrak{q}$ is given, we define $\Gamma(\mathfrak{q})$ to be the Γ in (5.88), and we define $\Pi'(\mathfrak{q})$ to be the set of simple roots in the linear span of $\Gamma(\mathfrak{q}) \cap -\Gamma(\mathfrak{q})$. Then $\mathfrak{q} \mapsto \Pi'(\mathfrak{q})$ is a map from parabolic subalgebras $\mathfrak{q}$ containing $\mathfrak{b}$ to subsets of simple roots. In the reverse direction, if Π' is given, we define $\Gamma(\Pi')$ to be the Γ in (5.89), and then $\mathfrak{q}(\Pi')$ is defined by means of (5.88). We have seen that $\mathfrak{q}(\Pi')$ is a subalgebra, and thus $\Pi' \mapsto \mathfrak{q}(\Pi')$ is a map from subsets of simple roots to parabolic subalgebras containing $\mathfrak{b}$.

To complete the proof we have to show that these two maps are inverse to one another. To see that $\Pi'(\mathfrak{q}(\Pi')) = \Pi'$, we observe that

$$\{\alpha \in \Delta(\mathfrak{g}, \mathfrak{h}) \mid \alpha \in \operatorname{span}(\Pi')\}$$

is closed under negatives. Therefore (5.89) gives

$$\begin{aligned}\Gamma(\Pi') \cap -\Gamma(\Pi') &= (\Delta^+(\mathfrak{g}, \mathfrak{h}) \cup \{\alpha \in \Delta(\mathfrak{g}, \mathfrak{h}) \mid \alpha \in \operatorname{span}(\Pi')\}) \\ &\qquad \cap (-\Delta^+(\mathfrak{g}, \mathfrak{h}) \cup \{\alpha \in \Delta(\mathfrak{g}, \mathfrak{h}) \mid \alpha \in \operatorname{span}(\Pi')\}) \\ &= (\Delta^+(\mathfrak{g}, \mathfrak{h}) \cap -\Delta^+(\mathfrak{g}, \mathfrak{h})) \cup \{\alpha \in \Delta(\mathfrak{g}, \mathfrak{h}) \mid \alpha \in \operatorname{span}(\Pi')\} \\ &= \{\alpha \in \Delta(\mathfrak{g}, \mathfrak{h}) \mid \alpha \in \operatorname{span}(\Pi')\}.\end{aligned}$$

The simple roots in the span of the right side are the members of Π', by the independence in Proposition 2.49, and it follows that $\Pi'(\mathfrak{q}(\Pi')) = \Pi'$.

To see that $\mathfrak{q}(\Pi'(\mathfrak{q})) = \mathfrak{q}$, we are to show that $\Gamma(\Pi'(\mathfrak{q})) = \Gamma(\mathfrak{q})$. Since $\Delta^+(\mathfrak{g}, \mathfrak{h}) \subseteq \Gamma(\mathfrak{q})$, the inclusion $\Gamma(\Pi'(\mathfrak{q})) \subseteq \Gamma(\mathfrak{q})$ will follow if we show that

$$\{\alpha \in \Delta(\mathfrak{g}, \mathfrak{h}) \mid \alpha \in \operatorname{span}(\Pi'(\mathfrak{q}))\} \subseteq \Gamma(\mathfrak{q}). \tag{5.91}$$

Since $\Gamma(\mathfrak{q}) = \Delta^+(\mathfrak{g}, \mathfrak{h}) \cup (\Gamma(\mathfrak{q}) \cap -\Gamma(\mathfrak{q}))$, the inclusion $\Gamma(\Pi'(\mathfrak{q})) \supseteq \Gamma(\mathfrak{q})$ will follow if we show that

$$\Gamma(\mathfrak{q}) \cap -\Gamma(\mathfrak{q}) \subseteq \Gamma(\Pi'(\mathfrak{q})). \tag{5.92}$$

Let us first prove (5.91). The positive members of the left side of (5.91) are elements of the right side since $\mathfrak{b} \subseteq \mathfrak{q}$. Any negative root in the left side is a negative-integer combination of members of $\Pi'(\mathfrak{q})$ by Proposition 2.49. Let $-\alpha$ be such a root, and expand α in terms of the simple roots $\Pi = \{\alpha_i\}_{i=1}^l$ as $\alpha = \sum_i n_i \alpha_i$. We prove by induction on the level $\sum n_i$ that a nonzero root vector $E_{-\alpha}$ for $-\alpha$ is in $\mathfrak{q}$. When the level is 1, this assertion is just the definition of $\Pi'(\mathfrak{q})$. When the level of α is > 1, we can choose positive roots β and γ with $\alpha = \beta + \gamma$. Then β and γ are positive integer combinations of members of $\Pi'(\mathfrak{q})$. By inductive hypothesis, $-\beta$ and $-\gamma$ are in $\Gamma(\mathfrak{q})$. Hence the corresponding root vectors $E_{-\beta}$ and $E_{-\gamma}$ are in $\mathfrak{q}$. By Corollary 2.35, $[E_{-\beta}, E_{-\gamma}]$ is a nonzero root vector for $-\alpha$. Since $\mathfrak{q}$ is a subalgebra, $-\alpha$ must be in $\Gamma(\mathfrak{q})$. This proves (5.91).

Finally let us prove (5.92). Let $-\alpha$ be a negative root in $\Gamma(\mathfrak{q})$, and expand α in terms of simple roots as $\alpha = \sum_i n_i \alpha_i$. The assertion is that each α_i for which $n_i > 0$ is in $\Pi'(\mathfrak{q})$, i.e., has $-\alpha_i \in \Gamma(\mathfrak{q})$. We prove this assertion by induction on the level $\sum n_i$, the case of level 1 being trivial. If the level of α is > 1, then $\alpha = \beta + \gamma$ with β and γ in $\Delta^+(\mathfrak{g}, \mathfrak{h})$. The root vectors $E_{-\alpha}$ and E_β are in $\mathfrak{q}$, and hence so is their bracket, which is a nonzero multiple of $E_{-\gamma}$ by Corollary 2.35. Similarly $E_{-\alpha}$ and E_γ are in $\mathfrak{q}$, and hence so is $E_{-\beta}$. Thus $-\gamma$ and $-\beta$ are in $\Gamma(\mathfrak{q})$. By induction the constituent simple roots of β and γ are in $\Pi'(\mathfrak{q})$, and thus the same thing is true of α. This proves (5.92) and completes the proof of the proposition.

Now define

$$\mathfrak{l} = \mathfrak{h} \oplus \bigoplus_{\alpha \in \Gamma \cap -\Gamma} \mathfrak{g}_\alpha \qquad \text{and} \qquad \mathfrak{u} = \bigoplus_{\substack{\alpha \in \Gamma, \\ \alpha \notin -\Gamma}} \mathfrak{g}_\alpha, \tag{5.93a}$$

so that

$$q = \mathfrak{l} \oplus \mathfrak{u}. \tag{5.93b}$$

Corollary 5.94. Relative to a parabolic subalgebra $\mathfrak{q}$ containing $\mathfrak{b}$,

(a) $\mathfrak{l}$ and $\mathfrak{u}$ are subalgebras of $\mathfrak{q}$, and $\mathfrak{u}$ is an ideal in $\mathfrak{q}$
(b) $\mathfrak{u}$ is nilpotent
(c) $\mathfrak{l}$ is reductive with center $\mathfrak{h}'' = \bigcap_{\alpha\in\Gamma\cap-\Gamma}\ker\alpha \subseteq \mathfrak{h}$ and with semisimple part $\mathfrak{l}_{ss}$ having root-space decomposition

$$\mathfrak{l}_{ss} = \mathfrak{h}' \oplus \bigoplus_{\alpha\in\Gamma\cap-\Gamma} \mathfrak{g}_\alpha,$$

where $\mathfrak{h}' = \sum_{\alpha\in\Gamma\cap-\Gamma}\mathbb{C}H_\alpha$.

PROOF. By Proposition 5.90 let $\mathfrak{q}$ be built from Π' by means of (5.89) and (5.88). Then (a) is clear. In (b), we have $\mathfrak{u} \subseteq \mathfrak{n}$, and hence $\mathfrak{u}$ is nilpotent.

Let us prove (c). Let $\mathfrak{h}_0$ be the real form of $\mathfrak{h}$ on which all roots are real-valued. Then $\mathfrak{h}_0' = \mathfrak{h}_0 \cap \mathfrak{h}'$ and $\mathfrak{h}_0'' = \mathfrak{h}_0 \cap \mathfrak{h}''$ are real forms of $\mathfrak{h}'$ and $\mathfrak{h}''$, respectively. The form B for $\mathfrak{g}$ has $B|_{\mathfrak{h}_0\times\mathfrak{h}_0}$ positive definite, and it is clear that $\mathfrak{h}_0'$ and $\mathfrak{h}_0''$ are othogonal complements of each other. Therefore $\mathfrak{h}_0 = \mathfrak{h}_0' \oplus \mathfrak{h}_0''$ and $\mathfrak{h} = \mathfrak{h}' \oplus \mathfrak{h}''$. Thus with $\mathfrak{l}_{ss}$ defined as in the statement of (c), $\mathfrak{l} = \mathfrak{h}'' \oplus \mathfrak{l}_{ss}$. Moreover it is clear that $\mathfrak{h}''$ and $\mathfrak{l}_{ss}$ are ideals in $\mathfrak{l}$ and that $\mathfrak{h}''$ is contained in the center. To complete the proof, it is enough to show that $\mathfrak{l}_{ss}$ is semisimple.

Thus let B' be the Killing form of $\mathfrak{l}_{ss}$. Relative to B', $\mathfrak{h}'$ is orthogonal to each $\mathfrak{g}_\alpha$ in $\mathfrak{l}$, and each $\mathfrak{g}_\alpha$ in $\mathfrak{l}$ is orthogonal to all $\mathfrak{g}_\beta$ in $\mathfrak{l}$ except $\mathfrak{g}_{-\alpha}$. For $\alpha \in \Gamma \cap -\Gamma$, choose root vectors E_α and $E_{-\alpha}$ with $B(E_\alpha, E_{-\alpha}) = 1$, so that $[E_\alpha, E_{-\alpha}] = H_\alpha$. We shall show that $B'(E_\alpha, E_{-\alpha}) > 0$ and that B' is positive definite on $\mathfrak{h}_0' \times \mathfrak{h}_0'$. Then it follows that B' is nondegenerate, and $\mathfrak{l}_{ss}$ is semisimple by Cartan's Criterion for Semisimplicity (Theorem 1.42).

In considering $B'(E_\alpha, E_{-\alpha})$, we observe from Corollary 2.37 that $\operatorname{ad} E_\alpha \operatorname{ad} E_{-\alpha}$ acts with eigenvalue ≥ 0 on any $\mathfrak{g}_\beta$. On $H \in \mathfrak{h}$, it gives $\alpha(H)H_\alpha$, which is a positive multiple of H_α if $H = H_\alpha$ and is 0 if H is in $\ker\alpha$. Thus $\operatorname{ad} E_\alpha \operatorname{ad} E_{-\alpha}$ has trace > 0 on $\mathfrak{h}$ and trace ≥ 0 on each $\mathfrak{g}_\beta$. Consequently $B'(E_\alpha, E_{-\alpha}) > 0$.

If H is in $\mathfrak{h}_0'$, then $B'(H, H) = \sum_{\alpha\in\Gamma\cap-\Gamma}\alpha(H)^2$, and each term is ≥ 0. To get 0, we must have $\alpha(H) = 0$ for all $\alpha \in \Gamma \cap -\Gamma$. This condition forces H to be in $\mathfrak{h}''$. Since $\mathfrak{h}' \cap \mathfrak{h}'' = 0$, we find that $H = 0$. Consequently B' is positive definite on $\mathfrak{h}_0' \times \mathfrak{h}_0'$, as asserted.

In the decomposition (5.93) of $\mathfrak{q}$, $\mathfrak{l}$ is called the **Levi factor** and $\mathfrak{u}$ is called the **nilpotent radical**. The nilpotent radical can be characterized solely in terms of $\mathfrak{q}$ as the radical of the symmetric bilinear form $B|_{\mathfrak{q}\times\mathfrak{q}}$, where B is the invariant form for $\mathfrak{g}$. But the Levi factor $\mathfrak{l}$ depends on $\mathfrak{h}$ as well as $\mathfrak{q}$.

Define

(5.95a) $$\mathfrak{u}^- = \bigoplus_{\substack{\alpha\in\Gamma,\\ \alpha\notin-\Gamma}} \mathfrak{g}_{-\alpha}.$$

and

(5.95b) $$\mathfrak{q}^- = \mathfrak{l}\oplus\mathfrak{u}^-,$$

(The subalgebra $\mathfrak{q}^-$ is a parabolic subalgebra containing the Borel subalgebra $\mathfrak{b}^- = \mathfrak{h}\oplus\mathfrak{n}^-$.) Then we have the important identities

(5.96) $$\mathfrak{l} = \mathfrak{q}\cap\mathfrak{q}^-$$

and

(5.97) $$\mathfrak{g} = \mathfrak{u}^-\oplus\mathfrak{l}\oplus\mathfrak{u}.$$

Now we shall examine parabolic subalgebras in terms of centralizers and eigenvalues. We begin with some notation. In the background will be our Cartan subalgebra $\mathfrak{h}$ and the Borel subalgebra $\mathfrak{b}$. We suppose that V is a finite-dimensional completely reducible representation of $\mathfrak{h}$, and we denote by $\Delta(V)$ the set of weights of $\mathfrak{h}$ in V. Some examples are

$$\begin{aligned}
\Delta(\mathfrak{g}) &= \Delta(\mathfrak{g},\mathfrak{h})\cup\{0\}\\
\Delta(\mathfrak{n}) &= \Delta^+(\mathfrak{g},\mathfrak{h})\\
\Delta(\mathfrak{q}) &= \Gamma\cup\{0\}\\
\Delta(\mathfrak{l}) &= (\Gamma\cap-\Gamma)\cup\{0\}\\
\Delta(\mathfrak{u}) &- \{\alpha\in\Gamma\mid -\alpha\notin\Gamma\}.
\end{aligned}$$

For each weight $\omega\in\Delta(V)$, let m_ω be the multiplicity of ω. We define

(5.98) $$\delta(V) = \tfrac{1}{2}\sum_{\omega\in\Delta(V)} m_\omega\omega,$$

half the sum of the weights with multiplicities counted. An example is that $\delta(\mathfrak{n}) = \delta$, with δ defined as in §II.6 and again in (5.8). The following result generalizes Proposition 2.69.

Proposition 5.99. Let V be a finite-dimensional representation of $\mathfrak{g}$, and let Λ be a subset of $\Delta(V)$. Suppose that α is a root such that $\lambda \in \Lambda$ and $\alpha+\lambda \in \Delta(V)$ together imply $\alpha+\lambda \in \Lambda$. Then $\left\langle \sum_{\lambda\in\Lambda} m_\lambda \lambda, \alpha \right\rangle \geq 0$. Strict inequality holds when the representation is the adjoint representation of $\mathfrak{g}$ on $V = \mathfrak{g}$ and α is in Λ and $-\alpha$ is not in Λ.

PROOF. Theorem 5.29 shows that V is completely reducible. If E_α and $E_{-\alpha}$ denote nonzero root vectors for α and $-\alpha$, V is therefore completely reducible under $\mathfrak{h} + \operatorname{span}\{H_\alpha, E_\alpha, E_{-\alpha}\}$. Let λ be in Λ, and suppose that $\langle \lambda, \alpha \rangle < 0$. Then the theory for $\mathfrak{sl}(2, \mathbb{C})$ shows that $\lambda, \lambda + \alpha, \dots, s_\alpha \lambda$ are in $\Delta(V)$, and the hypothesis forces all of these weights to be in Λ. In particular $s_\alpha\lambda$ is in Λ. Theorem 5.5e says that $m_\lambda = m_{s_\alpha \lambda}$. Therefore

$$\sum_{\lambda\in\Lambda} m_\lambda \lambda = \sum_{\substack{\lambda\in\Lambda,\\ \langle\lambda,\alpha\rangle<0}} m_\lambda(\lambda + s_\alpha\lambda) + \sum_{\substack{\lambda\in\Lambda,\\ \langle\lambda,\alpha\rangle=0}} m_\lambda\lambda + \sum_{\substack{\lambda\in\Lambda,\, s_\alpha\lambda\notin\Lambda,\\ \langle\lambda,\alpha\rangle>0}} m_\lambda\lambda.$$

The inner product of α with the first two sums on the right is 0, and the inner product of α with the third sum is term-by-term positive. This proves the first assertion. In the case of the adjoint representation, if $\alpha \in \Lambda$ and $-\alpha \notin \Lambda$, then α occurs in the third sum and gives a positive inner product. This proves the second assertion.

Corollary 5.100. Let $\mathfrak{q}$ be a parabolic subalgebra containing $\mathfrak{b}$. If α is in $\Delta^+(\mathfrak{g}, \mathfrak{h})$, then

$$\langle \delta(\mathfrak{u}), \alpha \rangle \quad \text{is} \quad \begin{cases} = 0 & \text{if } \alpha \in \Delta(\mathfrak{l}, \mathfrak{h}) \\ > 0 & \text{if } \alpha \in \Delta(\mathfrak{u}). \end{cases}$$

PROOF. In Proposition 5.99 let $V = \mathfrak{g}$ and $\Lambda = \Delta(\mathfrak{u})$. If α is in $\Delta(\mathfrak{l}, \mathfrak{h})$, the proposition applies to α and $-\alpha$ and gives $\langle \delta(\mathfrak{u}), \alpha \rangle = 0$. If α is in $\Delta(\mathfrak{u})$, then $-\alpha$ is not in Λ and the proposition gives $\langle \delta(\mathfrak{u}), \alpha \rangle > 0$.

Corollary 5.101. Let $\mathfrak{q} = \mathfrak{l} \oplus \mathfrak{u}$ be a parabolic subalgebra containing $\mathfrak{b}$. Then the element $H = H_{\delta(\mathfrak{u})}$ of $\mathfrak{h}$ has the property that all roots are real-valued on H and

$$\begin{aligned} \mathfrak{u} &= \text{sum of eigenspaces of } \operatorname{ad} H \text{ for positive eigenvalues} \\ \mathfrak{l} &= Z_{\mathfrak{g}}(H) = \text{eigenspace of } \operatorname{ad} H \text{ for eigenvalue } 0 \\ \mathfrak{u}^- &= \text{sum of eigenspaces of } \operatorname{ad} H \text{ for negative eigenvalues.} \end{aligned}$$

PROOF. This is immediate from Corollary 5.100.

We are ready to examine the role of parabolic subalgebras in finite-dimensional representations. The idea is to obtain a generalization of the Theorem of the Highest Weight (Theorem 5.5) in which $\mathfrak{h}$ and $\mathfrak{n}$ get replaced by $\mathfrak{l}$ and $\mathfrak{u}$.

The Levi factor $\mathfrak{l}$ of a parabolic subalgebra $\mathfrak{q}$ containing $\mathfrak{b}$ is reductive by Corollary 5.94c, but it is usually not semisimple. In the representations that we shall study, $\mathfrak{h}$ will act completely reducibly, and hence the subalgebra $\mathfrak{h}''$ in that corollary will act completely reducibly. Each simultaneous eigenspace of $\mathfrak{h}''$ will give a representation of $\mathfrak{l}_{ss}$, which will be completely reducible by Theorem 5.29. We summarize these remarks as follows.

Proposition 5.102. Let $\mathfrak{q}$ be a parabolic subalgebra containing $\mathfrak{b}$. In any finite-dimensional representation of $\mathfrak{l}$ for which $\mathfrak{h}$ acts completely reducibly, $\mathfrak{l}$ acts completely reducibly. This happens in particular when the action of a representation of $\mathfrak{g}$ is restricted to $\mathfrak{l}$.

Each irreducible constituent from Proposition 5.102 consists of a scalar action by $\mathfrak{h}''$ and an irreducible representation of $\mathfrak{l}_{ss}$, and the Theorem of the Highest Weight (Theorem 5.5) is applicable for the latter. Reassembling matters, we see that we can treat $\mathfrak{h}$ as a Cartan subalgebra of $\mathfrak{l}$ and treat $\Gamma \cap -\Gamma$ as the root system $\Delta(\mathfrak{l}, \mathfrak{h})$. The Theorem of the Highest Weight may then be reinterpreted as valid for $\mathfrak{l}$. Even though $\mathfrak{l}$ is merely reductive, we shall work with $\mathfrak{l}$ in this fashion without further special comment.

Let a finite-dimensional representation of $\mathfrak{g}$ be given on a space V, and fix a parabolic subalgebra $\mathfrak{q} = \mathfrak{l} \oplus \mathfrak{u}$ containing $\mathfrak{b}$. The key tool for our investigation will be the subspace of $\mathfrak{u}$ **invariants** given by

$$V^{\mathfrak{u}} = \{v \in V \mid Xv = 0 \text{ for all } X \in \mathfrak{u}\}.$$

This subspace carries a representation of $\mathfrak{l}$ since $H \in \mathfrak{l}$, $v \in V^{\mathfrak{u}}$, and $X \in \mathfrak{u}$ imply

$$X(Hv) = H(Xv) + [X, H]v = 0 + 0 = 0$$

by Corollary 5.94a. By Corollary 5.31c the representation of $\mathfrak{l}$ on $V^{\mathfrak{u}}$ is determined up to equivalence by the representation of $\mathfrak{h}$ on the space of $\mathfrak{l} \cap \mathfrak{n}$ invariants. But

$$(V^{\mathfrak{u}})^{\mathfrak{l}\cap\mathfrak{n}} = V^{\mathfrak{u}\oplus(\mathfrak{l}\cap\mathfrak{n})} = V^{\mathfrak{n}}, \tag{5.103}$$

and the right side is given by the Theorem of the Highest Weight for $\mathfrak{g}$. This fact allows us to treat the representation of $\mathfrak{l}$ on $V^{\mathfrak{u}}$ as a generalization of the highest weight of the representation of $\mathfrak{g}$ on V.

Theorem 5.104. Let $\mathfrak{g}$ be a complex semisimple Lie algebra, let $\mathfrak{h}$ be a Cartan subalgebra, let $\Delta^+(\mathfrak{g}, \mathfrak{h})$ be a positive system for the set of roots, and define $\mathfrak{n}$ by (5.8). Let $\mathfrak{q} = \mathfrak{l} \oplus \mathfrak{u}$ be a parabolic subalgebra containing the Borel subalgebra $\mathfrak{b} = \mathfrak{h} \oplus \mathfrak{n}$.

(a) If an irreducible finite-dimensional representation of $\mathfrak{g}$ is given on V, then the corresponding representation of $\mathfrak{l}$ on $V^{\mathfrak{u}}$ is irreducible. The highest weight of this representation of $\mathfrak{l}$ matches the highest weight of V and is therefore algebraically integral and dominant for $\Delta^+(\mathfrak{g}, \mathfrak{h})$.

(b) If irreducible finite-dimensional representations of $\mathfrak{g}$ are given on V_1 and V_2 such that the associated irreducible representations of $\mathfrak{l}$ on $V_1^{\mathfrak{u}}$ and $V_2^{\mathfrak{u}}$ are equivalent, then V_1 and V_2 are equivalent.

(c) If an irreducible finite-dimensional representation of $\mathfrak{l}$ on M is given whose highest weight is algebraically integral and dominant for $\Delta^+(\mathfrak{g}, \mathfrak{h})$, then there exists an irreducible finite-dimensional representation of $\mathfrak{g}$ on a space V such that $V^{\mathfrak{u}} \cong M$ as representations of $\mathfrak{l}$.

PROOF.

(a) By (5.103), $(V^{\mathfrak{u}})^{\mathfrak{l}\cap\mathfrak{n}} = V^{\mathfrak{n}}$. Parts (b) and (c) of Theorem 5.5 for $\mathfrak{g}$ say that $V^{\mathfrak{n}}$ is 1-dimensional. Hence the space of $\mathfrak{l}\cap\mathfrak{n}$ invariants for $V^{\mathfrak{u}}$ is 1-dimensional. Since $V^{\mathfrak{u}}$ is completely reducible under $\mathfrak{l}$ by Proposition 5.102, Theorem 5.5c for $\mathfrak{l}$ shows that $V^{\mathfrak{u}}$ is irreducible under $\mathfrak{l}$. If λ is the highest weight of V under $\mathfrak{g}$, then λ is the highest weight of $V^{\mathfrak{u}}$ under $\mathfrak{l}$ since $V_\lambda = V^{\mathfrak{n}} \subseteq V^{\mathfrak{u}}$. Then λ is algebraically integral and dominant for $\Delta^+(\mathfrak{g}, \mathfrak{h})$ by Theorem 5.5 for $\mathfrak{g}$.

(b) If $V_1^{\mathfrak{u}}$ and $V_2^{\mathfrak{u}}$ are equivalent under $\mathfrak{l}$, then $(V_1^{\mathfrak{u}})^{\mathfrak{l}\cap\mathfrak{n}}$ and $(V_2^{\mathfrak{u}})^{\mathfrak{l}\cap\mathfrak{n}}$ are equivalent under $\mathfrak{h}$. By (5.103), $V_1^{\mathfrak{n}}$ and $V_2^{\mathfrak{n}}$ are equivalent under $\mathfrak{h}$. By uniqueness in Theorem 5.5, V_1 and V_2 are equivalent under $\mathfrak{g}$.

(c) Let M have highest weight λ, which is assumed algebraically integral and dominant for $\Delta^+(\mathfrak{g}, \mathfrak{h})$. By Theorem 5.5 we can form an irreducible finite-dimensional representation of $\mathfrak{g}$ on a space V with highest weight λ. Then $V^{\mathfrak{u}}$ has highest weight λ by (a), and $V^{\mathfrak{u}} \cong M$ as representations of $\mathfrak{l}$ by uniqueness in Theorem 5.5 for $\mathfrak{l}$.

Proposition 5.105. Let $\mathfrak{g}$ be a complex semisimple Lie algebra, and let $\mathfrak{q} = \mathfrak{l} \oplus \mathfrak{u}$ be a parabolic subalgebra containing $\mathfrak{b}$. If V is any finite-dimensional $U(\mathfrak{g})$ module, then

(a) $V = V^{\mathfrak{u}} \oplus \mathfrak{u}^- V$

(b) the natural map $V^{\mathfrak{u}} \to V/(\mathfrak{u}^- V)$ is an isomorphism of $U(\mathfrak{l})$ modules

(c) the $U(\mathfrak{l})$ module $V^{\mathfrak{u}}$ determines the $U(\mathfrak{g})$ module V up to equivalence; the number of irreducible constituents of $V^{\mathfrak{u}}$ equals the number of irreducible constituents of V, and the multiplicity of an irreducible $U(\mathfrak{l})$ module in $V^{\mathfrak{u}}$ equals the multiplicity in V of the irreducible $U(\mathfrak{g})$ module with that same highest weight.

PROOF. We have seen that $V^{\mathfrak{u}}$ is a $U(\mathfrak{l})$ module, and similarly $\mathfrak{u}^- V$ is a $U(\mathfrak{l})$ module. Conclusion (b) is immediate from (a), and conclusion (c) is immediate from Theorems 5.29 and 5.104. Thus we are left with proving (a).

By Theorem 5.29, V is a direct sum of irreducible representations, and there is no loss of generality in assuming that V is irreducible, say of highest weight λ.

With V irreducible, we argue as in the proof of Corollary 5.31, using a Poincaré-Birkhoff-Witt basis of $U(\mathfrak{g})$ built from root vectors in $\mathfrak{u}^-$, root vectors in $\mathfrak{l}$ together with members of $\mathfrak{h}$, and root vectors in $\mathfrak{u}$. We may do so because of (5.97). Each such root vector is an eigenvector under ad $H_{\delta(\mathfrak{u})}$, and the eigenvalues are negative, zero, and positive in the three cases by Corollary 5.101. Using this eigenvalue as a substitute for "weight" in the proof of Corollary 5.31, we see that

$$V = U(\mathfrak{l})V_\lambda \oplus \mathfrak{u}^- V.$$

But $\mathfrak{l}$ acts irreducibly on $V^{\mathfrak{u}}$ by Theorem 5.104a, and $V_\lambda = V^{\mathfrak{n}} \subseteq V^{\mathfrak{u}}$. Hence $U(\mathfrak{l})V_\lambda = V^{\mathfrak{u}}$, and (a) is proved. This completes the proof of the proposition.

8. Application to Compact Lie Groups

As was mentioned in §1, one of the lines of motivation for studying finite-dimensional representations of complex semisimple Lie algebras is the representation theory of compact connected Lie groups. We now return to that theory in order to interpret the results of this chapter in that context.

Throughout this section we let G be a compact connected Lie group with Lie algebra $\mathfrak{g}_0$ and complexified Lie algebra $\mathfrak{g}$, and we let T be a maximal torus with Lie algebra $\mathfrak{t}_0$ and complexified Lie algebra $\mathfrak{t}$. The Lie algebra $\mathfrak{g}$ is reductive (Corollary 4.25), and we saw in §IV.4 how to interpret $\mathfrak{t}$ as a Cartan subalgebra and how the theory of roots extended from the semisimple case to this reductive case. Let $\Delta = \Delta(\mathfrak{g}, \mathfrak{t})$ be the set of roots, and let $W = W(\Delta)$ be the Weyl group.

Recall that a member λ of $\mathfrak{t}^*$ is **analytically integral** if it is the differential of a multiplicative character ξ_λ of T, i.e., if $\xi_\lambda(\exp H) = e^{\lambda(H)}$

for all $H \in \mathfrak{t}_0$. If λ is analytically integral, then λ takes purely imaginary values on $\mathfrak{t}_0$ by Proposition 4.58. Every root is analytically integral by Proposition 4.58. Every analytically integral member of $\mathfrak{t}^*$ is algebraically integral by Proposition 4.59.

Lemma 5.106. If Φ is a finite-dimensional representation of the compact connected Lie group G and if λ is a weight of the differential of Φ, then λ is analytically integral.

PROOF. We observed in §1 that $\Phi|_T$ is the direct sum of 1-dimensional invariant subspaces with $\Phi|_T$ acting in each by a multiplicative character ξ_{λ_j}. Then the weights are the various λ_j's. Since each weight is the differential of a multiplicative character of T, each weight is analytically integral.

Theorem 5.107. Let G be a simply connected compact semisimple Lie group, let T be a maximal torus, and let $\mathfrak{t}$ be the complexified Lie algebra of T. Then every algebraically integral member of $\mathfrak{t}^*$ is analytically integral.

PROOF. Let $\lambda \in \mathfrak{t}^*$ be algebraically integral. Then λ is real-valued on $i\mathfrak{t}_0$, and the real span of the roots is $(i\mathfrak{t}_0)^*$ by semisimplicity of $\mathfrak{g}$. Hence λ is in the real span of the roots. By Proposition 2.67 we can introduce a positive system $\Delta^+(\mathfrak{g}, \mathfrak{t})$ such that λ is dominant. By the Theorem of the Highest Weight (Theorem 5.5), there exists an irreducible finite-dimensional representation φ of $\mathfrak{g}$ with highest weight λ. Since G is simply connected, there exists an irreducible finite-dimensional representation Φ of G with differential $\varphi|_{\mathfrak{g}_0}$. By Lemma 5.106, λ is analytically integral.

Corollary 5.108. If G is a compact semisimple Lie group, then the order of the fundamental group of G equals the index of the group of analytically integral forms for G in the group of algebraically integral forms.

PROOF. Let $\tilde{G}$ be a simply connected covering group of G. By Weyl's Theorem (Theorem 4.69), $\tilde{G}$ is compact. Theorem 5.107 shows that the analytically integral forms for $\tilde{G}$ coincide with the algebraically integral forms. Then it follows from Proposition 4.67 that the index of the group of analytically integral forms for G in the group of algebraically integral forms equals the order of the kernel of the covering homomorphism $\tilde{G} \to G$. Since $\tilde{G}$ is simply connected, this kernel is isomorphic to the fundamental group of G.

EXAMPLE. Let $G = SO(2n+1)$ with $n \geq 1$ or $G = SO(2n)$ with $n \geq 2$. The analytically integral forms in standard notation are all expressions $\sum_{j=1}^{n} c_j e_j$ with all c_j in $\mathbb{Z}$. The algebraically integral forms are all expressions $\sum_{j=1}^{n} c_j e_j$ with all c_j in $\mathbb{Z}$ or all c_j in $\mathbb{Z}+\frac{1}{2}$. Corollary 5.108 therefore implies that the fundamental group of G has order 2.

Corollary 5.109. If G is a simply connected compact semisimple Lie group, then the order of the center Z_G of G equals the determinant of the Cartan matrix.

PROOF. Let G' be the adjoint group of G so that Z_G is the kernel of the covering map $G \to G'$. The analytically integral forms for G coincide with the algebraically integral forms by Theorem 5.107, and the analytically integral forms for G' coincide with the $\mathbb{Z}$ combinations of roots by Proposition 4.68. Thus the corollary follows by combining Propositions 4.64 and 4.67.

Now we give results that do not assume that G is semisimple. Since $\mathfrak{g}_0$ is reductive, we can write $\mathfrak{g}_0 = Z_{\mathfrak{g}_0} \oplus [\mathfrak{g}_0, \mathfrak{g}_0]$ with $[\mathfrak{g}_0, \mathfrak{g}_0]$ semisimple. Put $\mathfrak{t}_0' = \mathfrak{t}_0 \cap [\mathfrak{g}_0, \mathfrak{g}_0]$. The root-space decomposition of $\mathfrak{g}$ is then

$$\mathfrak{g} = \mathfrak{t} \oplus \bigoplus_{\alpha \in \Delta(\mathfrak{g},\mathfrak{t})} \mathfrak{g}_\alpha = Z_{\mathfrak{g}} \oplus \Big(\mathfrak{t}' \oplus \bigoplus_{\alpha \in \Delta(\mathfrak{g},\mathfrak{t})} \mathfrak{g}_\alpha\Big).$$

By Proposition 4.24 the compactness of G implies that there is an invariant inner product on the Lie algebra $\mathfrak{g}_0$, and we let B be its negative. (This form was used in Chapter IV, beginning in §5.) If we were assuming that $\mathfrak{g}_0$ is semisimple, then B could be taken to be the Killing form, according to Corollary 4.26. We extend B to be complex bilinear on $\mathfrak{g} \times \mathfrak{g}$. The restriction of B to $i\mathfrak{t}_0 \times i\mathfrak{t}_0$ is an inner product, which transfers to give an inner product on $(i\mathfrak{t}_0)^*$. Analytically integral forms are always in $(i\mathfrak{t}_0)^*$. If a positive system $\Delta^+(\mathfrak{g}, \mathfrak{t})$ is given for the roots, then the condition of dominance for the form depends only on the restriction of the form to $i\mathfrak{t}_0'$.

Theorem 5.110 (Theorem of the Highest Weight). Let G be a compact connected Lie group with complexified Lie algebra $\mathfrak{g}$, let T be a maximal torus with complexified Lie algebra $\mathfrak{t}$, and let $\Delta^+(\mathfrak{g}, \mathfrak{t})$ be a positive system for the roots. Apart from equivalence the irreducible finite-dimensional representations Φ of G stand in one-one correspondence with the dominant analytically integral linear functionals λ on $\mathfrak{t}$, the correspondence being that λ is the highest weight of Φ.

REMARK. The highest weight has the additional properties given in Theorem 5.5.

PROOF. Let notation be as above. If Φ is given, then the highest weight λ of Φ is analytically integral by Lemma 5.106. To see dominance, let φ be the differential of Φ. Extend φ complex linearly from $\mathfrak{g}_0$ to $\mathfrak{g}$, and restrict to $[\mathfrak{g}, \mathfrak{g}]$. The highest weight of φ on $[\mathfrak{g}, \mathfrak{g}]$ is the restriction of λ to $\mathfrak{t}'$, and this must be dominant by Theorem 5.5. Therefore λ is dominant.

By Theorem 4.29, G is a commuting product $G = (Z_G)_0 G_{ss}$ with G_{ss} compact semisimple. Suppose that Φ and Φ' are irreducible representations of G, both with highest weight λ. By Schur's Lemma (Corollary 4.9), $\Phi|_{(Z_G)_0}$ and $\Phi'|_{(Z_G)_0}$ are scalar, and the scalar is determined by the restriction of λ to the Lie algebra $Z_{\mathfrak{g}_0}$ of $(Z_G)_0$. Hence $\Phi|_{(Z_G)_0} = \Phi'|_{(Z_G)_0}$. On G_{ss}, the differentials φ and φ' give irreducible representations of $[\mathfrak{g}, \mathfrak{g}]$ with the same highest weight $\lambda|_{\mathfrak{t}'}$, and these are equivalent by Theorem 5.5. Then it follows that φ and φ' are equivalent as representations of $\mathfrak{g}$, and Φ and Φ' are equivalent as representations of G.

Finally if an analytically integral dominant λ is given, we shall produce a representation Φ of G with highest weight λ. The form λ is algebraically integral by Proposition 4.59. We construct an irreducible representation φ of $\mathfrak{g}$ with highest weight λ: This comes in two parts, with $\varphi|_{[\mathfrak{g},\mathfrak{g}]}$ equal to the representation in Theorem 5.5 corresponding to $\lambda|_{\mathfrak{t}'}$ and with $\varphi|_{Z_{\mathfrak{g}}}$ given by scalar operators equal to $\lambda|_{Z_{\mathfrak{g}}}$.

Let $\tilde{G}$ be the universal covering group of G. Since $\tilde{G}$ is simply connected, there exists an irreducible representation $\tilde{\Phi}$ of $\tilde{G}$ with differential $\varphi|_{\mathfrak{g}_0}$, hence with highest weight λ. To complete the proof, we need to show that $\tilde{\Phi}$ descends to a representation Φ of G.

Since $G = (Z_G)_0 G_{ss}$, $\tilde{G}$ is of the form $\mathbb{R}^n \times \tilde{G}_{ss}$, where $\tilde{G}_{ss}$ is the universal covering group of G_{ss}. Let Z be the discrete subgroup of the center $Z_{\tilde{G}}$ of $\tilde{G}$ such that $G \cong \tilde{G}/Z$. By Weyl's Theorem (Theorem 4.69), $\tilde{G}_{ss}$ is compact. Thus Corollary 4.47 shows that the center of $\tilde{G}_{ss}$ is contained in every maximal torus of $\tilde{G}_{ss}$. Since $Z_{\tilde{G}} \subseteq \mathbb{R}^n \times Z_{\tilde{G}_{ss}}$, it follows that $Z_{\tilde{G}} \subseteq \exp \mathfrak{t}_0$. Now λ is analytically integral for G, and consequently the corresponding multiplicative character ξ_λ on $\exp \mathfrak{t}_0 \subseteq \tilde{G}$ is trivial on Z. By Schur's Lemma, $\tilde{\Phi}$ is scalar on $Z_{\tilde{G}}$, and its scalar values must agree with those of ξ_λ since λ is a weight. Thus $\tilde{\Phi}$ is trivial on Z, and $\tilde{\Phi}$ descends to a representation Φ of G, as required.

Next we take up characters. Let Φ be an irreducible finite-dimensional representation of the compact connected Lie group G with highest weight λ, let V be the underlying vector space, and let φ be the differential, regarded as a representation of $\mathfrak{g}$. The Weyl Character Formula, as stated in Theorem 5.75, gives a kind of generating function for the weights of an irreducible Lie algebra representation in the semisimple case. Hence it is applicable to the semisimple Lie algebra $[\mathfrak{g}, \mathfrak{g}]$, the

Cartan subalgebra $\mathfrak{t}'$, the representation $\varphi|_{[\mathfrak{g},\mathfrak{g}]}$, and the highest weight $\lambda|_{\mathfrak{t}'}$. By Schur's Lemma, $\Phi|_{(Z_G)_0}$ is scalar, necessarily with differential $\varphi|_{Z_\mathfrak{g}} = \lambda|_{Z_\mathfrak{g}}$. Thus we can extend the Weyl Character Formula as stated in Theorem 5.75 to be meaningful for our reductive $\mathfrak{g}$ by extending all weights from $\mathfrak{t}'$ to $\mathfrak{t}$ with $\lambda|_{Z_\mathfrak{g}}$ as their values on $Z_\mathfrak{g}$. The formula looks the same:

$$\Big(e^\delta \prod_{\alpha\in\Delta^+} (1 - e^{-\alpha})\Big)\mathrm{char}(V) = \sum_{w\in W} \varepsilon(w) e^{w(\lambda+\delta)}. \tag{5.111}$$

We can apply the evaluation homomorphism ϵ_H to both sides for any $H \in \mathfrak{t}$, but we want to end up with an expression for char(V) as a function on the maximal torus T. This is a question of analytic integrality. The expressions char(V) and $\prod(1 - e^{-\alpha})$ give well defined functions on T since each weight and root is analytically integral. But e^δ need not give a well defined function on T since δ need not be analytically integral. (It is not analytically integral for $SO(3)$, for example.) Matters are resolved by the following lemma.

Lemma 5.112. For each $w \in W$, $\delta - w\delta$ is analytically integral. In fact, $\delta - w\delta$ is the sum of all posiitve roots β such that $w^{-1}\beta$ is negative.

PROOF. We write

$$\delta = \tfrac{1}{2}\sum\{\beta \mid \beta > 0,\ w^{-1}\beta > 0\} + \tfrac{1}{2}\sum\{\beta \mid \beta > 0,\ w^{-1}\beta < 0\}$$

and

$$\begin{aligned} w\delta &= \tfrac{1}{2}w\sum\{\alpha \mid \alpha > 0,\ w\alpha > 0\} + \tfrac{1}{2}w\sum\{\alpha \mid \alpha > 0,\ w\alpha < 0\} \\ &= \tfrac{1}{2}\sum\{w\alpha \mid \alpha > 0,\ w\alpha > 0\} + \tfrac{1}{2}\sum\{w\alpha \mid \alpha > 0,\ w\alpha < 0\} \\ &= \tfrac{1}{2}\sum\{\beta \mid w^{-1}\beta > 0,\ \beta > 0\} + \tfrac{1}{2}\sum\{\eta \mid w^{-1}\eta > 0,\ \eta < 0\} \\ &\qquad\qquad\qquad\qquad \text{under } \beta = w\alpha \text{ and } \eta = w\alpha \\ &= \tfrac{1}{2}\sum\{\beta \mid w^{-1}\beta > 0,\ \beta > 0\} - \tfrac{1}{2}\sum\{\beta \mid w^{-1}\beta < 0,\ \beta > 0\} \\ &\qquad\qquad\qquad\qquad \text{under } \beta = -\eta. \end{aligned}$$

Subtracting, we obtain

$$\delta - w\delta = \sum\{\beta \mid \beta > 0,\ w^{-1}\beta < 0\}$$

as required.

Theorem 5.113 (Weyl Character Formula). Let G be a compact connected Lie group, let T be a maximal torus, let $\Delta^+ = \Delta^+(\mathfrak{g}, \mathfrak{t})$ be a positive system for the roots, and let $\lambda \in \mathfrak{t}^*$ be analytically integral and dominant. Then the character χ_{Φ_λ} of the irreducible finite-dimensional representation Φ_λ of G with highest weight λ is given by

$$\chi_{\Phi_\lambda} = \frac{\sum_{w \in W} \varepsilon(w) \xi_{w(\lambda+\delta)-\delta}(t)}{\prod_{\alpha \in \Delta^+} (1 - \xi_{-\alpha}(t))}$$

at every $t \in T$ where no ξ_α takes the value 1 on t. If G is simply connected, then this formula can be rewritten as

$$\chi_{\Phi_\lambda} = \frac{\sum_{w \in W} \varepsilon(w) \xi_{w(\lambda+\delta)}(t)}{\xi_\delta(t) \prod_{\alpha \in \Delta^+} (1 - \xi_{-\alpha}(t))} = \frac{\sum_{w \in W} \varepsilon(w) \xi_{w(\lambda+\delta)}(t)}{\sum_{w \in W} \varepsilon(w) \xi_{w\delta}(t)}.$$

REMARK. Theorem 4.36 says that every member of G is conjugate to a member of T. Since characters are constant on conjugacy classes, the above formulas determine the characters everywhere on G.

PROOF. Theorem 5.110 shows that Φ_λ exists when λ is analytically integral and dominant. We apply Theorem 5.75 in the form of (5.111). When we divide (5.111) by e^δ, Lemma 5.112 says that all the exponentials yield well defined functions on T. The first formula follows. If G is simply connected, then G is semisimple as a consequence of Proposition 1.99. The linear functional δ is algebraically integral by Proposition 2.69, hence analytically integral by Theorem 5.107. Thus we can regroup the formula as indicated. The version of the formula with an alternating sum in the denominator uses Theorem 5.77 in place of Theorem 5.75.

Finally we discuss how parabolic subalgebras play a role in the representation theory of compact Lie groups. With G and T given, fix a positive system $\Delta^+(\mathfrak{g}, \mathfrak{t})$ for the roots, define $\mathfrak{n}$ as in (5.8), and let $\mathfrak{q} = \mathfrak{l} \oplus \mathfrak{u}$ be a parabolic subalgebra of $\mathfrak{g}$ containing $\mathfrak{b} = \mathfrak{h} \oplus \mathfrak{n}$. Corollary 5.101 shows that $\mathfrak{l} = Z_\mathfrak{g}(H_{\delta(\mathfrak{u})})$, and we can equally well write $\mathfrak{l} = Z_\mathfrak{g}(iH_{\delta(\mathfrak{u})})$. Since $iH_{\delta(\mathfrak{u})}$ is in $\mathfrak{t}_0 \subseteq \mathfrak{g}_0$, $\mathfrak{l}$ is the complexification of the subalgebra

$$\mathfrak{l}_0 = Z_{\mathfrak{g}_0}(iH_{\delta(\mathfrak{u})})$$

of $\mathfrak{g}_0$. Define

$$L = Z_G(iH_{\delta(\mathfrak{u})}).$$

This is a compact subgroup of G containing T. Since the closure of $\exp i\mathbb{R}H_{\delta(\mathfrak{u})}$ is a torus in G, L is the centralizer of a torus in G and is connected by Corollary 4.51. Thus we have an inclusion of compact connected Lie groups $T \subseteq L \subseteq G$, and T is a maximal torus in both L and G. Hence analytic integrality is the same for L as for G. Combining Theorems 5.104 and 5.110, we obtain the following result.

Theorem 5.114. Let G be a compact connected Lie group with maximal torus T, let $\mathfrak{g}_0$ and $\mathfrak{t}_0$ be the Lie algebras, and let $\mathfrak{g}$ and $\mathfrak{t}$ be the complexifications. Let $\Delta^+(\mathfrak{g}, \mathfrak{t})$ be a positive system for the roots, and define $\mathfrak{n}$ by (5.8). Let $\mathfrak{q} = \mathfrak{l} \oplus \mathfrak{u}$ be a parabolic subalgebra containing $\mathfrak{b} = \mathfrak{h} \oplus \mathfrak{n}$, let $\mathfrak{l}_0 = \mathfrak{l} \cap \mathfrak{g}_0$, and let L be the analytic subgroup of G with Lie algebra $\mathfrak{l}_0$.

(a) The subgroup L is compact connected, and T is a maximal torus in it.

(b) If an irreducible finite-dimensional representation of G is given on V, then the corresponding representation of L on $V^{\mathfrak{u}}$ is irreducible. The highest weight of this representation of L matches the highest weight of V and is therefore analytically integral and dominant for $\Delta^+(\mathfrak{g}, \mathfrak{h})$.

(c) If irreducible finite-dimensional representations of G are given on V_1 and V_2 such that the associated irreducible representations of L on $V_1^{\mathfrak{u}}$ and $V_2^{\mathfrak{u}}$ are equivalent, then V_1 and V_2 are equivalent.

(d) If an irreducible finite-dimensional representation of L on M is given whose highest weight is analytically integral and dominant for $\Delta^+(\mathfrak{g}, \mathfrak{h})$, then there exists an irreducible finite-dimensional representation of G on a space V such that $V^{\mathfrak{u}} \cong M$ as representations of L.

9. Problems

1. Let $\mathfrak{g}$ be a complex semisimple Lie algebra, and let φ be a finite-dimensional representation of $\mathfrak{g}$ on the space V. The contragredient φ^c is defined in (4.4).
   (a) Show that the weights of φ^c are the negatives of the weights of φ.
   (b) Let w_0 be the element of the Weyl group produced in Problem 18 of Chapter II such that $w_0\Delta^+ = -\Delta^+$. If φ is irreducible with highest weight λ, prove that φ^c is irreducible with highest weight $-w_0\lambda$.

2. As in Problems 8–13 of Chapter IV, let V_N be the space of polynomials in $x_1, \ldots, x_n$ homogeneous of degree N, and let H_N be the subspace of harmonic polynomials. The compact group $G = SO(n)$ acts on V_N, and hence so does the complexified Lie algebra $\mathfrak{so}(n, \mathbb{C})$. The subspace H_N is an invariant subspace. In the parts of this problem, it is appropriate to handle separately the cases of n odd and n even.
   (a) The weights of V_N are identified in §1. Check that Ne_1 is the highest weight, and conclude that Ne_1 is the highest weight of H_N.
   (b) Calculate the dimension of the irreducible representation of $\mathfrak{so}(n, \mathbb{C})$ with highest weight Ne_1, compare with the result of Problem 13 of Chapter IV, and conclude that $\mathfrak{so}(n, \mathbb{C})$ acts irreducibly on H_N.

3. As in Problems 14–16 of Chapter IV, let V_N be the space of polynomials in $z_1, \dots, z_n, \bar{z}_1, \dots, \bar{z}_n$ homogeneous of degree N, and let $V_{p,q}$ be the subspace of polynomials with p z-type factors and q $\bar{z}$-type factors. The compact group $G = SU(n)$ acts on V_N, and hence so does the complexified Lie algebra $\mathfrak{sl}(n, \mathbb{C})$. The subspace $H_{p,q}$ of harmonic polynomials in $V_{p,q}$ is an invariant subspace.
   (a) The weights of $V_{p,q}$ are identified in §1. Check that $qe_1 - pe_n$ is the highest weight, and conclude that $qe_1 - pe_n$ is the highest weight of $H_{p,q}$.
   (b) Calculate the dimension of the irreducible representation of $\mathfrak{sl}(n, \mathbb{C})$ with highest weight $qe_1 - pe_n$, compare with the result of Problem 16 of Chapter IV, and conclude that $\mathfrak{sl}(n, \mathbb{C})$ acts irreducibly on $H_{p,q}$.
4. For $\mathfrak{g} = \mathfrak{sl}(3, \mathbb{C})$, show that the space $\mathcal{H}^W$ of Weyl-group invariants contains a nonzero element homogeneous of degree 3.
5. Give an interpretation of the Weyl Denominator Formula for $\mathfrak{sl}(n, \mathbb{C})$ in terms of the evaluation of Vandermonde determinants.
6. Prove that the Kostant partition function $\mathcal{P}$ satisfies the recursion formula
$$\mathcal{P}(\lambda) = -\sum_{\substack{w \in W, \\ w \neq 1}} \varepsilon(w) \mathcal{P}(\lambda - (\delta - w\delta))$$
for $\lambda \neq 0$ in Q^+. Here $\mathcal{P}(\nu)$ is understood to be 0 if ν is not in Q^+.

Problems 7–10 address irreducibility of certain representations in spaces of alternating tensors.

7. Show that the representation of $\mathfrak{sl}(n, \mathbb{C})$ on $\bigwedge^l \mathbb{C}^n$ is irreducible by showing that the dimension of the irreducible representation with highest weight $\sum_{k=1}^l e_k$ is $\binom{n}{l}$.
8. Show that the representation of $\mathfrak{so}(2n+1, \mathbb{C})$ on $\bigwedge^l \mathbb{C}^{2n+1}$ is irreducible for $l \leq n$ by showing that the dimension of the irreducible representation with highest weight $\sum_{k=1}^l e_k$ is $\binom{2n+1}{l}$.
9. Show that the representation of $\mathfrak{so}(2n, \mathbb{C})$ on $\bigwedge^l \mathbb{C}^{2n}$ is irreducible for $l < n$ by showing that the dimension of the irreducible representation with highest weight $\sum_{k=1}^l e_k$ is $\binom{2n}{l}$.
10. Show that the representation of $\mathfrak{so}(2n, \mathbb{C})$ on $\bigwedge^n \mathbb{C}^{2n}$ is reducible, being the sum of two irreducible representations with respective highest weights $\left(\sum_{k=1}^{n-1} e_k\right) \pm e_n$.

Problems 11–13 concern Verma modules.

11. Prove for arbitrary λ and μ in $\mathfrak{h}^*$ that every nonzero $U(\mathfrak{g})$ linear map of $V(\mu)$ into $V(\lambda)$ is one-one.

12. Prove for arbitrary λ and μ in $\mathfrak{h}^*$ that if $V(\mu)$ is isomorphic to a $U(\mathfrak{g})$ submodule of $V(\lambda)$, then μ is in $\lambda - Q^+$ and is in the orbit of λ under the Weyl group.

13. Let λ be in $\mathfrak{h}^*$, and let M be an irreducible quotient of a $U(\mathfrak{g})$ submodule of $V(\lambda)$. Prove that M is isomorphic to the $U(\mathfrak{g})$ module $L(\mu)$ of Proposition 5.15 for some μ in $\lambda - Q^+$ such that μ is in the orbit of λ under the Weyl group.

Problems 14–20 deal with decomposing tensor products into irreducible representations. Let $\mathfrak{g}$ be a complex semisimple Lie algebra, and let notation be as in §2.

14. Let φ_λ and $\varphi_{\lambda'}$ be irreducible representations of $\mathfrak{g}$ with highest weights λ and λ', respectively. Prove that the weights of $\varphi_\lambda \otimes \varphi_{\lambda'}$ are all sums $\mu + \mu'$, where μ is a weight of φ_λ and μ' is a weight of $\varphi_{\lambda'}$. How is the multiplicity of $\mu + \mu'$ related to multiplicities in φ_λ and $\varphi_{\lambda'}$?

15. Let v_λ and $v_{\lambda'}$ be highest weight vectors in φ_λ and $\varphi_{\lambda'}$, respectively. Prove that $v_\lambda \otimes v_{\lambda'}$ is a highest weight vector in $\varphi_\lambda \otimes \varphi_{\lambda'}$. Conclude that $\varphi_{\lambda+\lambda'}$ occurs exactly once in $\varphi_\lambda \otimes \varphi_{\lambda'}$. (This occurrence is sometimes called the **Cartan composition** of φ_λ and $\varphi_{\lambda'}$.)

16. Let λ'' be any highest weight in $\varphi_\lambda \otimes \varphi_{\lambda'}$, i.e., the highest weight of some irreducible constituent. Prove that λ'' is of the form $\lambda'' = \lambda + \mu'$ for some weight μ' of $\varphi_{\lambda'}$.

17. Prove that if all weights of φ_λ have multiplicity one, then each irreducible constituent of $\varphi_\lambda \otimes \varphi_{\lambda'}$ has multiplicity one.

18. If λ is algebraically integral and if there exists $w_0 \neq 1$ in W fixing λ, prove that $\sum_{w \in W} (\det w) \xi_{w\lambda} = 0$.

19. Let $m_\lambda(\mu)$ be the multiplicity of the weight μ in φ_λ, and define $\operatorname{sgn}\mu$ by

$$\operatorname{sgn}\mu = \begin{cases} 0 & \text{if some } w \neq 1 \text{ in } W \text{ fixes } \mu \\ \det w & \text{otherwise, where } w \text{ is chosen in } W \text{ to make } w\mu \text{ dominant.} \end{cases}$$

Write the character of φ_λ as $\chi_\lambda = \sum m_\lambda(\lambda'')\xi_{\lambda''}$, write λ' as in the Weyl Character Formula, and multiply. With $\mu^\vee$ denoting the result of applying an element of W to μ to obtain something dominant, obtain the formula

$$\chi_\lambda \chi_{\lambda'} = \sum_{\lambda'' = \text{weight of } \varphi_\lambda} m_\lambda(\lambda'') \operatorname{sgn}(\lambda'' + \lambda' + \delta) \chi_{(\lambda''+\lambda'+\delta)^\vee - \delta}.$$

20. Let $-\mu$ be the lowest weight of φ_λ. Deduce from Problem 19 that if $\lambda' - \mu$ is dominant, then $\varphi_{\lambda'-\mu}$ occurs in $\varphi_\lambda \otimes \varphi_{\lambda'}$ with multiplicity one.

Problems 21–23 use Problem 19 to identify a particular constituent of a tensor product of irreducible representations, beyond the one in Problem 20. Let λ and λ' be dominant and algebraically integral. Let w be in W, and suppose that $\lambda' + w\lambda$ is dominant. The goal is to prove that $\varphi_{\lambda'+w\lambda}$ occurs in $\varphi_\lambda \otimes \varphi_{\lambda'}$ with multiplicity one.

21. Prove that $\lambda'' = w\lambda$ contributes $\chi_{\lambda'+w\lambda}$ to the right side of the formula in Problem 19.
22. To see that there is no other contribution of $\chi_{\lambda'+w\lambda}$, suppose that λ'' contributes. Then $(\lambda' + \delta + \lambda'')^\vee - \delta = \lambda' + w\lambda$. Solve for λ'', compute its length squared, and use the assumed dominance to obtain $|\lambda''|^2 \geq |w\lambda|^2$. Show how to conclude that $\lambda'' = w\lambda$.
23. Complete the proof that $\varphi_{\lambda'+w\lambda}$ occurs in $\varphi_\lambda \otimes \varphi_{\lambda'}$ with multiplicity one.

Problems 24–26 begin a construction of "spin representations." Let $u_1, \ldots, u_n$ be the standard orthonormal basis of $\mathbb{R}^n$. The **Clifford algebra** $\text{Cliff}(\mathbb{R}^n)$ is an associative algebra over $\mathbb{R}$ of dimension 2^n with a basis parametrized by subsets of $\{1, \ldots, n\}$ and given by

$$\{u_{i_1}u_{i_2}\cdots u_{i_k} \mid i_1 < i_2 < \cdots < i_k\}.$$

The generators multiply by the rules

$$u_i^2 = -1, \qquad u_iu_j = -u_ju_i \ \text{ if } i \neq j.$$

24. Verify that the Clifford algebra is associative.
25. The Clifford algebra, like any associative algebra, becomes a Lie algebra under the bracket operation $[x, y] = xy - yx$. Put

$$\mathfrak{q} = \sum_{i \neq j} \mathbb{R}u_iu_j.$$

Verify that $\mathfrak{q}$ is a Lie subalgebra of $\text{Cliff}(\mathbb{R}^n)$ isomorphic to $\mathfrak{so}(n)$, the isomorphism being $\varphi : \mathfrak{so}(n) \to \mathfrak{q}$ with

$$\varphi(E_{ji} - E_{ij}) = \tfrac{1}{2}u_iu_j.$$

26. With φ as in Problem 25, verify that

$$[\varphi(x), u_j] = xu_j \qquad \text{for all } x \in \mathfrak{so}(n).$$

Here the left side is a bracket in $\text{Cliff}(\mathbb{R}^n)$, and the right side is the product of the matrix x by the column vector u_j, the product being reinterpreted as a member of $\text{Cliff}(\mathbb{R}^n)$.

Problems 27–35 continue the construction of spin representations. We form the complexification $\mathrm{Cliff}^{\mathbb{C}}(\mathbb{R}^n)$ and denote left multiplication by c, putting $c(x)y = xy$. Then c is a representation of the associative algebra $\mathrm{Cliff}^{\mathbb{C}}((\mathbb{R}^n)$ on itself, hence also of the Lie algebra $\mathrm{Cliff}^{\mathbb{C}}(\mathbb{R}^n)$ on itself, hence also of the Lie subalgebra $\mathfrak{q}^{\mathbb{C}} \cong \mathfrak{so}(n, \mathbb{C})$ on $\mathrm{Cliff}^{\mathbb{C}}(\mathbb{R}^n)$. Let $n = 2m+1$ or $n = 2m$, according as n is odd or even. For $1 \leq j \leq m$, let

$$z_j = u_{2j-1} + iu_{2j} \qquad \text{and} \qquad \bar{z}_j = u_{2j-1} - iu_{2j}.$$

For each subset S of $\{1, \ldots, m\}$, define

$$z_S = \Big(\prod_{j \in S} z_j\Big)\Big(\prod_{j=1}^{m} \bar{z}_j\Big),$$

with each product arranged so that the indices are in increasing order. If n is odd, define also

$$z'_S = \Big(\prod_{j \in S} z_j\Big)\Big(\prod_{j=1}^{m} \bar{z}_j\Big)u_{2m+1}.$$

27. Check that

$$z_j^2 = \bar{z}_j^2 = 0 \qquad \text{and} \qquad \bar{z}_j z_j \bar{z}_j = -4z_j,$$

and deduce that

$$c(z_j)z_S = \begin{cases} \pm z_{S \cup \{j\}} & \text{if } j \notin S \\ 0 & \text{if } j \in S \end{cases}$$

$$c(\bar{z}_j)z_S = \begin{cases} 0 & \text{if } j \notin S \\ \pm 4 z_{S - \{j\}} & \text{if } j \in S. \end{cases}$$

28. When n is odd, check that $c(z_j)z'_S$ and $c(\bar{z}_j)z'_S$ are given by formulas similar to those in Problem 27, and compute also $c(u_{2m+1})z_S$ and $c(u_{2m+1})z'_S$, up to sign.

29. For n even let

$$\mathcal{S} = \sum_{S \subseteq \{1,\ldots,m\}} \mathbb{C} z_S,$$

of dimension 2^m. For n odd let

$$\mathcal{S} = \sum_{S \subseteq \{1,\ldots,m\}} \mathbb{C} z_S + \sum_{T \subseteq \{1,\ldots,m\}} \mathbb{C} z'_T,$$

of dimension 2^{m+1}. Prove that $c(\mathrm{Cliff}^{\mathbb{C}}(\mathbb{R}^n))$ carries $\mathcal{S}$ to itself, hence that $c(\mathfrak{q}^{\mathbb{C}})$ carries $\mathcal{S}$ to itself.

30. For n even, write $\mathcal{S} = \mathcal{S}^+ \oplus \mathcal{S}^-$, where $\mathcal{S}^+$ refers to sets S with an even number of elements and where $\mathcal{S}^-$ corresponds to sets S with an odd number of elements. Prove that $\mathcal{S}^+$ and $\mathcal{S}^-$ are invariant subspaces under $c(\mathfrak{q}^{\mathbb{C}})$, of dimension 2^{m-1}. (The representations $\mathcal{S}^+$ and $\mathcal{S}^-$ are the **spin representations** of $\mathfrak{so}(2m, \mathbb{C})$.)

31. For n odd, write $\mathcal{S} = \mathcal{S}^+ \oplus \mathcal{S}^-$, where $\mathcal{S}^+$ corresponds to sets S with an even number of elements and sets T with an odd number of elements and where $\mathcal{S}^-$ corresponds to sets S with an odd number of elements and sets T with an even number of elements. Prove that $\mathcal{S}^+$ and $\mathcal{S}^-$ are invariant subspaces under $c(\mathfrak{q}^{\mathbb{C}})$, of dimension 2^m, and that they are equivalent under right multiplication by u_{2m+1}. (The **spin representation** of $\mathfrak{so}(2m+1, \mathbb{C})$ is either of the equivalent representations $\mathcal{S}^+$ and $\mathcal{S}^-$.)

32. Let $\mathfrak{t}_0$ be the maximal abelian subspace of $\mathfrak{so}(n)$ in §IV.5. In terms of the isomorphism φ in Problem 25, check that the corresponding maximal abelian subspace of $\mathfrak{q}$ is $\varphi(\mathfrak{t}_0) = \sum \mathbb{R}u_{2j}u_{2j-1}$. In the notation of §II.1, check also that $\frac{1}{2}iu_{2j}u_{2j-1}$ is φ of the element of $\mathfrak{t}$ on which e_j is 1 and e_i is 0 for $i \neq j$.

33. In the notation of the previous problem, prove that

$$c(\varphi(h))z_S = \frac{1}{2}\Big(\sum_{j \notin S} e_j - \sum_{j \in S} e_j\Big)(h) z_S$$

for $h \in \mathfrak{t}$. Prove also that a similar formula holds for the action on z'_S when n is odd.

34. Suppose that n is even.
   (a) Conclude from Problem 33 that the weights of $\mathcal{S}^+$ are all expressions $\frac{1}{2}(\pm e_1 \pm \cdots \pm e_m)$ with an even number of minus signs, while the weights of $\mathcal{S}^-$ are all expressions $\frac{1}{2}(\pm e_1 \pm \cdots \pm e_m)$ with an odd number of minus signs.
   (b) Compute the dimensions of the irreducible representations with highest weights $\frac{1}{2}(e_1 + \cdots + e_{m-1} + e_m)$ and $\frac{1}{2}(e_1 + \cdots + e_{m-1} - e_m)$, and conclude that $\mathfrak{so}(2m, \mathbb{C})$ acts irreducibly on $\mathcal{S}^+$ and $\mathcal{S}^-$.

35. Suppose that n is odd.
   (a) Conclude from Problem 33 that the weights of $\mathcal{S}^+$ are all expressions $\frac{1}{2}(\pm e_1 \pm \cdots \pm e_m)$ and that the weights of $\mathcal{S}^-$ are the same.
   (b) Compute the dimension of the irreducible representation with highest weight $\frac{1}{2}(e_1 + \cdots + e_m)$, and conclude that $\mathfrak{so}(2m+1, \mathbb{C})$ acts irreducibly on $\mathcal{S}^+$ and $\mathcal{S}^-$.

Problems 36–41 concern fundamental representations. Let $\alpha_1, \dots, \alpha_l$ be the simple roots, and define $\varpi_1, \dots, \varpi_l$ by $2\langle \varpi_i, \alpha_j \rangle / |\alpha_j|^2 = \delta_{ij}$. The dominant algebraically integral linear functionals are then all expressions $\sum_i n_i \varpi_i$ with all n_i integers ≥ 0. We call ϖ_i the **fundamental weight** attached to the simple root α_i, and the corresponding irreducible representation is called the **fundamental representation** attached to that simple root.

36. Let $\mathfrak{g} = \mathfrak{sl}(n, \mathbb{C})$.
    (a) Verify that the fundamental weights are $\sum_{k=1}^{l} e_k$ for $1 \leq l \leq n-1$.
    (b) Using Problem 7, verify that the fundamental representations are the usual alternating-tensor representations.

37. Let $\mathfrak{g} = \mathfrak{so}(2n+1, \mathbb{C})$. Let $\alpha_i = e_i - e_{i+1}$ for $i < n$, and let $\alpha_i = e_n$.
    (a) Verify that the fundamental weights are $\varpi_l = \sum_{k=1}^{l} e_k$ for $1 \leq l \leq n-1$ and $\varpi_n = \frac{1}{2} \sum_{k=1}^{n} e_k$.
    (b) Using Problem 8, verify that the fundamental representations attached to simple roots other than the last one are alternating-tensor representations.
    (c) Using Problem 35, verify that the fundamental representation attached to the last simple root is the spin representation.

38. Let $\mathfrak{g} = \mathfrak{so}(2n, \mathbb{C})$. Let $\alpha_i = e_i - e_{i+1}$ for $i < n-1$, and let $\alpha_{n-1} = e_{n-1} - e_n$ and $\alpha_n = e_{n-1} + e_n$.
    (a) Verify that the fundamental weights are $\varpi_l = \sum_{k=1}^{l} e_k$ for $1 \leq l \leq n-2$, $\varpi_{n-1} = \frac{1}{2} \sum_{k=1}^{n} e_k$, and $\varpi_n = \frac{1}{2} \big(\sum_{k=1}^{n-1} e_k - e_n \big)$.
    (b) Using Problem 9, verify that the fundamental representations attached to simple roots other than the last two are alternating-tensor representations.
    (c) Using Problem 34, verify that the fundamental representations attached to the last two simple roots are the spin representations.

39. Let λ and λ' be dominant algebraically integral, and suppose that $\lambda - \lambda_i'$ is dominant and nonzero. Prove that the dimension of an irreducible representation with highest weight λ is greater than the dimension of an irreducible representation with highest weight λ'.

40. Given $\mathfrak{g}$, prove for each integer N that there are only finitely many irreducible representations of $\mathfrak{g}$, up to equivalence, of dimension $< N$.

41. Let $\mathfrak{g}$ be a complex simple Lie algebra of type G_2.
    (a) Using Problem 42 in Chapter II, construct a 7-dimensional nonzero representation of $\mathfrak{g}$.
    (b) Let α_1 be the long simple root, and let α_2 be the short simple root. Verify that $\varpi_1 = 2\alpha_1 + 3\alpha_2$ and that $\varpi_2 = \alpha_1 + 2\alpha_2$.

(c) Verify that the dimensions of the fundamental representations of $\mathfrak{g}$ are 7 and 14. Which one has dimension 7?

(d) Using Problem 39, conclude that the representation constructed in (a) is irreducible.

CHAPTER VI

Structure Theory of Semisimple Groups

Abstract. Every complex semisimple Lie algebra has a compact real form, as a consequence of a particular normalization of root vectors whose construction uses the Isomorphism Theorem of Chapter II. If $\mathfrak{g}_0$ is a real semisimple Lie algebra, then the use of a compact real form of $(\mathfrak{g}_0)^{\mathbb{C}}$ leads to the construction of a "Cartan involution" θ of $\mathfrak{g}_0$. This involution has the property that if $\mathfrak{g}_0 = \mathfrak{k}_0 \oplus \mathfrak{p}_0$ is the corresponding eigenspace decomposition or "Cartan decomposition," then $\mathfrak{k}_0 \oplus i\mathfrak{p}_0$ is a compact real form of $(\mathfrak{g}_0)^{\mathbb{C}}$. Any two Cartan involutions of $\mathfrak{g}_0$ are conjugate by an inner automorphism. The Cartan decomposition generalizes the decomposition of a classical matrix Lie algebra into its skew-Hermitian and Hermitian parts.

If G is a semisimple Lie group, then a Cartan decomposition $\mathfrak{g}_0 = \mathfrak{k}_0 \oplus \mathfrak{p}_0$ of its Lie algebra leads to a global decomposition $G = K \exp \mathfrak{p}_0$, where K is the analytic subgroup of G with Lie algebra $\mathfrak{g}_0$. This global decomposition generalizes the polar decomposition of matrices. The group K contains the center of G and, if the center of G is finite, is a maximal compact subgroup of G.

The Iwasawa decomposition $G = KAN$ exhibits closed subgroups A and N of G such that A is simply connected abelian, N is simply connected nilpotent, A normalizes N, and multiplication from $K \times A \times N$ to G is a diffeomorphism onto. This decomposition generalizes the Gram-Schmidt orthogonalization process. Any two Iwasawa decompositions of G are conjugate. The Lie algebra $\mathfrak{a}_0$ of A may be taken to be any maximal abelian subspace of $\mathfrak{p}_0$, and the Lie algebra of N is defined from a kind of root-space decomposition of $\mathfrak{g}_0$ with respect to $\mathfrak{a}_0$. The simultaneous eigenspaces are called "restricted roots," and the restricted roots form an abstract root system. The Weyl group of this system coincides with the quotient of normalizer by centralizer of $\mathfrak{a}_0$ in K.

A Cartan subalgebra of $\mathfrak{g}_0$ is a subalgebra whose complexification is a Cartan subalgebra of $(\mathfrak{g}_0)^{\mathbb{C}}$. One Cartan subalgebra of $\mathfrak{g}_0$ is obtained by adjoining to the above $\mathfrak{a}_0$ a maximal abelian subspace of the centralizer of $\mathfrak{a}_0$ in $\mathfrak{k}_0$. This Cartan subalgebra is θ stable. Any Cartan subalgebra of $\mathfrak{g}_0$ is conjugate by an inner automorphism to a θ stable one, and the subalgebra built from $\mathfrak{a}_0$ as above is maximally noncompact among all θ stable Cartan subalgebras. Any two maximally noncompact Cartan subalgebras are conjugate, and so are any two maximally compact ones. Cayley transforms allow one to pass between any two θ stable Cartan subalgebras, up to conjugacy.

A Vogan diagram of $\mathfrak{g}_0$ superimposes certain information about the real form $\mathfrak{g}_0$ on the Dynkin diagram of $(\mathfrak{g}_0)^{\mathbb{C}}$. The extra information involves a maximally compact θ stable Cartan subalgebra and an allowable choice of a positive system of roots. The effect of θ on simple roots is labeled, and imaginary simple roots are painted if they are "noncompact," left unpainted if they are "compact." Such a diagram is not unique for

$\mathfrak{g}_0$, but it determines $\mathfrak{g}_0$ up to isomorphism. Every diagram that looks formally like a Vogan diagram arises from some $\mathfrak{g}_0$.

Vogan diagrams lead quickly to a classification of all simple real Lie algebras, the only difficulty being eliminating the redundancy in the choice of positive system of roots. This difficulty is resolved by the Borel and de Siebenthal Theorem. Using a succession of Cayley transforms to pass from a maximally compact Cartan subalgebra to a maximally noncompact Cartan subalgebra, one readily identifies the restricted roots for each simple real Lie algebra.

1. Existence of a Compact Real Form

An important clue to the structure of semisimple Lie groups comes from the examples of the classical semisimple groups in §§I.8 and I.14. In each case the Lie algebra $\mathfrak{g}_0$ is a real Lie algebra of matrices over $\mathbb{R}$, $\mathbb{C}$, or $\mathbb{H}$ closed under conjugate transpose $(\cdot)^*$. This fact is the key ingredient used in Proposition 1.56 to detect semisimplicity of $\mathfrak{g}_0$.

Using the techniques at the end of §I.8, we can regard $\mathfrak{g}_0$ as a Lie algebra of matrices over $\mathbb{R}$ closed under transpose $(\cdot)^*$. Then $\mathfrak{g}_0$ is the direct sum of the set $\mathfrak{k}_0$ of its skew-symmetric members and the set $\mathfrak{p}_0$ of its symmetric members. The real vector space $\mathfrak{u}_0 = \mathfrak{k}_0 \oplus i\mathfrak{p}_0$ of complex matrices is closed under brackets and is a Lie subalgebra of skew-Hermitian matrices.

Meanwhile we can regard the complexification $\mathfrak{g}$ of $\mathfrak{g}_0$ as the Lie algebra of complex matrices $\mathfrak{g} = \mathfrak{g}_0 + i\mathfrak{g}_0$. Putting $\mathfrak{k} = (\mathfrak{k}_0)^{\mathbb{C}}$ and $\mathfrak{p} = (\mathfrak{p}_0)^{\mathbb{C}}$, we write $\mathfrak{g} = \mathfrak{k} \oplus \mathfrak{p}$ as vector spaces. The complexification of $\mathfrak{u}_0$ is the same set of matrices: $(\mathfrak{u}_0)^{\mathbb{C}} = \mathfrak{k} \oplus \mathfrak{p}$.

Since $\mathfrak{g}_0$ has been assumed semisimple, $\mathfrak{g}$ is semisimple by Corollary 1.50, and $\mathfrak{u}_0$ is semisimple by the same corollary. The claim is that $\mathfrak{u}_0$ is a compact Lie algebra in the sense of §IV.4. In fact, let us introduce the inner product $\langle X, Y\rangle = \operatorname{Re}\operatorname{Tr}(XY^*)$ on $\mathfrak{u}_0$. The proof of Proposition 1.56 shows that

$$\langle (\operatorname{ad} Y)X, Z\rangle = \langle X, (\operatorname{ad}(Y^*))Z\rangle$$

and hence

$$(\operatorname{ad} Y)^* = \operatorname{ad}(Y^*). \tag{6.1}$$

Since $Y^* = -Y$, $\operatorname{ad} Y$ is skew Hermitian. Thus $(\operatorname{ad} Y)^2$ has eigenvalues ≤ 0, and the Killing form $B_{\mathfrak{u}_0}$ of $\mathfrak{u}_0$ satisfies

$$B_{\mathfrak{u}_0}(Y, Y) = \operatorname{Tr}((\operatorname{ad} Y)^2) \le 0.$$

Since $\mathfrak{u}_0$ is semisimple, $B_{\mathfrak{u}_0}$ is nondegenerate (Theorem 1.42) and must be negative definite. By Proposition 4.27, $\mathfrak{u}_0$ is a compact Lie algebra.

In the terminology of real forms as in §I.3, the splitting of any of the classical semisimple Lie algebras $\mathfrak{g}_0$ in §I.8 is equivalent with associating to $\mathfrak{g}_0$ the compact Lie algebra $\mathfrak{u}_0$ that is a real form of the complexification of $\mathfrak{g}_0$. Once we have this splitting of $\mathfrak{g}_0$, the arguments in §I.14 allowed us to obtain a polar-like decomposition of the analytic group of matrices G with Lie algebra $\mathfrak{g}_0$. This polar-like decomposition was a first structure theorem for the classical groups, giving insight into the topology of G and underlining the importance of a certain compact subgroup K of G.

The idea for beginning an investigation of the structure of a general semisimple Lie group G, not necessarily classical, is to look for this same kind of structure. We start with the Lie algebra $\mathfrak{g}_0$ and seek a decomposition into skew-symmetric and symmetric parts. To get this decomposition, we look for the occurrence of a compact Lie algebra $\mathfrak{u}_0$ as a real form of the complexification $\mathfrak{g}$ of $\mathfrak{g}_0$.

Actually not just any $\mathfrak{u}_0$ of this kind will do. The real forms $\mathfrak{u}_0$ and $\mathfrak{g}_0$ must be aligned so that the skew-symmetric part $\mathfrak{k}_0$ and the symmetric part $\mathfrak{p}_0$ can be recovered as $\mathfrak{k}_0 = \mathfrak{g}_0 \cap \mathfrak{u}_0$ and $\mathfrak{p}_0 = \mathfrak{g}_0 \cap i\mathfrak{u}_0$. The condition of proper alignment for $\mathfrak{u}_0$ is that the conjugations of $\mathfrak{g}$ with respect to $\mathfrak{g}_0$ and to $\mathfrak{u}_0$ must commute with each other.

The first step will be to associate to a complex semisimple Lie algebra $\mathfrak{g}$ a real form $\mathfrak{u}_0$ that is compact. This construction will occupy us for the remainder of this section. In §2 we shall address the alignment question when $\mathfrak{g}$ is the complexification of a real semisimple Lie algebra $\mathfrak{g}_0$. The result will yield the desired Lie algebra decomposition $\mathfrak{g}_0 = \mathfrak{k}_0 \oplus \mathfrak{p}_0$, known as the "Cartan decomposition" of the Lie algebra. Then in §3 we shall pass from the Cartan decomposition of the Lie algebra to a "Cartan decomposition" of the Lie group that generalizes the polar-like decomposition in Proposition 1.122.

The argument in the present section for constructing a compact real form from a complex semisimple $\mathfrak{g}$ will be somewhat roundabout. We shall use the Isomorphism Theorem (Theorem 2.108) to show that root vectors can be selected so that the constants arising in the bracket products of root vectors are all real. More precisely this result gives us a real form of $\mathfrak{g}$ known as a "split real form." It is not a compact Lie algebra but in a certain sense is as noncompact as possible. When $\mathfrak{g}$ is $\mathfrak{sl}(2, \mathbb{C})$, the real subalgebra $\mathfrak{sl}(2, \mathbb{R})$ is a split form, and the desired real form that is compact is $\mathfrak{su}(2)$. In general we obtain the real form that is compact by taking suitable linear combinations of the root vectors that define the split real form.

For the remainder of this section, let $\mathfrak{g}$ be a complex semisimple Lie algebra, let $\mathfrak{h}$ be a Cartan subalgebra, let $\Delta = \Delta(\mathfrak{g}, \mathfrak{h})$ be the set of roots of $\mathfrak{g}$ with respect to $\mathfrak{h}$, and let B be the Killing form. (The Killing form has the property that it is invariant under all automorphisms of $\mathfrak{g}$, according to Proposition 1.96, and this property is not always shared by other forms. To take advantage of this property, we shall insist that B is the Killing form in §§1–3. After that, we shall allow more general forms in place of B.)

For each pair $\{\alpha, -\alpha\}$ in Δ, we fix $E_\alpha \in \mathfrak{g}_\alpha$ and $E_{-\alpha} \in \mathfrak{g}_{-\alpha}$ so that $B(E_\alpha, E_{-\alpha}) = 1$. Then $[E_\alpha, E_{-\alpha}] = H_\alpha$ by Proposition 2.18a. Let α and β be roots. If $\alpha + \beta$ is in Δ, define $C_{\alpha,\beta}$ by

$$[E_\alpha, E_\beta] = C_{\alpha,\beta} E_{\alpha+\beta}.$$

If $\alpha + \beta$ is not in Δ, put $C_{\alpha,\beta} = 0$.

Lemma 6.2. $C_{\alpha,\beta} = -C_{\beta,\alpha}$.

PROOF. This follows from the skew symmetry of the bracket.

Lemma 6.3. If α, β, and γ are in Δ and $\alpha + \beta + \gamma = 0$, then

$$C_{\alpha,\beta} = C_{\beta,\gamma} = C_{\gamma,\alpha}.$$

PROOF. By the Jacobi identity,

$$[[E_\alpha, E_\beta], E_\gamma] + [[E_\beta, E_\gamma], E_\alpha] + [[E_\gamma, E_\alpha], E_\beta] = 0.$$

Thus $$C_{\alpha,\beta}[E_{-\gamma}, E_\gamma] + C_{\beta,\gamma}[E_{-\alpha}, E_\alpha] + C_{\gamma,\alpha}[E_{-\beta}, E_\beta] = 0$$

and $$C_{\alpha,\beta} H_\gamma + C_{\beta,\gamma} H_\alpha + C_{\gamma,\alpha} H_\beta = 0.$$

Substituting $H_\gamma = -H_\alpha - H_\beta$ and using the linear independence of $\{H_\alpha, H_\beta\}$, we obtain the result.

Lemma 6.4. Let α, β, and $\alpha + \beta$ be in Δ, and let $\beta + n\alpha$, with $-p \le n \le q$, be the α string containing β. Then

$$C_{\alpha,\beta}, C_{-\alpha,-\beta} = -\tfrac{1}{2} q(1 + p)|\alpha|^2.$$

PROOF. By Corollary 2.37,

$$[E_{-\alpha}, [E_\alpha, E_\beta]] = \tfrac{1}{2} q(1 + p)|\alpha|^2 B(E_\alpha, E_{-\alpha}) E_\beta.$$

The left side is $C_{-\alpha,\alpha+\beta} C_{\alpha,\beta} E_\beta$, and $B(E_\alpha, E_{-\alpha}) = 1$ on the right side. Therefore

$$C_{-\alpha,\alpha+\beta} C_{\alpha,\beta} = \tfrac{1}{2} q(1 + p)|\alpha|^2. \tag{6.5}$$

Since $(-\alpha) + (\alpha + \beta) + (-\beta) = 0$, Lemmas 6.3 and 6.2 give

$$C_{-\alpha,\alpha+\beta} = C_{-\beta,-\alpha} = -C_{-\alpha,-\beta},$$

and the result follows by substituting this formula into (6.5).

Theorem 6.6. Let $\mathfrak{g}$ be a complex semisimple Lie algebra, let $\mathfrak{h}$ be a Cartan subalgebra, and let Δ be the set of roots. For each $\alpha \in \Delta$, it is possible to choose root vectors $X_\alpha \in \mathfrak{g}_\alpha$ such that, for all α and β in Δ,

$$\begin{aligned} [X_\alpha, X_{-\alpha}] &= H_\alpha \\ [X_\alpha, X_\beta] &= N_{\alpha,\beta} X_{\alpha+\beta} && \text{if } \alpha+\beta \in \Delta \\ [X_\alpha, X_\beta] &= 0 && \text{if } \alpha+\beta \neq 0 \text{ and } \alpha+\beta \notin \Delta \end{aligned}$$

with constants $N_{\alpha,\beta}$ that satisfy

$$N_{\alpha,\beta} = -N_{-\alpha,-\beta}.$$

For any such choice of the system $\{X_\alpha\}$ of root vectors, the constants $N_{\alpha,\beta}$ satisfy

$$N_{\alpha,\beta}^2 = \tfrac{1}{2} q(1+p)|\alpha|^2,$$

where $\beta + n\alpha$, with $-p \leq n \leq q$, is the α string containing β.

PROOF. The transpose of the linear map $\varphi : \mathfrak{h} \to \mathfrak{h}$ given by $\varphi(h) = -h$ carries Δ to Δ, and thus φ extends to an automorphism $\tilde{\varphi}$ of $\mathfrak{g}$, by the Isomorphism Theorem (Theorem 2.108). (See Example 3 at the end of §II.10.) Since $\tilde{\varphi}(E_\alpha)$ is in $\mathfrak{g}_{-\alpha}$, there exists a constant $c_{-\alpha}$ such that $\tilde{\varphi}(E_\alpha) = c_{-\alpha} E_{-\alpha}$. By Proposition 1.96,

$$B(\tilde{\varphi}X, \tilde{\varphi}Y) = B(X, Y) \qquad \text{for all } X \text{ and } Y \text{ in } \mathfrak{g}.$$

Applying this formula with $X = E_\alpha$ and $Y = E_{-\alpha}$, we obtain

$$c_{-\alpha} c_\alpha = c_{-\alpha} c_\alpha B(E_{-\alpha}, E_\alpha) = B(\tilde{\varphi}E_\alpha, \tilde{\varphi}E_{-\alpha}) = B(E_\alpha, E_{-\alpha}) = 1.$$

Thus $c_{-\alpha} c_\alpha = 1$. Because of this relation we can choose a_α for each $\alpha \in \Delta$ such that

$$a_\alpha a_{-\alpha} = +1 \tag{6.7a}$$

$$a_\alpha^2 = -c_\alpha. \tag{6.7b}$$

For example, fix a pair $\{\alpha, -\alpha\}$, and write $c_\alpha = re^{i\theta}$ and $c_{-\alpha} = r^{-1}e^{-i\theta}$; then we can take $a_\alpha = r^{1/2} i e^{i\theta/2}$ and $a_{-\alpha} = -r^{-1/2} i e^{-i\theta/2}$.

With the choices of the a_α's in place so that (6.7) holds, define $X_\alpha = a_\alpha E_\alpha$. The root vectors X_α satisfy

$$[X_\alpha, X_{-\alpha}] = a_\alpha a_{-\alpha}[E_\alpha, E_{-\alpha}] = H_\alpha \qquad \text{by (6.7a)}$$

and

$$\begin{aligned}\tilde{\varphi}(X_\alpha) = a_\alpha\tilde{\varphi}(E_\alpha) &= a_\alpha c_{-\alpha}E_{-\alpha} \\ &= a_{-\alpha}^{-1}c_{-\alpha}E_{-\alpha} && \text{by (6.7a)} \\ &= -a_{-\alpha}E_{-\alpha} && \text{by (6.7b)} \\ &= -X_{-\alpha}. && (6.8)\end{aligned}$$

Define constants $N_{\alpha,\beta}$ relative to the root vectors X_γ in the same way that the constants $C_{\alpha,\beta}$ are defined relative to the root vectors E_γ. Then (6.8) gives

$$\begin{aligned}-N_{\alpha,\beta}X_{-\alpha-\beta} &= \tilde{\varphi}(N_{\alpha,\beta}X_{\alpha+\beta}) = \tilde{\varphi}[X_\alpha, X_\beta] \\ &= [\tilde{\varphi}X_\alpha, \tilde{\varphi}X_\beta] = [-X_{-\alpha}, -X_{-\beta}] = N_{-\alpha,-\beta}X_{-\alpha-\beta},\end{aligned}$$

and we find that $N_{\alpha,\beta} = -N_{-\alpha,-\beta}$. The formula for $N_{\alpha,\beta}^2$ follows by substituting into Lemma 6.4, and the proof is complete.

Theorem 6.6 has an interpretation in terms of real forms of the complex Lie algebra $\mathfrak{g}$. With notation as in Theorem 6.6, define

$$\mathfrak{h}_0 = \{H \in \mathfrak{h} \mid \alpha(H) \in \mathbb{R} \text{ for all } \alpha \in \Delta\}, \tag{6.9}$$

and put

$$\mathfrak{g}_0 = \mathfrak{h}_0 \oplus \bigoplus_{\alpha\in\Delta} \mathbb{R}X_\alpha.$$

The formula $N_{\alpha,\beta}^2 = \frac{1}{2}q(1+p)|\alpha|^2$ shows that $N_{\alpha,\beta}$ is real. Therefore $\mathfrak{g}_0$ is a subalgebra of $\mathfrak{g}^{\mathbb{R}}$. Since it is clear that $\mathfrak{g}^{\mathbb{R}} = \mathfrak{g}_0 \oplus i\mathfrak{g}_0$ as real vector spaces, $\mathfrak{g}_0$ is a real form of $\mathfrak{g}$. A real form of $\mathfrak{g}$ that contains $\mathfrak{h}_0$ as in (6.9) for some Cartan subalgebra $\mathfrak{h}$ is called a **split real form** of $\mathfrak{g}$. We summarize the above remarks as follows.

Corollary 6.10. Any complex semisimple Lie algebra contains a split real form.

EXAMPLES. It is clear from the computations in §II.1 that $\mathfrak{sl}(n,\mathbb{R})$ and $\mathfrak{sp}(n,\mathbb{R})$ are split real forms of $\mathfrak{sl}(n,\mathbb{C})$ and $\mathfrak{sp}(n,\mathbb{C})$, respectively. We shall see in §4 that $\mathfrak{so}(n+1,n)$ and $\mathfrak{so}(n,n)$ are isomorphic to split real forms of $\mathfrak{so}(2n+1,\mathbb{C})$ and $\mathfrak{so}(2n,\mathbb{C})$, respectively.

As we indicated at the beginning of this section, we shall study real semisimple Lie algebras by relating them to other real forms that are compact Lie algebras. A real form of the complex semisimple Lie algebra $\mathfrak{g}$ that is a compact Lie algebra is called a **compact real form** of $\mathfrak{g}$.

Theorem 6.11. If $\mathfrak{g}$ is a complex semisimple Lie algebra, then $\mathfrak{g}$ has a compact real form $\mathfrak{u}_0$.

REMARKS.

1) The compact real forms of the classical complex semisimple Lie algebras are already familiar. For $\mathfrak{sl}(n,\mathbb{C})$, $\mathfrak{so}(n,\mathbb{C})$, and $\mathfrak{sp}(n,\mathbb{C})$, they are $\mathfrak{su}(n)$, $\mathfrak{so}(n)$, and $\mathfrak{sp}(n)$, respectively. In the case of $\mathfrak{sp}(n,\mathbb{C})$, this fact uses the isomorphism $\mathfrak{sp}(n) \cong \mathfrak{sp}(n,\mathbb{C}) \cap \mathfrak{u}(2n)$ proved in §I.8.

2) We denote the compact real forms of the complex Lie algebras of types E_6, E_7, E_8, F_4, and G_2 by $\mathfrak{e}_6$, $\mathfrak{e}_7$, $\mathfrak{e}_8$, $\mathfrak{f}_4$, and $\mathfrak{g}_2$, respectively. Corollary 6.20 will show that these compact real forms are well defined up to isomorphism.

PROOF. Let $\mathfrak{h}$ be a Cartan subalgebra, and define root vectors X_α as in Theorem 6.6. Let

$$\mathfrak{u}_0 = \sum_{\alpha\in\Delta} \mathbb{R}(iH_\alpha) + \sum_{\alpha\in\Delta} \mathbb{R}(X_\alpha - X_{-\alpha}) + \sum_{\alpha\in\Delta} \mathbb{R}i(X_\alpha + X_{-\alpha}). \tag{6.12}$$

It is clear that $\mathfrak{g}^{\mathbb{R}} = \mathfrak{u}_0 \oplus i\mathfrak{u}_0$ as real vector spaces. Let us see that $\mathfrak{u}_0$ is closed under brackets. The term $\sum \mathbb{R}(iH_\alpha)$ on the right side of (6.12) is abelian, and we have

$$\begin{aligned}
[iH_\alpha, (X_\alpha - X_{-\alpha})] &= |\alpha|^2 i(X_\alpha + X_{-\alpha}) \\
[iH_\alpha, i(X_\alpha + X_{-\alpha})] &= -|\alpha|^2 (X_\alpha - X_{-\alpha}).
\end{aligned}$$

Therefore the term $\sum \mathbb{R}(iH_\alpha)$ brackets $\mathfrak{u}_0$ into $\mathfrak{u}_0$. For the other brackets of elements of $\mathfrak{u}_0$, we recall from Theorem 6.6 that $N_{\alpha,\beta} = -N_{-\alpha,-\beta}$, and we compute for $\beta \neq \pm\alpha$ that

$$\begin{aligned}
&[(X_\alpha - X_{-\alpha}), (X_\beta - X_{-\beta})] \\
&\quad = N_{\alpha,\beta}X_{\alpha+\beta} + N_{-\alpha,-\beta}X_{-\alpha-\beta} - N_{-\alpha,\beta}X_{-\alpha+\beta} - N_{\alpha,-\beta}X_{\alpha-\beta} \\
&\quad = N_{\alpha,\beta}(X_{\alpha+\beta} - X_{-(\alpha+\beta)}) - N_{-\alpha,\beta}(X_{-\alpha+\beta} - X_{-(-\alpha+\beta)})
\end{aligned}$$

and similarly that

$$\begin{aligned}
&[(X_\alpha - X_{-\alpha}), i(X_\beta + X_{-\beta})] \\
&\quad = N_{\alpha,\beta}i(X_{\alpha+\beta} + X_{-(\alpha+\beta)}) - N_{-\alpha,\beta}i(X_{-\alpha+\beta} + X_{-(-\alpha+\beta)})
\end{aligned}$$

and

$$\begin{aligned}
&[i(X_\alpha + X_{-\alpha}), i(X_\beta + X_{-\beta})] \\
&\quad = -N_{\alpha,\beta}(X_{\alpha+\beta} - X_{-(\alpha+\beta)}) - N_{-\alpha,\beta}(X_{-\alpha+\beta} - X_{-(-\alpha+\beta)}).
\end{aligned}$$

Finally

$$[(X_\alpha - X_{-\alpha}), i(X_\alpha + X_{-\alpha})] = 2iH_\alpha,$$

and therefore $\mathfrak{u}_0$ is closed under brackets. Consequently $\mathfrak{u}_0$ is a real form.

To show that $\mathfrak{u}_0$ is a compact Lie algebra, it is enough, by Proposition 4.27, to show that the Killing form of $\mathfrak{u}_0$ is negative definite. The Killing forms $B_{\mathfrak{u}_0}$ of $\mathfrak{u}_0$ and B of $\mathfrak{g}$ are related by $B_{\mathfrak{u}_0} = B|_{\mathfrak{u}_0\times\mathfrak{u}_0}$, according to (1.20). The first term on the right side of (6.12) is orthogonal to the other two terms by Proposition 2.17a, and B is positive on $\sum \mathbb{R}H_\alpha$ by Corollary 2.38. Hence B is negative on $\sum \mathbb{R}iH_\alpha$. Next we use Proposition 2.17a to observe for $\beta \neq \pm\alpha$ that

$$\begin{aligned} B((X_\alpha - X_{-\alpha}), (X_\beta - X_{-\beta})) &= 0 \\ B((X_\alpha - X_{-\alpha}), i(X_\beta + X_{-\beta})) &= 0 \\ B(i(X_\alpha + X_{-\alpha}), i(X_\beta + X_{-\beta})) &= 0. \end{aligned}$$

Finally we have

$$\begin{aligned} B((X_\alpha - X_{-\alpha}), (X_\alpha - X_{-\alpha})) &= -2B(X_\alpha, X_{-\alpha}) = -2 \\ B(i(X_\alpha + X_{-\alpha}), i(X_\alpha + X_{-\alpha})) &= -2B(X_\alpha, X_{-\alpha}) = -2, \end{aligned}$$

and therefore $B|_{\mathfrak{u}_0\times\mathfrak{u}_0}$ is negative definite.

2. Cartan Decomposition on the Lie Algebra Level

To detect semisimplicity of some specific Lie algebras of matrices in §I.8, we made critical use of the conjugate transpose mapping $X \mapsto X^*$. Slightly better is the map $\theta(X) = -X^*$, which is actually an **involution**, i.e., an automorphism of the Lie algebra with square equal to the identity. To see that θ respects brackets, we just write

$$\theta[X, Y] = -[X, Y]^* = -[Y^*, X^*] = [-X^*, -Y^*] = [\theta(X), \theta(Y)].$$

Let B be the Killing form. The involution θ has the property that $B_\theta(X, Y) = -B(X, \theta Y)$ is symmetric and positive definite because Proposition 1.96 gives

$$\begin{aligned} B_\theta(X, Y) &= -B(X, \theta Y) = -B(\theta X, \theta^2 Y) \\ &= -B(\theta X, Y) = -B(Y, \theta X) = B_\theta(Y, X) \end{aligned}$$

and (6.1) gives

$$\begin{aligned} B_\theta(X, Y) &= -B(X, \theta Y) = -\mathrm{Tr}((\mathrm{ad}\, X)(\mathrm{ad}\, \theta Y)) \\ &= \mathrm{Tr}((\mathrm{ad}\, X)(\mathrm{ad}\, X^*)) = \mathrm{Tr}(\mathrm{ad}\, X)(\mathrm{ad}\, X)^*) \geq 0. \end{aligned}$$

An involution θ of a real semisimple Lie algebra $\mathfrak{g}_0$ such that the symmetric bilinear form

$$B_\theta(X, Y) = -B(X, \theta Y) \tag{6.13}$$

is positive definite is called a **Cartan involution**. We shall see that any real semisimple Lie algebra has a Cartan involution and that the Cartan involution is unique up to inner automorphism. As a consequence of the proof, we shall obtain a converse to the arguments of §I.8: Every real semisimple Lie algebra can be realized as a Lie algebra of real matrices closed under transpose.

Theorem 6.11 says that any complex semisimple Lie algebra $\mathfrak{g}$ has a compact real form. According to the next proposition, it follows that $\mathfrak{g}^{\mathbb{R}}$ has a Cartan involution.

Proposition 6.14. Let $\mathfrak{g}$ be a complex semisimple Lie algebra, let $\mathfrak{u}_0$ be a compact real form of $\mathfrak{g}$, and let τ be the conjugation of $\mathfrak{g}$ with respect to $\mathfrak{u}_0$. If $\mathfrak{g}$ is regarded as a real Lie algebra $\mathfrak{g}^{\mathbb{R}}$, then τ is a Cartan involution of $\mathfrak{g}^{\mathbb{R}}$.

REMARK. The real Lie algebra $\mathfrak{g}^{\mathbb{R}}$ is semisimple by (1.58).

PROOF. It is clear that τ is an involution. The Killing forms $B_{\mathfrak{g}}$ of $\mathfrak{g}$ and $B_{\mathfrak{g}^{\mathbb{R}}}$ of $\mathfrak{g}^{\mathbb{R}}$ are related by

$$B_{\mathfrak{g}^{\mathbb{R}}}(Z_1, Z_2) = 2\operatorname{Re} B_{\mathfrak{g}}(Z_1, Z_2),$$

according to (1.57). Write $Z \in \mathfrak{g}$ as $Z = X + iY$ with X and Y in $\mathfrak{u}_0$. Then

$$\begin{aligned} B_{\mathfrak{g}}(Z, \tau Z) &= B_{\mathfrak{g}}(X + iY, X - iY) \\ &= B_{\mathfrak{g}}(X, X) + B_{\mathfrak{g}}(Y, Y) \\ &= B_{\mathfrak{u}_0}(X, X) + B_{\mathfrak{u}_0}(Y, Y), \end{aligned}$$

and the right side is < 0 unless $Z = 0$. In the notation of (6.13), it follows that

$$(B_{\mathfrak{g}^{\mathbb{R}}})_\tau(Z_1, Z_2) = -B_{\mathfrak{g}^{\mathbb{R}}}(Z_1, \tau Z_2) = -2\operatorname{Re} B_{\mathfrak{g}}(Z_1, \tau Z_2)$$

is positive definite on $\mathfrak{g}^{\mathbb{R}}$, and therefore τ is a Cartan involution of $\mathfrak{g}^{\mathbb{R}}$.

Now we address the problem of aligning a compact real form properly when we start with a real semisimple Lie algebra $\mathfrak{g}_0$ and obtain $\mathfrak{g}$ by complexification. Corollaries give the existence and uniqueness (up to conjugacy) of Cartan involutions.

Lemma 6.15. Let $\mathfrak{g}_0$ be a real finite-dimensional Lie algebra, and let ρ be an automorphism of $\mathfrak{g}_0$ that is diagonable with positive eigenvalues $d_1, \ldots, d_m$ and corresponding eigenspaces $(\mathfrak{g}_0)_{d_j}$. For $-\infty < r < \infty$, define ρ^r to be the linear transformation on $\mathfrak{g}_0$ that is d_j^r on $(\mathfrak{g}_0)_{d_j}$. Then $\{\rho^r\}$ is a one-parameter group in $\operatorname{Aut}\mathfrak{g}_0$. If $\mathfrak{g}_0$ is semisimple, then ρ^r lies in $\operatorname{Int}\mathfrak{g}_0$.

PROOF. If X is in $(\mathfrak{g}_0)_{d_i}$ and Y is in $(\mathfrak{g}_0)_{d_j}$, then

$$\rho[X, Y] = [\rho X, \rho Y] = d_i d_j [X, Y]$$

since ρ is an automorphism. Hence $[X, Y]$ is in $(\mathfrak{g}_0)_{d_i d_j}$, and we obtain

$$\rho^r[X, Y] = (d_i d_j)^r [X, Y] = [d_i^r X, d_j^r Y] = [\rho^r X, \rho^r Y].$$

Consequently ρ^r is an automorphism. Therefore $\{\rho^r\}$ is a one-parameter group in $\operatorname{Aut}\mathfrak{g}_0$, hence in the identity component $(\operatorname{Aut}\mathfrak{g}_0)_0$. If $\mathfrak{g}_0$ is semisimple, then Propositions 1.97 and 1.98 show that $(\operatorname{Aut}\mathfrak{g}_0)_0 = \operatorname{Int}\mathfrak{g}_0$, and the lemma follows.

Theorem 6.16. Let $\mathfrak{g}_0$ be a real semisimple Lie algebra, let θ be a Cartan involution, and let σ be any involution. Then there exists $\varphi \in \operatorname{Int}\mathfrak{g}_0$ such that $\varphi\theta\varphi^{-1}$ commutes with σ.

PROOF. Since θ is given as a Cartan involution, B_θ is an inner product for $\mathfrak{g}_0$. Put $\omega = \sigma\theta$. This is an automorphism of $\mathfrak{g}_0$, and Proposition 1.96 shows that it leaves B invariant. From $\sigma^2 = \theta^2 = 1$, we therefore have

$$B(\omega X, \theta Y) = B(X, \omega^{-1}\theta Y) = B(X, \theta\omega Y)$$

and hence
$$B_\theta(\omega X, Y) = B_\theta(X, \omega Y).$$

Thus ω is symmetric, and its square $\rho = \omega^2$ is positive definite. Write ρ^r for the positive-definite r^{th} power of ρ, $-\infty < \rho < \infty$. Lemma 6.15 shows that ρ^r is a one-parameter group in $\operatorname{Int}\mathfrak{g}_0$.

Now

$$\rho\theta = \omega^2\theta = \sigma\theta\sigma\theta\theta = \sigma\theta\sigma = \theta\theta\sigma\theta\sigma = \theta\omega^{-2} = \theta\rho^{-1}.$$

In terms of a basis of $\mathfrak{g}_0$ that diagonalizes ρ, the matrix form of this equation is

$$\rho_{ii}\theta_{ij} = \theta_{ij}\rho_{jj}^{-1} \qquad \text{for all } i \text{ and } j.$$

Considering separately the cases $\theta_{ij} = 0$ and $\theta_{ij} \neq 0$, we see that

$$\rho_{ii}^r\theta_{ij} = \theta_{ij}\rho_{jj}^{-r}$$

and therefore that

$$\rho^r\theta = \theta\rho^{-r}. \tag{6.17}$$

Put $\varphi = \rho^{1/4}$. Then two applications of (6.17) give

$$\begin{aligned}(\varphi\theta\varphi^{-1})\sigma &= \rho^{1/4}\theta\rho^{-1/4}\sigma = \rho^{1/2}\theta\sigma \\ &= \rho^{1/2}\omega^{-1} = \rho^{-1/2}\rho\omega^{-1} \\ &= \rho^{-1/2}\omega = \omega\rho^{-1/2} \\ &= \sigma\theta\rho^{-1/2} = \sigma\rho^{1/4}\theta\rho^{-1/4} = \sigma(\varphi\theta\varphi^{-1}),\end{aligned}$$

as required.

Corollary 6.18. If $\mathfrak{g}_0$ is a real semisimple Lie algebra, then $\mathfrak{g}_0$ has a Cartan involution.

PROOF. Let $\mathfrak{g}$ be the complexification of $\mathfrak{g}_0$, and choose by Theorem 6.11 a compact real form $\mathfrak{u}_0$ of $\mathfrak{g}$. Let σ and τ be the conjugations of $\mathfrak{g}$ with respect to $\mathfrak{g}_0$ and $\mathfrak{u}_0$. If we regard $\mathfrak{g}$ as a real Lie algebra $\mathfrak{g}^{\mathbb{R}}$, then σ and τ are involutions of $\mathfrak{g}^{\mathbb{R}}$, and Proposition 6.14 shows that τ is a Cartan involution. By Theorem 6.16 we can find $\varphi \in \operatorname{Int}(\mathfrak{g}^{\mathbb{R}}) = \operatorname{Int}\mathfrak{g}$ such that $\varphi\tau\varphi^{-1}$ commutes with σ.

Here $\varphi\tau\varphi^{-1}$ is the conjugation of $\mathfrak{g}$ with respect to $\varphi(\mathfrak{u}_0)$, which is another compact real form of $\mathfrak{g}$. Thus

$$(B_{\mathfrak{g}^{\mathbb{R}}})_{\varphi\tau\varphi^{-1}}(Z_1, Z_2) = -2\operatorname{Re} B_{\mathfrak{g}}(Z_1, \varphi\tau\varphi^{-1}Z_2)$$

is positive definite on $\mathfrak{g}^{\mathbb{R}}$.

The Lie algebra $\mathfrak{g}_0$ is characterized as the fixed set of σ. If $\sigma X = X$, then

$$\sigma(\varphi\tau\varphi^{-1}X) = \varphi\tau\varphi^{-1}\sigma X = \varphi\tau\varphi^{-1}X.$$

Hence $\varphi\tau\varphi^{-1}$ restricts to an involution θ of $\mathfrak{g}_0$. We have

$$B_\theta(X, Y) = -B_{\mathfrak{g}_0}(X, \theta Y) = -B_{\mathfrak{g}}(X, \varphi\tau\varphi^{-1}Y) = \tfrac{1}{2}(B_{\mathfrak{g}^{\mathbb{R}}})_{\varphi\tau\varphi^{-1}}(X, Y).$$

Thus B_θ is positive definite on $\mathfrak{g}_0$, and θ is a Cartan involution.

Corollary 6.19. If $\mathfrak{g}_0$ is a real semisimple Lie algebra, then any two Cartan involutions of $\mathfrak{g}_0$ are conjugate via $\operatorname{Int}\mathfrak{g}_0$.

PROOF. Let θ and θ' be two Cartan involutions. Taking $\sigma = \theta'$ in Theorem 6.16, we can find $\varphi \in \operatorname{Int} \mathfrak{g}_0$ such that $\varphi\theta\varphi^{-1}$ commutes with θ'. Here $\varphi\theta\varphi^{-1}$ is another Cartan involution of $\mathfrak{g}_0$. So we may as well assume that θ and θ' commute from the outset. We shall prove that $\theta = \theta'$.

Since θ and θ' commute, they have compatible eigenspace decompositions into $+1$ and -1 eigenspaces. By symmetry it is enough to show that no nonzero $X \in \mathfrak{g}_0$ is in the $+1$ eigenspace for θ and the -1 eigenspace for θ'. Assuming the contrary, suppose that $\theta X = X$ and $\theta' X = -X$. Then we have

$$0 < B_\theta(X, X) = -B(X, \theta X) = -B(X, X)$$

$$0 < B_{\theta'}(X, X) = -B(X, \theta' X) = +B(X, X),$$

contradiction. We conclude that $\theta = \theta'$, and the proof is complete.

Corollary 6.20. If $\mathfrak{g}$ is a complex semisimple Lie algebra, then any two compact real forms of $\mathfrak{g}$ are conjugate via $\operatorname{Int} \mathfrak{g}$.

PROOF. Each compact real form has an associated conjugation that determines it, and this conjugation is a Cartan involution of $\mathfrak{g}^{\mathbb{R}}$, by Proposition 6.14. Applying Corollary 6.19 to $\mathfrak{g}^{\mathbb{R}}$, we see that the two conjugations are conjugate by a member of $\operatorname{Int}(\mathfrak{g}^{\mathbb{R}})$. Since $\operatorname{Int}(\mathfrak{g}^{\mathbb{R}}) = \operatorname{Int} \mathfrak{g}$, the corollary follows.

Corollary 6.21. If $A = (A_{ij})_{i,j=1}^{l}$ is an abstract Cartan matrix, then there exists, up to isomorphism, one and only one compact semisimple Lie algebra $\mathfrak{g}_0$ whose complexification $\mathfrak{g}$ has a root system with A as Cartan matrix.

PROOF. Existence of $\mathfrak{g}$ is given in Theorem 2.111, and uniqueness of $\mathfrak{g}$ is given in Example 1 of §II.10. The passage from $\mathfrak{g}$ to $\mathfrak{g}_0$ is accomplished by Theorem 6.11 and Corollary 6.20.

Corollary 6.22. If $\mathfrak{g}$ is a complex semisimple Lie algebra, then the only Cartan involutions of $\mathfrak{g}^{\mathbb{R}}$ are the conjugations with respect to the compact real forms of $\mathfrak{g}$.

PROOF. Theorem 6.11 and Proposition 6.14 produce a Cartan involution of $\mathfrak{g}^{\mathbb{R}}$ that is conjugation with respect to some compact real form of $\mathfrak{g}$. Any other Cartan involution is conjugate to this one, according to Corollary 6.19, and hence is also the conjugation with respect to a compact real form of $\mathfrak{g}$.

A Cartan involution θ of $\mathfrak{g}_0$ yields an eigenspace decomposition

$$\mathfrak{g}_0 = \mathfrak{k}_0 \oplus \mathfrak{p}_0 \tag{6.23}$$

of $\mathfrak{g}_0$ into $+1$ and -1 eigenspaces, and these must bracket according to the rules

$$[\mathfrak{k}_0, \mathfrak{k}_0] \subseteq \mathfrak{k}_0, \quad [\mathfrak{k}_0, \mathfrak{p}_0] \subseteq \mathfrak{p}_0, \quad [\mathfrak{p}_0, \mathfrak{p}_0] \subseteq \mathfrak{k}_0 \tag{6.24}$$

since θ is an involution. From (6.23) and (6.24) it follows that

$$\mathfrak{k}_0 \text{ and } \mathfrak{p}_0 \text{ are orthogonal under } B_{\mathfrak{g}_0} \text{ and under } B_\theta \tag{6.25}$$

In fact, if X is in $\mathfrak{k}_0$ and Y is in $\mathfrak{p}_0$, then $\operatorname{ad} X \operatorname{ad} Y$ carries $\mathfrak{k}_0$ to $\mathfrak{p}_0$ and $\mathfrak{p}_0$ to $\mathfrak{k}_0$. Thus it has trace 0, and $B_{\mathfrak{g}_0}(X, Y) = 0$; since $\theta Y = -Y$, $B_\theta(X, Y) = 0$ also.

Since B_θ is positive definite, the eigenspaces $\mathfrak{k}_0$ and $\mathfrak{p}_0$ in (6.23) have the property that

$$B_{\mathfrak{g}_0} \text{ is } \begin{cases} \text{negative definite on } \mathfrak{k}_0 \\ \text{positive definite on } \mathfrak{p}_0. \end{cases} \tag{6.26}$$

A decomposition (6.23) of $\mathfrak{g}_0$ that satisfies (6.24) and (6.26) is called a **Cartan decomposition** of $\mathfrak{g}_0$.

Conversely a Cartan decomposition determines a Cartan involution θ by the formula

$$\theta = \begin{cases} +1 & \text{on } \mathfrak{k}_0 \\ -1 & \text{on } \mathfrak{p}_0. \end{cases}$$

Here (6.24) shows that θ respects brackets, and (6.25) and (6.26) show that B_θ is positive definite. (B_θ is symmetric by Proposition 1.96 since θ has order 2.)

If $\mathfrak{g}_0 = \mathfrak{k}_0 \oplus \mathfrak{p}_0$ is a Cartan decomposition of $\mathfrak{g}_0$, then $\mathfrak{k}_0 \oplus i\mathfrak{p}_0$ is a compact real form of $\mathfrak{g} = (\mathfrak{g}_0)^{\mathbb{C}}$. Conversely if $\mathfrak{h}_0$ and $\mathfrak{q}_0$ are the $+1$ and -1 eigenspaces of an involution σ, then σ is a Cartan involution only if the real form $\mathfrak{h}_0 \oplus i\mathfrak{q}_0$ of $\mathfrak{g} = (\mathfrak{g}_0)^{\mathbb{C}}$ is compact.

If $\mathfrak{g}$ is a complex semisimple Lie algebra, then it follows from Corollary 6.22 that the most general Cartan decomposition of $\mathfrak{g}^{\mathbb{R}}$ is $\mathfrak{g}^{\mathbb{R}} = \mathfrak{u}_0 \oplus i\mathfrak{u}_0$, where $\mathfrak{u}_0$ is a compact real form of $\mathfrak{g}_0$.

Corollaries 6.18 and 6.19 have shown for an arbitrary real semisimple Lie algebra $\mathfrak{g}_0$ that Cartan decompositions exist and are unique up to conjugacy by $\operatorname{Int} \mathfrak{g}_0$. Let us see as a consequence that every real semsimple Lie algebra can be realized as a Lie algebra of real matrices closed under transpose.

Lemma 6.27. If $\mathfrak{g}_0$ is a real semisimple Lie algebra and θ is a Cartan involution, then

$$(\operatorname{ad} X)^* = -\operatorname{ad} \theta X \qquad \text{for all } X \in \mathfrak{g}_0,$$

where adjoint $(\,\cdot\,)^*$ is defined relative to the inner product B_θ.

PROOF. We have

$$\begin{aligned} B_\theta((\operatorname{ad} \theta X)Y, Z) &= -B([\theta X, Y], \theta Z) \\ &= B(Y, [\theta X, \theta Z]) = B(Y, \theta[X, Z]) \\ &= -B_\theta(Y, (\operatorname{ad} X)Z) = -B_\theta((\operatorname{ad} X)^* Y, Z). \end{aligned}$$

Proposition 6.28. If $\mathfrak{g}_0$ is a real semisimple Lie algebra, then $\mathfrak{g}_0$ is isomorphic to a Lie algebra of real matrices that is closed under transpose. If a Cartan involution θ of $\mathfrak{g}_0$ has been specified, then the isomorphism may be chosen so that θ is carried to negative transpose.

PROOF. Let θ be a Cartan involution of $\mathfrak{g}_0$ (existence by Corollary 6.18), and define the inner product B_θ on $\mathfrak{g}_0$ as in (6.13). Since $\mathfrak{g}_0$ is semisimple, $\mathfrak{g}_0 \cong \operatorname{ad} \mathfrak{g}_0$. The matrices of $\operatorname{ad} \mathfrak{g}_0$ in an orthonormal basis relative to B_θ will be the required Lie algebra of matrices. We have only to show that $\operatorname{ad} \mathfrak{g}_0$ is closed under adjoint. But this follows from Lemma 6.27 and the fact that $\mathfrak{g}_0$ is closed under θ.

Corollary 6.29. If $\mathfrak{g}_0$ is a real semisimple Lie algebra and θ is a Cartan involution, then any θ stable subalgebra $\mathfrak{s}_0$ of $\mathfrak{g}_0$ is reductive.

PROOF. Proposition 6.28 allows us to regard $\mathfrak{g}_0$ as a real Lie algebra of real matrices closed under transpose, and θ becomes negative transpose. Then $\mathfrak{s}_0$ is a Lie subalgebra of matrices closed under transpose, and the result follows from Proposition 1.56.

3. Cartan Decomposition on the Lie Group Level

In this section we turn to a consideration of groups. Let G be a semisimple Lie group, and let $\mathfrak{g}_0$ be its Lie algebra. The results of §2 established that $\mathfrak{g}_0$ has a Cartan involution and that any two Cartan involutions are conjugate by an inner automorphism. The theorem in this section lifts the corresponding Cartan decomposition $\mathfrak{g}_0 = \mathfrak{k}_0 \oplus \mathfrak{p}_0$ given in (6.23) to a decomposition of G.

In the course of the proof, we shall consider $\operatorname{Ad}(G)$ first, proving the theorem in this special case. Then we shall use the result for $\operatorname{Ad}(G)$ to obtain the theorem for G. The following proposition clarifies one detail about this process.

Proposition 6.30. If G is a semisimple Lie group and Z is its center, then G/Z has trivial center.

REMARK. The center Z is discrete, being a closed subgroup of G whose Lie algebra is 0.

PROOF. Let $\mathfrak{g}_0$ be the Lie algebra of G. For $x \in G$, $\operatorname{Ad}(x)$ is the differential of conjugation by x and is 1 if and only if x is in Z. Thus $G/Z \cong \operatorname{Ad}(G)$. If $g \in \operatorname{Ad}(G)$ is central, we have $g\operatorname{Ad}(x) = \operatorname{Ad}(x)g$ for all $x \in G$. Differentiation gives $g(\operatorname{ad} X) = (\operatorname{ad} X)g$ for $X \in \mathfrak{g}_0$, and application of both sides of this equation to $Y \in \mathfrak{g}_0$ gives $g([X, Y]) = [X, gY]$. Replacing Y by $g^{-1}Y$, we obtain $[gX, Y] = [X, Y]$. Interchanging X and Y gives $[X, gY] = [X, Y]$ and hence $g([X, Y]) = [X, Y]$. Since $[\mathfrak{g}_0, \mathfrak{g}_0] = \mathfrak{g}_0$ by Corollary 1.52, the linear transformation g is 1 on all of $\mathfrak{g}_0$, i.e., $g = 1$. Thus $\operatorname{Ad}(G)$ has trivial center.

Theorem 6.31. Let G be a semisimple Lie group, let θ be a Cartan involution of its Lie algebra $\mathfrak{g}_0$, let $\mathfrak{g}_0 = \mathfrak{k}_0 \oplus \mathfrak{p}_0$ be the corresponding Cartan decomposition, and let K be the analytic subgroup of G with Lie algebra $\mathfrak{k}_0$. Then

- (a) there exists a Lie group automorphism Θ of G with differential θ, and Θ has $\Theta^2 = 1$
- (b) the subgroup of G fixed by Θ is K
- (c) the mapping $K \times \mathfrak{p}_0 \to G$ given by $(k, X) \mapsto k \exp X$ is a diffeomorphism onto
- (d) K is closed
- (e) K contains the center Z of G
- (f) K is compact if and only if Z is finite
- (g) when Z is finite, K is a maximal compact subgroup of G.

REMARKS.

1) This theorem generalizes and extends Proposition 1.122, where (c) reduces to the polar decomposition of matrices. Proposition 1.122 therefore points to a host of examples of the theorem.

2) The automorphism Θ of the theorem will be called the **global Cartan involution**, and (c) is the **global Cartan decomposition**. Many authors follow the convention of writing θ for Θ, using the same symbol for the involution of G as for the involution of $\mathfrak{g}_0$, but we shall use distinct symbols for the two kinds of involution.

PROOF. Let $\bar{G} = \operatorname{Ad}(G)$. We shall prove the theorem for $\bar{G}$ and then deduce as a consequence the theorem for G. For the case of $\bar{G}$, we begin by constructing Θ as in (a), calling it $\bar{\Theta}$. Then we define $\bar{K}^\#$ to be the subgroup fixed by $\bar{\Theta}$, and we prove (c) with K replaced by $\bar{K}^\#$. The rest of the proof of the theorem for $\bar{G}$ is then fairly easy.

For $\bar{G}$, the Lie algebra is $\operatorname{ad}\mathfrak{g}_0$, and the Cartan involution $\bar{\theta}$ is $+1$ on $\operatorname{ad}_{\mathfrak{g}_0}(\mathfrak{k}_0)$ and -1 on $\operatorname{ad}_{\mathfrak{g}_0}(\mathfrak{p}_0)$. Let us write members of $\operatorname{ad}\mathfrak{g}_0$ with bars over them. Define the inner product B_θ on $\mathfrak{g}_0$ by (6.13), and let adjoint $(\cdot)^*$ be defined for linear maps of $\mathfrak{g}_0$ into itself by means of B_θ. Lemma 6.27 says that

$$(\operatorname{ad} W)^* = -\operatorname{ad}\theta W \qquad \text{for all } W \in \mathfrak{g}_0, \tag{6.32}$$

and therefore

$$\bar{\theta}\bar{W} = -\bar{W}^* \qquad \text{for all } \bar{W} \in \operatorname{ad}\mathfrak{g}_0. \tag{6.33}$$

If g is in $\operatorname{Aut}\mathfrak{g}_0$, we shall prove that g^* is in $\operatorname{Aut}\mathfrak{g}_0$. Since B_θ is definite, we are to prove that

$$B_\theta([g^*X, g^*Y], Z) \stackrel{?}{=} B_\theta(g^*[X, Y], Z) \tag{6.34}$$

for all $X, Y, Z \in \mathfrak{g}_0$. Using (6.32) three times, we have

$$\begin{aligned}
B_\theta([g^*X, g^*Y], Z) &= -B_\theta(g^*Y, [\theta g^*X, Z]) = -B_\theta(Y, [g\theta g^*X, gZ]) \\
&= B_\theta((\operatorname{ad} gZ)g\theta g^*X, Y) = -B_\theta(g\theta g^*X, [\theta gZ, Y]) \\
&= B(g\theta g^*X, [gZ, \theta Y]) = -B_\theta(g^*X, g^{-1}[gZ, \theta Y]) \\
&= -B_\theta(X, [gZ, \theta Y]) = B_\theta(X, (\operatorname{ad}\theta Y)gZ) \\
&= B_\theta([X, Y], gZ) = B_\theta(g^*[X, Y], Z),
\end{aligned}$$

and (6.34) is established.

We apply this fact when $g = \bar{x}$ is in $\operatorname{Ad}(G) = \bar{G}$. Then $\bar{x}^*\bar{x}$ is a positive definite element in $\operatorname{Aut}\mathfrak{g}_0$. By Lemma 6.15 the positive definite r^{th} power, which we write as $(\bar{x}^*\bar{x})^r$, is in $\operatorname{Int}\mathfrak{g}_0 = \operatorname{Ad}(G) = \bar{G}$ for every real r. Hence

$$(\bar{x}^*\bar{x})^r = \exp r\bar{X} \tag{6.35}$$

for some $\bar{X} \in \operatorname{ad}\mathfrak{g}_0$. Differentiating with respect to r and putting $r = 0$, we see that $\bar{X}^* = \bar{X}$. By (6.32), $\bar{X}$ is in $\operatorname{ad}_{\mathfrak{g}_0}(\mathfrak{p}_0)$.

Specializing to the case $r = 1$, we see that $\bar{G}$ is closed under adjoint. Hence we may define $\bar{\Theta}(\bar{x}) = (\bar{x}^*)^{-1}$, and $\bar{\Theta}$ is an automorphism of $\bar{G}$ with $\bar{\Theta}^2 = 1$. The differential of $\bar{\Theta}$ is $\bar{Y} \mapsto -\bar{Y}^*$, and (6.33) shows that this is $\bar{\theta}$. This proves (a) for $\bar{G}$.

The fixed group for $\bar{\Theta}$ is a closed subgroup of $\bar{G}$ that we define to be $\bar{K}^\#$. The members $\bar{k}$ of $\bar{K}^\#$ have $(\bar{k}^*)^{-1} = \bar{k}$ and hence are in the orthogonal group on $\mathfrak{g}_0$. Since $\bar{G} = \operatorname{Int}\mathfrak{g}_0$ and since Propositions 1.97

and 1.98 show that $\text{Int}\,\mathfrak{g}_0 = (\text{Aut}\,\mathfrak{g}_0)_0$, $\bar{K}^{\#}$ is closed in $GL(\mathfrak{g}_0)$. Since $\bar{K}^{\#}$ is contained in the orthogonal group, $\bar{K}^{\#}$ is compact. The Lie algebra of $\bar{K}^{\#}$ is the subalgebra of all $\bar{T} \in \text{ad}\,\mathfrak{g}_0$ where $\bar{\theta}(\bar{T}) = \bar{T}$, and this is just $\text{ad}_{\mathfrak{g}_0}(\mathfrak{k}_0)$.

Consider the smooth mapping $\varphi_{\bar{G}} : \bar{K}^{\#} \times \text{ad}_{\mathfrak{g}_0}(\mathfrak{p}_0) \to \bar{G}$ given by $\varphi_{\bar{G}}(\bar{k}, \bar{S}) = \bar{k} \exp \bar{S}$. Let us prove that $\varphi_{\bar{G}}$ maps onto $\bar{G}$. Given $\bar{x} \in \bar{G}$, define $\bar{X} \in \text{ad}_{\mathfrak{g}_0}(\mathfrak{p}_0)$ by (6.35), and put $\bar{p} = \exp \frac{1}{2}\bar{X}$. The element $\bar{p}$ is in $\text{Ad}(G)$, and $\bar{p}^* = \bar{p}$. Put $\bar{k} = \bar{x}\bar{p}^{-1}$, so that $\bar{x} = \bar{k}\bar{p}$. Then $\bar{k}^*\bar{k} = (\bar{p}^{-1})^*\bar{x}^*\bar{x}\bar{p}^{-1} = (\exp -\frac{1}{2}\bar{X})(\exp \bar{X})(\exp -\frac{1}{2}\bar{X}) = 1$, and hence $\bar{k}^* = \bar{k}^{-1}$. Consequently $\bar{\Theta}(\bar{k}) = (\bar{k}^*)^{-1} = \bar{k}$, and we conclude that $\varphi_{\bar{G}}$ is onto.

Let us see that $\varphi_{\bar{G}}$ is one-one. If $\bar{x} = \bar{k} \exp \bar{X}$, then $\bar{x}^* = (\exp \bar{X}^*)\bar{k}^* = (\exp \bar{X})\bar{k}^* = (\exp \bar{X})\bar{k}^{-1}$. Hence $\bar{x}^*\bar{x} = \exp 2\bar{X}$. The two sides of this equation are equal positive definite linear transformations. Their positive definite r^{th} powers must be equal for all real r, necessarily to $\exp 2r\bar{X}$. Differentiating $(\bar{x}^*\bar{x})^r = \exp 2r\bar{X}$ with respect to r and putting $r = 0$, we see that $\bar{x}$ determines $\bar{X}$. Hence $\bar{x}$ determines also $\bar{k}$, and $\varphi_{\bar{G}}$ is one-one.

To complete the proof of (c) (but with K replaced by $\bar{K}^{\#}$), we are to show that the inverse map is smooth. It is enough to prove that the corresponding inverse map in the case of all n-by-n real nonsingular matrices is smooth, where $n = \dim \mathfrak{g}_0$. In fact, the given inverse map is a restriction of the inverse map for all matrices, and we recall from §I.10 that if M is an analytic subgroup of a Lie group M', then a smooth map into M' with image in M is smooth into M.

Thus we are to prove smoothness of the inverse for the case of matrices. The forward map is $O(n) \times \mathfrak{p}(n, \mathbb{R}) \to GL(n, \mathbb{R})$ with $(k, X) \mapsto ke^X$, where $\mathfrak{p}(n, \mathbb{R})$ denotes the vector space of real symmetric matrices. It is enough to prove local invertibility of this mapping near $(1, X_0)$. Thus we examine the differential at $k = 1$ and $X = X_0$ of $(k, X) \mapsto ke^Xe^{-X_0}$, identifying tangent spaces as follows: At $k = 1$, we use the linear Lie algebra of $O(n)$, which is the space $\mathfrak{so}(n)$ of skew-symmetric real matrices. Near $X = X_0$, write $X = X_0 + S$, and use $\{S\} = \mathfrak{p}(n, \mathbb{R})$ as tangent space. In $GL(n, \mathbb{R})$, we use the linear Lie algebra, which consists of all real matrices.

To compute the differential, we consider restrictions of the forward map with each coordinate fixed in turn. The differential of $(k, X_0) \mapsto k$ is $(T, 0) \mapsto T$ for $T \in \mathfrak{so}(n)$. The map $(1, X) \mapsto e^Xe^{-X_0}$ has derivative at $t = 0$ along the curve $X = X_0 + tS$ equal to

$$\frac{d}{dt} e^{X_0+tS}e^{-X_0}|_{t=0}.$$

Thus we ask whether it is possible to have

(6.36a)

$$\begin{aligned} 0 &\overset{?}{=} T + \frac{d}{dt} e^{X_0 + tS} e^{-X_0}|_{t=0} \\ &= T + \frac{d}{dt}(1 + (X_0 + tS) + \tfrac{1}{2!}(X_0 + tS)^2 + \cdots) e^{-X_0}|_{t=0} \\ &= T + \big(S + \tfrac{1}{2!}(SX_0 + X_0 S) + \cdots + \tfrac{1}{(n+1)!} \sum_{k=0}^{n} X_0^k S X_0^{n-k} + \cdots\big) e^{-X_0}. \end{aligned}$$

We left-bracket by X_0, noting that

$$\big[X_0, \sum_{k=0}^{n} X_0^k S X_0^{n-k}\big] = X_0^{n+1} S - S X_0^{n+1}.$$

Then we have

(6.36b)

$$\begin{aligned} 0 &\overset{?}{=} [X_0, T] + \big((X_0 S - S X_0) + \tfrac{1}{2!}(X_0^2 S - S X_0^2) \\ &\qquad + \cdots + \tfrac{1}{(n+1)!}(X_0^{n+1} S - S X_0^{n+1}) + \cdots\big) e^{-X_0} \\ &= [X_0, T] + (e^{X_0} S - S e^{X_0}) e^{-X_0} \\ &= [X_0, T] + (e^{X_0} S e^{-X_0} - S). \end{aligned}$$

Since $[\mathfrak{p}(n, \mathbb{R}), \mathfrak{so}(n)] \subseteq \mathfrak{p}(n, \mathbb{R})$, we conclude that $e^{X_0} S e^{-X_0} - S$ is symmetric. Let v be an eigenvector, and let λ be the eigenvalue for v. Let $\langle \cdot, \cdot \rangle$ denote ordinary dot product on $\mathbb{R}^n$. Since e^{X_0} and S are symmetric, $e^{X_0} S - S e^{X_0}$ is skew symmetric, and we have

$$\begin{aligned} 0 &= \langle (e^{X_0} S - S e^{X_0}) e^{-X_0} v,\ e^{-X_0} v \rangle \\ &= \langle (e^{X_0} S e^{-X_0} - S) v,\ e^{-X_0} v \rangle \\ &= \lambda \langle v, e^{-X_0} v \rangle. \end{aligned}$$

But e^{-X_0} is positive definite, and hence $\lambda = 0$. Thus

$$e^{X_0} S e^{-X_0} = S. \tag{6.37}$$

This equation forces

$$X_0 S = S X_0. \tag{6.38}$$

In fact, there is no loss of generality is assuming that X_0 is diagonal with diagonal entries d_i. Then (6.37) implies $e^{d_i} S_{ij} = S_{ij} e^{d_j}$. Considering the two cases $S_{ij} = 0$ and $S_{ij} \neq 0$ separately, we deduce that $d_i S_{ij} = S_{ij} d_j$, and (6.38) is the result. Because of (6.37), (6.36a) collapses to

$$0 \stackrel{?}{=} T + S,$$

and we conclude that $T = S = 0$. Thus the differential is everywhere an isomorphism, and the proof of local invertibility of the forward map is complete. This completes the proof of (c) for $\bar{G}$, but with K replaced by $\bar{K}^{\#}$.

The homeomorphism $\bar{K}^{\#} \times \mathrm{ad}_{\mathfrak{g}_0}(\mathfrak{p}_0) \xrightarrow{\sim} \bar{G}$ of (c) forces $\bar{K}^{\#}$ to be connected. Thus $\bar{K}^{\#}$ is the analytic subgroup of $\bar{G}$ with Lie algebra $\mathrm{ad}_{\mathfrak{g}_0}(\mathfrak{k}_0)$, which we denote $\bar{K}$. This proves (c) for $\bar{K}$ and also (b).

To complete the proof for the adjoint group $\bar{G}$, we need to verify (d) through (g) with $\bar{K}$ in place of K. Since $\bar{K}$ is compact, (d) is immediate. Proposition 6.30 shows that $\bar{G}$ has trivial center, and then (e) and (f) follow.

For (g) suppose on the contrary that $\bar{K} \subsetneqq \bar{K}_1$ with $\bar{K}_1$ compact. Let $\bar{x}$ be in $\bar{K}_1$ but not $\bar{K}$, and write $\bar{x} = \bar{k} \exp \bar{X}$ as in (c). Then $\exp \bar{X}$ is in $\bar{K}_1$ and is not 1. The powers of $\exp \bar{X}$ have unbounded eigenvalues, and this fact contradicts the compactness of $\bar{K}_1$. Thus (g) follows, and the proof of the theorem is complete for $\bar{G}$.

Now we shall prove the theorem for G. Write $e : G \to \bar{G}$ for the covering homomorphism $\mathrm{Ad}_{\mathfrak{g}_0}(\,\cdot\,)$. Let $\bar{K}$ be the analytic subgroup of $\bar{G}$ with Lie algebra $\mathrm{ad}\,\mathfrak{k}_0$, and let $K = e^{-1}(\bar{K})$. The subgroup K is closed in G since $\bar{K}$ is closed in $\bar{G}$.

From the covering homomorphism e, we obtain a smooth mapping $\psi : G/K \to \bar{G}/\bar{K}$ by defining $\psi(gK) = e(g)\bar{K}$. The definition of K makes ψ one-one, and e onto makes ψ onto. Let us see that ψ^{-1} is continuous. Let $\lim \bar{g}_n = \bar{g}$ in $\bar{G}$, and choose g_n and g in G with $e(g_n) = \bar{g}_n$ and $e(g) = \bar{g}$. Then $e(g^{-1}g_n) = \bar{g}^{-1}\bar{g}_n$ tends to 1. Fix an open neighborhood N of 1 in $\bar{G}$ that is evenly covered by e. Then we can write $g^{-1}g_n = v_n z_n$ with $v_n \in N$ and $z_n \in Z$, and we have $\lim v_n = 1$. Since $Z \subseteq K$ by definition of K, $g_n K = g v_n K$ tends to gK. Therefore ψ^{-1} is continuous.

Hence G/K is homeomorphic with $\bar{G}/\bar{K}$. Conclusion (c) for $\bar{G}$ shows that $\bar{G}/\bar{K}$ is simply connected. Hence G/K is simply connected, and it follows that K is connected. Thus K is the analytic subgroup of G with Lie algebra $\mathfrak{k}_0$. This proves (d) and (e) for G. Since $Z \subseteq K$, the map $e|_K : K \to \bar{K}$ has kernel Z, and hence K is compact if and only if Z is finite. This proves (f) for G.

Now let us prove (c) for G. Define $\varphi_G : K \times \mathfrak{p}_0 \to G$ by $\varphi_G(k, X) = k \exp_G X$. From (1.84) we have

$$e\varphi_G(k, X) = e(k)e(\exp_G X) = e(k)\exp_{\bar{G}}(\mathrm{ad}_{\mathfrak{g}_0}(X)) = \varphi_{\bar{G}}(e(k), \mathrm{ad}_{\mathfrak{g}_0}(X)),$$

and therefore the diagram

$$\begin{array}{ccc} K \times \mathfrak{p}_0 & \xrightarrow{\varphi_G} & G \\ {\scriptstyle e|_K \times \mathrm{ad}_{\mathfrak{g}_0}} \downarrow & & \downarrow {\scriptstyle e} \\ \bar{K} \times \mathrm{ad}_{\mathfrak{g}_0}(\mathfrak{p}_0) & \xrightarrow{\varphi_{\bar{G}}} & \bar{G} \end{array}$$

commutes. The maps on the sides are covering maps since K is connected, and $\varphi_{\bar{G}}$ is a diffeomorphism by (c) for $\bar{G}$. If we show that φ_G is one-one onto, then it follows that φ_G is a diffeomorphism, and (c) is proved for G.

First let us check that φ_G is one-one. Suppose $k \exp_G X = k' \exp_G X'$. Applying e, we have $e(k) \exp_{\bar{G}}(\mathrm{ad}_{\mathfrak{g}_0}(X)) = e(k') \exp_{\bar{G}}(\mathrm{ad}_{\mathfrak{g}_0}(X'))$. Then $X = X'$ from (c) for $\bar{G}$, and consequently $k = k'$.

Second let us check that φ_G is onto. Let $x \in G$ be given. Write $e(x) = \bar{k} \exp_{\bar{G}}(\mathrm{ad}_{\mathfrak{g}_0}(X))$ by (c) for $\bar{G}$, and let k be any member of $e^{-1}(\bar{k})$. Then $e(x) = e(k \exp_G X)$, and we see that $x = zk \exp_G X$ for some $z \in Z$. Since $Z \subseteq K$, $x = (zk) \exp_G X$ is the required decomposition. This completes the proof of (c) for G.

The next step is to construct Θ. Let $\tilde{G}$ be a simply connected covering group of G, let $\tilde{K}$ be the analytic subgroup of $\tilde{G}$ with Lie algebra $\mathfrak{k}_0$, let $\tilde{Z}$ be the center of $\tilde{G}$, and let $\tilde{e} : \tilde{G} \to G$ be the covering homomorphism. Since $\tilde{G}$ is simply connected, there exists a unique involution $\tilde{\Theta}$ of $\tilde{G}$ with differential θ. Since θ is 1 on $\mathfrak{k}_0$, $\tilde{\Theta}$ is 1 on $\tilde{K}$. By (c) for $\tilde{G}$, $\tilde{Z} \subseteq \tilde{K}$. Therefore $\ker \tilde{e} \subseteq \tilde{K}$, and $\tilde{\Theta}$ descends to an involution Θ of G with differential θ. This proves (a) for G.

Suppose that x is a member of G with $\Theta(x) = x$. Using (c), we can write $x = k \exp_G X$ and see that

$$k(\exp_G X)^{-1} = k \exp_G \theta X = k\Theta(\exp_G X) = \Theta(x) = x = k \exp_G X.$$

Then $\exp_G 2X = 1$, and it follows from (c) that $X = 0$. Thus x is in K, and (b) is proved for G.

Finally we are to prove (g) for G. Suppose that K is compact and that $K \subseteq K_1$ with K_1 compact. Applying e, we obtain a compact subgroup $e(K_1)$ of $\bar{G}$ that contains $\bar{K}$. By (g) for $\bar{G}$, $e(K_1) = e(K)$. Therefore $K_1 \subseteq ZK = K$, and we must have $K_1 = K$. This completes the proof of the theorem.

The Cartan decomposition on the Lie algebra level led in Proposition 6.28 to the conclusion that any real semisimple Lie algebra can be realized as a Lie algebra of real matrices closed under transpose. There is no corresponding proposition about realizing a semisimple Lie group as a group of real matrices. It is true that a semisimple Lie group of matrices is necessarily closed, and we shall prove this fact in Chapter VII. But the following example shows that a semisimple Lie group need not be realizable as a group of matrices.

EXAMPLE. By Proposition 1.122 the group $SL(2,\mathbb{R})$ has the same fundamental group as $SO(2)$, namely $\mathbb{Z}$, while $SL(2,\mathbb{C})$ has the same fundamental group as $SU(2)$, namely $\{1\}$. Then $SL(2,\mathbb{R})$ has a two-fold covering group G that is unique up to isomorphism. Let us see that G is not isomorphic to a group of n-by-n real matrices. If it were, then its linear Lie algebra $\mathfrak{g}_0$ would have the matrix Lie algebra $\mathfrak{g} = \mathfrak{g}_0 + i\mathfrak{g}_0$ as complexification. Let $G^{\mathbb{C}}$ be the analytic subgroup of $GL(n,\mathbb{C})$ with Lie algebra $\mathfrak{g}$. The diagram

$$\begin{array}{ccc} G & \longrightarrow & G^{\mathbb{C}} \\ \downarrow & & \uparrow \\ SL(2,\mathbb{R}) & \longrightarrow & SL(2,\mathbb{C}) \end{array} \tag{6.39}$$

has inclusions at the top and bottom, a two-fold covering map on the left, and a homomorphism on the right that exists since $SL(2,\mathbb{C})$ is simply connected and has Lie algebra isomorphic to $\mathfrak{g}$. The corresponding diagram of Lie algebras commutes, and hence so does the diagram (6.39) of Lie groups. However, the top map of (6.39) is one-one, while the composition of left, bottom, and right maps is not one-one. We have a contradiction, and we conclude that G is not isomorphic to a group of real matrices.

4. Iwasawa Decomposition

The Iwasawa decomposition is a second global decomposition of a semisimple Lie group. Unlike with the Cartan decomposition, the factors in the Iwasawa decomposition are closed subgroups. The prototype is the Gram-Schmidt orthogonalization process in linear algebra.

EXAMPLE. Let $G = SL(m,\mathbb{C})$. The group K from Proposition 1.122 or the global Cartan decomposition (Theorem 6.31) is $SU(m)$. Let A be the subgroup of G of diagonal matrices with positive diagonal entries, and let N be the upper-triangular group with 1 in each diagonal entry. The Iwasawa decomposition is $G = KAN$ in the sense that multiplication $K\times A\times N \to G$ is a diffeomorphism onto. To see that this decomposition of $SL(m, \mathbb{C})$ amounts to the Gram-Schmidt orthogonalization process, let $\{e_1, \ldots, e_m\}$ be the standard basis of $\mathbb{C}^m$, let $g \in G$ be given, and form the basis $\{ge_1, \ldots, ge_m\}$. The Gram-Schmidt process yields an orthonormal basis $v_1, \ldots, v_m$ such that

$$\text{span}\{ge_1, \ldots, ge_j\} = \text{span}\{v_1, \ldots, v_j\}$$

$$v_j \in \mathbb{R}^+(ge_j) + \text{span}\{v_1, \ldots, v_{j-1}\}$$

for $1 \leq j \leq m$. Define a matrix $k \in U(m)$ by $k^{-1}v_j = e_j$. Then $k^{-1}g$ is upper triangular with positive diagonal entries. Since g has determinant 1 and k has determinant of modulus 1, k must have determinant 1. Then k is in $K = SU(m)$, $k^{-1}g$ is in AN, and $g = k(k^{-1}g)$ exhibits g as in $K(AN)$. This proves that $K \times A \times N \to G$ is onto. It is one-one since $K \cap AN = \{1\}$, and the inverse is smooth because of the explicit formulas for the Gram-Schmidt process.

The decomposition in the example extends to all semisimple Lie groups. To prove such a theorem, we first obtain a Lie algebra decomposition, and then we lift the result to the Lie group.

Throughout this section, G will denote a semisimple Lie group. Changing notation from earlier sections of this chapter, we write $\mathfrak{g}$ for the Lie algebra of G. (We shall have relatively little use for the complexification of the Lie algebra in this section and write $\mathfrak{g}$ in place of $\mathfrak{g}_0$ to make the notation less cumbersome.) Let θ be a Cartan involution of $\mathfrak{g}$ (Corollary 6.18), let $\mathfrak{g} = \mathfrak{k} \oplus \mathfrak{p}$ be the corresponding Cartan decomposition (6.23), and let K be the analytic subgroup of G with Lie algebra $\mathfrak{k}_0$.

Insistence on using the Killing form as our nondegenerate symmetric invariant bilinear form on $\mathfrak{g}$ will turn out to be inconvenient later when we want to compare the form on $\mathfrak{g}$ with a corresponding form on a semisimple subalgebra of $\mathfrak{g}$. Thus we shall allow some flexibility in choosing a form B. For now it will be enough to let B be any nondegenerate symmetric invariant bilinear form on $\mathfrak{g}$ such that $B(\theta X, \theta Y) = B(X, Y)$ for all X and Y in $\mathfrak{g}$ and such that the form B_θ defined in terms of B by (6.13) is positive definite. Then it follows that B is negative definite on the compact real form $\mathfrak{k} \oplus i\mathfrak{p}$. Therefore B is negative definite on a maximal abelian subspace of $\mathfrak{k} \oplus i\mathfrak{p}$, and we conclude as in the remarks with Corollary 2.38 that, for any Cartan subalgebra of $\mathfrak{g}^{\mathbb{C}}$, B is positive definite on the real subspace where all the roots are real-valued.

The Killing form is one possible choice for B, but there are others. In any event, B_θ is an inner product on $\mathfrak{g}$, and we use it to define orthogonality and adjoints.

Let $\mathfrak{a}$ be a maximal abelian subspace of $\mathfrak{p}$. This exists by finite-dimensionality. Since $(\operatorname{ad} X)^* = -\operatorname{ad} \theta X$ by Lemma 6.27, the set

$\{\text{ad } H \mid H \in \mathfrak{a}\}$ is a commuting family of self-adjoint transformations of $\mathfrak{g}$. Then $\mathfrak{g}$ is the orthogonal direct sum of simultaneous eigenspaces, all the eigenvalues being real. If we fix such an eigenspace and if λ_H is the eigenvalue of ad H, then the equation $(\text{ad } H)X = \lambda_H X$ shows that λ_H is linear in H. Hence the simultaneous eigenvalues are members of the dual space $\mathfrak{a}^*$. For $\lambda \in \mathfrak{a}^*$, we write

$$\mathfrak{g}_\lambda = \{X \in \mathfrak{g} \mid (\text{ad } H)X = \lambda(H)X \text{ for all } H \in \mathfrak{a}\}.$$

If $\mathfrak{g}_\lambda \neq 0$ and $\lambda \neq 0$, we call λ a **restricted root** of $\mathfrak{g}$ or a **root** of $(\mathfrak{g}, \mathfrak{a})$. The set of restricted roots is denoted Σ. Any nonzero $\mathfrak{g}_\lambda$ is called a **restricted-root space**, and each member of $\mathfrak{g}_\lambda$ is called a **restricted-root vector** for the restricted root λ.

Proposition 6.40. The restricted roots and the restricted-root spaces have the following properties:

(a) $\mathfrak{g}$ is the orthogonal direct sum $\mathfrak{g} = \mathfrak{g}_0 \oplus \bigoplus_{\lambda \in \Sigma} \mathfrak{g}_\lambda$
(b) $[\mathfrak{g}_\lambda, \mathfrak{g}_\mu] \subseteq \mathfrak{g}_{\lambda+\mu}$
(c) $\theta\mathfrak{g}_\lambda = \mathfrak{g}_{-\lambda}$, and hence $\lambda \in \Sigma$ implies $-\lambda \in \Sigma$
(d) $\mathfrak{g}_0 = \mathfrak{a} \oplus \mathfrak{m}$ orthogonally, where $\mathfrak{m} = Z_{\mathfrak{k}}(\mathfrak{a})$.

REMARK. The decomposition in (a) is called the **restricted-root space decomposition** of $\mathfrak{g}$.

PROOF. We saw (a) in the course of the construction of restricted-root spaces, and (b) follows from the Jacobi identity. For (c) let X be in $\mathfrak{g}_\lambda$; then $[H, \theta X] = \theta[\theta H, X] = -\theta[H, X] = -\lambda(H)\theta X$.

In (d) we have $\theta\mathfrak{g}_0 = \mathfrak{g}_0$ by (c). Hence $\mathfrak{g}_0 = (\mathfrak{k} \cap \mathfrak{g}_0) \oplus (\mathfrak{p} \cap \mathfrak{g}_0)$. Since $\mathfrak{a} \subseteq \mathfrak{p} \cap \mathfrak{g}_0$ and $\mathfrak{a}$ is maximal abelian in $\mathfrak{p}$, $\mathfrak{a} = \mathfrak{p} \cap \mathfrak{g}_0$. Also $\mathfrak{k} \cap \mathfrak{g}_0 = Z_{\mathfrak{k}}(\mathfrak{a})$. This proves (d).

EXAMPLES.

1) Let $G = SL(n, \mathbb{K})$, where $\mathbb{K}$ is $\mathbb{R}$, $\mathbb{C}$, or $\mathbb{H}$. The Lie algebra is $\mathfrak{g} = \mathfrak{sl}(n, \mathbb{K})$ in the sense of §I.8. For a Cartan decomposition we can take $\mathfrak{k}$ to consist of the skew-Hermitian members of $\mathfrak{g}$ and $\mathfrak{p}$ to consist of the Hermitian members. The space of real diagonal matrices of trace 0 is a maximal abelian subspace of $\mathfrak{p}$, and we use it as $\mathfrak{a}$. Note that $\dim \mathfrak{a} = n - 1$. The restricted-root space decomposition of $\mathfrak{g}$ is rather similar to Example 1 in §II.1. Let f_i be evaluation of the ith diagonal entry of members of $\mathfrak{a}$. Then the restricted roots are all linear functionals $f_i - f_j$ with $i \neq j$, and $\mathfrak{g}_{f_i - f_j}$ consists of all matrices with all entries other than the $(i, j)^{\text{th}}$ equal to 0. The dimension of each restricted-root space is 1, 2, or 4 when $\mathbb{K}$ is $\mathbb{R}$, $\mathbb{C}$, or $\mathbb{H}$. The subalgebra $\mathfrak{m}$ of Proposition 6.40d

consists of all skew-Hermitian diagonal matrices in $\mathfrak{g}$. For $\mathbb{K} = \mathbb{R}$ this is 0, and for $\mathbb{K} = \mathbb{C}$ it is all purely imaginary matrices of trace 0 and has dimension $n-1$. For $\mathbb{K} = \mathbb{H}$, $\mathfrak{m}$ consists of all diagonal matrices whose diagonal entries x_j have $\bar{x}_j = -x_j$ and is isomorphic to the direct sum of n copies of $\mathfrak{su}(2)$; its dimension is $3n$.

2) Let $G = SU(p,q)$ with $p \geq q$. We can write the Lie algebra in block form as

$$\mathfrak{g} = \begin{matrix} & \begin{matrix} p & q \end{matrix} & \\ & \begin{pmatrix} a & b \\ b^* & d \end{pmatrix} & \begin{matrix} p \\ q \end{matrix} \end{matrix} \tag{6.41}$$

with all entries complex, with a and d skew Hermitian, and with $\operatorname{Tr} a + \operatorname{Tr} d = 0$. We take $\mathfrak{k}$ to be all matrices in $\mathfrak{g}$ with $b = 0$, and we take $\mathfrak{p}$ to be all matrices in $\mathfrak{g}$ with $a = 0$ and $d = 0$. One way of forming a maximal abelian subspace $\mathfrak{a}$ of $\mathfrak{p}$ is to allow b to have nonzero real entries only in the lower-left entry and the entries extending diagonally up from that one:

$$\mathfrak{a} = \left\{ \begin{pmatrix} \vdots & \cdots & \vdots \\ 0 & \cdots & 0 \\ 0 & \cdots & a_q \\ & \cdot^{\cdot^{\cdot}} & \\ a_1 & \cdots & 0 \end{pmatrix} \right\}, \tag{6.42}$$

with $p - q$ rows of 0's at the top. Let f_i be the member of $\mathfrak{a}^*$ whose value on the matrix in (6.42) is a_i. Then the restricted roots include all linear functionals $\pm f_i \pm f_j$ with $i \neq j$ and $\pm 2f_i$ for all i. Also the $\pm f_i$ are restricted roots if $p \neq q$. The restricted-root spaces are described as follows: Let $i < j$, and let $J(z)$, $I_{2,0}$, and $I_{1,1}$ be the 2-by-2 matrices

$$J(z) = \begin{pmatrix} 0 & z \\ -\bar{z} & 0 \end{pmatrix}, \qquad I_{2,0} = \begin{pmatrix} 1 & 0 \\ 0 & 1 \end{pmatrix}, \qquad I_{1,1} = \begin{pmatrix} 1 & 0 \\ 0 & -1 \end{pmatrix}.$$

Here z is any complex number. The restricted-root spaces for $\pm f_i \pm f_j$ are 2-dimensional and are nonzero only in the 16 entries corresponding to row and column indices $p-j+1, p-i+1, p+i, p+j$, where they are

$$\mathfrak{g}_{f_i - f_j} = \left\{ \begin{pmatrix} J(z) & -zI_{2,0} \\ -zI_{2,0} & -J(z) \end{pmatrix} \right\}, \qquad \mathfrak{g}_{-f_i + f_j} = \left\{ \begin{pmatrix} J(z) & zI_{2,0} \\ zI_{2,0} & -J(z) \end{pmatrix} \right\},$$

$$\mathfrak{g}_{f_i + f_j} = \left\{ \begin{pmatrix} J(z) & -zI_{1,1} \\ -zI_{1,1} & J(z) \end{pmatrix} \right\}, \qquad \mathfrak{g}_{-f_i - f_j} = \left\{ \begin{pmatrix} J(z) & zI_{1,1} \\ zI_{1,1} & J(z) \end{pmatrix} \right\}.$$

The restricted-root spaces for $\pm 2f_i$ have dimension 1 and are nonzero only in the 4 entries corresponding to row and column indices $p-i+1$ and $p+i$, where they are

$$\mathfrak{g}_{2f_i} = i\mathbb{R}\begin{pmatrix} -1 & 1 \\ 1 & 1 \end{pmatrix} \quad \text{and} \quad \mathfrak{g}_{2f_i} = i\mathbb{R}\begin{pmatrix} 1 & 1 \\ 1 & -1 \end{pmatrix}.$$

The restricted-root spaces for $\pm f_i$ have dimension $2(p-q)$ and are nonzero only in the entries corresponding to row and column indices 1 to $p-q$, $p-i+1$, and $p+i$, where they are

$$\mathfrak{g}_{f_i} = \left\{\begin{pmatrix} 0 & v & -v \\ -v^* & 0 & 0 \\ -v^* & 0 & 0 \end{pmatrix}\right\} \quad \text{and} \quad \mathfrak{g}_{-f_i} = \left\{\begin{pmatrix} 0 & v & v \\ -v^* & 0 & 0 \\ v^* & 0 & 0 \end{pmatrix}\right\}.$$

Here v is any member of $\mathbb{C}^{p-q}$. The subalgebra $\mathfrak{m}$ of Proposition 6.40d consists of all skew-Hermitian matrices of trace 0 that are arbitrary in the upper left block of size $p-q$, are otherwise diagonal, and have the $(p-i+1)^{\text{st}}$ diagonal entry equal to the $(p+i)^{\text{th}}$ diagonal entry for $1 \le i \le q$; thus $\mathfrak{m} \cong \mathfrak{su}(p-q) \oplus \mathbb{R}^q$. In the next section we shall see that Σ is an abstract root system; this example shows that this root system need not be reduced.

3) Let $G = SO(p,q)_0$ with $p \ge q$. We can write the Lie algebra in block form as in (6.41) but with all entries real and with a and d skew symmetric. As in Example 2, we take $\mathfrak{k}$ to be all matrices in $\mathfrak{g}$ with $b = 0$, and we take $\mathfrak{p}$ to be all matrices in $\mathfrak{g}$ with $a = 0$ and $d = 0$. We again choose $\mathfrak{a}$ as in (6.42). Let f_i be the member whose value on the matrix in (6.42) is a_i. Then the restricted roots include all linear functionals $\pm f_i \pm f_j$ with $i \ne j$. Also the $\pm f_i$ are restricted roots if $p \ne q$. The restricted-root spaces are the intersections with $\mathfrak{so}(p,q)$ of the restricted-root spaces in Example 2. Then the restricted-root spaces for $\pm f_i \pm f_j$ are 1-dimensional, and the restricted-root spaces for $\pm f_i$ have dimension $p-q$. The linear functionals $\pm 2f_i$ are no longer restricted roots. The subalgebra $\mathfrak{m}$ of Proposition 6.40d consists of all skew-symmetric matrices that are nonzero only in the upper left block of size $p-q$; thus $\mathfrak{m} \cong \mathfrak{so}(p-q)$.

Choose a notion of positivity for $\mathfrak{a}^*$ in the manner of §II.5, as for example by using a lexicographic ordering. Let Σ^+ be the set of positive roots, and define $\mathfrak{n} = \bigoplus_{\lambda \in \Sigma^+} \mathfrak{g}_\lambda$. By Proposition 6.40b, $\mathfrak{n}$ is a Lie subalgebra of $\mathfrak{g}$ and is nilpotent.

Proposition 6.43 (Iwasawa decomposition of Lie algebra). With notation as above, $\mathfrak{g}$ is a vector-space direct sum $\mathfrak{g} = \mathfrak{k} \oplus \mathfrak{a} \oplus \mathfrak{n}$. Here $\mathfrak{a}$ is abelian, $\mathfrak{n}$ is nilpotent, $\mathfrak{a} \oplus \mathfrak{n}$ is a solvable Lie subalgebra of $\mathfrak{g}$, and $[\mathfrak{a} \oplus \mathfrak{n}, \mathfrak{a} \oplus \mathfrak{n}] = \mathfrak{n}$.

PROOF. We know that $\mathfrak{a}$ is abelian and that $\mathfrak{n}$ is nilpotent. Since $[\mathfrak{a}, \mathfrak{g}_\lambda] = \mathfrak{g}_\lambda$ for each $\lambda \neq 0$, we see that $[\mathfrak{a}, \mathfrak{n}] = \mathfrak{n}$ and that $\mathfrak{a} \oplus \mathfrak{n}$ is a solvable subalgebra with $[\mathfrak{a} \oplus \mathfrak{n}, \mathfrak{a} \oplus \mathfrak{n}] = \mathfrak{n}$.

To prove that $\mathfrak{k} + \mathfrak{a} + \mathfrak{n}$ is a direct sum, let X be in $\mathfrak{k} \cap (\mathfrak{a} \oplus \mathfrak{n})$. Then $\theta X = X$ with $\theta X \in \mathfrak{a} \oplus \theta\mathfrak{n}$. Since $\mathfrak{a} \oplus \mathfrak{n} \oplus \theta\mathfrak{n}$ is a direct sum (by (a) and (c) in Proposition 6.40), X is in $\mathfrak{a}$. But then X is in $\mathfrak{k} \cap \mathfrak{p} = 0$.

The sum $\mathfrak{k} \oplus \mathfrak{a} \oplus \mathfrak{n}$ is all of $\mathfrak{g}$ because we can write any $X \in \mathfrak{g}$, using some $H \in \mathfrak{a}$, some $X_0 \in \mathfrak{m}$, and elements $X_\lambda \in \mathfrak{g}_\lambda$, as

$$\begin{aligned} X &= H + X_0 + \sum_{\lambda \in \Sigma} X_\lambda \\ &= \big(X_0 + \sum_{\lambda \in \Sigma^+} (X_{-\lambda} + \theta X_{-\lambda})\big) + H + \big(\sum_{\lambda \in \Sigma^+} (X_\lambda - \theta X_{-\lambda})\big), \end{aligned}$$

and the right side is in $\mathfrak{k} \oplus \mathfrak{a} \oplus \mathfrak{n}$.

To prepare to prove a group decomposition, we prove two lemmas.

Lemma 6.44. Let H be an analytic group with Lie algebra $\mathfrak{h}$, and suppose that $\mathfrak{h}$ is a vector-space direct sum of Lie subalgebras $\mathfrak{h} = \mathfrak{s} \oplus \mathfrak{t}$. If S and T denote the analytic subgroups of H corresponding to $\mathfrak{s}$ and $\mathfrak{t}$, then the multiplication map $\Phi(s, t) = st$ of $S \times T$ into H is everywhere regular.

PROOF. The tangent space at (s_0, t_0) in $S \times T$ can be identified by left translation within S and within T with $\mathfrak{s} \oplus \mathfrak{t} = \mathfrak{h}$, and the tangent space at $s_0 t_0$ in H can be identified by left translation within H with $\mathfrak{h}$. With these identifications we compute the differential $d\Phi$ at (s_0, t_0). Let X be in $\mathfrak{s}$ and Y be in $\mathfrak{t}$. Then

$$\Phi(s_0 \exp rX, t_0) = s_0 \exp(rX) t_0 = s_0 t_0 \exp(\operatorname{Ad}(t_0^{-1}) rX)$$

and

$$\Phi(s_0, t_0 \exp rY) = s_0 t_0 \exp rY,$$

from which it follows that

$$d\Phi(X) = \operatorname{Ad}(t_0^{-1}) X$$

and

$$d\Phi(Y) = Y.$$

In matrix form, $d\Phi$ is therefore block triangular, and hence

$$\det d\Phi = \frac{\det \operatorname{Ad}_{\mathfrak{h}}(t_0^{-1})}{\det \operatorname{Ad}_{\mathfrak{t}}(t_0^{-1})} = \frac{\det \operatorname{Ad}_{\mathfrak{t}}(t_0)}{\det \operatorname{Ad}_{\mathfrak{h}}(t_0)}.$$

This is nonzero, and hence Φ is regular.

Lemma 6.45. There exists a basis of $\{X_i\}$ of $\mathfrak{g}$ such that the matrices representing $\operatorname{ad}\mathfrak{g}$ have the following properties:

(a) the matrices of $\operatorname{ad}\mathfrak{k}$ are skew symmetric
(b) the matrices of $\operatorname{ad}\mathfrak{a}$ are diagonal with real entries
(c) the matrices of $\operatorname{ad}\mathfrak{n}$ are upper triangular with 0's on the diagonal.

PROOF. Let $\{X_i\}$ be an orthonormal basis of $\mathfrak{g}$ compatible with the orthogonal decomposition of $\mathfrak{g}$ in Proposition 6.40a and having the property that $X_i \in \mathfrak{g}_{\lambda_i}$ and $X_j \in \mathfrak{g}_{\lambda_j}$ with $i < j$ implies $\lambda_i \geq \lambda_j$. For $X \in \mathfrak{k}$, we have $(\operatorname{ad} X)^* = -\operatorname{ad}\theta X = -\operatorname{ad} X$ from Lemma 6.27, and this proves (a). Since each X_i is a restricted-root vector or is in $\mathfrak{g}_0$, the matrices of $\operatorname{ad}\mathfrak{a}$ are diagonal, necessarily with real entries. This proves (b). Conclusion (c) follows from Proposition 6.40b.

Theorem 6.46 (Iwasawa decomposition). Let G be a semisimple Lie group, let $\mathfrak{g} = \mathfrak{k} \oplus \mathfrak{a} \oplus \mathfrak{n}$ be an Iwasawa decomposition of the Lie algebra $\mathfrak{g}$ of G, and let A and N be the analytic subgroups of G with Lie algebras $\mathfrak{a}$ and $\mathfrak{n}$. Then the multiplication map $K \times A \times N \to G$ given by $(k, a, n) \mapsto kan$ is a diffeomorphism onto. The groups A and N are simply connected.

PROOF. Let $\bar{G} = \operatorname{Ad}(G)$, regarded as the closed subgroup $(\operatorname{Aut}\mathfrak{g})_0$ of $GL(\mathfrak{g})$ (Propositions 1.97 and 1.98). We shall prove the theorem for $\bar{G}$ and then lift the result to G.

We impose the inner product B_θ on $\mathfrak{g}$ and write matrices for elements of $\bar{G}$ and $\operatorname{ad}\mathfrak{g}$ relative to the basis in Lemma 6.45. Let $\bar{K} = \operatorname{Ad}_{\mathfrak{g}}(K)$, $\bar{A} = \operatorname{Ad}_{\mathfrak{g}}(A)$, and $\bar{N} = \operatorname{Ad}_{\mathfrak{g}}(N)$. Lemma 6.45 shows that the matrices of $\bar{K}$ are rotation matrices, those for $\bar{A}$ are diagonal with positive entries on the diagonal, and those for $\bar{N}$ are upper triangular with 1's on the diagonal. We know that $\bar{K}$ is compact (Proposition 6.30 and Theorem 6.31f). The diagonal subgroup of $GL(\mathfrak{g})$ with positive diagonal entries is simply connected abelian, and $\bar{A}$ is an analytic subgroup of it. By Corollary 1.111, $\bar{A}$ is closed in $GL(\mathfrak{g})$ and hence closed in $\bar{G}$. Similarly the upper-triangular subgroup of $GL(\mathfrak{g})$ with 1's on the diagonal is simply connected nilpotent, and $\bar{N}$ is an analytic subgroup of it. By Corollary 1.111, $\bar{N}$ is closed in $GL(\mathfrak{g})$ and hence closed in $\bar{G}$.

The map $\bar{A} \times \bar{N}$ into $GL(\mathfrak{g})$ given by $(\bar{a}, \bar{n}) \mapsto \bar{a}\bar{n}$ is one-one since we can recover $\bar{a}$ from the diagonal entries, and it is onto a subgroup $\bar{A}\bar{N}$ since $\bar{a}_1\bar{n}_1\bar{a}_2\bar{n}_2 = \bar{a}_1\bar{a}_2(\bar{a}_2^{-1}\bar{n}_1\bar{a}_2)\bar{n}_2$ and $(\bar{a}\bar{n})^{-1} = \bar{n}^{-1}\bar{a}^{-1} = \bar{a}^{-1}(\bar{a}\bar{n}\bar{a}^{-1})$. This subgroup is closed. In fact, if $\lim \bar{a}_m\bar{n}_m = x$, let $\bar{a}$ be the diagonal matrix with the same diagonal entries as x. Then $\lim \bar{a}_m = \bar{a}$, and $\bar{a}$ must be in $\bar{A}$ since $\bar{A}$ is closed in $GL(\mathfrak{g})$. Also $\bar{n}_m = \bar{a}_m^{-1}(\bar{a}_m\bar{n}_m)$ has limit $\bar{a}^{-1}x$, which has to be in $\bar{N}$ since $\bar{N}$ is closed in $\bar{G}$. Thus $\lim \bar{a}_m\bar{n}_m$ is in $\bar{A}\bar{N}$, and $\bar{A}\bar{N}$ is closed.

Clearly the closed subgroup $\bar{A}\bar{N}$ has Lie algebra $\mathfrak{a} \oplus \mathfrak{n}$. By Lemma 6.44, $\bar{A} \times \bar{N} \to \bar{A}\bar{N}$ is a diffeomorphism.

The subgroup $\bar{K}$ is compact, and thus the image of $\bar{K} \times \bar{A} \times \bar{N} \to \bar{K} \times \bar{A}\bar{N} \to \bar{G}$ is the product of a compact set and a closed set and is closed. Also the image is open since the map is everywhere regular (Lemma 6.45) and since the equality $\mathfrak{g} = \mathfrak{k} \oplus \mathfrak{a} \oplus \mathfrak{n}$ shows that the dimensions add properly. Since the image of $\bar{K} \times \bar{A} \times \bar{N}$ is open and closed and since $\bar{G}$ is connected, the image is all of $\bar{G}$.

Thus the multiplication map is smooth, regular, and onto. Finally $\bar{K} \cap \bar{A}\bar{N} = \{1\}$ since a rotation matrix with positive eigenvalues is 1. Since $\bar{A} \times \bar{N} \to \bar{A}\bar{N}$ is one-one, it follows that $\bar{K} \times \bar{A} \times \bar{N} \to \bar{G}$ is one-one. This completes the proof for the adjoint group $\bar{G}$.

We now lift the above result to G. Let $e : G \to \bar{G} = \mathrm{Ad}(G)$ be the covering homomorphism. Using a locally defined inverse of e, we can write the map $(k, a, n) \mapsto kan$ locally as

$$(k, a, n) \mapsto (e(k), e(a), e(n)) \mapsto e(k)e(a)e(n) = e(kan) \mapsto kan,$$

and therefore the multiplication map is smooth and everywhere regular. Since A and N are connected, $e|_A$ and $e|_N$ are covering maps to $\bar{A}$ and $\bar{N}$, respectively. Since $\bar{A}$ and $\bar{N}$ are simply connected, it follows that e is one-one on A and on N and that A and N are simply connected.

Let us prove that the multiplication map is onto G. If $g \in G$ is given, write $e(g) = \bar{k}\bar{a}\bar{n}$. Put $a = (e|_A)^{-1}(\bar{a}) \in A$ and $n = (e|_N)^{-1}(\bar{N}) \in N$. Let k be in $e^{-1}(\bar{k})$. Then $e(kan) = \bar{k}\bar{a}\bar{n}$, so that $e(g(kan)^{-1}) = 1$. Thus $g(kan)^{-1} = z$ is in the center of G. By Theorem 6.31e, z is in K. Therefore $g = (zk)an$ exhibits g as in the image of the multiplication map.

Finally we show that the multiplication map is one-one. Since $\bar{A} \times \bar{N} \to \bar{A}\bar{N}$ is one-one, so is $A \times N \to AN$. The set of products AN is a group, just as in the adjoint case, and therefore it is enough to prove that $K \cap AN = \{1\}$. If x is in $K \cap AN$, then $e(x)$ is in $\bar{K} \cap \bar{A}\bar{N} = \{1\}$. Hence $e(x) = 1$. Write $x = an \in AN$. Then $1 = e(x) = e(an) = e(a)e(n)$, and the result for the adjoint case implies that $e(a) = e(n) = 1$. Since e is one-one on A and on N, $a = n = 1$. Thus $x = 1$. This completes the proof.

Recall from §IV.5 that a subalgebra $\mathfrak{h}$ of $\mathfrak{g}$ is called a **Cartan subalgebra** if $\mathfrak{h}^{\mathbb{C}}$ is a Cartan subalgebra of $\mathfrak{g}^{\mathbb{C}}$. The **rank** of $\mathfrak{g}$ is the dimension of any Cartan subalgebra; this is well defined since Proposition 2.15 shows that any two Cartan subalgebras of $\mathfrak{g}^{\mathbb{C}}$ are conjugate via $\mathrm{Int}\,\mathfrak{g}^{\mathbb{C}}$.

Proposition 6.47. If $\mathfrak{t}$ is a maximal abelian subspace of $\mathfrak{m} = Z_{\mathfrak{k}}(\mathfrak{a})$, then $\mathfrak{h} = \mathfrak{a} \oplus \mathfrak{t}$ is a Cartan subalgebra of $\mathfrak{g}$.

PROOF. By Corollary 2.13 it is enough to show that $\mathfrak{h}^{\mathbb{C}}$ is maximal abelian in $\mathfrak{g}^{\mathbb{C}}$ and that $\operatorname{ad}_{\mathfrak{g}^{\mathbb{C}}} \mathfrak{h}^{\mathbb{C}}$ is simultaneously diagonable.

Certainly $\mathfrak{h}^{\mathbb{C}}$ is abelian. Let us see that it is maximal abelian. If $Z = X + iY$ commutes with $\mathfrak{h}^{\mathbb{C}}$, then so do X and Y. Thus there is no loss in generality in considering only X. The element X commutes with $\mathfrak{h}^{\mathbb{C}}$, hence commutes with $\mathfrak{a}$, and hence is in $\mathfrak{a} \oplus \mathfrak{m}$. The same thing is true of θX. Then $X + \theta X$, being in $\mathfrak{k}$, is in $\mathfrak{m}$ and commutes with $\mathfrak{t}$, hence is in $\mathfrak{t}$, while $X - \theta X$ is in $\mathfrak{a}$. Thus X is in $\mathfrak{a} \oplus \mathfrak{t}$, and we conclude that $\mathfrak{h}^{\mathbb{C}}$ is maximal abelian.

In the basis of Lemma 6.45, the matrices representing $\operatorname{ad} \mathfrak{t}$ are skew symmetric and hence are diagonable over $\mathbb{C}$, while the matrices representing $\operatorname{ad} \mathfrak{a}$ are already diagonal. Since all the matrices in question form a commuting family, the members of $\operatorname{ad} \mathfrak{h}^{\mathbb{C}}$ are diagonable.

With notation as in Proposition 6.47, $\mathfrak{h} = \mathfrak{a} \oplus \mathfrak{t}$ is a Cartan subalgebra of $\mathfrak{g}$, and it is meaningful to speak of the set $\Delta = \Delta(\mathfrak{g}^{\mathbb{C}}, \mathfrak{h}^{\mathbb{C}})$ of roots of $\mathfrak{g}^{\mathbb{C}}$ with respect to $\mathfrak{h}^{\mathbb{C}}$. We can write the corresponding root-space decomposition as

$$\mathfrak{g}^{\mathbb{C}} = \mathfrak{h}^{\mathbb{C}} \oplus \bigoplus_{\alpha \in \Delta} (\mathfrak{g}^{\mathbb{C}})_{\alpha}. \tag{6.48a}$$

Then it is clear that

$$\mathfrak{g}_{\lambda} = \mathfrak{g} \cap \bigoplus_{\substack{\alpha \in \Delta, \\ \alpha|_{\mathfrak{a}} = \lambda}} (\mathfrak{g}^{\mathbb{C}})_{\alpha} \tag{6.48b}$$

and

$$\mathfrak{m}^{\mathbb{C}} = \mathfrak{t}^{\mathbb{C}} \oplus \bigoplus_{\substack{\alpha \in \Delta, \\ \alpha|_{\mathfrak{a}} = 0}} (\mathfrak{g}^{\mathbb{C}})_{\alpha}. \tag{6.48c}$$

That is, the restricted roots are the nonzero restrictions to $\mathfrak{a}$ of the roots, and $\mathfrak{m}$ arises from the roots that restrict to 0 on $\mathfrak{a}$.

Corollary 6.49. If $\mathfrak{t}$ is a maximal abelian subspace of $\mathfrak{m} = Z_{\mathfrak{k}}(\mathfrak{a})$, then the Cartan subalgebra $\mathfrak{h} = \mathfrak{a} \oplus \mathfrak{t}$ of $\mathfrak{g}$ has the property that all of the roots are real on $\mathfrak{a} \oplus i\mathfrak{t}$. If $\mathfrak{m} = 0$, then $\mathfrak{g}$ is a split real form of $\mathfrak{g}^{\mathbb{C}}$.

PROOF. In view of (6.48) the values of the roots on a member H of $\operatorname{ad} \mathfrak{h}$ are the eigenvalues of $\operatorname{ad} H$. For $H \in \mathfrak{a}$, these are real since $\operatorname{ad} H$ is self adjoint. For $H \in \mathfrak{t}$, they are purely imaginary since $\operatorname{ad} H$ is skew adjoint. The first assertion follows.

If $\mathfrak{m} = 0$, then $\mathfrak{t} = 0$. So the roots are real on $\mathfrak{h} = \mathfrak{a}$. Thus $\mathfrak{g}$ contains the real subspace of a Cartan subalgebra $\mathfrak{h}^{\mathbb{C}}$ of $\mathfrak{g}^{\mathbb{C}}$ where all the roots are real, and $\mathfrak{g}$ is a split real form of $\mathfrak{g}^{\mathbb{C}}$.

EXAMPLE. Corollary 6.49 shows that the Lie algebras $\mathfrak{so}(n+1,n)$ and $\mathfrak{so}(n,n)$ are split real forms of their complexifications, since Example 3 earlier in this section showed that $\mathfrak{m}=0$ in each case. For any p and q, the complexification of $\mathfrak{so}(p,q)$ is conjugate to $\mathfrak{so}(p+q,\mathbb{C})$ by a diagonal matrix whose diagonal consists of p entries i and then q entries 1. Consequently $\mathfrak{so}(n+1,n)$ is isomorphic to a split real form of $\mathfrak{so}(2n+1,\mathbb{C})$, and $\mathfrak{so}(n,n)$ is isomorphic to a split real form of $\mathfrak{so}(2n,\mathbb{C})$.

With Δ as above, we can impose a positive system on Δ so that Δ^+ extends Σ^+. Namely we just take $\mathfrak{a}$ before $i\mathfrak{t}$ in forming a lexicographic ordering of $(\mathfrak{a}+i\mathfrak{t})^*$. If $\alpha\in\Delta$ is nonzero on $\mathfrak{a}$, then the positivity of α depends only on the $\mathfrak{a}$ part, and thus positivity for Σ has been extended to Δ.

5. Uniqueness Properties of the Iwasawa Decomposition

We continue with G as a semisimple Lie group, with $\mathfrak{g}$ as the Lie algebra of G, and with other notation as in §4. In this section we shall show that an Iwasawa decomposition of $\mathfrak{g}$ is unique up to conjugacy by $\operatorname{Int}\mathfrak{g}$; therefore an Iwasawa decomposition of G is unique up to inner automorphism.

We already know from Corollary 6.19 that any two Cartan decompositions are conjugate via $\operatorname{Int}\mathfrak{g}$. Hence $\mathfrak{k}$ is unique up to conjugacy. Next we show that with $\mathfrak{k}$ fixed, $\mathfrak{a}$ is unique up to conjugacy. Finally with $\mathfrak{k}$ and $\mathfrak{a}$ fixed, we show that the various possibilities for $\mathfrak{n}$ are conjugate.

Lemma 6.50. If $H\in\mathfrak{a}$ has $\lambda(H)\neq0$ for all $\lambda\in\Sigma$, then $Z_{\mathfrak{g}}(H)=\mathfrak{m}\oplus\mathfrak{a}$. Hence $Z_{\mathfrak{p}}(H)=\mathfrak{a}$.

PROOF. Let X be in $Z_{\mathfrak{g}}(H)$, and use Proposition 6.40 to write $X=H_0+X_0+\sum_{\lambda\in\Sigma}X_\lambda$ with $H_0\in\mathfrak{a}$, $X_0\in\mathfrak{m}$, and $X_\lambda\in\mathfrak{g}_\lambda$. Then $0=[H,X]=\sum\lambda(H)X_\lambda$, and hence $\lambda(H)X_\lambda=0$ for all λ. Since $\lambda(H)\neq0$ by assumption, $X_\lambda=0$.

Theorem 6.51. If $\mathfrak{a}$ and $\mathfrak{a}'$ are two maximal abelian subspaces of $\mathfrak{p}$, then there is a member k of K with $\operatorname{Ad}(k)\mathfrak{a}'=\mathfrak{a}$. Consequently the space $\mathfrak{p}$ satisfies $\mathfrak{p}=\bigcup_{k\in K}\operatorname{Ad}(k)\mathfrak{a}$.

REMARKS.

1) In the case of $SL(m,\mathbb{C})$, this result amounts to the Spectral Theorem for Hermitian matrices.

2) The proof should be compared with the proof of Theorem 4.34.

PROOF. There are only finitely many restricted roots relative to $\mathfrak{a}$, and the union of their kernels therefore cannot exhaust $\mathfrak{a}$. By Lemma 6.50 we can find $H \in \mathfrak{a}$ such that $Z_\mathfrak{p}(H) = \mathfrak{a}$. Similarly we can find $H' \in \mathfrak{a}'$ such that $Z_\mathfrak{p}(H') = \mathfrak{a}'$. Choose by compactness of $\text{Ad}(K)$ a member $k = k_0$ of K that minimizes $B(\text{Ad}(k)H', H)$. For any $Z \in \mathfrak{k}$, $r \mapsto B(\text{Ad}(\exp rZ)\text{Ad}(k_0)H', H)$ is then a smooth function of r that is minimized for $r = 0$. Differentiating and setting $r = 0$, we obtain

$$0 = B((\text{ad}\, Z)\text{Ad}(k_0)H', H) = B(Z, [\text{Ad}(k_0)H', H]).$$

Here $[\text{Ad}(k_0)H', H]$ is in $\mathfrak{k}$, and Z is arbitrary in $\mathfrak{k}$. Since $B(\mathfrak{k}, \mathfrak{p}) = 0$ by (6.25) and since B is nondegenerate, $[\text{Ad}(k_0)H', H] = 0$. Thus $\text{Ad}(k_0)H'$ is in $Z_\mathfrak{p}(H) = \mathfrak{a}$. Since $\mathfrak{a}$ is abelian, this means

$$\mathfrak{a} \subseteq Z_\mathfrak{p}(\text{Ad}(k_0)H') = \text{Ad}(k_0)Z_\mathfrak{p}(H') = \text{Ad}(k_0)\mathfrak{a}'.$$

Equality must hold since $\mathfrak{a}$ is maximal abelian in $\mathfrak{p}$. Thus $\mathfrak{a} = \text{Ad}(k_0)\mathfrak{a}'$.

If X is any member of $\mathfrak{p}$, then we can extend $\mathbb{R}X$ to a maximal abelian subspace $\mathfrak{a}'$ of $\mathfrak{p}$. As above, we can write $\mathfrak{a}' = \text{Ad}(k)\mathfrak{a}$, and hence X is in $\bigcup_{k \in K} \text{Ad}(k)\mathfrak{a}$. Therefore $\mathfrak{p} = \bigcup_{k \in K} \text{Ad}(k)\mathfrak{a}$.

Now we think of $\mathfrak{k}$ and $\mathfrak{a}$ as fixed and consider the various possibilities for $\mathfrak{n}$. The inner product B_θ on $\mathfrak{g}$ can be restricted to $\mathfrak{a}$ and transferred to $\mathfrak{a}^*$ to give an inner product and norm denoted by $\langle \cdot, \cdot \rangle$ and $|\cdot|$, respectively. We write H_λ for the element of $\mathfrak{a}$ that corresponds to $\lambda \in \mathfrak{a}^*$.

Proposition 6.52. Let λ be a restricted root, and let E_λ be a nonzero restricted-root vector for λ.

(a) $[E_\lambda, \theta E_\lambda] = B(E_\lambda, \theta E_\lambda)H_\lambda$, and $B(E_\lambda, \theta E_\lambda) < 0$.

(b) $\mathbb{R}H_\lambda \oplus \mathbb{R}E_\lambda \oplus \mathbb{R}\theta E_\lambda$ is a Lie subalgebra of $\mathfrak{g}$ isomorphic to $\mathfrak{sl}(2, \mathbb{R})$, and the isomorphism can be defined so that the vector $H'_\lambda = 2|\lambda|^{-2}H_\lambda$ corresponds to $h = \begin{pmatrix} 1 & 0 \\ 0 & -1 \end{pmatrix}$.

(c) If E_λ is normalized so that $B(E_\lambda, \theta E_\lambda) = -2/|\lambda|^2$, then $k = \exp \frac{\pi}{2}(E_\lambda + \theta E_\lambda)$ is a member of the normalizer $N_K(\mathfrak{a})$, and $\text{Ad}(k)$ acts as the reflection s_λ on $\mathfrak{a}^*$.

PROOF.

(a) By Proposition 6.40 the vector $[E_\lambda, \theta E_\lambda]$ is in $[\mathfrak{g}_\lambda, \mathfrak{g}_{-\lambda}] \subseteq \mathfrak{g}_0 = \mathfrak{a} \oplus \mathfrak{m}$, and $\theta[E_\lambda, \theta E_\lambda] = [\theta E_\lambda, E_\lambda] = -[E_\lambda, \theta E_\lambda]$. Thus $[E_\lambda, \theta E_\lambda]$ is in $\mathfrak{a}$. Then $H \in \mathfrak{a}$ gives

$$\begin{aligned} B([E_\lambda, \theta E_\lambda], H) &= B(E_\lambda, [\theta E_\lambda, H]) = \lambda(H)B(E_\lambda, \theta E_\lambda) \\ &= B(H_\lambda, H)B(E_\lambda, \theta E_\lambda) = B(B(E_\lambda, \theta E_\lambda)H_\lambda, H). \end{aligned}$$

By nondegeneracy of B on $\mathfrak{a}$, $[E_\lambda, \theta E_\lambda] = B(E_\lambda, \theta E_\lambda) H_\lambda$. Finally $B(E_\lambda, \theta E_\lambda) = -B_\theta(E_\lambda, E_\lambda) < 0$ since B_θ is positive definite.

(b) Put

$$H'_\lambda = \frac{2}{|\lambda|^2} H_\lambda, \quad E'_\lambda = \frac{2}{|\lambda|^2} E_\lambda, \quad E'_{-\lambda} = \theta E_\lambda.$$

Then (a) shows that

$$[H'_\lambda, E'_\lambda] = 2E'_\lambda, \quad [H'_\lambda, E'_{-\lambda}] = -2E'_{-\lambda}, \quad [E'_\lambda, E'_{-\lambda}] = H'_\lambda,$$

and (b) follows.

(c) Note from (a) that the normalization $B(E_\lambda, \theta E_\lambda) = -2/|\lambda|^2$ is allowable. If $\lambda(H) = 0$, then

$$\begin{aligned} \operatorname{Ad}(k)H &= \operatorname{Ad}(\exp \tfrac{\pi}{2}(E_\lambda + \theta E_\lambda))H \\ &= (\exp \operatorname{ad} \tfrac{\pi}{2}(E_\lambda + \theta E_\lambda))H \\ &= \sum_{n=0}^{\infty} \tfrac{1}{n!}(\operatorname{ad} \tfrac{\pi}{2}(E_\lambda + \theta E_\lambda))^n H \\ &= H. \end{aligned}$$

On the other hand, for the element H'_λ, we first calculate that

$$(\operatorname{ad} \tfrac{\pi}{2}(E_\lambda + \theta E_\lambda))H'_\lambda = \pi(\theta E_\lambda - E_\lambda)$$

and

$$(\operatorname{ad} \tfrac{\pi}{2}(E_\lambda + \theta E_\lambda))^2 H'_\lambda = -\pi^2 H'_\lambda.$$

Therefore

$$\begin{aligned} \operatorname{Ad}(k)H'_\lambda &= \sum_{n=0}^{\infty} \tfrac{1}{n!}(\operatorname{ad} \tfrac{\pi}{2}(E_\lambda + \theta E_\lambda))^n H'_\lambda \\ &= \sum_{m=0}^{\infty} \tfrac{1}{(2m)!}((\operatorname{ad} \tfrac{\pi}{2}(E_\lambda + \theta E_\lambda))^2)^m H'_\lambda \\ &\quad + \sum_{m=0}^{\infty} \tfrac{1}{(2m+1)!}(\operatorname{ad} \tfrac{\pi}{2}(E_\lambda + \theta E_\lambda))((\operatorname{ad} \tfrac{\pi}{2}(E_\lambda + \theta E_\lambda))^2)^m H'_\lambda \\ &= \sum_{m=0}^{\infty} \tfrac{1}{(2m)!}(-\pi^2)^m H'_\lambda + \sum_{m=0}^{\infty} \tfrac{1}{(2m+1)!}(-\pi^2)^m \pi(\theta E_\lambda - E_\lambda) \\ &= (\cos \pi) H'_\lambda + (\sin \pi)(E_\lambda - \theta E_\lambda) \\ &= -H'_\lambda, \end{aligned}$$

and (c) follows.

Corollary 6.53. Σ is an abstract root system in $\mathfrak{a}^*$.

REMARKS. Examples of Σ appear in §4 after Proposition 6.40. The example of $SU(p,q)$ for $p > q$ shows that the abstract root system Σ need not be reduced.

PROOF. We verify that Σ satisfies the axioms for an abstract root system. To see that Σ spans $\mathfrak{a}^*$, let $\lambda(H) = 0$ for all $\lambda \in \Sigma$. Then $[H, \mathfrak{g}_\lambda] = 0$ for all λ and hence $[H, \mathfrak{g}] = 0$. But $\mathfrak{g}$ has 0 center, and therefore $H = 0$. Thus Σ spans $\mathfrak{a}^*$.

Let us show that $2\langle\mu, \lambda\rangle/|\lambda|^2$ is an integer whenever μ and λ are in Σ. Consider the subalgebra of Proposition 6.52b, calling it $\mathfrak{sl}_\lambda$. This acts by ad on $\mathfrak{g}$ and hence on $\mathfrak{g}^{\mathbb{C}}$. Complexifying, we obtain a representation of $(\mathfrak{sl}_\lambda)^{\mathbb{C}} \cong \mathfrak{sl}(2, \mathbb{C})$ on $\mathfrak{g}^{\mathbb{C}}$. We know from Corollary 1.69 that the element $H'_\lambda = 2|\lambda|^{-2}H_\lambda$, which corresponds to h, has to act diagonably with integer eigenvalues. The action of H'_λ on $\mathfrak{g}_\mu$ is by the scalar $\mu(2|\lambda|^{-2}H_\lambda) = 2\langle\mu, \lambda\rangle/|\lambda|^2$. Hence $2\langle\mu, \lambda\rangle/|\lambda|^2$ is an integer.

Finally we are to show that $s_\lambda(\mu)$ is in Σ whenever μ and λ are in Σ. Define k as in Proposition 6.52c, let H be in $\mathfrak{a}$, and let X be in $\mathfrak{g}_\mu$. Then we have

$$\begin{aligned}[H, \mathrm{Ad}(k)X] &= \mathrm{Ad}(k)[\mathrm{Ad}(k)^{-1}H, X] = \mathrm{Ad}(k)[s_\lambda^{-1}(H), X] \\ &= \mu(s_\lambda^{-1}(H))\mathrm{Ad}(k)X = (s_\lambda\mu)(H)\mathrm{Ad}(k)X,\end{aligned} \tag{6.54}$$

and hence $\mathfrak{g}_{s_\lambda\mu}$ is not 0. This completes the proof.

The possibilities for the subalgebra $\mathfrak{n}$ are given by all possible Σ^+'s resulting from different orderings of $\mathfrak{a}^*$, and it follows from Corollary 6.53 that the Σ^+'s correspond to all possible simple systems for Σ. Any two such simple systems are conjugate by the Weyl group $W(\Sigma)$ of Σ, and it follows from Proposition 6.52c that the conjugation can be achieved by a member of $N_K(\mathfrak{a})$. The same computation as in (6.54) shows that if $k \in N_K(\mathfrak{a})$ represents the member s of $W(\Sigma)$, then $\mathrm{Ad}(k)\mathfrak{g}_\lambda = \mathfrak{g}_{s\lambda}$. We summarize this discussion in the following corollary.

Corollary 6.55. Any two choices of $\mathfrak{n}$ are conjugate by Ad of a member of $N_K(\mathfrak{a})$.

This completes our discussion of the conjugacy of different Iwasawa decompositions.

We now examine $N_K(\mathfrak{a})$ further. Define

$$W(G, A) = N_K(\mathfrak{a})/Z_K(\mathfrak{a}).$$

This is a group of linear transformations of $\mathfrak{a}$, telling all possible ways that members of k can act on $\mathfrak{a}$ by Ad. We have already seen that $W(\Sigma) \subseteq W(G, A)$, and we are going to prove that $W(\Sigma) = W(G, A)$.

We write M for the group $Z_K(\mathfrak{a})$. This is a compact group (being a closed subgroup of K) with Lie algebra $Z_{\mathfrak{k}}(\mathfrak{a}) = \mathfrak{m}$. After Proposition 6.40 we saw examples of restricted-root space decompositions and the associated Lie algebras $\mathfrak{m}$. The following examples continue that discussion.

EXAMPLES.

1) Let $G = SL(n, \mathbb{K})$, where $\mathbb{K}$ is $\mathbb{R}, \mathbb{C}$, or $\mathbb{H}$. The subgroup M consists of all diagonal members of K. When $\mathbb{K} = \mathbb{R}$, the diagonal entries are ± 1, but there are only $n-1$ independent signs since the determinant is 1. Thus M is finite abelian and is the product of $n-1$ groups of order 2. When $\mathbb{K} = \mathbb{C}$, the diagonal entries are complex numbers of modulus 1, and again the determinant is 1. Thus M is a torus of dimension $n-1$. When $\mathbb{K} = \mathbb{H}$, the diagonal entries are quaternions of absolute value 1, and there is no restriction on the determinant. Thus M is the product of n copies of $SU(2)$.

2) Let $G = SU(p, q)$ with $p \geq q$. The group M consists of all unitary matrices of determinant 1 that are arbitrary in the upper left block of size $p-q$, are otherwise diagonal, and have the $(p-i+1)^{\text{st}}$ diagonal entry equal to the $(p+i)^{\text{th}}$ diagonal entry for $1 \leq i \leq q$. Let us abbreviate such a matrix as

$$m = \operatorname{diag}(\omega, e^{i\theta_q}, \ldots, e^{i\theta_1}, e^{i\theta_1}, \ldots, e^{i\theta_q}),$$

where ω is the upper left block of size $p-q$. When $p = q$, the condition that the determinant be 1 says that $\sum_{j=1}^{q} \theta_j \in \pi\mathbb{Z}$. Thus we can take $\theta_1, \ldots, \theta_{q-1}$ to be arbitrary and use $e^{i\theta_q} = \pm e^{-i(\theta_1+\cdots+\theta_{q-1})}$. Consequently M is the product of a torus of dimension $q-1$ and a 2-element group. When $p > q$, M is connected. In fact, the homomorphism that maps the above matrix m to the $2q$-by-$2q$ diagonal matrix

$$\operatorname{diag}(e^{i\theta_q}, \ldots, e^{i\theta_1}, e^{i\theta_1}, \ldots, e^{i\theta_q})$$

has a (connected) q-dimensional torus as image, and the kernel is isomorphic to the connected group $SU(p-q)$; thus M itself is connected.

3) Let $G = SO(p, q)_0$ with $p \geq q$. The subgroup M for this example is the intersection of $SO(p) \times SO(q)$ with the M of the previous example. Thus M here consists of matrices that are orthogonal matrices of total determinant 1, are arbitrary in the upper left block of size $p-q$, are

otherwise diagonal, have q diagonal entries ± 1 after the upper left block, and then have those q diagonal entries ± 1 repeated in reverse order. For the lower right q entries to yield a matrix in $SO(q)$, the product of the q entries ± 1 must be 1. For the upper left p entries to yield a matrix in $SO(p)$, the orthogonal matrix in the upper left block of size $p-q$ must have determinant 1. Therefore M is isomorphic to the product of $SO(p-q)$ and the product of $q-1$ groups of order 2.

Lemma 6.56. The Lie algebra of $N_K(\mathfrak{a})$ is $\mathfrak{m}$. Therefore $W(G, A)$ is a finite group.

PROOF. The second conclusion follows from the first, since the first conclusion implies that $W(G, A)$ is 0-dimensional and compact, hence finite. For the first conclusion, the Lie algebra in question is $N_{\mathfrak{k}}(\mathfrak{a})$. Let $X = H_0 + X_0 + \sum_{\lambda\in\Sigma} X_\lambda$ be a member of $N_{\mathfrak{k}}(\mathfrak{a})$, with $H_0 \in \mathfrak{a}$, $X_0 \in \mathfrak{m}$, and $X_\lambda \in \mathfrak{g}_\lambda$. Since X is to be in $\mathfrak{k}$, θ must fix X, and we see that X may be rewritten as $X = X_0 + \sum_{\lambda\in\Sigma^+}(X_\lambda + \theta X_\lambda)$. When we apply $\operatorname{ad} H$ for $H \in \mathfrak{a}$, we obtain $[H, X] = \sum_{\lambda\in\Sigma^+} \lambda(H)(X_\lambda - \theta X_\lambda)$. This element is supposed to be in $\mathfrak{a}$, since we started with X in the normalizer of $\mathfrak{a}$, and that means $[H, X]$ is 0. But then $X_\lambda = 0$ for all λ, and X reduces to the member X_0 of $\mathfrak{m}$.

Theorem 6.57. The group $W(G, A)$ coincides with $W(\Sigma)$.

REMARK. This theorem should be compared with Theorem 4.54.

PROOF. Let us observe that $W(G, A)$ permutes the restricted roots. In fact, let k be in $N_K(\mathfrak{a})$, let λ be in Σ, and let E_λ be in $\mathfrak{g}_\lambda$. Then

$$\begin{aligned}[H, \operatorname{Ad}(k)E_\lambda] &= \operatorname{Ad}(k)[\operatorname{Ad}(k)^{-1}H, E_\lambda] = \operatorname{Ad}(k)(\lambda(\operatorname{Ad}(k)^{-1}H)E_\lambda) \\ &= \lambda(\operatorname{Ad}(k)^{-1}H)\operatorname{Ad}(k)E_\lambda = (k\lambda)(H)\operatorname{Ad}(k)E_\lambda\end{aligned}$$

shows that $k\lambda$ is in Σ and that $\operatorname{Ad}(k)E_\lambda$ is a restricted-root vector for $k\lambda$. Thus $W(G, A)$ permutes the restricted roots.

We have seen that $W(\Sigma) \subseteq W(G, A)$. Fix a simple system Σ^+ for Σ. In view of Theorem 2.63, it suffices to show that if $k \in N_K(\mathfrak{a})$ has $\operatorname{Ad}(k)\Sigma^+ = \Sigma^+$, then k is in $Z_K(\mathfrak{a})$.

The element $\operatorname{Ad}(k) = w$ acts as a permutation of Σ^+. Let 2δ denote the sum of the reduced members of Σ^+, so that w fixes δ. If λ_i is a simple restricted root, then Proposition 2.69 shows that $2\langle\delta, \lambda_i\rangle/|\lambda_i|^2 = 1$. Therefore $\langle\delta, \lambda\rangle > 0$ for all $\lambda \in \Sigma^+$.

Let $\mathfrak{u} = \mathfrak{k} \oplus i\mathfrak{p}$ be the compact real form of $\mathfrak{g}^{\mathbb{C}}$ associated to θ, and let U be the adjoint group of $\mathfrak{u}$. Then $\operatorname{Ad}_{\mathfrak{g}^{\mathbb{C}}}(K) \subseteq U$, and in particular $\operatorname{Ad}(k)$ is a member of U. Form $S = \overline{\{\exp ir \operatorname{ad} H_\delta\}} \subseteq U$. Here S is a torus in U, and we let $\mathfrak{s}$ be the Lie algebra of S. The element $\mathrm{Ad}(k)$ is in $Z_U(S)$, and the claim is that every member of $Z_U(S)$ centralizes $\mathfrak{a}$. If so, then $\mathrm{Ad}(k)$ is 1 on $\mathfrak{a}$, and k is in $Z_K(\mathfrak{a})$, as required.

By Corollary 4.51 we can verify that $Z_U(S)$ centralizes $\mathfrak{a}$ by showing that $Z_\mathfrak{u}(\mathfrak{s})$ centralizes $\mathfrak{a}$. Here

$$Z_\mathfrak{u}(\mathfrak{s}) = \mathfrak{u} \cap Z_{\mathfrak{g}^\mathbb{C}}(\mathfrak{s}) = \mathfrak{u} \cap Z_{\mathfrak{g}^\mathbb{C}}(H_\delta).$$

To evaluate the right side, we complexify the statement of Lemma 6.50. Since $\langle \lambda, \delta \rangle \neq 0$, the centralizer $Z_{\mathfrak{g}^\mathbb{C}}(H_\delta)$ is just $\mathfrak{a}^\mathbb{C} \oplus \mathfrak{m}^\mathbb{C}$. Therefore

$$Z_\mathfrak{u}(\mathfrak{s}) = \mathfrak{u} \cap (\mathfrak{a}^\mathbb{C} \oplus \mathfrak{m}^\mathbb{C}) = i\mathfrak{a} \cap \mathfrak{m}.$$

Every member of the right side centralizes $\mathfrak{a}$, and the proof is complete.

6. Cartan Subalgebras

Proposition 6.47 showed that every real semisimple Lie algebra has a Cartan subalgebra. But as we shall see shortly, not all Cartan subalgebras are conjugate. In this section and the next we investigate the conjugacy classes of Cartan subalgebras and some of their relationships to each other.

We revert to the use of subscripted Gothic letters for real Lie algebras and to unsubscripted letters for complexifications. Let $\mathfrak{g}_0$ be a real semisimple Lie algebra, let θ be a Cartan involution, and let $\mathfrak{g}_0 = \mathfrak{k}_0 \oplus \mathfrak{p}_0$ be the corresponding Cartan decomposition. Let $\mathfrak{g}$ be the complexification of $\mathfrak{g}_0$, and write $\mathfrak{g} = \mathfrak{k} \oplus \mathfrak{p}$ for the complexification of the Cartan decomposition. Let B be any nondegenerate symmetric invariant bilinear form on $\mathfrak{g}_0$ such that $B(\theta X, \theta Y) = B(X, Y)$ and such that B_θ, defined by (6.13), is positive definite.

All Cartan subalgebras of $\mathfrak{g}_0$ have the same dimension, since their complexifications are Cartan subalgebras of $\mathfrak{g}$ and are conjugate via $\mathrm{Int}\,\mathfrak{g}$, according to Theorem 2.15.

Let $K = \mathrm{Int}_{\mathfrak{g}_0}(\mathfrak{k}_0)$. This subgroup of $\mathrm{Int}\,\mathfrak{g}_0$ is compact.

EXAMPLE. Let $G = SL(2, \mathbb{R})$ and $\mathfrak{g}_0 = \mathfrak{sl}(2, \mathbb{R})$. A Cartan subalgebra $\mathfrak{h}_0$ complexifies to a Cartan subalgebra of $\mathfrak{sl}(2, \mathbb{C})$ and therefore has dimension 1. Therefore let us consider which 1-dimensional subspaces $\mathbb{R}X$ of $\mathfrak{sl}(2, \mathbb{R})$ are Cartan subalgebras. The matrix X has trace 0, and we divide matters into cases according to the sign of $\det X$. If $\det X < 0$, then X has real eigenvalues μ and $-\mu$, and X is conjugate via $SL(2, \mathbb{R})$ to a diagonal matrix. Thus, for some $g \in SL(2, \mathbb{R})$,

$$\mathbb{R}X = \{\mathrm{Ad}(g)\mathbb{R}h\}.$$

where $h = \begin{pmatrix} 1 & 0 \\ 0 & -1 \end{pmatrix}$ as usual. The subspace $\mathbb{R}h$ is maximal abelian in $\mathfrak{g}_0$ and $\operatorname{ad} h$ acts diagonably on $\mathfrak{g}$ with eigenvectors h, e, f. Since (1.84) gives

$$\operatorname{ad}(\operatorname{Ad}(g)h) = \operatorname{Ad}(g)(\operatorname{ad} h)\operatorname{Ad}(g)^{-1},$$

$\operatorname{ad}(\operatorname{Ad}(g)h)$ acts diagonably with eigenvectors $\operatorname{Ad}(g)h$, $\operatorname{Ad}(g)e$, $\operatorname{Ad}(g)f$. Therefore $\mathbb{R}X$ is a Cartan subalgebra when $\det X < 0$, and it is conjugate via $\operatorname{Int} \mathfrak{g}_0$ to $\mathbb{R}h$.

If $\det X > 0$, then X has purely imaginary eigenvalues μ and $-\mu$, and X is conjugate via $SL(2, \mathbb{R})$ to a real multiple of ih_B, where

$$h_B = \begin{pmatrix} 0 & i \\ -i & 0 \end{pmatrix}. \tag{6.58a}$$

Thus, for some $g \in SL(2, \mathbb{R})$,

$$\mathbb{R}X = \{\operatorname{Ad}(g)\mathbb{R}ih_B\}.$$

The subspace $\mathbb{R}ih_B$ is maximal abelian in $\mathfrak{g}_0$ and $\operatorname{ad} ih_B$ acts diagonably on $\mathfrak{g}$ with eigenvectors h_B, e_B, f_B, where

$$e_B = \frac{1}{2}\begin{pmatrix} 1 & -i \\ -i & -1 \end{pmatrix} \quad \text{and} \quad f_B = \frac{1}{2}\begin{pmatrix} 1 & i \\ i & -1 \end{pmatrix}. \tag{6.58b}$$

Then $\operatorname{ad}(\operatorname{Ad}(g)ih_B)$ acts diagonably with eigenvectors $\operatorname{Ad}(g)h_B$, $\operatorname{Ad}(g)e_B, \operatorname{Ad}(g)f_B$. Therefore $\mathbb{R}X$ is a Cartan subalgebra when $\det X > 0$, and it is conjugate via $\operatorname{Int} \mathfrak{g}_0$ to $\mathbb{R}ih_B$.

If $\det X = 0$, then X has both eigenvalues equal to 0, and X is conjugate via $SL(2, \mathbb{R})$ to a real multiple of $e = \begin{pmatrix} 0 & 1 \\ 0 & 0 \end{pmatrix}$. Thus, for some $g \in SL(2, \mathbb{R})$,

$$\mathbb{R}X = \{\operatorname{Ad}(g)\mathbb{R}e\}.$$

The subspace $\mathbb{R}e$ is maximal abelian in $\mathfrak{g}_0$, but the element $\operatorname{ad} e$ does not act diagonably on $\mathfrak{g}$. It follows that $\operatorname{ad}(\operatorname{Ad}(g)e)$ does not act diagonably. Therefore $\mathbb{R}X$ is not a Cartan subalgebra when $\det X = 0$.

In the above example every Cartan subalgebra is conjugate either to $\mathbb{R}h$ or to $\mathbb{R}ih_B$, and these two are θ stable. We shall see in Proposition 6.59 that this kind of conjugacy remains valid for all real semisimple Lie algebras $\mathfrak{g}_0$.

Another feature of the above example is that the two Cartan subalgebras $\mathbb{R}h$ and $\mathbb{R}ih_B$ are not conjugate. In fact, h has nonzero real eigenvalues, and ih_B has nonzero purely imaginary eigenvalues, and thus the two cannot be conjugate.

Proposition 6.59. Any Cartan subalgebra $\mathfrak{h}_0$ of $\mathfrak{g}_0$ is conjugate via $\operatorname{Int}\mathfrak{g}_0$ to a θ stable Cartan subalgebra.

PROOF. Let $\mathfrak{h}$ be the complexification of $\mathfrak{h}_0$, and let σ be the conjugation of $\mathfrak{g}$ with respect to $\mathfrak{g}_0$. Let $\mathfrak{u}_0$ be the compact real form constructed from $\mathfrak{h}$ and other data in Theorem 6.11, and let τ be the conjugation of $\mathfrak{g}$ with respect to $\mathfrak{u}_0$. The construction of $\mathfrak{u}_0$ has the property that $\tau(\mathfrak{h}) = \mathfrak{h}$.

The conjugations σ and τ are involutions of $\mathfrak{g}^{\mathbb{R}}$, and τ is a Cartan involution by Proposition 6.14. Theorem 6.16 shows that the element φ of $\operatorname{Int}\mathfrak{g}^{\mathbb{R}} = \operatorname{Int}\mathfrak{g}$ given by $\varphi = ((\sigma\tau)^2)^{1/4}$ has the property that the Cartan involution $\tilde{\eta} = \varphi\tau\varphi^{-1}$ of $\mathfrak{g}^{\mathbb{R}}$ commutes with σ. Since $\sigma(\mathfrak{h}) = \mathfrak{h}$ and $\tau(\mathfrak{h}) = \mathfrak{h}$, it follows that $\varphi(\mathfrak{h}) = \mathfrak{h}$. Therefore $\tilde{\eta}(\mathfrak{h}) = \mathfrak{h}$.

Since $\tilde{\eta}$ and σ commute, it follows that $\tilde{\eta}(\mathfrak{g}_0) = \mathfrak{g}_0$. Since $\mathfrak{h}_0 = \mathfrak{h} \cap \mathfrak{g}_0$, we obtain $\tilde{\eta}(\mathfrak{h}_0) = \mathfrak{h}_0$.

Put $\eta = \tilde{\eta}|_{\mathfrak{g}_0}$, so that $\eta(\mathfrak{h}_0) = \mathfrak{h}_0$. Since $\tilde{\eta}$ is the conjugation of $\mathfrak{g}$ with respect to the compact real form $\varphi(\mathfrak{u}_0)$, the proof of Corollary 6.18 shows that η is a Cartan involution of $\mathfrak{g}_0$. Corollary 6.19 shows that η and θ are conjugate via $\operatorname{Int}\mathfrak{g}_0$, say $\theta = \psi\eta\psi^{-1}$ with $\psi \in \operatorname{Int}\mathfrak{g}_0$. Then $\psi(\mathfrak{h}_0)$ is a Cartan subalgebra of $\mathfrak{g}_0$, and

$$\theta(\psi(\mathfrak{h}_0)) = \psi\eta\psi^{-1}\psi(\mathfrak{h}_0) = \psi(\eta\mathfrak{h}_0) = \psi(\mathfrak{h}_0),$$

shows that it is θ stable.

Thus it suffices to study θ stable Cartan subalgebras. When $\mathfrak{h}_0$ is θ stable, we can write $\mathfrak{h}_0 = \mathfrak{t}_0 \oplus \mathfrak{a}_0$ with $\mathfrak{t}_0 \subseteq \mathfrak{k}_0$ and $\mathfrak{a}_0 \subseteq \mathfrak{p}_0$. By the same argument as for Corollary 6.49, roots of $(\mathfrak{g}, \mathfrak{h})$ are real-valued on $\mathfrak{a}_0 \oplus i\mathfrak{t}_0$. Consequently the **compact dimension** $\dim \mathfrak{t}_0$ and the **noncompact dimension** $\dim \mathfrak{a}_0$ of $\mathfrak{h}_0$ are unchanged when $\mathfrak{h}_0$ is conjugated via $\operatorname{Int}\mathfrak{g}_0$ to another θ stable Cartan subalgebra.

We say that a θ stable Cartan subalgebra $\mathfrak{h}_0 = \mathfrak{t}_0 \oplus \mathfrak{a}_0$ is **maximally noncompact** if its noncompact dimension is as large as possible, **maximally compact** if its compact dimension is as large as possible. In $\mathfrak{sl}(2, \mathbb{R})$, $\mathbb{R}h$ is maximally noncompact, and $\mathbb{R}ih_B$ is maximally compact. In any case $\mathfrak{a}_0$ is an abelian subspace of $\mathfrak{p}_0$, and thus Proposition 6.47 implies that $\mathfrak{h}_0$ is maximally noncompact if and only if $\mathfrak{a}_0$ is a maximal abelian subspace of $\mathfrak{p}_0$.

Proposition 6.60. Let $\mathfrak{t}_0$ be a maximal abelian subspace of $\mathfrak{k}_0$. Then $\mathfrak{h}_0 = Z_{\mathfrak{g}_0}(\mathfrak{t}_0)$ is a θ stable Cartan subalgebra of $\mathfrak{g}_0$ of the form $\mathfrak{h}_0 = \mathfrak{t}_0 \oplus \mathfrak{a}_0$ with $\mathfrak{a}_0 \subseteq \mathfrak{p}_0$.

PROOF. The subalgebra $\mathfrak{h}_0$ is θ stable and hence is a vector-space direct sum $\mathfrak{h}_0 = \mathfrak{t}_0 \oplus \mathfrak{a}_0$, where $\mathfrak{a}_0 = \mathfrak{h}_0 \cap \mathfrak{p}_0$. Since $\mathfrak{h}_0$ is θ stable, Proposition 6.29 shows that it is reductive. By Corollary 1.53, $[\mathfrak{h}_0, \mathfrak{h}_0]$ is semisimple.

We have $[\mathfrak{h}_0, \mathfrak{h}_0] = [\mathfrak{a}_0, \mathfrak{a}_0]$, and $[\mathfrak{a}_0, \mathfrak{a}_0] \subseteq \mathfrak{t}_0$ since $\mathfrak{a}_0 \subseteq \mathfrak{p}_0$ and $\mathfrak{h}_0 \cap \mathfrak{k}_0 = \mathfrak{t}_0$. Thus the semisimple Lie algebra $[\mathfrak{h}_0, \mathfrak{h}_0]$ is abelian and must be 0. Consequently $\mathfrak{h}_0$ is abelian.

It is clear that $\mathfrak{h} = (\mathfrak{h}_0)^{\mathbb{C}}$ is maximal abelian in $\mathfrak{g}$, and $\operatorname{ad} \mathfrak{h}_0$ is certainly diagonable on $\mathfrak{g}$ since the members of $\operatorname{ad}_{\mathfrak{g}_0}(\mathfrak{t}_0)$ are skew adjoint, the members of $\operatorname{ad}_{\mathfrak{g}_0}(\mathfrak{a}_0)$ are self adjoint, and $\mathfrak{t}_0$ commutes with $\mathfrak{a}_0$. By Corollary 2.13, $\mathfrak{h}$ is a Cartan subalgebra of $\mathfrak{g}$, and hence $\mathfrak{h}_0$ is a Cartan subalgebra of $\mathfrak{g}_0$.

With any θ stable Cartan subalgebra $\mathfrak{h}_0 = \mathfrak{t}_0 \oplus \mathfrak{a}_0$, $\mathfrak{t}_0$ is an abelian subspace of $\mathfrak{k}_0$, and thus Proposition 6.60 implies that $\mathfrak{h}_0$ is maximally compact if and only if $\mathfrak{t}_0$ is a maximal abelian subspace of $\mathfrak{k}_0$.

Proposition 6.61. Among θ stable Cartan subalgebras $\mathfrak{h}_0$ of $\mathfrak{g}_0$, the maximally noncompact ones are all conjugate via K, and the maximally compact ones are all conjugate via K.

PROOF. Let $\mathfrak{h}_0$ and $\mathfrak{h}_0'$ be given Cartan subalgebras. In the first case, as we observed above, $\mathfrak{h}_0 \cap \mathfrak{p}_0$ and $\mathfrak{h}_0' \cap \mathfrak{p}_0$ are maximal abelian in $\mathfrak{p}_0$, and Theorem 6.51 shows that there is no loss of generality in assuming that $\mathfrak{h}_0 \cap \mathfrak{p}_0 = \mathfrak{h}_0' \cap \mathfrak{p}_0$. Thus $\mathfrak{h}_0 = \mathfrak{t}_0 \oplus \mathfrak{a}_0$ and $\mathfrak{h}_0' = \mathfrak{t}_0' \oplus \mathfrak{a}_0$, where $\mathfrak{a}_0$ is maximal abelian in $\mathfrak{p}_0$. Define $\mathfrak{m}_0 = Z_{\mathfrak{k}_0}(\mathfrak{a}_0)$. Then $\mathfrak{t}_0$ and $\mathfrak{t}_0'$ are in $\mathfrak{m}_0$ and are maximal abelian there. Let $M = Z_K(\mathfrak{a}_0)$. This is a compact subgroup of K with Lie algebra $\mathfrak{m}_0$, and we let M_0 be its identity component. Theorem 4.34 says that $\mathfrak{t}_0$ and $\mathfrak{t}_0'$ are conjugate via M_0, and this conjugacy clearly fixes $\mathfrak{a}_0$. Hence $\mathfrak{h}_0$ and $\mathfrak{h}_0'$ are conjugate via K.

In the second case, as we observed above, $\mathfrak{h}_0 \cap \mathfrak{k}_0$ and $\mathfrak{h}_0' \cap \mathfrak{k}_0$ are maximal abelian in $\mathfrak{k}_0$, and Theorem 4.34 shows that there is no loss of generality in assuming that $\mathfrak{h}_0 \cap \mathfrak{k}_0 = \mathfrak{h}_0' \cap \mathfrak{k}_0$. Then Proposition 6.60 shows that $\mathfrak{h}_0 = \mathfrak{h}_0'$, and the proof is complete.

If we examine the proof of the first part of Proposition 6.61 carefully, we find that we can adjust it to obtain root data that determine a Cartan subalgebra up to conjugacy. As a consequence there are only finitely many conjugacy classes of Cartan subalgebras.

Lemma 6.62. Let $\mathfrak{h}_0$ and $\mathfrak{h}_0'$ be θ stable Cartan subalgebras of $\mathfrak{g}_0$ such that $\mathfrak{h}_0 \cap \mathfrak{p}_0 = \mathfrak{h}_0' \cap \mathfrak{p}_0$. Then $\mathfrak{h}_0$ and $\mathfrak{h}_0'$ are conjugate via K.

PROOF. Since the $\mathfrak{p}_0$ parts of the two Cartan subalgebras are the same and since Cartan subalgebras are abelian, the $\mathfrak{k}_0$ parts $\mathfrak{h}_0 \cap \mathfrak{k}_0$ and $\mathfrak{h}_0' \cap \mathfrak{k}_0$ are both contained in $\tilde{\mathfrak{m}}_0 = Z_{\mathfrak{k}_0}(\mathfrak{h}_0 \cap \mathfrak{p}_0)$. The Cartan subalgebras are maximal abelian in $\mathfrak{g}_0$, and therefore $\mathfrak{h}_0 \cap \mathfrak{k}_0$ and $\mathfrak{h}_0' \cap \mathfrak{k}_0$ are both maximal abelian in $\tilde{\mathfrak{m}}_0$. Let $\tilde{M} = Z_K(\mathfrak{h}_0 \cap \mathfrak{p}_0)$. This is a compact Lie group with

Lie algebra $\tilde{\mathfrak{m}}_0$, and we let $\tilde{M}_0$ be its identity component. Theorem 4.34 says that $\mathfrak{h}_0 \cap \mathfrak{k}_0$ and $\mathfrak{h}'_0 \cap \mathfrak{k}_0$ are conjugate via $\tilde{M}_0$, and this conjugacy clearly fixes $\mathfrak{h}_0 \cap \mathfrak{p}_0$. Hence $\mathfrak{h}_0$ and $\mathfrak{h}'_0$ are conjugate via K.

Lemma 6.63. Let $\mathfrak{a}_0$ be a maximal abelian subspace of $\mathfrak{p}_0$, and let Σ be the set of restricted roots of $(\mathfrak{g}_0, \mathfrak{a}_0)$. Suppose that $\mathfrak{h}_0$ is a θ stable Cartan subalgebra such that $\mathfrak{h}_0 \cap \mathfrak{p}_0 \subseteq \mathfrak{a}_0$. Let $\Sigma' = \{\lambda \in \Sigma \mid \lambda(\mathfrak{h}_0 \cap \mathfrak{p}_0) = 0\}$. Then $\mathfrak{h}_0 \cap \mathfrak{p}_0$ is the common kernel of all $\lambda \in \Sigma'$.

PROOF. Let $\mathfrak{a}'_0$ be the common kernel of all $\lambda \in \Sigma'$. Then $\mathfrak{h}_0 \cap \mathfrak{p}_0 \subseteq \mathfrak{a}'_0$, and we are to prove that equality holds. Since $\mathfrak{h}_0$ is maximal abelian in $\mathfrak{g}_0$, it is enough to prove that $\mathfrak{h}_0 + \mathfrak{a}'_0$ is abelian.

Let $\mathfrak{g}_0 = \mathfrak{a}_0 \oplus \mathfrak{m}_0 \oplus \bigoplus_{\lambda \in \Sigma} (\mathfrak{g}_0)_\lambda$ be the restricted-root space decomposition of $\mathfrak{g}_0$, and let $X = H_0 + X_0 + \sum_{\lambda \in \Sigma} X_\lambda$ be an element of $\mathfrak{g}_0$ that centralizes $\mathfrak{h}_0 \cap \mathfrak{p}_0$. Bracketing the formula for X with $H \in \mathfrak{h}_0 \cap \mathfrak{p}_0$, we obtain $0 = \sum_{\lambda \in \Sigma - \Sigma'} \lambda(H) X_\lambda$, from which we conclude that $\lambda(H) X_\lambda = 0$ for all $H \in \mathfrak{h}_0 \cap \mathfrak{p}_0$ and all $\lambda \in \Sigma - \Sigma'$. Since the λ's in $\Sigma - \Sigma'$ have $\lambda(\mathfrak{h}_0 \cap \mathfrak{p}_0)$ not identically 0, we see that $X_\lambda = 0$ for all $\lambda \in \Sigma - \Sigma'$. Thus any X that centralizes $\mathfrak{h}_0 \cap \mathfrak{p}_0$ is of the form

$$X = H_0 + X_0 + \sum_{\lambda \in \Sigma'} X_\lambda.$$

Since $\mathfrak{h}_0$ is abelian, the elements $X \in \mathfrak{h}_0$ are of this form, and $\mathfrak{a}'_0$ commutes with any X of this form. Hence $\mathfrak{h}_0 + \mathfrak{a}'_0$ is abelian, and the proof is complete.

Proposition 6.64. Up to conjugacy by $\operatorname{Int} \mathfrak{g}_0$, there are only finitely many Cartan subalgebras of $\mathfrak{g}_0$.

PROOF. Fix a maximal abelian subspace $\mathfrak{a}_0$ of $\mathfrak{p}_0$. Let $\mathfrak{h}_0$ be a Cartan subalgebra. Proposition 6.59 shows that we may assume that $\mathfrak{h}_0$ is θ stable, and Theorem 6.51 shows that we may assume that $\mathfrak{h}_0 \cap \mathfrak{p}_0$ is contained in $\mathfrak{a}_0$. Lemma 6.63 associates to $\mathfrak{h}_0$ a subset of the set Σ of restricted roots that determines $\mathfrak{h}_0 \cap \mathfrak{p}_0$, and Lemma 6.62 shows that $\mathfrak{h}_0 \cap \mathfrak{p}_0$ determines $\mathfrak{h}_0$ up to conjugacy. Hence the number of conjugacy classes of Cartan subalgebras is bounded by the number of subsets of Σ.

7. Cayley Transforms

The classification of real semisimple Lie algebras later in this chapter will use maximally compact Cartan subalgebras, but much useful information about a semisimple Lie algebra $\mathfrak{g}_0$ comes about from a maximally noncompact Cartan subalgebra. To correlate this information, we need to be able to track down the conjugacy via $\mathfrak{g} = (\mathfrak{g}_0)^{\mathbb{C}}$ of a maximally compact Cartan subalgebra and a maximally noncompact one.

Cayley transforms are one-step conjugacies of θ stable Cartan subalgebras whose iterates explicitly relate any θ stable Cartan subalgebra with any other. We develop Cayley transforms in this section and show that in favorable circumstances we can see past the step-by-past process to understand the composite conjugation all at once.

There are two kinds of Cayley transforms, essentially inverse to each other. They are modeled on what happens in $\mathfrak{sl}(2,\mathbb{R})$. In the case of $\mathfrak{sl}(2,\mathbb{R})$, we start with the standard basis h, e, f for $\mathfrak{sl}(2,\mathbb{C})$ as in (1.5), as well as the members h_B, e_B, f_B of $\mathfrak{sl}(2,\mathbb{C})$ defined in (6.58). The latter elements satisfy the familiar bracket relations

$$[h_B, e_B] = 2e_B, \quad [h_B, f_B] = -2f_B, \quad [e_B, f_B] = h_B.$$

The definitions of e_B and f_B make $e_B + f_B$ and $i(e_B - f_B)$ be in $\mathfrak{sl}(2,\mathbb{R})$, while $i(e_B + f_B)$ and $e_B - f_B$ are in $\mathfrak{su}(2)$. The first kind of Cayley transform within $\mathfrak{sl}(2,\mathbb{C})$ is the mapping

$$\operatorname{Ad}\left(\frac{\sqrt{2}}{2}\begin{pmatrix} 1 & i \\ i & 1 \end{pmatrix}\right) = \operatorname{Ad}(\exp \tfrac{\pi}{4}(f_B - e_B)),$$

which carries h_B, e_B, f_B to the multiple $-i$ of h, e, f and carries the Cartan subalgebra $\mathbb{R}\begin{pmatrix} 0 & 1 \\ -1 & 0 \end{pmatrix}$ to $i\mathbb{R}\begin{pmatrix} 1 & 0 \\ 0 & -1 \end{pmatrix}$. When generalized below, this Cayley transform will be called $\mathbf{c}_\beta$.

The second kind of Cayley transform within $\mathfrak{sl}(2,\mathbb{C})$ is the mapping

$$\operatorname{Ad}\left(\frac{\sqrt{2}}{2}\begin{pmatrix} 1 & -i \\ -i & 1 \end{pmatrix}\right) = \operatorname{Ad}(\exp i\tfrac{\pi}{4}(-f - e)),$$

which carries h, e, f to the multiple i of h_B, e_B, f_B and carries the Cartan subalgebra $\mathbb{R}\begin{pmatrix} 1 & 0 \\ 0 & -1 \end{pmatrix}$ to $i\mathbb{R}\begin{pmatrix} 0 & 1 \\ -1 & 0 \end{pmatrix}$. In view of the explicit formula for the matrices of the Cayley transforms, the two transforms are inverse to one another. When generalized below, this second Cayley transform will be called $\mathbf{d}_\alpha$.

The idea is to embed each of these constructions into constructions in the complexification of our underlying semisimple algebra that depend upon a single root of a special kind, leaving fixed the part of the Cartan subalgebra that is orthogonal to the embedded copy of $\mathfrak{sl}(2,\mathbb{C})$.

Turning to the case of a general real semisimple Lie algebra, we continue with the notation of the previous section. We extend the inner product B_θ on $\mathfrak{g}_0$ to a Hermitian inner product on $\mathfrak{g}$ by the definition

$$B_\theta(Z_1, Z_2) = -B(Z_1, \theta \bar{Z}_2),$$

where bar denotes the conjugation of $\mathfrak{g}$ with respect to $\mathfrak{g}_0$. In this expression θ and bar commute.

If $\mathfrak{h}_0 = \mathfrak{t}_0 \oplus \mathfrak{a}_0$ is a θ stable Cartan subalgebra of $\mathfrak{g}_0$, we have noted that roots of $(\mathfrak{g}, \mathfrak{h})$ are imaginary on $\mathfrak{t}_0$ and real on $\mathfrak{a}_0$. A root is **real** if it takes on real values on $\mathfrak{h}_0$ (i.e., vanishes on $\mathfrak{t}_0$), **imaginary** if it takes on purely imaginary values on $\mathfrak{h}_0$ (i.e., vanishes on $\mathfrak{a}_0$), and **complex** otherwise.

For any root α, $\theta\alpha$ is the root $\theta\alpha(H) = \alpha(\theta^{-1}H)$. To see that $\theta\alpha$ is a root, we let E_α be a nonzero root vector for α, and we calculate

$$[H, \theta E_\alpha] = \theta[\theta^{-1}H, E_\alpha] = \alpha(\theta^{-1}H)\theta E_\alpha = (\theta\alpha)(H)\theta E_\alpha.$$

If α is imaginary, then $\theta\alpha = \alpha$. Thus $\mathfrak{g}_\alpha$ is θ stable, and we have $\mathfrak{g}_\alpha = (\mathfrak{g}_\alpha \cap \mathfrak{k}) \oplus (\mathfrak{g}_\alpha \cap \mathfrak{p})$. Since $\mathfrak{g}_\alpha$ is 1-dimensional, $\mathfrak{g}_\alpha \subseteq \mathfrak{k}$ or $\mathfrak{g}_\alpha \subseteq \mathfrak{p}$. We call an imaginary root α **compact** if $\mathfrak{g}_\alpha \subseteq \mathfrak{k}$, **noncompact** if $\mathfrak{g}_\alpha \subseteq \mathfrak{p}$.

We introduce two kinds of Cayley transforms, starting from a given θ stable Cartan subalgebra:

(i) Using an imaginary noncompact root β, we construct a new Cartan subalgebra whose intersection with $\mathfrak{p}_0$ goes up by 1 in dimension.

(ii) Using a real root α, we construct a new Cartan subalgebra whose intersection with $\mathfrak{p}_0$ goes down by 1 in dimension.

First we give the construction that starts from a Cartan subalgebra $\mathfrak{h}_0$ and uses an imaginary noncompact root β. Let E_β be a nonzero root vector. Since β is imaginary, $\overline{E_\beta}$ is in $\mathfrak{g}_{-\beta}$. Since β is noncompact, we have

$$0 < B_\theta(E_\beta, E_\beta) = -B(E_\beta, \overline{\theta E_\beta}) = B(E_\beta, \overline{E_\beta}).$$

Thus we are allowed to normalize E_β to make $B(E_\beta, \overline{E_\beta})$ be any positive constant. We choose to make $B(E_\beta, \overline{E_\beta}) = 2/|\beta|^2$. From Lemma 2.18a we have

$$[E_\beta, \overline{E_\beta}] = B(E_\beta, \overline{E_\beta})H_\beta = 2|\beta|^{-2}H_\beta.$$

Put $H'_\beta = 2|\beta|^{-2}H_\beta$. Then we have the bracket relations

$$[H'_\beta, E_\beta] = 2E_\beta, \quad [H'_\beta, \overline{E_\beta}] = -2\overline{E_\beta}, \quad [E_\beta, \overline{E_\beta}] = H'_\beta.$$

Also the elements $E_\beta + \overline{E_\beta}$ and $i(E_\beta - \overline{E_\beta})$ are fixed by bar and hence are in $\mathfrak{g}_0$. In terms of our discussion above of $\mathfrak{sl}(2, \mathbb{C})$, the correspondence is

$$H'_\beta \leftrightarrow \begin{pmatrix} 0 & i \\ -i & 0 \end{pmatrix}$$

$$E_\beta \leftrightarrow \tfrac{1}{2}\begin{pmatrix} 1 & -i \\ -i & -1 \end{pmatrix}$$

$$\overline{E_\beta} \leftrightarrow \tfrac{1}{2}\begin{pmatrix} 1 & i \\ i & -1 \end{pmatrix}$$

$$\overline{E_\beta} - E_\beta \leftrightarrow \begin{pmatrix} 0 & i \\ i & 0 \end{pmatrix}.$$

Define

$$\mathbf{c}_\beta = \operatorname{Ad}(\exp \tfrac{\pi}{4}(\overline{E_\beta} - E_\beta)) \tag{6.65a}$$

and

$$\mathfrak{h}'_0 = \mathfrak{g}_0 \cap \mathbf{c}_\beta(\mathfrak{h}) = \ker(\beta|_{\mathfrak{h}_0}) \oplus \mathbb{R}(E_\beta + \overline{E_\beta}). \tag{6.65b}$$

Note that E_β is not uniquely determined by the conditions on it, and both formulas (6.65) depend on the particular choice we make for E_β. To see that (6.65b) is valid, we can use infinite series to calculate that

$$\mathbf{c}_\beta(H'_\beta) = E_\beta + \overline{E_\beta} \tag{6.66a}$$

$$\mathbf{c}_\beta(E_\beta - \overline{E_\beta}) = E_\beta - \overline{E_\beta} \tag{6.66b}$$

$$\mathbf{c}_\beta(E_\beta + \overline{E_\beta}) = -H'_\beta. \tag{6.66c}$$

Then (6.66a) implies (6.65b).

Next we give the construction that starts from a Cartan subalgebra $\mathfrak{h}'_0$ and uses a real root α. Let E_α be a nonzero root vector. Since α is real, $\overline{E_\alpha}$ is in $\mathfrak{g}_\alpha$. Adjusting E_α, we may therefore assume that E_α is in $\mathfrak{g}_0$. Since α is real, θE_α is in $\mathfrak{g}_{-\alpha}$, and we know from Proposition 6.52a that $[E_\alpha, \theta E_\alpha] = B(E_\alpha, \theta E_\alpha)H_\alpha$ with $B(E_\alpha, \theta E_\alpha) < 0$. We normalize E_α by a real constant to make $B(E_\alpha, \theta E_\alpha) = -2/|\alpha|^2$, and put $H'_\alpha = 2|\alpha|^{-2}H_\alpha$. Then we have the bracket relations

$$[H'_\alpha, E_\alpha] = 2E_\alpha, \quad [H'_\alpha, \theta E_\alpha] = -2\theta E_\alpha, \quad [E_\alpha, \theta E_\alpha] = -H'_\alpha.$$

In terms of our discussion above of $\mathfrak{sl}(2, \mathbb{C})$, the correspondence is

$$H'_\alpha \leftrightarrow \begin{pmatrix} 1 & 0 \\ 0 & -1 \end{pmatrix}$$

$$E_\alpha \leftrightarrow \begin{pmatrix} 0 & 1 \\ 0 & 0 \end{pmatrix}$$

$$\theta E_\alpha \leftrightarrow \begin{pmatrix} 0 & 0 \\ -1 & 0 \end{pmatrix}$$

$$i(\theta E_\alpha - E_\alpha) \leftrightarrow \begin{pmatrix} 0 & -i \\ -i & 0 \end{pmatrix}.$$

Define

$$\mathbf{d}_\alpha = \mathrm{Ad}(\exp i \tfrac{\pi}{4}(\theta E_\alpha - E_\alpha)) \tag{6.67a}$$

and

$$\mathfrak{h}_0 = \mathfrak{g}_0 \cap \mathbf{d}_\alpha(\mathfrak{h}') = \ker(\alpha|_{\mathfrak{h}'_0}) \oplus \mathbb{R}(E_\alpha + \theta E_\alpha). \tag{6.67b}$$

To see that (6.67b) is valid, we can use infinite series to calculate that

$$\mathbf{d}_\alpha(H'_\alpha) = i(E_\alpha + \theta E_\alpha) \tag{6.68a}$$

$$\mathbf{d}_\alpha(E_\alpha - \theta E_\alpha) = E_\alpha - \theta E_\alpha \tag{6.68b}$$

$$\mathbf{d}_\alpha(E_\alpha + \theta E_\alpha) = i H'_\alpha. \tag{6.68c}$$

Then (6.68a) implies (6.67b).

Proposition 6.69. The two kinds of Cayley transforms are essentially inverse to each other in the following senses:

(a) If β is a noncompact imaginary root, then in the computation of $\mathbf{d}_{\mathbf{c}_\beta(\beta)} \circ \mathbf{c}_\beta$ the root vector $E_{\mathbf{c}_\beta(\beta)}$ can be taken to be $-i\mathbf{c}_\beta(E_\beta)$ and this choice makes the composition the identity.

(b) If α is a real root, then in the the computation of $\mathbf{c}_{\mathbf{d}_\alpha(\alpha)} \circ \mathbf{d}_\alpha$ the root vector $E_{\mathbf{d}_\alpha(\alpha)}$ can be taken to be $-i\mathbf{d}_\alpha(E_\alpha)$ and this choice makes the composition the identity.

PROOF.

(a) By (6.66),

$$\mathbf{c}_\beta(E_\beta) = \tfrac{1}{2}\mathbf{c}_\beta(E_\beta + \overline{E_\beta}) - \tfrac{1}{2}\mathbf{c}_\beta(E_\beta - \overline{E_\beta}) = -\tfrac{1}{2}H'_\beta - \tfrac{1}{2}(E_\beta - \overline{E_\beta}).$$

Both terms on the right side are in $i\mathfrak{g}_0$, and hence $-i\mathbf{c}_\beta(E_\beta)$ is in $\mathfrak{g}_0$. Since H'_β is in $\mathfrak{k}$ while E_β and $\overline{E_\beta}$ are in $\mathfrak{p}$,

$$\theta\mathbf{c}_\beta(E_\beta) = -\tfrac{1}{2}H'_\beta + \tfrac{1}{2}(E_\beta - \overline{E_\beta}).$$

Put $E_{\mathbf{c}_\beta(\beta)} = -i\mathbf{c}_\beta(E_\beta)$. From $B(E_\beta, \overline{E_\beta}) = 2/|\beta|^2$, we obtain

$$B(E_{\mathbf{c}_\beta(\beta)}, \theta E_{\mathbf{c}_\beta(\beta)}) = -2/|\beta|^2 = -2/|\mathbf{c}_\beta(\beta)|^2.$$

Thus $E_{\mathbf{c}_\beta(\beta)}$ is properly normalized. Then $\mathbf{d}_{\mathbf{c}_\beta(\beta)}$ becomes

$$\begin{aligned}\mathbf{d}_{\mathbf{c}_\beta(\beta)} &= \operatorname{Ad}(\exp i\tfrac{\pi}{4}(\theta E_{\mathbf{c}_\beta(\beta)} - E_{\mathbf{c}_\beta(\beta)}))\\ &= \operatorname{Ad}(\exp \tfrac{\pi}{4}(\theta\mathbf{c}_\beta(E_\beta) - \mathbf{c}_\beta(E_\beta)))\\ &= \operatorname{Ad}(\exp \tfrac{\pi}{4}(E_\beta - \overline{E_\beta})),\end{aligned}$$

and this is the inverse of

$$\mathbf{c}_\beta = \operatorname{Ad}(\exp \tfrac{\pi}{4}(\overline{E_\beta} - E_\beta)).$$

(b) By (6.68),

$$\mathbf{d}_\alpha(E_\alpha) = \tfrac{1}{2}\mathbf{d}_\alpha(E_\alpha + \theta E_\alpha) + \tfrac{1}{2}\mathbf{d}_\alpha(E_\alpha - \theta E_\alpha) = \tfrac{1}{2}iH'_\alpha + \tfrac{1}{2}(E_\alpha - \theta E_\alpha).$$

Since H'_α, E_α, and θE_α are in $\mathfrak{g}_0$,

$$\overline{\mathbf{d}_\alpha(E_\alpha)} = -\tfrac{1}{2}iH'_\alpha + \tfrac{1}{2}(E_\alpha - \theta E_\alpha).$$

Put $E_{\mathbf{d}_\alpha(\alpha)} = -i\mathbf{d}_\alpha(E_\alpha)$. From $B(E_\alpha, \theta E_\alpha) = -2/|\alpha|^2$, we obtain

$$B(E_{\mathbf{d}_\alpha(\alpha)}, \overline{E_{\mathbf{d}_\alpha(\alpha)}}) = 2/|\alpha|^2 = 2/|\mathbf{d}_\alpha(\alpha)|^2.$$

Thus $E_{\mathbf{d}_\alpha(\alpha)}$ is properly normalized. Then $\mathbf{c}_{\mathbf{d}_\alpha(\alpha)}$ becomes

$$\begin{aligned}\mathbf{c}_{\mathbf{d}_\alpha(\alpha)} &= \operatorname{Ad}(\exp \tfrac{\pi}{4}(\overline{E_{\mathbf{d}_\alpha(\alpha)}} - E_{\mathbf{d}_\alpha(\alpha)}))\\ &= \operatorname{Ad}(\exp i\tfrac{\pi}{4}(\mathbf{d}_\alpha(E_\alpha) + \overline{\mathbf{d}_\alpha(E_\alpha)}))\\ &= \operatorname{Ad}(\exp i\tfrac{\pi}{4}(E_\alpha - \theta E_\alpha)),\end{aligned}$$

and this is the inverse of

$$\mathbf{d}_\alpha = \operatorname{Ad}(\exp i\tfrac{\pi}{4}(\theta E_\alpha - E_\alpha)).$$

Proposition 6.70. Let $\mathfrak{h}_0$ be a θ stable Cartan subalgebra of $\mathfrak{g}_0$. Then there are no noncompact imaginary roots if and only if $\mathfrak{h}_0$ is maximally noncompact, and there are no real roots if and only if $\mathfrak{h}_0$ is maximally compact.

PROOF. The Cayley transform construction $\mathbf{c}_\beta$ tells us that if $\mathfrak{h}_0$ has a noncompact imaginary root β, then $\mathfrak{h}_0$ is not maximally noncompact. Similarly the Cayley transform construction $\mathbf{d}_\alpha$ tells us that if $\mathfrak{h}_0$ has a real root α, then $\mathfrak{h}_0$ is not maximally compact.

For the converses write $\mathfrak{h}_0 = \mathfrak{t}_0 \oplus \mathfrak{a}_0$, and let $\Delta = \Delta(\mathfrak{g}, \mathfrak{h})$ be the set of roots. Form the expansion

$$\mathfrak{g} = \mathfrak{h} \oplus \bigoplus_{\alpha \in \Delta} \mathfrak{g}_\alpha. \tag{6.71}$$

Suppose there are no noncompact imaginary roots. Then

$$Z_\mathfrak{g}(\mathfrak{a}_0) = \mathfrak{h} \oplus \bigoplus_{\substack{\alpha\in\Delta, \\ \alpha \text{ imaginary}}} \mathfrak{g}_\alpha = \mathfrak{h} \oplus \bigoplus_{\substack{\alpha\in\Delta, \\ \alpha \text{ compact} \\ \text{imaginary}}} \mathfrak{g}_\alpha$$

and $$\mathfrak{p}_0 \cap Z_\mathfrak{g}(\mathfrak{a}_0) = \mathfrak{g}_0 \cap (\mathfrak{p} \cap Z_\mathfrak{g}(\mathfrak{a}_0)) = \mathfrak{g}_0 \cap (\mathfrak{p} \cap \mathfrak{h}) = \mathfrak{a}_0.$$

Hence $\mathfrak{a}_0$ is maximal abelian in $\mathfrak{p}_0$, and $\mathfrak{h}_0$ is maximally noncompact.

Suppose there are no real roots. From the expansion (6.71) we obtain

$$Z_\mathfrak{g}(\mathfrak{t}_0) = \mathfrak{h} \oplus \bigoplus_{\substack{\alpha\in\Delta, \\ \alpha \text{ real}}} \mathfrak{g}_\alpha = \mathfrak{h}$$

and $\mathfrak{k}_0 \cap Z_\mathfrak{g}(\mathfrak{t}_0) = \mathfrak{k}_0 \cap \mathfrak{h} = \mathfrak{t}_0$. Therefore $\mathfrak{t}_0$ is maximal abelian in $\mathfrak{k}_0$, and $\mathfrak{h}_0$ is maximally compact.

The Cayley transforms and the above propositions give us a method of finding all Cartan subalgebras up to conjugacy. In fact, if we start with a θ stable Cartan subalgebra, we can apply various Cayley transforms $\mathbf{c}_\beta$ as long as there are noncompact imaginary roots, and we know that the resulting Cartan subalgebra will be maximally noncompact when we are done. Consequently if we apply various Cayley transforms $\mathbf{d}_\alpha$ in the reverse order, starting from a maximally noncompact Cartan subalgebra, we obtain all Cartan subalgebras up to conjugacy.

Alternatively if we start with a θ stable Cartan subalgebra, we can apply various Cayley transforms $\mathbf{d}_\alpha$ as long as there are real roots, and we know that the resulting Cartan subalgebra will be maximally compact when we are done. Consequently if we apply various Cayley transforms $\mathbf{c}_\beta$ in the reverse order, starting from a maximally compact Cartan subalgebra, we obtain all Cartan subalgebras up to conjugacy.

EXAMPLE. Let $\mathfrak{g}_0 = \mathfrak{sp}(2, \mathbb{R})$ with θ given by negative transpose. We can take the Iwasawa $\mathfrak{a}_0$ to be the diagonal subalgebra

$$\mathfrak{a}_0 = \{\operatorname{diag}(s, t, -s, -t)\}.$$

Let f_1 and f_2 be the linear functionals on $\mathfrak{a}_0$ that give s and t on the indicated matrix. For this example, $\mathfrak{m}_0 = 0$. Thus Proposition 6.47 shows that $\mathfrak{a}_0$ is a maximally noncompact Cartan subalgebra. The roots are $\pm 2f_1, \pm 2f_2, \pm(f_1 + f_2), \pm(f_1 - f_2)$. All of them are real. We begin with a $\mathbf{d}_\alpha$ type Cayley transform, noting that $\pm\alpha$ give the same thing. The data for $2f_1$ and $2f_2$ are conjugate within $\mathfrak{g}_0$, and so are the data for $f_1 + f_2$ and $f_1 - f_2$. So there are only two essentially different first steps, say $\mathbf{d}_{2f_2}$ and $\mathbf{d}_{f_1-f_2}$. After $\mathbf{d}_{2f_2}$, the only real roots are $\pm 2f_1$ (or more precisely $\mathbf{d}_{2f_2}(\pm 2f_1)$). A second Cayley transform $\mathbf{d}_{2f_1}$ leads to all roots imaginary, hence to a maximally compact Cartan subalgebra, and we can go no further. Similarly after $\mathbf{d}_{f_1-f_2}$, the only real roots are $\pm(f_1 + f_2)$, and the second Cayley transform $\mathbf{d}_{f_1+f_2}$ leads to all roots imaginary. A little computation shows that we have produced

$$\begin{pmatrix} s & 0 & 0 & 0 \\ 0 & t & 0 & 0 \\ 0 & 0 & -s & 0 \\ 0 & 0 & 0 & -t \end{pmatrix}, \quad \begin{pmatrix} s & 0 & 0 & 0 \\ 0 & 0 & 0 & \theta \\ 0 & 0 & -s & 0 \\ 0 & -\theta & 0 & 0 \end{pmatrix},$$

$$\begin{pmatrix} t & \theta & 0 & 0 \\ -\theta & t & 0 & 0 \\ 0 & 0 & -t & \theta \\ 0 & 0 & -\theta & -t \end{pmatrix}, \quad \begin{pmatrix} 0 & 0 & \theta_1 & 0 \\ 0 & 0 & 0 & \theta_2 \\ -\theta_1 & 0 & 0 & 0 \\ 0 & -\theta_2 & 0 & 0 \end{pmatrix}.$$

The second Cartan subalgebra results from the first by applying $\mathbf{d}_{2f_2}$, the third results from the first by applying $\mathbf{d}_{f_1-f_2}$, and the fourth results from the first by applying $\mathbf{d}_{2f_1}\mathbf{d}_{2f_2}$.

As in the example, when we pass from $\mathfrak{h}_0'$ to $\mathfrak{h}_0$ by $\mathbf{d}_\alpha$, we can anticipate what roots will be real for $\mathfrak{h}_0$. What we need in order to do a succession of such Cayley transforms is a sequence of real roots that become imaginary one at a time. In other words, we can do a succession of such Cayley transforms with ease if we have an orthogonal sequence of real roots.

Similarly when we apply $\mathbf{c}_\alpha$ to pass from $\mathfrak{h}_0$ to $\mathfrak{h}_0'$, we can anticipate what roots will be imaginary for $\mathfrak{h}_0'$. But a further condition on a root beyond "imaginary" is needed to do a Cayley transform $\mathbf{c}_\alpha$; we need the imaginary root to be noncompact. The following proposition tells how to anticipate which imaginary roots are noncompact after a Cayley transform.

Proposition 6.72. Let α be a noncompact imaginary root. Let β be a root orthogonal to α, so that the α string containing β is symmetric about β. Let E_α and E_β be nonzero roots vectors for α and β, and normalize E_α as in the definition of the Cayley transform $\mathbf{c}_\alpha$.

(a) If $\beta \pm \alpha$ are not roots, then $\mathbf{c}_\alpha(E_\beta) = E_\beta$. Thus if β is imaginary, then β is compact if and only if $\mathbf{c}_\alpha(\beta)$ is compact.

(b) If $\beta \pm \alpha$ are roots, then $\mathbf{c}_\alpha(E_\beta) = \frac{1}{2}([\overline{E_\alpha}, E_\beta] - [E_\alpha, E_\beta])$. Thus if β is imaginary, then β is compact if and only if $\mathbf{c}_\alpha(\beta)$ is noncompact.

PROOF. Recall that $\mathbf{c}_\alpha = \mathrm{Ad}(\exp \frac{\pi}{4}(\overline{E_\alpha} - E_\alpha))$ with $[E_\alpha, \overline{E_\alpha}] = H'_\alpha$.

(a) In this case $\mathbf{c}_\alpha(E_\beta) = E_\beta$ clearly. If β is imaginary, then the equal vectors $\mathbf{c}_\alpha(E_\beta)$ and E_β are both in $\mathfrak{k}$ or both in $\mathfrak{p}$.

(b) Here we use Corollary 2.37 and Proposition 2.48g to calculate that

$$\begin{aligned} \mathrm{ad}\, \tfrac{\pi}{4}(\overline{E_\alpha} - E_\alpha)E_\beta &= \tfrac{\pi}{4}([\overline{E_\alpha}, E_\beta] - [E_\alpha, E_\beta]) \\ \mathrm{ad}^2(\tfrac{\pi}{4}(\overline{E_\alpha} - E_\alpha))E_\beta &= -(\tfrac{\pi}{4})^2([E_\alpha, [\overline{E_\alpha}, E_\beta]] + [\overline{E_\alpha}, [E_\alpha, E_\beta]]) \\ &= -(\tfrac{\pi}{4})^2(2E_\beta + 2E_\beta) \\ &= -(\tfrac{\pi}{2})^2 E_\beta. \end{aligned}$$

Then we have

$$\begin{aligned} \mathbf{c}_\alpha(E_\beta) &= \sum_{n=0}^{\infty} \tfrac{1}{(2n)!}\mathrm{ad}^{2n}(\tfrac{\pi}{4}(\overline{E_\alpha} - E_\alpha))E_\beta \\ &\quad + \sum_{n=0}^{\infty} \tfrac{1}{(2n+1)!}\mathrm{ad}(\tfrac{\pi}{4}(\overline{E_\alpha} - E_\alpha))\mathrm{ad}^{2n}(\tfrac{\pi}{4}(\overline{E_\alpha} - E_\alpha))E_\beta \\ &= \sum_{n=0}^{\infty} \tfrac{1}{(2n)!}(-1)^n(\tfrac{\pi}{2})^{2n}E_\beta \\ &\quad + \sum_{n=0}^{\infty} \tfrac{1}{(2n+1)!}(-1)^n(\tfrac{\pi}{2})^{2n}(\tfrac{\pi}{4})([\overline{E_\alpha}, E_\beta] - [E_\alpha, E_\beta]) \\ &= (\cos \tfrac{\pi}{2})E_\beta + \tfrac{1}{2}(\sin \tfrac{\pi}{2})([\overline{E_\alpha}, E_\beta] - [E_\alpha, E_\beta]) \\ &= \tfrac{1}{2}([\overline{E_\alpha}, E_\beta] - [E_\alpha, E_\beta]). \end{aligned}$$

If β is imaginary, then $\mathbf{c}_\alpha(E_\beta)$ is in $\mathfrak{k}$ if and only if E_β is in $\mathfrak{p}$ since E_α and $\overline{E_\alpha}$ are in $\mathfrak{p}$.

We say that two orthogonal roots α and β are **strongly orthogonal** if $\beta \pm \alpha$ are not roots. Proposition 6.72 indicates that we can do a succession of Cayley transforms $\mathbf{c}_\beta$ with ease if we have a strongly orthogonal sequence of noncompact imaginary roots.

If α and β are orthogonal but not strongly orthogonal, then

$$|\beta \pm \alpha|^2 = |\beta|^2 + |\alpha|^2 \tag{6.73}$$

shows that there are at least two root lengths. Actually we must have $|\beta|^2 = |\alpha|^2$, since otherwise (6.73) would produce three root lengths, which is forbidden within a simple component of a reduced root system. Thus (6.73) becomes $|\beta \pm \alpha|^2 = 2|\alpha|^2$, and the simple component of the root system containing α and β has a double line in its Dynkin diagram. In other words, whenever the Dynkin diagram of the root system has no double line, then orthogonal roots are automatically strongly orthogonal.

8. Vogan Diagrams

To a real semisimple Lie algebra $\mathfrak{g}_0$, in the presence of some other data, we shall associate a diagram consisting of the Dynkin diagram of $\mathfrak{g} = (\mathfrak{g}_0)^{\mathbb{C}}$ with some additional information superimposed. This diagram will be called a "Vogan diagram." We shall see that the same Vogan diagram cannot come from two nonisomorphic $\mathfrak{g}_0$'s and that every diagram that looks formally like a Vogan diagram comes from some $\mathfrak{g}_0$. Thus Vogan diagrams give us a handle on the problem of classification, and all we need to do is to sort out which Vogan diagrams come from the same $\mathfrak{g}_0$.

Let $\mathfrak{g}_0$ be a real semisimple Lie algebra, let $\mathfrak{g}$ be its complexification, let θ be a Cartan involution, let $\mathfrak{g}_0 = \mathfrak{k}_0 \oplus \mathfrak{p}_0$ be the corresponding Cartan decomposition, and let B be as in §§6–7. We introduce a maximally compact θ stable Cartan subalgebra $\mathfrak{h}_0 = \mathfrak{t}_0 \oplus \mathfrak{a}_0$ of $\mathfrak{g}_0$, with complexification $\mathfrak{h} = \mathfrak{t} \oplus \mathfrak{a}$, and we let $\Delta = \Delta(\mathfrak{g}, \mathfrak{h})$ be the set of roots. By Proposition 6.70 there are no real roots, i.e., no roots that vanish on $\mathfrak{t}$.

Choose a positive system Δ^+ for Δ that takes $i\mathfrak{t}_0$ before $\mathfrak{a}$. For example, Δ^+ can be defined in terms of a lexicographic ordering built from a basis of $i\mathfrak{t}_0$ followed by a basis of $\mathfrak{a}_0$. Since θ is $+1$ on $\mathfrak{t}_0$ and -1 on $\mathfrak{a}_0$ and since there are no real roots, $\theta(\Delta^+) = \Delta^+$. Therefore θ permutes the simple roots. It must fix the simple roots that are imaginary and permute in 2-cycles the simple roots that are complex.

By the **Vogan diagram** of the triple $(\mathfrak{g}_0, \mathfrak{h}_0, \Delta^+)$, we mean the Dynkin diagram of Δ^+ with the 2-element orbits under θ so labeled and with the 1-element orbits painted or not, according as the corresponding imaginary simple root is noncompact or compact.

For example if $\mathfrak{g}_0 = \mathfrak{su}(3, 3)$, let us take θ to be negative conjugate transpose, $\mathfrak{h}_0 = \mathfrak{t}_0$ to be the diagonal subalgebra, and Δ^+ to be determined by the conditions $e_1 \geq e_2 \geq e_4 \geq e_5 \geq e_3 \geq e_6$. The Dynkin diagram is of type A_5, and all simple roots are imaginary since $\mathfrak{a}_0 = 0$. In particular, θ acts as the identity in the Dynkin diagram. The compact roots $e_i - e_j$ are those with i and j in the same set $\{1, 2, 3\}$ or $\{4, 5, 6\}$, while the noncompact roots are those with i and j in opposite sets. Then among the simple roots, $e_1 - e_2$ is compact, $e_2 - e_4$ is noncompact, $e_4 - e_5$ is compact, $e_5 - e_3$ is noncompact, and $e_3 - e_6$ is noncompact. Hence the Vogan diagram is

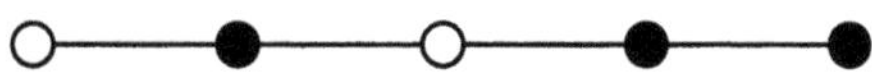

Here are two infinite classes of examples.

EXAMPLES.

1) Let $\mathfrak{g}_0 = \mathfrak{su}(p, q)$ with negative conjugate transpose as Cartan involution. We take $\mathfrak{h}_0 = \mathfrak{t}_0$ to be the diagonal subalgebra. Then θ is 1 on all the roots. We use the standard ordering, so that the positive roots are $e_i - e_j$ with $i < j$. A positive root is compact if i and j are both in $\{1, \ldots, p\}$ or both in $\{p+1, \ldots, p+q\}$. It is noncompact if i is in $\{1, \ldots, p\}$ and j is in $\{p+1, \ldots, p+q\}$. Thus among the simple roots $e_i - e_{i+1}$, the root $e_p - e_{p+1}$ is noncompact, and the others are compact. The Vogan diagram is

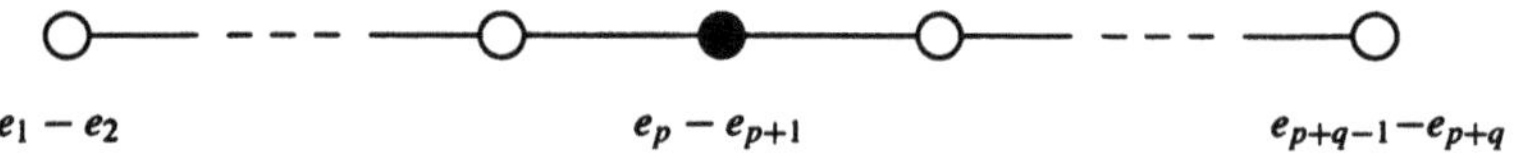

2) Let $\mathfrak{g}_0 = \mathfrak{sl}(2n, \mathbb{R})$ with negative transpose as Cartan involution, and define

$$\mathfrak{h}_0 = \left\{ \begin{pmatrix} x_1 & \theta_1 & & & \\ -\theta_1 & x_1 & & & \\ & & \ddots & & \\ & & & x_n & \theta_n \\ & & & -\theta_n & x_n \end{pmatrix} \right\}.$$

The matrices here are understood to be built from 2-by-2 blocks and to have $\sum_{j=1}^n x_j = 0$. The subspace $\mathfrak{t}_0$ corresponds to the θ_j part, $1 \leq j \leq n$, i.e., it is the subspace where all x_j are 0. The subspace $\mathfrak{a}_0$ similarly corresponds to the x_j part, $1 \leq j \leq n$. We define linear functionals e_j and f_j to depend only on the jth block, the dependence being

$$e_j \begin{pmatrix} x_j & -iy_j \\ iy_j & x_j \end{pmatrix} = y_j \quad \text{and} \quad f_j \begin{pmatrix} x_j & -iy_j \\ iy_j & x_j \end{pmatrix} = x_j.$$

Computation shows that

$$\Delta = \{\pm e_j \pm e_k \pm (f_j - f_k) \mid j \neq k\} \cup \{\pm 2e_l \mid 1 \leq l \leq n\}.$$

Roots that involve only e_j's are imaginary, those that involve only f_j's are real, and the remainder are complex. It is apparent that there are no real roots, and therefore $\mathfrak{h}_0$ is maximally compact. The involution θ acts as $+1$ on the e_j's and -1 on the f_j's. We define a lexicographic ordering by using the spanning set

$$e_1, \ldots, e_n, f_1, \ldots, f_n,$$

and we obtain

$$\Delta^+ = \begin{cases} e_j + e_k \pm (f_j - f_k), & \text{all } j \neq k \\ e_j - e_k \pm (f_j - f_k), & j < k \\ 2e_j, & 1 \leq l \leq n. \end{cases}$$

The Vogan diagram is

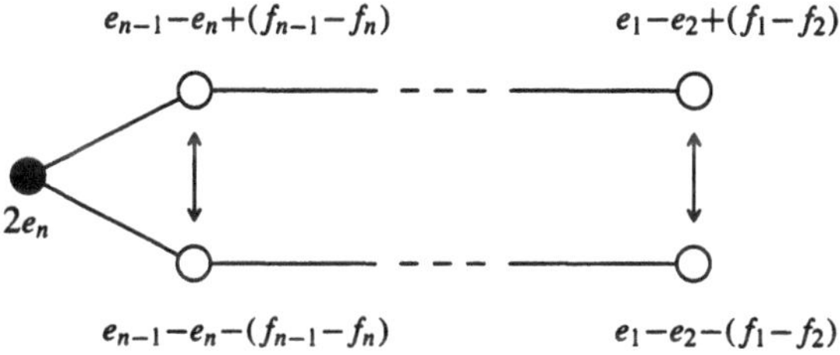

Theorem 6.74. Let $\mathfrak{g}_0$ and $\mathfrak{g}_0'$ be real semisimple Lie algebras. With notation as above, if two triples $(\mathfrak{g}_0, \mathfrak{h}_0, \Delta^+)$ and $(\mathfrak{g}_0', \mathfrak{h}_0', (\Delta')^+)$ have the same Vogan diagram, then $\mathfrak{g}_0$ and $\mathfrak{g}_0'$ are isomorphic.

REMARK. This theorem is an analog for real semisimple Lie algebras of the Isomorphism Theorem (Theorem 2.108) for complex semisimple Lie algebras.

PROOF. Since the Dynkin diagrams are the same, the Isomorphism Theorem (Theorem 2.108) shows that there is no loss of generality in assuming that $\mathfrak{g}_0$ and $\mathfrak{g}_0'$ have the same complexification $\mathfrak{g}$. Let $\mathfrak{u}_0 = \mathfrak{k}_0 \oplus i\mathfrak{p}_0$ and $\mathfrak{u}_0' = \mathfrak{k}_0' \oplus i\mathfrak{p}_0'$ be the associated compact real forms of $\mathfrak{g}$. By Corollary 6.20, there exists $x \in \operatorname{Int}\mathfrak{g}$ such that $x\mathfrak{u}_0' = \mathfrak{u}_0$. The real form $x\mathfrak{g}_0'$ of $\mathfrak{g}$ is isomorphic to $\mathfrak{g}_0'$ and has Cartan decomposition $x\mathfrak{g}_0' = x\mathfrak{k}_0' \oplus x\mathfrak{p}_0'$. Since $x\mathfrak{k}_0' \oplus ix\mathfrak{p}_0' = x\mathfrak{u}_0' = \mathfrak{u}_0$, there is no loss of generality in assuming that $\mathfrak{u}_0' = \mathfrak{u}_0$ from the outset. Then

$$\theta(\mathfrak{u}_0) = \mathfrak{u}_0 \quad \text{and} \quad \theta'(\mathfrak{u}_0) = \mathfrak{u}_0. \tag{6.75}$$

Let us write the effect of the Cartan decompositions on the Cartan subalgebras as $\mathfrak{h}_0 = \mathfrak{t}_0 \oplus \mathfrak{a}_0$ and $\mathfrak{h}'_0 = \mathfrak{t}'_0 \oplus \mathfrak{a}'_0$. Then $\mathfrak{t}_0 \oplus i\mathfrak{a}_0$ and $\mathfrak{t}'_0 \oplus i\mathfrak{a}'_0$ are maximal abelian subspaces of $\mathfrak{u}_0$. By Theorem 4.34 there exists $k \in \operatorname{Int} \mathfrak{u}_0$ with $k(\mathfrak{t}'_0 \oplus i\mathfrak{a}'_0) = \mathfrak{t}_0 \oplus i\mathfrak{a}_0$. Replacing $\mathfrak{g}'_0$ by $k\mathfrak{g}'_0$ and arguing as above, we may assume that $\mathfrak{t}'_0 \oplus i\mathfrak{a}'_0 = \mathfrak{t}_0 \oplus i\mathfrak{a}_0$ from the outset. Therefore $\mathfrak{h}_0$ and $\mathfrak{h}'_0$ have the same complexification, which we denote $\mathfrak{h}$. The space

$$\mathfrak{u}_0 \cap \mathfrak{h} = \mathfrak{t}_0 \oplus i\mathfrak{a}_0 = \mathfrak{t}'_0 \oplus i\mathfrak{a}'_0$$

is a maximal abelian subspace of $\mathfrak{u}_0$.

Now that the complexifications $\mathfrak{g}$ and $\mathfrak{h}$ have been aligned, the root systems are the same. Let the positive systems given in the respective triples be Δ^+ and $\Delta^{+\prime}$. By Theorems 4.54 and 2.63 there exists $k' \in \operatorname{Int} \mathfrak{u}_0$ normalizing $\mathfrak{u}_0 \cap \mathfrak{h}$ with $k'\Delta^{+\prime} = \Delta^+$. Replacing $\mathfrak{g}'_0$ by $k'\mathfrak{g}'_0$ and arguing as above, we may assume that $\Delta^{+\prime} = \Delta^+$ from the outset.

The next step is to choose normalizations of root vectors relative to $\mathfrak{h}$. For this proof let B be the Killing form of $\mathfrak{g}$. We start with root vectors X_α produced from $\mathfrak{h}$ as in Theorem 6.6. Using (6.12), we construct a compact real form $\tilde{\mathfrak{u}}_0$ of $\mathfrak{g}$. The subalgebra $\tilde{\mathfrak{u}}_0$ contains the real subspace of $\mathfrak{h}$ where the roots are imaginary, which is just $\mathfrak{u}_0 \cap \mathfrak{h}$. By Corollary 6.20, there exists $g \in \operatorname{Int} \mathfrak{g}$ such that $g\tilde{\mathfrak{u}}_0 = \mathfrak{u}_0$. Then $g\tilde{\mathfrak{u}}_0 = \mathfrak{u}_0$ is built by (6.12) from $g(\mathfrak{u}_0 \cap \mathfrak{h})$ and the root vectors gX_α. Since $\mathfrak{u}_0 \cap \mathfrak{h}$ and $g(\mathfrak{u}_0 \cap \mathfrak{h})$ are maximal abelian in $\mathfrak{u}_0$, Theorem 4.34 produces $u \in \operatorname{Int} \mathfrak{u}_0$ with $ug(\mathfrak{u}_0 \cap \mathfrak{h}) = \mathfrak{u}_0 \cap \mathfrak{h}$. Then $\mathfrak{u}_0$ is built by (6.12) from $ug(\mathfrak{u}_0 \cap \mathfrak{h})$ and the root vectors ugX_α. For $\alpha \in \Delta$, put $Y_\alpha = ugX_\alpha$. Then we have established that

$$\mathfrak{u}_0 = \sum_{\alpha \in \Delta} \mathbb{R}(iH_\alpha) + \sum_{\alpha \in \Delta} \mathbb{R}(Y_\alpha - Y_{-\alpha}) + \sum_{\alpha \in \Delta} \mathbb{R}i(Y_\alpha + Y_{-\alpha}). \tag{6.76}$$

We have not yet used the information that is superimposed on the Dynkin diagram of Δ^+. Since the automorphisms of Δ^+ defined by θ and θ' are the same, θ and θ' have the same effect on $\mathfrak{h}^*$. Thus

$$\theta(H) = \theta'(H) \qquad \text{for all } H \in \mathfrak{h}. \tag{6.77}$$

If α is an imaginary simple root, then

$$\theta(Y_\alpha) = Y_\alpha = \theta'(Y_\alpha) \qquad \text{if } \alpha \text{ is unpainted,} \tag{6.78a}$$

$$\theta(Y_\alpha) = -Y_\alpha = \theta'(Y_\alpha) \qquad \text{if } \alpha \text{ is painted.} \tag{6.78b}$$

We still have to deal with the complex simple roots. For $\alpha \in \Delta$, write $\theta Y_\alpha = a_\alpha Y_\alpha$. From (6.75) we know that

$$\theta(\mathfrak{u}_0 \cap \operatorname{span}\{Y_\alpha, Y_{-\alpha}\}) \subseteq \mathfrak{u}_0 \cap \operatorname{span}\{Y_{\theta\alpha}, Y_{-\theta\alpha}\}.$$

In view of (6.76) this inclusion means that

$$\theta(\mathbb{R}(Y_\alpha - Y_{-\alpha}) + \mathbb{R}i(Y_\alpha + Y_{-\alpha})) \subseteq \theta(\mathbb{R}(Y_{\theta\alpha} - Y_{-\theta\alpha}) + \mathbb{R}i(Y_{\theta\alpha} + Y_{-\theta\alpha})).$$

If x and y are real and if $z = x + yi$, then we have

$$x(Y_\alpha - Y_{-\alpha}) + yi(Y_\alpha + Y_{-\alpha}) = zY_\alpha - \bar{z}Y_{-\alpha}.$$

Thus the expression $\theta(zY_\alpha - \bar{z}Y_{-\alpha}) = za_\alpha Y_{\theta\alpha} - \bar{z}a_{-\alpha}Y_{-\theta\alpha}$ must be of the form $wY_{\theta\alpha} - \bar{w}Y_{-\theta\alpha}$, and we conclude that

$$a_{-\alpha} = \overline{a_\alpha}. \tag{6.79}$$

Meanwhile $a_\alpha a_{-\alpha} = B(a_\alpha Y_{\theta\alpha}, a_{-\alpha}Y_{-\theta\alpha}) = B(\theta Y_\alpha, \theta Y_{-\alpha}) = B(Y_\alpha, Y_{-\alpha}) = 1$ shows that

$$a_\alpha a_{-\alpha} = 1. \tag{6.80}$$

Combining (6.79) and (6.80), we see that

$$|a_\alpha| = 1. \tag{6.81}$$

Next we observe that

$$a_\alpha a_{\theta\alpha} = 1 \tag{6.82}$$

since $Y_\alpha = \theta^2 Y_\alpha = \theta(a_\alpha Y_{\theta\alpha}) = a_\alpha a_{\theta\alpha} Y_\alpha$.

For each pair of complex simple roots α and $\theta\alpha$, choose square roots $a_\alpha^{1/2}$ and $a_{\theta\alpha}^{1/2}$ so that

$$a_\alpha^{1/2} a_{\theta\alpha}^{1/2} = 1. \tag{6.83}$$

This is possible by (6.82).

Similarly write $\theta' Y_\alpha = b_\alpha Y_{\theta\alpha}$ with

$$|b_\alpha| = 1, \tag{6.84}$$

and define $b_\alpha^{1/2}$ and $b_{\theta\alpha}^{1/2}$ for α and $\theta\alpha$ simple so that

$$b_\alpha^{1/2} b_{\theta\alpha}^{1/2} = 1. \tag{6.85}$$

By (6.81) and (6.84), we can define H and H' in $\mathfrak{u}_0 \cap \mathfrak{h}$ by the conditions that $\alpha(H) = \alpha(H') = 0$ for α imaginary simple and

$$\exp(\tfrac{1}{2}\alpha(H)) = a_\alpha^{1/2}, \quad \exp(\tfrac{1}{2}\theta\alpha(H)) = a_{\theta\alpha}^{1/2},$$
$$\exp(\tfrac{1}{2}\alpha(H')) = b_\alpha^{1/2}, \quad \exp(\tfrac{1}{2}\theta\alpha(H')) = b_{\theta\alpha}^{1/2}$$

for α and $\theta\alpha$ complex simple.

We shall show that

$$\theta' \circ \mathrm{Ad}(\exp \tfrac{1}{2}(H - H')) = \mathrm{Ad}(\exp \tfrac{1}{2}(H - H')) \circ \theta. \tag{6.86}$$

In fact, the two sides of (6.86) are equal on $\mathfrak{h}$ and also on each X_α for α imaginary simple, by (6.77) and (6.78), since the Ad factor drops out from each side. If α is complex simple, then

$$\begin{aligned}
\theta' \circ \mathrm{Ad}(\exp \tfrac{1}{2}(H - H'))Y_\alpha &= \theta'(e^{\frac{1}{2}\alpha(H-H')}Y_\alpha) \\
&= b_\alpha a_\alpha^{1/2} b_\alpha^{-1/2} Y_{\theta\alpha} \\
&= b_\alpha^{1/2} a_\alpha^{-1/2} \theta Y_\alpha \\
&= b_{\theta\alpha}^{-1/2} a_{\theta\alpha}^{1/2} \theta Y_\alpha \quad \text{by (6.83) and (6.85)} \\
&= \mathrm{Ad}(\exp \tfrac{1}{2}(H - H')) \circ \theta Y_\alpha .
\end{aligned}$$

This proves (6.86).

Applying (6.86) to $\mathfrak{k}$ and then to $\mathfrak{p}$, we see that

$$\begin{aligned}
&\mathrm{Ad}(\exp \tfrac{1}{2}(H - H'))(\mathfrak{k}) \subseteq \mathfrak{k}' \\
&\mathrm{Ad}(\exp \tfrac{1}{2}(H - H'))(\mathfrak{p}) \subseteq \mathfrak{p}',
\end{aligned} \tag{6.87}$$

and then equality must hold in each line of (6.87). Since the element $\mathrm{Ad}(\exp \frac{1}{2}(H-H'))$ carries $\mathfrak{u}_0$ to itself, it must carry $\mathfrak{k}_0 = \mathfrak{u}_0 \cap \mathfrak{k}$ to $\mathfrak{k}_0' = \mathfrak{u}_0 \cap \mathfrak{k}'$ and $\mathfrak{p}_0 = \mathfrak{u}_0 \cap \mathfrak{p}$ to $\mathfrak{p}_0' = \mathfrak{u}_0 \cap \mathfrak{p}'$. Hence it must carry $\mathfrak{g}_0 = \mathfrak{k}_0 \oplus \mathfrak{p}_0$ to $\mathfrak{g}_0' = \mathfrak{k}_0' \oplus \mathfrak{p}_0'$. This completes the proof.

Now let us address the question of existence. We define an **abstract Vogan diagram** to be an abstract Dynkin diagram with two pieces of additional structure indicated: One is an automorphism of order 1 or 2 of the diagram, which is to be indicated by labeling the 2-element orbits. The other is a subset of the 1-element orbits, which is to be indicated by painting the vertices corresponding to the members of the subset. Every Vogan diagram is of course an abstract Vogan diagram.

Theorem 6.88. If an abstract Vogan diagram is given, then there exist a real semisimple Lie algebra $\mathfrak{g}_0$, a Cartan involution θ, a maximally compact θ stable Cartan subalgebra $\mathfrak{h}_0 = \mathfrak{t}_0 \oplus \mathfrak{a}_0$, and a positive system Δ^+ for $\Delta = \Delta(\mathfrak{g}, \mathfrak{h})$ that takes $\mathfrak{t}_0$ before $i\mathfrak{a}_0$ such that the given diagram is the Vogan diagram of $(\mathfrak{g}_0, \mathfrak{h}_0, \Delta^+)$.

REMARK. Briefly the theorem says that any abstract Vogan diagram comes from some $\mathfrak{g}_0$. Thus the theorem is an analog for real semisimple Lie algebras of the Existence Theorem (Theorem 2.111) for complex semisimple Lie algebras.

PROOF. By the Existence Theorem (Theorem 2.111) let $\mathfrak{g}$ be a complex semisimple Lie algebra with the given abstract Dynkin diagram as its Dynkin diagram, and let $\mathfrak{h}$ be a Cartan subalgebra (Theorem 2.9). Put $\Delta = \Delta(\mathfrak{g}, \mathfrak{h})$, and let Δ^+ be the positive system determined by the given data. Introduce root vectors X_α normalized as in Theorem 6.6, and define a compact real form $\mathfrak{u}_0$ of $\mathfrak{g}$ in terms of $\mathfrak{h}$ and the X_α by (6.12). The formula for $\mathfrak{u}_0$ is

$$\mathfrak{u}_0 = \sum_{\alpha\in\Delta} \mathbb{R}(iH_\alpha) + \sum_{\alpha\in\Delta} \mathbb{R}(X_\alpha - X_{-\alpha}) + \sum_{\alpha\in\Delta} \mathbb{R}i(X_\alpha + X_{-\alpha}). \tag{6.89}$$

The given data determine an automorphism θ of the Dynkin diagram, which extends linearly to $\mathfrak{h}^*$ and is isometric. Let us see that $\theta(\Delta) = \Delta$. It is enough to see that $\theta(\Delta^+) \subseteq \Delta$. We prove that $\theta(\Delta^+) \subseteq \Delta$ by induction on the level $\sum n_i$ of a positive root $\alpha = \sum n_i \alpha_i$. If the level is 1, then the root α is simple and we are given that $\theta\alpha$ is a simple root. Let $n > 1$, and assume inductively that $\theta\alpha$ is in Δ if $\alpha \in \Delta^+$ has level $< n$. Let α have level n. If we choose α_i simple with $\langle \alpha, \alpha_i \rangle > 0$, then $s_{\alpha_i}(\alpha)$ is a positive root β with smaller level than α. By inductive hypothesis, $\theta\beta$ and $\theta\alpha_i$ are in Δ. Since θ is isometric, $\theta\alpha = s_{\theta\alpha_i}(\theta\beta)$, and therefore $\theta\alpha$ is in Δ. This completes the induction. Thus $\theta(\Delta) = \Delta$.

We can then transfer θ to $\mathfrak{h}$, retaining the same name θ. Define θ on the root vectors X_α for simple roots by

$$\theta X_\alpha = \begin{cases} X_\alpha & \text{if } \alpha \text{ is unpainted and forms a 1-element orbit} \\ -X_\alpha & \text{if } \alpha \text{ is painted and forms a 1-element orbit} \\ X_{\theta\alpha} & \text{if } \alpha \text{ is in a 2-element orbit.} \end{cases}$$

By the Isomorphism Theorem (Theorem 2.108), θ extends to an automorphism of $\mathfrak{g}$ consistently with these definitions on $\mathfrak{h}$ and on the X_α's for α simple. The uniqueness in Theorem 2.108 implies that $\theta^2 = 1$.

The main step is to prove that $\theta\mathfrak{u}_0 = \mathfrak{u}_0$. Let B be the Killing form of $\mathfrak{g}$. For $\alpha \in \Delta$, define a constant a_α by $\theta X_\alpha = a_\alpha X_{\theta\alpha}$. Then $a_\alpha a_{-\alpha} = B(a_\alpha X_{\theta\alpha}, a_{-\alpha} X_{-\theta\alpha}) = B(\theta X_\alpha, \theta X_{-\alpha}) = B(X_\alpha, X_{-\alpha}) = 1$ shows that

$$a_\alpha a_{-\alpha} = 1. \tag{6.90}$$

We shall prove that

$$a_\alpha = \pm 1 \qquad \text{for all } \alpha \in \Delta. \tag{6.91}$$

To prove (6.91), it is enough because of (6.90) to prove the result for $\alpha \in \Delta^+$. We do so by induction on the level of α. If the level is 1, then $a_\alpha = \pm 1$ by definition. Thus it is enough to prove that if (6.91) holds for positive roots α and β and if $\alpha+\beta$ is a root, then it holds for $\alpha+\beta$. In the notation of Theorem 6.6, we have

$$\begin{aligned}\theta X_{\alpha+\beta} &= N_{\alpha,\beta}^{-1}\theta[X_\alpha, X_\beta] = N_{\alpha,\beta}^{-1}[\theta X_\alpha, \theta X_\beta] \\ &= N_{\alpha,\beta}^{-1}a_\alpha a_\beta[X_{\theta\alpha}, X_{\theta\beta}] = N_{\alpha,\beta}^{-1}N_{\theta\alpha,\theta\beta}a_\alpha a_\beta X_{\theta\alpha+\theta\beta}.\end{aligned}$$

Therefore

$$a_{\alpha+\beta} = N_{\alpha,\beta}^{-1}N_{\theta\alpha,\theta\beta}a_\alpha a_\beta.$$

Here $a_\alpha a_\beta = \pm 1$ by assumption, while Theorem 6.6 and the fact that θ is an automorphism of Δ say that $N_{\alpha,\beta}$ and $N_{\theta\alpha,\theta\beta}$ are real with

$$N_{\alpha,\beta}^2 = \tfrac{1}{2}q(1+p)|\alpha|^2 = \tfrac{1}{2}q(1+p)|\theta\alpha|^2 = N_{\theta\alpha,\theta\beta}^2.$$

Hence $a_{\alpha+\beta} = \pm 1$, and (6.91) is proved.

Let us see that

(6.92)
$$\theta(\mathbb{R}(X_\alpha - X_{-\alpha}) + \mathbb{R}i(X_\alpha + X_{-\alpha})) \subseteq \mathbb{R}(X_{\theta\alpha} - X_{-\theta\alpha}) + \mathbb{R}i(X_{\theta\alpha} + X_{-\theta\alpha}).$$

If x and y are real and if $z = x + yi$, then we have

$$x(X_\alpha - X_{-\alpha}) + yi(X_\alpha + X_{-\alpha}) = zX_\alpha - \bar{z}X_{-\alpha}.$$

Thus (6.92) amounts to the assertion that the expression

$$\theta(zX_\alpha - \bar{z}X_{-\alpha}) = za_\alpha X_{\theta\alpha} - \bar{z}a_{-\alpha}X_{-\theta\alpha}$$

is of the form $wX_{\theta\alpha} - \bar{w}X_{-\theta\alpha}$, and this follows from (6.91) and (6.90). Since θ carries roots to roots,

$$\theta\Big(\sum_{\alpha\in\Delta}\mathbb{R}(iH_\alpha)\Big) = \sum_{\alpha\in\Delta}\mathbb{R}(iH_\alpha). \tag{6.93}$$

Combining (6.92) and (6.93) with (6.89), we see that $\theta\mathfrak{u}_0 = \mathfrak{u}_0$.

Let $\mathfrak{k}$ and $\mathfrak{p}$ be the $+1$ and -1 eigenspaces for θ in $\mathfrak{g}$, so that $\mathfrak{g} = \mathfrak{k} \oplus \mathfrak{p}$. Since $\theta\mathfrak{u}_0 = \mathfrak{u}_0$, we have

$$\mathfrak{u}_0 = (\mathfrak{u}_0 \cap \mathfrak{k}) \oplus (\mathfrak{u}_0 \cap \mathfrak{p}).$$

Define $k_0 = \mathfrak{u}_0 \cap \mathfrak{k}$ and $\mathfrak{p}_0 = i(\mathfrak{u}_0 \cap \mathfrak{p})$, so that

$$\mathfrak{u}_0 = \mathfrak{k}_0 \oplus i\mathfrak{p}_0.$$

Since $\mathfrak{u}_0$ is a real form of $\mathfrak{g}$ as a vector space, so is

$$\mathfrak{g}_0 = \mathfrak{k}_0 \oplus \mathfrak{p}_0.$$

Since $\theta \mathfrak{u}_0 = \mathfrak{u}_0$ and since θ is an involution, we have the bracket relations

$$[\mathfrak{k}_0, \mathfrak{k}_0] \subseteq \mathfrak{k}_0, \quad [\mathfrak{k}_0, \mathfrak{p}_0] \subseteq \mathfrak{p}_0, \quad [\mathfrak{p}_0, \mathfrak{p}_0] \subseteq \mathfrak{k}_0.$$

Therefore $\mathfrak{g}_0$ is closed under brackets and is a real form of $\mathfrak{g}$ as a Lie algebra. The involution θ is $+1$ on $\mathfrak{k}_0$ and is -1 on $\mathfrak{p}_0$; it is a Cartan involution of $\mathfrak{g}_0$ by the remarks following (6.26), since $\mathfrak{k}_0 \oplus i\mathfrak{p}_0 = \mathfrak{u}_0$ is compact.

Formula (6.93) shows that θ maps $\mathfrak{u}_0 \cap \mathfrak{h}$ to itself, and therefore

$$\begin{aligned} \mathfrak{u}_0 \cap \mathfrak{h} &= (\mathfrak{u}_0 \cap \mathfrak{k} \cap \mathfrak{h}) \oplus (\mathfrak{u}_0 \cap \mathfrak{p} \cap \mathfrak{h}) \\ &= (\mathfrak{k}_0 \cap \mathfrak{h}) \oplus (i\mathfrak{p}_0 \cap \mathfrak{h}) \\ &= (\mathfrak{k}_0 \cap \mathfrak{h}) \oplus i(\mathfrak{p}_0 \cap \mathfrak{h}). \end{aligned}$$

The abelian subspace $\mathfrak{u}_0 \cap \mathfrak{h}$ is a real form of $\mathfrak{h}$, and hence so is

$$\mathfrak{h}_0 = (\mathfrak{k}_0 \cap \mathfrak{h}) \oplus (\mathfrak{p}_0 \cap \mathfrak{h}).$$

The subspace $\mathfrak{h}_0$ is contained in $\mathfrak{g}_0$, and it is therefore a θ stable Cartan subalgebra of $\mathfrak{g}_0$.

A real root α relative to $\mathfrak{h}_0$ has the property that $\theta\alpha = -\alpha$. Since θ preserves positivity relative to Δ^+, there are no real roots. By Proposition 6.70, $\mathfrak{h}_0$ is maximally compact.

Let us verify that Δ^+ results from a lexicographic ordering that takes $i(\mathfrak{k}_0 \cap \mathfrak{h})$ before $\mathfrak{p}_0 \cap \mathfrak{h}$. Let $\{\beta_i\}_{i=1}^l$ be the set of simple roots of Δ^+ in 1-element orbits under θ, and let $\{\gamma_i, \theta\gamma_i\}_{i=1}^m$ be the set of simple roots of Δ^+ in 2-element orbits. Relative to basis $\{\alpha_i\}_{i=1}^{l+2m}$ consisting of all simple roots, let $\{\omega_i\}$ be the dual basis defined by $\langle \omega_i, \alpha_j \rangle = \delta_{ij}$. We shall write ω_{β_j} or ω_{γ_j} or $\omega_{\theta\gamma_j}$ in place of ω_i in what follows. We define a lexicographic ordering by using inner products with the ordered basis

$$\omega_{\beta_1}, \dots, \omega_{\beta_l}, \omega_{\gamma_1} + \omega_{\theta\gamma_1}, \dots, \omega_{\gamma_m} + \omega_{\theta\gamma_m}, \omega_{\gamma_1} - \omega_{\theta\gamma_1}, \dots, \omega_{\gamma_m} - \omega_{\theta\gamma_m},$$

which takes $i(\mathfrak{k}_0 \cap \mathfrak{h})$ before $\mathfrak{p}_0 \cap \mathfrak{h}$. Let α be in Δ^+, and write

$$\alpha = \sum_{i=1}^{l} n_i\beta_i + \sum_{j=1}^{m} r_j\gamma_j + \sum_{j=1}^{m} s_j\theta\gamma_j.$$

Then

$$\langle \alpha, \omega_{\beta_j} \rangle = n_j \geq 0$$

and

$$\langle \alpha, \omega_{\gamma_j} + \omega_{\theta\gamma_j} \rangle = r_j + s_j \geq 0.$$

If all these inner products are 0, then all coefficients of α are 0, contradiction. Thus α has positive inner product with the first member of our ordered basis for which the inner product is nonzero, and the lexicographic ordering yields Δ^+ as positive system. Consequently $(\mathfrak{g}_0, \mathfrak{h}_0, \Delta^+)$ is a triple.

Our definitions of θ on $\mathfrak{h}^*$ and on the X_α for α simple make it clear that the Vogan diagram of $(\mathfrak{g}_0, \mathfrak{h}_0, \Delta^+)$ coincides with the given data. This completes the proof.

9. Complexification of a Simple Real Lie Algebra

This section deals with some preliminaries for the classification of simple real Lie algebras. Our procedure in the next section is to start from a complex semisimple Lie algebra and pass to all possible real forms that are simple. In order to use this method effectively, we need to know what complex semisimple Lie algebras can arise in this way.

Theorem 6.94. Let $\mathfrak{g}_0$ be a simple Lie algebra over $\mathbb{R}$, and let $\mathfrak{g}$ be its complexification. Then there are just two possibilities:

(a) $\mathfrak{g}_0$ is complex, i.e., is of the form $\mathfrak{s}^{\mathbb{R}}$ for some complex $\mathfrak{s}$, and then $\mathfrak{g}$ is $\mathbb{C}$ isomorphic to $\mathfrak{s} \oplus \mathfrak{s}$

(b) $\mathfrak{g}_0$ is not complex, and then $\mathfrak{g}$ is simple over $\mathbb{C}$.

PROOF.

(a) Let J be multiplication by $\sqrt{-1}$ in $\mathfrak{g}_0$, and define an $\mathbb{R}$ linear map $L : \mathfrak{g} \to \mathfrak{s} \oplus \mathfrak{s}$ by $L(X + iY) = (X + JY, X - JY)$ for X and Y in $\mathfrak{g}_0$. We readily check that L is one-one and respects brackets. Since the domain and range have the same real dimension, L is an $\mathbb{R}$ isomorphism.

Moreover L satisfies

$$\begin{aligned} L(i(X + iY)) &= L(-Y + iX) \\ &= (-Y + JX, -Y - JX) \\ &= (J(X + JY), -J(X - JY)). \end{aligned}$$

This equation exhibits L as a $\mathbb{C}$ isomorphism of $\mathfrak{g}$ with $\mathfrak{s} \oplus \bar{\mathfrak{s}}$, where $\bar{\mathfrak{s}}$ is the same real Lie algebra as $\mathfrak{g}_0$ but where the multiplication by $\sqrt{-1}$ is defined as multiplication by $-i$.

To complete the proof of (a), we show that $\bar{\mathfrak{s}}$ is $\mathbb{C}$ isomorphic to $\mathfrak{s}$. By Theorem 6.11, $\mathfrak{s}$ has a compact real form $\mathfrak{u}_0$. The conjugation τ of $\mathfrak{s}$ with respect to $\mathfrak{u}_0$ is $\mathbb{R}$ linear and respects brackets, and the claim is that τ is a $\mathbb{C}$ isomorphism of $\mathfrak{s}$ with $\bar{\mathfrak{s}}$. In fact, if U and V are in $\mathfrak{u}_0$, then

$$\tau(J(U{+}JV)) = \tau(-V{+}JU) = -V - JU = -J(U{-}JV) = -J\tau(U{+}JV),$$

and (a) follows.

(b) Let bar denote conjugation of $\mathfrak{g}$ with respect to $\mathfrak{g}_0$. If $\mathfrak{a}$ is a simple ideal in $\mathfrak{g}$, then $\mathfrak{a} \cap \bar{\mathfrak{a}}$ and $\mathfrak{a} + \bar{\mathfrak{a}}$ are ideals in $\mathfrak{g}$ invariant under conjugation and hence are complexifications of ideals in $\mathfrak{g}_0$. Thus they are 0 or $\mathfrak{g}$. Since $\mathfrak{a} \neq 0$, $\mathfrak{a} + \bar{\mathfrak{a}} = \mathfrak{g}$.

If $\mathfrak{a} \cap \bar{\mathfrak{a}} = 0$, then $\mathfrak{g} = \mathfrak{a} \oplus \bar{\mathfrak{a}}$. The inclusion of $\mathfrak{g}_0$ into $\mathfrak{g}$, followed by projection to $\mathfrak{a}$, is an $\mathbb{R}$ homomorphism φ of Lie algebras. If $\ker \varphi$ is nonzero, then $\ker \varphi$ must be $\mathfrak{g}_0$. In this case $\mathfrak{g}_0$ is contained in $\bar{\mathfrak{a}}$. But conjugation fixes $\mathfrak{g}_0$, and thus $\mathfrak{g}_0 \subseteq \mathfrak{a} \cap \bar{\mathfrak{a}} = 0$, contradiction. We conclude that φ is one-one. A count of dimensions shows that φ is an $\mathbb{R}$ isomorphism of $\mathfrak{g}_0$ onto $\mathfrak{a}$. But then $\mathfrak{g}_0$ is complex, contradiction.

We conclude that $\mathfrak{a} \cap \bar{\mathfrak{a}} = \mathfrak{g}$ and hence $\mathfrak{a} = \mathfrak{g}$. Therefore $\mathfrak{g}$ is simple, as asserted.

Proposition 6.95. If $\mathfrak{g}$ is a complex Lie algebra simple over $\mathbb{C}$, then $\mathfrak{g}^{\mathbb{R}}$ is simple over $\mathbb{R}$.

PROOF. Suppose that $\mathfrak{a}$ is an ideal in $\mathfrak{g}^{\mathbb{R}}$. Since $\mathfrak{g}^{\mathbb{R}}$ is semisimple, $[\mathfrak{a}, \mathfrak{g}^{\mathbb{R}}] \subseteq \mathfrak{a} = [\mathfrak{a}, \mathfrak{a}] \subseteq [\mathfrak{a}, \mathfrak{g}^{\mathbb{R}}]$. Therefore $\mathfrak{a} = [\mathfrak{a}, \mathfrak{g}^{\mathbb{R}}]$. Let X be in $\mathfrak{a}$, and write $X = \sum_j [X_j, Y_j]$ with $X_j \in \mathfrak{a}$ and $Y_j \in \mathfrak{g}$. Then

$$iX = \sum_j i[X_j, Y_j] = \sum [X_j, iY_j] \in [\mathfrak{a}, \mathfrak{g}^{\mathbb{R}}] = \mathfrak{a}.$$

So $\mathfrak{a}$ is a complex ideal in $\mathfrak{g}$. Since $\mathfrak{g}$ is complex simple, $\mathfrak{a} = 0$ or $\mathfrak{a} = \mathfrak{g}$. Thus $\mathfrak{g}^{\mathbb{R}}$ is simple over $\mathbb{R}$.

10. Classification of Simple Real Lie Algebras

Before taking up the problem of classification, a word of caution is in order. The virtue of classification is that it provides a clear indication of the scope of examples in the subject. It is rarely a sound idea to prove a theorem by proving it case-by-case for all simple real Lie algebras. Instead the important thing about classification is the techniques that are involved. Techniques that are subtle enough to identify all the examples are probably subtle enough to help in investigating all semisimple Lie algebras simultaneously.

Theorem 6.94 divided the simple real Lie algebras into two kinds, and we continue with that distinction in this section.

The first kind is a complex simple Lie algebra that is regarded as a real Lie algebra and remains simple when regarded that way. Proposition 6.95 shows that every complex simple Lie algebra may be used for this purpose. In view of the results of Chapter II, the classification of this kind is complete. We obtain complex Lie algebras of the usual types A_n through G_2. Matrix realizations of the complex Lie algebras of the classical types A_n through D_n are listed in (2.43).

The other kind is a noncomplex simple Lie algebra $\mathfrak{g}_0$, and its complexification is then simple over $\mathbb{C}$. Since the complexification is simple, any Vogan diagram for $\mathfrak{g}_0$ will have its underlying Dynkin diagram connected. Conversely any real semisimple Lie algebra $\mathfrak{g}_0$ with a Vogan diagram having connected Dynkin diagram has $(\mathfrak{g}_0)^{\mathbb{C}}$ simple, and therefore $\mathfrak{g}_0$ has to be simple. We know from Theorem 6.74 that the same Vogan diagram cannot come from nonisomorphic $\mathfrak{g}_0$'s, and we know from Theorem 6.88 that every abstract Vogan diagram is a Vogan diagram. Therefore the classification of this type of simple real Lie algebra comes down to classifying abstract Vogan diagrams whose underlying Dynkin diagram is connected.

Thus we want to eliminate the redundancy in connected Vogan diagrams. There is no redundancy from the automorphism. The only connected Dynkin diagrams admitting nontrivial automorphisms of order 2 are A_n, D_n, and E_6. In these cases a nontrivial automorphism of order 2 of the Dynkin diagram is unique up to an automorphism of the diagram (and is absolutely unique except in D_4). A Vogan diagram for $\mathfrak{g}_0$ incorporates a nontrivial automorphism of order 2 if and only if there exist complex roots, and this condition depends only on $\mathfrak{g}_0$.

The redundancy comes about through having many allowable choices for the positive system Δ^+. The idea, partly but not completely, is that we can always change Δ^+ so that at most one imaginary simple root is painted.

Theorem 6.96 (Borel and de Siebenthal Theorem). Let $\mathfrak{g}_0$ be a noncomplex simple real Lie algebra, and let the Vogan diagram of $\mathfrak{g}_0$ be given that corresponds to the triple $(\mathfrak{g}_0, \mathfrak{h}_0, \Delta^+)$. Then there exists a simple system Π' for $\Delta = \Delta(\mathfrak{g}, \mathfrak{h})$, with corresponding positive system $\Delta^{+\prime}$, such that $(\mathfrak{g}_0, \mathfrak{h}_0, \Delta^{+\prime})$ is a triple and there is at most one painted simple root in its Vogan diagram. Furthermore suppose that the automorphism associated with the Vogan diagram is the identity, that $\Pi' = \{\alpha_1, \dots, \alpha_l\}$, and that $\{\omega_1, \dots, \omega_l\}$ is the dual basis given by $\langle \omega_j, \alpha_k \rangle = \delta_{jk}$. Then the single painted simple root α_i may be chosen so that there is no i' with $\langle \omega_i - \omega_{i'}, \omega_{i'} \rangle > 0$.

REMARKS.

1) The proof will be preceded by two lemmas. The main conclusion of the theorem is that we can arrange that at most one simple root is painted. The second conclusion (concerning ω_i and therefore limiting which simple root can be painted) is helpful only when the Dynkin diagram is exceptional (E_6, E_7, E_8, F_4, or G_2).

2) The proof simplifies somewhat when the automorphism marked as part of the Vogan diagram is the identity. This is the case that $\mathfrak{h}_0$ is contained in $\mathfrak{k}_0$, and most examples will turn out to have this property.

Lemma 6.97. Let Δ be an irreducible abstract reduced root system in a real vector space V, let Π be a simple system, and let ω and ω' be nonzero members of V that are dominant relative to Π. Then $\langle \omega, \omega' \rangle > 0$.

PROOF. The first step is to show that in the expansion $\omega = \sum_{\alpha \in \Pi} a_\alpha \alpha$, all the a_α are ≥ 0. Let us enumerate Π as $\alpha_1, \dots, \alpha_l$ so that

$$\omega = \sum_{i=1}^{r} a_i \alpha_i - \sum_{i=r+1}^{s} b_i \alpha_i = \omega^+ - \omega^-$$

with all $a_i \geq 0$ and all $b_i > 0$. We shall show that $\omega^- = 0$. Since $\omega^- = \omega^+ - \omega$, we have

$$0 \leq |\omega^-|^2 = \langle \omega^+, \omega^- \rangle - \langle \omega^-, \omega \rangle = \sum_{i=1}^{r} \sum_{j=r+1}^{s} a_i b_j \langle \alpha_i, \alpha_j \rangle - \sum_{j=r+1}^{l} b_j \langle \omega, \alpha_j \rangle.$$

The first term on the right side is ≤ 0 by Lemma 2.51, and the second term on the right side (with the minus sign included) is term-by-term ≤ 0 by hypothesis. Therefore the right side is ≤ 0, and we conclude that $\omega^- = 0$.

Thus we can write $\omega = \sum_{j=1}^{l} a_j \alpha_j$ with all $a_j \geq 0$. The next step is to show from the irreducibility of Δ that $a_j > 0$ for all j. Assuming the contrary, suppose that $a_i = 0$. Then

$$0 \leq \langle \omega, \alpha_i \rangle = \sum_{j \neq i} a_j \langle \alpha_j, \alpha_i \rangle,$$

and every term on the right side is ≤ 0 by Lemma 2.51. Thus $a_j = 0$ for every α_j such that $\langle \alpha_j, \alpha_i \rangle < 0$, i.e., for all neighbors of α_i in the Dynkin diagram. Since the Dynkin diagram is connected (Proposition 2.54), iteration of this argument shows that all coefficients are 0 once one of them is 0.

Now we can complete the proof. For at least one index i, $\langle \alpha_i, \omega' \rangle > 0$, since $\omega' \neq 0$. Then

$$\langle \omega, \omega' \rangle = \sum_{j} a_j \langle \alpha_j, \omega' \rangle \geq a_i \langle \alpha_i, \omega' \rangle,$$

and the right side is > 0 since $a_i > 0$. This proves the lemma.

Lemma 6.98. Let $\mathfrak{g}_0$ be a noncomplex simple real Lie algebra, and let the Vogan diagram of $\mathfrak{g}_0$ be given that corresponds to the triple $(\mathfrak{g}_0, \mathfrak{h}_0, \Delta^+)$. Write $\mathfrak{h}_0 = \mathfrak{t}_0 \oplus \mathfrak{a}_0$ as usual. Let V be the span of the simple roots that are imaginary, let Δ_0 be the root system $\Delta \cap V$, let $\mathcal{H}$ be the subset of $i\mathfrak{t}_0$ paired with V, and let Λ be the subset of $\mathcal{H}$ where all roots of Δ_0 take integer values and all noncompact roots of Δ_0 take odd-integer values. Then Λ is nonempty. In fact, if $\alpha_1, \dots, \alpha_m$ is any simple system for Δ_0 and if $\omega_1, \dots, \omega_m$ in V are defined by $\langle \omega_j, \alpha_k \rangle = \delta_{jk}$, then the element

$$\omega = \sum_{\substack{i \text{ with } \alpha_i \\ \text{noncompact}}} \omega_i .$$

is in Λ.

PROOF. Fix a simple system $\alpha_1, \dots, \alpha_m$ for Δ_0, and let Δ_0^+ be the set of positive roots of Δ_0. Define $\omega_1, \dots, \omega_m$ by $\langle \omega_j, \alpha_k \rangle = \delta_{jk}$. If $\alpha = \sum_{i=1}^m n_i \alpha_i$ is a positive root of Δ_0, then $\langle \omega, \alpha \rangle$ is the sum of the n_i for which α_i is noncompact. This is certainly an integer.

We shall prove by induction on the level $\sum_{i=1}^m n_i$ that $\langle \omega, \alpha \rangle$ is even if α is compact, odd if α is noncompact. When the level is 1, this assertion is true by definition. In the general case, let α and β be in Δ_0^+ with $\alpha + \beta$ in Δ, and suppose that the assertion is true for α and β. Since the sum of the n_i for which α_i is noncompact is additive, we are to prove that imaginary roots satisfy

$$\begin{aligned} \text{compact} + \text{compact} &= \text{compact} \\ \text{compact} + \text{noncompact} &= \text{noncompact} \\ \text{noncompact} + \text{noncompact} &= \text{compact}. \end{aligned} \tag{6.99}$$

But this is immediate from Corollary 2.35 and the bracket relations (6.24).

PROOF OF THEOREM 6.96. Define V, Δ_0, and Λ as in Lemma 6.98. Before we use Lemma 6.97, it is necessary to observe that the Dynkin diagram of Δ_0 is connected, i.e., that the roots in the Dynkin diagram of Δ fixed by the given automorphism form a connected set. There is no problem when the automorphism is the identity, and we observe the connectednes in the other cases one at a time by inspection.

Let $\Delta_0^+ = \Delta^+ \cap V$. The set Λ is discrete, being a subset of a lattice, and Lemma 6.98 has just shown that it is nonempty. Let H_0 be a member of Λ with norm as small as possible. By Proposition 2.67 we can choose a new positive system $\Delta_0^{+\prime}$ for Δ_0 that makes H_0 dominant. The main step is to show that

$$\text{at most one simple root of } \Delta_0^{+\prime} \text{ is painted.} \tag{6.100}$$

Suppose $H_0 = 0$. If α is in Δ_0, then $\langle H_0, \alpha \rangle$ is 0 and is not an odd integer. By definition of Λ, α is compact. Thus all roots of Δ_0 are compact, and (6.100) is true.

Now suppose $H_0 \neq 0$. Let $\alpha_1, \dots, \alpha_m$ be the simple roots of Δ_0 relative to $\Delta_0^{+\prime}$, and define $\omega_1, \dots, \omega_m$ by $\langle \omega_j, \alpha_k \rangle = \delta_{jk}$. We can write $H_0 = \sum_{j=1}^m n_j \omega_j$ with $n_j = \langle H_0, \alpha_j \rangle$. The number n_j is an integer since H_0 is in Λ, and it is ≥ 0 since H_0 is dominant relative to $\Delta_0^{+\prime}$.

Since $H_0 \neq 0$, we have $n_i > 0$ for some i. Then $H_0 - \omega_i$ is dominant relative to $\Delta_0^{+\prime}$, and Lemma 6.97 shows that $\langle H_0 - \omega_i, \omega_i \rangle \geq 0$ with equality only if $H_0 = \omega_i$. If strict inequality holds, then the element $H_0 - 2\omega_i$ is in Λ and satisfies

$$|H_0 - 2\omega_i|^2 = |H_0|^2 - 4\langle H_0 - \omega_i, \omega_i \rangle < |H_0|^2,$$

in contradiction with the minimal-norm condition on H_0. Hence equality holds, and $H_0 = \omega_i$.

Since H_0 is in Λ, a simple root α_j in $\Delta_0^{+\prime}$ is noncompact only if $\langle H_0, \alpha_j \rangle$ is an odd integer. Since $\langle H_0, \alpha_j \rangle = 0$ for $j \neq i$, the only possible noncompact simple root in $\Delta_0^{+\prime}$ is α_i. This proves (6.100).

If the automorphism associated with the Vogan diagram is the identity, then (6.100) proves the first conclusion of the theorem. For the second conclusion we are assuming that $H_0 = \omega_i$; then an inequality $\langle \omega_i - \omega_{i'}, \omega_{i'} \rangle > 0$ would imply that

$$|H_0 - 2\omega_{i'}|^2 = |H_0|^2 - 4\langle \omega_i - \omega_{i'}, \omega_{i'} \rangle < |H_0|^2,$$

in contradiction with the minimal-norm condition on H_0.

To complete the proof of the theorem, we have to prove the first conclusion when the automorphism associated with the Vogan diagram is not the identity. Choose by Theorem 2.63 an element $s \in W(\Delta_0)$ with $\Delta_0^{+\prime} = s\Delta_0^+$, and define $\Delta^{+\prime} = s\Delta^+$. With $\mathfrak{h}_0 = \mathfrak{t}_0 \oplus \mathfrak{a}_0$ as usual, the element s maps $i\mathfrak{t}_0$ to itself. Since Δ^+ is defined by an ordering that takes $i\mathfrak{t}_0$ before $\mathfrak{a}_0$, so is $\Delta^{+\prime}$. Let the simple roots of Δ^+ be $\beta_1, \dots, \beta_l$ with $\beta_1, \dots, \beta_m$ in Δ_0. Then the simple roots of $\Delta^{+\prime}$ are $s\beta_1, \dots, s\beta_l$. Among these, $s\beta_1, \dots, s\beta_m$ are the simple roots $\alpha_1, \dots, \alpha_m$ of $\Delta_0^{+\prime}$ considered above, and (6.100) says that at most one of them is noncompact. The roots $s\beta_{m+1}, \dots, s\beta_l$ are complex since $\beta_{m+1}, \dots, \beta_l$ are complex and s carries complex roots to complex roots. Thus $\Delta^{+\prime}$ has at most one simple root that is noncompact imaginary. This completes the proof.

Now we can mine the consequences of the theorem. To each connected abstract Vogan diagram that survives the redundancy tests of Theorem 6.96, we associate a noncomplex simple real Lie algebra. If the underlying Dynkin diagram is classical, we find a known Lie algebra of matrices with that Vogan diagram, and we identify any isomorphisms among the Lie algebras obtained. If the underlying Dynkin digram is exceptional, we give the Lie algebra a name, and we eliminate any remaining redundancy.

As we shall see, the data at hand from a Vogan diagram for $\mathfrak{g}_0$ readily determine the Lie subalgebra $\mathfrak{k}_0$ in the Cartan decomposition $\mathfrak{g}_0 = \mathfrak{k}_0 \oplus \mathfrak{p}_0$. This fact makes it possible to decide which of the Lie algebras obtained are isomorphic to one another.

First suppose that the automorphism of the underlying Dynkin diagram is trivial. When no simple root is painted, then $\mathfrak{g}_0$ is a compact real form. For the classical Dynkin diagrams, the compact real forms are as follows:

(6.101)

Diagram	Compact Real Form
A_n	$\mathfrak{su}(n+1)$
B_n	$\mathfrak{so}(2n+1)$
C_n	$\mathfrak{sp}(n)$
D_n	$\mathfrak{so}(2n)$

For the situation in which one simple root is painted, we treat the classical Dynkin diagrams separately from the exceptional ones. Let us begin with the classical cases. For each classical Vogan diagram with just one simple root painted, we attach a known Lie algebra of matrices to that diagram. The result is that we are associating a Lie algebra of matrices to each simple root of each classical Dynkin diagram. We can assemble all the information for one Dynkin diagram in one picture by labeling each root of the Dynkin diagram with the associated Lie algebra of matrices. Those results are in Figure 6.1.

Verification of the information in Figure 6.1 is easy for the most part. For A_n, Example 1 in §8 gives the outcome, which is that $\mathfrak{su}(p,q)$ results when $p+q=n+1$ and the pth simple root from the left is painted.

For B_n, suppose that $p+q=2n+1$ and that p is even. Represent $\mathfrak{so}(p,q)$ by real matrices $\begin{pmatrix} a & b \\ b^* & d \end{pmatrix}$ with a and d skew symmetric. For $\mathfrak{h}_0$, we use block-diagonal matrices whose first n blocks are $\mathbb{R}\begin{pmatrix} 0 & 1 \\ -1 & 0 \end{pmatrix}$ of size 2-by-2 and whose last block is of size 1-by-1. With linear functionals on $(\mathfrak{h}_0)^{\mathbb{C}}$ as in Example 2 of §II.1 and with the positive system as in that example, the Vogan diagram is as indicated by Figure 6.1.

For C_n, the analysis for the first $n-1$ simple roots uses $\mathfrak{sp}(p,q)$ with $p+q=n$ in the same way that the analysis for A_n uses $\mathfrak{su}(p,q)$ with $p+q=n+1$. The analysis for the last simple root is different. For

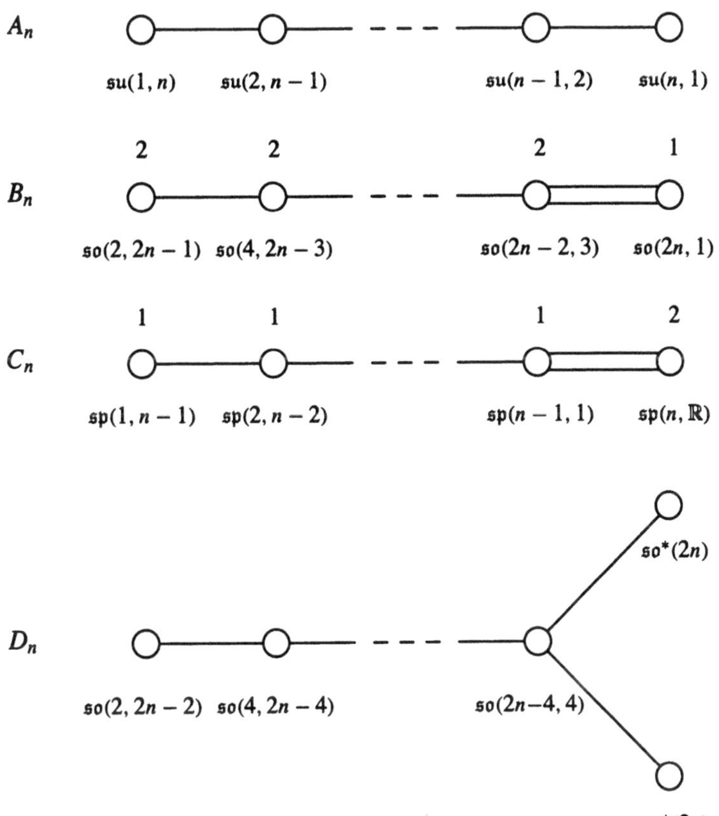

FIGURE 6.1. Association of classical matrix algebras to Vogan diagrams with the trivial automorphism

this case we take the Lie algebra to be $\mathfrak{sp}(n,\mathbb{R})$. Actually it is more convenient to use the isomorphic Lie algebra $\mathfrak{g}_0 = \mathfrak{su}(n,n) \cap \mathfrak{sp}(n,\mathbb{C})$, which is conjugate to $\mathfrak{sp}(n,\mathbb{R})$ by the matrix given in block form as $\frac{\sqrt{2}}{2}\begin{pmatrix} 1 & i \\ i & 1 \end{pmatrix}$. Within $\mathfrak{g}_0$, we take

$$\mathfrak{h}_0 = \{\operatorname{diag}(iy_1, \ldots, iy_n, -iy_1, \ldots, -iy_n)\}. \tag{6.102}$$

If we define e_j of the indicated matrix to be iy_j, then the roots are those of type C_n on (2.43), and we choose as positive system the customary one given in (2.50). The roots $e_i - e_j$ are compact, and the roots $\pm(e_i + e_j)$

and $\pm 2e_j$ are noncompact. Thus $2e_n$ is the unique noncompact simple root.

For D_n, the analysis for the first $n-2$ simple roots uses $\mathfrak{so}(p,q)$ with p and q even and $p+q=2n$. It proceeds in the same way as with B_n. The analysis for either of the last two simple roots is different. For one of the two simple roots we take $\mathfrak{g}_0 = \mathfrak{so}^*(2n)$. We use the same $\mathfrak{h}_0$ and e_j as in (6.102). Then the roots are those of type D_n in (2.43), and we introduce the customary positive system (2.50). The roots $e_i - e_j$ are compact, and the roots $\pm(e_i + e_j)$ are noncompact. Thus $e_{n-1} + e_n$ is the unique noncompact simple root. The remaining Vogan diagram is isomorphic to the one we have just considered, and hence it too must correspond to $\mathfrak{so}^*(2n)$.

For the exceptional Dynkin diagrams we make use of the additional conclusion in Theorem 6.96; this says that we can disregard the case in which α_i is the unique simple noncompact root if $\langle \omega_i - \omega_{i'}, \omega_{i'} \rangle > 0$ for some i'. First let us see how to apply this test in practice. Write $\alpha_i = \sum_k d_{ik}\omega_k$. Taking the inner product with α_j shows that $d_{ij} = \langle \alpha_i, \alpha_j \rangle$. If we put $\omega_j = \sum_l c_{lj}\alpha_l$, then

$$\delta_{ij} = \langle \alpha_i, \omega_j \rangle = \sum_{k,l} d_{ik} c_{lj} \langle \omega_k, \alpha_l \rangle = \sum_k d_{ik} c_{kj}.$$

Thus the matrix (c_{ij}) is the inverse of the matrix (d_{ij}). Finally the quantity of interest is just $\langle \omega_j, \omega_{j'} \rangle = c_{j'j}$.

The Cartan matrix will serve as (d_{ij}) if all roots have the same length because we can assume that $|\alpha_i|^2 = 2$ for all i; then the coefficients c_{ij} are obtained by inverting the Cartan matrix. When there are two root lengths, (d_{ij}) is a simple modification of the Cartan matrix.

Appendix C gives all the information necessary to make the computations quickly. Let us indicate details for E_6. Let the simple roots be $\alpha_1, \ldots, \alpha_6$ as in (2.86c). Then Appendix C gives

$$\begin{aligned}
\omega_1 &= \tfrac{1}{3}(4\alpha_1 + 3\alpha_2 + 5\alpha_3 + 6\alpha_4 + 4\alpha_5 + 2\alpha_6) \\
\omega_2 &= 1\alpha_1 + 2\alpha_2 + 2\alpha_3 + 3\alpha_4 + 2\alpha_5 + 1\alpha_6 \\
\omega_3 &= \tfrac{1}{3}(5\alpha_1 + 6\alpha_2 + 10\alpha_3 + 12\alpha_4 + 8\alpha_5 + 4\alpha_6) \\
\omega_4 &= 2\alpha_1 + 3\alpha_2 + 4\alpha_3 + 6\alpha_4 + 4\alpha_5 + 2\alpha_6 \\
\omega_5 &= \tfrac{1}{3}(4\alpha_1 + 6\alpha_2 + 8\alpha_3 + 12\alpha_4 + 10\alpha_5 + 5\alpha_6) \\
\omega_6 &= \tfrac{1}{3}(2\alpha_1 + 3\alpha_2 + 4\alpha_3 + 6\alpha_4 + 5\alpha_5 + 4\alpha_6).
\end{aligned}$$

Let us use Theorem 6.96 to rule out $i = 3, 4$, and 5. For $i = 3$, we take $i' = 1$; we have $\langle \omega_3, \omega_1 \rangle = \frac{5}{3}$ and $\langle \omega_1, \omega_1 \rangle = \frac{4}{3}$, so that $\langle \omega_3 - \omega_1, \omega_1 \rangle > 0$.

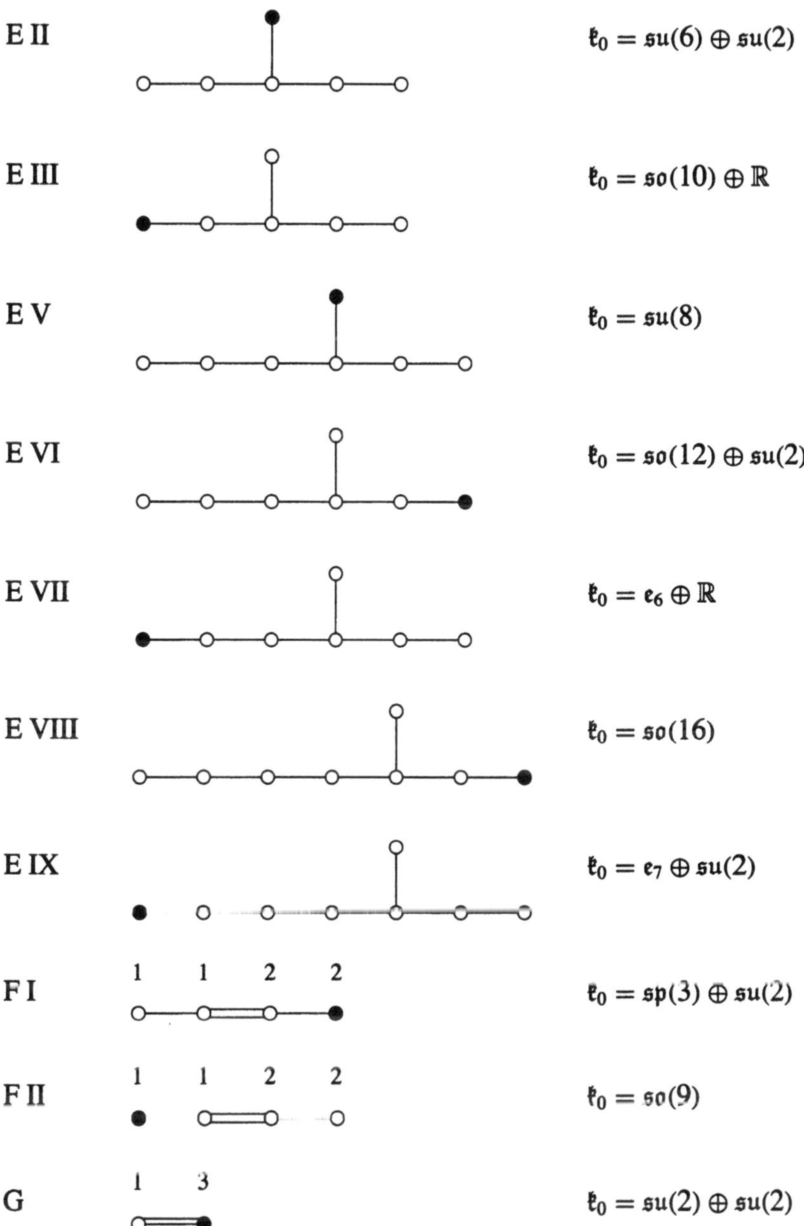

FIGURE 6.2. Noncompact noncomplex exceptional simple real Lie algebras with the trivial automorphism in the Vogan diagram

For $i = 4$, we take $i' = 1$; we have $\langle\omega_4, \omega_1\rangle = 2$ and $\langle\omega_1, \omega_1\rangle = \frac{4}{3}$, so that $\langle\omega_4 - \omega_1, \omega_1\rangle > 0$. For $i = 5$, we take $i' = 6$; we have $\langle\omega_5, \omega_6\rangle = \frac{5}{3}$ and $\langle\omega_6, \omega_6\rangle = \frac{4}{3}$, so that $\langle\omega_5 - \omega_6, \omega_6\rangle > 0$. Although there are six abstract Vogan diagrams of E_6 with trivial automorphism and with one noncompact simple root, Theorem 6.96 says that we need to consider only the three where the simple root is α_1, α_2, or α_6. Evidently α_6 yields a result isomorphic to that for α_1 and may be disregarded.

By similar computations for the other exceptional Dynkin diagrams, we find that we may take α_i to be an endpoint vertex of the Dynkin diagram. Moreover, in G_2, α_i may be taken to be the long simple root, while in E_8, we do not have to consider α_2 (the endpoint vertex on the short branch). Thus we obtain the 10 Vogan diagrams in Figure 6.2. We have given each of them its name from the Cartan listing [1927a]. Computing $\mathfrak{k}_0$ is fairly easy. As a Lie algebra, $\mathfrak{k}_0$ is reductive by Corollary 4.25. The root system of its semisimple part is the system of compact roots, which we can compute from the Vogan diagram if we remember (6.99) and use the tables in Appendix C that tell which combinations of simple roots are roots. Then we convert the result into a compact Lie algebra using (6.101), and we add $\mathbb{R}$ as center if necessary to make the dimension of the Cartan subalgebra work out correctly. Notice in Figure 6.2 that when the Vogan diagrams for two $\mathfrak{g}_0$'s have the same underlying Dynkin diagram, then the $\mathfrak{k}_0$'s are different. By Corollary 6.19 the $\mathfrak{g}_0$'s are nonisomorphic.

Now we suppose that the automorphism of the underlying Dynkin diagram is nontrivial. We already observed that the Dynkin diagram has to be of type A_n, D_n, or E_6.

For type A_n, we distinguish n even from n odd. For n even there is just one abstract Vogan diagram, namely

It must correspond to $\mathfrak{sl}(n+1, \mathbb{R})$ since we have not yet found a Vogan diagram for $\mathfrak{sl}(n+1, \mathbb{R})$ and since the equality $\mathfrak{sl}(n+1, \mathbb{R})^{\mathbb{C}} = \mathfrak{sl}(n+1, \mathbb{C})$ determines the underlying Dynkin diagram as being A_n.

For A_n with n odd, there are two abstract Vogan diagrams, namely

and

The first of these, according to Example 2 in §8, comes from $\mathfrak{sl}(n+1, \mathbb{R})$. The second one comes from $\mathfrak{sl}(\frac{1}{2}(n+1), \mathbb{H})$. In the latter case we take

$$\mathfrak{h}_0 = \left\{\operatorname{diag}(x_1 + iy_1, \dots, x_{\frac{1}{2}(n+1)} + iy_{\frac{1}{2}(n+1)}) \mid \sum x_m = 0\right\}.$$

If e_m and f_m on the indicated member of $\mathfrak{h}_0$ are iy_m and x_m, respectively, then Δ is the same as in Example 2 of §8. The imaginary roots are the $\pm 2e_m$, and they are compact. (The root vectors for $\pm 2e_m$ generate the complexification of the $\mathfrak{su}(2)$ in the jth diagonal entry formed by the skew-Hermitian quaternions there.)

For type D_n, the analysis uses $\mathfrak{so}(p,q)$ with p and q odd and with $p+q = 2n$. Represent $\mathfrak{so}(p,q)$ by real matrices $\begin{pmatrix} a & b \\ b^* & d \end{pmatrix}$ with a and d skew symmetric. For $\mathfrak{h}_0$, we use block-diagonal matrices with all blocks of size 2-by-2. The first $\frac{1}{2}(p-1)$ and the last $\frac{1}{2}(q-1)$ blocks are $\mathbb{R}\begin{pmatrix} 0 & 1 \\ -1 & 0 \end{pmatrix}$, and the remaining one is $\mathbb{R}\begin{pmatrix} 0 & 1 \\ 1 & 0 \end{pmatrix}$. The blocks $\mathbb{R}\begin{pmatrix} 0 & 1 \\ -1 & 0 \end{pmatrix}$ contribute to $\mathfrak{t}_0$, while $\mathbb{R}\begin{pmatrix} 0 & 1 \\ 1 & 0 \end{pmatrix}$ contributes to $\mathfrak{a}_0$. The linear functionals e_j for $j \neq \frac{1}{2}(p+1)$ are as in Example 4 of §II.1, and $e_{\frac{1}{2}(p+1)}$ on the embedded $\begin{pmatrix} 0 & t \\ t & 0 \end{pmatrix} \in \mathbb{R}\begin{pmatrix} 0 & 1 \\ 1 & 0 \end{pmatrix}$ is just t. The roots are $\pm e_i \pm e_j$ with $i \neq j$, and those involving index $\frac{1}{2}(p+1)$ are complex.

Suppose $q = 1$. Then the standard ordering takes $i\mathfrak{t}_0$ before $\mathfrak{a}_0$. The simple roots as usual are

$$e_1 - e_2, \; \dots, \; e_{n-2} - e_{n-1}, \; e_{n-1} - e_n, \; e_{n-1} + e_n.$$

The last two are complex, and the others are compact imaginary. Similarly if $p = 1$, we can use the reverse of the standard ordering and conclude that all imaginary roots are compact.

Now suppose $p > 1$ and $q > 1$. In this case we cannot use the standard ordering. To have $i\mathfrak{t}_0$ before $\mathfrak{a}_0$ in defining positivity, we take $\frac{1}{2}(p+1)$ last, and the simple roots are

$$e_1 - e_2, \; \dots, \; e_{\frac{1}{2}(p-1)-1} - e_{\frac{1}{2}(p-1)}, \; e_{\frac{1}{2}(p-1)} - e_{\frac{1}{2}(p+1)+1},$$
$$e_{\frac{1}{2}(p+1)+1} - e_{\frac{1}{2}(p+1)+2}, \; \dots, \; e_{n-1} - e_n, \; e_n - e_{\frac{1}{2}(p+1)}, \; e_n + e_{\frac{1}{2}(p+1)}.$$

The last two are complex, and the others are imaginary. Among the imaginary simple roots, $e_{\frac{1}{2}(p-1)} - e_{\frac{1}{2}(p+1)+1}$ is the unique noncompact simple root.

We can assemble our results for D_n in a diagram like that in Figure 6.1. As we observed above, the situation with all imaginary roots unpainted corresponds to $\mathfrak{so}(1, 2n-1) \cong \mathfrak{so}(2n-1, 1)$. If one imaginary root is painted, the associated matrix algebra may be seen from the diagram

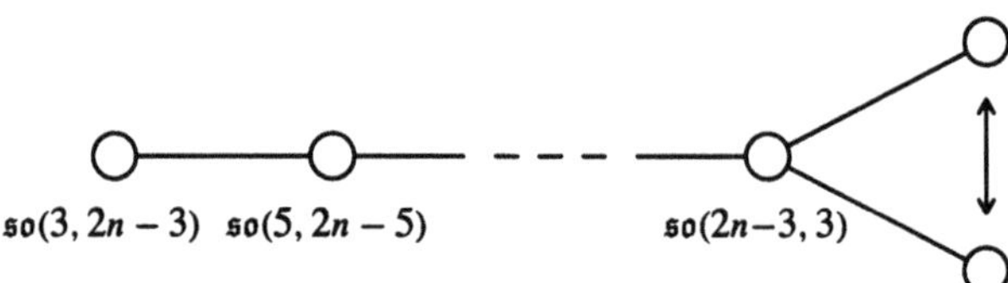

For type E_6, Theorem 6.96 gives us three diagrams to consider. As in (2.86c) let α_2 be the simple root corresponding to the endpoint vertex of the short branch in the Dynkin diagram, and let α_4 correspond to the triple point. The Vogan diagram in which α_4 is painted gives the same $\mathfrak{g}_0$ (up to isomorphism) as the Vogan diagram with α_2 painted. In fact, the Weyl group element $s_{\alpha_4}s_{\alpha_2}$ carries the one with α_2 painted to the one with α_4 painted. Thus there are only two Vogan diagrams that need to be considered, and they are in Figure 6.3. The figure also gives the names of the Lie algebras $\mathfrak{g}_0$ in the Cartan listing [1927a] and identifies $\mathfrak{k}_0$.

To compute $\mathfrak{k}_0$ for each case of Figure 6.3, we regroup the root-space decomposition of $\mathfrak{g}$ as

$$\begin{aligned} \mathfrak{g} = \Big(\mathfrak{t} \oplus \bigoplus_{\substack{\alpha \text{ imaginary} \\ \text{compact}}} \mathfrak{g}_\alpha \oplus \bigoplus_{\substack{\text{complex pairs} \\ \{\alpha, \theta\alpha\}}} (X_\alpha + \theta X_\alpha)\Big) \\ \oplus \Big(\mathfrak{a} \oplus \bigoplus_{\substack{\alpha \text{ imaginary} \\ \text{noncompact}}} \mathfrak{g}_\alpha \oplus \bigoplus_{\substack{\text{complex pairs} \\ \{\alpha, \theta\alpha\}}} (X_\alpha - \theta X_\alpha)\Big), \end{aligned} \tag{6.103}$$

and it is clear that the result is $\mathfrak{g} = \mathfrak{k} \oplus \mathfrak{p}$. Therefore the roots in $\Delta(\mathfrak{k}, \mathfrak{t})$ are the restrictions to $\mathfrak{t}$ of the imaginary compact roots in $\Delta(\mathfrak{g}, \mathfrak{h})$, together with the restrictions to $\mathfrak{t}$ of each pair $\{\alpha, \theta\alpha\}$ of complex roots in $\Delta(\mathfrak{g}, \mathfrak{h})$. Also the dimension of $\mathfrak{a}_0$ is the number of 2-element orbits in the Vogan diagram and is therefore 2 in each case.

We can tell which roots are complex, and we need to know how to decide which imaginary roots are compact. This determination can be carried out by induction on the level in the expansion in terms of simple roots. Thus suppose that α and β are positive roots with β simple, and

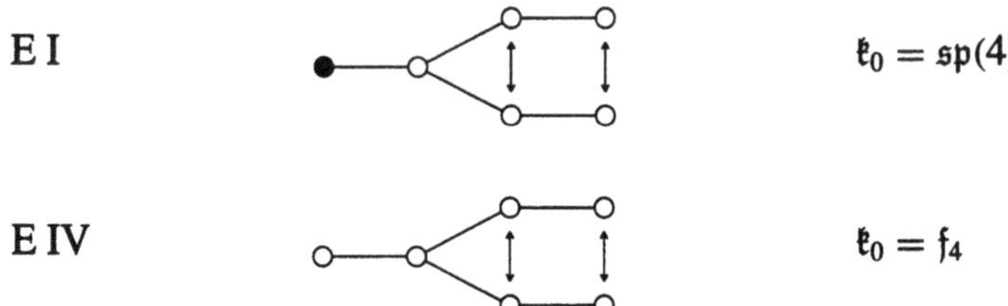

FIGURE 6.3. Noncompact noncomplex exceptional simple real Lie algebras with a nontrivial automorphism in the Vogan diagram

suppose $\alpha+\beta$ is an imaginary root. If β is imaginary, then (6.99) settles matters. Otherwise β is complex simple, and Figure 6.3 shows that $\langle\beta,\theta\beta\rangle = 0$. Therefore the following proposition settles matters for $\mathfrak{g}_0$ as in Figure 6.3 and allows us to complete the induction.

Proposition 6.104. For a connected Vogan diagram involving a nontrivial automorphism, suppose that α and β are positive roots, that β is complex simple, that β is orthogonal to $\theta\beta$, and that $\alpha+\beta$ is an imaginary root. Then $\alpha-\theta\beta$ is an imaginary root, and $\alpha-\theta\beta$ and $\alpha+\beta$ have the same type, compact or noncompact.

PROOF. Taking the common length of all roots to be 2, we have

$$\begin{aligned}1 = 2-1 &= \langle\beta,\beta\rangle + \langle\beta,\alpha\rangle = \langle\beta,\alpha+\beta\rangle \\ &= \langle\theta\beta,\theta(\beta+\alpha)\rangle = \langle\theta\beta,\alpha+\beta\rangle = \langle\theta\beta,\alpha\rangle + \langle\theta\beta,\beta\rangle = \langle\theta\beta,\alpha\rangle.\end{aligned}$$

Thus $\alpha-\theta\beta$ is a root, and we have

$$\alpha+\beta = \theta\beta + (\alpha-\theta\beta) + \beta.$$

Since $\alpha+\beta$ is imaginary, $\alpha-\theta\beta$ is imaginary. Therefore we can write $\theta X_{\alpha-\theta\beta} = sX_{\alpha-\theta\beta}$ with $s=\pm1$. Write $\theta X_\beta = tX_{\theta\beta}$ and $\theta X_{\theta\beta} = tX_\beta$ with $t=\pm1$. Then we have

$$\begin{aligned}\theta[[X_{\theta\beta}, X_{\alpha-\theta\beta}], X_\beta] &= [[\theta X_{\theta\beta}, \theta X_{\alpha-\theta\beta}], \theta X_\beta] \\ &= st^2[[X_\beta, X_{\alpha-\theta\beta}], X_{\theta\beta}] \\ &= -s[[X_{\alpha-\theta\beta}, X_{\theta\beta}], X_\beta] - s[[X_{\theta\beta}, X_\beta], X_{\alpha-\theta\beta}] \\ &= -s[[X_{\alpha-\theta\beta}, X_{\theta\beta}], X_\beta] \\ &= s[[X_{\theta\beta}, X_{\alpha-\theta\beta}], X_\beta],\end{aligned}$$

and the proof is complete.

Let us summarize our results.

Theorem 6.105 (classification). Up to isomorphism every simple real Lie algebra is in the following list, and everything in the list is a simple real Lie algebra:

(a) the Lie algebra $\mathfrak{g}^{\mathbb{R}}$, where $\mathfrak{g}$ is complex simple of type A_n for $n \geq 1$, B_n for $n \geq 2$, C_n for $n \geq 3$, D_n for $n \geq 4$, E_6, E_7, E_8, F_4, or G_2

(b) the compact real form of any $\mathfrak{g}$ as in (a)

(c) the classical matrix algebras

$\mathfrak{su}(p,q)$	with	$p \geq q > 0,\ p+q \geq 2$
$\mathfrak{so}(p,q)$	with	$p > q > 0,\ p+q$ odd, $p+q \geq 3$
	or with	$p \geq q > 0,\ p+q$ even, $p+q \geq 8$
$\mathfrak{sp}(p,q)$	with	$p \geq q > 0,\ p+q \geq 3$
$\mathfrak{sp}(n,\mathbb{R})$	with	$n \geq 3$
$\mathfrak{so}^*(2n)$	with	$n \geq 4$
$\mathfrak{sl}(n,\mathbb{R})$	with	$n \geq 3$
$\mathfrak{sl}(n,\mathbb{H})$	with	$n \geq 2$

(d) the 12 exceptional noncomplex noncompact simple Lie algebras given in Figures 6.2 and 6.3.

The only isomorphism among Lie algebras in the above list is $\mathfrak{so}^*(8) \cong \mathfrak{so}(6,2)$.

REMARKS. The restrictions on rank in (a) prevent coincidences in Dynkin diagrams. These restrictions are maintained in (b) and (c) for the same reason. Note for $\mathfrak{sl}(n,\mathbb{R})$ and $\mathfrak{sl}(n,\mathbb{H})$ that the restrictions on n force the automorphism to be nontrivial. In (c) there are no isomorphisms within a series because the $\mathfrak{k}_0$'s are different. To have an isomorphism between members of two series, we need at least two series with the same Dynkin diagram and automorphism. Then we examine the possibilities and are led to compare $\mathfrak{so}^*(8)$ with $\mathfrak{so}(6,2)$. The standard Vogan diagrams for these two Lie algebras are isomorphic, and hence the Lie algebras are isomorphic by Theorem 6.74.

11. Restricted Roots in the Classification

Additional information about the simple real Lie algebras of §10 comes by switching from a maximally compact Cartan subalgebra to a maximally noncompact Cartan subalgebra. The switch exposes the system of restricted roots, which governs the Iwasawa decomposition and some further structure theory that will be developed in Chapter VII.

According to §7 the switch in Cartan subalgebra is best carried out when we can find a maximal strongly orthogonal sequence of noncompact imaginary roots such that, after application of the Cayley transforms, no noncompact imaginary roots remain. If $\mathfrak{g}_0$ is a noncomplex simple real Lie algebra and if we have a Vogan diagram for $\mathfrak{g}_0$ as in Theorem 6.96, such a sequence is readily at hand by an inductive construction. We start with a noncompact imaginary simple root, form the set of roots orthogonal to it, label their compactness or noncompactness by means of Proposition 6.72, and iterate the process.

EXAMPLE. Let $\mathfrak{g}_0 = \mathfrak{su}(p, n-p)$ with $p \le n-p$. The distinguished Vogan diagram is of type A_{n-1} with $e_p - e_{p+1}$ as the unique noncompact imaginary simple root. Since the Dynkin diagram does not have a double line, orthogonality implies strong orthogonality. The above process yields the sequence of noncompact imaginary roots

$$\begin{aligned} 2f_1 &= e_p - e_{p+1} \\ 2f_2 &= e_{p-1} - e_{p+2} \\ &\vdots \\ 2f_p &= e_1 - e_{2p}. \end{aligned} \tag{6.106}$$

We do a Cayley transform with respect to each of these. The order is irrelevant; since the roots are strongly orthogonal, the individual Cayley transforms commute. It is helpful to use the same names for roots before and after Cayley transform but always to remember what Cartan subalgebra is being used. After Cayley transform the remaining imaginary roots are those roots involving only indices $2p+1, \dots, n$, and such roots are compact. Thus a maximally compact Cartan subalgebra has noncompact dimension p. The restricted roots are obtained by projecting all $e_k - e_l$ on the linear span of (6.106). If $1 \le k < l \le p$, we have

$$\begin{aligned} e_k - e_l &= \tfrac{1}{2}(e_k - e_{2p+1-k}) - \tfrac{1}{2}(e_l - e_{2p+1-l}) + (\text{orthogonal to (6.106)}) \\ &= (f_k - f_l) + (\text{orthogonal to (6.106)}). \end{aligned}$$

Thus $f_k - f_l$ is a restricted root. For the same k and l, $e_k - e_{2p+1-l}$ restricts to $f_k + f_l$. In addition, if $k + l = 2p+1$, then $e_k - e_l$ restricts to $2f_k$, while if $k \le p$ and $l > 2p$, then $e_k - e_l$ restricts to f_k. Consequently the set of restricted roots is

$$\Sigma = \begin{cases} \{\pm f_k \pm f_l\} \cup \{\pm 2f_k\} \cup \{\pm f_k\} & \text{if } 2p < n \\ \{\pm f_k \pm f_l\} \cup \{\pm 2f_k\} & \text{if } 2p = n. \end{cases}$$

Thus Σ is of type $(BC)_p$ if $2p < n$ and of type C_p if $2p = n$.

We attempt to repeat the construction in the above example for all of the classical matrix algebras and exceptional algebras in Theorem 6.105, parts (c) and (d). There is no difficulty when the automorphism in the Vogan diagram is trivial. However, the cases where the automorphism is nontrivial require special comment. Except for $\mathfrak{sl}(2n+1,\mathbb{R})$, which we can handle manually, each of these Lie algebras has β orthogonal to $\theta\beta$ whenever β is a complex simple root. Then it follows from Proposition 6.104 that any positive imaginary root is the sum of imaginary simple roots and a number of pairs $\beta, \theta\beta$ of complex simple roots and that the complex simple roots can be disregarded in deciding compactness or noncompactness. In particular, $\mathfrak{sl}(n,\mathbb{H})$ and E IV have no noncompact imaginary roots.

EXAMPLE. Let $\mathfrak{g}_0 = \text{E I}$. The Vogan diagram is

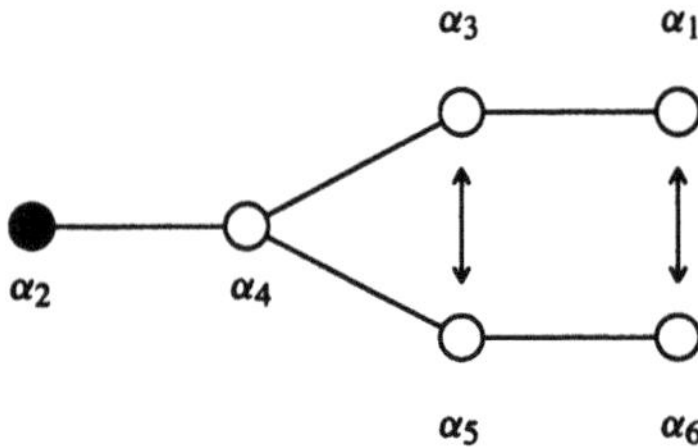

Let α_2 be the first member in the orthogonal sequence of imaginary noncompact roots. From the theory for D_4, a nonobvious root orthogonal to α_2 is $\alpha_0 = \alpha_2 + 2\alpha_4 + \alpha_3 + \alpha_5$. This root is imaginary, and no smaller imaginary root is orthogonal to α_2. We can disregard the complex pair α_3, α_5 in deciding compactness or noncompactness (Proposition 6.104), and we see that α_0 is noncompact. Following our algorithm, we can expand our list to α_2, α_0. The Vogan diagram of the system orthogonal to α_2 is

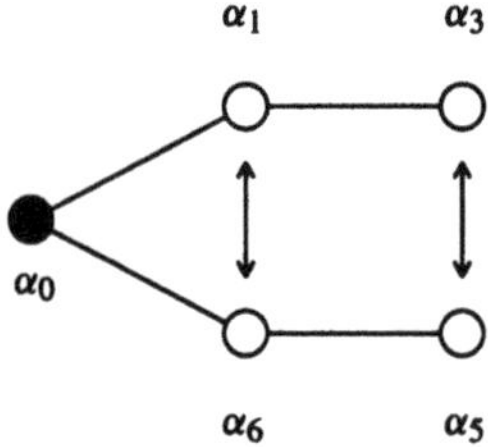

This is the Vogan diagram of $\mathfrak{sl}(6,\mathbb{R})$, and we therefore know that the list extends to

$$\alpha_2,\ \alpha_0,\ \alpha_1 + \alpha_0 + \alpha_6,\ \alpha_3 + (\alpha_1 + \alpha_0 + \alpha_6) + \alpha_5.$$

Thus the Cayley transforms increase the noncompact dimension of the Cartan subalgebra by 4 from 2 to 6, and it follows that E I is a split real form.

It is customary to refer to the noncompact dimension of a maximal noncompact Cartan subalgebra of $\mathfrak{g}_0$ as the **real rank** of $\mathfrak{g}_0$. We are led to the following information about restricted roots. In the case of the classical matrix algebras, the results are

(6.107)

$\mathfrak{g}_0$	Condition	Real Rank	Restricted Roots
$\mathfrak{su}(p,q)$	$p \geq q$	q	$(BC)_q$ if $p > q$, C_q if $p = q$
$\mathfrak{so}(p,q)$	$p \geq q$	q	B_q if $p > q$, D_q if $p = q$
$\mathfrak{sp}(p,q)$	$p \geq q$	q	$(BC)_q$ if $p > q$, C_q if $p = q$
$\mathfrak{sp}(n,\mathbb{R})$		n	C_n
$\mathfrak{so}^*(2n)$		$[\frac{n}{2}]$	$C_{\frac{1}{2}n}$ if n even, $(BC)_{\frac{1}{2}(n-1)}$ if n odd
$\mathfrak{sl}(n,\mathbb{R})$		$n-1$	A_{n-1}
$\mathfrak{sl}(n,\mathbb{H})$		$n-1$	A_{n-1}

For the exceptional Lie algebras the results are

(6.108)

$\mathfrak{g}_0$	Real Rank	Restricted Roots
E I	6	E_6
E II	4	F_4
E III	2	$(BC)_2$
E IV	2	A_2
E V	7	E_7
E VI	4	F_4
E VII	3	C_3
E VIII	8	E_8
E IX	4	F_4
F I	4	F_4
F II	1	$(BC)_1$
G	2	G_2

For the Lie algebras in Theorem 6.105a, the above analysis simplifies. Here $\mathfrak{g}$ is complex simple, and we take $\mathfrak{g}_0 = \mathfrak{g}^{\mathbb{R}}$. Let J be multiplication by $\sqrt{-1}$ within $\mathfrak{g}^{\mathbb{R}}$. If θ is a Cartan involution of $\mathfrak{g}^{\mathbb{R}}$, then Corollary 6.22 shows that θ comes from conjugation of $\mathfrak{g}$ with respect to a compact real form $\mathfrak{u}_0$. In other words, $\mathfrak{g}^{\mathbb{R}} = \mathfrak{u}_0 \oplus J\mathfrak{u}_0$ with $\theta(X + JY) = X - JY$. Let $\mathfrak{h}_0 = \mathfrak{t}_0 \oplus \mathfrak{a}_0$ be a θ stable Cartan subalgebra of $\mathfrak{g}^{\mathbb{R}}$. Since $\mathfrak{t}_0$ commutes with $\mathfrak{a}_0$, $\mathfrak{t}_0$ commutes with $J\mathfrak{a}_0$. Also $\mathfrak{a}_0$ commutes with $J\mathfrak{a}_0$. Since $\mathfrak{h}_0$ is maximal abelian, $J\mathfrak{a}_0 \subseteq \mathfrak{t}_0$. Similarly $J\mathfrak{t}_0 \subseteq \mathfrak{a}_0$. Therefore $J\mathfrak{t}_0 = \mathfrak{a}_0$, and $\mathfrak{h}_0$ is actually a complex subalgebra of $\mathfrak{g}$. By Proposition 2.7, $\mathfrak{h}_0$ is a (complex) Cartan subalgebra of $\mathfrak{g}$. Let

$$\mathfrak{g} = \mathfrak{h}_0 \oplus \bigoplus_{\alpha \in \Delta} \mathfrak{g}_\alpha$$

be the root-space decomposition. Here each α is complex-linear on the complex vector space $\mathfrak{h}_0$. Thus distinct α's have distinct restrictions to $\mathfrak{a}_0$. Hence

$$\mathfrak{g}^{\mathbb{R}} = \mathfrak{a}_0 \oplus \mathfrak{t}_0 \oplus \bigoplus_{\alpha \in \Delta} \mathfrak{g}_\alpha$$

is the restricted-root space decomposition, each restricted-root space being 2-dimensional over $\mathbb{R}$. Consequently the real rank of $\mathfrak{g}^{\mathbb{R}}$ equals the rank of $\mathfrak{g}$, and the system of restricted roots of $\mathfrak{g}^{\mathbb{R}}$ is canonically identified (by restriction or complexification) with the system of roots of $\mathfrak{g}$. In particular the system Σ of restricted roots is of the same type (A_n through G_2) as the system Δ of roots.

The simple real Lie algebras of real-rank one will play a special role in Chapter VII. From Theorem 6.105 and our determination above of the real rank of each example, the full list of such Lie algebras is

$$\begin{array}{lll} & \mathfrak{su}(p,1) & \text{with } p \geq 1 \\ (6.109) & \mathfrak{so}(p,1) & \text{with } p \geq 3 \\ & \mathfrak{sp}(p,1) & \text{with } p \geq 2 \\ & \text{F II} & \end{array}$$

Low-dimensional isomorphisms show that other candidates are redundant:

$$\begin{aligned} \mathfrak{sl}(2,\mathbb{C}) &\cong \mathfrak{so}(3,1) \\ \mathfrak{so}(2,1) &\cong \mathfrak{su}(1,1) \\ \mathfrak{sp}(1,1) &\cong \mathfrak{so}(4,1) \\ \mathfrak{sp}(1,\mathbb{R}) &\cong \mathfrak{su}(1,1) \\ \mathfrak{so}^*(4) &\cong \mathfrak{su}(2) \oplus \mathfrak{su}(1,1) \\ \mathfrak{so}^*(6) &\cong \mathfrak{su}(3,1) \\ \mathfrak{sl}(2,\mathbb{R}) &\cong \mathfrak{su}(1,1) \\ \mathfrak{sl}(2,\mathbb{H}) &\cong \mathfrak{so}(5,1). \end{aligned} \tag{6.110}$$

12. Problems

1. Prove that if $\mathfrak{g}$ is a complex semisimple Lie algebra, then any two split real forms of $\mathfrak{g}$ are conjugate via $\operatorname{Aut}\mathfrak{g}$.

2. Let $\mathfrak{g}_0 = \mathfrak{k}_0 \oplus \mathfrak{p}_0$ be a Cartan decomposition of a real semisimple Lie algebra. Prove that $\mathfrak{k}_0$ is compactly embedded in $\mathfrak{g}_0$ and that it is maximal with respect to this property.

3. Let G be semisimple, let $\mathfrak{g}_0 = \mathfrak{k}_0 \oplus \mathfrak{p}_0$ be a Cartan decomposition of the Lie algebra, and let X and Y be in $\mathfrak{p}_0$. Prove that $\exp X \exp Y \exp X$ is in $\exp \mathfrak{p}_0$.

4. Let $g \in SL(m, \mathbb{C})$ be positive definite. Prove that g can be decomposed as $g = lu$, where l is lower triangular and u is upper triangular.

5. In the development of the Iwasawa decomposition for $SO(p, 1)_0$ and $SU(p, 1)$, make particular choices of a positive system for the restricted roots, and compute N in each case.

6. (a) Prove that $\mathfrak{g}_0 = \mathfrak{so}^*(2n)$ consists in block form of all complex matrices $\begin{pmatrix} a & b \\ -\bar{b} & \bar{a} \end{pmatrix}$ with a skew Hermitian and b skew symmetric.
   (b) In $\mathfrak{g}_0$, let $\mathfrak{h}_0$ be the Cartan subalgebra in (6.102). Assuming that the roots are $\pm e_i \pm e_j$, find the root vectors. Show that $e_i - e_j$ is compact and $e_i + e_j$ is noncompact.
   (c) Show that a choice of maximal abelian subspace of $\mathfrak{p}_0$ is to take a to be 0 and take b to be block diagonal and real with blocks of sizes $2, \ldots, 2$ if n is even and $1, 2, \ldots, 2$ if n is odd.
   (d) Find the restricted-root space decomposition of $\mathfrak{g}_0$ relative to the maximal abelian subspace of $\mathfrak{p}_0$ given in (c).

7. Let $\mathfrak{h}_0 = \mathfrak{t}_0 \oplus \mathfrak{a}_0$ be a maximally noncompact θ stable Cartan subalgebra, and let $\mathfrak{h} = \mathfrak{t} \oplus \mathfrak{a}$ be the complexification. Fix a positive system Σ^+ for the restricted roots, and introduce a positive system Δ^+ for the roots so that a nonzero restriction to $\mathfrak{a}_0$ of a member of Δ^+ is always in Σ^+.
   (a) Prove that every simple restricted root for Σ^+ is the restriction of a simple root for Δ^+.
   (b) Let V be the span of the imaginary simple roots. Prove for each simple α_i not in V that $-\theta\alpha_i$ is in $\alpha_{i'} + V$ for a unique simple $\alpha_{i'}$, so that $\alpha_i \mapsto \alpha_{i'}$ defines a permutation of order 2 on the simple roots not in V.
   (c) For each orbit $\{i, i'\}$ of one or two simple roots not in V, define an element $H = H_{\{i,i'\}} \in \mathfrak{h}$ by $\alpha_i(H) = \alpha_{i'}(H) = 1$ and $\alpha_j(H) = 0$ for all other j. Prove that H is in $\mathfrak{a}$.

(d) Using the elements constructed in (c), prove that the linear span of the restrictions to $\mathfrak{a}_0$ of the simple roots has dimension equal to the number of orbits.

(e) Conclude from (d) that the nonzero restriction to $\mathfrak{a}_0$ of a simple root for Δ^+ is simple for Σ^+.

8. The group K for $G = SL(3, \mathbb{R})$ is $K = SO(3)$, which has a double cover $\tilde{K}$. Therefore G itself has a double cover $\tilde{G}$. The group $M = Z_K(A)$ is known from Example 1 of §5 to be the direct sum of two 2-element groups. Prove that $\tilde{M} = Z_{\tilde{K}}(A)$ is isomorphic to the subgroup $\{\pm 1, \pm i, \pm j, \pm k\}$ of the unit quaternions.

9. Suppose that D and D' are Vogan diagrams corresponding to $\mathfrak{g}_0$ and $\mathfrak{g}_0'$, respectively. Prove that an inclusion $D \subseteq D'$ induces a one-one Lie algebra homomorphism $\mathfrak{g}_0 \to \mathfrak{g}_0'$.

10. Let G be a semisimple Lie group with Lie algebra $\mathfrak{g}_0$. Fix a Cartan involution θ and Cartan decomposition $\mathfrak{g}_0 = \mathfrak{k}_0 \oplus \mathfrak{p}_0$, and let K be the analytic subgroup of G with Lie algebra $\mathfrak{k}_0$. Suppose that $\mathfrak{g}_0$ has a Cartan subalgebra contained in $\mathfrak{k}_0$.

    (a) Prove that there exists $k \in K$ such that $\theta = \mathrm{Ad}(k)$.

    (b) Prove that if Σ is the system of restricted roots of $\mathfrak{g}_0$, then -1 is in the Weyl group of Σ.

11. Let G be a semisimple Lie group with Lie algebra $\mathfrak{g}_0$. Fix a Cartan involution θ and Cartan decomposition $\mathfrak{g}_0 = \mathfrak{k}_0 \oplus \mathfrak{p}_0$, and let K be the analytic subgroup of G with Lie algebra $\mathfrak{k}_0$. Prove that if $\mathfrak{g}_0$ does not have a Cartan subalgebra contained in $\mathfrak{k}_0$, then there does not exist $k \in K$ such that $\theta = \mathrm{Ad}(k)$.

12. Let $\mathfrak{t}_0 \oplus \mathfrak{a}_0$ be a maximally noncompact θ stable Cartan subalgebra. Prove that if α is a root, then $\alpha + \theta\alpha$ is not a root.

13. For $\mathfrak{g}_0 = \mathfrak{sl}(2n, \mathbb{R})$, let $\mathfrak{h}_0^{(i)}$ consist of all block-diagonal matrices whose first i blocks are of size 2 of the form $\left\{\begin{pmatrix} t_j & \theta_j \\ -\theta_j & t_j \end{pmatrix}\right\}$, for $1 \le j \le i$, and whose remaining blocks are $2(n-i)$ blocks of size 1.

    (a) Prove that the $\mathfrak{h}_0^{(i)}$, $0 \le i \le n$, form a complete set of nonconjugate Cartan subalgebras of $\mathfrak{g}_0$.

    (b) Relate $\mathfrak{h}_0^{(i)}$ to the maximally compact θ stable Cartan subalgebra of Example 2 in §8, using Cayley transforms.

    (c) Relate $\mathfrak{h}_0^{(i)}$ to the maximally noncompact θ stable Cartan subalgebra of diagonal matrices, using Cayley transforms.

14. The example in §7 constructs four Cartan subalgebras for $\mathfrak{sp}(2, \mathbb{R})$. The first one $\mathfrak{h}_0$ is maximally noncompact, and the last one $\mathfrak{h}'_0$ is maximally compact. The second one has noncompact part contained in $\mathfrak{h}_0$ and compact part contained in $\mathfrak{h}'_0$, but the third one does not. Show that the third one is not even conjugate to a Cartan subalgebra whose noncompact part is contained in $\mathfrak{h}_0$ and whose compact part is contained in $\mathfrak{h}'_0$.

15. Let a $(2n)$-by-$(2n)$ matrix be given in block form by $\frac{\sqrt{2}}{2}\begin{pmatrix} 1 & i \\ i & 1 \end{pmatrix}$. Define a mapping $X \mapsto Y$ of the set of $(2n)$-by-$(2n)$ complex matrices to itself by $Y = \begin{pmatrix} 1 & i \\ i & 1 \end{pmatrix} X \begin{pmatrix} 1 & i \\ i & 1 \end{pmatrix}^{-1}$.
   (a) Prove that the map carries $\mathfrak{su}(n, n)$ to an image whose members Y are characterized by $\operatorname{Tr} Y = 0$ and $JY + Y^*J = 0$, where J is as in Example 2 of §I.8.
   (b) Prove that the mapping exhibits $\mathfrak{su}(n, n) \cap \mathfrak{sp}(n, \mathbb{C})$ as isomorphic with $\mathfrak{sp}(n, \mathbb{R})$.
   (c) Within $\mathfrak{g}_0 = \mathfrak{su}(n, n) \cap \mathfrak{sp}(n, \mathbb{C})$, let θ be negative conjugate transpose. Define $\mathfrak{h}_0$ to be the Cartan subalgebra in (6.102). Referring to Example 3 in §II.1, find all root vectors and identify which are compact and which are noncompact. Interpret the above mapping on $(\mathfrak{g}_0)^{\mathbb{C}}$ as a product of Cayley transforms $\mathbf{c}_\beta$. Which roots β are involved?

16. (a) Prove that every element of $SL(2, \mathbb{R})$ is conjugate to at least one matrix of the form

$$\begin{pmatrix} a & 0 \\ 0 & a^{-1} \end{pmatrix} \text{ with } a \neq 0, \quad \begin{pmatrix} 1 & t \\ 0 & 1 \end{pmatrix}, \quad \begin{pmatrix} -1 & t \\ 0 & -1 \end{pmatrix},$$

$$\text{or} \quad \begin{pmatrix} \cos\theta & \sin\theta \\ -\sin\theta & \cos\theta \end{pmatrix}.$$

   (b) Prove that the exponential map from $\mathfrak{sl}(2, \mathbb{R})$ into $SL(2, \mathbb{R})$ has image

$$\{X \mid \operatorname{Tr} X > -2\} \cup \{-1\}.$$

17. Let $\mathfrak{g}$ be a simple complex Lie algebra. Describe the Vogan diagram of $\mathfrak{g}^{\mathbb{R}}$.

18. This problem examines the effect on the painting in a Vogan diagram when the positive system is changed from Δ^+ to $s_\alpha \Delta^+$, where α is an imaginary simple root.
   (a) Show that the new diagram is a Vogan diagram with the same Dynkin diagram and automorphism and with the painting unchanged at the position of α and at all positions not adjacent to α.

(b) If α is compact, show that there is no change in the painting of imaginary roots in positions adjacent to α.
(c) If α is noncompact, show that the painting of an imaginary root at a position adjacent to α is reversed unless the root is connected by a double line to α and is long, in which case it is unchanged.
(d) Devise an algorithm for a Vogan diagram of type A_n for a step-by-step change of positive system so that ultimately at most one simple root is painted (as is asserted to be possible by Theorem 6.96).

19. In the Vogan diagram from Theorem 6.96 for the Lie algebra F II of §10, the simple root $\frac{1}{2}(e_1 - e_2 - e_3 - e_4)$ is noncompact, and the simple roots $e_2 - e_3, e_3 - e_4$, and e_4 are compact.
(a) Verify that $\frac{1}{2}(e_1 - e_2 + e_3 + e_4)$ is noncompact.
(b) The roots $\frac{1}{2}(e_1 - e_2 - e_3 - e_4)$ and $\frac{1}{2}(e_1 - e_2 + e_3 + e_4)$ are orthogonal and noncompact, yet (6.108) says that F II has real rank one. Explain.

20. The Vogan diagram of F I, as given by Theorem 6.96, has $e_2 - e_3$ as its one and only noncompact simple root. What strongly orthogonal set of noncompact roots is produced by the algorithm of §11?

21. Verify the assertion in (6.108) that E VII has real rank 3 and restricted roots of type C_3.

Problems 22–24 give further information about the Cartan decomposition $\mathfrak{g}_0 = \mathfrak{k}_0 \oplus \mathfrak{p}_0$ of a real semisimple Lie algebra. Let B be the Killing form of $\mathfrak{g}_0$.

22. Let $\mathfrak{p}_0'$ be an $\operatorname{ad} \mathfrak{k}_0$ invariant subspace of $\mathfrak{p}_0$, and let

$$\mathfrak{p}_0'^{\perp} = \{X \in \mathfrak{p}_0 \mid B(X, \mathfrak{p}_0') = 0\}.$$

Prove that $B([\mathfrak{p}_0', \mathfrak{p}_0'^{\perp}], \mathfrak{k}_0) = 0$ and conclude that $[\mathfrak{p}_0', \mathfrak{p}_0'^{\perp}] = 0$.

23. If $\mathfrak{p}_0'$ is an $\operatorname{ad} \mathfrak{k}_0$ invariant subspace of $\mathfrak{p}_0$, prove that $[\mathfrak{p}_0', \mathfrak{p}_0] \oplus \mathfrak{p}_0'$ is an ideal in $\mathfrak{g}_0$.

24. Under the additional assumption that $\mathfrak{g}_0$ is simple but not compact, prove that
(a) $[\mathfrak{p}_0, \mathfrak{p}_0] = \mathfrak{k}_0$
(b) $\mathfrak{k}_0$ is a maximal proper Lie subalgebra of $\mathfrak{g}_0$.

Problems 25–27 deal with low-dimensional isomorphisms.

25. Establish the following isomorphisms by using Vogan diagrams:
(a) the isomorphisms in (6.110)
(b) $\mathfrak{sl}(4, \mathbb{R}) \cong \mathfrak{so}(3, 3), \mathfrak{su}(2, 2) \cong \mathfrak{so}(4, 2), \mathfrak{sp}(2, \mathbb{R}) \cong \mathfrak{so}(3, 2)$
(c) $\mathfrak{sp}(2) \cong \mathfrak{so}(5), \mathfrak{su}(4) \cong \mathfrak{so}(6), \mathfrak{su}(2) \oplus \mathfrak{su}(2) \cong \mathfrak{so}(4)$.

26. (a) Prove that the mapping of Problem 36 of Chapter II gives an isomorphism of $\mathfrak{sl}(4, \mathbb{R})$ onto $\mathfrak{so}(3, 3)$.
    (b) Prove that the mapping of Problem 38 of Chapter II gives an isomorphism of $\mathfrak{sp}(2, \mathbb{R})$ onto $\mathfrak{so}(3, 2)$.
27. Prove that the Lie algebra isomorphisms of Problem 25b induce Lie group homomorphisms $SL(4, \mathbb{R}) \to SO(3, 3)_0$, $SU(2, 2) \to SO(4, 2)_0$, and $Sp(2, \mathbb{R}) \to SO(3, 2)_0$. What is the kernel in each case?

CHAPTER VII

Advanced Structure Theory

Abstract. The first main results are that simply connected compact semisimple Lie groups are in one-one correspondence with abstract Cartan matrices and their associated Dynkin diagrams and that the outer automorphisms of such a group correspond exactly to automorphisms of the Dynkin diagram. The remainder of the first section prepares for the definition of a reductive Lie group: A compact connected Lie group has a complexification that is unique up to holomorphic isomorphism. A semisimple Lie group of matrices is topologically closed and has finite center.

Reductive Lie groups G are defined as 4-tuples (G, K, θ, B) satisfying certain compatibility conditions. Here G is a Lie group, K is a compact subgroup, θ is an involution of the Lie algebra $\mathfrak{g}_0$ of G, and B is a bilinear form on $\mathfrak{g}_0$. Examples include semisimple Lie groups with finite center, any connected closed linear group closed under conjugate transpose, and the centralizer in a reductive group of a θ stable abelian subalgebra of the Lie algebra. The involution θ, which is called the "Cartan involution" of the Lie algebra, is the differential of a global Cartan involution Θ of G. In terms of Θ, G has a global Cartan decomposition that generalizes the polar decomposition of matrices.

A number of properties of semisimple Lie groups with finite center generalize to reductive Lie groups. Among these are the conjugacy of the maximal abelian subspaces of the -1 eigenspace $\mathfrak{p}_0$ of θ, the theory of restricted roots, the Iwasawa decomposition, and properties of Cartan subalgebras. The chapter addresses also some properties not discussed in Chapter VI, such as the $KA_{\mathfrak{p}}K$ decomposition and the Bruhat decomposition. Here $A_{\mathfrak{p}}$ is the analytic subgroup corresponding to a maximal abelian subspace of $\mathfrak{p}_0$.

The degree of disconnectedness of the subgroup $M_{\mathfrak{p}} = Z_K(A_{\mathfrak{p}})$ controls the disconnectedness of many other subgroups of G. The most complete description of $M_{\mathfrak{p}}$ is in the case that G has a complexification, and then serious results from Chapter V about representation theory play a decisive role.

Parabolic subgroups are closed subgroups containing a conjugate of $M_{\mathfrak{p}}A_{\mathfrak{p}}N_{\mathfrak{p}}$. They are parametrized up to conjugacy by subsets of simple restricted roots. A Cartan subgroup is defined to be the centralizer of a Cartan subalgebra. It has only finitely many components, and each regular element of G lies in one and only one Cartan subgroup of G. When G has a complexification, the component structure of Cartan subgroups can be identified in terms of the elements that generate $M_{\mathfrak{p}}$.

A reductive Lie group G that is semisimple has the property that G/K admits a complex structure with G acting holomorphically if and only if the centralizer in $\mathfrak{g}_0$ of the center of the Lie algebra $\mathfrak{k}_0$ of K is just $\mathfrak{k}_0$. In this case, G/K may be realized as a bounded domain in some $\mathbb{C}^n$ by means of the Harish-Chandra decomposition. The proof of the Harish-Chandra decomposition uses facts about parabolic subgroups. The spaces G/K of this kind may be classified easily by inspection of the classification of simple real Lie algebras in Chapter VI.

1. Further Properties of Compact Real Forms

Some aspects of compact real forms of complex semisimple Lie algebras were omitted in Chapter VI in order to move more quickly toward the classification of simple real Lie algebras. We take up these aspects now in order to prepare for the more advanced structure theory to be discussed in this chapter. The topics in this section are classification of compact semisimple Lie algebras and simply connected compact semisimple Lie groups, complex structures on semisimple Lie groups whose Lie algebras are complex, automorphisms of complex semisimple Lie algebras, and properties of connected linear groups with reductive Lie algebra. Toward the end of this section we discuss Weyl's unitary trick.

Proposition 7.1. The isomorphism classes of compact semisimple Lie algebras $\mathfrak{g}_0$ and the isomorphism classes of complex semisimple Lie algebras $\mathfrak{g}$ are in one-one correspondence, the correspondence being that $\mathfrak{g}$ is the complexification of $\mathfrak{g}_0$ and $\mathfrak{g}_0$ is a compact real form of $\mathfrak{g}$. Under this correspondence simple Lie algebras correspond to simple Lie algebras.

REMARK. The proposition implies that the complexification of a compact simple Lie algebra is simple. It then follows from Theorem 6.94 that a compact simple Lie algebra is never complex.

PROOF. If a compact semisimple $\mathfrak{g}_0$ is given, we know that its complexification $\mathfrak{g}$ is complex semisimple. In the reverse direction Theorem 6.11 shows that any complex semisimple $\mathfrak{g}$ has a compact real form, and Corollary 6.20 shows that the compact real form is unique up to isomorphism. This proves the correspondence. If a complex $\mathfrak{g}$ is simple, then it is trivial that any real form is simple.

Conversely suppose that $\mathfrak{g}_0$ is compact simple. Arguing by contradiction, suppose that the complexification $\mathfrak{g}$ is semisimple but not simple. Write $\mathfrak{g}$ as the direct sum of simple ideals $\mathfrak{g}_i$ by Theorem 1.51, and let $(\mathfrak{g}_i)_0$ be a compact real form of $\mathfrak{g}_i$ as in Theorem 6.11. The Killing forms of distinct $\mathfrak{g}_i$'s are orthogonal, and it follows that the Killing form of the direct sum of the $(\mathfrak{g}_i)_0$'s is negative definite. By Proposition 4.27, the direct sum of the $(\mathfrak{g}_i)_0$'s is a compact real form of $\mathfrak{g}$. By Corollary 6.20 the direct sum of the $(\mathfrak{g}_i)_0$'s is isomorphic to $\mathfrak{g}_0$ and exhibits $\mathfrak{g}_0$ as semisimple but not simple, contradiction.

Proposition 7.2. The isomorphism classes of simply connected compact semisimple Lie groups are in one-one correspondence with the isomorphism classes of compact semisimple Lie algebras by passage from a Lie group to its Lie algebra.

PROOF. The Lie algebra of a compact semisimple group is compact semisimple by Proposition 4.23. Conversely if a compact semisimple Lie algebra $\mathfrak{g}_0$ is given, then the Killing form of $\mathfrak{g}_0$ is negative definite by Corollary 4.26 and Cartan's Criterion for Semisimplicity (Theorem 1.42). Consequently $\operatorname{Int}\mathfrak{g}_0$ is a subgroup of a compact orthogonal group. On the other hand, Propositions 1.97 and 1.98 show that $\operatorname{Int}\mathfrak{g}_0 \cong (\operatorname{Aut}\mathfrak{g}_0)_0$ and hence that $\operatorname{Int}\mathfrak{g}_0$ is closed. Thus $\operatorname{Int}\mathfrak{g}_0$ is a compact connected Lie group with Lie algebra $\operatorname{ad}\mathfrak{g}_0 \cong \mathfrak{g}_0$. By Weyl's Theorem (Theorem 4.69) a universal covering group of $\operatorname{Int}\mathfrak{g}_0$ is a simply connected compact semisimple group with Lie algebra $\mathfrak{g}_0$. Since two simply connected analytic groups with isomorphic Lie algebras are isomorphic, the proposition follows.

Corollary 7.3. The isomorphism classes of

(a) simply connected compact semisimple Lie groups,
(b) compact semisimple Lie algebras,
(c) complex semisimple Lie algebras,
(d) reduced abstract root systems,
(e) abstract Cartan matrices and their associated Dynkin diagrams

are in one-one correspondence by passage from a Lie group to its Lie algebra, then to the complexification of the Lie algebra, and then to the underlying root system.

PROOF. The correspondence of (a) to (b) is addressed by Proposition 7.2, that of (b) to (c) is addressed by Proposition 7.1, and that of (c) to (d) to (e) is addressed by Chapter II.

Proposition 7.4. A semisimple Lie group whose Lie algebra is complex admits uniquely the structure of a complex Lie group in such a way that the exponential mapping is holomorphic.

REMARK. Consequently we may speak unambiguously of a **complex semisimple Lie group** as being a semisimple Lie group whose Lie algebra is complex.

PROOF. Let G be the Lie group, and let $\mathfrak{g}$ be the Lie algebra. Since $\mathfrak{g}$ is complex, the analytic group $\operatorname{Ad}(G)$ is an analytic subgroup of the complex Lie group $GL(\mathfrak{g})$ with Lie algebra closed under multiplication by i. By the remarks at the end of §I.10, $\operatorname{Ad}(G)$ is a complex Lie group in a unique way such that the exponential map is holomorphic. Since

$g \mapsto \mathrm{Ad}(g)$ is a covering map, the complex structure on $\mathrm{Ad}(G)$ lifts uniquely to a complex structure on G such that the covering map is holomorphic and regular. This lift is the unique lift making the exponential mapping holomorphic, and the result follows.

Proposition 7.5. A complex semisimple Lie group necessarily has finite center. Let G and G' be complex semisimple Lie groups, and let K and K' be the subgroups fixed by the respective global Cartan involutions of G and G'. Then K and K' are compact, and a homomorphism of K into K' as Lie groups induces a holomorphic homomorphism of G into G'. If the homomorphism $K \to K'$ is an isomorphism, then the holomorphic homomorphism $G \to G'$ is a holomorphic isomorphism.

PROOF. If G has Lie algebra $\mathfrak{g}$, then the most general Cartan decomposition of $\mathfrak{g}^{\mathbb{R}}$ is $\mathfrak{g}^{\mathbb{R}} = \mathfrak{g}_0 \oplus i\mathfrak{g}_0$, where $\mathfrak{g}_0$ is a compact real form of $\mathfrak{g}$ by Proposition 6.14 and Corollary 6.19. The Lie algebra $\mathfrak{g}_0$ is compact semisimple, and Weyl's Theorem (Theorem 4.69) shows that the corresponding analytic subgroup K is compact. Theorem 6.31f then shows that G has finite center.

In a similar fashion let $\mathfrak{g}'$ be the Lie algebra of G'. We may suppose that there is a Cartan decomposition $\mathfrak{g}'^{\mathbb{R}} = \mathfrak{g}'_0 \oplus i\mathfrak{g}'_0$ of $\mathfrak{g}'^{\mathbb{R}}$ such that K' is the analytic subgroup of G' with Lie algebra $\mathfrak{g}'_0$. As with K, K' is compact.

A homomorphism φ of K into K' yields a homomorphism $d\varphi$ of $\mathfrak{g}_0$ into $\mathfrak{g}'_0$, and this extends uniquely to a complex-linear homomorphism, also denoted $d\varphi$, of $\mathfrak{g}$ into $\mathfrak{g}'$. Let $\tilde{G}$ be a universal covering group of G, let $e : \tilde{G} \to G$ be the covering homomorphism, and let $\tilde{K}$ be the analytic subgroup of $\tilde{G}$ with Lie algebra $\mathfrak{g}_0$. Since $\tilde{G}$ is simply connected, $d\varphi$ lifts to a smooth homomorphism $\tilde{\varphi}$ of $\tilde{G}$ into G'.

We want to see that $\tilde{\varphi}$ descends to a homomorphism of G into G'. To see this, we show that $\tilde{\varphi}$ is 1 on the kernel of e. The restriction $\tilde{\varphi}|_{\tilde{K}}$ and the composition $\varphi \circ (e|_{\tilde{K}})$ both have $d\varphi$ as differential. Therefore they are equal, and $\tilde{\varphi}$ is 1 on the kernel of $e|_{\tilde{K}}$. Theorem 6.31e shows that the kernel of e in $\tilde{G}$ is contained in $\tilde{K}$, and it follows that $\tilde{\varphi}$ descends to a homomorphism of G into G' with differential $d\varphi$. Let us call this homomorphism φ. Then φ is a homomorphism between complex Lie groups, and its differential is complex linear. By remarks near the end of §I.10, φ is holomorphic.

If the given homomorphism is an isomorphism, then we can reverse the roles of G and G', obtaining a holomorphic homomorphism $\psi : G' \to G$ with differential the inverse of $d\varphi$. Since $\psi \circ \varphi$ and $\varphi \circ \psi$ have differential the identity, φ and ψ are inverses. Therefore φ is a holomorphic isomorphism.

Corollary 7.6. If G is a complex semisimple Lie group, then G is holomorphically isomorphic to a complex Lie group of matrices.

PROOF. Let $\mathfrak{g}$ be the Lie algebra of G, let $\mathfrak{g}^{\mathbb{R}} = \mathfrak{g}_0 \oplus i\mathfrak{g}_0$ be a Cartan decomposition of $\mathfrak{g}^{\mathbb{R}}$, and let K be the analytic subgroup of G with Lie algebra $\mathfrak{g}_0$. By Corollary 4.22, K is isomorphic to a closed linear group, say K', and there is no loss of generality in assuming that the members of K' are in $GL(V)$ for a real vector space V. Let $\mathfrak{g}_0'$ be the linear Lie algebra of K', and write the complexification $\mathfrak{g}'$ of $\mathfrak{g}_0'$ as a Lie algebra of complex endomorphisms of $V^{\mathbb{C}}$. If G' is the analytic subgroup of $GL(V^{\mathbb{C}})$ with Lie algebra $\mathfrak{g}'$, then G' is a complex Lie group since $GL(V^{\mathbb{C}})$ is complex and $\mathfrak{g}'$ is closed under multiplication by i. Applying Proposition 7.5, we can extend the isomorphism of K onto K' to a holomorphic isomorphism of G onto G'. Thus G' provides the required complex Lie group of matrices.

Let G be a semisimple Lie group, and suppose that $G^{\mathbb{C}}$ is a complex semisimple Lie group such that G is an analytic subgroup of $G^{\mathbb{C}}$ and the Lie algebra of $G^{\mathbb{C}}$ is the complexification of the Lie algebra of G. Then we say that $G^{\mathbb{C}}$ is a **complexification** of G and that G has a complexification $G^{\mathbb{C}}$. For example, $SU(n)$ and $SL(n, \mathbb{R})$ both have $SL(n, \mathbb{C})$ as complexification. Because of Corollary 7.6 it will follow from Proposition 7.9 below that if G has a complexification $G^{\mathbb{C}}$, then G is necessarily closed in $G^{\mathbb{C}}$. Not every semisimple Lie group has a complexification; because of Corollary 7.6, the example at the end of §VI.3 shows that a double cover of $SL(2, \mathbb{R})$ has no complexification. If G has a complexification, then the complexification is not necessarily unique up to isomorphism. However, Proposition 7.5 shows that the complexification is unique if G is compact.

We now use the correspondence of Corollary 7.3 to investigate automorphisms of complex semisimple Lie algebras.

Lemma 7.7. Let G be a complex semisimple Lie group with Lie algebra $\mathfrak{g}$, let $\mathfrak{h}$ be a Cartan subalgebra of $\mathfrak{g}$, and let $\Delta^+(\mathfrak{g}, \mathfrak{h})$ be a positive system for the roots. If H denotes the analytic subgroup of G with Lie algebra $\mathfrak{h}$, then any member of $\operatorname{Int}\mathfrak{g}$ carrying $\mathfrak{h}$ to itself and $\Delta^+(\mathfrak{g}, \mathfrak{h})$ to itself is in $\operatorname{Ad}_{\mathfrak{g}}(H)$.

PROOF. The construction of Theorem 6.11 produces a compact real form $\mathfrak{g}_0$ of $\mathfrak{g}$ such that $\mathfrak{g}_0 \cap \mathfrak{h} = \mathfrak{h}_0$ is a maximal abelian subspace of $\mathfrak{g}_0$. The decomposition $\mathfrak{g}^{\mathbb{R}} = \mathfrak{g}_0 \oplus i\mathfrak{g}_0$ is a Cartan decomposition of $\mathfrak{g}^{\mathbb{R}}$ by Proposition 6.14, and we let θ be the Cartan involution. Let K be the analytic subgroup of G with Lie algebra $\mathfrak{g}_0$. The subgroup K is compact by Proposition 7.5. If T is the analytic subgroup of K with Lie algebra $\mathfrak{h}_0$, then T is a maximal torus of K.

Let g be in G, and suppose that $\mathrm{Ad}(g)$ carries $\mathfrak{h}$ to itself and $\Delta^+(\mathfrak{g}, \mathfrak{h})$ to itself. By Theorem 6.31 we can write $g = k \exp X$ with $k \in K$ and $X \in i\mathfrak{g}_0$. The map $\mathrm{Ad}(\Theta g)$ is the differential at 1 of $g \mapsto (\Theta g)x(\Theta g)^{-1} = \Theta(g(\Theta x)g^{-1})$, hence is $\theta \mathrm{Ad}(g)\theta$. Since $\theta\mathfrak{h} = \mathfrak{h}$, $\mathrm{Ad}(\Theta g)$ carries $\mathfrak{h}$ to itself. Therefore so does $\mathrm{Ad}((\Theta g)^{-1}g) = \mathrm{Ad}(\exp 2X)$.

The linear transformation $\mathrm{Ad}(\exp 2X)$ is diagonable on $\mathfrak{g}^{\mathbb{R}}$ with positive eigenvalues. Since it carries $\mathfrak{h}$ to $\mathfrak{h}$, there exists a real subspace $\mathfrak{h}'$ of $\mathfrak{g}^{\mathbb{R}}$ carried to itself by $\mathrm{Ad}(\exp 2X)$ such that $\mathfrak{g}^{\mathbb{R}} = \mathfrak{h} \oplus \mathfrak{h}'$. The transformation $\mathrm{Ad}(\exp 2X)$ has a unique diagonable logarithm with real eigenvalues, and there are two candidates for this logarithm. One is $\mathrm{ad}\, 2X$, and the other is the sum of the logarithms on $\mathfrak{h}$ and $\mathfrak{h}'$ separately. By uniqueness we conclude that $\mathrm{ad}\, 2X$ carries $\mathfrak{h}$ to $\mathfrak{h}$. By Proposition 2.7, X is in $\mathfrak{h}$.

Therefore $\exp X$ is in H, and it is enough to show that k is in T. Here k is a member of K such that $\mathrm{Ad}(k)$ leaves $\mathfrak{h}_0$ stable and $\Delta^+(\mathfrak{g}, \mathfrak{h})$ stable. Since $\mathrm{Ad}(k)$ leaves $\mathfrak{h}_0$ stable, Theorem 4.54 says that $\mathrm{Ad}(k)$ is in the Weyl group $W(\mathfrak{g}, \mathfrak{h})$. Since $\mathrm{Ad}(k)$ leaves $\Delta^+(\mathfrak{g}, \mathfrak{h})$ stable, Theorem 2.63 says that $\mathrm{Ad}(k)$ is the identity element in $W(\mathfrak{g}, \mathfrak{h})$. Therefore $\mathrm{Ad}(k)$ is 1 on $\mathfrak{h}$, and k commutes with T. By Corollary 4.52, k is in T.

Theorem 7.8. If $\mathfrak{g}_0$ is a compact semisimple Lie algebra and $\mathfrak{g}$ is its complexification, then the following three groups are canonically isomorphic:

(a) $\mathrm{Aut}_{\mathbb{R}}\, \mathfrak{g}_0 / \mathrm{Int}\, \mathfrak{g}_0$
(b) $\mathrm{Aut}_{\mathbb{C}}\, \mathfrak{g} / \mathrm{Int}\, \mathfrak{g}$
(c) the group of automorphisms of the Dynkin diagram of $\mathfrak{g}$.

PROOF. By Proposition 7.4 let G be a simply connected complex Lie group with Lie algebra $\mathfrak{g}$, for example a universal covering group of $\mathrm{Int}\, \mathfrak{g}$. The analytic subgroup K with Lie algebra $\mathfrak{g}_0$ is simply connected by Theorem 6.31, and K is compact by Proposition 7.5.

Fix a maximal abelian subspace $\mathfrak{h}_0$ of $\mathfrak{g}_0$, let $\Delta^+(\mathfrak{g}, \mathfrak{h})$ be a positive system of roots, and let T be the maximal torus of K with Lie algebra $\mathfrak{h}_0$. Let D be the Dynkin diagram of $\mathfrak{g}$, and let $\mathrm{Aut}\, D$ be the group of automorphisms of D. Any member of $\mathrm{Aut}_{\mathbb{R}}\, \mathfrak{g}_0$ extends by complexifying to a member of $\mathrm{Aut}_{\mathbb{C}}\, \mathfrak{g}$, and members of $\mathrm{Int}\, \mathfrak{g}_0$ yield members of $\mathrm{Int}\, \mathfrak{g}$. Thus we obtain a group homomorphism $\Phi : \mathrm{Aut}_{\mathbb{R}} \mathfrak{g}_0 / \mathrm{Int}\, \mathfrak{g}_0 \to \mathrm{Aut}_{\mathbb{C}}\, \mathfrak{g} / \mathrm{Int}\, \mathfrak{g}$.

Let us observe that Φ is onto. In fact, if a member φ of $\mathrm{Aut}_{\mathbb{C}}\, \mathfrak{g}$ is given, then $\varphi(\mathfrak{g}_0)$ is a compact real form of $\mathfrak{g}$. By Corollary 6.20 we can adjust φ by a member of $\mathrm{Int}\, \mathfrak{g}$ so that φ carries $\mathfrak{g}_0$ into itself. Thus some automorphism of $\mathfrak{g}_0$ is carried to the coset of φ under Φ.

We shall construct a group homomorphism $\Psi : \mathrm{Aut}_{\mathbb{C}}\, \mathfrak{g} / \mathrm{Int}\, \mathfrak{g} \to \mathrm{Aut}\, D$. Let $\varphi \in \mathrm{Aut}_{\mathbb{C}}\, \mathfrak{g}$ be given. Since $\mathfrak{h}$ is a Cartan subalgebra of $\mathfrak{g}$ (by Corollary

2.13), $\varphi(\mathfrak{h})$ is another Cartan subalgebra. By Theorem 2.15 there exists $\psi_1 \in \operatorname{Int}\mathfrak{g}$ with $\psi_1\varphi(\mathfrak{h}) = \mathfrak{h}$. Then $\psi_1\varphi$ maps $\Delta(\mathfrak{g},\mathfrak{h})$ to itself and carries $\Delta^+(\mathfrak{g},\mathfrak{h})$ to another positive system $(\Delta^+)'(\mathfrak{g},\mathfrak{h})$. By Theorem 2.63 there exists a unique member w of the Weyl group $W(\mathfrak{g},\mathfrak{h})$ carrying $(\Delta^+)'(\mathfrak{g},\mathfrak{h})$ to $\Delta^+(\mathfrak{g},\mathfrak{h})$. Theorem 4.54 shows that w is implemented by a member of $\operatorname{Ad}(K)$, hence by a member ψ_2 of $\operatorname{Int}\mathfrak{g}$. Then $\psi_2\psi_1\varphi$ maps $\Delta^+(\mathfrak{g},\mathfrak{h})$ to itself and yields an automorphism of the Dynkin diagram.

Let us see the effect of the choices we have made. With different choices, we would be led to some $\psi_2'\psi_1'\varphi$ mapping $\Delta^+(\mathfrak{g},\mathfrak{h})$ to itself, and the claim is that we get the same member of $\operatorname{Aut} D$. In fact the composition $\psi = (\psi_2'\psi_1'\varphi)\circ(\psi_2\psi_1\varphi)^{-1}$ is in $\operatorname{Int}\mathfrak{g}$. Lemma 7.7 shows that ψ acts as the identity on $\mathfrak{h}$, and hence the automorphism of the Dynkin diagram corresponding to ψ is the identity. Therefore $\psi_2\psi_1\varphi$ and $\psi_2'\psi_1'\varphi$ lead to the same member of $\operatorname{Aut} D$.

Consequently the above construction yields a well defined function $\Psi : \operatorname{Aut}_{\mathbb{C}}\mathfrak{g}/\operatorname{Int}\mathfrak{g} \to \operatorname{Aut} D$. Since we can adjust any $\varphi \in \operatorname{Aut}_{\mathbb{C}}\mathfrak{g}$ by a member of $\operatorname{Int}\mathfrak{g}$ so that $\mathfrak{h}$ maps to itself and $\Delta^+(\mathfrak{g},\mathfrak{h})$ maps to itself, it is clear that Ψ is a homomorphism.

Let us prove that $\Psi\circ\Phi$ is one-one. Thus let $\varphi \in \operatorname{Aut}_{\mathbb{R}}\mathfrak{g}_0$ lead to the identity element of $\operatorname{Aut} D$. Write φ also for the corresponding complex-linear automorphism on $\mathfrak{g}$. Theorem 4.34 shows that we may adjust φ by a member of $\operatorname{Int}\mathfrak{g}_0$ so that φ carries $\mathfrak{h}_0$ to itself, and Theorems 2.63 and 4.54 show that we may adjust φ further by a member of $\operatorname{Int}\mathfrak{g}_0$ so that φ carries $\Delta^+(\mathfrak{g},\mathfrak{h})$ to itself. Let E_{α_i} be root vectors for the simple roots $\alpha_1,\dots,\alpha_l$ of $\mathfrak{g}$. Since φ is the identity on $\mathfrak{h}$, $\varphi(E_{\alpha_i}) = c_iE_{\alpha_i}$ for nonzero constants $c_1,\dots,c_l$. For each j, let x_j be any complex number with $e^{x_j} = c_j$. Choose, for $1 \le i \le l$, members h_j of $\mathfrak{h}$ with $\alpha_i(h_j) = \delta_{ij}$, and put $g = \exp\left(\sum_{j=1}^l x_jh_j\right)$. The element g is in H. Then $\operatorname{Ad}(g)(E_{\alpha_i}) = c_iE_{\alpha_i}$ for each i. Consequently $\operatorname{Ad}(g)$ is a member of $\operatorname{Int}\mathfrak{g}$ that agrees with φ on $\mathfrak{h}$ and on each E_{α_i}. By the Isomorphism Theorem (Theorem 2.108), $\varphi = \operatorname{Ad}(g)$.

To complete the proof that $\Psi\circ\Phi$ is one-one, we show that g is in T. We need to see that $|c_j| = 1$ for all j, so that x_j can be chosen purely imaginary. First we show that $\overline{E}_{\alpha_j}$ is a root vector for $-\alpha_j$ if bar denotes the conjugation of $\mathfrak{g}$ with respect to $\mathfrak{g}_0$. In fact, write $E_{\alpha_j} = X_j + iY_j$ with X_j and Y_j in $\mathfrak{g}_0$. If h is in $\mathfrak{h}_0$, then $\alpha_j(h)$ is purely imaginary. Since $[\mathfrak{h}_0,\mathfrak{g}_0] \subseteq \mathfrak{g}_0$, it follows from the equality

$$[h, X_j] + i[h, Y_j] = [h, E_{\alpha_j}] = \alpha_j(h)E_{\alpha_j} = i\alpha_j(h)Y_j + \alpha_j(h)X_j$$

that $[h, X_j] = i\alpha_j(h)Y_j$ and $i[h, Y_j] = \alpha_j(h)X_j$. Subtracting these two formulas gives

$$[h, X_j - iY_j] = i\alpha_j(h)Y_j - \alpha_j(h)X_j = -\alpha_j(h)(X_j - iY_j)$$

and shows that $\overline{E}_{\alpha_j}$ is indeed a root vector for $-\alpha_j$. Hence we find that $[E_{\alpha_j}, \overline{E}_{\alpha_j}]$ is in $\mathfrak{h}$. Since φ is complex linear and carries $\mathfrak{g}_0$ to itself, φ respects bar. Therefore $\varphi(\overline{E}_{\alpha_j}) = \bar{c}_j \overline{E}_{\alpha_j}$. Since φ fixes every element of $\mathfrak{h}$, φ fixes $[E_{\alpha_j}, \overline{E}_{\alpha_j}]$, and it follows that $c_j\bar{c}_j = 1$. We conclude that g is in T and that $\Psi \circ \Phi$ is one-one.

Since Φ is onto and $\Psi \circ \Phi$ is one-one, both Φ and Ψ are one-one. The fact that Ψ is onto is a consequence of the Isomorphism Theorem (Theorem 2.108) and is worked out in detail in the second example at the end of §II.10. This completes the proof of the theorem.

Now we take up some properties of Lie groups of matrices to prepare for the definition of "reductive Lie group" in the next section.

Proposition 7.9. Let G be an analytic subgroup of real or complex matrices whose Lie algebra $\mathfrak{g}_0$ is semisimple. Then G has finite center and is a closed linear group.

PROOF. Without loss of generality we may assume that G is an analytic subgroup of $GL(V)$ for a real vector space V. Let $\mathfrak{g}_0$ be the linear Lie algebra of G, and write the complexification $\mathfrak{g}$ of $\mathfrak{g}_0$ as a Lie algebra of complex endomorphisms of $V^{\mathbb{C}}$. Let $\mathfrak{g}_0 = \mathfrak{k}_0 \oplus \mathfrak{p}_0$ be a Cartan decomposition, and let K be the analytic subgroup of G with Lie algebra $\mathfrak{k}_0$. The Lie subalgebra $\mathfrak{u}_0 = \mathfrak{k}_0 \oplus i\mathfrak{p}_0$ of $\operatorname{End}_{\mathbb{C}} V$ is a compact semisimple Lie algebra, and we let U be the analytic subgroup of $GL(V^{\mathbb{C}})$ with Lie algebra $\mathfrak{u}_0$. Proposition 7.2 implies that the universal covering group $\tilde{U}$ of U is compact, and it follows that U is compact. Since U has discrete center, the center Z_U of U must be finite.

The center Z_G of G is contained in K by Theorem 6.31e, and $K \subseteq U$ since $\mathfrak{k}_0 \subseteq \mathfrak{u}_0$. Since $\operatorname{Ad}_{\mathfrak{g}}(Z_G)$ acts as 1 on $\mathfrak{u}_0$, we conclude that $Z_G \subseteq Z_U$. Therefore Z_G is finite. This proves the first conclusion. By Theorem 6.31f, K is compact.

Since U is compact, Proposition 4.6 shows that $V^{\mathbb{C}}$ has a Hermitian inner product preserved by U. Then U is contained in the unitary group $U(V^{\mathbb{C}})$. Let $\mathfrak{p}(V^{\mathbb{C}})$ be the vector space of Hermitian transformations of $V^{\mathbb{C}}$ so that $GL(V^{\mathbb{C}})$ has the polar decomposition $GL(V^{\mathbb{C}}) = U(V^{\mathbb{C}}) \exp \mathfrak{p}(V^{\mathbb{C}})$. The members of $\mathfrak{u}_0$ are skew Hermitian, and hence the members of $\mathfrak{k}_0$ are skew Hermitian and the members of $\mathfrak{p}_0$ are Hermitian. Therefore the global Cartan decomposition $G = K \exp \mathfrak{p}_0$ of G that is given in Theorem 6.31c is compatible with the polar decomposition of $GL(V^{\mathbb{C}})$.

We are to prove that G is closed in $GL(V^{\mathbb{C}})$. Let $g_n = k_n \exp X_n$ tend to $g \in GL(V^{\mathbb{C}})$. Using the compactness of K and passing to a subsequence, we may assume that k_n tends to $k \in K$. Therefore $\exp X_n$

converges. Since the polar decomposition of $GL(V^{\mathbb{C}})$ is a homeomorphism, it follows that $\exp X_n$ has limit $\exp X$ for some $X \in \mathfrak{p}(V^{\mathbb{C}})$. Since $\mathfrak{p}_0$ is closed in $\mathfrak{p}(V^{\mathbb{C}})$, X is in $\mathfrak{p}_0$. Therefore $g = k \exp X$ exhibits g as in G, and G is closed.

Corollary 7.10. Let G be an analytic subgroup of real or complex matrices whose Lie algebra $\mathfrak{g}_0$ is reductive, and suppose that the identity component of the center of G is compact. Then G is a closed linear group.

REMARK. In this result and some to follow, we shall work with analytic groups whose Lie algebras are direct sums. If G is an analytic group whose Lie algebra $\mathfrak{g}_0$ is a direct sum $\mathfrak{g}_0 = \mathfrak{a}_0 \oplus \mathfrak{b}_0$ of ideals and if A and B are the analytic subgroups corresponding to $\mathfrak{a}_0$ and $\mathfrak{b}_0$, then G is a commuting product $G = AB$. This fact follows from Proposition 1.99 or may be derived directly, as in the proof of Theorem 4.29.

PROOF. Write $\mathfrak{g}_0 = Z_{\mathfrak{g}_0} \oplus [\mathfrak{g}_0, \mathfrak{g}_0]$ by Corollary 1.53. The analytic subgroup of G corresponding to $Z_{\mathfrak{g}_0}$ is $(Z_G)_0$, and we let G_{ss} be the analytic subgroup corresponding to $[\mathfrak{g}_0, \mathfrak{g}_0]$. By the remarks before the proof, G is the commuting product $(Z_G)_0 G_{ss}$. The group G_{ss} is closed as a group of matrices by Proposition 7.1, and $(Z_G)_0$ is compact by assumption. Hence the set of products, which is G, is closed.

Corollary 7.11. Let G be a connected closed linear group whose Lie algebra $\mathfrak{g}_0$ is reductive. Then the analytic subgroup G_{ss} of G with Lie algebra $[\mathfrak{g}_0, \mathfrak{g}_0]$ is closed, and G is the commuting product $G = (Z_G)_0 G_{ss}$.

PROOF. The subgroup G_{ss} is closed by Proposition 7.1, and G is the commuting product $(Z_G)_0 G_{ss}$ by the remarks with Corollary 7.10.

Proposition 7.12. Let G be a compact connected linear Lie group, and let $\mathfrak{g}_0$ be its linear Lie algebra. Then the complex analytic group $G^{\mathbb{C}}$ of matrices with linear Lie algebra $\mathfrak{g} = \mathfrak{g}_0 \oplus i\mathfrak{g}_0$ is a closed linear group.

REMARKS. If G is a compact connected Lie group, then Corollary 4.22 implies that G is isomorphic to a closed linear group. If G is realized as a closed linear group in two different ways, then this proposition in principle produces two different groups $G^{\mathbb{C}}$. However, Proposition 7.5 shows that the two groups $G^{\mathbb{C}}$ are isomorphic. Therefore with no reference to linear groups, we can speak of the complexification $G^{\mathbb{C}}$ of a compact connected Lie group G, and $G^{\mathbb{C}}$ is unique up to isomorphism. Proposition 7.5 shows that a homomorphism between two such groups G and G' induces a holomorphic homomorphism between their complexifications.

PROOF. By Theorem 4.29 let us write $G = (Z_G)_0 G_{ss}$ with G_{ss} compact semisimple. Proposition 4.6 shows that we may assume without loss of generality that G is a connected closed subgroup of a unitary group $U(n)$ for some n, and Corollary 4.7 shows that we may take $(Z_G)_0$ to be diagonal.

Let us complexify the decomposition $\mathfrak{g}_0 = Z_{\mathfrak{g}_0} \oplus [\mathfrak{g}_0, \mathfrak{g}_0]$ to obtain $\mathfrak{g}^{\mathbb{R}} = Z_{\mathfrak{g}_0} \oplus iZ_{\mathfrak{g}_0} \oplus [\mathfrak{g}, \mathfrak{g}]$. The analytic subgroup corresponding to $Z_{\mathfrak{g}_0}$ is $G_1 = (Z_G)_0$ and is compact. Since $iZ_{\mathfrak{g}_0}$ consists of real diagonal matrices, Corollary 1.111 shows that its corresponding analytic subgroup G_2 is closed. In addition the analytic subgroup G_3 with Lie algebra $[\mathfrak{g}, \mathfrak{g}]$ is closed by Proposition 7.9. By the remarks with Corollary 7.10, the group $G^{\mathbb{C}}$ is the commuting product of these three subgroups, and we are to show that the product is closed.

For G_3, negative conjugate transpose is a Cartan involution of its Lie algebra, and therefore conjugate transpose inverse is a global Cartan involution of G_3. Consequently G_3 has a global Cartan decomposition $G_3 = G_{ss} \exp(\mathfrak{p}_3)_0$, where $(\mathfrak{p}_3)_0 = i[\mathfrak{g}_0, \mathfrak{g}_0]$. Since $iZ_{\mathfrak{g}_0}$ commutes with $(\mathfrak{p}_3)_0$ and since the polar decomposition of all matrices is a homeomorphism, it follows that the product G_2G_3 is closed. Since G_1 is compact, $G^{\mathbb{C}} = G_1G_2G_3$ is closed.

Lemma 7.13. On matrices let Θ be conjugate transpose inverse, and let θ be negative conjugate transpose. Let G be a connected abelian closed linear group that is stable under Θ, and let $\mathfrak{g}_0$ be its linear Lie algebra, stable under θ. Let $\mathfrak{g}_0 = \mathfrak{k}_0 \oplus \mathfrak{p}_0$ be the decomposition of $\mathfrak{g}_0$ into $+1$ and -1 eigenspaces under θ, and let $K = \{x \in G \mid \Theta x = x\}$. Then the map $K \times \mathfrak{p}_0 \to G$ given by $(k, X) \mapsto k \exp X$ is a Lie group isomorphism.

PROOF. The group K is a closed subgroup of the unitary group and is compact with Lie algebra $\mathfrak{k}_0$. Since $\mathfrak{p}_0$ is abelian, $\exp \mathfrak{p}_0$ is the analytic subgroup of G with Lie algebra $\mathfrak{p}_0$. By the remarks with Corollary 7.10, $G = K \exp \mathfrak{p}_0$. The smooth map $K \times \mathfrak{p}_0 \to G$ is compatible with the polar decomposition of matrices and is therefore one-one. It is a Lie group homomorphism since G and $\mathfrak{p}_0$ are abelian. Its inverse is smooth since the inverse of the polar decomposition of matrices is smooth (by an argument in the proof of Theorem 6.31).

Proposition 7.14. On matrices let Θ be conjugate transpose inverse, and let θ be negative conjugate transpose. Let G be a connected closed linear group that is stable under Θ, and let $\mathfrak{g}_0$ be its linear Lie algebra, stable under θ. Let $\mathfrak{g}_0 = \mathfrak{k}_0 \oplus \mathfrak{p}_0$ be the decomposition of $\mathfrak{g}_0$ into $+1$ and -1 eigenspaces under θ, and let $K = \{x \in G \mid \Theta x = x\}$. Then the map $K \times \mathfrak{p}_0 \to G$ given by $(k, X) \mapsto k \exp X$ is a diffeomorphism onto.

PROOF. By Proposition 1.56, $\mathfrak{g}_0$ is reductive. Therefore Corollary 1.53 allows us to write $\mathfrak{g}_0 = Z_{\mathfrak{g}_0} \oplus [\mathfrak{g}_0, \mathfrak{g}_0]$ with $[\mathfrak{g}_0, \mathfrak{g}_0]$ semisimple. The analytic subgroup of G with Lie algebra $Z_{\mathfrak{g}_0}$ is $(Z_G)_0$, and we let G_{ss} be the analytic subgroup of G with Lie algebra $[\mathfrak{g}_0, \mathfrak{g}_0]$. By Corollary 7.11, $(Z_G)_0$ and G_{ss} are closed, and $G = (Z_G)_0 G_{ss}$. It is clear that $Z_{\mathfrak{g}_0}$ and $[\mathfrak{g}_0, \mathfrak{g}_0]$ are stable under θ, and hence $(Z_G)_0$ and G_{ss} are stable under Θ.

Because of the polar decomposition of matrices, the map $K \times \mathfrak{p}_0 \to G$ is smooth and one-one. The parts of this map associated with $(Z_G)_0$ and G_{ss} are onto by Lemma 7.13 and Theorem 6.31, respectively. Since $(Z_G)_0$ and G_{ss} commute with each other, it follows that $K \times \mathfrak{p}_0 \to G$ is onto. The inverse is smooth since the inverse of the polar decomposition of marices is smooth (by an argument in the proof of Theorem 6.31).

Proposition 7.15 (Weyl's unitary trick). Let G be an analytic subgroup of complex matrices whose linear Lie algebra $\mathfrak{g}_0$ is semisimple and is stable under the map θ given by negative conjugate transpose. Let $\mathfrak{g}_0 = \mathfrak{k}_0 \oplus \mathfrak{p}_0$ be the Cartan decomposition of $\mathfrak{g}_0$ defined by θ, and suppose that $\mathfrak{k}_0 \cap i\mathfrak{p}_0 = 0$. Let U and $G^{\mathbb{C}}$ be the analytic subgroups of matrices with respective Lie algebras $\mathfrak{u}_0 = \mathfrak{k}_0 \oplus i\mathfrak{p}_0$ and $\mathfrak{g} = (\mathfrak{k}_0 \oplus \mathfrak{p}_0)^{\mathbb{C}}$. The group U is compact. Suppose that U is simply connected. If V is any finite-dimensional complex vector space, then a representation of any of the following kinds on V leads, via the formula

$$\mathfrak{g} = \mathfrak{g}_0 \oplus i\mathfrak{g}_0 = \mathfrak{u}_0 \oplus i\mathfrak{u}_0, \tag{7.16}$$

to a representation of each of the other kinds. Under this correspondence invariant subspaces and equivalences are preserved:

(a) a representation of G on V
(b) a representation of U on V
(c) a holomorphic representation of $G^{\mathbb{C}}$ on V
(d) a representation of $\mathfrak{g}_0$ on V
(e) a representation of $\mathfrak{u}_0$ on V
(f) a complex-linear representation of $\mathfrak{g}$ on V.

PROOF. The groups G, U, and $G^{\mathbb{C}}$ are closed linear groups by Proposition 7.9, and U is compact, being a closed subgroup of the unitary group. Since U is simply connected and its Lie algebra is a compact real form of $\mathfrak{g}$, $G^{\mathbb{C}}$ is simply connected.

We can pass from (c) to (a) or (b) by restriction. Since continuous homomorphisms between Lie groups are smooth, we can pass from (a) or (b) to (d) or (e) by taking differentials. Formula (7.16) allows us to pass from (d) or (e) to (f). Since $G^{\mathbb{C}}$ is simply connected, a Lie algebra homomorphism as in (f) lifts to a group homomorphism, and the group homomorphism must be holomorphic since the Lie algebra homomorphism is assumed complex linear. Thus we can pass from (f) to (c). If we follow the steps all the way around, starting from (c), we end up with the original representation, since the differential at the identity uniquely determines a homomorphism of Lie groups. Thus invariant subspaces and equivalence are preserved.

EXAMPLE. Weyl's unitary trick gives us a new proof of the fact that finite-dimensional complex-linear representations of complex semisimple Lie algebras are completely reducible (Theorem 5.29); the crux of the new proof is the existence of a compact real form (Theorem 6.11). For the argument let the Lie algebra $\mathfrak{g}$ be given, and let G be a simply connected complex semisimple group with Lie algebra $\mathfrak{g}$. Corollary 7.6 allows us to regard G as a subgroup of $GL(V^{\mathbb{C}})$ for some finite-dimensional complex vector space $V^{\mathbb{C}}$. Let $\mathfrak{u}_0$ be a compact real form of $\mathfrak{g}$, so that $\mathfrak{g}^{\mathbb{R}} = \mathfrak{u}_0 \oplus i\mathfrak{u}_0$, and let U be the analytic subgroup of G with Lie algebra $\mathfrak{u}_0$. Proposition 7.15 notes that U is compact. By Proposition 4.6 we can introduce a Hermitian inner product into $V^{\mathbb{C}}$ so that U is a subgroup of the unitary group. If a complex-linear representation of $\mathfrak{g}$ is given, we can use the passage (f) to (b) in Proposition 7.15 to obtain a representation of U. This is completely reducible by Corollary 4.7, and the complete reducibility of the given representation of $\mathfrak{g}$ follows.

The final proposition shows how to recognize a Cartan decomposition of a real semisimple Lie algebra in terms of a bilinear form other than the Killing form.

Proposition 7.17. Let $\mathfrak{g}_0$ be a real semisimple Lie algebra, let θ be an involution of $\mathfrak{g}_0$, and let B be a nondegnerate symmetric invariant bilinear form on $\mathfrak{g}_0$ such that $B(\theta X, \theta Y) = B(X, Y)$ for all X and Y in $\mathfrak{g}_0$. If the form $B_\theta(X, Y) = -B(X, \theta Y)$ is positive definite, then θ is a Cartan involution of $\mathfrak{g}_0$.

PROOF. Let $\mathfrak{g}_0 = \mathfrak{k}_0 \oplus \mathfrak{p}_0$ be the decomposition of $\mathfrak{g}_0$ into $+1$ and -1 eigenspaces under θ, and extend B to be complex bilinear on the complexification $\mathfrak{g}$ of $\mathfrak{g}_0$. Since θ is an involution, $\mathfrak{u}_0 = \mathfrak{k}_0 \oplus i\mathfrak{p}_0$ is a Lie subalgebra of $\mathfrak{g} = (\mathfrak{g}_0)^{\mathbb{C}}$, necessarily a real form. Here $\mathfrak{g}$ is semisimple, and then so is $\mathfrak{u}_0$. Since B_θ is positive definite, B is negative definite on $\mathfrak{k}_0$ and on $i\mathfrak{p}_0$. Also $\mathfrak{k}_0$ and $i\mathfrak{p}_0$ are orthogonal since $X \in \mathfrak{k}_0$ and $Y \in i\mathfrak{p}_0$ implies

$$B(X, Y) = B(\theta X, \theta Y) = B(X, -Y) = -B(X, Y).$$

Hence B is real-valued and negative definite on $\mathfrak{u}_0$.

By Propositions 1.97 and 1.98, $\operatorname{Int}\mathfrak{u}_0 = (\operatorname{Aut}_{\mathbb{R}}\mathfrak{u}_0)_0$. Consequently $\operatorname{Int}\mathfrak{u}_0$ is a closed subgroup of $GL(\mathfrak{u}_0)$. On the other hand, we have just seen that $-B$ is an inner product on $\mathfrak{u}_0$, and in this inner product every member of $\operatorname{ad}\mathfrak{u}_0$ is skew symmetric. Therefore the corresponding analytic subgroup $\operatorname{Int}\mathfrak{u}_0$ of $GL(\mathfrak{u}_0)$ acts by orthogonal transformations. Since $\operatorname{Int}\mathfrak{u}_0$ is then exhibited as a closed subgroup of the orthogonal group, $\operatorname{Int}\mathfrak{u}_0$ is compact. Hence $\mathfrak{u}_0$ is a compact real form of $\mathfrak{g}$. By the remarks preceding Lemma 6.27, θ is a Cartan involution of $\mathfrak{g}_0$.

2. Reductive Lie Groups

We are ready to define the class of groups that will be the objects of study in this chapter. The intention is to study semisimple groups, but, as was already the case in Chapters IV and VI, we shall often have to work with centralizers of abelian analytic subgroups invariant under a Cartan involution, and these centralizers may be disconnected and may have positive-dimensional center. To be able to use arguments that take advantage of such subgroups and proceed by induction on the dimension, we are forced to enlarge the class of groups under study. Groups in the enlarged class are always called "reductive," but their characterizing properties vary from author to author. We shall use the following definition.

A **reductive Lie group** is actually a 4-tuple (G, K, θ, B) consisting of a Lie group G, a compact subgroup K of G, a Lie algebra involution θ of the Lie algebra $\mathfrak{g}_0$ of G, and a nondegenerate, $\operatorname{Ad}(G)$ invariant, θ invariant, bilinear form B on $\mathfrak{g}_0$ such that

- (i) $\mathfrak{g}_0$ is a reductive Lie algebra,
- (ii) the decomposition of $\mathfrak{g}_0$ into $+1$ and -1 eigenspaces under θ is $\mathfrak{g}_0 = \mathfrak{k}_0 \oplus \mathfrak{p}_0$, where $\mathfrak{k}_0$ is the Lie algebra of K,
- (iii) $\mathfrak{k}_0$ and $\mathfrak{p}_0$ are orthogonal under B, and B is positive definite on $\mathfrak{p}_0$ and negative definite on $\mathfrak{k}_0$,
- (iv) multiplication, as a map from $K \times \exp\mathfrak{p}_0$ into G, is a diffeomorphism onto, and
- (v) every automorphism $\operatorname{Ad}(g)$ of $\mathfrak{g} = (\mathfrak{g}_0)^{\mathbb{C}}$ is **inner** for $g \in G$, i.e., is given by some x in $\operatorname{Int}\mathfrak{g}$.

When informality permits, we shall refer to the reductive Lie group simply as G. Then θ will be called the **Cartan involution,** $\mathfrak{g}_0 = \mathfrak{k}_0 \oplus \mathfrak{p}_0$ will be called the **Cartan decomposition** of $\mathfrak{g}_0$, K will be called the associated **maximal compact subgroup** (see Proposition 7.19a below), and B will be called the **invariant bilinear form**.

The idea is that a reductive Lie group G is a Lie group whose Lie algebra is reductive, whose center is not too wild, and whose disconnectedness is not too wild. The various properties make precise the notion "not too wild." Note that property (iv) and the compactness of K say that G has only finitely many components.

We write G_{ss} for the semisimple analytic subgroup of G with Lie algebra $[\mathfrak{g}_0, \mathfrak{g}_0]$. The decomposition of G is property (iv) is called the **global Cartan decomposition**. Sometimes one assumes about a reductive Lie group that also

(vi) G_{ss} has finite center.

In this case the reductive group will be said to be in the **Harish-Chandra class** because of the use of axioms equivalent with (i) through (vi) by Harish-Chandra. Reductive groups in the Harish-Chandra class have often been the groups studied in representation theory.

EXAMPLES.

1) G is any semisimple Lie group with finite center, θ is a Cartan involution, K is the analytic subgroup with Lie algebra $\mathfrak{k}_0$, and B is the Killing form. Property (iv) and the compactness of K follow from Theorem 6.31. Property (v) is automatic since G connected makes $\mathrm{Ad}(G) = \mathrm{Int}\,\mathfrak{g}_0 \subseteq \mathrm{Int}\,\mathfrak{g}$. Property (vi) has been built into the definition for this example.

2) G is any connected closed linear group of real or complex matrices closed under conjugate transpose inverse, θ is negative conjugate transpose, K is the intersection of G with the unitary group, and $B(X, Y)$ is $\mathrm{Re}\,\mathrm{Tr}(XY)$. The compactness of K follows since K is the intersection of the unitary group with the closed group of matrices G. Property (iv) follows from Proposition 7.14, and property (v) is automatic since G is connected. Property (vi) is automatic for any linear group by Proposition 7.9.

3) G is any compact Lie group satisfying property (v). Then $K = G$, $\theta = 1$, and B is the negative of an inner product constructed as in Proposition 4.24. Properties (i) through (iv) are trivial, and property (vi) follows from Theorem 4.21. Every finite group G is trivially an example where property (v) holds. Property (v) is satisfied by the orthogonal group $O(n)$ if n is odd but not by $O(n)$ if n is even.

4) G is any closed linear group of real or complex matrices closed under conjugate transpose inverse, given as the common zero locus of some set of real-valued polynomials in the real and imaginary parts of the matrix entries, and satisfying property (v). Here θ is negative conjugate transpose, K is the intersection of G with the unitary group, and $B(X, Y)$ is $\mathrm{Re}\,\mathrm{Tr}(XY)$. The compactness of K follows since K is the intersection

of the unitary group with the closed group of matrices G. Properties (iv) and (vi) follow from Propositions 1.122 and 7.9, respectively. The closed linear group of real matrices of determinant ± 1 satisfies property (v) since

$$\mathrm{Ad}(\mathrm{diag}(-1, 1, \ldots, 1)) = \mathrm{Ad}(\mathrm{diag}(e^{i\pi(n-1)/n}, e^{-i\pi/n}, \ldots, e^{-i\pi/n})).$$

But as noted in Example 3, the orthogonal group $O(n)$ does not satisfy property (v) if n is even.

5) G is the centralizer in a reductive group $\tilde{G}$ of a θ stable abelian subalgebra of the Lie algebra of $\tilde{G}$. Here K is obtained by intersection, and θ and B are obtained by restriction. The verification that G is a reductive Lie group will be given below in Proposition 7.21.

If G is semisimple with finite center and if K, θ, and B are specified so that G is considered as a reductive group, then θ is forced to be a Cartan involution in the sense of Chapter VI. This is the content of Proposition 7.17. Hence the new terms "Cartan involution" and "Cartan decomposition" are consistent with the terminology of Chapter VI in the case that G is semisimple.

An alternate way of saying (iii) is that the symmetric bilinear form

$$B_\theta(X, Y) = -B(X, \theta Y) \tag{7.18}$$

is positive definite on $\mathfrak{g}_0$.

We use the notation $\mathfrak{g}$, $\mathfrak{k}$, $\mathfrak{p}$, etc., to denote the complexifications of $\mathfrak{g}_0, \mathfrak{k}_0, \mathfrak{p}_0$, etc. Using complex linearity, we extend θ from $\mathfrak{g}_0$ to $\mathfrak{g}$ and B from $\mathfrak{g}_0 \times \mathfrak{g}_0$ to $\mathfrak{g} \times \mathfrak{g}$.

Proposition 7.19. If G is a reductive Lie group, then

(a) K is a maximal compact subgroup of G
(b) K meets every component of G, i.e., $G = KG_0$
(c) each member of $\mathrm{Ad}(K)$ leaves $\mathfrak{k}_0$ and $\mathfrak{p}_0$ stable and therefore commutes with θ
(d) $(\mathrm{ad}\, X)^* = -\mathrm{ad}\, \theta X$ relative to B_θ if X is in $\mathfrak{g}_0$
(e) θ leaves $Z_{\mathfrak{g}_0}$ and $[\mathfrak{g}_0, \mathfrak{g}_0]$ stable, and the restriction of θ to $[\mathfrak{g}_0, \mathfrak{g}_0]$ is a Cartan decomposition
(f) the identity component G_0 is a reductive Lie group (with maximal compact subgroup obtained by intersection and with Cartan involution and invariant form unchanged).

PROOF. For (a) assume the contrary, and let K_1 be a compact subgroup of G properly containing K. If k_1 is in K_1 but not K, write $k_1 = k\exp X$ according to (iv). Then $\exp X$ is in K_1. By compactness of K_1, $(\exp X)^n = \exp nX$ has a convergent subsequence in G, but this contradicts the homeomorphism in (iv).

Conclusion (b) is clear from (iv). In (c), $\mathrm{Ad}(K)(\mathfrak{k}_0) \subseteq \mathfrak{k}_0$ since K has Lie algebra $\mathfrak{k}_0$. Since B is $\mathrm{Ad}(K)$ invariant, $\mathrm{Ad}(K)$ leaves stable the subspace of $\mathfrak{g}_0$ orthogonal to $\mathfrak{k}_0$, and this is just $\mathfrak{p}_0$.

For (d) we have

$$\begin{aligned} B_\theta((\mathrm{ad}\,X)Y, Z) &= -B((\mathrm{ad}\,X)Y, \theta Z) \\ &= B(Y, [X, \theta Z]) = B(Y, \theta[\theta X, Z]) = B_\theta(Y, -(\mathrm{ad}\,\theta X)Z), \end{aligned}$$

and (d) is proved. Conclusion (e) follows from the facts that θ is an involution and B_θ is positive definite, and conclusion (f) is trivial.

Proposition 7.20. If G is a reductive Lie group in the Harish-Chandra class, then

(a) G_{ss} is a closed subgroup
(b) any semisimple analytic subgroup of G_{ss} has finite center.

REMARK. Because of (b), in checking whether a particular subgroup of G is reductive in the Harish-Chandra class, property (vi) is automatic for the subgroup if it holds for G.

PROOF.

(a) Write the global Cartan decomposition of Theorem 6.31c for G_{ss} as $G_{ss} = K_{ss}\exp(\mathfrak{p}_0 \cap [\mathfrak{g}_0, \mathfrak{g}_0])$. This is compatible with the decomposition in (iv). By (vi) and Theorem 6.31f, K_{ss} is compact. Hence $K_{ss} \times (\mathfrak{p}_0 \cap [\mathfrak{g}_0, \mathfrak{g}_0])$ is closed in $K \times \mathfrak{p}_0$, and (iv) implies that G_{ss} is closed in G.

(b) Let S be a semisimple analytic subgroup of G_{ss} with Lie algebra $\mathfrak{s}_0$. The group $\mathrm{Ad}_\mathfrak{g}(S)$ is a semisimple analytic subgroup of the linear group $GL(\mathfrak{g})$ and has finite center by Proposition 7.9. Under $\mathrm{Ad}_\mathfrak{g}$, Z_S maps into the center of $\mathrm{Ad}_\mathfrak{g}(S)$. Hence the image of Z_S is finite. The kernel of $\mathrm{Ad}_\mathfrak{g}$ on S consists of certain members x of G_{ss} for which $\mathrm{Ad}_\mathfrak{g}(x) = 1$. These x's are in $Z_{G_{ss}}$, and the kernel is then finite by property (vi) for G. Consequently Z_S is finite.

Proposition 7.21. If G is a reductive Lie group, then the function $\Theta : G \to G$ defined by

$$\Theta(k\exp X) = k\exp(-X) \qquad \text{for } k \in K \text{ and } X \in \mathfrak{p}_0$$

is an automorphism of G and its differential is θ.

REMARK. As in the semisimple case, Θ is called the **global Cartan involution**.

PROOF. The function Θ is a well defined diffeomorphism by (iv). First consider its restriction to the analytic subgroup G_{ss} with Lie algebra $[\mathfrak{g}_0, \mathfrak{g}_0]$. By Proposition 7.19e the Lie algebra $[\mathfrak{g}_0, \mathfrak{g}_0]$ has a Cartan decomposition

$$[\mathfrak{g}_0, \mathfrak{g}_0] = ([\mathfrak{g}_0, \mathfrak{g}_0] \cap \mathfrak{k}_0) \oplus ([\mathfrak{g}_0, \mathfrak{g}_0] \cap \mathfrak{p}_0).$$

If K_{ss} denotes the analytic subgroup of G_{ss} whose Lie algebra is the first summand on the right side, then Theorem 6.31 shows that G_{ss} consists exactly of the elements in $K_{ss} \exp([\mathfrak{g}_0, \mathfrak{g}_0] \cap \mathfrak{p}_0)$ and that Θ is an automorphism on G_{ss} with differential θ.

Next consider the restriction of Θ to the analytic subgroup $(Z_{G_0})_0$. By Proposition 7.19e the Lie algebra of this abelian group decomposes as

$$Z_{\mathfrak{g}_0} = (Z_{\mathfrak{g}_0} \cap \mathfrak{k}_0) \oplus (Z_{\mathfrak{g}_0} \cap \mathfrak{p}_0).$$

Since all the subalgebras in question are abelian, the exponential mappings in question are onto, and $(Z_{G_0})_0$ is a commuting product

$$(Z_{G_0})_0 = \exp(Z_{\mathfrak{g}_0} \cap \mathfrak{k}_0) \exp(Z_{\mathfrak{g}_0} \cap \mathfrak{p}_0)$$

contained in $K \exp \mathfrak{p}_0$. Thus Θ on $(Z_{G_0})_0$ is the lift to the group of θ on the Lie algebra and hence is an automorphism of the subgroup $(Z_{G_0})_0$.

The subgroups G_{ss} and $(Z_{G_0})_0$ commute, and hence Θ is an automorphism of their commuting product, which is G_0 by the remarks with Corollary 7.10.

Now consider Θ on all of G, where it is given consistently by $\Theta(kg_0) = k\Theta(g_0)$ for $k \in K$ and $g_0 \in G_0$. By Proposition 7.19c we have $\theta \operatorname{Ad}(k) = \operatorname{Ad}(k)\theta$ on $\mathfrak{g}_0$, from which we obtain $\Theta(k \exp X\, k^{-1}) = k\Theta(\exp X)k^{-1}$ for $k \in K$ and $X \in \mathfrak{g}_0$. Therefore

$$\Theta(kg_0k^{-1}) = k\Theta(g_0)k^{-1} \qquad \text{for } k \in K \text{ and } g \in G_0.$$

On the product of two general elements kg_0 and $k'g_0'$ of G, we therefore have

$$\begin{aligned}\Theta(kg_0k'g_0') &= \Theta(kk'k'^{-1}g_0k'g_0') = kk'\Theta(k'^{-1}g_0k'g_0') \\ &= kk'\Theta(k'^{-1}g_0k')\Theta(g_0') = k\Theta(g_0)k'\Theta(g_0') = \Theta(kg_0)\Theta(k'g_0'),\end{aligned}$$

as required.

Lemma 7.22. Let G be a reductive Lie group, and let $g = k \exp X$ be the global Cartan decomposition of an element g of G. If $\mathfrak{s}_0$ is a θ stable subspace of $\mathfrak{g}_0$ such that $\mathrm{Ad}(g)$ normalizes $\mathfrak{s}_0$, then $\mathrm{Ad}(k)$ and $\mathrm{ad}\, X$ each normalize $\mathfrak{s}_0$. If $\mathrm{Ad}(g)$ centralizes $\mathfrak{s}_0$, then $\mathrm{Ad}(k)$ and $\mathrm{ad}\, X$ each centralize $\mathfrak{s}_0$.

PROOF. For $x \in G$, we have $(\Theta g)x(\Theta g)^{-1} = \Theta(g(\Theta x)g^{-1})$. Differentiating at $x = 1$, we obtain

$$\mathrm{Ad}(\Theta g) = \theta \mathrm{Ad}(g) \theta. \tag{7.23}$$

Therefore $\mathrm{Ad}(\Theta g)$ normalizes $\mathfrak{s}_0$. Since $\Theta g = k \exp(-X)$, it follows that Ad of $(\Theta g)^{-1} g = \exp 2X$ normalizes $\mathfrak{s}_0$. Because of Proposition 7.19d, $\mathrm{Ad}(\exp 2X)$ is positive definite relative to B_θ, hence diagonable. Then there exists a vector subspace $\mathfrak{s}_0'$ of $\mathfrak{g}_0$ invariant under $\mathrm{Ad}(\exp 2X)$ such that $\mathfrak{g}_0 = \mathfrak{s}_0 \oplus \mathfrak{s}_0'$. The transformation $\mathrm{Ad}(\exp 2X)$ has a unique logarithm with real eigenvalues, and $\mathrm{ad}\, 2X$ is a candidate for it. Another candidate is the logarithm on each subspace, which normalizes $\mathfrak{s}_0$ and $\mathfrak{s}_0'$. These two candidates must be equal, and therefore $\mathrm{ad}\, 2X$ normalizes $\mathfrak{s}_0$ and $\mathfrak{s}_0'$. Hence the same thing is true of $\mathrm{ad}\, X$. Then $\mathrm{Ad}(\exp X)$ and $\mathrm{Ad}(g)$ both normalize $\mathfrak{s}_0$ and $\mathfrak{s}_0'$, and the same thing must be true of $\mathrm{Ad}(k)$.

If $\mathrm{Ad}(g)$ centralizes $\mathfrak{s}_0$, we can go over the above argument to see that $\mathrm{Ad}(k)$ and $\mathrm{ad}\, X$ each centralize $\mathfrak{s}_0$. In fact, $\mathrm{Ad}(\exp 2X)$ must centralize $\mathfrak{s}_0$, the unique real logarithm must be 0 on $\mathfrak{s}_0$, and $\mathrm{ad}\, X$ must be 0 on $\mathfrak{s}_0$. The lemma follows.

Lemma 7.24. Let G be a reductive Lie group, and let $\mathfrak{u}_0 = \mathfrak{k}_0 \oplus i\mathfrak{p}_0$. Then $\mathrm{Ad}_\mathfrak{g}(K)$ is contained in $\mathrm{Int}_\mathfrak{g}(\mathfrak{u}_0)$.

PROOF. The group $\mathrm{Int}\, \mathfrak{g}$ is complex semisimple with Lie algebra $\mathrm{ad}_\mathfrak{g}(\mathfrak{g})$. If bar denotes the conjugation of $\mathfrak{g}$ with respect to $\mathfrak{g}_0$, then the extension $B_\theta(Z_1, Z_2) = -B(Z_1, \theta \bar{Z}_2)$ is a Hermitian inner product on $\mathfrak{g}$, and the compact real form $\mathrm{ad}_\mathfrak{g}(\mathfrak{u}_0)$ of $\mathrm{ad}_\mathfrak{g}(\mathfrak{g})$ consists of skew Hermitian transformations. Hence $\mathrm{Int}_\mathfrak{g}(\mathfrak{u}_0)$ consists of unitary transformations and $\mathrm{ad}_\mathfrak{g}(i\mathfrak{u}_0)$ consists of Hermitian transformations. Therefore the global Cartan decomposition of $\mathrm{Int}\, \mathfrak{g}$ given in Theorem 6.31c is compatible with the polar decomposition relative to B_θ, and every unitary member of $\mathrm{Int}\, \mathfrak{g}$ is in the compact real form $\mathrm{Int}_\mathfrak{g}(\mathfrak{u}_0)$.

Let k be in K. The transformation $\mathrm{Ad}_\mathfrak{g}(k)$ is in $\mathrm{Int}\, \mathfrak{g}$ by property (v) for G, and $\mathrm{Ad}_\mathfrak{g}(k)$ is unitary since B is $\mathrm{Ad}(k)$ invariant and since $\mathrm{Ad}(k)$ commutes with bar and θ (Proposition 7.19c). From the result of the previous paragraph, we conclude that $\mathrm{Ad}_\mathfrak{g}(k)$ is in $\mathrm{Int}_\mathfrak{g}(\mathfrak{u}_0)$.

Proposition 7.25. If G is a reductive Lie group and $\mathfrak{h}_0$ is a θ stable abelian subalgebra of its Lie algebra, then $Z_G(\mathfrak{h}_0)$ is a reductive Lie group.

Here the maximal compact subgroup of $Z_G(\mathfrak{h}_0)$ is given by intersection, and the Cartan involution and invariant form are given by restriction.

REMARK. The hypothesis "abelian" will be used only in the proof of property (v) for $Z_G(\mathfrak{h}_0)$, and we shall make use of this fact in Corollary 7.26 below.

PROOF. The group $Z_G(\mathfrak{h}_0)$ is closed, hence Lie. Its Lie algebra is $Z_{\mathfrak{g}_0}(\mathfrak{h}_0)$, which is θ stable. Then it follows, just as in the proof of Corollary 6.29, that $Z_{\mathfrak{g}_0}(\mathfrak{h}_0)$ is reductive. This proves property (i) of a reductive Lie group. Since $Z_{\mathfrak{g}_0}(\mathfrak{h}_0)$ is θ stable, we have

$$Z_{\mathfrak{g}_0}(\mathfrak{h}_0) = (Z_{\mathfrak{g}_0}(\mathfrak{h}_0) \cap \mathfrak{k}_0) \oplus (Z_{\mathfrak{g}_0}(\mathfrak{h}_0) \cap \mathfrak{p}_0),$$

and the first summand on the right side is the Lie algebra of $Z_G(\mathfrak{h}_0) \cap K$. This proves property (ii), and property (iii) is trivial.

In view of property (iv) for G, what needs proof in (iv) for $Z_G(\mathfrak{h}_0)$ is that $Z_K(\mathfrak{h}_0) \times (Z_{\mathfrak{g}_0}(\mathfrak{h}_0) \cap \mathfrak{p}_0)$ maps onto $Z_G(\mathfrak{h}_0)$. That is, we need to see that if $g = k \exp X$ is the global Cartan decomposition of a member g of $Z_G(\mathfrak{h}_0)$, then k is in $Z_G(\mathfrak{h}_0)$ and X is in $Z_{\mathfrak{g}_0}(\mathfrak{h}_0)$. But this is immediate from Lemma 7.22, and (iv) follows.

For property (v) we are to show that $\mathrm{Ad}_{Z_{\mathfrak{g}}(\mathfrak{h})}$ carries $Z_G(\mathfrak{h}_0)$ into $\mathrm{Int}\, Z_{\mathfrak{g}}(\mathfrak{h})$. If $x \in Z_G(\mathfrak{h}_0)$ is given, then property (iv) allows us to write $x = k \exp X$ with $k \in Z_K(\mathfrak{h}_0)$ and $X \in Z_{\mathfrak{g}_0}(\mathfrak{h}_0) \cap \mathfrak{p}_0$. Then $\mathrm{Ad}_{Z_{\mathfrak{g}}(\mathfrak{h})}(\exp X)$ is in $\mathrm{Int}\, Z_{\mathfrak{g}}(\mathfrak{h})$, and it is enough to treat k. By Lemma 7.24, $\mathrm{Ad}_{\mathfrak{g}}(k)$ is in the subgroup $\mathrm{Int}_{\mathfrak{g}}(\mathfrak{u}_0)$, which is compact by Proposition 7.9.

The element $\mathrm{Ad}_{\mathfrak{g}}(k)$ centralizes $\mathfrak{h}_0$, hence centralizes the variant $(\mathfrak{h}_0 \cap \mathfrak{k}_0) \oplus i(\mathfrak{h}_0 \cap \mathfrak{p}_0)$. Since $(\mathfrak{h}_0 \cap \mathfrak{k}_0) \oplus i(\mathfrak{h}_0 \cap \mathfrak{p}_0)$ is an abelian subalgebra of $\mathfrak{g}$, the centralizer of $\mathfrak{h}_0$ in $\mathrm{Int}_{\mathfrak{g}}(\mathfrak{u}_0)$ is the centralizer of a torus, which is connected by Corollary 4.51. Therefore $\mathrm{Ad}_{\mathfrak{g}}(k)$ is in the analytic subgroup of $\mathrm{Int}\,\mathfrak{g}$ with Lie algebra $Z_{\mathfrak{u}_0}((\mathfrak{h}_0 \cap \mathfrak{k}_0) \oplus i(\mathfrak{h}_0 \cap \mathfrak{p}_0))$. By Corollary 4.48 we can write $\mathrm{Ad}_{\mathfrak{g}}(k) = \exp \mathrm{ad}_{\mathfrak{g}} Y$ with Y in this Lie algebra. Then $\mathrm{Ad}_{Z_{\mathfrak{g}}(\mathfrak{h})}(k) = \exp \mathrm{ad}_{Z_{\mathfrak{g}}(\mathfrak{h})} Y$, and Y is in $Z_{\mathfrak{g}}(\mathfrak{h})$. Then $\mathrm{Ad}_{Z_{\mathfrak{g}}(\mathfrak{h})}(k)$ is in $\mathrm{Int}\, Z_{\mathfrak{g}}(\mathfrak{h})$, and (v) is proved.

Corollary 7.26. If G is a reductive Lie group, then

(a) $(Z_{G_0})_0 \subseteq Z_G$

(b) Z_G is a reductive Lie group (with maximal compact subgroup given by intersection and with Cartan involution and invariant form given by restriction).

PROOF. Property (v) for G gives $\mathrm{Ad}_{\mathfrak{g}}(G) \subseteq \mathrm{Int}\,\mathfrak{g}$, and $\mathrm{Int}\,\mathfrak{g}$ acts trivially on $Z_{\mathfrak{g}}$. Hence $\mathrm{Ad}(G)$ acts trivially on $Z_{\mathfrak{g}_0}$, and G centralizes $(Z_{G_0})_0$. This proves (a).

From (a) it follows that Z_G has Lie algebra $Z_{\mathfrak{g}_0}$, which is also the Lie algebra of $Z_G(\mathfrak{g}_0)$. Therefore property (v) is trivial for both Z_G and $Z_G(\mathfrak{g}_0)$. Propositon 7.25 and its remark show that $Z_G(\mathfrak{g}_0)$ is reductive, and consequently only property (iv) needs proof for Z_G. We need to see that if $z \in Z_G$ decomposes in G under (iv) as $z = k \exp X$, then k is in $Z_G \cap K$ and X is in $Z_{\mathfrak{g}_0}$. By Lemma 7.22 we know that k is in $Z_G(\mathfrak{g}_0)$ and X is in $Z_{\mathfrak{g}_0}$. Then $\exp X$ is in $(Z_{G_0})_0$, and (a) shows that $\exp X$ is in Z_G. Since z and $\exp X$ are in Z_G, so is k. This completes the proof of (iv), and (b) follows.

Let G be reductive. Since $\operatorname{ad}_{\mathfrak{g}} \mathfrak{g}$ carries $[\mathfrak{g}, \mathfrak{g}]$ to itself, $\operatorname{Int} \mathfrak{g}$ carries $[\mathfrak{g}, \mathfrak{g}]$ to itself. By (v), $\operatorname{Ad}(G)$ normalizes $[\mathfrak{g}_0, \mathfrak{g}_0]$. Consequently ${}^0G = KG_{ss}$ is a subgroup of G.

The vector subspace $\mathfrak{p}_0 \cap Z_{\mathfrak{g}_0}$ is an abelian subspace of $\mathfrak{g}_0$, and therefore $Z_{vec} = \exp(\mathfrak{p}_0 \cap Z_{\mathfrak{g}_0})$ is an analytic subgroup of G.

Proposition 7.27. If G is a reductive Lie group, then

(a) ${}^0G = K \exp(\mathfrak{p}_0 \cap [\mathfrak{g}_0, \mathfrak{g}_0])$, and ${}^0G$ is a closed subgroup
(b) the Lie algebra ${}^0\mathfrak{g}_0$ of ${}^0G$ is $\mathfrak{k}_0 \oplus (\mathfrak{p}_0 \cap [\mathfrak{g}_0, \mathfrak{g}_0])$
(c) ${}^0G$ is reductive (with maximal compact subgroup K and with Cartan involution and invariant form given by restriction)
(d) the center of ${}^0G$ is a compact subgroup of K
(e) Z_{vec} is closed, is isomorphic to the additive group of a Euclidean space, and is contained in the center of G
(f) the multiplication map exhibits ${}^0G \times Z_{vec}$ as isomorphic to G.

REMARK. The closed subgroup Z_{vec} is called the **split component** of G.

PROOF.
(a) If we write the global Cartan decomposition of G_{ss} as $G_{ss} = K_{ss} \exp(\mathfrak{p}_0 \cap [\mathfrak{g}_0, \mathfrak{g}_0])$, then ${}^0G = K \exp(\mathfrak{p}_0 \cap [\mathfrak{g}_0, \mathfrak{g}_0])$, and we see from property (iv) that ${}^0G$ is closed.

(b) Because of (a), ${}^0G$ is a Lie subgroup. Since ${}^0G$ contains K and G_{ss}, its Lie algebra must contain $\mathfrak{k}_0 \oplus (\mathfrak{p}_0 \cap [\mathfrak{g}_0, \mathfrak{g}_0])$. From property (iv) for G, the formula ${}^0G = K \exp(\mathfrak{p}_0 \cap [\mathfrak{g}_0, \mathfrak{g}_0])$ shows that $\dim {}^0\mathfrak{g}_0 = \dim \mathfrak{k}_0 + \dim(\mathfrak{p}_0 \cap [\mathfrak{g}_0, \mathfrak{g}_0])$. So ${}^0\mathfrak{g}_0 = \mathfrak{k}_0 \oplus (\mathfrak{p}_0 \cap [\mathfrak{g}_0, \mathfrak{g}_0])$.

(c) From (b) we see that ${}^0\mathfrak{g}_0$ is θ stable. From this fact all the properties of a reductive group are clear except properties (iv) and (v). Property (iv) follows from (a). For property (v) we know that any $\operatorname{Ad}_{\mathfrak{g}}(g)$ for $g \in {}^0G$ is in $\operatorname{Int} \mathfrak{g}$. Therefore we can write $\operatorname{Ad}_{\mathfrak{g}}(g)$ as a product of elements $\exp \operatorname{ad}_{\mathfrak{g}}(X_j)$ with X_j in $[\mathfrak{g}, \mathfrak{g}]$ or $Z_{\mathfrak{g}}$. When X_j is in $Z_{\mathfrak{g}}$, $\exp \operatorname{ad}_{\mathfrak{g}}(X_j)$ is trivial. Therefore $\operatorname{Ad}_{\mathfrak{g}}(g)$ agrees with a product of elements $\exp \operatorname{ad}_{\mathfrak{g}}(X_j)$

with X_j in $[\mathfrak{g}, \mathfrak{g}]$. Restricting the action to $[\mathfrak{g}, \mathfrak{g}]$, we see that $\mathrm{Ad}_{[\mathfrak{g},\mathfrak{g}]}(g)$ is in $\mathrm{Int}\,[\mathfrak{g}, \mathfrak{g}]$.

(d) Conclusion (c) and Corollary 7.26 show that the center of ${}^0G$ is reductive. The intersection of the Lie algebra of the center with $\mathfrak{p}_0$ is 0, and hence property (iv) shows that the center is contained in K.

(e) Since $\mathfrak{p}_0 \cap Z_{\mathfrak{g}_0}$ is a closed subspace of $\mathfrak{p}_0$, property (iv) implies that Z_{vec} is closed and that Z_{vec} is isomorphic to the additive group of a Euclidean space. Since $\mathrm{Int}\,\mathfrak{g}$ acts trivially on $Z_{\mathfrak{g}}$, property (v) implies that $\mathrm{Ad}(g) = 1$ on $\mathfrak{p}_0 \cap Z_{\mathfrak{g}_0}$ for every $g \in G$. Hence Z_{vec} is contained in the center of G.

(f) Multiplication is a diffeomorphism, as we see by combining (a), property (iv), and the formula $\exp(X + Y) = \exp X \exp Y$ for $X \in \mathfrak{p}_0 \cap [\mathfrak{g}_0, \mathfrak{g}_0]$ and $Y \in \mathfrak{p}_0 \cap Z_{\mathfrak{g}_0}$. Multiplication is a homomorphism since, by (e), Z_{vec} is contained in the center of G.

Reductive Lie groups are supposed to have all the essential structure-theoretic properties of semisimple groups and to be closed under various operations that allow us to prove theorems by induction on the dimension of the group. The remainder of this section will be occupied with reviewing the structure theory developed in Chapter VI to describe how the results should be interpreted for reductive Lie groups.

The first remarks concern the Cartan decomposition. The decomposition on the Lie algebra level is built into the definition of reductive Lie group, and the properties of the global Cartan decomposition (generalizing Theorem 6.31) are given partly in property (iv) of the definition and partly in Proposition 7.21.

It might look as if property (iv) would be a hard thing to check for a particular candidate for a reductive group. It is possible to substitute various axioms concerning the component structure of G that are easier to state, but it is often true that ones gets at the component structure by first proving (iv). Proposition 1.122 and Lemma 7.22 provide examples of this order of events; the global Cartan decomposition in those cases implies that the number of components of the group under study is finite. Thus property (iv) is the natural property to include in the definition even though its statement is complicated.

The other two general structure-theoretic topics in Chapter VI are the Iwasawa decomposition and Cartan subalgebras. Let us first extend the notion of an Iwasawa decomposition to the context of reductive Lie groups. Let G be a reductive Lie group, and write its Lie algebra as $\mathfrak{g}_0 = Z_{\mathfrak{g}_0} \oplus [\mathfrak{g}_0, \mathfrak{g}_0]$. Let $\mathfrak{a}_0$ be a maximal abelian subspace of $\mathfrak{p}_0$. Certainly $\mathfrak{a}_0$ contains $\mathfrak{p}_0 \cap Z_{\mathfrak{g}_0}$, and therefore $\mathfrak{a}_0$ is of the form

$$\mathfrak{a}_0 = (\mathfrak{p}_0 \cap Z_{\mathfrak{g}_0}) \oplus (\mathfrak{a}_0 \cap [\mathfrak{g}_0, \mathfrak{g}_0]), \tag{7.28}$$

where $\mathfrak{a}_0 \cap [\mathfrak{g}_0, \mathfrak{g}_0]$ is a maximal abelian subspace of $\mathfrak{p}_0 \cap [\mathfrak{g}_0, \mathfrak{g}_0]$. Theorem 6.51 shows that any two maximal abelian subspaces of $\mathfrak{p}_0 \cap [\mathfrak{g}_0, \mathfrak{g}_0]$ are conjugate via $\mathrm{Ad}(K)$, and it follows from (7.28) that this result extends to our reductive $\mathfrak{g}_0$.

Proposition 7.29. Let G be a reductive Lie group. If $\mathfrak{a}_0$ and $\mathfrak{a}_0'$ are two maximal abelian subspaces of $\mathfrak{p}_0$, then there is a member k of K with $\mathrm{Ad}(k)\mathfrak{a}_0' = \mathfrak{a}_0$. The member k of K can be taken to be in $K \cap G_{ss}$. Hence $\mathfrak{p}_0 = \bigcup_{k \in K_{ss}} \mathrm{Ad}(k)\mathfrak{a}_0$.

Relative to $\mathfrak{a}_0$, we can form restricted roots just as in §VI.4. A **restricted root** of $\mathfrak{g}_0$, also called a **root** of $(\mathfrak{g}_0, \mathfrak{a}_0)$, is a nonzero $\lambda \in \mathfrak{a}_0^*$ such that the space

$$(\mathfrak{g}_0)_\lambda = \{X \in \mathfrak{g}_0 \mid (\mathrm{ad}\, H)X = \lambda(H)X \text{ for all } H \in \mathfrak{a}_0\}$$

is nonzero. It is apparent that such a restricted root is obtained by taking a restricted root for $[\mathfrak{g}_0, \mathfrak{g}_0]$ and extending it from $\mathfrak{a}_0 \cap [\mathfrak{g}_0, \mathfrak{g}_0]$ to $\mathfrak{a}_0$ by making it be 0 on $\mathfrak{p}_0 \cap Z_{\mathfrak{g}_0}$. The restricted-root space decomposition for $[\mathfrak{g}_0, \mathfrak{g}_0]$ gives us a restricted-root space decomposition for $\mathfrak{g}_0$. We define $\mathfrak{m}_0 = Z_{\mathfrak{k}_0}(\mathfrak{a}_0)$, so that the centralizer of $\mathfrak{a}_0$ in $\mathfrak{g}_0$ is $\mathfrak{m}_0 \oplus \mathfrak{a}_0$.

The set of restricted roots is denoted Σ. Choose a notion of positivity for $\mathfrak{a}_0^*$ in the manner of §II.5, as for example by using a lexicographic ordering. Let Σ^+ be the set of positive restricted roots, and define $\mathfrak{n}_0 = \bigoplus_{\lambda \in \Sigma^+} (\mathfrak{g}_0)_\lambda$. Then $\mathfrak{n}_0$ is a nilpotent Lie subalgebra of $\mathfrak{g}_0$, and we have an Iwasawa decomposition

$$\mathfrak{g}_0 = \mathfrak{k}_0 \oplus \mathfrak{a}_0 \oplus \mathfrak{n}_0 \tag{7.30}$$

with all the properties in Proposition 6.43.

Proposition 7.31. Let G be a reductive Lie group, let (7.30) be an Iwasawa decomposition of $\mathfrak{g}_0$ of G, and let A and N be the analytic subgroups of G with Lie algebras $\mathfrak{a}_0$ and $\mathfrak{n}_0$. Then the multiplication map $K \times A \times N \to G$ given by $(k, a, n) \mapsto kan$ is a diffeomorphism onto. The groups A and N are simply connected.

PROOF. Multiplication is certainly smooth, and it is regular by Lemma 6.44. To see that it is one-one, it is enough, as in the proof of Theorem 6.46, to see that we cannot have $kan = 1$ nontrivially. The identity $kan = 1$ would force the orthogonal transformation $\mathrm{Ad}(k)$ to be upper triangular with positive diagonal entries in the matrix realization of Lemma 6.45, and consequently we may assume that $\mathrm{Ad}(k) = \mathrm{Ad}(a) = \mathrm{Ad}(n) = 1$. Thus k, a, and n are in $Z_G(\mathfrak{g}_0)$. By Lemma 7.22, a is the exponential of something in $Z_{\mathfrak{g}_0}(\mathfrak{g}_0) = Z_{\mathfrak{g}_0}$. Hence a is in Z_{vec}. By construction n is in G_{ss}, and hence k and n are in ${}^0G$. By Proposition 7.27f, $a = 1$ and $kn = 1$. But then the identity $kn = 1$ is valid in G_{ss}, and Theorem 6.46 implies that $k = n = 1$.

To see that multiplication is onto G, we observe from Theorem 6.46 that $\exp(\mathfrak{p}_0 \cap [\mathfrak{g}_0, \mathfrak{g}_0])$ is in the image. By Proposition 7.27a, the image contains ${}^0G$. Also Z_{vec} is in the image (of $1 \times A \times 1$), and Z_{vec} commutes with ${}^0G$. Hence the image contains ${}^0GZ_{vec}$. This is all of G by Proposition 7.27f.

We define $\mathfrak{n}_0^- = \bigoplus_{\lambda\in\Sigma^+}(\mathfrak{g}_0)_{-\lambda}$. Then $\mathfrak{n}_0^-$ is a nilpotent Lie subalgebra of $\mathfrak{g}_0$, and we let N^- be the corresponding analytic subgroup. Since $-\Sigma^+$ is the set of positive restricted roots for another notion of positivity on $\mathfrak{a}_0^*$, $\mathfrak{g}_0 = \mathfrak{k}_0 \oplus \mathfrak{a}_0 \oplus \mathfrak{n}_0$ is another Iwasawa decomposition of $\mathfrak{g}_0$ and $G = KAN^-$ is another Iwasawa decomposition of G. The identity $\theta(\mathfrak{g}_0)_\lambda = (\mathfrak{g}_0)_{-\lambda}$ given in Proposition 6.40c implies that $\theta\mathfrak{n}_0 = \mathfrak{n}_0^-$. By Proposition 7.21, $\Theta N = N^-$.

We write M for the group $Z_K(\mathfrak{a}_0)$. This is a compact subgroup since it is closed in K, and its Lie algebra is $Z_{\mathfrak{k}_0}(\mathfrak{a}_0)$. This subgroup normalizes each $(\mathfrak{g}_0)_\lambda$ since

$$\begin{aligned}\operatorname{ad}(H)(\operatorname{Ad}(m)X_\lambda) &= \operatorname{Ad}(m)\operatorname{ad}(\operatorname{Ad}(m)^{-1}H)X_\lambda \\ &= \operatorname{Ad}(m)\operatorname{ad}(H)X_\lambda = \lambda(H)\operatorname{Ad}(m)X_\lambda\end{aligned}$$

for $m \in M$, $H \in \mathfrak{a}_0$, and $X_\lambda \in (\mathfrak{g}_0)_\lambda$. Consequently M normalizes $\mathfrak{n}_0$. Thus M centralizes A and normalizes N. Since M is compact and AN is closed, MAN is a closed subgroup.

Reflections in the restricted roots generate a group $W(\Sigma)$, which we call the **Weyl group** of Σ. The elements of $W(\Sigma)$ are nothing more than the elements of the Weyl group for the restricted roots of $[\mathfrak{g}_0, \mathfrak{g}_0]$, with each element extended to $\mathfrak{a}_0^*$ by being defined to be the identity on $\mathfrak{p}_0 \cap Z_{\mathfrak{g}_0}$.

We define $W(G, A) = N_K(\mathfrak{a}_0)/Z_K(\mathfrak{a}_0)$. By the same proof as for Lemma 6.56, the Lie algebra of $N_K(\mathfrak{a}_0)$ is $\mathfrak{m}_0$. Therefore $W(G, A)$ is a finite group.

Proposition 7.32. If G is a reductive Lie group, then the group $W(G, A)$ coincides with $W(\Sigma)$.

PROOF. Just as with the corresponding result in the semisimple case (Theorem 6.57), we know that $W(\Sigma) \subseteq W(G, A)$. Fix a simple system Σ^+ for Σ. As in the proof of Theorem 6.57, it suffices to show that if $k \in N_K(\mathfrak{a}_0)$ has $\operatorname{Ad}(k)\Sigma^+ = \Sigma^+$, then k is in $Z_K(\mathfrak{a}_0)$. By Lemma

7.24, $\mathrm{Ad}_{\mathfrak{g}}(k)$ is in the compact semisimple Lie group $\mathrm{Int}_{\mathfrak{g}}(\mathfrak{u}_0)$, where $\mathfrak{u}_0 = \mathfrak{k}_0 \oplus i\mathfrak{p}_0$. The connectedness of $\mathrm{Int}_{\mathfrak{g}}(\mathfrak{u}_0)$ is the key, and the remainder of the proof of Theorem 6.57 is applicable to this situation.

Proposition 7.33. If G is a reductive Lie group, then M meets every component of K, hence every component of G.

PROOF. Let $k \in K$ be given. Since $\mathrm{Ad}(k)^{-1}(\mathfrak{a}_0)$ is maximal abelian in $\mathfrak{p}_0$, Proposition 7.28 gives us $k_0 \in K_0$ with $\mathrm{Ad}(k_0^{-1}k^{-1})(\mathfrak{a}_0) = \mathfrak{a}_0$. Thus $k_0^{-1}k^{-1}$ normalizes $\mathfrak{a}_0$. Comparison of Proposition 7.32 and Theorem 6.57 produces $k_1^{-1} \in K_0$ so that $k_1^{-1}k_0^{-1}k^{-1}$ centralizes $\mathfrak{a}_0$. Then kk_0k_1 is in M, and k is in MK_0.

Next let us extend the notion of Cartan subalgebras to the context of reductive Lie groups. We recall from §IV.5 that a Lie subalgebra $\mathfrak{h}_0$ of $\mathfrak{g}_0$ is a **Cartan subalgebra** if its complexification $\mathfrak{h}$ is a Cartan subalgebra of $\mathfrak{g} = (\mathfrak{g}_0)^{\mathbb{C}}$. Since $\mathfrak{h}$ must equal its own normalizer (Proposition 2.7), it follows that $Z_{\mathfrak{g}} \subseteq \mathfrak{h}$. Therefore $\mathfrak{h}_0$ must be of the form

$$\mathfrak{h}_0 = Z_{\mathfrak{g}_0} \oplus (\mathfrak{h}_0 \cap [\mathfrak{g}_0, \mathfrak{g}_0]), \tag{7.34}$$

where $\mathfrak{h}_0 \cap [\mathfrak{g}_0, \mathfrak{g}_0]$ is a Cartan subalgebra of the semisimple Lie algebra $[\mathfrak{g}_0, \mathfrak{g}_0]$. By Corollary 2.13 a sufficient condition for $\mathfrak{h}_0$ to be a Cartan subalgebra of $\mathfrak{g}_0$ is that $\mathfrak{h}_0$ is maximal abelian in $\mathfrak{g}_0$ and $\mathrm{ad}_{\mathfrak{g}} \mathfrak{h}_0$ is simultaneously diagonable.

As in the special case (4.31), we can form a set of roots $\Delta(\mathfrak{g}, \mathfrak{h})$, which amount to the roots of $[\mathfrak{g}, \mathfrak{g}]$ with respect to $\mathfrak{h} \cap [\mathfrak{g}, \mathfrak{g}]$, extended to $\mathfrak{h}$ by being defined to be 0 on $Z_{\mathfrak{g}}$. We can form also a Weyl group $W(\mathfrak{g}, \mathfrak{h})$ generated by the reflections in the members of Δ; $W(\mathfrak{g}, \mathfrak{h})$ consists of the members of $W([\mathfrak{g}, \mathfrak{g}], \mathfrak{h} \cap [\mathfrak{g}, \mathfrak{g}])$ extended to $\mathfrak{g}$ by being defined to be the identity on $Z_{\mathfrak{g}}$.

Because of the form (7.34) of Cartan subalgebras of $\mathfrak{g}_0$, Proposition 6.59 implies that any Cartan subalgebra is conjugate via $\mathrm{Int}\,\mathfrak{g}_0$ to a θ stable Cartan subalgebra. There are only finitely many conjugacy classes (Proposition 6.64), and these can be related by Cayley transforms.

The maximally noncompact θ stable Cartan subalgebras are obtained by adjoining to an Iwasawa $\mathfrak{a}_0$ a maximal abelian subspace of $\mathfrak{m}_0$. As in Proposition 6.61, all such Cartan subalgebras are conjugate via K. The restricted roots relative to $\mathfrak{a}_0$ are the nonzero restrictions to $\mathfrak{a}_0$ of the roots relative to this Cartan subalgebra.

Any maximally compact θ stable Cartan subalgebra is obtained as the centralizer of a maximal abelian subspace of $\mathfrak{k}_0$. As in Proposition 6.61, all such Cartan subalgebras are conjugate via K.

Proposition 7.35. Let G be a reductive Lie group. If two θ stable Cartan subalgebras of $\mathfrak{g}_0$ are conjugate via G, then they are conjugate via G_{ss} and in fact by $K \cap G_{ss}$.

PROOF. Let $\mathfrak{h}_0$ and $\mathfrak{h}_0'$ be θ stable Cartan subalgebras, and suppose that $\text{Ad}(g)(\mathfrak{h}_0) = \mathfrak{h}_0'$. By (7.23), $\text{Ad}(\Theta g)(\mathfrak{h}_0) = \mathfrak{h}_0'$. If $g = k \exp X$ with $k \in K$ and $X \in \mathfrak{p}_0$, then it follows that Ad of $(\Theta g)^{-1} g = \exp 2X$ normalizes $\mathfrak{h}_0$. Applying Lemma 7.22 to $\exp 2X$, we see that $[X, \mathfrak{h}_0] \subseteq \mathfrak{h}_0$. Therefore $\exp X$ normalizes $\mathfrak{h}_0$, and $\text{Ad}(k)$ carries $\mathfrak{h}_0$ to $\mathfrak{h}_0'$.

Since $\text{Ad}(k)$ commutes with θ, $\text{Ad}(k)$ carries $\mathfrak{h}_0 \cap \mathfrak{p}_0$ to $\mathfrak{h}_0' \cap \mathfrak{p}_0$. Let $\mathfrak{a}_0$ be a maximal abelian subspace of $\mathfrak{p}_0$ containing $\mathfrak{h}_0 \cap \mathfrak{p}_0$, and choose $k_0 \in K_0$ by Proposition 7.29 so that $\text{Ad}(k_0 k)(\mathfrak{a}_0) = \mathfrak{a}_0$. Comparing Proposition 7.32 and Theorem 6.57, we can find $k_1 \in K_0$ so that $k_1 k_0 k$ centralizes $\mathfrak{a}_0$. Then $\text{Ad}(k)|_{\mathfrak{a}_0} = \text{Ad}(k_0^{-1} k_1^{-1})|_{\mathfrak{a}_0}$, and the element $k' = k_0^{-1} k_1^{-1}$ of K_0 has the property that $\text{Ad}(k')(\mathfrak{h}_0 \cap \mathfrak{p}_0) = \mathfrak{h}_0' \cap \mathfrak{p}_0$. The θ stable Cartan subalgebras $\mathfrak{h}_0$ and $\text{Ad}(k')^{-1}(\mathfrak{h}_0')$ therefore have the same $\mathfrak{p}_0$ part, and Lemma 6.62 shows that they are conjugate via $K \cap G_{ss}$.

3. KAK Decomposition

Throughout this section we let G be a reductive Lie group, and we let other notation be as in §2.

From the global Cartan decomposition $G = K \exp \mathfrak{p}_0$ and from the equality $\mathfrak{p}_0 = \bigoplus_{k \in K} \text{Ad}(k)\mathfrak{a}_0$ of Proposition 7.29, it is immediate that $G = KAK$ in the sense that every element of G can be decomposed as a product of an element of K, an element of A, and a second element of K. In this section we shall examine the degree of nonuniqueness of this decomposition.

Lemma 7.36. If X is in $\mathfrak{p}_0$, then $Z_G(\exp X) = Z_G(\mathbb{R}X)$.

PROOF. Certainly $Z_G(\mathbb{R}X) \subseteq Z_G(\exp X)$. In the reverse direction if g is in $Z_G(\exp X)$, then $\text{Ad}(g)\text{Ad}(\exp X) = \text{Ad}(\exp X)\text{Ad}(g)$. By Proposition 7.19d, $\text{Ad}(\exp X)$ is positive definite on $\mathfrak{g}_0$, thus diagonable. Consequently $\text{Ad}(g)$ carries each eigenspace of $\text{Ad}(\exp X)$ to itself, and it follows that $\text{Ad}(g)\text{ad}(X) = \text{ad}(X)\text{Ad}(g)$. By Lemma 1.95,

$$\text{ad}(\text{Ad}(g)X) = \text{ad}(X). \tag{7.37}$$

Write $X = Y + Z$ with $Y \in Z_{\mathfrak{g}_0}$ and $Z \in [\mathfrak{g}_0, \mathfrak{g}_0]$. By property (v) of a reductive group, $\text{Ad}(g)Y = Y$. Comparing this equality with (7.37), we see that $\text{ad}(\text{Ad}(g)Z) = \text{ad}(Z)$, hence that $\text{Ad}(g)Z - Z$ is in the center of $\mathfrak{g}_0$. Since it is in $[\mathfrak{g}_0, \mathfrak{g}_0]$ also, it is 0. Therefore $\text{Ad}(g)X = X$, and g is in the centralizer of $\mathbb{R}X$.

Lemma 7.38. If k is in K and if a and a' are in A with $kak^{-1} = a'$, then there exists k_0 in $N_K(\mathfrak{a}_0)$ with $k_0ak_0^{-1} = a'$.

PROOF. The subgroup $Z_G(a')$ is reductive by Lemma 7.36 and Proposition 7.25, and its Lie algebra is $Z_{\mathfrak{g}_0}(a') = \{X \in \mathfrak{g}_0 \mid \mathrm{Ad}(a')X = X\}$. Now $\mathfrak{a}_0$ and $\mathrm{Ad}(k)\mathfrak{a}_0$ are two maximal abelian subspaces of $Z_{\mathfrak{g}_0}(a') \cap \mathfrak{p}_0$ since $kak^{-1} = a'$. By Proposition 7.29 there exists k_1 in $K \cap Z_G(a')$ with $\mathrm{Ad}(k_1)\mathrm{Ad}(k)\mathfrak{a}_0 = \mathfrak{a}_0$. Then $k_0 = k_1k$ is in $N_K(\mathfrak{a}_0)$, and

$$k_0ak_0^{-1} = k_1(kak^{-1})k_1^{-1} = k_1a'k_1^{-1} = a'.$$

Theorem 7.39 (KAK decomposition). Every element in G has a decomposition as k_1ak_2 with $k_1, k_2 \in K$ and $a \in A$. In this decomposition, a is uniquely determined up to conjugation by a member of $W(G, A)$. If a is fixed as $\exp H$ with $H \in \mathfrak{a}_0$ and if $\lambda(H) \neq 0$ for all $\lambda \in \Sigma$, then k_1 is unique up to right multiplication by a member of M.

PROOF. Existence of the decomposition was noted at the beginning of the section. For uniqueness suppose $k_1'a'k_2' = k_1''ak_2''$. If $k' = k_1''^{-1}k_1'$ and $k = k_2'k_2''^{-1}$, then $k'a'k = a$ and hence $(k'k)(k^{-1}a'k) = a$. By the uniqueness of the global Cartan decomposition, $k'k = 1$ and $k^{-1}a'k = a$. Lemma 7.38 then shows that a' and a are conjugate via $N_K(\mathfrak{a}_0)$.

Now let $a = a' = \exp H$ with $H \in \mathfrak{a}_0$ and $\lambda(H) \neq 0$ for all $\lambda \in \Sigma$. We have seen that $k^{-1}ak = a$. By Lemma 7.36, $\mathrm{Ad}(k)^{-1}H = H$. Since $\lambda(H) \neq 0$ for all $\lambda \in \Sigma$, Lemma 6.50 shows that $Z_{\mathfrak{g}_0}(H) = \mathfrak{a}_0 \oplus \mathfrak{m}_0$. Hence the centralizer of H in $\mathfrak{p}_0$ is $\mathfrak{a}_0$, and the centralizer of $\mathrm{Ad}(k)^{-1}H$ in $\mathfrak{p}_0$ is $\mathrm{Ad}(k)^{-1}\mathfrak{a}_0$. But $\mathrm{Ad}(k)^{-1}H = H$ implies that these centralizers are the same: $\mathrm{Ad}(k)^{-1}\mathfrak{a}_0 = \mathfrak{a}_0$. Thus k is in $N_K(\mathfrak{a}_0)$.

By Proposition 7.32, $\mathrm{Ad}(k)$ is given by an element w of the Weyl group $W(\Sigma)$. Since $\lambda(H) \neq 0$ for all $\lambda \in \Sigma$, we can define a lexicographic ordering so that the positive restricted roots are positive on H. Then $\mathrm{Ad}(k)H = H$ says that w permutes the positive restricted roots. By Theorem 2.63, $w = 1$. Therefore $\mathrm{Ad}(k)$ centralizes $\mathfrak{a}_0$, and k is in M.

From $k'k = 1$, we see that k' is in M. Then $k' = k_1''^{-1}k_1'$ shows that k_1' and k_1'' differ by an element of M on the right.

4. Bruhat Decomposition

We continue to assume that G is a reductive Lie group and that other notation is as in §2.

We know that the subgroup $M = Z_K(\mathfrak{a}_0)$ of K is compact, and we saw in §2 that MAN is a closed subgroup of G. It follows from the Iwasawa decomposition that the multiplication map $M \times A \times N \to MAN$ is a diffeomorphism onto.

The Bruhat decomposition describes the double coset decomposition $MAN\backslash G/MAN$ of G with respect to MAN. Here is an example.

EXAMPLE. Let $G = SL(2,\mathbb{R})$. Here $MAN = \left\{\begin{pmatrix} a & b \\ 0 & b^{-1}\end{pmatrix}\right\}$. The normalizer $N_K(\mathfrak{a}_0)$ consists of the four matrices $\pm\begin{pmatrix} 1 & 0 \\ 0 & 1\end{pmatrix}$ and $\pm\begin{pmatrix} 0 & 1 \\ -1 & 0\end{pmatrix}$, while the centralizer $Z_K(\mathfrak{a}_0)$ consists of the two matrices $\pm\begin{pmatrix} 1 & 0 \\ 0 & 1\end{pmatrix}$. Thus $|W(G,A)| = 2$, and $\tilde{w} = \begin{pmatrix} 0 & -1 \\ 1 & 0\end{pmatrix}$ is a representative of the nontrivial element of $W(G,A)$. Let $g = \begin{pmatrix} a & b \\ c & d\end{pmatrix}$ be given in G. If $c=0$, then g is in MAN. If $c \neq 0$, then

$$\begin{pmatrix} 0 & 1 \\ -1 & 0\end{pmatrix}\begin{pmatrix} a & b \\ c & d\end{pmatrix} = \begin{pmatrix} c & d \\ -a & -b\end{pmatrix} = \begin{pmatrix} 1 & 0 \\ -ac^{-1} & 1\end{pmatrix}\begin{pmatrix} c & d \\ 0 & c^{-1}\end{pmatrix}$$

$$= \begin{pmatrix} 0 & 1 \\ -1 & 0\end{pmatrix}\begin{pmatrix} 1 & ac^{-1} \\ 0 & 1\end{pmatrix}\begin{pmatrix} 0 & -1 \\ 1 & 0\end{pmatrix}\begin{pmatrix} c & d \\ 0 & c^{-1}\end{pmatrix}.$$

Hence

$$\begin{pmatrix} a & b \\ c & d\end{pmatrix} = \begin{pmatrix} 1 & ac^{-1} \\ 0 & 1\end{pmatrix}\begin{pmatrix} 0 & -1 \\ 1 & 0\end{pmatrix}\begin{pmatrix} c & d \\ 0 & c^{-1}\end{pmatrix}$$

exhibits $\begin{pmatrix} a & b \\ c & d\end{pmatrix}$ as in $MAN\tilde{w}MAN$. Thus the double-coset space $MAN\backslash G/MAN$ consists of two elements, with 1 and $\tilde{w}$ as representatives.

Theorem 7.40 (Bruhat decomposition). The double cosets of $MAN\backslash G/MAN$ are parametrized in a one-one fashion by $W(G,A)$, the double coset corresponding to $w \in W(G,A)$ being $MAN\tilde{w}MAN$, where $\tilde{w}$ is any representative of w in $N_K(\mathfrak{a}_0)$.

PROOF OF UNIQUENESS. Suppose that w_1 and w_2 are in $W(G,A)$, with $\tilde{w}_1$ and $\tilde{w}_2$ as representatives, and that x_1 and x_2 in MAN have

$$x_1\tilde{w}_1 = \tilde{w}_2 x_2. \tag{7.41}$$

Now $\mathrm{Ad}(N) = \exp(\mathrm{ad}(\mathfrak{n}_0))$ by Theorem 1.104, and hence $\mathrm{Ad}(N)$ carries $\mathfrak{a}_0$ to $\mathfrak{a}_0 \oplus \mathfrak{n}_0$ while leaving the $\mathfrak{a}_0$ component unchanged. Meanwhile under Ad, $N_K(\mathfrak{a}_0)$ permutes the restricted-root spaces and thus carries $\mathfrak{m}_0 \oplus \bigoplus_{\lambda\in\Sigma}(\mathfrak{g}_0)_\lambda$ to itself. Apply Ad of both sides of (7.41) to an element $H \in \mathfrak{a}_0$ and project to $\mathfrak{a}_0$ along $\mathfrak{m}_0 \oplus \bigoplus_{\lambda\in\Sigma}(\mathfrak{g}_0)_\lambda$. The resulting left side is in $\mathfrak{a}_0 \oplus \mathfrak{n}_0$ with $\mathfrak{a}_0$ component $\mathrm{Ad}(\tilde{w}_1)H$, while the right side is in $\mathrm{Ad}(\tilde{w}_2)H + \mathrm{Ad}(\tilde{w}_2)(\mathfrak{m}_0 \oplus \mathfrak{n}_0)$. Hence $\mathrm{Ad}(\tilde{w}_1)H = \mathrm{Ad}(\tilde{w}_2)H$. Since H is arbitrary, $\tilde{w}_2^{-1}\tilde{w}_1$ centralizes $\mathfrak{a}_0$. Therefore $w_1 = w_2$.

The proof of existence in Theorem 7.40 will be preceded by three lemmas.

Lemma 7.42. Let $H \in \mathfrak{a}_0$ be such that $\lambda(H) \neq 0$ for all $\lambda \in \Sigma$. Then the mapping $\varphi : N \to \mathfrak{g}_0$ given by $n \mapsto \operatorname{Ad}(n)H - H$ carries N onto $\mathfrak{n}_0$.

PROOF. Write $\mathfrak{n}_0 = \bigoplus (\mathfrak{g}_0)_\lambda$ as a sum of restricted-root spaces, and regard the restricted roots as ordered lexicographically. For any restricted root α, the subspace $\mathfrak{n}_\alpha = \bigoplus_{\lambda \geq \alpha} (\mathfrak{g}_0)_\lambda$ is an ideal, and we prove by induction downward on α that φ carries $N_\alpha = \exp \mathfrak{n}_\alpha$ onto $\mathfrak{n}_\alpha$. This conclusion for α equal to the smallest positive restricted root gives the lemma.

If α is given, we can write $\mathfrak{n}_\alpha = (\mathfrak{g}_0)_\alpha \oplus \mathfrak{n}_\beta$ with $\beta > \alpha$. Let X be given in $\mathfrak{n}_\alpha$, and write X as $X_1 + X_2$ with $X_1 \in (\mathfrak{g}_0)_\alpha$ and $X_2 \in \mathfrak{n}_\beta$. Since $\alpha(H) \neq 0$, we can choose $Y_1 \in (\mathfrak{g}_0)_\alpha$ with $[H, Y_1] = X_1$. Then

$$\operatorname{Ad}(\exp Y_1)H - H = H + [Y_1, H] + \tfrac{1}{2}(\operatorname{ad} Y_1)^2 H + \cdots - H = -X_1 + (\mathfrak{n}_\beta \text{ terms}),$$

and hence $\operatorname{Ad}(\exp Y_1)(H + X) - H$ is in $\mathfrak{n}_\beta$. By inductive hypothesis we can find $n \in N_\beta$ with

$$\operatorname{Ad}(n)H - H = \operatorname{Ad}(\exp Y_1)(H + X) - H.$$

Then $\operatorname{Ad}((\exp Y_1)^{-1} n)H - H = X$, and the element $(\exp Y_1)^{-1} n$ of N_α is the required element to complete the induction.

Lemma 7.43. Let $\mathfrak{s}_0 = \mathfrak{m}_0 \oplus \mathfrak{a}_0 \oplus \mathfrak{n}_0$. Then

(a) $\mathfrak{n}_0 \oplus Z_{\mathfrak{g}_0} = \{Z \in \mathfrak{s}_0 \mid \operatorname{ad}_{\mathfrak{g}}(Z) \text{ is nilpotent}\}$

(b) $\mathfrak{a}_0 \oplus \mathfrak{n}_0 \oplus (\mathfrak{m}_0 \cap Z_{\mathfrak{g}_0}) = \{Z \in \mathfrak{s}_0 \mid \operatorname{ad}_{\mathfrak{g}}(Z) \text{ has all eigenvalues real}\}$.

PROOF. Certainly the left sides in (a) and (b) are contained in the right sides. For the reverse containments write $Z \in \mathfrak{s}_0$ as $Z = X_0 + H + X$ with $X_0 \in \mathfrak{m}_0$, $H \in \mathfrak{a}_0$, and $X \in \mathfrak{n}_0$. Extend $\mathbb{R}X_0$ to a maximal abelian subspace $\mathfrak{t}_0$ of $\mathfrak{m}_0$, so that $\mathfrak{a}_0 \oplus \mathfrak{t}_0$ is a Cartan subalgebra of $\mathfrak{g}_0$. Extending the ordering of $\mathfrak{a}_0$ to one of $\mathfrak{a}_0 \oplus i\mathfrak{t}_0$ so that $\mathfrak{a}_0$ is taken before $i\mathfrak{t}_0$, we obtain a positive system Δ^+ for $\Delta(\mathfrak{g}, (\mathfrak{a} \oplus \mathfrak{t}))$ such that Σ^+ arises as the set of nonzero restrictions of members of Δ^+. Arrange the members of Δ^+ in decreasing order and form the matrix of $\operatorname{ad} Z$ in a corresponding basis of root vectors (with vectors from $\mathfrak{a} \oplus \mathfrak{t}$ used at the appropriate place in the middle). The matrix is upper triangular. The diagonal entries in the positions corresponding to the root vectors are $\alpha(X_0 + H) = \alpha(X_0) + \alpha(H)$ for $\alpha \in \Delta$, and the diagonal entries are 0 in the positions corresponding to basis vectors in $\mathfrak{a} \oplus \mathfrak{t}$. Here $\alpha(X_0)$ is imaginary, and $\alpha(H)$ is real. To have $\operatorname{ad} Z$ nilpotent, we must get 0 for all α. Thus the component of $X_0 + H$ in $[\mathfrak{g}_0, \mathfrak{g}_0]$ is 0. This proves (a). To have $\operatorname{ad} Z$ have real eigenvalues, we must have $\alpha(X_0) = 0$ for all $X \in \Delta$. Thus the component of X_0 in $[\mathfrak{g}_0, \mathfrak{g}_0]$ is 0. This proves (b).

Lemma 7.44. For each $g \in G$, put $\mathfrak{s}_0^g = \mathfrak{s}_0 \cap \operatorname{Ad}(g)\mathfrak{s}_0$. Then

$$\mathfrak{s}_0 = \mathfrak{s}_0^g + \mathfrak{n}_0.$$

PROOF. Certainly $\mathfrak{s}_0 \supseteq \mathfrak{s}_0^g + \mathfrak{n}_0$, and therefore it is enough to show that $\dim(\mathfrak{s}_0^g + \mathfrak{n}_0) = \dim \mathfrak{s}_0$. Since $G = KAN$, there is no loss of generality in assuming that g is in K. Write $k = g$. Let $(\cdot)^\perp$ denote orthogonal complement within $\mathfrak{g}_0$ relative to B_θ. From $\theta(\mathfrak{g}_0)_\lambda = (\mathfrak{g}_0)_{-\lambda}$, we have $\mathfrak{s}_0^\perp = \theta\mathfrak{n}_0$. Since $\operatorname{Ad}(k)$ acts in an orthogonal fashion,

$$(7.45)\qquad \begin{aligned}(\mathfrak{s}_0 + \operatorname{Ad}(k)\mathfrak{s}_0)^\perp &= \mathfrak{s}_0^\perp \cap (\operatorname{Ad}(k)\mathfrak{s}_0)^\perp = \theta\mathfrak{n}_0 \cap \operatorname{Ad}(k)\mathfrak{s}_0^\perp \\ &= \theta\mathfrak{n}_0 \cap \operatorname{Ad}(k)\theta\mathfrak{n}_0 = \theta(\mathfrak{n}_0 \cap \operatorname{Ad}(k)\mathfrak{n}_0).\end{aligned}$$

Let X be in $\mathfrak{s}_0 \cap \operatorname{Ad}(k)\mathfrak{s}_0$ and in $\mathfrak{n}_0$. Then $\operatorname{ad}_\mathfrak{g}(X)$ is nilpotent by Lemma 7.43a. Since $\operatorname{ad}_\mathfrak{g}(\operatorname{Ad}(k)^{-1}X)$ and $\operatorname{ad}_\mathfrak{g}(X)$ have the same eigenvalues, $\operatorname{ad}_\mathfrak{g}(\operatorname{Ad}(k)^{-1}X)$ is nilpotent. By Lemma 7.43a, $\operatorname{Ad}(k)^{-1}X$ is in $\mathfrak{n}_0 \oplus Z_{\mathfrak{g}_0}$. Since $\operatorname{Ad}(k)$ fixes $Z_{\mathfrak{g}_0}$ (by property (v)), $\operatorname{Ad}(k)^{-1}X$ is in $\mathfrak{n}_0$. Therefore X is in $\operatorname{Ad}(k)\mathfrak{n}_0$, and we obtain

$$(7.46)\qquad \mathfrak{n}_0 \cap \operatorname{Ad}(k)\mathfrak{n}_0 = \mathfrak{n}_0 \cap (\mathfrak{s}_0 \cap \operatorname{Ad}(k)\mathfrak{s}_0) = \mathfrak{n}_0 \cap \mathfrak{s}_0^k.$$

Consequently

$$\begin{aligned}2\dim \mathfrak{s}_0 - \dim \mathfrak{s}_0^k &= \dim(\mathfrak{s}_0 + \operatorname{Ad}(k)\mathfrak{s}_0) \\ &= \dim \mathfrak{g}_0 - \dim(\mathfrak{n}_0 \cap \operatorname{Ad}(k)\mathfrak{n}_0) && \text{by (7.45)} \\ &= \dim \mathfrak{g}_0 - \dim(\mathfrak{n}_0 \cap \mathfrak{s}_0^k) && \text{by (7.46)} \\ &= \dim \mathfrak{g}_0 + \dim(\mathfrak{n}_0 + \mathfrak{s}_0^k) - \dim \mathfrak{n}_0 - \dim \mathfrak{s}_0^k,\end{aligned}$$

and we conclude that

$$\dim \mathfrak{g}_0 + \dim(\mathfrak{n}_0 + \mathfrak{s}_0^k) - \dim \mathfrak{n}_0 = 2\dim \mathfrak{s}_0.$$

Since $\dim \mathfrak{n}_0 + \dim \mathfrak{s}_0 = \dim \mathfrak{g}_0$, we obtain $\dim(\mathfrak{n}_0 + \mathfrak{s}_0^k) = \dim \mathfrak{s}_0$, as required.

PROOF OF EXISTENCE IN THEOREM 7.40. Fix $H \in \mathfrak{a}_0$ with $\lambda(H) \neq 0$ for all $\lambda \in \Sigma$. Let $x \in G$ be given. Since $\mathfrak{a}_0 \subseteq \mathfrak{s}_0$, Lemma 7.44 allows us to write $H = X + Y$ with $X \in \mathfrak{n}_0$ and $Y \in \mathfrak{s}_0^x$. By Lemma 7.42 we can choose $n_1 \in N$ with $\operatorname{Ad}(n_1)H - H = -X$. Then

$$\operatorname{Ad}(n_1)H = H - X = Y \in \mathfrak{s}_0^x \subseteq \operatorname{Ad}(x)\mathfrak{s}_0.$$

So $Z = \operatorname{Ad}(x^{-1}n_1)H$ is in $\mathfrak{s}_0$. Since $\operatorname{ad}_\mathfrak{g} Z$ and $\operatorname{ad}_\mathfrak{g} H$ have the same eigenvalues, Lemma 7.43b shows that Z is in $\mathfrak{a}_0 \oplus \mathfrak{n}_0 \oplus (\mathfrak{m}_0 \cap Z_{\mathfrak{g}_0})$. Since $\operatorname{Ad}(x^{-1}n_1)^{-1}$ fixes $Z_{\mathfrak{g}_0}$ (by property (v)), Z is in $\mathfrak{a}_0 \oplus \mathfrak{m}_0$. Write $Z = H' + X'$ correspondingly. Here $\operatorname{ad} H$ and $\operatorname{ad} H'$ have the same eigenvalues, so that $\lambda(H') \neq 0$ for all $\lambda \in \Sigma$. By Lemma 7.42 there exists $n_2 \in N$ with $\operatorname{Ad}(n_2)^{-1}H' - H' = X'$. Then $\operatorname{Ad}(n_2)^{-1}H' = H' + X' = Z$, and

$$H' = \operatorname{Ad}(n_2)Z = \operatorname{Ad}(n_2x^{-1}n_1)H.$$

The centralizers of H' and H are both $\mathfrak{a}_0 \oplus \mathfrak{m}_0$ by Lemma 6.50. Thus

$$\operatorname{Ad}(n_2x^{-1}n_1)(\mathfrak{a}_0 \oplus \mathfrak{m}_0) = \mathfrak{a}_0 \oplus \mathfrak{m}_0. \tag{7.47}$$

If X is in $\mathfrak{a}_0$, then $\operatorname{ad}_\mathfrak{g}(X)$ has real eigenvalues by Lemma 7.43b. Since $\operatorname{ad}_\mathfrak{g}(\operatorname{Ad}(n_2x^{-1}n_1)X)$ and $\operatorname{ad}_\mathfrak{g}(X)$ have the same eigenvalues, Lemma 7.43b shows that $\operatorname{Ad}(n_2x^{-1}n_1)X$ is in $\mathfrak{a}_0 \oplus (\mathfrak{m}_0 \cap Z_{\mathfrak{g}_0})$. Since $\operatorname{Ad}(n_2x^{-1}n_1)^{-1}$ fixes $Z_{\mathfrak{g}_0}$ (by property (v)), $\operatorname{Ad}(n_2x^{-1}n_1)X$ is in $\mathfrak{a}_0$. We conclude that $n_2x^{-1}n_1$ is in $N_G(\mathfrak{a}_0)$.

Let $n_2x^{-1}n_1 = u \exp X_0$ be the global Cartan decomposition of $n_2x^{-1}n_1$. By Lemma 7.22, u is in $N_K(\mathfrak{a}_0)$ and X_0 is in $N_{\mathfrak{g}_0}(\mathfrak{a}_0)$. By the same argument as in Lemma 6.56, $N_{\mathfrak{g}_0}(\mathfrak{a}_0) = \mathfrak{a}_0 \oplus \mathfrak{m}_0$. Since X_0 is in $\mathfrak{p}_0$, X_0 is in $\mathfrak{a}_0$. Therefore u is in $N_K(\mathfrak{a}_0)$ and $\exp X_0$ is in A. In other words, $n_2^{-1}n_1$ is in uA, and x is in the same MAN double coset as the member u^{-1} of $N_K(\mathfrak{a}_0)$.

5. Structure of M

We continue to assume that G is a reductive Lie group and that other notation is as in §2. The fundamental source of disconnectedness in the structure theory of semisimple groups is the behavior of the subgroup $M = Z_K(\mathfrak{a}_0)$. We shall examine M in this section, paying particular attention to its component structure. For the first time we shall make serious use of results from Chapter V.

Proposition 7.48. *M is a reductive Lie group.*

PROOF. Proposition 7.25 shows that $Z_G(\mathfrak{a}_0)$ is a reductive Lie group, necessarily of the form $Z_K(\mathfrak{a}_0)\exp(Z_{\mathfrak{g}_0}(\mathfrak{a}_0) \cap \mathfrak{p}_0) = MA$. By Proposition 7.27, ${}^0(MA) = M$ is a reductive Lie group.

Proposition 7.33 already tells us that M meets every component of G. But M can be disconnected even when G is disconnected. (Recall from the examples in §VI.5 that M is disconnected when $G = SL(n, \mathbb{R})$.) Choose and fix a maximal abelian subspace $\mathfrak{t}_0$ of $\mathfrak{m}_0$. Then $\mathfrak{a}_0 \oplus \mathfrak{t}_0$ is a Cartan subalgebra of $\mathfrak{g}_0$.

Proposition 7.49. Every component of M contains a member of M that centralizes $\mathfrak{t}_0$, so that $M = Z_M(\mathfrak{t}_0)M_0$.

REMARK. The proposition says that we may focus our attention on $Z_M(\mathfrak{t}_0)$. After this proof we shall study $Z_M(\mathfrak{t}_0)$ by considering it as a subgroup of $Z_K(\mathfrak{t}_0)$.

PROOF. If $m \in M$ is given, then $\text{Ad}(m)\mathfrak{t}_0$ is a maximal abelian subspace of $\mathfrak{m}_0$. By Theorem 4.34 (applied to M_0), there exists $m_0 \in M_0$ such that $\text{Ad}(m_0)\text{Ad}(m)\mathfrak{t}_0 = \mathfrak{t}_0$. Then m_0m is in $N_M(\mathfrak{m}_0)$. Introduce a positive system Δ^+ for the root system $\Delta = \Delta(\mathfrak{m}, \mathfrak{t})$. Then $\text{Ad}(m_0m)\Delta^+$ is a positive system for Δ, and Theorems 4.54 and 2.63 together say that we can find $m_1 \in M_0$ such that $\text{Ad}(m_1m_0m)$ maps Δ^+ to itself. By Proposition 7.48, M satisfies property (v) of reductive Lie groups. Therefore $\text{Ad}_{\mathfrak{m}}(m_1m_0m)$ is in $\text{Int}\,\mathfrak{m}$. Then $\text{Ad}_{\mathfrak{m}}(m_1m_0m)$ must be induced by in $\text{Int}_{\mathfrak{m}}\,[\mathfrak{m}, \mathfrak{m}]$, and Theorem 7.8 says that this element fixes each member of Δ^+. Therefore m_1m_0m centralizes $\mathfrak{t}_0$, and the result follows.

Suppose that the root α in $\Delta(\mathfrak{g}, \mathfrak{a}\oplus\mathfrak{t})$ is real, i.e., α vanishes on $\mathfrak{t}$. As in the discussion following (6.66), the root space $\mathfrak{g}_\alpha$ in $\mathfrak{g}$ is invariant under the conjugation of $\mathfrak{g}$ with respect to $\mathfrak{g}_0$. Since $\dim_{\mathbb{C}} \mathfrak{g}_\alpha = 1$, $\mathfrak{g}_\alpha$ contains a nonzero root vector E_α that is in $\mathfrak{g}_0$. Also as in the discussion following (6.66), we may normalize E_α by a real constant so that $B(E_\alpha, \theta E_\alpha) = -2/|\alpha|^2$. Put $H'_\alpha = 2|\alpha|^{-2}H_\alpha$. Then $\{H'_\alpha, E_\alpha, \theta E_\alpha\}$ spans a copy of $\mathfrak{sl}(2,\mathbb{R})$ with

$$H'_\alpha \leftrightarrow h, \qquad E_\alpha \leftrightarrow e, \qquad \theta E_\alpha \leftrightarrow -f. \tag{7.50}$$

Let us write $(\mathfrak{g}_0)_\alpha$ for $\mathbb{R}E_\alpha$ and $(\mathfrak{g}_0)_{-\alpha}$ for $\mathbb{R}\theta E_\alpha$.

Proposition 7.51. The subgroup $Z_G(\mathfrak{t}_0)$ of G

(a) is reductive with global Cartan decomposition

$$Z_G(\mathfrak{t}_0) = Z_K(\mathfrak{t}_0)\exp(\mathfrak{p}_0 \cap Z_{\mathfrak{g}_0}(\mathfrak{t}_0))$$

(b) has Lie algebra

$$Z_{\mathfrak{g}_0}(\mathfrak{t}_0) = \mathfrak{t}_0 \oplus \mathfrak{a}_0 \oplus \bigoplus_{\substack{\alpha\in\Delta(\mathfrak{g},\mathfrak{a}\oplus\mathfrak{t}),\\ \alpha \text{ real}}} (\mathfrak{g}_0)_\alpha,$$

which is the direct sum of its center with a real semisimple Lie algebra that is a split real form of its complexification

(c) is such that the component groups of G, K, $Z_G(\mathfrak{t}_0)$, and $Z_K(\mathfrak{t}_0)$ are all isomorphic.

PROOF. Conclusion (a) is immediate from Proposition 7.25. For (b) it is clear that

$$Z_{\mathfrak{g}}(\mathfrak{t}_0) = \mathfrak{t} \oplus \mathfrak{a} \oplus \bigoplus_{\substack{\alpha \in \Delta(\mathfrak{g}, \mathfrak{a} \oplus \mathfrak{t}), \\ \alpha \text{ real}}} \mathfrak{g}_\alpha .$$

The conjugation of $\mathfrak{g}$ with respect to $\mathfrak{g}_0$ carries every term of the right side into itself, and therefore we obtain the formula of (b). Here $\mathfrak{a}_0$ is maximal abelian in $\mathfrak{p}_0 \cap Z_{\mathfrak{g}_0}(\mathfrak{t}_0)$, and therefore this decomposition is the restricted-root space decomposition of $\mathfrak{g}_0$. Applying Corollary 6.49 to $[\mathfrak{g}_0, \mathfrak{g}_0]$, we obtain (b). In (c), G and K have isomorphic component groups as a consequence of the global Cartan decomposition, and $Z_G(\mathfrak{t}_0)$ and $Z_K(\mathfrak{t}_0)$ have the same component groups as a consequence of (a). Consider the natural homomorphism

$$Z_K(\mathfrak{t}_0)/Z_K(\mathfrak{t}_0)_0 \to K/K_0$$

induced by inclusion. Propositions 7.49 and 7.33 show that this map is onto, and Corollary 4.51 shows that it is one-one. This proves (c).

We cannot expect to say much about the disconnectedness of M that results from the disconnectedness of G. Thus we shall assume for the remainder of this section that G is connected. Proposition 7.51c notes that $Z_G(\mathfrak{t}_0)$ is connected. To study $Z_G(\mathfrak{t}_0)$, we shall work with the analytic subgroup of $Z_G(\mathfrak{t}_0)$ whose Lie algebra is $[Z_{\mathfrak{g}_0}(\mathfrak{t}_0), Z_{\mathfrak{g}_0}(\mathfrak{t}_0)]$. This is the subgroup that could be called $Z_G(\mathfrak{t}_0)_{ss}$ in the notation of §2. It is semisimple, and its Lie algebra is a split real form. We call the subgroup the **associated split semisimple subgroup**, and we introduce the notation G_{split} for it in order to emphasize that its Lie algebra is split.

Let T be the maximal torus of M_0 with Lie algebra $\mathfrak{t}_0$. Under the assumption that G is connected, it follows from Proposition 7.51b that $Z_G(\mathfrak{t}_0)$ is a commuting product

$$Z_G(\mathfrak{t}_0) = TAG_{\text{split}}.$$

By Proposition 7.27,

$$^0Z_G(\mathfrak{t}_0) = TG_{\text{split}}$$

is a reductive Lie group.

The group G_{split} need not have finite center, but the structure theory of Chapter VI is available to describe it. Let K_{split} and A_{split} be the analytic subgroups with Lie algebras given as the intersections of $\mathfrak{k}_0$ and $\mathfrak{a}_0$ with $[Z_{\mathfrak{g}_0}(\mathfrak{t}_0), Z_{\mathfrak{g}_0}(\mathfrak{t}_0)]$. Let $F = M_{\text{split}}$ be the centralizer of A_{split} in K_{split}. The subgroup F will play a key role in the analysis of M. It centralizes both T and A.

Corollary 7.52. The subgroup F normalizes M_0, and $M = FM_0$.

PROOF. Since F centralizes A and is a subgroup of K, it is a subgroup of M. Therefore F normalizes M_0, and FM_0 is a group. We know from Proposition 7.49 that $M = Z_M(\mathfrak{t}_0)M_0$. Since $T \subseteq M_0$, it is enough to prove that $Z_M(\mathfrak{t}_0) = TF$. The subgroup $Z_M(\mathfrak{t}_0)$ is contained in $Z_K(\mathfrak{t}_0)$, which in turn is contained in ${}^0Z_G(\mathfrak{t}_0) = TG_{\text{split}}$. Since $Z_M(\mathfrak{t}_0)$ is contained in K, it is therefore contained in TK_{split}. Decompose a member m of $Z_M(\mathfrak{t}_0)$ in a corresponding fashion as $m = tk$. Since m and t centralize A, so does k. Therefore k is in $F = M_{\text{split}}$, and the result follows.

Without additional hypotheses we cannot obtain further nontrivial results about F, and accordingly we recall the following definition from §1.

A semisimple group G has a **complexification** $G^{\mathbb{C}}$ if $G^{\mathbb{C}}$ is a connected complex Lie group with Lie algebra $\mathfrak{g}$ such that G is the analytic subgroup corresponding to the real form $\mathfrak{g}_0$ of $\mathfrak{g}$. By Corollary 7.6, $G^{\mathbb{C}}$ is isomorphic to a matrix group, and hence the same thing is true of G and G_{split}. By Proposition 7.9, each of G and G_{split} has finite center. Therefore we may consider G and G_{split} in the context of reductive Lie groups.

Fix K, θ, and B for G. If the Cartan decomposition of $\mathfrak{g}_0$ is $\mathfrak{g}_0 = \mathfrak{k}_0 \oplus \mathfrak{p}_0$, then

$$\mathfrak{g} = (\mathfrak{k}_0 \oplus i\mathfrak{p}_0) \oplus (\mathfrak{p}_0 \oplus i\mathfrak{k}_0)$$

is a Cartan decomposition of $\mathfrak{g}$, and the corresponding Cartan involution of $\mathfrak{g}$ is bar $\circ\, \theta$, where bar is the conjugation of $\mathfrak{g}$ with respect to $\mathfrak{g}_0$. The Lie algebra $\mathfrak{u}_0 = \mathfrak{k}_0 \oplus i\mathfrak{p}_0$ is compact semisimple, and it follows from Proposition 7.9 that the corresponding analytic subgroup U of $G^{\mathbb{C}}$ is compact. Then the tuple $(G^{\mathbb{C}}, U, \text{bar} \circ \theta, B)$ makes $G^{\mathbb{C}}$ into a reductive Lie group. Whenever a semisimple Lie group G has a complexification $G^{\mathbb{C}}$ and we consider G as a reductive Lie group (G, K, θ, B), we may consider $G^{\mathbb{C}}$ as the reductive Lie group $(G^{\mathbb{C}}, U, \text{bar} \circ \theta, B)$.

Under the assumption that the semisimple group G has a complexification $G^{\mathbb{C}}$, $\exp i\mathfrak{a}_0$ is well defined as an analytic subgroup of U.

Theorem 7.53. Suppose that the reductive Lie group G is semisimple and has a complexification $G^{\mathbb{C}}$. Then

(a) $F = K_{\text{split}} \cap \exp i\mathfrak{a}_0$
(b) F is contained in the center of M
(c) M is the commuting product $M = FM_0$
(d) F is finite abelian, and every element $f \neq 1$ in F has order 2.

PROOF.

(a) Every member of $K_{\text{split}} \cap \exp i\mathfrak{a}_0$ centralizes $\mathfrak{a}_0$ and lies in K_{split}, hence lies in F. For the reverse inclusion we have $F \subseteq K_{\text{split}}$ by definition. To see that $F \subseteq \exp i\mathfrak{a}_0$, let U_{split} be the analytic subgroup of $G^{\mathbb{C}}$ with Lie algebra the intersection of $\mathfrak{u}_0$ with the Lie algebra $[Z_{\mathfrak{g}}(\mathfrak{t}_0), Z_{\mathfrak{g}}(\mathfrak{t}_0)]$. Then U_{split} is compact, and $i\mathfrak{a}_0 \cap [Z_{\mathfrak{g}}(\mathfrak{t}_0), Z_{\mathfrak{g}}(\mathfrak{t}_0)]$ is a maximal abelian subspace of its Lie algebra. By Corollary 4.52 the corresponding torus is its own centralizer. Hence the centralizer of $\mathfrak{a}_0$ in U_{split} is contained in $\exp i\mathfrak{a}_0$. Since $K_{\text{split}} \subseteq U_{\text{split}}$, it follows that $F \subseteq \exp i\mathfrak{a}_0$.

(b, c) Corollary 7.52 says that $M = FM_0$. By (a), every element of F commutes with any element that centralizes $\mathfrak{a}_0$. Hence F is central in M, and (b) and (c) follow.

(d) Since G_{split} has finite center, F is compact. Its Lie algebra is 0, and thus it is finite. By (b), F is abelian. We still have to prove that every element $f \neq 1$ in F has order 2.

Since G has a complexification, so does G_{split}. Call this group $G^{\mathbb{C}}_{\text{split}}$, let $\tilde{G}^{\mathbb{C}}_{\text{split}}$ be a simply connected covering group, and let φ be the covering map. Let $\tilde{G}_{\text{split}}$ be the analytic subgroup with the same Lie algebra as for G_{split}, and form the subgroups $\tilde{K}_{\text{split}}$ and $\tilde{F}$ of $\tilde{G}_{\text{split}}$. The subgroup $\tilde{F}$ is the complete inverse image of F under φ. Let $\tilde{U}_{\text{split}}$ play the same role for $\tilde{G}^{\mathbb{C}}_{\text{split}}$ that U plays for G. The automorphism θ of the Lie algebra of G_{split} complexifies and lifts to an automorphism $\tilde{\theta}$ of $\tilde{G}^{\mathbb{C}}_{\text{split}}$ that carries $\tilde{U}_{\text{split}}$ into itself. The automorphism $\tilde{\theta}$ acts as $x \mapsto x^{-1}$ on $\exp i\mathfrak{a}_0$ and as the identity on $\tilde{K}_{\text{split}}$. The elements of $\tilde{F}$ are the elements of the intersection, by (a), and hence $\tilde{f}^{-1} = \tilde{f}$ for every element $\tilde{f}$ of $\tilde{F}$. That is $\tilde{f}^2 = 1$. Applying φ and using the fact that φ maps $\tilde{F}$ onto F, we conclude that every element $f \neq 1$ in F has order 2.

EXAMPLE. When G does not have a complexification, the subgroup F need not be abelian. For an example we observe that the group K for $SL(3, \mathbb{R})$ is $SO(3)$, which has $SU(2)$ as a 2-sheeted simply connected covering group. Thus $SL(3, \mathbb{R})$ has a 2-sheeted simply connected covering group, and we take this covering group as G. We already noted in §VI.5 that the group M for $SL(3, \mathbb{R})$ consists of the diagonal matrices with diagonal entries ± 1 and determinant 1. Thus M is the direct sum of two 2-element groups. The subgroup F of G is the complete inverse image of M under the covering map and thus has order 8. Moreover it is a subgroup of $SU(2)$, which has only one element of order 2. Thus F is a group of order 8 with only one element of order 2 and no element of order 8. Of the five abstract groups of order 8, only the 8-element subgroup $\{\pm 1, \pm i, \pm j, \pm k\}$ of the quaternions has this property. This group is nonabelian, and hence F is nonabelian.

Let α be a real root of $\Delta(\mathfrak{g}, \mathfrak{a} \oplus \mathfrak{t})$. From (7.50) we obtain a one-one homomorphism $\mathfrak{sl}(2, \mathbb{R}) \to \mathfrak{g}_0$ whose only ambiguity is a sign in the definition of E_α. This homomorphism carries $\mathfrak{so}(2)$ to $\mathfrak{k}_0$ and complexifies to a homomorphism $\mathfrak{sl}(2, \mathbb{C}) \to \mathfrak{g}$. Under the assumption that G is semisimple and has a complexification $G^{\mathbb{C}}$, we can form the analytic subgroup of $G^{\mathbb{C}}$ with Lie algebra $\mathfrak{sl}(2, \mathbb{C})$. This will be a homomorphic image of $SL(2, \mathbb{C})$ since $SL(2, \mathbb{C})$ is simply connected. We let γ_α be the image of $\begin{pmatrix} -1 & 0 \\ 0 & -1 \end{pmatrix}$. This element is evidently in the image of $SO(2) \subseteq SL(2, \mathbb{R})$ and hence lies in K_{split}. Clearly it does not depend upon the choice of the ambiguous sign in the definition of E_α. A formula for γ_α is

$$\gamma_\alpha = \exp 2\pi i |\alpha|^{-2} H_\alpha. \tag{7.54}$$

Theorem 7.55. Suppose that the reductive Lie group G is semisimple and has a complexification $G^{\mathbb{C}}$. Then F is generated by all elements γ_α for all real roots α.

PROOF. Our construction of γ_α shows that γ_α is in both K_{split} and $\exp i\mathfrak{a}_0$. By Theorem 7.53a, γ_α is in F. In the reverse direction we use the construction in the proof of Theorem 7.53d, forming a simply connected cover $\tilde{G}^{\mathbb{C}}_{\text{split}}$ of the complexification $G^{\mathbb{C}}_{\text{split}}$ of G_{split}. We form also the groups $\tilde{K}_{\text{split}}$, $\tilde{F}$, and $\tilde{U}_{\text{split}}$. The elements γ_α are well defined in $\tilde{F}$ via (7.54), and we show that they generate $\tilde{F}$. Then the theorem will follow by applying the covering map $\tilde{G}^{\mathbb{C}}_{\text{split}} \to G^{\mathbb{C}}_{\text{split}}$, since $\tilde{F}$ maps onto F.

Let $\tilde{H}$ be the maximal torus of $\tilde{U}_{\text{split}}$ with Lie algebra $i\mathfrak{a}_0$. We know from Theorem 7.53 that $\tilde{F}$ is a finite subgroup of $\tilde{H}$. Arguing by contradiction, suppose that the elements γ_α generate a proper subgroup $\tilde{F}_0$ of $\tilde{F}$. Let $\tilde{f}$ be an element of $\tilde{F}$ not in $\tilde{F}_0$. Applying the Peter-Weyl Theorem (Theorem 4.20) to $\tilde{H}/\tilde{F}_0$, we can obtain a multiplicative character χ_ν of $\tilde{H}$ that is 1 on $\tilde{F}_0$ and is $\neq 1$ on $\tilde{f}$. Here ν is the imaginary-valued linear functional on $i\mathfrak{a}_0$ such that $\chi_\nu(\exp ih) = e^{\nu(ih)}$ for $h \in \mathfrak{a}_0$. The roots for $\tilde{U}_{\text{split}}$ are the real roots for $\mathfrak{g}_0$, and our assumption is that each such real root α has

$$1 = \chi_\nu(\gamma_\alpha) = \chi(\exp 2\pi i |\alpha|^{-2} H_\alpha) = e^{\nu(2\pi i |\alpha|^{-2} H_\alpha)} = e^{\pi i (2\langle \nu, \alpha \rangle / |\alpha|^2)}.$$

That is $2\langle \nu, \alpha \rangle / |\alpha|^2$ is an even integer for all α. Hence $\frac{1}{2}\nu$ is algebraically integral.

Since $\tilde{U}_{\text{split}}$ is simply connected, Theorem 5.107 shows that $\frac{1}{2}\nu$ is analytically integral. Thus the multiplicative character $\chi_{\frac{1}{2}\nu}$ of $\tilde{H}$ given by $\chi_{\frac{1}{2}\nu}(\exp ih) = e^{\frac{1}{2}\nu(ih)}$ is well defined. Theorem 7.53d says that $\tilde{f}^2 = 1$, and therefore $\chi_{\frac{1}{2}\nu}(\tilde{f}) = \pm 1$. Since $\chi_\nu = (\chi_{\frac{1}{2}\nu})^2$, we obtain $\chi_\nu(\tilde{f}) = 1$, contradiction. We conclude that $\tilde{F}_0$ equals $\tilde{F}$, and the proof is complete.

It is sometimes handy to enlarge the collection of elements γ_α. Let β be any restricted root, and let X_β be any restricted-root vector corresponding to β. Then θX_β is a restricted-root vector for the restricted root $-\beta$ by Proposition 6.40c. Proposition 6.52 shows that we can normalize X_β so that $[X_\beta, \theta X_\beta] = -2|\beta|^{-2}H_\beta$, and then the correspondence

$$(7.56) \qquad h \leftrightarrow 2|\beta|^{-2}H_\beta, \qquad e \leftrightarrow X_\beta, \qquad f \leftrightarrow -\theta X_\beta$$

is an isomorphism of $\mathfrak{sl}(2,\mathbb{R})$ with the real span of $H_\beta, X_\beta, \theta X_\beta$ in $\mathfrak{g}_0$. Once again this homomorphism carries $\mathfrak{so}(2) = \mathbb{R}(e-f)$ to $\mathfrak{k}_0$ and complexifies to a homomorphism $\mathfrak{sl}(2,\mathbb{C}) \to \mathfrak{g}$. Under the assumption that G is semisimple and has a complexification $G^{\mathbb{C}}$, we can form the analytic subgroup of $G^{\mathbb{C}}$ with Lie algebra $\mathfrak{sl}(2,\mathbb{C})$. This will be a homomorphic image of $SL(2,\mathbb{C})$ since $SL(2,\mathbb{C})$ is simply connected. We let γ_β be the image of $\begin{pmatrix} -1 & 0 \\ 0 & -1 \end{pmatrix}$, namely

$$(7.57) \qquad \gamma_\beta = \exp 2\pi i |\beta|^{-2} H_\beta.$$

This element is evidently in the image of $SO(2) \subseteq SL(2,\mathbb{R})$ and hence lies in K. Formula (7.57) makes it clear that γ_β does not depend on the choice of X_β, except for the normalization, and also (7.57) shows that γ_β commutes with $\mathfrak{a}_0$. Hence

$$(7.58) \qquad \gamma_\beta \text{ is in } M \text{ for each restricted root } \beta.$$

Since $\begin{pmatrix} -1 & 0 \\ 0 & -1 \end{pmatrix}$ has square the identity, it follows that

$$(7.59) \qquad \gamma_\beta^2 = 1 \qquad \text{for each restricted root } \beta.$$

In the special case that β extends to a real root α of $\Delta(\mathfrak{g}, \mathfrak{a} \oplus \mathfrak{t})$ when set equal to 0 on $\mathfrak{t}$, γ_β equals the element γ_α defined in (7.54). The more general elements (7.57) are not needed for the description of F in Theorem 7.55, but they will play a role in Chapter VIII.

6. Real-Rank-One Subgroups

We continue to assume that G is a reductive Lie group, and we use the other notation of §2. In addition, we use the notation F of §5.

The **real rank** of G is the dimension of a maximal abelian subspace of $\mathfrak{p}_0$. Proposition 7.29 shows that real rank is well defined. Since any maximal abelian subspace of $\mathfrak{p}_0$ contains $\mathfrak{p}_0 \cap Z_{\mathfrak{g}_0}$, it follows that

$$\text{real rank}(G) = \text{real rank}({}^0G) + \dim Z_{vec}. \tag{7.60}$$

Our objective in this section is to identify some subgroups of G of real rank one and illustrate how information about these subgroups can give information about G.

"Real rank" is meaningful for a real semisimple Lie algebra outside the context of reductive Lie groups (G, K, θ, B), since Cartan decompositions exist and all are conjugate. But it is not meaningful for a reductive Lie algebra by itself, since the splitting of $Z_{\mathfrak{g}_0}$ into its $\mathfrak{k}_0$ part and its $\mathfrak{p}_0$ part depends upon the choice of θ.

The Lie subalgebra $[\mathfrak{g}_0, \mathfrak{g}_0]$ of $\mathfrak{g}_0$, being semisimple, is uniquely the sum of simple ideals. These ideals are orthogonal with respect to B, since if $\mathfrak{g}_i$ and $\mathfrak{g}_j$ are distinct ideals, then

$$B(\mathfrak{g}_i, \mathfrak{g}_j) = B([\mathfrak{g}_i, \mathfrak{g}_i], \mathfrak{g}_j) = B(\mathfrak{g}_i, [\mathfrak{g}_i, \mathfrak{g}_j]) = B(\mathfrak{g}_i, 0) = 0. \tag{7.61}$$

Since $[\mathfrak{g}_0, \mathfrak{g}_0]$ is invariant under θ, θ permutes these simple ideals, necessarily in orbits of one or two ideals. But actually there are no 2-ideal orbits since if X and θX are nonzero elements of distinct ideals, then (7.61) gives

$$0 < B_\theta(X, X) = -B(X, \theta X) = 0,$$

contradiction. Hence each simple ideal is invariant under θ, and it follows that $\mathfrak{p}_0$ is the direct sum of its components in each simple ideal and its component in $Z_{\mathfrak{g}_0}$.

We would like to conclude that the real rank of G is the sum of the real ranks from the components and from the center. But to do so, we need either to define real rank for triples $(\mathfrak{g}_0, \theta, B)$ or to lift the setting from Lie algebras to Lie groups. Following the latter procedure, assume that G is in the Harish-Chandra class; this condition is satisfied automatically if G is semisimple. If G_i is the analytic subgroup of G whose Lie algebra is one of the various simple ideals of G, then Proposition 7.20b shows that G_i has finite center. Consequently G_i is a reductive group. Also in this case the subgroup K_i of G_i fixed by Θ is compact, and it follows from property (iv) that G_i is closed in G. We summarize as follows.

Proposition 7.62. Let the reductive Lie group G be in the Harish-Chandra class, and let $G_1, \ldots, G_n$ be the analytic subgroups of G whose Lie algebra are the simple ideals of $\mathfrak{g}_0$. Then $G_1, \ldots, G_n$ are reductive Lie groups, they are closed in G, and the sum of the real ranks of the G_i's, together with the dimension of Z_{vec}, equals the real rank of $\mathfrak{g}_0$.

With the maximal abelian subspace $\mathfrak{a}_0$ of $\mathfrak{p}_0$ fixed, let λ be a restricted root. Denote by $H_\lambda^\perp$ the orthogonal complement of $\mathbb{R}H_\lambda$ in $\mathfrak{a}_0$ relative to B_θ. Propositions 7.25 and 7.27 show that $Z_G(H_\lambda^\perp)$ and ${}^0Z_G(H_\lambda^\perp)$ are reductive Lie groups. All of $\mathfrak{a}_0$ is in $Z_G(H_\lambda^\perp)$, and therefore $Z_G(H_\lambda^\perp)$ has the same real rank as G. The split component of $Z_G(H_\lambda^\perp)$ is $H_\lambda^\perp$, and it follows from (7.60) that ${}^0Z_G(H_\lambda^\perp)$ is a reductive Lie group of real rank one.

The subgroup ${}^0Z_G(H_\lambda^\perp)$ is what is meant by the real-rank-one reductive subgroup of G corresponding to the restricted root λ. A maximal abelian subspace of the $\mathfrak{p}_0$ for ${}^0Z_G(H_\lambda^\perp)$ is $\mathbb{R}H_\lambda$, and the restricted roots for this group are those nonzero multiples of λ that provide restricted roots for $\mathfrak{g}_0$. In other words the restricted-root space decomposition of the Lie algebra of ${}^0Z_G(H_\lambda^\perp)$ is

$$\mathbb{R}H_\lambda \oplus \mathfrak{m}_0 \oplus \bigoplus_{c\neq 0} (\mathfrak{g}_0)_{c\lambda}. \tag{7.63}$$

Sometimes it is desirable to associate to λ a real-rank-one subgroup whose Lie algebra is simple. To do so, let us assume that G is in the Harish-Chandra class. Then so is ${}^0Z_G(H_\lambda^\perp)$. Since this group has compact center, Proposition 7.62 shows that the sum of the real ranks of the subgroups G_i of ${}^0Z_G(H_\lambda^\perp)$ corresponding to the simple ideals of the Lie algebra is 1. Hence exactly one G_i has real rank one, and that is the real-rank-one reductive subgroup that we can use. The part of (7.63) that is being dropped to get a simple Lie algebra is contained in $\mathfrak{m}_0$.

In the case that the reductive group G is semisimple and has a complexification, the extent of the disconnectedness of M can be investigated with the help of the real-rank-one subgroups ${}^0Z_G(H_\lambda^\perp)$. The result that we use about the real-rank-one case is given in Theorem 7.66 below.

Lemma 7.64. $N^- \cap MAN = \{1\}$.

PROOF. Let $x \neq 1$ be in $N^- = \Theta N$. By Theorem 1.104 write $x = \exp X$ with X in $\mathfrak{n}_0^- = \theta\mathfrak{n}_0$. Recall from Proposition 6.40c that $\theta(\mathfrak{g}_0)_\lambda = (\mathfrak{g}_0)_{-\lambda}$, let $X = \sum_{\mu\in\Sigma} X_\mu$ be the decomposition of X into restricted-root vectors, and choose $\mu = \mu_0$ as large as possible so that $X_\mu \neq 0$. If we take any

$H \in \mathfrak{a}_0$ such that $\lambda(H) \neq 0$ for all $\lambda \in \Sigma$, then

$$\begin{aligned}\mathrm{Ad}(x)H - H &= e^{\mathrm{ad}\,X}H - H \\ &= [X, H] + \tfrac{1}{2}[X, [X, H]] + \cdots \\ &= [X_{\mu_0}, H] + \text{terms for lower restricted roots.}\end{aligned}$$

In particular, $\mathrm{Ad}(x)H - H$ is in $\mathfrak{n}_0^-$ and is not 0. On the other hand, if x is in MAN, then $\mathrm{Ad}(x)H - H$ is in $\mathfrak{n}_0$. Since $\mathfrak{n}_0^- \cap \mathfrak{n}_0 \neq 0$, we must have $N^- \cap MAN = \{1\}$.

Lemma 7.65. The map $K/M \to G/MAN$ induced by inclusion is a diffeomorphism.

PROOF. The given map is certainly smooth. If $\kappa(g)$ denotes the K component of g in the Iwasawa decomposition $G = KAN$ of Proposition 7.31, then $g \mapsto \kappa(g)$ is smooth, and the map $gMAN \mapsto \kappa(g)M$ is a two-sided inverse to the given map.

Theorem 7.66. Suppose that the reductive Lie group G is semisimple, is of real rank one, and has a complexification $G^{\mathbb{C}}$. Then M is connected unless $\dim \mathfrak{n}_0 = 1$.

REMARKS. Since G is semisimple, it is in the Harish-Chandra class. The above remarks about simple components are therefore applicable. The condition $\dim \mathfrak{n}_0 = 1$ is the same as the condition that the simple component of $\mathfrak{g}_0$ containing $\mathfrak{a}_0$ is isomorphic to $\mathfrak{sl}(2, \mathbb{R})$. In fact, if $\dim \mathfrak{n}_0 = 1$, then $\mathfrak{n}_0$ is of the form $\mathbb{R}X$ for some X. Then X, θX, and $[X, \theta X]$ span a copy of $\mathfrak{sl}(2, \mathbb{R})$, and we obtain $\mathfrak{g}_0 \cong \mathfrak{sl}(2, \mathbb{R}) \oplus \mathfrak{m}_0$. The Lie subalgebra $\mathfrak{m}_0$ must centralize X, θX, and $[X, \theta X]$ and hence must be an ideal in $\mathfrak{g}_0$. The complementary ideal is $\mathfrak{sl}(2, \mathbb{R})$, as asserted.

PROOF. The multiplication map $N^- \times M_0AN \to G$ is smooth and everywhere regular by Lemma 6.44. Hence the map $N^- \to G/M_0AN$ induced by inclusion is smooth and regular, and so is the map

$$N^- \to G/MAN, \tag{7.67}$$

which is the composition of $N^- \to G/M_0AN$ and a covering map. Also the map (7.67) is one-one by Lemma 7.64. Therefore (7.67) is a diffeomorphism onto an open set.

Since G is semisimple and has real rank 1, the Weyl group $W(\Sigma)$ has two elements. By Proposition 7.32, $W(G, A)$ has two elements. Let $\tilde{w} \in N_K(\mathfrak{a}_0)$ represent the nontrivial element of $W(G, A)$. By the Bruhat decomposition (Theorem 7.40),

$$G = MAN \cup MAN\tilde{w}MAN = MAN \cup N\tilde{w}MAN. \tag{7.68}$$

Since $\mathrm{Ad}(\tilde{w})^{-1}$ acts as -1 on $\mathfrak{a}_0$, it sends the positive restricted roots to the negative restricted roots, and it follows from Proposition 6.40c that $\mathrm{Ad}(\tilde{w})^{-1}\mathfrak{n}_0 = \mathfrak{n}_0^-$. Therefore $\tilde{w}^{-1}N\tilde{w} = N^-$. Multiplying (7.68) on the left by $\tilde{w}^{-1}$, we obtain

$$G = \tilde{w}MAN \cup N^-MAN.$$

Hence G/MAN is the disjoint union of the single point $\tilde{w}MAN$ and the image of the map (7.67).

We have seen that (7.67) is a diffeomorphism onto an open subset of G/MAN. Lemma 7.65 shows that G/MAN is diffeomorphic to K/M. Since Theorem 1.104 shows that N^- is diffeomorphic to Euclidean space, K/M is a one-point compactification of a Euclidean space, hence a sphere. Since K is connected, M must be connected whenever K/M is simply connected, i.e., whenever $\dim K/M > 1$. Since $\dim K/M = \dim \mathfrak{n}_0$, M is connected unless $\dim \mathfrak{n}_0 = 1$.

Corollary 7.69. Suppose that the reductive Lie group G is semisimple and has a complexification $G^{\mathbb{C}}$. Let $\alpha \in \Delta(\mathfrak{g}, \mathfrak{a} \oplus \mathfrak{t})$ be a real root. If the positive multiples of the restricted root $\alpha|_{\mathfrak{a}_0}$ have combined restricted-root multiplicity greater than one, then γ_α is in M_0.

PROOF. The element γ_α is in the homomorphic image of $SL(2, \mathbb{R})$ associated to the root α, hence is in the subgroup $G' = {}^0Z_G(H_\alpha^\perp)_0$. Consequently it is in the M subgroup of G'. The subgroup G' satisfies the hypotheses of Theorem 7.66, and its $\mathfrak{n}_0$ has dimension >1 by hypothesis. By Theorem 7.66 its M subgroup is connected. Hence γ_α is in the identity component of the M subgroup for G.

7. Parabolic Subgroups

In this section G will denote a reductive Lie group, and we shall use the other notation of §2 concerning the Cartan decomposition. But we shall abandon the use of $\mathfrak{a}_0$ as a maximal abelian subspace of $\mathfrak{p}_0$, as well as the other notation connected with the Iwasawa decomposition. Instead of using the symbols $\mathfrak{a}_0$, $\mathfrak{n}_0$, $\mathfrak{m}_0$, $\mathfrak{a}$, $\mathfrak{n}$, $\mathfrak{m}$, A, N, and M for these objects, we shall use the symbols $\mathfrak{a}_{\mathfrak{p},0}$, $\mathfrak{n}_{\mathfrak{p},0}$, $\mathfrak{m}_{\mathfrak{p},0}$, $\mathfrak{a}_\mathfrak{p}$, $\mathfrak{n}_\mathfrak{p}$, $\mathfrak{m}_\mathfrak{p}$, $A_\mathfrak{p}$, $N_\mathfrak{p}$, and $M_\mathfrak{p}$.

Our objective is to define and characterize "parabolic subgroups" of G, first working with "parabolic subalgebras" of $\mathfrak{g}_0$. Each parabolic subgroup Q will have a canonical decomposition in the form $Q = MAN$, known as the "Langlands decomposition" of Q. As we suggested at the start of §2, a number of arguments with reductive Lie groups are carried out by induction on the dimension of the group. One way of implementing this idea is to reduce proofs from G to the M of some parabolic subgroup. For such a procedure to succeed, we build into the definition of M the fact that M is a reductive Lie group.

In developing our theory, one approach would be to define a parabolic subalgebra of $\mathfrak{g}_0$ to be a subalgebra whose complexification is a parabolic subalgebra of $\mathfrak{g}$. Then we could deduce properties of parabolic subalgebras of $\mathfrak{g}_0$ from the theory in §V.7. But it will be more convenient to work with parabolic subalgebras of $\mathfrak{g}_0$ directly, proving results by imitating the theory of §V.7, rather than by applying it.

A **minimal parabolic subalgebra** of $\mathfrak{g}_0$ is any subalgebra of $\mathfrak{g}_0$ that is conjugate to $\mathfrak{q}_{\mathfrak{p},0} = \mathfrak{m}_{\mathfrak{p},0} \oplus \mathfrak{a}_{\mathfrak{p},0} \oplus \mathfrak{n}_{\mathfrak{p},0}$ via $\text{Ad}(G)$. Because of the Iwasawa decomposition $G = KA_{\mathfrak{p}}N_{\mathfrak{p}}$, we may as well assume that the conjugacy is via $\text{Ad}(K)$. The subalgebra $\mathfrak{q}_{\mathfrak{p},0}$ contains the maximally noncompact θ stable Cartan subalgebra $\mathfrak{a}_{\mathfrak{p},0} \oplus \mathfrak{t}_{\mathfrak{p},0}$, where $\mathfrak{t}_{\mathfrak{p},0}$ is any maximal abelian subspace of $\mathfrak{m}_{\mathfrak{p},0}$, and $\text{Ad}(k)$ sends any such Cartan subalgebra into another such Cartan subalgebra if k is in K. Hence every minimal parabolic subalgebra of $\mathfrak{g}_0$ contains a maximally noncompact θ stable Cartan subalgebra of $\mathfrak{g}_0$. A **parabolic subalgebra** $\mathfrak{q}_0$ of $\mathfrak{g}_0$ is a Lie subalgebra containing some minimal parabolic subalgebra. A parabolic subalgebra must contain a maximally noncompact θ stable Cartan subalgebra of $\mathfrak{g}_0$.

Therefore there is no loss of generality in assuming that $\mathfrak{q}_0$ contains a minimal parabolic subalgebra of the form $\mathfrak{m}_{\mathfrak{p},0} \oplus \mathfrak{a}_{\mathfrak{p},0} \oplus \mathfrak{n}_{\mathfrak{p},0}$, where $\mathfrak{a}_{\mathfrak{p},0}$ is maximal abelian in $\mathfrak{p}_0$, and $\mathfrak{m}_{\mathfrak{p},0}$ and $\mathfrak{n}_{\mathfrak{p},0}$ are constructed are usual. Let Σ denote the set of restricted roots of $\mathfrak{g}_0$ relative to $\mathfrak{a}_{\mathfrak{p},0}$. The restricted roots contributing to $\mathfrak{n}_{\mathfrak{p},0}$ are taken to be the positive ones.

We can obtain examples of parabolic subalgebras as follows. Let Π be the set of simple restricted roots, fix a subset Π' of Π, and let

$$\Gamma = \Sigma^+ \cup \{\beta \in \Sigma \mid \beta \in \text{span}(\Pi')\}. \tag{7.70}$$

Then

$$\mathfrak{q}_0 = \mathfrak{a}_{\mathfrak{p},0} \oplus \mathfrak{m}_{\mathfrak{p},0} \oplus \bigoplus_{\beta \in \Gamma} (\mathfrak{g}_0)_\beta \tag{7.71}$$

is a parabolic subalgebra of $\mathfrak{g}_0$ containing $\mathfrak{m}_{\mathfrak{p},0} \oplus \mathfrak{a}_{\mathfrak{p},0} \oplus \mathfrak{n}_{\mathfrak{p},0}$. This construction is an analog of the corresponding construction of parabolic subalgebras of $\mathfrak{g}$ given in (5.88) and (5.89), and Proposition 7.76 will show that every parabolic subalgebra of $\mathfrak{g}_0$ is of the form given in (7.70) and (7.71). But the proof requires more preparation than in the situation with (5.88) and (5.89).

EXAMPLES.

1) Let $G = SL(n, \mathbb{K})$, where $\mathbb{K}$ is $\mathbb{R}$, $\mathbb{C}$, or $\mathbb{H}$. When $\mathfrak{g}_0$ is realized as matrices, the Lie subalgebra of upper-triangular matrices is a minimal parabolic subalgebra $\mathfrak{q}_{\mathfrak{p},0}$. The other examples of parabolic subalgebras $\mathfrak{q}_0$ containing $\mathfrak{q}_{\mathfrak{p},0}$ and written as in (7.70) and (7.71) are the Lie subalgebras of block upper-triangular matrices, one subalgebra for each arrangement of blocks.

2) Let G have compact center and be of real rank one. The examples as in (7.70) and (7.71) are the minimal parabolic subalgebras and $\mathfrak{g}_0$ itself.

We shall work with a vector X in the restricted-root space $(\mathfrak{g}_0)_\gamma$ and with θX in $(\mathfrak{g}_0)_{-\gamma}$. (See Proposition 6.40c.) Proposition 6.52 shows that $B(X, \theta X)$ is a negative multiple of H_γ. Normalizing, we may assume that $B(X, \theta X) = -2/|\gamma|^2$. Put $H'_\gamma = 2|\gamma|^{-2} H_\gamma$. Then the linear span $\mathfrak{sl}_X$ of $\{X, \theta X, H'_\gamma\}$ is isomorphic to $\mathfrak{s}(2, \mathbb{R})$ under the isomorphism

$$H'_\gamma \leftrightarrow h, \qquad X \leftrightarrow e, \qquad \theta X \leftrightarrow -f. \tag{7.72}$$

We shall make use of the copy $\mathfrak{sl}_X$ of $\mathfrak{sl}(2, \mathbb{R})$ in the same way as in the proof of Corollary 6.53. This subalgebra of $\mathfrak{g}_0$ acts by ad on $\mathfrak{g}_0$ and hence acts on $\mathfrak{g}$. We know from Theorem 1.64 that the resulting representation of $\mathfrak{sl}_X$ is completely reducible, and we know the structure of each irreducible subspace from Theorem 1.63.

Lemma 7.73. Let γ be a restricted root, and let $X \neq 0$ be in $(\mathfrak{g}_0)_\gamma$. Then

(a) $\operatorname{ad} X$ carries $(\mathfrak{g}_0)_\gamma$ onto $(\mathfrak{g}_0)_{2\gamma}$

(b) $(\operatorname{ad} \theta X)^2$ carries $(\mathfrak{g}_0)_\gamma$ onto $(\mathfrak{g}_0)_{-\gamma}$

(c) $(\operatorname{ad} \theta X)^4$ carries $(\mathfrak{g}_0)_{2\gamma}$ onto $(\mathfrak{g}_0)_{-2\gamma}$.

PROOF. Without loss of generality, we may assume that X is normalized as in (7.72). The complexification of $\bigoplus_{c \in \mathbb{Z}} (\mathfrak{g}_0)_{c\gamma}$ is an invariant subspace of $\mathfrak{g}$ under the representation ad of $\mathfrak{sl}_X$. Using Theorem 1.64, we decompose it as the direct sum of irreducible representations. Each member of $(\mathfrak{g}_0)_{c\gamma}$ is an eigenvector for ad H'_γ with eigenvalue $2c$, and H'_γ corresponds to the member h of $\mathfrak{sl}(2, \mathbb{R})$. From Theorem 1.63 we see that the only possibilities for irreducible subspaces are 5-dimensional subspaces consisting of one dimension each from

$$(\mathfrak{g}_0)_{2\gamma},\ (\mathfrak{g}_0)_\gamma,\ \mathfrak{m}_0,\ (\mathfrak{g}_0)_{-\gamma},\ (\mathfrak{g}_0)_{-2\gamma};$$

3-dimensional subspaces consisting of one dimension each from

$$(\mathfrak{g}_0)_\gamma,\ \mathfrak{m}_0,\ (\mathfrak{g}_0)_{-\gamma};$$

and 1-dimensional subspaces consisting of one dimension each from $\mathfrak{m}_0$. In any 5-dimensional such subspace, $\operatorname{ad} X$ carries a nonzero vector of eigenvalue 2 to a nonzero vector of eigenvalue 4. This proves (a). Also in any 5-dimensional such subspace, $(\operatorname{ad}\theta X)^4$ carries a nonzero vector of eigenvalue 4 to a nonzero vector of eigenvalue -4. This proves (c). Finally in any 5-dimensional such subspace or 3-dimensional such subspace, $(\operatorname{ad}\theta X)^2$ carries a nonzero vector of eigenvalue 2 to a nonzero vector of eigenvalue -2. This proves (b).

Lemma 7.74. Every parabolic subalgebra $\mathfrak{q}_0$ of $\mathfrak{g}_0$ containing $\mathfrak{m}_{\mathfrak{p},0} \oplus \mathfrak{a}_{\mathfrak{p},0} \oplus \mathfrak{n}_{\mathfrak{p},0}$ is of the form

$$\mathfrak{q}_0 = \mathfrak{a}_{\mathfrak{p},0} \oplus \mathfrak{m}_{\mathfrak{p},0} \oplus \bigoplus_{\beta\in\Gamma} (\mathfrak{g}_0)_\beta$$

for some subset Γ of Σ that contains Σ^+.

PROOF. Since $\mathfrak{q}_0$ contains $\mathfrak{a}_{\mathfrak{p},0} \oplus \mathfrak{m}_{\mathfrak{p},0}$ and is invariant under $\operatorname{ad}(\mathfrak{a}_{\mathfrak{p},0})$, it is of the form

$$\mathfrak{q}_0 = \mathfrak{a}_{\mathfrak{p},0} \oplus \mathfrak{m}_{\mathfrak{p},0} \oplus \bigoplus_{\beta\in\Sigma} ((\mathfrak{g}_0)_\beta \cap \mathfrak{q}_0).$$

Thus we are to show that if $\mathfrak{q}_0$ contains one nonzero vector Y of $(\mathfrak{g}_0)_\beta$, then it contains all of $(\mathfrak{g}_0)_\beta$. Since $\mathfrak{q}_0$ contains $\mathfrak{n}_{\mathfrak{p},0}$, we may assume that β is negative. We apply Lemma 7.73b with $X = \theta Y$ and $\gamma = -\beta$. The lemma says that $(\operatorname{ad} Y)^2$ carries $(\mathfrak{g}_0)_{-\beta}$ onto $(\mathfrak{g}_0)_\beta$. Since Y and $(\mathfrak{g}_0)_{-\beta}$ are contained in $\mathfrak{q}_0$, so is $(\mathfrak{g}_0)_\beta$.

Lemma 7.75. If β, γ, and $\beta+\gamma$ are restricted roots and X is a nonzero member of $(\mathfrak{g}_0)_\gamma$, then $[X, (\mathfrak{g}_0)_\beta]$ is a nonzero subspace of $(\mathfrak{g}_0)_{\beta+\gamma}$.

PROOF. Without loss of generality, we may assume that X is normalized as in (7.72). The complexification of $\bigoplus_{c\in\mathbb{Z}} (\mathfrak{g}_0)_{\beta+c\gamma}$ is an invariant subspace of $\mathfrak{g}$ under the representation ad of $\mathfrak{sl}_X$. Using Theorem 1.64, we decompose it as the direct sum of irreducible representations. Each member of $(\mathfrak{g}_0)_{\beta+c\gamma}$ is an eigenvector for $\operatorname{ad} H'_\gamma$ with eigenvalue $\frac{2\langle\beta,\gamma\rangle}{|\gamma|^2}+2c$, and H'_γ corresponds to the member h of $\mathfrak{sl}(2,\mathbb{R})$. We apply Theorem 1.63 and divide matters into cases according to the sign of $\frac{2\langle\beta,\gamma\rangle}{|\gamma|^2}$. If the sign is < 0, then $\operatorname{ad} X$ is one-one on $(\mathfrak{g}_0)_\beta$, and the lemma follows. If the sign is ≥ 0, then $\operatorname{ad}\theta X$ and $\operatorname{ad} X \operatorname{ad}\theta X$ are one-one on $(\mathfrak{g}_0)_\beta$, and hence $\operatorname{ad} X$ is nonzero on the member $[\theta X, Y]$ if Y is nonzero in $(\mathfrak{g}_0)_{\beta+\gamma}$.

Proposition 7.76. The parabolic subalgebras $\mathfrak{q}_0$ containing $\mathfrak{m}_{\mathfrak{p},0}\oplus\mathfrak{a}_{\mathfrak{p},0}\oplus\mathfrak{n}_{\mathfrak{p},0}$ are parametrized by the set of subsets of simple restricted roots; the one corresponding to a subset Π' is of the form (7.71) with Γ as in (7.70).

PROOF. Lemma 7.74 establishes that any $\mathfrak{q}_0$ is of the form (7.71) for some subset Γ. We can now go over the proof of Proposition 5.90 to see that it applies. What is needed is a substitute for Corollary 2.35, which says that $[\mathfrak{g}_\beta, \mathfrak{g}_\gamma] = \mathfrak{g}_{\beta+\gamma}$ if β, γ, and $\beta+\gamma$ are all roots. Lemma 7.75 provides the appropriate substitute, and the proposition follows.

In the notation of the proposition, $\Gamma \cap -\Gamma$ consists of all restricted roots in the span of Π', and the other members of Γ are all positive and have expansions in terms of simple restricted roots that involve a simple restricted root not in Π'. Define

$$\begin{aligned}
\mathfrak{a}_0 &= \bigcap_{\beta\in\Gamma\cap-\Gamma} \ker\beta \subseteq \mathfrak{a}_{\mathfrak{p},0} \\
\mathfrak{a}_{M,0} &= \mathfrak{a}_0^{\perp} \subseteq \mathfrak{a}_{\mathfrak{p},0} \\
\mathfrak{m}_0 &= \mathfrak{a}_{M,0} \oplus \mathfrak{m}_{\mathfrak{p},0} \oplus \bigoplus_{\beta\in\Gamma\cap-\Gamma} (\mathfrak{g}_0)_\beta \\
\mathfrak{n}_0 &= \bigoplus_{\substack{\beta\in\Gamma,\\ \beta\notin-\Gamma}} (\mathfrak{g}_0)_\beta \\
\mathfrak{n}_{M,0} &= \mathfrak{n}_{\mathfrak{p},0} \cap \mathfrak{m}_0,
\end{aligned} \tag{7.77a}$$

so that

$$\mathfrak{q}_0 = \mathfrak{m}_0 \oplus \mathfrak{a}_0 \oplus \mathfrak{n}_0. \tag{7.77b}$$

The decomposition (7.77b) is called the **Langlands decomposition** of $\mathfrak{q}_0$.

EXAMPLE. Let $G = SU(2,2)$. The Lie algebra $\mathfrak{g}_0$ consists of all 4-by-4 complex matrices of the block form

$$\begin{pmatrix} X_{11} & X_{12} \\ X_{12}^* & X_{22} \end{pmatrix}$$

with X_{11} and X_{22} skew Hermitian and the total trace equal to 0. We take the Cartan involution to be negative conjugate transpose, so that

$$\mathfrak{k}_0 = \left\{\begin{pmatrix} X_{11} & 0 \\ 0 & X_{22} \end{pmatrix}\right\} \qquad \text{and} \qquad \mathfrak{p}_0 = \left\{\begin{pmatrix} 0 & X_{12} \\ X_{12}^* & 0 \end{pmatrix}\right\}.$$

Let us take

$$\mathfrak{a}_{\mathfrak{p},0} = \left\{ \begin{pmatrix} 0 & 0 & s & 0 \\ 0 & 0 & 0 & t \\ s & 0 & 0 & 0 \\ 0 & t & 0 & 0 \end{pmatrix} \;\middle|\; s \text{ and } t \text{ in } \mathbb{R} \right\}.$$

Define linear functionals f_1 and f_2 on $\mathfrak{a}_{\mathfrak{p},0}$ by saying that f_1 of the above matrix is s and f_2 of the matrix is t. Then

$$\Sigma = \{\pm f_1 \pm f_2,\ \pm 2f_1,\ \pm 2f_2\},$$

which is a root system of type C_2. Here $\pm f_1 \pm f_2$ have multiplicity 2, and the others have multiplicity one. In the obvious ordering, Σ^+ consists of $f_1 \pm f_2$ and $2f_1$ and $2f_2$, and the simple restricted roots are $f_1 - f_2$ and $2f_2$. Then

$$\mathfrak{m}_{\mathfrak{p},0} = \{\operatorname{diag}(ir, -ir, ir, -ir)\}$$

$$\mathfrak{n}_{\mathfrak{p},0} = \bigoplus_{\beta \in \Sigma^+} (\mathfrak{g}_0)_\beta \quad \text{with } \dim \mathfrak{n}_{\mathfrak{p},0} = 6.$$

Our minimal parabolic subalgebra is $\mathfrak{q}_{\mathfrak{p},0} = \mathfrak{m}_{\mathfrak{p},0} \oplus \mathfrak{a}_{\mathfrak{p},0} \oplus \mathfrak{n}_{\mathfrak{p},0}$, and this is reproduced as $\mathfrak{q}_0$ by (7.70) and (7.71) with $\Pi' = \emptyset$. When $\Pi' = \{f_1 - f_2, 2f_2\}$, then $\mathfrak{q}_0 = \mathfrak{g}_0$. The two intermediate cases are as follows. If $\Pi' = \{f_1 - f_2\}$, then

$$\mathfrak{a}_0 = \{H \in \mathfrak{a}_{\mathfrak{p},0} \mid (f_1 - f_2)(H) = 0\} \qquad (s = t \text{ in } \mathfrak{a}_{\mathfrak{p},0})$$

$$\mathfrak{m}_0 = \left\{ \begin{pmatrix} ir & w & x & z \\ -\bar{w} & -ir & \bar{z} & -x \\ x & z & ir & w \\ \bar{z} & -x & -\bar{w} & -ir \end{pmatrix} \;\middle|\; x, r \in \mathbb{R} \text{ and } w, z \in \mathbb{C} \right\}$$

$$\mathfrak{n}_0 = (\mathfrak{g}_0)_{2f_1} \oplus (\mathfrak{g}_0)_{f_1+f_2} \oplus (\mathfrak{g}_0)_{2f_2}.$$

If $\Pi' = \{2f_2\}$, then

$$\mathfrak{a}_0 = \{H \in \mathfrak{a}_{\mathfrak{p},0} \mid 2f_2(H) = 0\} \qquad (t = 0 \text{ in } \mathfrak{a}_{\mathfrak{p},0})$$

$$\mathfrak{m}_0 = \mathfrak{m}_{\mathfrak{p},0} \oplus \left\{ \begin{pmatrix} 0 & 0 & 0 & 0 \\ 0 & is & 0 & z \\ 0 & 0 & 0 & 0 \\ 0 & \bar{z} & 0 & -is \end{pmatrix} \;\middle|\; s \in \mathbb{R} \text{ and } z \in \mathbb{C} \right\}$$

$$\mathfrak{n}_0 = (\mathfrak{g}_0)_{2f_1} \oplus (\mathfrak{g}_0)_{f_1+f_2} \oplus (\mathfrak{g}_0)_{f_1-f_2}.$$

Proposition 7.76 says that there are no other parabolic subalgebras $\mathfrak{q}_0$ containing $\mathfrak{q}_{\mathfrak{p},0}$.

Proposition 7.78. A parabolic subalgebra $\mathfrak{q}_0$ containing the minimal parabolic subalgebra $\mathfrak{m}_{\mathfrak{p},0} \oplus \mathfrak{a}_{\mathfrak{p},0} \oplus \mathfrak{n}_{\mathfrak{p},0}$ has the properties that

(a) $\mathfrak{m}_0$, $\mathfrak{a}_0$, and $\mathfrak{n}_0$ are Lie subalgebras, and $\mathfrak{n}_0$ is an ideal in $\mathfrak{q}_0$
(b) $\mathfrak{a}_0$ is abelian, and $\mathfrak{n}_0$ is nilpotent
(c) $\mathfrak{a}_0 \oplus \mathfrak{m}_0$ is the centralizer of $\mathfrak{a}_0$ in $\mathfrak{g}_0$
(d) $\mathfrak{q}_0 \cap \theta\mathfrak{q}_0 = \mathfrak{a}_0 \oplus \mathfrak{m}_0$, and $\mathfrak{a}_0 \oplus \mathfrak{m}_0$ is reductive
(e) $\mathfrak{a}_{\mathfrak{p},0} = \mathfrak{a}_0 \oplus \mathfrak{a}_{M,0}$
(f) $\mathfrak{n}_{\mathfrak{p},0} = \mathfrak{n}_0 \oplus \mathfrak{n}_{M,0}$ as vector spaces
(g) $\mathfrak{g}_0 = \mathfrak{a}_0 \oplus \mathfrak{m}_0 \oplus \mathfrak{n}_0 \oplus \theta\mathfrak{n}_0$ orthogonally with respect to θ
(h) $\mathfrak{m}_0 = \mathfrak{m}_{\mathfrak{p},0} \oplus \mathfrak{a}_{M,0} \oplus \mathfrak{n}_{M,0} \oplus \theta\mathfrak{n}_{M,0}$.

PROOF.

(a, b, e, f) All parts of these are clear.

(c) The centralizer of $\mathfrak{a}_0$ is spanned by $\mathfrak{a}_{\mathfrak{p},0}$, $\mathfrak{m}_{\mathfrak{p},0}$, and all the restricted root spaces for restricted roots vanishing on $\mathfrak{a}_0$. The sum of these is $\mathfrak{a}_0 \oplus \mathfrak{m}_0$.

(d) Since $\theta(\mathfrak{g}_0)_\beta = (\mathfrak{g}_0)_{-\beta}$ by Proposition 6.40c, $\mathfrak{q}_0 \cap \theta\mathfrak{q}_0 = \mathfrak{a}_0 \oplus \mathfrak{m}_0$. Then $\mathfrak{a}_0 \oplus \mathfrak{m}_0$ is reductive by Corollary 1.53.

(g, h) These follow from Proposition 6.40.

Proposition 7.79. Among the parabolic subalgebras containing $\mathfrak{q}_{\mathfrak{p},0}$, let $\mathfrak{q}_0$ be the one corresponding to the subset Π' of simple restricted roots. For $\eta \neq 0$ in $\mathfrak{a}_0^*$, let

$$(\mathfrak{g}_0)_{(\eta)} = \bigoplus_{\substack{\beta \in \mathfrak{a}_{\mathfrak{p},0}^*, \\ \beta|_{\mathfrak{a}_0} = \eta_0}} (\mathfrak{g}_0)_\beta .$$

Then $(\mathfrak{g}_0)_{(\eta)} \subseteq \mathfrak{n}_0$ or $(\mathfrak{g}_0)_{(\eta)} \subseteq \theta\mathfrak{n}_0$.

PROOF. We have

$$\mathfrak{a}_{M,0} = \mathfrak{a}_0^\perp = \Big(\bigcap_{\beta \in \Gamma \cap -\Gamma} \ker \beta\Big)^\perp = \Big(\bigcap_{\beta \in \Gamma \cap -\Gamma} H_\beta^\perp\Big)^\perp = \sum_{\beta \in \Gamma \cap -\Gamma} \mathbb{R} H_\beta = \sum_{\beta \in \Pi'} \mathbb{R} H_\beta .$$

Let β and β' be restricted roots with a common nonzero restriction η to members of $\mathfrak{a}_0$. Then $\beta - \beta'$ is 0 on $\mathfrak{a}_0$, and $H_\beta - H_{\beta'}$ is in $\mathfrak{a}_{M,0}$. From the formula for $\mathfrak{a}_{M,0}$, the expansion of $\beta - \beta'$ in terms of simple restricted roots involves only the members of Π'. Since $\eta \neq 0$, the individual expansions of β and β' involve nonzero coefficients for at least one simple restricted root other than the ones in Π'. The coefficients for this other simple restricted root must be equal and in particular of the same sign. By Proposition 2.49, β and β' are both positive or both negative, and the result follows.

Motivated by Proposition 7.79, we define, for $\eta \in \mathfrak{a}_0^*$,

$$(\mathfrak{g}_0)_{(\eta)} = \{X \in \mathfrak{g}_0 \mid [H, X] = \eta(H)X \text{ for all } H \in \mathfrak{a}_0\}. \tag{7.80}$$

We say that η is an $\mathfrak{a}_0$ **root**, or root of $(\mathfrak{g}_0, \mathfrak{a}_0)$, if $\eta \neq 0$ and $(\mathfrak{g}_0)_{(\eta)} \neq 0$. In this case we call $(\mathfrak{g}_0)_{(\eta)}$ the corresponding $\mathfrak{a}_0$ **root space**. The proposition says that $\mathfrak{n}_0$ is the sum of $\mathfrak{a}_0$ root spaces, and so is $\theta\mathfrak{n}_0$. We call an $\mathfrak{a}_0$ root **positive** if it contributes to $\mathfrak{n}_0$, otherwise **negative**. The set of $\mathfrak{a}_0$ roots does not necessarily form an abstract root system, but the notion of an $\mathfrak{a}_0$ root is still helpful.

Corollary 7.81. The normalizer of $\mathfrak{a}_0$ in $\mathfrak{g}_0$ is $\mathfrak{a}_0 \oplus \mathfrak{m}_0$.

PROOF. The normalizer contains $\mathfrak{a}_0 \oplus \mathfrak{m}_0$ by Proposition 7.78c. In the reverse direction let X be in the normalizer, and write

$$X = H_0 + X_0 + \sum_{\substack{\eta\neq 0,\\ \eta\in\mathfrak{a}_0^*}} X_\eta \qquad \text{with } H_0 \in \mathfrak{a}_0,\ X_0 \in \mathfrak{m}_0,\ X_\eta \in (\mathfrak{g}_0)_{(\eta)}.$$

If H is in $\mathfrak{a}_0$, then $[X, H] = -\sum_\eta \eta(H)X_\eta$, and this can be in $\mathfrak{a}_0$ for all such H only if $X_\eta = 0$ for all η. Therefore $X = H_0 + X_0$ is in $\mathfrak{a}_0 \oplus \mathfrak{m}_0$.

Now let A and N be the analytic subgroups of G with Lie algebras $\mathfrak{a}_0$ and $\mathfrak{n}_0$, and define $M = {}^0Z_G(\mathfrak{a}_0)$. We shall see in Proposition 7.83 below that $Q = MAN$ is the normalizer of $\mathfrak{m}_0 \oplus \mathfrak{a}_0 \oplus \mathfrak{n}_0$ in G, and we define it to be the **parabolic subgroup** associated to the parabolic subalgebra $\mathfrak{q}_0 = \mathfrak{m}_0 \oplus \mathfrak{a}_0 \oplus \mathfrak{n}_0$. The decomposition of elements of Q according to MAN will be seen to be unique, and $Q = MAN$ is called the **Langlands decomposition** of Q. When $\mathfrak{q}_0$ is a minimal parabolic subalgebra, the corresponding Q is called a **minimal parabolic subgroup**. We write $N^- = \Theta N$.

Let A_M and N_M be the analytic subgroups of $\mathfrak{g}_0$ with Lie algebras $\mathfrak{a}_{M,0}$ and $\mathfrak{n}_{M,0}$, and let $M_M = Z_{K\cap M}(\mathfrak{a}_{M,0})$. Define $K_M = K \cap M$. Recall the subgroup F of G that is the subject of Corollary 7.52.

Proposition 7.82. The subgroups M, A, N, K_M, M_M, A_M, and N_M have the properties that

(a) $MA = Z_G(\mathfrak{a}_0)$ is reductive, $M = {}^0(MA)$ is reductive, and A is Z_{vec} for MA
(b) M has Lie algebra $\mathfrak{m}_0$
(c) $M_M = M_\mathfrak{p}$, $M_{\mathfrak{p},0}A_M N_M$ is a minimal parabolic subgroup of M, and $M = K_M A_M N_M$
(d) $M = FM_0$ if G is connected
(e) $A_\mathfrak{p} = AA_M$ as a direct product
(f) $N_\mathfrak{p} = NN_M$ as a semidirect product with N normal.

PROOF.

(a, b) The subgroups $Z_G(\mathfrak{a}_0)$ and ${}^0Z_G(\mathfrak{a}_0)$ are reductive by Propositions 7.25 and 7.27. By Proposition 7.78, $Z_{\mathfrak{g}_0}(\mathfrak{a}_0) = \mathfrak{a}_0 \oplus \mathfrak{m}_0$. Thus the space Z_{vec} for the group $Z_G(\mathfrak{a}_0)$ is the analytic subgroup corresponding to the intersection of $\mathfrak{p}_0$ with the center of $\mathfrak{a}_0 \oplus \mathfrak{m}_0$. From the definition of $\mathfrak{m}_0$, the center of $Z_{\mathfrak{g}_0}(\mathfrak{a}_0)$ has to be contained in $\mathfrak{a}_{\mathfrak{p},0} \oplus \mathfrak{m}_{\mathfrak{p},0}$, and the $\mathfrak{p}_0$ part of this is $\mathfrak{a}_{\mathfrak{p},0}$. The part of $\mathfrak{a}_{\mathfrak{p},0}$ that commutes with $\mathfrak{m}_0$ is $\mathfrak{a}_0$ by definition of $\mathfrak{m}_0$. Therefore $Z_{vec} = \exp \mathfrak{a}_0 = A$, and $Z_G(\mathfrak{a}_0) = ({}^0Z_G(\mathfrak{a}_0))A$ by Proposition 7.27. Then (a) and (b) follow.

(c) By (a), M is reductive. It is clear that $\mathfrak{a}_{M,0}$ is a maximal abelian subspace of $\mathfrak{p}_0 \cap \mathfrak{m}_0$, since $\mathfrak{m}_0 \cap \mathfrak{a}_0 = 0$. The restricted roots of $\mathfrak{m}_0$ relative to $\mathfrak{a}_{M,0}$ are then the members of $\Gamma \cap -\Gamma$, and the sum of the restricted-root spaces for the positive such restricted roots is $\mathfrak{n}_{M,0}$. Therefore the minimal parabolic subgroup in question for M is $M_M A_M N_M$. The computation

$$\begin{aligned} M_M = Z_{K \cap M}(\mathfrak{a}_{M,0}) &= MA \cap Z_K(\mathfrak{a}_{M,0}) \\ &= Z_G(\mathfrak{a}_0) \cap Z_K(\mathfrak{a}_{M,0}) = Z_K(\mathfrak{a}_{\mathfrak{p},0}) = M_{\mathfrak{p}} \end{aligned}$$

identifies M_M, and $M = K_M A_M N_M$ by the Iwasawa decomposition for M (Proposition 7.31).

(d) By (a), M is reductive. Hence $M = M_M M_0$ by Proposition 7.33. But (c) shows that $M_M = M_{\mathfrak{p}}$, and Corollary 7.52 shows that $M_{\mathfrak{p}} = F(M_{\mathfrak{p}})_0$. Hence $M = FM_0$.

(e) This follows from Proposition 7.78e and the simple connectivity of $A_{\mathfrak{p}}$.

(f) This follows from Proposition 7.78f, Theorem 1.102, and the simple connectivity of $N_{\mathfrak{p}}$.

Proposition 7.83. The subgroups M, A, and N have the properties that

(a) MA normalizes N, so that $Q = MAN$ is a group
(b) $Q = N_G(\mathfrak{m}_0 \oplus \mathfrak{a}_0 \oplus \mathfrak{n}_0)$, and hence Q is a closed subgroup
(c) Q has Lie algebra $\mathfrak{q}_0 = \mathfrak{m}_0 \oplus \mathfrak{a}_0 \oplus \mathfrak{n}_0$
(d) multiplication $M \times A \times N \to Q$ is a diffeomorphism
(e) $N^- \cap Q = \{1\}$
(f) $G = KQ$.

PROOF.

(a) Let z be in $MA = Z_G(\mathfrak{a}_0)$, and fix $(\mathfrak{g}_0)_{(\eta)} \subseteq \mathfrak{n}_0$ as in (7.80). If X is in $(\mathfrak{g}_0)_{(\mathfrak{n}_0)}$ and H is in $\mathfrak{a}_0$, then

$$[H, \mathrm{Ad}(z)X] = [\mathrm{Ad}(z)H, \mathrm{Ad}(z)X] = \mathrm{Ad}(z)[H, X] = \eta(H)\mathrm{Ad}(z)X.$$

Hence $\text{Ad}(z)X$ is in $(\mathfrak{g}_0)_{(\eta)}$, and $\text{Ad}(z)$ maps $(\mathfrak{g}_0)_{(\eta)}$ into itself. Since $\mathfrak{n}_0$ is the sum of such spaces, $\text{Ad}(z)\mathfrak{n}_0 \subseteq \mathfrak{n}_0$. Therefore MA normalizes N.

(b) The subgroup MA normalizes its Lie algebra $\mathfrak{m}_0 \oplus \mathfrak{a}_0$, and it normalizes $\mathfrak{n}_0$ by (a). The subgroup N normalizes $\mathfrak{q}_0$ because it is connected with a Lie algebra that normalizes $\mathfrak{q}_0$ by Proposition 7.78a. Hence MAN normalizes $\mathfrak{q}_0$. In the reverse direction let x be in $N_G(\mathfrak{q}_0)$. We are to prove that x is in MAN. Let us write x in terms of the Iwasawa decomposition $G = KA_\mathfrak{p}N_\mathfrak{p}$. Here $A_\mathfrak{p} = AA_M$ by Proposition 7.82e, and A and A_M are both contained in MA. Also $N_\mathfrak{p} = NN_M$ by Proposition 7.82f, and N and N_M are both contained in MN. Thus we may assume that x is in $N_K(\mathfrak{q}_0)$. By (7.23), $\text{Ad}(\Theta x) = \theta\text{Ad}(x)\theta$, and thus $\text{Ad}(\Theta x)$ normalizes $\theta\mathfrak{q}_0$. But $\Theta x = x$ since x is in K, and therefore $\text{Ad}(x)$ normalizes both $\mathfrak{q}_0$ and $\theta\mathfrak{q}_0$. By Proposition 7.78d, $\text{Ad}(x)$ normalizes $\mathfrak{a}_0 \oplus \mathfrak{m}_0$. Since $\mathfrak{a}_0$ is the $\mathfrak{p}_0$ part of the center of $\mathfrak{a}_0 \oplus \mathfrak{m}_0$, $\text{Ad}(x)$ normalizes $\mathfrak{a}_0$ and $\mathfrak{m}_0$ individually. Let η be an $\mathfrak{a}_0$ root contributing to $\mathfrak{n}_0$. If X is in $(\mathfrak{g}_0)_\eta$ and H is in $\mathfrak{a}_0$, then

$$\begin{aligned}[H, \text{Ad}(x)X] &= \text{Ad}(x)[\text{Ad}(x)^{-1}H, X] \\ &= \eta(\text{Ad}(x)^{-1}H)\text{Ad}(x)X = (\text{Ad}(x)\eta)(H)\text{Ad}(x)X.\end{aligned}$$

In other words, $\text{Ad}(x)$ carries $(\mathfrak{g}_0)_{(\eta)}$ to $(\mathfrak{g}_0)_{(\text{Ad}(x)\eta)}$. So whenever η is the restriction to $\mathfrak{a}_0$ of a positive restricted root, so is $\text{Ad}(x)\eta$. Meanwhile, $\text{Ad}(x)$ carries $\mathfrak{a}_{M,0}$ to a maximal abelian subspace of $\mathfrak{p}_0 \cap \mathfrak{m}_0$, and Proposition 7.29 allows us to adjust it by some $\text{Ad}(k) \in \text{Ad}(K \cap M)$ so that $\text{Ad}(kx)\mathfrak{a}_{M,0} = \mathfrak{a}_{M,0}$. Taking Proposition 7.32 and Theorem 2.63 into account, we can choose $k' \in K \cap M$ so that $\text{Ad}(k'kx)$ is the identity on $\mathfrak{a}_{M,0}$. Then $\text{Ad}(k'kx)$ sends Σ^+ to itself. By Proposition 7.32 and Theorem 2.63, $\text{Ad}(k'kx)$ is the identity on $\mathfrak{a}_{\mathfrak{p},0}$ and in particular on $\mathfrak{a}_0$. Hence $k'kx$ is in M, and so is x. We conclude that $MAN = N_G(\mathfrak{q}_0)$, and consequently MAN is closed.

(c) By (b), Q is closed, hence Lie. The Lie algebra of Q is $N_{\mathfrak{g}_0}(\mathfrak{q}_0)$, which certainly contains $\mathfrak{q}_0$. In the reverse direction let $X \in \mathfrak{g}_0$ normalize $\mathfrak{q}_0$. Since $\mathfrak{a}_{\mathfrak{p},0}$ and $\mathfrak{n}_{\mathfrak{p},0}$ are contained in $\mathfrak{q}_0$, the Iwasawa decomposition on the Lie algebra level allows us to assume that X is in $\mathfrak{k}_0$. Since X normalizes $\mathfrak{q}_0$, θX normalizes $\theta\mathfrak{q}_0$. But $X = \theta X$, and hence X normalizes $\mathfrak{q}_0 \cap \theta\mathfrak{q}_0$, which is $\mathfrak{a}_0 \oplus \mathfrak{m}_0$ by Proposition 7.78d. Since $\mathfrak{a}_0$ is the $\mathfrak{p}_0$ part of the center of $\mathfrak{a}_0 \oplus \mathfrak{m}_0$, X normalizes $\mathfrak{a}_0$ and $\mathfrak{m}_0$ individually. By Corollary 7.81, X is in $\mathfrak{a}_0 \oplus \mathfrak{m}_0$.

(d) Use of Lemma 6.44 twice shows that the smooth map $M \times A \times N \to Q$ is regular on $M_0 \times A \times N$, and translation to M shows that it is regular everywhere. We are left with showing that it is one-one. Since $A \subseteq A_\mathfrak{p}$ and $N \subseteq N_\mathfrak{p}$, the uniqueness for the Iwasawa decomposition of G (Proposition 7.31) shows that it is enough to prove that $M \cap AN = \{1\}$. Given $m \in M$, let the Iwasawa decomposition of m

according to $M = K_M A_M N_M$ be $m = k_M a_M n_M$. If this element is to be in AN, then $k_M = 1$, a_M is in $A_M \cap A$, and n_M is in $N_M \cap N$, by uniqueness of the Iwasawa decomposition in G. But $A_M \cap A = \{1\}$ and $N_M \cap N = \{1\}$ by (e) and (f) of Proposition 7.82. Therefore $m = 1$, and we conclude that $M \cap AN = \{1\}$.

(e) This is proved in the same way as Lemma 7.64, which is stated for a minimal parabolic subgroup.

(f) Since $Q \supseteq A_\mathfrak{p} N_\mathfrak{p}$, $G = KQ$ by the Iwasawa decomposition for G (Proposition 7.31).

Although the set of $\mathfrak{a}_0$ roots does not necessarily form an abstract root system, it is still meaningful to define

$$W(G, A) = N_K(\mathfrak{a}_0)/Z_K(\mathfrak{a}_0), \tag{7.84a}$$

just as we did in the case that $\mathfrak{a}_0$ is maximal abelian in $\mathfrak{p}_0$. Corollary 7.81 and Proposition 7.78c show that $N_K(\mathfrak{a}_0)$ and $Z_K(\mathfrak{a}_0)$ both have $\mathfrak{k}_0 \cap \mathfrak{m}_0$ as Lie algebra. Hence $W(G, A)$ is a compact 0-dimensional group, and we conclude that $W(G, A)$ is finite. An alternate formula for $W(G, A)$ is

$$W(G, A) = N_G(\mathfrak{a}_0)/Z_G(\mathfrak{a}_0). \tag{7.84b}$$

The equality of the right sides of (7.84a) and (7.84b) is an immediate consequence of Lemma 7.22 and Corollary 7.81. To compute $N_K(\mathfrak{a}_0)$, it is sometimes handy to use the following proposition.

Proposition 7.85. Every element of $N_K(\mathfrak{a}_0)$ decomposes as a product zn, where n is in $N_K(\mathfrak{a}_{\mathfrak{p},0})$ and z is in $Z_K(\mathfrak{a}_0)$.

PROOF. Let k be in $N_K(\mathfrak{a}_0)$ and form $\mathrm{Ad}(k)\mathfrak{a}_{M,0}$. Since $\mathfrak{a}_{M,0}$ commutes with $\mathfrak{a}_0$, $\mathrm{Ad}(k)\mathfrak{a}_{M,0}$ commutes with $\mathrm{Ad}(k)\mathfrak{a}_0 = \mathfrak{a}_0$. By Proposition 7.78c, $\mathrm{Ad}(k)\mathfrak{a}_{M,0}$ is contained in $\mathfrak{a}_0 \oplus \mathfrak{m}_0$. Since $\mathfrak{a}_{M,0}$ is orthogonal to $\mathfrak{a}_0$ under B_θ, $\mathrm{Ad}(k)\mathfrak{a}_{M,0}$ is orthogonal to $\mathrm{Ad}(k)\mathfrak{a}_0 = \mathfrak{a}_0$. Hence $\mathrm{Ad}(k)\mathfrak{a}_{M,0}$ is contained in $\mathfrak{m}_0$ and therefore in $\mathfrak{p}_0 \cap \mathfrak{m}_0$. By Proposition 7.29 there exists z in $K \cap M$ with $\mathrm{Ad}(z)^{-1}\mathrm{Ad}(k)\mathfrak{a}_{M,0} = \mathfrak{a}_{M,0}$. Then $n = z^{-1}k$ is in $N_K(\mathfrak{a}_0)$ and in $N_K(\mathfrak{a}_{M,0})$, hence in $N_K(\mathfrak{a}_{\mathfrak{p},0})$.

EXAMPLE. Let $G = SL(3, \mathbb{R})$. Take $\mathfrak{a}_{\mathfrak{p},0}$ to be the diagonal subalgebra, and let $\Sigma^+ = \{f_1 - f_2,\ f_2 - f_3,\ f_1 - f_3\}$ in the notation of Example 1 of §VI.4. Define a parabolic subalgebra $\mathfrak{q}_0$ by using $\Pi' = \{f_1 - f_2\}$. The corresponding parabolic subgroup is the block upper-triangular group with blocks of sizes 2 and 1, respectively. The subalgebra $\mathfrak{a}_0$ equals $\{\mathrm{diag}(r, r, -2r)\}$. Suppose that w is in $W(G, A)$. Proposition 7.85 says that w extends to a member of $W(G, A_\mathfrak{p})$ leaving $\mathfrak{a}_0$ and $\mathfrak{a}_{M,0}$ individually stable. Here $W(G, A_\mathfrak{p}) = W(\Sigma)$, and the only member of $W(\Sigma)$ sending $\mathfrak{a}_0$ to itself is the identity. So $W(G, A) = \{1\}$.

The members of $W(G, A)$ act on set of the $\mathfrak{a}_0$ roots, and we have the following substitute for Theorem 2.63.

Proposition 7.86. The only member of $W(G, A)$ that leaves stable the set of positive $\mathfrak{a}_0$ roots is the identity.

PROOF. Let k be in $N_K(\mathfrak{a}_0)$. By assumption $\mathrm{Ad}(k)\mathfrak{n}_0 = \mathfrak{n}_0$. The centralizer of $\mathfrak{a}_0$ in $\mathfrak{g}_0$ is $\mathfrak{a}_0 \oplus \mathfrak{m}_0$ by Proposition 7.78c. If X is in this centralizer and if H is arbitrary in $\mathfrak{a}_0$, then

$$[H, \mathrm{Ad}(k)X] = \mathrm{Ad}(k)[\mathrm{Ad}(k)^{-1}H, X] = 0$$

shows that $\mathrm{Ad}(k)X$ is in the centralizer. Hence $\mathrm{Ad}(k)(\mathfrak{a}_0 \oplus \mathfrak{m}_0) = \mathfrak{a}_0 \oplus \mathfrak{m}_0$. By Proposition 7.83b, k is in MAN. By Proposition 7.82c and the uniqueness of the Iwasawa decomposition for G, k is in M. Therefore k is in $Z_K(\mathfrak{a}_0)$.

A parabolic subalgebra $\mathfrak{q}_0$ of $\mathfrak{g}_0$ and the corresponding parabolic subgroup $Q = MAN$ of G are said to be **cuspidal** if $\mathfrak{m}_0$ has a θ stable compact Cartan subalgebra, say $\mathfrak{t}_0$. In this case, $\mathfrak{h}_0 = \mathfrak{t}_0 \oplus \mathfrak{a}_0$ is a θ stable Cartan subalgebra of $\mathfrak{g}_0$. The restriction of a root in $\Delta(\mathfrak{g}, \mathfrak{h})$ to $\mathfrak{a}_0$ is an $\mathfrak{a}_0$ root if it is not 0, and we can identify $\Delta(\mathfrak{m}, \mathfrak{t})$ with the set of roots in $\Delta(\mathfrak{g}, \mathfrak{h})$ that vanish on $\mathfrak{a}$. Let us choose a positive system $\Delta^+(\mathfrak{m}, \mathfrak{t})$ for $\mathfrak{m}$ and extend it to a positive system $\Delta^+(\mathfrak{g}, \mathfrak{h})$ by saying that a root $\alpha \in \Delta(\mathfrak{g}, \mathfrak{h})$ with nonzero restriction to $\mathfrak{a}_0$ is positive if $\alpha|_{\mathfrak{a}_0}$ is a positive $\mathfrak{a}_0$ root. Let us decompose members α of $\mathfrak{h}^*$ according to their projections on $\mathfrak{a}^*$ and $\mathfrak{t}^*$ as $\alpha = \alpha_\mathfrak{a} + \alpha_\mathfrak{t}$. Now $\theta\alpha = -\alpha_\mathfrak{a} + \alpha_\mathfrak{t}$, and θ carries roots to roots. Hence if $\alpha_\mathfrak{a} + \alpha_\mathfrak{t}$ is a root, so is $\alpha_\mathfrak{a} - \alpha_\mathfrak{t}$.

The positive system $\Delta^+(\mathfrak{g}, \mathfrak{h})$ just defined is given by a lexicographic ordering that takes $\mathfrak{a}_0$ before $i\mathfrak{t}_0$. In fact, write the half sum of positive roots as $\delta = \delta_\mathfrak{a} + \delta_\mathfrak{t}$. The claim is that positivity is determined by inner products with the ordered set $\{\delta_\mathfrak{a}, \delta_\mathfrak{t}\}$ and that $\delta_\mathfrak{t}$ is equal to the half sum of the members of $\Delta^+(\mathfrak{m}, \mathfrak{t})$. To see this, let $\alpha = \alpha_\mathfrak{a} + \alpha_\mathfrak{t}$ be in $\Delta^+(\mathfrak{g}, \mathfrak{h})$. If $\alpha_\mathfrak{a} \neq 0$, then $\alpha_\mathfrak{a} - \alpha_\mathfrak{t}$ is in $\Delta^+(\mathfrak{g}, \mathfrak{h})$, and

$$\langle \alpha, \delta_\mathfrak{a} \rangle = \langle \alpha_\mathfrak{a}, \delta_\mathfrak{a} \rangle = \langle \alpha_\mathfrak{a}, \delta \rangle = \tfrac{1}{2}\langle \alpha_\mathfrak{a} + \alpha_\mathfrak{t}, \delta \rangle + \tfrac{1}{2}\langle \alpha_\mathfrak{a} - \alpha_\mathfrak{t}, \delta \rangle > 0.$$

Since the positive roots with nonzero restriction to $\mathfrak{a}$ cancel in pairs when added, we see that $\delta_\mathfrak{t}$ equals half the sum of the members of $\Delta^+(\mathfrak{m}, \mathfrak{t})$. Finally if $\alpha_\mathfrak{a} = 0$, then $\langle \alpha, \delta_\mathfrak{a} \rangle = 0$ and $\langle \alpha, \delta_\mathfrak{t} \rangle > 0$. Hence $\Delta^+(\mathfrak{g}, \mathfrak{h})$ is indeed given by a lexicographic ordering of the type described.

The next proposition gives a converse that tells a useful way to construct cuspidal parabolic subalgebras of $\mathfrak{g}_0$ directly.

Proposition 7.87. Let $\mathfrak{h}_0 = \mathfrak{t}_0 \oplus \mathfrak{a}_0$ be the decomposition of a θ stable Cartan subalgebra according to θ, and suppose that a lexicographic ordering taking $\mathfrak{a}_0$ before $i\mathfrak{t}_0$ is used to define a positive system $\Delta^+(\mathfrak{g}, \mathfrak{h})$. Define

$$\mathfrak{m}_0 = \mathfrak{g}_0 \cap \Big(\mathfrak{t} \oplus \bigoplus_{\substack{\alpha\in\Delta(\mathfrak{g},\mathfrak{h}),\\ \alpha|_{\mathfrak{a}}=0}} \mathfrak{g}_\alpha\Big)$$

and

$$\mathfrak{n}_0 = \mathfrak{g}_0 \cap \Big(\bigoplus_{\substack{\alpha\in\Delta^+(\mathfrak{g},\mathfrak{h}),\\ \alpha|_{\mathfrak{a}}\neq 0}} \mathfrak{g}_\alpha\Big).$$

Then $\mathfrak{q}_0 = \mathfrak{m}_0 \oplus \mathfrak{a}_0 \oplus \mathfrak{n}_0$ is the Langlands decomposition of a cuspidal parabolic subgroup of $\mathfrak{g}_0$.

PROOF. In view of the definitions, we have to relate $\mathfrak{q}_0$ to a minimal parabolic subalgebra. Let bar denote conjugation of $\mathfrak{g}$ with respect to $\mathfrak{g}_0$. If $\alpha = \alpha_\mathfrak{a} + \alpha_\mathfrak{t}$ is a root, let $\bar\alpha = -\theta\alpha = \alpha_\mathfrak{a} - \alpha_\mathfrak{t}$. Then $\overline{\mathfrak{g}_\alpha} = \mathfrak{g}_{\bar\alpha}$, and it follows that

$$\mathfrak{m} = \mathfrak{t} \oplus \bigoplus_{\substack{\alpha\in\Delta(\mathfrak{g},\mathfrak{h}),\\ \alpha|_{\mathfrak{a}}=0}} \mathfrak{g}_\alpha \qquad \text{and} \qquad \mathfrak{n} = \bigoplus_{\substack{\alpha\in\Delta^+(\mathfrak{g},\mathfrak{h}),\\ \alpha|_{\mathfrak{a}}\neq 0}} \mathfrak{g}_\alpha. \tag{7.88}$$

In particular, $\mathfrak{m}_0$ is θ stable, hence reductive. Let $\mathfrak{h}_{M,0} = \mathfrak{t}_{M,0} \oplus \mathfrak{a}_{M,0}$ be the decomposition of a maximally noncompact θ stable Cartan subalgebra of $\mathfrak{m}_0$ according to θ. Since Theorem 2.15 shows that $\mathfrak{h}_M$ is conjugate to $\mathfrak{t}$ via $\operatorname{Int}\mathfrak{m}$, $\mathfrak{h}' = \mathfrak{a} \oplus \mathfrak{h}_M$ is conjugate to $\mathfrak{h} = \mathfrak{a} \oplus \mathfrak{t}$ via a member of $\operatorname{Int}\mathfrak{g}$ that fixes $\mathfrak{a}_0$. In particular, $\mathfrak{h}_0' = \mathfrak{a}_0 \oplus \mathfrak{h}_{M,0}$ is a Cartan subalgebra of $\mathfrak{g}_0$. Applying our constructed member of $\operatorname{Int}\mathfrak{g}$ to (7.88), we obtain

$$\mathfrak{m} = \mathfrak{h}_M \oplus \bigoplus_{\substack{\alpha\in\Delta(\mathfrak{g},\mathfrak{h}'),\\ \alpha|_{\mathfrak{a}}=0}} \mathfrak{g}_\alpha \qquad \text{and} \qquad \mathfrak{n} = \bigoplus_{\substack{\alpha\in\Delta^+(\mathfrak{g},\mathfrak{h}'),\\ \alpha|_{\mathfrak{a}}\neq 0}} \mathfrak{g}_\alpha \tag{7.89}$$

for the positive system $\Delta^+(\mathfrak{g}, \mathfrak{h}')$ obtained by transferring positivity from $\Delta^+(\mathfrak{g}, \mathfrak{h})$.

Let us note that $\mathfrak{a}_{\mathfrak{p},0} = \mathfrak{a}_0 \oplus \mathfrak{a}_{M,0}$ is a maximal abelian subspace of $\mathfrak{p}_0$. In fact, the centralizer of $\mathfrak{a}_0$ in $\mathfrak{g}_0$ is $\mathfrak{a}_0 \oplus \mathfrak{m}_0$, and $\mathfrak{a}_{M,0}$ is maximal abelian in $\mathfrak{m}_0 \cap \mathfrak{p}_0$; hence the assertion follows. We introduce a lexicographic ordering for $\mathfrak{h}_0'$ that is as before on $\mathfrak{a}_0$, takes $\mathfrak{a}_0$ before $\mathfrak{a}_{M,0}$, and takes $\mathfrak{a}_{M,0}$ before $i\mathfrak{t}_{M,0}$. Then we obtain a positive system $\Delta^{+\prime}(\mathfrak{g}, \mathfrak{h}')$ with the property that a root α with $\alpha|_{\mathfrak{a}_0} \neq 0$ is positive if and only if $\alpha|_{\mathfrak{a}_0}$ is the restriction to $\mathfrak{a}_0$ of a member of $\Delta^+(\mathfrak{g}, \mathfrak{h})$. Consequently we can replace $\Delta^+(\mathfrak{g}, \mathfrak{h}')$ in (7.89) by $\Delta^{+\prime}(\mathfrak{g}, \mathfrak{h}')$. Then it is apparent that

$\mathfrak{m} \oplus \mathfrak{a} \oplus \mathfrak{n}$ contains $\mathfrak{m}_\mathfrak{p} \oplus \mathfrak{a}_\mathfrak{p} \oplus \mathfrak{n}_\mathfrak{p}$ defined relative to the positive restricted roots obtained from $\Delta^{+\prime}(\mathfrak{g}, \mathfrak{h}')$, and hence $\mathfrak{q}_0$ is a parabolic subalgebra. Referring to (7.77), we see that $\mathfrak{q}_0 = \mathfrak{m}_0 \oplus \mathfrak{a}_0 \oplus \mathfrak{n}_0$ is the Langlands decomposition. Finally $\mathfrak{t}_0$ is a Cartan subalgebra of $\mathfrak{m}_0$ by Corollary 2.13, and hence $\mathfrak{q}_0$ is cuspidal.

8. Cartan Subgroups

We continue to assume that G is a reductive Lie group and to use the notation of §2 concerning the Cartan decomposition. A **Cartan subgroup** of G is the centralizer in G of a Cartan subalgebra. We know from §§VI.6 and VII.2 that any Cartan subalgebra is conjugate via Int $\mathfrak{g}_0$ to a θ stable Cartan subalgebra and that there are only finitely many conjugacy classes of Cartan subalgebras. Consequently any Cartan subgroup of G is conjugate via G to a Θ stable Cartan subgroup, and there are only finitely many conjugacy classes of Cartan subgroups. A Θ stable Cartan subgroup is a reductive Lie group by Proposition 7.25.

When G is compact connected and T is a maximal torus, every element of G is conjugate to a member of T, according to Theorem 4.36. In particular every member of G lies in a Cartan subgroup. This statement does not extend to noncompact groups, as the following example shows.

EXAMPLE. Let $G = SL(2, \mathbb{R})$. We saw in §VI.6 that every Cartan subalgebra is conjugate to one of

$$\left\{\begin{pmatrix} r & 0 \\ 0 & -r \end{pmatrix}\right\} \quad \text{and} \quad \left\{\begin{pmatrix} 0 & r \\ -r & 0 \end{pmatrix}\right\},$$

and the corresponding Cartan subgroups are

$$\left\{\pm\begin{pmatrix} e^r & 0 \\ 0 & e^{-r} \end{pmatrix}\right\} \quad \text{and} \quad \left\{\begin{pmatrix} \cos r & \sin r \\ -\sin r & \cos r \end{pmatrix}\right\}.$$

Some features of these subgroups are worth noting. The first Cartan subgroup is disconnected; disconnectedness is common among Cartan subgroups for general G. Also every member of either Cartan subgroup is diagonable over $\mathbb{C}$. Hence $\begin{pmatrix} 1 & 1 \\ 0 & 1 \end{pmatrix}$ lies in no Cartan subgroup.

Although the union of the Cartan subgroups of G need not exhaust G, it turns out that the union exhausts almost all of G. This fact is the most important conclusion about Cartan subgroups to be derived in this section and appears below as Theorem 7.108. When we treat integration in Chapter VIII, this fact will permit integration of functions on G by integrating over the conjugates of a finite set of Cartan subgroups; the resulting formula, known as the "Weyl Integration Formula," is an important tool for harmonic analysis on G.

Before coming to this main result, we give a proposition about the component structure of Cartan subgroups and we introduce a finite group $W(G, H)$ for each Cartan subgroup analogous to the groups $W(G, A)$ considered in §7.

Proposition 7.90. Let H be a Cartan subgroup of G.

(a) If H is maximally noncompact, then H meets every component of G.

(b) If H is maximally compact and if G is connected, then H is connected.

REMARKS. The modifiers "maximally noncompact" and "maximally compact" are to be interpreted in terms of the Lie algebras. If $\mathfrak{h}_0$ is a Cartan subalgebra, $\mathfrak{h}_0$ is conjugate to a θ stable Cartan subalgebra $\mathfrak{h}_0'$, and we defined "maximally noncompact" and "maximally compact" for $\mathfrak{h}_0'$ in §§VI.6 and VII.2. Proposition 7.35 says that any two candidates for $\mathfrak{h}_0'$ are conjugate via K, and hence it is meaningful to say that $\mathfrak{h}_0$ is maximally noncompact or maximally compact if $\mathfrak{h}_0'$ is.

PROOF. Let $\mathfrak{h}_0$ be the Lie algebra of H. We may assume that $\mathfrak{h}_0$ is θ stable. Let $\mathfrak{h}_0 = \mathfrak{t}_0 \oplus \mathfrak{a}_0$ be the decomposition of $\mathfrak{h}_0$ into $+1$ and -1 eigenspaces under θ.

(a) If $\mathfrak{h}_0$ is maximally noncompact, then $\mathfrak{a}_0$ is a maximal abelian subspace of $\mathfrak{p}_0$. The group H contains the subgroup F introduced before Corollary 7.52, and Corollary 7.52 and Proposition 7.33 show that F meets every component of G.

(b) If $\mathfrak{h}_0$ is maximally compact, then $\mathfrak{t}_0$ is a maximal abelian subspace of $\mathfrak{k}_0$. Since K is connected, the subgroup $Z_K(\mathfrak{t}_0)$ is connected by Corollary 4.51, and $Z_K(\mathfrak{t}_0) \exp \mathfrak{a}_0$ is therefore a connected closed subgroup of G with Lie algebra $\mathfrak{h}_0$. On the other hand, Proposition 7.25 implies that

$$H = Z_K(\mathfrak{h}_0) \exp \mathfrak{a}_0 \subseteq Z_K(\mathfrak{t}_0) \exp \mathfrak{a}_0.$$

Since H and $Z_K(\mathfrak{t}_0) \exp \mathfrak{a}_0$ are closed subgroups with the same Lie algebra and since $Z_K(\mathfrak{t}_0) \exp \mathfrak{a}_0$ is connected, it follows that $H = Z_K(\mathfrak{t}_0) \exp \mathfrak{a}_0$.

Corollary 7.91. If a maximally noncompact Cartan subgroup H of G is abelian, then $Z_{G_0} \subseteq Z_G$.

PROOF. By Proposition 7.90a, $G = G_0 H$. If z is in Z_{G_0}, then $\operatorname{Ad}(z) = 1$ on $\mathfrak{h}_0$, and hence z is in $Z_G(\mathfrak{h}_0) = H$. Let $g \in G$ be given, and write $g = g_0 h$ with $g \in G_0$ and $h \in H$. Then $zg_0 = g_0 z$ since z commutes with members of G_0, and $zh = hz$ since z is in H and H is abelian. Hence $zg = gz$, and z is in Z_G.

If H is a Cartan subgroup of G with Lie algebra $\mathfrak{h}_0$, we define

$$W(G, H) = N_G(\mathfrak{h}_0)/Z_G(\mathfrak{h}_0). \tag{7.92a}$$

Here $Z_G(\mathfrak{h}_0)$ is nothing more than H itself, by definition. When $\mathfrak{h}_0$ is θ stable, an alternate formula for $W(G, H)$ is

$$W(G, H) = N_K(\mathfrak{h}_0)/Z_K(\mathfrak{h}_0). \tag{7.92b}$$

The equality of the right sides of (7.92a) and (7.92b) is an immediate consequence of Lemma 7.22 and Proposition 2.7. Proposition 2.7 shows that $N_K(\mathfrak{h}_0)$ and $Z_K(\mathfrak{h}_0)$ both have $\mathfrak{k}_0 \cap \mathfrak{h}_0 = \mathfrak{t}_0$ as Lie algebra. Hence $W(G, H)$ is a compact 0-dimensional group, and we conclude that $W(G, H)$ is finite.

Each member of $N_G(\mathfrak{h}_0)$ sends roots of $\Delta = \Delta(\mathfrak{g}, \mathfrak{h})$ to roots, and the action of $N_G(\mathfrak{h}_0)$ on Δ descends to $W(G, H)$. It is clear that only the identity in $W(G, H)$ acts as the identity on Δ. Since $\operatorname{Ad}_{\mathfrak{g}}(G) \subseteq \operatorname{Int} \mathfrak{g}$, it follows from Theorem 7.8 that

$$W(G, H) \subseteq W(\Delta(\mathfrak{g}, \mathfrak{h})). \tag{7.93}$$

EXAMPLE. Let $G = SL(2, \mathbb{R})$. For any $\mathfrak{h}$, $W(\mathfrak{g}, \mathfrak{h})$ has order 2. When $\mathfrak{h}_0 = \left\{\begin{pmatrix} r & 0 \\ 0 & -r \end{pmatrix}\right\}$, $W(G, H)$ has order 2, a representative of the nontrivial coset being $\begin{pmatrix} 0 & 1 \\ -1 & 0 \end{pmatrix}$. When $\mathfrak{h}_0 = \left\{\begin{pmatrix} 0 & r \\ -r & 0 \end{pmatrix}\right\}$, $W(G, H)$ has order 1.

Now we begin to work toward the main result of this section, that the union of all Cartan subgroups of G exhausts almost all of G. We shall use the notion of a "regular element" of G. Recall that in Chapter II we introduced regular elements in the complexified Lie algebra $\mathfrak{g}$. Let $\dim \mathfrak{g} = n$. For $X \in \mathfrak{g}$, we formed the characteristic polynomial

$$\det(\lambda 1 - \operatorname{ad} X) = \lambda^n + \sum_{j=0}^{n-1} d_j(X)\lambda^j. \tag{7.94}$$

Here each d_j is a holomorphic polynomial function on $\mathfrak{g}$. The **rank** of $\mathfrak{g}$ is the minimum index l such that $d_l(X) \not\equiv 0$, and the **regular elements** of $\mathfrak{g}$ are those elements X such that $d_l(X) \neq 0$. For such an X, Theorem 2.9′ shows that the generalized eigenspace of $\operatorname{ad} X$ for eigenvalue 0 is a Cartan subalgebra of $\mathfrak{g}$. Because $\mathfrak{g}$ is reductive, the Cartan subalgebra acts completely reducibly on $\mathfrak{g}$, and hence the generalized eigenspace of $\operatorname{ad} X$ for eigenvalue 0 is nothing more than the centralizer of X in $\mathfrak{g}$.

Within $\mathfrak{g}$, let $\mathfrak{h}$ be a Cartan subalgebra, and let $\Delta = \Delta(\mathfrak{g}, \mathfrak{h})$. For $X \in \mathfrak{h}$, $d_l(X) = \prod_{\alpha\in\Delta} \alpha(X)$, so that $X \in \mathfrak{h}$ is regular if and only if no root vanishes on X. If $\mathfrak{h}_0$ is a Cartan subalgebra of our real form $\mathfrak{g}_0$, then we can find $X \in \mathfrak{h}_0$ so that $\alpha(X) \neq 0$ for all $\alpha \in \Delta$.

On the level of Lie algebras, we have concentrated on eigenvalue 0 for $\operatorname{ad} X$. On the level of reductive Lie groups, the analogous procedure is to concentrate on eigenvalue 1 for $\operatorname{Ad}(x)$. Thus for $x \in G$, we define

$$D(x, \lambda) = \det((\lambda + 1)1 - \operatorname{Ad}(x)) = \lambda^n + \sum_{j=0}^{n-1} D_j(x)\lambda^j.$$

Here each $D_j(x)$ is real analytic on G and descends to a real analytic function on $\operatorname{Ad}(G)$. But $\operatorname{Ad}(G) \subseteq \operatorname{Int} \mathfrak{g}$ by property (v) for reductive Lie groups, and the formula for $D_j(x)$ extends to be valid on $\operatorname{Int} \mathfrak{g}$ and to define a holomorphic function on $\operatorname{Int} \mathfrak{g}$. Let l' be the minimum index such that $D_{l'}(x) \not\equiv 0$ (on G or equivalently on $\operatorname{Int} \mathfrak{g}$). We shall observe shortly that $l' = l$. With this understanding the **regular elements** of G are those elements x such that $D_l(x) \neq 0$. Elements that are not regular are **singular**. The set of regular elements is denoted G'. Note that

$$D(yxy^{-1}, \lambda) = D(x, \lambda), \tag{7.95}$$

from which it follows that G' is stable under group conjugation. It is almost but not quite true that the centralizer of a regular element of G is a Cartan subgroup. Here is an example of how close things get in a complex group.

EXAMPLE. Let $G = SL(2, \mathbb{C})/\{\pm 1\}$. We work with elements of G as 2-by-2 matrices identified when they differ only by a sign. The element $\begin{pmatrix} z & 0 \\ 0 & z^{-1} \end{pmatrix}$, with $z \neq 0$, is regular if $z \neq \pm 1$. For most other values of z, the centralizer of $\begin{pmatrix} z & 0 \\ 0 & z^{-1} \end{pmatrix}$ is the diagonal subgroup, which is a Cartan subgroup. But for $z = \pm i$, the centralizer is generated by the diagonal subgroup and $\begin{pmatrix} 0 & 1 \\ -1 & 0 \end{pmatrix}$; thus the Cartan subgroup has index 2 in the centralizer.

Now, as promised, we prove that $l = l'$, i.e., the minimum index l such that $d_l(X) \not\equiv 0$ equals the minimum index l' such that $D_{l'}(x) \not\equiv 0$. Let

$\operatorname{ad} X$ have generalized eigenvalue 0 exactly l times. For sufficiently small r, $\operatorname{ad} X$ has all eigenvalues $< 2\pi$ in absolute value, and it follows for such X that $\operatorname{Ad}(\exp X)$ has generalized eigenvalue 1 exactly l times. Thus $l' \leq l$. In the reverse direction suppose $D_{l'}(x) \not\equiv 0$. Since $D_{l'}$ extends holomorphically to the connected complex group $\operatorname{Int} \mathfrak{g}$, $D_{l'}$ cannot be identically 0 in any neighborhood of the identity in $\operatorname{Int} \mathfrak{g}$. Hence $D_{l'}(x)$ cannot be identically 0 in any neighborhood of $x = 1$ in G. Choose a neighborhood U of X's in $\mathfrak{g}_0$ about 0 such that all $\operatorname{ad} X$ have all eigenvalues $< 2\pi$ in absolute value and such that exp is a diffeomorphism onto a neighborhood of 1 in G. Under these conditions the multiplicity of 0 as a generalized eigenvalue for $\operatorname{ad} X$ equals the multiplicity of 1 as a generalized eigenvalue for $\operatorname{Ad}(\exp X)$. Thus if $D_{l'}(x)$ is somewhere nonzero on $\exp U$, then $d_l(X)$ is somewhere nonzero on U. Thus $l \leq l'$, and we conclude that $l = l'$.

To understand the relationship between regular elements and Cartan subgroups, we shall first study the case of a complex group (which in practice will usually be $\operatorname{Int} \mathfrak{g}$). The result in this case is Theorem 7.101 below. We establish notation for this theorem after proving three lemmas.

Lemma 7.96. Let Z be a connected complex manifold, and let $f : Z \to \mathbb{C}^n$ be a holomorphic function not identically 0. Then the subset of Z where f is not 0 is connected.

PROOF. Lemma 2.14 proves this result for the case that $Z = \mathbb{C}^m$ and f is a polynomial. But the same proof works if Z is a bounded polydisc $\Pi_{j=1}^m \{|z_j| < r_j\}$ and f is a holomorphic function on a neighborhood of the closure of the polydisc. We shall piece together local results of this kind to handle general Z.

Thus let the manifold structure of Z be specified by compatible charts $(V_\alpha, \varphi_\alpha)$ with $\varphi_\alpha : V_\alpha \to \mathbb{C}^m$ holomorphic onto a bounded polydisc. There is no loss of generality in assuming that there are open subsets U_α covering Z such that $\varphi_\alpha(U_\alpha)$ is an open polydisc whose closure is contained in $\varphi_\alpha(V_\alpha)$. For any subset S of Z, let S' denote the subset of S where f is not 0. The result of the previous paragraph implies that U'_α is connected for each α, and we are to prove that Z' is connected. Also U'_α is dense in U_α since the subset of a connected open set where a nonzero holomorphic function takes on nonzero values is dense.

Fix $U = U_0$. To each point $z \in Z$, we can find a chain of U_α's of the form $U = U_0, U_1, \ldots, U_k$ such that z is in U_k and $U_{i-1} \cap U_i \neq \emptyset$ for $1 \leq i \leq k$. In fact, the set of z's for which this assertion is true is nonempty open closed and hence is all of Z.

Now let $z \in Z'$ be given, and form the chain $U = U_0, U_1, \ldots, U_k$. Here z is in U'_k. We readily see by induction on $m \leq k$ that $U'_0 \cup \cdots \cup U'_m$ is connected, hence that $U'_0 \cup \cdots \cup U'_k$ is connected. Thus each $z \in Z'$ lies in a connected open set containing U'_0, and it follows that the union of these connected open sets is connected. The union is Z', and hence Z' is connected.

Lemma 7.97. Let N be a simply connected nilpotent Lie group with Lie algebra $\mathfrak{n}_0$, and let $\mathfrak{n}'_0$ be an ideal in $\mathfrak{n}_0$. If X is in $\mathfrak{n}_0$ and Y is in $\mathfrak{n}'_0$, then $\exp(X+Y) = \exp X \exp Y'$ for some Y' in $\mathfrak{n}'_0$.

PROOF. If N' is the analytic subgroup corresponding to $\mathfrak{n}'_0$, then N' is certainly normal, and N' is closed as a consequence of Theorem 1.104. Let $\varphi : N \to N/N'$ be the quotient homomorphism, and let $d\varphi$ be its differential. Since $d\varphi(Y) = 0$, we have

$$\begin{aligned}\varphi((\exp(X+Y))(\exp X)^{-1}) &= \varphi(\exp(X+Y))\varphi(\exp X)^{-1} \\ &= \exp(d\varphi(X) + d\varphi(Y))(\exp d\varphi(X))^{-1} \\ &= \exp(d\varphi(X))(\exp d\varphi(X))^{-1} = 1.\end{aligned}$$

Therefore $(\exp(X+Y))(\exp X)^{-1}$ is in N', and Theorem 1.104 shows that it is of the form $\exp Y'$ for some $Y' \in \mathfrak{n}'_0$.

Lemma 7.98. Let $G = KAN$ be an Iwasawa decomposition of the reductive group G, let $M = Z_K(A)$, and let $\mathfrak{n}_0$ be the Lie algebra of N. If $h \in MA$ has the property that $\text{Ad}(h)$ acts as a scalar on each restricted-root space and $\text{Ad}(h)^{-1} - 1$ is nonsingular on $\mathfrak{n}_0$, then the map $\varphi : N \to N$ given by $\varphi(n) = h^{-1}nhn^{-1}$ is onto N.

REMARK. This lemma may be regarded as a Lie group version of the Lie algebra result given as Lemma 7.42.

PROOF. Write $\mathfrak{n}_0 = \bigoplus (\mathfrak{g}_0)_\lambda$ as a sum of restricted-root spaces, and regard the restricted roots as ordered lexicographically. For any restricted root α, the subspace $\mathfrak{n}_\alpha = \bigoplus_{\lambda \geq \alpha} (\mathfrak{g}_0)_\lambda$ is an ideal, and we prove by induction downward on α that φ carries $\exp \mathfrak{n}_\alpha$ onto itself. This conclusion when α is equal to the smallest positive restricted root gives the lemma since exp carries $\mathfrak{n}_0$ onto N (Theorem 1.104).

If α is given, we can write $\mathfrak{n}_\alpha = (\mathfrak{g}_0)_\alpha \oplus \mathfrak{n}_\beta$ with $\beta > \alpha$. Let X be given in $\mathfrak{n}_\alpha$, and write X as $X_1 + X_2$ with $X_1 \in (\mathfrak{g}_0)_\alpha$ and $X_2 \in \mathfrak{n}_\beta$. Since $\text{Ad}(h)^{-1} - 1$ is nonsingular on $(\mathfrak{g}_0)_\alpha$, we can choose $Y_1 \in (\mathfrak{g}_0)_\alpha$ with $X_1 = (\text{Ad}(h)^{-1} - 1)Y_1$. Put $n_1 = \exp Y_1$. Since $\text{Ad}(h)^{-1}Y_1$ is a multiple of Y_1, $\text{Ad}(h)^{-1}Y_1$ commutes with Y_1. Therefore

$$\begin{aligned}h^{-1}n_1hn_1^{-1} &= (\exp \text{Ad}(h)^{-1}Y_1)(\exp Y_1)^{-1} \\ &= \exp((\text{Ad}(h)^{-1} - 1)Y_1) = \exp X_1.\end{aligned} \tag{7.99}$$

Thus

$$\begin{aligned}
\exp X &= \exp(X_1 + X_2) \\
&= \exp X_1 \exp X_2' && \text{by Lemma 7.97} \\
&= h^{-1}n_1hn_1^{-1} \exp X_2' && \text{by (7.99)} \\
&= h^{-1}n_1h \exp X_2'' n_1^{-1} && \text{with } X_2'' \in \mathfrak{n}_\beta.
\end{aligned}$$

By induction $\exp X_2'' = h^{-1}n_2hn_2^{-1}$. Hence $\exp X = h^{-1}(n_1n_2)h(n_1n_2)^{-1}$, and the induction is complete.

Now we are ready for the main result about Cartan subgroups in the complex case. Let G_c be a complex semisimple Lie group (which will usually be $\operatorname{Int}\mathfrak{g}$ when we return to our reductive Lie group G). Proposition 7.5 shows that G_c is a reductive Lie group. Let $G_c = UAN$ be an Iwasawa decomposition of G_c, and let $M = Z_U(A)$. We denote by $\mathfrak{g}$, $\mathfrak{u}_0$, $\mathfrak{a}_0$, $\mathfrak{n}_0$, and $\mathfrak{m}_0$ the respective Lie algebras. Here $\mathfrak{m}_0 = i\mathfrak{a}_0$, $\mathfrak{m}_0$ is maximal abelian in $\mathfrak{u}_0$, and $\mathfrak{h} = \mathfrak{a}_0 \oplus \mathfrak{m}_0$ is a Cartan subalgebra of $\mathfrak{g}$. The corresponding Cartan subgroup of G_c is of the form $H_c = MA$ since Proposition 7.25 shows that H_c is a reductive Lie group. Since

$$M = Z_U(\mathfrak{a}_0) = Z_U(i\mathfrak{a}_0) = Z_U(\mathfrak{m}_0),$$

Corollary 4.52 shows that M is connected. Therefore

$$H_c \text{ is connected.} \tag{7.100}$$

Let G_c' denote the regular set in G_c.

Theorem 7.101. For the complex semisimple Lie group G_c, the regular set G_c' is connected and satisfies $G_c' \subseteq \bigcup_{x\in G_c} xH_cx^{-1}$. If X_0 is any regular element in $\mathfrak{h}$, then $Z_{G_c}(X_0) = H_c$.

PROOF. We may regard $D_l(x)$ as a holomorphic function on G_c. The regular set G_c' is the set where $D_l(x) \neq 0$, and Lemma 7.96 shows that G_c' is connected.

Let $H_c' = H_c \cap G_c'$, and define $V' = \bigcup_{x\in G_c} xH_c'x^{-1}$. Then $V' \subseteq G_c'$ by (7.95). If $X_0 \in \mathfrak{h}$ is chosen so that no root in $\Delta(\mathfrak{g}, \mathfrak{h})$ vanishes on X_0, then we have seen that $\exp rX_0$ is in H_c' for all sufficiently small $r > 0$. Hence V' is nonempty. We shall prove that V' is open and closed in G_c', and then it follows that $G_c' = V'$, hence that $G_c' \subseteq \bigcup_{x\in G_c} xH_cx^{-1}$.

To prove that V' is closed in G_c', we observe that H_cN is closed in G_c, being the minimal parabolic subgroup MAN. Since U is compact, it follows that

$$V = \bigcup_{u\in U} uH_cNu^{-1}$$

is closed in G_c. By (7.95),

$$V \cap G'_c = \bigcup_{u \in U} u(H_cN)'u^{-1},$$

where $(H_cN)' = H_cN \cap G'_c$. If h is in H_c and n is in N, then $\mathrm{Ad}(hn)$ has the same generalized eigenvalues as $\mathrm{Ad}(h)$. Hence $(H_cN)' = H'_cN$. If h is in H'_c, then $\mathrm{Ad}(h)$ is scalar on each restricted root space contributing to $\mathfrak{n}_0$, and $\mathrm{Ad}(h) - 1$ is nonsingular on $\mathfrak{n}_0$. By Lemma 7.98 such an h has the property that $n \mapsto h^{-1}nhn^{-1}$ carries N onto N. Let $n_0 \in N$ be given, and write $n_0 = h^{-1}nhn^{-1}$. Then $hn_0 = nhn^{-1}$, and we see that every element of hN is an N conjugate of h. Since every N conjugate of h is certainly in hN, we obtain

$$H'_cN = \bigcup_{n \in N} nH'_cn^{-1}.$$

Therefore

$$V \cap G'_c = \bigcup_{u \in U} \bigcup_{n \in N} (un)H'_c(un)^{-1}.$$

Since $aH'_ca^{-1} = H'_c$ for $a \in A$ and since $G_c = UAN = UNA$, we obtain $V \cap G'_c = V'$. Thus V' is exhibited as the intersection of G'_c with the closed set V, and V' is therefore closed in G'_c.

To prove that V' is open in G'_c, it is enough to prove that the map $\psi : G_c \times H_c \to G_c$ given by $\psi(y, x) = yxy^{-1}$ has differential mapping onto at every point of $G_c \times H'_c$. The argument imitates part of the proof of Theorem 4.36. Let us abbreviate yxy^{-1} as x^y. Fix $y \in G_c$ and $x \in H'_c$. We identify the tangent spaces at y, x, and x^y with $\mathfrak{g}$, $\mathfrak{h}$, and $\mathfrak{g}$ by left translation. First let Y be in $\mathfrak{g}$. To compute $(d\psi)_{(y,x)}(Y, 0)$, we observe from (1.90) that

$$x^{y \exp rY} = x^y \exp(r\mathrm{Ad}(yx^{-1})Y)\exp(-r\mathrm{Ad}(y)Y). \tag{7.102}$$

We know from Lemma 1.92a that

$$\exp rX' \exp rY' = \exp\{r(X' + Y') + O(r^2)\} \qquad \text{as } r \to 0.$$

Hence the right side of (7.102) is

$$= x^y \exp(r\mathrm{Ad}(y)(\mathrm{Ad}(x^{-1}) - 1)Y + O(r^2)),$$

and

$$d\psi(Y, 0) = \mathrm{Ad}(y)(\mathrm{Ad}(x^{-1}) - 1)Y. \tag{7.103}$$

Next if X is in $\mathfrak{h}$, then (1.90) gives

$$(x \exp rX)^y = x^y \exp(r\operatorname{Ad}(y)X),$$

and hence

$$d\psi(0, X) = \operatorname{Ad}(y)X. \tag{7.104}$$

Combining (7.103) and (7.104), we obtain

$$d\psi(Y, X) = \operatorname{Ad}(y)((\operatorname{Ad}(x^{-1}) - 1)Y + X). \tag{7.105}$$

Since x is in H_c', $\operatorname{Ad}(x^{-1}) - 1$ is invertible on the sum of the restricted-root spaces, and thus the set of all $(\operatorname{Ad}(x^{-1}) - 1)Y$ contains this sum. Since X is arbitrary in $\mathfrak{h}$, the set of all $(\operatorname{Ad}(x^{-1}) - 1)Y + X$ is all of $\mathfrak{g}$. But $\operatorname{Ad}(y)$ is invertible, and thus (7.105) shows that $d\psi$ is onto $\mathfrak{g}$. This completes the proof that V' is open in G_c'.

We are left with proving that any regular element X_0 of $\mathfrak{h}$ has $Z_{G_c}(X_0) = H_c$. Let $x \in G_c$ satisfy $\operatorname{Ad}(x)X_0 = X_0$. Since the centralizer of X_0 in $\mathfrak{g}$ is $\mathfrak{h}$, $\operatorname{Ad}(x)\mathfrak{h} = \mathfrak{h}$. If $x = u \exp X$ is the global Cartan decomposition of x, then Lemma 7.22 shows that $\operatorname{Ad}(u)\mathfrak{h} = \mathfrak{h}$ and $(\operatorname{ad} X)\mathfrak{h} = \mathfrak{h}$. By Proposition 2.7, X is in $\mathfrak{h}$. Thus $\operatorname{Ad}(u)X_0 = X_0$, and it is enough to prove that u is in M. Write $X_0 = X_1 + iX_2$ with X_1 and X_2 in $\mathfrak{m}_0$. Since $\operatorname{Ad}(u)\mathfrak{u}_0 = \mathfrak{u}_0$, we must have $\operatorname{Ad}(u)X_1 = X_1$. The centralizer of the torus $\overline{\exp \mathbb{R}X_1}$ in U is connected, by Corollary 4.51, and must lie in the analytic subgroup of U with Lie algebra $Z_{\mathfrak{u}_0}(X_1)$. Since X_1 is regular, Lemma 4.33 shows that $Z_{\mathfrak{u}_0}(X_1) = \mathfrak{m}_0$. Therefore u is in M, and the proof is complete.

Corollary 7.106. For the complex semisimple Lie group G_c, let H_x denote the centralizer in G_c of a regular element x of G_c. Then the identity component of H_x is a Cartan subgroup $(H_x)_0$ of G_c, and H_x lies in the normalizer $N_{G_c}((H_x)_0)$. Consequently H_x has only a finite number of connected components.

REMARK. Compare this conclusion with the example of $SL(2, \mathbb{C})/\{\pm 1\}$ given after (7.95).

PROOF. Theorem 7.101 shows that we can choose $y \in G_c$ with $h = y^{-1}xy$ in H_c. Since x is regular, so is h. Therefore $\operatorname{Ad}(h)$ has 1 as a generalized eigenvalue with multiplicity $l = \dim_{\mathbb{C}} \mathfrak{h}$. Since $\operatorname{Ad}(h)$ acts as the identity on $\mathfrak{h}$, it follows that $\mathfrak{h}$ is the centralizer of h in $\mathfrak{g}$. Hence $\operatorname{Ad}(y)\mathfrak{h}$ is the centralizer of $x = yhy^{-1}$ in $\mathfrak{g}$, and $\operatorname{Ad}(y)\mathfrak{h}$ is therefore the Lie algebra of H_x. Then $(H_x)_0 = yH_cy^{-1}$ is a Cartan subgroup of G_c by (7.100).

Next any element of a Lie group normalizes its identity component, and hence H_x lies in the normalizer $N_{G_c}((H_x)_0)$. By (7.93), H_x has a finite number of components.

Corollary 7.107. For the complex semisimple Lie group G_c, the centralizer in $\mathfrak{g}$ of a regular element of G_c is a Cartan subalgebra of $\mathfrak{g}$.

PROOF. This follows from the first conclusion of Corollary 7.106.

We return to the general reductive Lie group G. The relationship between the regular set in G and the Cartan subgroups of G follows quickly from Corollary 7.107.

Theorem 7.108. For the reductive Lie group G, let $(\mathfrak{h}_1)_0, \dots, (\mathfrak{h}_r)_0$ be a maximal set of nonconjugate θ stable Cartan subalgebras of $\mathfrak{g}_0$, and let $H_1, \dots, H_r$ be the corresponding Cartan subgroups of G. Then

(a) $G' \subseteq \bigcup_{i=1}^r \bigcup_{x \in G} x H_i x^{-1}$

(b) each member of G' lies in just one Cartan subgroup of G

(c) each H_i is abelian if G is semisimple and has a complexification.

PROOF.

(a) We apply Corollary 7.107 with $G_c = \operatorname{Int} \mathfrak{g}$. Property (v) of reductive Lie groups says that $\operatorname{Ad}(G) \subseteq G_c$, and the regular elements of G are exactly the elements x of G for which $\operatorname{Ad}(x)$ is regular in G_c. If x is in G', then Corollary 7.107 shows that $Z_{\mathfrak{g}}(x)$ is a Cartan subalgebra of $\mathfrak{g}$. Since x is in G, $Z_{\mathfrak{g}}(x)$ is the complexification of $Z_{\mathfrak{g}_0}(x)$, and hence $Z_{\mathfrak{g}_0}(x)$ is a Cartan subalgebra of $\mathfrak{g}_0$. Therefore $Z_{\mathfrak{g}_0}(x) = \operatorname{Ad}(y)(\mathfrak{h}_i)_0$ for some $y \in G$ and some i with $1 \le i \le r$. Write $\tilde{\mathfrak{h}}_0$ for $Z_{\mathfrak{g}_0}(x)$, and let $\tilde{H} = Z_G(\tilde{\mathfrak{h}}_0)$ be the corresponding Cartan subgroup. By definition, x is in $\tilde{H}$. Since $\tilde{\mathfrak{h}}_0 = \operatorname{Ad}(y)(\mathfrak{h}_i)_0$, it follows that $\tilde{H} = y H_i y^{-1}$. Therefore x is in $y H_i y^{-1}$, and (a) is proved.

(b) We again apply Corollary 7.107 with $G_c = \operatorname{Int} \mathfrak{g}$. If $x \in G'$ lies in two distinct Cartan subgroups, then it centralizes two distinct Cartan subalgebras of $\mathfrak{g}_0$ and also their complexifications in $\mathfrak{g}$. Hence the centralizer of x in $\mathfrak{g}$ contains the sum of the two Cartan subalgebras in $\mathfrak{g}$, in contradiction with Corollary 7.107.

(c) This time we regard G_c as the complexification of G. Let $\mathfrak{h}_0$ be a Cartan subalgebra of $\mathfrak{g}_0$, and let H be the corresponding Cartan subgroup of G. The centralizer H_c of $\mathfrak{h}$ in G_c is connected by (7.100), and H is a subgroup of this group. Since H_c has abelian Lie algebra, it is abelian. Hence H is abelian.

Now we return to the component structure of Cartan subgroups, but we shall restrict attention to the case that the reductive Lie group G is semisimple and has a complexification $G^{\mathbb{C}}$. Let $\mathfrak{h}_0 = \mathfrak{t}_0 \oplus \mathfrak{a}_0$ be the decomposition into $+1$ and -1 eigenspaces under θ of a θ stable Cartan subalgebra $\mathfrak{h}_0$. Let H be the Cartan subgroup $Z_G(\mathfrak{h}_0)$, let $T = \exp \mathfrak{t}_0$, and let $A = \exp \mathfrak{a}_0$. Here T is closed in K since otherwise the Lie algebra of its closure would form with $\mathfrak{a}_0$ an abelian subspace larger than $\mathfrak{h}_0$. Hence T is a torus. If α is a real root in $\Delta(\mathfrak{g}, \mathfrak{h})$, then the same argument as for (7.54) shows that

$$\gamma_\alpha = \exp 2\pi i|\alpha|^{-2}H_\alpha \tag{7.109}$$

is an element of K with $\gamma_\alpha^2 = 1$. As α varies, the elements γ_α commute. Define $F(T)$ to be the subgroup of K generated by all the elements γ_α for α real. Theorem 7.55 identifies $F(T)$ in the special case that $\mathfrak{h}_0$ is maximally noncompact; the theorem says that $F(T) = F$ in this case.

Proposition 7.110. Let G be semisimple with a complexification $G^{\mathbb{C}}$, and let $\mathfrak{h}_0$ be a θ stable Cartan subalgebra. Then the corresponding Cartan subgroup is $H = ATF(T)$.

PROOF. By Proposition 7.25, $Z_G(\mathfrak{t}_0)$ is a reductive Lie group, and then it satisfies $Z_G(\mathfrak{t}_0) = Z_K(\mathfrak{t}_0)\exp(\mathfrak{p}_0 \cap Z_{\mathfrak{g}_0}(\mathfrak{t}_0))$. By Corollary 4.51, $Z_K(\mathfrak{t}_0)$ is connected. Therefore $Z_G(\mathfrak{t}_0)$ is connected.

Consequently $Z_G(\mathfrak{t}_0)$ is the analytic subgroup corresponding to

$$Z_{\mathfrak{g}_0}(\mathfrak{t}_0) = \mathfrak{g}_0 \cap \big(\mathfrak{h} + \sum_{\alpha \text{ real}} \mathfrak{g}_\alpha\big) = \mathfrak{h}_0 + \big(\sum_{\alpha \text{ real}} \mathbb{R}H_\alpha + \sum_{\alpha \text{ real}} (\mathfrak{g}_\alpha \cap \mathfrak{g}_0)\big).$$

The grouped term on the right is a split semisimple Lie algebra $\mathfrak{s}_0$. Let S be the corresponding analytic subgroup, so that $Z_G(\mathfrak{t}_0) = (\exp \mathfrak{h}_0)S = ATS$. Since the subspace $\mathfrak{a}_0' = \sum_{\alpha \text{ real}} \mathbb{R}H_\alpha$ of $\mathfrak{s}$ is a maximal abelian subspace of $\mathfrak{s}_0 \cap \mathfrak{p}_0$, Theorem 7.55 shows that the corresponding F group is just $F(T)$. By Theorem 7.53c, $Z_S(\mathfrak{a}_0') = (\exp \mathfrak{a}_0')F(T)$. Then

$$Z_G(\mathfrak{h}_0) = Z_{ATS}(\mathfrak{a}_0) = ATZ_S(\mathfrak{a}_0) = ATZ_S(\mathfrak{a}_0') = ATF(T).$$

Corollary 7.111. Let G be semisimple with a complexification $G^{\mathbb{C}}$, and let $Q = MAN$ be the Langlands decomposition of a cuspidal parabolic subgroup. Let $\mathfrak{t}_0$ be a θ stable compact Cartan subalgebra of $\mathfrak{m}_0$, and let $\mathfrak{h}_0 = \mathfrak{t}_0 \oplus \mathfrak{a}_0$ be the corresponding θ stable Cartan subalgebra of $\mathfrak{g}_0$. Define T and $F(T)$ from $\mathfrak{t}_0$. Then

(a) $Z_M(\mathfrak{t}_0) = TF(T)$
(b) $Z_{M_0} = Z_M \cap T$
(c) $Z_M = (Z_M \cap T)F(T) = Z_{M_0}F(T)$
(d) $M_0Z_M = M_0F(T)$.

REMARK. When Q is a minimal parabolic subgroup, the subgroup M_0Z_M is all of M. But for general Q, M_0Z_M need not exhaust M. For some purposes in representation theory, M_0Z_M plays an intermediate role in passing from representations of M_0 to representations of M.

PROOF.

(a) Proposition 7.110 gives $Z_M(\mathfrak{t}_0) = {}^0Z_G(\mathfrak{t}_0 \oplus \mathfrak{a}_0) = {}^0(ATF(T)) = TF(T)$.

(b) Certainly $Z_M \cap T \subseteq Z_{M_0}$. In the reverse direction, Z_{M_0} is contained in $K \cap M_0$, hence is contained in the center of $K \cap M_0$. The center of a compact connected Lie group is contained in every maximal torus (Corollary 4.47), and thus $Z_{M_0} \subseteq T$. To complete the proof of (b), we show that $Z_{M_0} \subseteq Z_M$. The sum of $\mathfrak{a}_0$ and a maximally noncompact Cartan subalgebra of $\mathfrak{m}_0$ is a Cartan subalgebra of $\mathfrak{g}_0$, and the corresponding Cartan subgroup of G is abelian by Proposition 7.110. The intersection of this Cartan subgroup with M is a maximal noncompact Cartan subgroup of M and is abelian. By Corollary 7.91, $Z_{M_0} \subseteq Z_M$.

(c) The subgroup $F(T)$ is contained in Z_M since it is in $K \cap \exp i\mathfrak{a}_0$. Therefore $Z_M = Z_M \cap Z_M(\mathfrak{t}_0) = Z_M \cap (TF(T)) = (Z_M \cap T)F(T)$, which proves the first equality of (c). The second equality follows from (b).

(d) By (c), $M_0 Z_M = M_0 Z_{M_0} F(T) = M_0 F(T)$.

9. Harish-Chandra Decomposition

For $G = SU(1,1) = \left\{ \begin{pmatrix} \alpha & \beta \\ \bar\beta & \bar\alpha \end{pmatrix} \,\middle|\, |\alpha|^2 - |\beta|^2 = 1 \right\}$, the subgroup K can be taken to be $K = \left\{ \begin{pmatrix} e^{i\theta} & 0 \\ 0 & e^{-i\theta} \end{pmatrix} \right\}$, and G/K may be identified with the disc $\{|z| < 1\}$ by $gK \leftrightarrow \beta/\bar\alpha$. If $g' = \begin{pmatrix} \alpha' & \beta' \\ \bar\beta' & \bar\alpha' \end{pmatrix}$ is given, then the equality $g'g = \begin{pmatrix} \alpha'\alpha + \beta'\bar\beta & \alpha'\beta + \beta'\bar\alpha \\ \bar\beta'\alpha + \bar\alpha'\bar\beta & \bar\beta'\beta + \bar\alpha'\bar\alpha \end{pmatrix}$ implies that

$$g'(gK) \leftrightarrow \frac{\alpha'\beta + \beta'\bar\alpha}{\bar\beta'\beta + \bar\alpha'\bar\alpha} = \frac{\alpha'(\beta/\bar\alpha) + \beta'}{\bar\beta'(\beta/\bar\alpha) + \bar\alpha'}.$$

In other words, under this identification, g' acts by the associated linear fractional transformation $z \mapsto \dfrac{\alpha' z + \beta'}{\bar\beta' z + \bar\alpha'}$. The transformations by which G acts on G/K are thus holomorphic once we have imposed a suitable complex-manifold structure on G/K.

If G is a semisimple Lie group, then we say that G/K is **Hermitian** if G/K admits a complex-manifold structure such that G acts by holomorphic transformations. In this section we shall classify the semisimple groups G for which G/K is Hermitian. Since the center of G is contained in K (Theorem 6.31e), we could assume, if we wanted, that G is an adjoint group. At any rate there is no loss of generality in assuming that

G is linear and hence has a complexification. We begin with a more complicated example.

EXAMPLE. Let $n \geq m$, let $M_{nm}(\mathbb{C})$ be the complex vector space of all n-by-m complex matrices, and let 1_m be the m-by-m identity matrix. Define

$$\Omega = \{Z \in M_{nm}(\mathbb{C}) \mid 1_m - Z^*Z \text{ is positive definite}\}.$$

We shall identify Ω with a quotient G/K, taking $G = SU(n,m)$ and

$$\begin{aligned} K &= S(U(n) \times U(m)) \\ &= \left\{ \begin{pmatrix} A & 0 \\ 0 & D \end{pmatrix} \,\middle|\, A \in U(n),\ D \in U(m),\ \det A \det D = 1 \right\}. \end{aligned}$$

The group action of G on Ω will be by

$$g(Z) = (AZ+B)(CZ+D)^{-1} \qquad \text{if } g = \begin{pmatrix} A & B \\ C & D \end{pmatrix}. \tag{7.112}$$

To see that (7.112) defines an action of G on Ω, we shall verify that $(CZ+D)^{-1}$ is defined in (7.112) and that $g(Z)$ is in Ω if Z is in Ω. To do so, we write

$$\begin{aligned} &(AZ+B)^*(AZ+B) - (CZ+D)^*(CZ+D) \\ &\qquad = (Z^* \quad 1_m)\, g^* \begin{pmatrix} 1_n & 0 \\ 0 & -1_m \end{pmatrix} g \begin{pmatrix} Z \\ 1_m \end{pmatrix} \\ &\qquad = (Z^* \quad 1_m) \begin{pmatrix} 1_n & 0 \\ 0 & -1_m \end{pmatrix} \begin{pmatrix} Z \\ 1_m \end{pmatrix} \qquad \text{since } g \text{ is in } SU(n,m) \\ &\qquad = Z^*Z - 1_m. \end{aligned} \tag{7.113}$$

Let $(CZ+D)v = 0$. Unless $v = 0$, we see from (7.113) that

$$0 \leq v^*(AZ+B)^*(AZ+B)v = v^*(Z^*Z - 1_m)v < 0,$$

a contradiction. Hence $(CZ+D)^{-1}$ exists, and then (7.113) gives

$$g(Z)^*g(Z) - 1_m = (CZ+D)^{*-1}(Z^*Z - 1_m)(CZ+D)^*.$$

The right side is negative definite, and hence $g(Z)$ is in Ω.

The isotropy subgroup at $Z = 0$ is the subgroup with $B = 0$, and this subgroup reduces to K. Let us see that G acts transitively on Ω. Let $Z \in M_{nm}(\mathbb{C})$ be given. The claim is that Z decomposes as

$$(7.114) \qquad Z = udv \qquad \text{with } u \in U(n),\ v \in U(m),$$

and d of the form $d = \begin{pmatrix} d_0 \\ 0 \end{pmatrix}$, where $d_0 = \operatorname{diag}(\lambda_1, \ldots, \lambda_m)$ with all $\lambda_j \geq 0$ and where 0 is of size $(n-m)$-by-m. To prove (7.114), we extend Z to a square matrix $(\,Z \quad 0\,)$ of size n-by-n and let the polar decomposition of $(\,Z \quad 0\,)$ be $(\,Z \quad 0\,) = u_1 p$ with $u_1 \in U(n)$ and p positive semidefinite. Since $(\,Z \quad 0\,)$ is 0 in the last $n-m$ columns, u_1 gives 0 when applied to the last $n-m$ columns of p. The matrix u_1 is nonsingular, and thus the last $n-m$ columns of p are 0. Since p is Hermitian, $p = \begin{pmatrix} p' & 0 \\ 0 & 0 \end{pmatrix}$ with p' positive semidefinite of size m-by-m. By the finite-dimensional Spectral Theorem, write $p' = u_2 d_0 u_2^{-1}$ with $u_2 \in U(m)$ and $d_0 = \operatorname{diag}(\lambda_1, \ldots, \lambda_m)$. Then (7.114) holds with $u = u_1 \begin{pmatrix} u_2 & 0 \\ 0 & 1_{n-m} \end{pmatrix}$, $d = \begin{pmatrix} d_0 \\ 0 \end{pmatrix}$, and $v = u_2^{-1}$.

With Z as in (7.114), the matrix $Z^*Z = v^*d^*dv$ has the same eigenvalues as d^*d, which has eigenvalues $\lambda_1^2, \ldots, \lambda_m^2$. Thus Z is in Ω if and only if $0 \leq \lambda_j < 1$ for $1 \leq j \leq m$. In the formula (7.114) there is no loss of generality in assuming that $(\det u)(\det v)^{-1} = 1$, so that $\begin{pmatrix} u & 0 \\ 0 & v^{-1} \end{pmatrix}$ is in K. Let a be the member of $SU(n,m)$ that is $\begin{pmatrix} \cosh t_j & \sinh t_j \\ \sinh t_j & \cosh t_j \end{pmatrix}$ in the jth and $(n+j)^{\text{th}}$ rows and columns for $1 \leq j \leq m$ and is otherwise the identity. Then $a(0) = d$, and $\begin{pmatrix} u & 0 \\ 0 & v^{-1} \end{pmatrix}(d) = udv = Z$. Hence $g = \begin{pmatrix} u & 0 \\ 0 & v^{-1} \end{pmatrix} a$ maps 0 to Z, and the action of G on Ω is transitive.

Throughout this section we let G be a semisimple Lie group with a complexification $G^{\mathbb{C}}$. We continue with the usual notation for G as a reductive Lie group. Let $\mathfrak{c}_0$ be the center of $\mathfrak{k}_0$. We shall see that a necessary and sufficient condition for G/K to be Hermitian is that $Z_{\mathfrak{g}_0}(\mathfrak{c}_0) = \mathfrak{k}_0$. In this case we shall exhibit G/K as holomorphically equivalent to a bounded domain in $\mathbb{C}^n$ for a suitable n. The explicit realization of G/K as a bounded domain is achieved through the "Harish-Chandra decomposition" of a certain open dense subset of $G^{\mathbb{C}}$.

First we shall prove that if G/K is Hermitian, then $Z_{\mathfrak{g}_0}(\mathfrak{c}_0) = \mathfrak{k}_0$. Before stating the precise theorem, it will be helpful to have a little detail available concerning the "multiplication-by-i" mapping mentioned in connection with holomorphic mappings at the end of §I.10. Let M and N be open sets in $\mathbb{C}^m$ and $\mathbb{C}^n$, and let $\Phi : M \to N$ be smooth. Let $(z_1, \ldots, z_m)$ and $(w_1, \ldots, w_n)$ be coordinates for M and N, and write $z_j = x_j + iy_j$ and $w_j = u_j + iv_j$. We think of Φ as given by expressions $w_i = w_i(z_1, \ldots, z_m)$, $1 \le i \le n$. If we put $x_{j+m} = y_j$ for $1 \le j \le m$ and $u_{i+n} = v_i$ for $1 \le i \le n$, then the derivative matrix of Φ at a point is $(\Phi') = \left(\frac{\partial u_i}{\partial x_j}\right)$. The Cauchy-Riemann equations for each w_i in each variable z_j say that (Φ') if of the form $\begin{pmatrix} a & b \\ -b & a \end{pmatrix}$, or equivalently that (Φ') satisfies the commutativity relation

$$(J_n)(\Phi') = (\Phi')(J_m), \tag{7.115a}$$

where $J_m = \begin{pmatrix} 0 & -1_m \\ 1_m & 0 \end{pmatrix}$ and where J_n is defined similarly. Here J_m is the matrix of the linear transformation (also denoted J_m) that is defined by $J_m\left(\frac{\partial}{\partial x_j}\right) = \left(\frac{\partial}{\partial y_j}\right)$ and $J_m\left(\frac{\partial}{\partial y_j}\right) = -\left(\frac{\partial}{\partial x_j}\right)$.

If Φ is the identity, then (7.115a) for arbitrary coordinates says that the linear transformation corresponding to J_m is well defined independently of basis. Then the above remarks make sense when Φ is a smooth mapping between complex manifolds. At each point p of M, we obtain a linear map J_p of the tangent space $T_p(M)$ to itself with $J_p^2 = -1$, and we similarly have such a map J'_q of the tangent space $T_q(N)$ to itself. If Φ is holomorphic, then the Cauchy-Riemann equations yield

$$J'_{\Phi(p)} \circ d\Phi_p = d\Phi_p \circ J_p \tag{7.115b}$$

for all p. The map J_p is the **multiplication-by-*i*** mapping at p, and the system of all J_p as p varies is the **almost-complex structure** on M associated to the complex structure.

Now let us consider the case that $M = N = G/K$ and p is the identity coset. If G/K is Hermitian, then each left translation L_k by $k \in K$ (defined by $L_k(k') = kk'$) is holomorphic and fixes the identity coset. If J denotes the multiplication-by-i mapping at the identity coset, then (7.115) gives

$$J \circ dL_k = dL_k \circ J.$$

We may identify the tangent space at the identity coset with $\mathfrak{p}_0$, and then $dL_k = \operatorname{Ad}(k)|_{\mathfrak{p}_0}$. Differentiating, we obtain

$$J \circ (\operatorname{ad} X)|_{\mathfrak{p}_0} = (\operatorname{ad} X)|_{\mathfrak{p}_0} \circ J \qquad \text{for all } X \in \mathfrak{k}_0. \tag{7.116}$$

Theorem 7.117. If G/K is Hermitian, then the multiplication-by-i mapping $J : \mathfrak{p}_0 \to \mathfrak{p}_0$ at the identity coset is of the form $J = (\operatorname{ad} X_0)|_{\mathfrak{p}_0}$ for some $X_0 \in \mathfrak{k}_0$. This element X_0 is in $\mathfrak{c}_0$ and satisfies $Z_{\mathfrak{g}_0}(X_0) = \mathfrak{k}_0$. Hence $Z_{\mathfrak{g}_0}(\mathfrak{c}_0) = \mathfrak{k}_0$.

PROOF. Since $J^2 = -1$ on $\mathfrak{p}_0$, the complexification $\mathfrak{p}$ is the direct sum of its $+i$ and $-i$ eigenspaces $\mathfrak{p}^+$ and $\mathfrak{p}^-$. The main step is to prove that

$$[X, Y] = 0 \qquad \text{if } X \in \mathfrak{p}^+ \text{ and } Y \in \mathfrak{p}^+. \tag{7.118}$$

Let B be the bilinear form on $\mathfrak{g}_0$ and $\mathfrak{g}$ that is part of the data of a reductive group, and define a bilinear form C on $\mathfrak{p}$ by

$$C(X, Y) = B(X, Y) + B(JX, JY).$$

Since B is positive definite on $\mathfrak{p}_0$, so is C. Hence C is nondegenerate on $\mathfrak{p}$. Let us prove that

$$C([[X, Y], Z], T) = C([[Z, T], X], Y) \tag{7.119}$$

for X, Y, Z, T in $\mathfrak{p}$. When X, Y, Z are in $\mathfrak{p}$, the bracket $[Y, Z]$ is in $\mathfrak{k}$, and therefore (7.116) implies that

$$J[X, [Y, Z]] = [JX, [Y, Z]]. \tag{7.120}$$

Using the Jacobi identity and (7.120) repeatedly, together with the invariance of B, we compute

$$\begin{aligned} B(J[[X, Y], Z], JT) &= B(J[X, [Y, Z]], JT) - B(J[Y, [X, Z]], JT) \\ &= B([JX, [Y, Z]], JT) - B([JY, [X, Z]], JT) \\ &= -B([JT, [Y, Z]], JX) + B([JT, [X, Z]], JY) \\ &= -B(J[T, [Y, Z]], JX) + B(J[T, [X, Z]], JY). \end{aligned} \tag{7.121}$$

Using the result (7.121) with Z and T interchanged, we obtain

$$\begin{aligned} B(J[[X, Y], Z], JT) &= B([[X, Y], JZ], JT) \\ &= -B([[X, Y], JT], JZ) \\ &= -B(J[[X, Y], T], JZ) \\ &= B(J[Z, [Y, T]], JX) - B(J[Z, [X, T]], JY). \end{aligned} \tag{7.122}$$

The sum of (7.121) and (7.122) is

$$\begin{aligned} 2B(J[[X, Y], Z], JT) &= -B(J[T, [Y, Z]], JX) + B(J[T, [X, Z]], JY) \\ &\quad + B(J[Z, [Y, T]], JX) - B(J[Z, [X, T]], JY) \\ &= B(J[Y, [Z, T]], JX) - B(J[X, [Z, T]], JY) \end{aligned}$$

$$
\begin{aligned}
&= B([JY,[Z,T]],JX) - B([JX,[Z,T]],JY) \\
&= 2B([Z,T],[JX,JY]) \\
&= 2B([[Z,T],JX],JY) \\
&= 2B(J[[Z,T],X],JY). \qquad (7.123)
\end{aligned}
$$

The calculation that leads to (7.123) remains valid if J is dropped throughout. If we add the results with J present and with J absent, we obtain (7.119). To prove (7.118), suppose that X and Y are in $\mathfrak{p}^+$, so that $JX = iX$ and $JY = iY$. Then

$$
\begin{aligned}
C([[Z,T],X],Y) &= C(J[[Z,T],X],JY) \\
&= C([[Z,T],JX],JY) \\
&= -C([[Z,T],X],Y)
\end{aligned}
$$

says $C([[Z,T],X],Y) = 0$. By (7.119), $C([[X,Y],Z],T) = 0$. Since T is arbitrary and C is nondegenerate,

$$
[[X,Y],Z] = 0 \qquad \text{for all } Z \in \mathfrak{p}. \qquad (7.124)
$$

If bar denotes conjugation of $\mathfrak{g}$ with respect to $\mathfrak{g}_0$, then $B(W,\overline{W}) < 0$ for all $W \neq 0$ in $\mathfrak{k}$. For $W = [X,Y]$, we have

$$
B([X,Y],\overline{[X,Y]}) = B([X,Y],[\overline{X},\overline{Y}]) = B([[X,Y],\overline{X}],\overline{Y}),
$$

and the right side is 0 by (7.124). Therefore $[X,Y] = 0$, and (7.118) is proved.

Let us extend J to a linear map $\tilde{J}$ defined on $\mathfrak{g}$, putting $\tilde{J} = 0$ on $\mathfrak{k}$. We shall deduce from (7.118) that $\tilde{J}$ is a derivation of $\mathfrak{g}_0$, i.e., that

$$
\tilde{J}[X,Y] = [\tilde{J}X,Y] + [X,\tilde{J}Y] \qquad \text{for } X,Y \in \mathfrak{g}_0. \qquad (7.125)
$$

If X and Y are in $\mathfrak{k}_0$, all terms are 0, and (7.125) is automatic. If X is in $\mathfrak{k}_0$ and Y is in $\mathfrak{p}_0$, then $[\tilde{J}X,Y] = 0$ since $\tilde{J}X = 0$, and (7.125) reduces to (7.116). Thus suppose X and Y are in $\mathfrak{p}_0$. The element $X - iJX$ is in $\mathfrak{p}^+$ since

$$
J(X - iJX) = JX - iJ^2X = JX + iX = i(X - iJX),
$$

and similarly $Y - iJY$ is in $\mathfrak{p}^+$. By (7.118),

$$
0 = [X - iJX, Y - iJY] = ([X,Y] - [JX,JY]) - i([JX,Y] + [X,JY]).
$$

The real and imaginary parts must each be 0. Since the imaginary part is 0, the right side of (7.125) is 0. The left side of (7.125) is 0 since $\tilde{J}$ is 0 on $\mathfrak{k}_0$. Hence $\tilde{J}$ is a derivation of $\mathfrak{g}_0$.

By Proposition 1.98, $\tilde{J} = \operatorname{ad} X_0$ for some $X_0 \in \mathfrak{g}_0$. Let $Y \in \mathfrak{p}_0$ be given. Since $J^2 = -1$ on $\mathfrak{p}_0$, the element $Y' = -JY$ of $\mathfrak{p}_0$ has $JY' = Y$. Then

$$B(X_0, Y) = B(X_0, JY') = B(X_0, [X_0, Y']) = B([X_0, X_0], Y') = 0.$$

Hence X_0 is orthogonal to $\mathfrak{p}_0$, and X_0 must be in $\mathfrak{k}_0$. Since $\tilde{J} = \operatorname{ad} X_0$ is 0 on $\mathfrak{k}_0$, X_0 is in $\mathfrak{c}_0$.

If Y is in $Z_{\mathfrak{g}_0}(X_0)$, then the $\mathfrak{k}_0$ component of Y already commutes with X_0 since X_0 is in $\mathfrak{c}_0$. Thus we may assume that Y is in $\mathfrak{p}_0$. But then $[X_0, Y] = JY$. Since J is nonsingular on $\mathfrak{p}_0$, $0 = [X_0, Y]$ implies $Y = 0$. We conclude that $Z_{\mathfrak{g}_0}(X_0) = \mathfrak{k}_0$. Finally we have

$$\mathfrak{k}_0 \subseteq Z_{\mathfrak{g}_0}(\mathfrak{c}_0) \subseteq Z_{\mathfrak{g}_0}(X_0) = \mathfrak{k}_0,$$

and equality must hold throughout. Therefore $Z_{\mathfrak{g}_0}(\mathfrak{c}_0) = \mathfrak{k}_0$.

For the converse we assume that $Z_{\mathfrak{g}_0}(\mathfrak{c}_0) = \mathfrak{k}_0$, and we shall exhibit a complex structure on G/K such that G operates by holomorphic transformations. Fix a maximal abelian subspace $\mathfrak{t}_0$ of $\mathfrak{k}_0$. Then $\mathfrak{c}_0 \subseteq \mathfrak{t}_0$, so that $Z_{\mathfrak{g}_0}(\mathfrak{t}_0) \subseteq Z_{\mathfrak{g}_0}(\mathfrak{c}_0) = \mathfrak{k}_0$. Consequently $\mathfrak{t}_0$ is a compact Cartan subalgebra of $\mathfrak{g}_0$. The corresponding Cartan subgroup T is connected by Proposition 7.90b, hence is a torus.

Every root in $\Delta = \Delta(\mathfrak{g}, \mathfrak{t})$ is imaginary, hence compact or noncompact in the sense of §VI.7. If Δ_K and Δ_n denote the sets of compact and noncompact roots, then we have

$$(7.126) \qquad \mathfrak{k} = \mathfrak{t} \oplus \bigoplus_{\alpha \in \Delta_K} \mathfrak{g}_\alpha \qquad \text{and} \qquad \mathfrak{p} = \bigoplus_{\alpha \in \Delta_n} \mathfrak{g}_\alpha,$$

just as in (6.103).

Lemma 7.127. A root α is compact if and only if α vanishes on the center $\mathfrak{c}$ of $\mathfrak{k}$.

PROOF. If α is in Δ, then $\alpha(\mathfrak{c}) = 0$ if and only if $[\mathfrak{c}, \mathfrak{g}_\alpha] = 0$, if and only if $\mathfrak{g}_\alpha \subseteq Z_{\mathfrak{g}}(\mathfrak{c})$, if and only if $\mathfrak{g}_\alpha \subseteq \mathfrak{k}$, if and only if α is compact.

By a **good ordering** for $i\mathfrak{t}_0$, we mean a system of positivity in which every noncompact positive root is larger than every compact root. A good ordering always exists; we can, for instance, use a lexicographic ordering that takes $i\mathfrak{c}_0$ before its orthogonal complement in $i\mathfrak{t}_0$. Fixing a good ordering, let Δ^+, Δ_K^+, and Δ_n^+ be the sets of positive roots in Δ, Δ_K, and Δ_n. Define

$$\mathfrak{p}^+ = \bigoplus_{\alpha\in\Delta_n^+} \mathfrak{g}_\alpha \qquad \text{and} \qquad \mathfrak{p}^- = \bigoplus_{\alpha\in\Delta_n^+} \mathfrak{g}_{-\alpha},$$

so that $\mathfrak{p} = \mathfrak{p}^+ \oplus \mathfrak{p}^-$.

In the example of $SU(n,m)$ earlier in this section, we have

$$i\mathfrak{c}_0 = \mathbb{R}\operatorname{diag}(\tfrac{1}{n},\dots,\tfrac{1}{n},-\tfrac{1}{m},\dots,-\tfrac{1}{m})$$

with n entries $\frac{1}{n}$ and m entries $-\frac{1}{m}$, and we may take $\mathfrak{t}_0$ to be the diagonal subalgebra. If roots $e_i - e_j$ that are positive on

$$\operatorname{diag}(\tfrac{1}{n},\dots,\tfrac{1}{n},-\tfrac{1}{m},\dots,-\tfrac{1}{m})$$

are declared to be positive, then $\mathfrak{p}^+$ has the block form $\begin{pmatrix} 0 & * \\ 0 & 0 \end{pmatrix}$ and $\mathfrak{p}^-$ has the block form $\begin{pmatrix} 0 & 0 \\ * & 0 \end{pmatrix}$.

Lemma 7.128. The subspaces $\mathfrak{p}^+$ and $\mathfrak{p}^-$ are abelian subspaces of $\mathfrak{p}$, and $[\mathfrak{k},\mathfrak{p}^+] \subseteq \mathfrak{p}^+$ and $[\mathfrak{k},\mathfrak{p}^-] \subseteq \mathfrak{p}^-$.

PROOF. Let α, β, and $\alpha+\beta$ be in Δ with α compact and β noncompact. Then $[\mathfrak{g}_\alpha,\mathfrak{g}_\beta] \subseteq \mathfrak{g}_{\alpha+\beta}$, and β and $\alpha+\beta$ are both positive or both negative because the ordering is good. Summing on α and β, we see that $[\mathfrak{k},\mathfrak{p}^+] \subseteq \mathfrak{p}^+$ and $[\mathfrak{k},\mathfrak{p}^-] \subseteq \mathfrak{p}^-$.

If α and β are in Δ_n^+, then $\alpha+\beta$ cannot be a root since it would have to be a compact root larger than the noncompact positive root α. Summing on α and β, we obtain $[\mathfrak{p}^+,\mathfrak{p}^+] = 0$. Similarly $[\mathfrak{p}^-,\mathfrak{p}^-] = 0$.

Let $\mathfrak{b}$ be the Lie subalgebra

$$\mathfrak{b} = \mathfrak{t} \oplus \bigoplus_{\alpha\in\Delta^+} \mathfrak{g}_{-\alpha}$$

of $\mathfrak{g}$, and let P^+, $K^{\mathbb{C}}$, P^-, and B be the analytic subgroups of $G^{\mathbb{C}}$ with Lie algebras $\mathfrak{p}^+$, $\mathfrak{k}$, $\mathfrak{p}^-$, and $\mathfrak{b}$. Since $G^{\mathbb{C}}$ is complex and $\mathfrak{p}^+$, $\mathfrak{k}$, $\mathfrak{p}^-$, $\mathfrak{b}$ are closed under multiplication by i, all the groups P^+, $K^{\mathbb{C}}$, P^-, B are complex subgroups.

Theorem 7.129 (Harish-Chandra decomposition). Let G be semisimple with a complexification $G^{\mathbb{C}}$, and suppose that the center $\mathfrak{c}_0$ of $\mathfrak{k}_0$ has $Z_{\mathfrak{g}_0}(\mathfrak{c}_0) = \mathfrak{k}_0$. Then multiplication from $P^+ \times K^{\mathbb{C}} \times P^-$ into $G^{\mathbb{C}}$ is one-one, holomorphic, and regular (with image open in $G^{\mathbb{C}}$), GB is open in $G^{\mathbb{C}}$, and there exists a bounded open subset $\Omega \subseteq P^+$ such that

$$GB = GK^{\mathbb{C}}P^- = \Omega K^{\mathbb{C}}P^-.$$

Moreover, G/K is Hermitian. In fact, the map $G \to \Omega$ given by $g \mapsto$ (P^+ component of g) exhibits G/K and Ω as diffeomorphic, and G acts holomorphically on Ω by $g(\omega) =$ (P^+ component of $g\omega$).

REMARKS.

1) We shall see in the proof that the complex group P^+ is holomorphically isomorphic with some $\mathbb{C}^n$, and the theorem asserts that Ω is a bounded open subset when regarded as in $\mathbb{C}^n$ in this fashion.

2) When $G = SU(n,m)$, $G^{\mathbb{C}}$ may be taken as $SL(n+m,\mathbb{C})$. The decomposition of an open subset of $G^{\mathbb{C}}$ as $P^+ \times K^{\mathbb{C}} \times P^-$ is

(7.130)
$$\begin{pmatrix} A & B \\ C & D \end{pmatrix} = \begin{pmatrix} 1 & BD^{-1} \\ 0 & 1 \end{pmatrix} \begin{pmatrix} A - BD^{-1}C & 0 \\ 0 & D \end{pmatrix} \begin{pmatrix} 1 & 0 \\ D^{-1}C & 1 \end{pmatrix},$$

valid whenever D is nonsingular. Whatever Ω is in the theorem, if $\omega = \begin{pmatrix} 1 & Z \\ 0 & 1 \end{pmatrix}$ is in Ω and $g = \begin{pmatrix} A & B \\ C & D \end{pmatrix}$ is in G, then $g\omega = \begin{pmatrix} A & AZ+B \\ C & CZ+D \end{pmatrix}$; hence (7.130) shows that the P^+ component of $g\omega$ is

$$\begin{pmatrix} 1 & (AZ+B)(CZ+D)^{-1} \\ 0 & 1 \end{pmatrix}.$$

So the action is

$$\text{(7.131)} \qquad \begin{pmatrix} A & B \\ C & D \end{pmatrix}\left(\begin{pmatrix} 1 & Z \\ 0 & 1 \end{pmatrix}\right) = \begin{pmatrix} 1 & (AZ+B)(CZ+D)^{-1} \\ 0 & 1 \end{pmatrix}.$$

We know from the example earlier in this section that the image of $Z = 0$ under $Z \mapsto (AZ+B)(CZ+D)^{-1}$ for all $\begin{pmatrix} A & B \\ C & D \end{pmatrix}$ in $SU(n,m)$ is all Z with $1_m - Z^*Z$ positive definite. Therefore Ω consists of all $\begin{pmatrix} 1 & Z \\ 0 & 1 \end{pmatrix}$ such that $1_m - Z^*Z$ is positive definite, and the action (7.131) corresponds to the action by linear fractional transformations in the example.

3) The proof will reduce matters to two lemmas, which we shall consider separately.

PROOF. Define

$$\mathfrak{n} = \bigoplus_{\alpha\in\Delta^+} \mathfrak{g}_\alpha, \qquad \mathfrak{n}^- = \bigoplus_{\alpha\in\Delta^+} \mathfrak{g}_{-\alpha}, \qquad \mathfrak{b}_K = \mathfrak{t} \oplus \bigoplus_{\alpha\in\Delta_K^+} \mathfrak{g}_{-\alpha},$$

$$N,\ N^-,\ B_K = \text{corresponding analytic subgroups of } G^{\mathbb{C}}.$$

Let $H_{\mathbb{R}}$ and H be the analytic subgroups of $G^{\mathbb{C}}$ with Lie algebras $i\mathfrak{t}_0$ and $\mathfrak{t}$, so that $H = TH_{\mathbb{R}}$ as a direct product. By (7.100) a Cartan subgroup of a complex semisimple Lie group is connected, and therefore H is a Cartan subgroup. The involution $\theta \circ$ bar, where bar is the conjugation of $\mathfrak{g}$ with respect to $\mathfrak{g}_0$, is a Cartan involution of $\mathfrak{g}$, and $i\mathfrak{t}_0$ is a maximal abelian subspace of the -1 eigenspace. The $+1$ eigenspace is $\mathfrak{k}_0 \oplus i\mathfrak{p}_0$, and the corresponding analytic subgroup of $G^{\mathbb{C}}$ we call U. Then

$$Z_U(i\mathfrak{t}_0) = Z_U(\mathfrak{t}) = U \cap Z_{G^{\mathbb{C}}}(\mathfrak{t}) = U \cap H = T.$$

So the $M_{\mathfrak{p}}$ group is just T. By Proposition 7.82 the M of every parabolic subgroup of $G^{\mathbb{C}}$ is connected.

The restricted roots of $\mathfrak{g}^{\mathbb{R}}$ relative to $i\mathfrak{t}_0$ are evidently the restrictions from $\mathfrak{t}$ to $i\mathfrak{t}_0$ of the roots. Therefore $\mathfrak{b} = \mathfrak{t} \oplus \mathfrak{n}^-$ is a minimal parabolic subalgebra of $\mathfrak{g}^{\mathbb{R}}$. Since parabolic subgroups of $G^{\mathbb{C}}$ are closed (by Proposition 7.83b) and connected, B is closed.

The subspace $\mathfrak{k}\oplus\mathfrak{p}^-$ is a Lie subalgebra of $\mathfrak{g}^{\mathbb{R}}$ containing $\mathfrak{b}$ and hence is a parabolic subalgebra. Then Proposition 7.83 shows that $K^{\mathbb{C}}$ and P^- are closed, $K^{\mathbb{C}}P^-$ is closed, and multiplication $K^{\mathbb{C}} \times P^-$ is a diffeomorphism onto. Similarly P^+ is closed.

Moreover the Lie algebra $\mathfrak{k} \oplus \mathfrak{p}^-$ of $K^{\mathbb{C}}P^-$ is complex, and hence $K^{\mathbb{C}}P^-$ is a complex manifold. Then multiplication $K^{\mathbb{C}} \times P^-$ is evidently holomorphic and has been observed to be one-one and regular. Since $\mathfrak{p}^+\oplus(\mathfrak{k}\oplus\mathfrak{p}^-) = \mathfrak{g}$, Lemma 6.44 shows that the holomorphic multiplication map $P^+ \times (K^{\mathbb{C}}P^-) \to G^{\mathbb{C}}$ is everywhere regular. It is one-one by Proposition 7.83e. Hence $P^+ \times K^{\mathbb{C}} \times P^- \to G$ is one-one, holomorphic, and regular.

Next we shall show that GB is open in $G^{\mathbb{C}}$. First let us observe that

$$\mathfrak{g}_0 \cap (i\mathfrak{t}_0 \oplus \mathfrak{n}^-) = 0. \tag{7.132}$$

In fact, since roots are imaginary on $\mathfrak{t}_0$, we have $\overline{\mathfrak{g}_\alpha} = \mathfrak{g}_{-\alpha}$. Thus if h is in $i\mathfrak{t}_0$ and $X_{-\alpha}$ is in $\mathfrak{n}^-$, then

$$\overline{h + \sum_{\alpha\in\Delta^+} X_{-\alpha}} = -h + \sum_{\alpha\in\Delta^+} \overline{X_{-\alpha}} \in -h + \mathfrak{n},$$

and (7.132) follows since members of $\mathfrak{g}_0$ equal their own conjugates. The real dimension of $i\mathfrak{t}_0 \oplus \mathfrak{n}^-$ is half the real dimension of $\mathfrak{t} \oplus \mathfrak{n} \oplus \mathfrak{n}^- = \mathfrak{g}$, and hence

$$\dim_{\mathbb{R}}(\mathfrak{g}_0 \oplus (i\mathfrak{t}_0 \oplus \mathfrak{n}^-)) = \dim_{\mathbb{R}} \mathfrak{g}. \tag{7.133}$$

Combining (7.132) and (7.133), we see that

$$\mathfrak{g} = \mathfrak{g}_0 \oplus (i\mathfrak{t}_0 \oplus \mathfrak{n}^-). \tag{7.134}$$

The subgroup $H_{\mathbb{R}}N^-$ of $G^{\mathbb{C}}$ is closed by Proposition 7.83, and hence $H_{\mathbb{R}}N^-$ is an anlytic subgroup, necessarily with Lie algebra $i\mathfrak{t}_0 \oplus \mathfrak{n}^-$. By Lemma 6.44 it follows from (7.134) that multiplication $G \times H_{\mathbb{R}}N^- \to G^{\mathbb{C}}$ is everywhere regular. The dimension relation (7.133) therefore implies that $GH_{\mathbb{R}}N^-$ is open in $G^{\mathbb{C}}$. Since $B = TH_{\mathbb{R}}N^-$ and $T \subseteq G$, GB equals $GH_{\mathbb{R}}N^-$ and is open in $G^{\mathbb{C}}$.

The subgroups P^+ and P^- are the N groups of parabolic subalgebras, and their Lie algebras are abelian by Lemma 7.128. Hence P^+ and P^- are Euclidean groups. Then $\exp : \mathfrak{p}^+ \to P^+$ is biholomorphic, and P^+ is biholomorphic with $\mathbb{C}^n$ for some n. Similarly P^- is biholomorphic with $\mathbb{C}^n$.

The subgroup $K^{\mathbb{C}}$ is a reductive group, being connected and having bar as a Cartan involution for its Lie algebra. It is the product of the identity component of its center by a complex semisimple Lie group, and our above considerations show that its parabolic subgroups are connected. Then B_K is a parabolic subgroup, and

$$K^{\mathbb{C}} = KB_K \tag{7.135}$$

by Proposition 7.83f.

Let A denote a specific $A_{\mathfrak{p}}$ component for the Iwasawa decomposition of G, to be specified in Lemma 7.143 below. We shall show in Lemma 7.145 that this A satisfies

$$A \subseteq P^+K^{\mathbb{C}}P^- \tag{7.136a}$$

and

(7.136b) $\qquad P^+$ components of members of A are bounded.

Theorem 7.39 shows that $G = KAK$. Since $\mathfrak{b} \subseteq \mathfrak{k} \oplus \mathfrak{p}^-$, we have $B \subseteq K^{\mathbb{C}}P^-$. Since Lemma 7.128 shows that $K^{\mathbb{C}}$ normalizes P^+ and P^-, (7.136a) gives

$$\begin{aligned} GB &\subseteq GK^{\mathbb{C}}P^- \subseteq KAKK^{\mathbb{C}}P^- \\ &\subseteq KP^+K^{\mathbb{C}}P^-K^{\mathbb{C}}P^- = P^+K^{\mathbb{C}}P^-. \end{aligned} \tag{7.137}$$

By (7.135) we have

$$GK^{\mathbb{C}}P^- = GKB_KP^- \subseteq GB_KP^- \subseteq GB. \tag{7.138}$$

Inclusions (7.137) and (7.138) together imply that

$$GB = GK^{\mathbb{C}}P^- \subseteq P^+K^{\mathbb{C}}P^-.$$

Since GB is open,

$$GB = GK^{\mathbb{C}}P^- = \Omega K^{\mathbb{C}}P^- \tag{7.139}$$

for some open set Ω in P^+.

Let us write $p^+(\cdot)$ for the P^+ component. For $gb \in GB$, we have $p^+(gb) = p^+(g)$, and thus p^+ restricts to a smooth map carrying G onto Ω. From (7.139) it follows that the map $G \times \Omega \to \Omega$ given by

$$(g, \omega) \mapsto p^+(g\omega) \tag{7.140}$$

is well defined. For fixed g, this is holomorphic since left translation by g is holomorphic on $G^{\mathbb{C}}$ and since p^+ is holomorphic from $P^+K^{\mathbb{C}}P^-$ to P^+. To see that (7.140) is a group action, we use that $K^{\mathbb{C}}P^-$ is a subgroup. Let g_1 and g_2 be given, and write $g_2\omega = p^+(g_2\omega)k_2p_2^-$ and $g_1g_2\omega = p^+(g_1g_2\omega)k^{\mathbb{C}}p^-$. Then

$$g_1p^+(g_2\omega) = g_1g_2\omega(k_2p_2^-)^{-1} = p^+(g_1g_2\omega)(k^{\mathbb{C}}p^-)(k_2p_2^-)^{-1}.$$

Since $(k^{\mathbb{C}}p^-)(k_2p_2^-)^{-1}$ is in $K^{\mathbb{C}}P^-$, $p^+(g_1p^+(g_2\omega)) = p^+(g_1g_2\omega)$. Therefore (7.140) is a group action. The action is evidently smooth, and we have seen that it is transitive.

If g is in G and k is in K, we can regard 1 as in Ω and write

$$p^+(gk) = p^+(gk1) = p^+(gp^+(k1)) = p^+(g1)$$

since $k1$ is in $K \subseteq K^{\mathbb{C}}$ and has P^+ component 1. Therefore $p^+ : G \to \Omega$ descends to a smooth map of G/K onto Ω. Let us see that it is one-one. If $p^+(g_1) = p^+(g_2)$, then $g_1 = g_2k^{\mathbb{C}}p^-$ since $K^{\mathbb{C}}P^-$ is a group, and hence $g_2^{-1}g_1 = k^{\mathbb{C}}p^-$. Thus the map $G/K \to \Omega$ will be one-one if we show that

$$G \cap K^{\mathbb{C}}P^- = K. \tag{7.141}$$

To prove (7.141), we note that $\supseteq$ is clear. Then we argue in the same way as for (7.131) that

$$\mathfrak{g}_0 \cap (\mathfrak{k} \oplus \mathfrak{p}^-) = \mathfrak{k}_0. \tag{7.142}$$

Since G and $K^{\mathbb{C}}P^-$ are closed in $G^{\mathbb{C}}$, their intersection is a closed subgroup of G with Lie algebra $\mathfrak{k}_0$. Let $g = k \exp X$ be the global Cartan decomposition of an element g of $G \cap K^{\mathbb{C}}P^-$. Then $\mathrm{Ad}(g)\mathfrak{k}_0 = \mathfrak{k}_0$, and Lemma 7.22 implies that $(\mathrm{ad}\, X)\mathfrak{k}_0 \subseteq \mathfrak{k}_0$. Since $\mathrm{ad}\, X$ is skew symmetric relative to B, $(\mathrm{ad}\, X)\mathfrak{p}_0 \subseteq \mathfrak{p}_0$. But $X \in \mathfrak{p}_0$ implies that $(\mathrm{ad}\, X)\mathfrak{k}_0 \subseteq \mathfrak{p}_0$ and $(\mathrm{ad}\, X)\mathfrak{p}_0 \subseteq \mathfrak{k}_0$. Hence $\mathrm{ad}\, X = 0$ and $X = 0$. This proves (7.141).

To see that $G/K \to \Omega$ is everywhere regular, it is enough, since (7.140) is a smooth group action, to show that the differential of $p^+ : G \to \Omega$ at the identity is one-one on $\mathfrak{p}_0$. But dp^+ complexifies to the projection of $\mathfrak{g} = \mathfrak{p}^+ \oplus \mathfrak{k} \oplus \mathfrak{p}^-$ on $\mathfrak{p}^+$, and (7.142) shows that the kernel of this projection meets $\mathfrak{p}_0$ only in 0. Therefore the map G/K is a diffeomorphism.

To see that Ω is bounded, we need to see that $p^+(g)$ remains bounded as g varies in G. If $g \in G$ is given, write $g = k_1 a k_2$ according to $G = KAK$. Then $p^+(g) = p^+(k_1 a) = k_1 p^+(a)$ since it follows from (7.139) that $K\Omega = \Omega$. Therefore it is enough to prove that $\|\log p^+(a)\|$ remains bounded, and this is just (7.136b). Thus the theorem reduces to proving (7.136), which we do in Lemmas 7.143 and 7.145 below.

Lemma 7.143. Inductively define $\gamma_1, \ldots, \gamma_s$ in Δ_n^+ as follows: γ_1 is the largest member of Δ_n^+, and γ_j is the largest member of Δ_n^+ orthogonal to $\gamma_1, \ldots, \gamma_{j-1}$. For $1 \le j \le s$, let E_{γ_j} be a nonzero root vector for γ_j. Then the roots $\gamma_1, \ldots, \gamma_s$ are strongly orthogonal, and

$$\mathfrak{a}_0 = \bigoplus_{j=1}^{s} \mathbb{R}(E_{\gamma_j} + \overline{E_{\gamma_j}})$$

is a maximal abelian subspace of $\mathfrak{p}_0$.

PROOF. We make repeated use of the fact that if E_β is in $\mathfrak{g}_\beta$, then $\overline{E_\beta}$ is in $\mathfrak{g}_{-\beta}$. Since $[\mathfrak{p}^+, \mathfrak{p}^+] = 0$ by Lemma 7.128, $\gamma_j + \gamma_i$ is never a root, and the γ_j's are strongly orthogonal. Then it follows that $\mathfrak{a}_0$ is abelian.

To see that $\mathfrak{a}_0$ is maximal abelian in $\mathfrak{p}_0$, let X be a member of $\mathfrak{p}_0$ commuting with $\mathfrak{a}_0$. By (7.126) we can write $X = \sum_{\beta \in \Delta_n} X_\beta$ with $X_\beta \in \mathfrak{g}_\beta$. Without loss of generality, we may assume that X is orthogonal to $\mathfrak{a}_0$, and then we are to prove that $X = 0$. Assuming that $X \neq 0$, let β_0 be the largest member of Δ_n such that $X_{\beta_0} \neq 0$. Since $X = \overline{X}$, $X_{-\beta_0} \neq 0$ also; thus β_0 is positive. Choose j as small as possible so that β_0 is not orthogonal to γ_j.

First suppose that $\beta_0 \neq \gamma_j$. Since $[\mathfrak{p}^+, \mathfrak{p}^+] = 0$, $\beta_0 + \gamma_j$ is not a root. Therefore $\beta_0 - \gamma_j$ is a root. The root β_0 is orthogonal to $\gamma_1, \ldots, \gamma_{j-1}$, and γ_j is the largest noncompact root orthogonal to $\gamma_1, \ldots, \gamma_{j-1}$. Thus $\beta_0 < \gamma_j$, and $\beta_0 - \gamma_j$ is negative. We have

$$0 = [X, E_{\gamma_j} + \overline{E_{\gamma_j}}] = \sum_{\beta \in \Delta_n} ([X_\beta, E_{\gamma_j}] + [X_\beta, \overline{E_{\gamma_j}}]), \tag{7.144}$$

and $[X_{\beta_0}, \overline{E_{\gamma_j}}]$ is not 0, by Corollary 2.35. Thus there is a compensating term $[X_\beta, E_{\gamma_j}]$, i.e., there exists $\beta \in \Delta_n$ with $\beta + \gamma_j = \beta_0 - \gamma_j$ and with $X_\beta \neq 0$. Since $X = \overline{X}$, $X_{-\beta} \neq 0$. By maximality of β_0, $\beta_0 > -\beta$. Since $\gamma_j - \beta_0$ is positive, $\gamma_j > \beta_0 > -\beta$. Therefore $\beta + \gamma_j$ is positive. But $\beta + \gamma_j = \beta_0 - \gamma_j$, and the right side is negative, contradiction.

Next suppose that $\beta_0 = \gamma_j$. Then $[X_{\gamma_j}, \overline{E_{\gamma_j}}] \neq 0$, and (7.144) gives

$$[X_{-\gamma_j}, E_{\gamma_j}] + [X_{\gamma_j}, \overline{E_{\gamma_j}}] = 0.$$

Define scalars c^+ and c^- by $X_{\gamma_j} = c^+ E_{\gamma_j}$ and $X_{-\gamma_j} = c^- \overline{E_{\gamma_j}}$. Substituting, we obtain

$$-c^-[E_{\gamma_j}, \overline{E_{\gamma_j}}] + c^+[E_{\gamma_j}, \overline{E_{\gamma_j}}] = 0,$$

and therefore $c^+ = c^-$. Consequently $X_{\gamma_j} + X_{-\gamma_j} = c^+(E_{\gamma_j} + \overline{E_{\gamma_j}})$ makes a contribution to X that is nonorthogonal to $E_{\gamma_j} + \overline{E_{\gamma_j}}$. Since the other terms of X are orthogonal to $E_{\gamma_j} + \overline{E_{\gamma_j}}$, we have a contradiction. We conclude that $X = 0$ and hence that $\mathfrak{a}_0$ is maximal abelian in $\mathfrak{p}_0$.

Lemma 7.145. With notation as in Lemma 7.143 and with the E_{γ_j}'s normalized so that $[E_{\gamma_j}, \overline{E_{\gamma_j}}] = 2|\gamma_j|^{-2} H_{\gamma_j}$, let $Z = \sum_{j=1}^{s} t_j (E_{\gamma_j} + \overline{E_{\gamma_j}})$ be in $\mathfrak{a}_0$. Then

$$\exp Z = \exp X_0 \exp H_0 \exp Y_0 \tag{7.146}$$

with

$$X_0 = \sum (\tanh t_j) E_{\gamma_j} \in \mathfrak{p}^+, \qquad Y_0 = \sum (\tanh t_j) \overline{E_{\gamma_j}} \in \mathfrak{p}^-,$$
$$H_0 = \sum (\log \cosh t_j)[E_{\gamma_j}, \overline{E_{\gamma_j}}] \in i\mathfrak{t}_0 \subseteq \mathfrak{k}.$$

Moreover the P^+ components $\exp X_0$ of $\exp Z$ remain bounded as Z varies through $\mathfrak{a}_0$.

REMARK. The given normalization is the one used with Cayley transforms in §VI.7 and in particular is permissible.

PROOF. For the special case that $G = SU(1,1) \subseteq SL(2, \mathbb{C})$, (7.146) is just the identity

$$\begin{pmatrix} \cosh t & \sinh t \\ \sinh t & \cosh t \end{pmatrix} = \begin{pmatrix} 1 & \tanh t \\ 0 & 1 \end{pmatrix} \begin{pmatrix} (\cosh t)^{-1} & 0 \\ 0 & \cosh t \end{pmatrix} \begin{pmatrix} 1 & 0 \\ \tanh t & 1 \end{pmatrix}.$$

Here we are using $E_\gamma = \begin{pmatrix} 0 & 1 \\ 0 & 0 \end{pmatrix}$ and $\overline{E_\gamma} = \begin{pmatrix} 0 & 0 \\ 1 & 0 \end{pmatrix}$.

We can embed the special case in the general case for each γ_j, $1 \le j \le s$, since the inclusion

$$\mathfrak{sl}(2,\mathbb{C}) = \mathbb{C}H_{\gamma_j} + \mathbb{C}E_{\gamma_j} + \mathbb{C}\overline{E_{\gamma_j}} \subseteq \mathfrak{g}$$

induces a homomorphism $SL(2,\mathbb{C}) \to G^{\mathbb{C}}$, $SL(2,\mathbb{C})$ being simply connected. This embedding handles each of the s terms of Z separately. Since the γ_j's are strongly orthogonal, the contributions to X_0, Y_0, and H_0 for γ_i commute with those for γ_j when $i \ne j$, and (7.146) follows for general Z.

Finally in the expression for X_0, the coefficients of each E_{γ_j} lie between -1 and $+1$ for all Z. Hence $\exp X_0$ remains bounded in P^+.

This completes the proof of Theorem 7.129. Let us see what it means in examples. First suppose that $\mathfrak{g}_0$ is simple. For $\mathfrak{c}_0$ to be nonzero, $\mathfrak{g}_0$ must certainly be noncompact. Consider the Vogan diagram of $\mathfrak{g}_0$ in a good ordering. Lemma 7.128 rules out having the sum of two positive noncompact roots be a root. Since the sum of any connected set of simple roots in a Dynkin diagram is a root, it follows that there cannot be two or more noncompact simple roots in the Vogan diagram. Hence there is just one noncompact simple root, and the Vogan diagram is one of those considered in §VI.10. Since there is just one noncompact simple root and that root cannot occur twice in any positive root, every positive noncompact root has the same restriction to $\mathfrak{c}_0$. In particular, $\dim \mathfrak{c}_0 = 1$.

To see the possibilities, we can refer to the classification in §VI.10 and see that $\mathfrak{c}_0 \ne 0$ for the following cases and only these up to isomorphism:

(7.147)

$\mathfrak{g}_0$	$\mathfrak{k}_0$
$\mathfrak{su}(p,q)$	$\mathfrak{su}(p) \oplus \mathfrak{su}(q) \oplus \mathbb{R}$
$\mathfrak{so}(2,n)$	$\mathfrak{so}(n) \oplus \mathbb{R}$
$\mathfrak{sp}(n,\mathbb{R})$	$\mathfrak{su}(n) \oplus \mathbb{R}$
$\mathfrak{so}^*(2n)$	$\mathfrak{su}(n) \oplus \mathbb{R}$
E III	$\mathfrak{so}(10) \oplus \mathbb{R}$
E VII	$\mathfrak{e}_6 \oplus \mathbb{R}$

Conversely each of these cases corresponds to a group G satisfying the condition $Z_{\mathfrak{g}_0}(\mathfrak{c}_0) = \mathfrak{k}_0$, and hence G/K is Hermitian in each case.

If $\mathfrak{g}_0$ is merely semisimple, then the condition $Z_{\mathfrak{g}_0}(\mathfrak{c}_0) = \mathfrak{k}_0$ forces the center of the component of $\mathfrak{k}_0$ in each noncompact simple component of $\mathfrak{g}_0$ to be nonzero. The corresponding G/K is then the product of spaces obtained in the preceding paragraph.

10. Problems

1. Prove that the orthogonal group $O(2n)$ does not satisfy property (v) of a reductive Lie group.

2. Let $\widetilde{SL}(2,\mathbb{R})$ be the universal covering group of $SL(2,\mathbb{R})$, and let φ be the covering homomorphism. Let $\tilde{K}$ be the subgroup of $\widetilde{SL}(2,\mathbb{R})$ fixed by the global Cartan involution Θ. Parametrize $\tilde{K} \cong \mathbb{R}$ so that $\ker\varphi = \mathbb{Z}$. Define $\tilde{G} = \widetilde{SL}(2,\mathbb{R}) \times \mathbb{R}$, and extend Θ to $\tilde{G}$ so as to be 1 in the second factor. Within the subgroup $\mathbb{R}\times\mathbb{R}$ where Θ is 1, let D be the discrete subgroup generated by $(0,1)$ and $(1,\sqrt{2})$, so that D is central in $\tilde{G}$. Define $G = \tilde{G}/D$.
   (a) Prove that G is a connected reductive Lie group with ${}^0G = G$.
   (b) Prove that G_{ss} has infinite center and is not closed in G.

3. In $G = SL(n,\mathbb{R})$, take $M_\mathfrak{p} A_\mathfrak{p} N_\mathfrak{p}$ to be the upper-triangular subgroup.
   (a) Follow the prescription of Proposition 7.76 to see that the proposition leads to all possible block upper-triangular subgroups of $SL(n,\mathbb{R})$.
   (b) Give a direct proof for $SL(n,\mathbb{R})$ that the only closed subgroups containing $M_\mathfrak{p} A_\mathfrak{p} N_\mathfrak{p}$ are the block upper-triangular subgroups.
   (c) Give a direct proof for $SL(n,\mathbb{R})$ that no two distinct block upper-triangular subgroups are conjugate within $SL(n,\mathbb{R})$.

4. In the notation for $G = SL(4,\mathbb{R})$ as in §VI.4, form the parabolic subgroup MAN containing the upper-triangular group and corresponding to the subset $\{f_3 - f_4\}$ of simple restricted roots.
   (a) Prove that the $\mathfrak{a}_0$ roots are $\pm(f_1 - f_2)$, $\pm(f_1 - \frac{1}{2}(f_3 + f_4))$, and $\pm(f_2 - \frac{1}{2}(f_3 + f_4))$.
   (b) Prove that the $\mathfrak{a}_0$ roots do not all have the same length and do not form a root system.

5. Show that a maximal proper parabolic subgroup MAN of $SL(3,\mathbb{R})$ is cuspidal and that $M \neq M_0 Z_M$.

6. For G equal to split G_2, show that there is a cuspidal maximal proper parabolic subgroup MAN such that the set of $\mathfrak{a}_0$ roots is of the form $\{\pm\eta, \pm 2\eta, \pm 3\eta\}$.

7. The group $G = Sp(2,\mathbb{R})$ has at most four nonconjugate Cartan subalgebras, according to §VI.7, and a representative of each conjugacy class is given in that section.
   (a) For each of the four, construct the MA of an associated cuspidal parabolic subgroup as in Proposition 7.87.
   (b) Use the result of (a) to show that the two Cartan subalgebras of noncompact dimension one are not conjugate.

8. Let G be $SO(n,2)_0$.
   (a) Show that $G^{\mathbb{C}} \cong SO(n+2,\mathbb{C})$.
   (b) Show that $Z_{\mathfrak{g}_0}(\mathfrak{c}_0) = \mathfrak{k}_0$.
   (c) The isomorphism in (a) identifies the root system of $SO(n,2)$ as of type $B_{(n+1)/2}$ if n is odd and of type $D_{(n+2)/2}$ if n is even. Identify which roots are compact and which are noncompact.
   (d) Decide on some particular good ordering in the sense of §9, and identify the positive roots.

Problems 9–12 concern a reductive Lie group G. Notation is as in §2.

9. Let $\mathfrak{a}_0$ be maximal abelian in $\mathfrak{p}_0$. The natural inclusion $N_K(\mathfrak{a}_0) \subseteq N_G(\mathfrak{a}_0)$ induces a homomorphism $N_K(\mathfrak{a}_0)/Z_K(\mathfrak{a}_0) \to N_G(\mathfrak{a}_0)/Z_G(\mathfrak{a}_0)$. Prove that this homomorphism is an isomorphism.

10. Let $\mathfrak{t}_0 \oplus \mathfrak{a}_0$ be a maximally noncompact θ stable Cartan subalgebra of $\mathfrak{g}_0$. Prove that every element of $N_K(\mathfrak{a}_0)$ decomposes as a product zn, where n is in $N_K(\mathfrak{t}_0 \oplus \mathfrak{a}_0)$ and z is in $Z_K(\mathfrak{a}_0)$.

11. Let H be a Cartan subgroup of G, and let s_α be a root reflection in $W(\mathfrak{g},\mathfrak{h})$.
   (a) Prove that s_α is in $W(G,H)$ if α is real or α is compact imaginary.
   (b) Prove that if H is compact and G is connected, then s_α is not in $W(G,H)$ when α is noncompact imaginary.
   (c) Give an example of a reductive Lie group G with a compact Cartan subgroup H such that s_α is in $W(G,H)$ for some noncompact imaginary root α.

12. Let $H = TA$ be the global Cartan decomposition of a Θ stable Cartan subgroup of G. Let $W(G,A) = N_G(\mathfrak{a}_0)/Z_G(\mathfrak{a}_0)$, and let $M = {}^0Z_G(\mathfrak{a}_0)$. Let $W_1(G,H)$ be the subgroup of $W(G,H)$ of elements normalizing $i\mathfrak{t}_0$ and $\mathfrak{a}_0$ separately.
   (a) Show that restriction to $\mathfrak{a}_0$ defines a homomorphism of $W_1(G,H)$ into $W(G,A)$.
   (b) Prove that the homomorphism in (a) is onto.
   (c) Prove that the kernel of the homomorphism in (a) may be identified with $W(M,T)$.

Problems 13–21 concern a reductive Lie group G that is semisimple. Notation is as in §2.

13. Let $\mathfrak{t}_0 \oplus \mathfrak{a}_0$ be a maximally noncompact θ stable Cartan subalgebra of $\mathfrak{g}_0$, impose an ordering on the roots that takes $\mathfrak{a}_0$ before $i\mathfrak{t}_0$, let $\mathfrak{b}$ be a Borel subalgebra of $\mathfrak{g}$ containing $\mathfrak{t} \oplus \mathfrak{a}$ and built from that ordering, and let bar denote the conjugation of $\mathfrak{g}$ with respect to $\mathfrak{g}_0$. Prove that the smallest Lie subalgebra of $\mathfrak{g}$ containing $\mathfrak{b}$ and $\bar{\mathfrak{b}}$ is the complexification of a minimal parabolic subalgebra of $\mathfrak{g}_0$.

14. Prove that $N_{\mathfrak{g}_0}(\mathfrak{k}_0) = \mathfrak{k}_0$.

15. Let G have a complexification $G^{\mathbb{C}}$. Prove that the normalizer of $\mathfrak{g}_0$ in $G^{\mathbb{C}}$ is a reductive Lie group.

16. Let G have a complexification $G^{\mathbb{C}}$, let $U \subseteq G^{\mathbb{C}}$ be the analytic subgroup with Lie algebra $\mathfrak{k}_0 \oplus i\mathfrak{p}_0$, and let $\mathfrak{h}_0 = \mathfrak{t}_0 \oplus \mathfrak{a}_0$ be the decomposition into $+1$ and -1 eigenspaces of a θ stable Cartan subalgebra of $\mathfrak{g}_0$. Prove that $\exp i\mathfrak{a}_0$ is closed in U.

17. Give an example of a semisimple G with complexification $G^{\mathbb{C}}$ such that $K \cap \exp i\mathfrak{a}_0$ strictly contains $K_{\text{split}} \cap \exp i\mathfrak{a}_0$. Here $\mathfrak{a}_0$ is assumed maximal abelian in $\mathfrak{p}_0$.

18. Suppose that G has a complexification $G^{\mathbb{C}}$ and that $\operatorname{rank} G = \operatorname{rank} K$. Prove that $Z_{G^{\mathbb{C}}} = Z_G$.

19. Suppose that $\operatorname{rank} G = \operatorname{rank} K$. Prove that any two complexifications of G are holomorphically isomorphic.

20. Show that the conclusions of Problems 18 and 19 are false for $G = SL(3, \mathbb{R})$.

21. Suppose that G/K is Hermitian and that $\mathfrak{g}_0$ is simple. Show that there are only two ways to impose a G invariant complex structure on G/K.

Problems 22–24 compare the integer span of the roots with the integer span of the compact roots. It is assumed that G is a reductive Lie group with $\operatorname{rank} G = \operatorname{rank} K$.

22. Fix a positive system Δ^+. Attach to each simple noncompact root the integer 1 and to each simple compact root the integer 0; extend additively to the group generated by the roots, obtaining a function $\gamma \to n(\gamma)$. Arguing as in Lemma 6.98, prove that $n(\gamma)$ is odd when γ is a positive noncompact root and is even when γ is a positive compact root.

23. Making use of the function $\gamma \to (-1)^{n(\gamma)}$, prove that a noncompact root can never be an integer combination of compact roots.

24. Suppose that G is semisimple, that $\mathfrak{g}_0$ is simple, and that G/K is not Hermitian. Prove that the lattice generated by the compact roots has index 2 in the lattice generated by all the roots.

Problems 25–29 give further properties of semisimple groups with $\operatorname{rank} G = \operatorname{rank} K$. Let $\mathfrak{t}_0 \subseteq \mathfrak{k}_0$ be a Cartan subalgebra of $\mathfrak{g}_0$, and form roots, compact and noncompact.

25. K acts on $\mathfrak{p}$ via the adjoint representation. Identify the weights as the noncompact roots, showing in particular that 0 is not a weight.

26. Show that the subalgebras of $\mathfrak{g}$ containing $\mathfrak{k}$ are of the form $\mathfrak{k} \oplus \bigoplus_{\alpha \in E} \mathfrak{g}_\alpha$ for some subset E of noncompact roots.

27. Suppose that $\mathfrak{k} \oplus \bigoplus_{\alpha\in E} \mathfrak{g}_\alpha$ is a subalgebra of $\mathfrak{g}$. Prove that

$$\mathfrak{k} \oplus \sum_{\alpha\in E} (\mathfrak{g}_\alpha \oplus \mathfrak{g}_{-\alpha}) \qquad \text{and} \qquad \mathfrak{k} \oplus \bigoplus_{\alpha\in(E\cap(-E))} \mathfrak{g}_\alpha$$

are subalgebras of $\mathfrak{g}$ that are the complexifications of subalgebras of $\mathfrak{g}_0$.

28. Suppose that $\mathfrak{g}_0$ is simple. Prove that the adjoint representation of K on $\mathfrak{p}$ splits into at most two irreducible pieces.

29. Suppose that $\mathfrak{g}_0$ is simple, and suppose that the adjoint representation of K on $\mathfrak{p}$ is reducible (necessarily into two pieces, according to Problem 28). Show that the center $\mathfrak{c}_0$ of $\mathfrak{k}_0$ is nonzero, that $Z_{\mathfrak{g}_0}(\mathfrak{c}_0) = \mathfrak{k}_0$, and that the irreducible pieces are $\mathfrak{p}^+$ and $\mathfrak{p}^-$.

Problems 30–33 concern the group $G = SU(n,n) \cap Sp(n,\mathbb{C})$. In the notation of §9, let Ω be the set of all $Z \in M_{nn}(\mathbb{C})$ such that $1_n - Z^*Z$ is positive definite and $Z = Z^t$.

30. Using Problem 15b from Chapter VI, prove that $G \cong Sp(n,\mathbb{R})$.

31. With the members of G written in block form, show that (7.112) defines an action of G on Ω by holomorphic transformations.

32. Identify the isotropy subgroup of G at 0 with

$$K = \left\{ \begin{pmatrix} A & 0 \\ 0 & \bar{A} \end{pmatrix} \,\middle|\, A \in U(n) \right\}.$$

33. The diagonal subalgebra of $\mathfrak{g}_0$ is a compact Cartan subalgebra. Exhibit a good ordering such that $\mathfrak{p}^+$ consists of block strictly upper-triangular matrices.

Problems 34–36 concern the group $G = SO^*(2n)$. In the notation of §9, let Ω be the set of all $Z \in M_{nn}(\mathbb{C})$ such that $1_n - Z^*Z$ is positive definite and $Z = -Z^t$.

34. With the members of G written in block form, show that (7.112) defines an action of G on Ω by holomorphic transformations.

35. Identify the isotropy subgroup of G at 0 with

$$K = \left\{ \begin{pmatrix} A & 0 \\ 0 & \bar{A} \end{pmatrix} \,\middle|\, A \in U(n) \right\}.$$

36. The diagonal subalgebra of $\mathfrak{g}_0$ is a compact Cartan subalgebra. Exhibit a good ordering such that $\mathfrak{p}^+$ consists of block strictly upper-triangular matrices.

Problems 37–41 concern the restricted roots in cases when G is semisimple and G/K is Hermitian.

37. In the example of §9 with $G = SU(n, m)$,
    (a) show that the roots γ_j produced in Lemma 7.143 are $\gamma_1 = e_1 - e_{n+m}$, $\gamma_2 = e_2 - e_{n+m-1}, \dots, \gamma_m = e_m - e_{m+1}$.
    (b) show that the restricted roots (apart from Cayley transform) always include all $\pm\gamma_j$ and all $\frac{1}{2}(\pm\gamma_i \pm \gamma_j)$. Show that there are no other restricted roots if $m = n$ and that $\pm\frac{1}{2}\gamma_i$ are the only other restricted roots if $m < n$.
38. In the example of Problems 30–33 with $G = SU(n, n) \cap Sp(n, \mathbb{C}) \cong Sp(n, \mathbb{R})$,
    (a) show that the roots γ_j produced in Lemma 7.143 are $\gamma_1 = 2e_1, \dots, \gamma_n = 2e_n$.
    (b) show that the restricted roots (apart from Cayley transform) are all $\pm\gamma_j$ and all $\frac{1}{2}(\pm\gamma_i \pm \gamma_j)$.
39. In the example of Problem 6 of Chapter VI and Problems 34–36 above with $G = SO^*(2n)$,
    (a) show that the roots γ_j produced in Lemma 7.143 are $\gamma_1 = e_1 + e_n$, $\gamma_2 = e_2 + e_{n-1}, \dots, \gamma_{[n/2]} = e_{[n/2]} + e_{n-[n/2]+1}$.
    (b) find the restricted roots apart from Cayley transform.
40. For general G with G/K Hermitian, suppose that α, β, and γ are roots with α compact and with β and γ positive noncompact in a good ordering. Prove that $\alpha + \beta$ and $\alpha + \beta + \gamma$ cannot both be roots.
41. Let the expansion of a root in terms of Lemma 7.143 be $\gamma = \sum_{i=1}^{s} c_i\gamma_i + \gamma'$ with γ' orthogonal to $\gamma_1, \dots, \gamma_s$.
    (a) Prove for each i that $2c_i$ is an integer with $|2c_i| \le 3$.
    (b) Rule out $c_i = -\frac{3}{2}$ by using Problem 40 and the γ_i string containing γ, and rule out $c_i = +\frac{3}{2}$ by applying this conclusion to $-\gamma$.
    (c) Rule out $c_i = \pm 1$ for some $j \ne i$ by a similar argument.
    (d) Show that $c_i \ne 0$ for at most two indices i by a similar argument.
    (e) Deduce that each restricted root, apart from Cayley transform, is of one of the forms $\pm\gamma_i$, $\frac{1}{2}(\pm\gamma_i \pm \gamma_j)$, or $\pm\frac{1}{2}\gamma_i$.
    (f) If $\mathfrak{g}_0$ is simple, conclude that the restricted root system is of type $(BC)_s$ or C_s.

Problems 42–44 yield a realization of G/K, in the Hermitian case, as a particularly nice unbounded open subset Ω' of P^+. Let notation be as in §9.

42. In the special case that $G = SU(1,1)$, let u be the Cayley transform matrix $\frac{1}{\sqrt{2}}\begin{pmatrix} 1 & i \\ i & 1 \end{pmatrix}$, let $G' = SL(2,\mathbb{R})$, and let

$$\Omega' = \left\{ \begin{pmatrix} 1 & z \\ 0 & 1 \end{pmatrix} \middle| \operatorname{Im} z > 0 \right\}.$$

It is easily verified that $uGu^{-1} = G'$. Prove that $uGB = G'uB = \Omega' K^{\mathbb{C}} P^-$ and that G' acts on Ω' by the usual action of $SL(2,\mathbb{R})$ on the upper half plane.

43. In the general case as in §9, let $\gamma_1, \ldots, \gamma_s$ be constructed as in Lemma 7.143. For each j, construct an element u_j in $G^{\mathbb{C}}$ that behaves for the 3-dimensional group corresponding to γ_j like the element u of Problem 42. Put $u = \prod_{j=1}^{s} u_j$.
   (a) Exhibit u as in $P^+ K^{\mathbb{C}} P^-$.
   (b) Let $\mathfrak{a}_0$ be the maximal abelian subspace of $\mathfrak{p}_0$ constructed in Lemma 7.143, and let $A_\mathfrak{p} = \exp \mathfrak{a}_0$. Show that $uA_\mathfrak{p}u^{-1} \subseteq K^{\mathbb{C}}$.
   (c) Show for a particular ordering on $\mathfrak{a}_0^*$ that $uN_\mathfrak{p}u^{-1} \subseteq P^+ K^{\mathbb{C}}$ if $N_\mathfrak{p}$ is built from the positive restricted roots.
   (d) Writing $G = N_\mathfrak{p} A_\mathfrak{p} K$ by the Iwasawa decomposition, prove that $uGB \subseteq P^+ K^{\mathbb{C}} P^-$.

44. Let $G' = uGu^{-1}$. Prove that $G'uB = \Omega' K^{\mathbb{C}} P^-$ for some open subset Ω' of P^+. Prove also that the resulting action of G' on Ω' is holomorphic and transitive, and identify Ω' with G/K.

CHAPTER VIII

Integration

Abstract. An m-dimensional manifold M that is oriented admits a notion of integration $f \mapsto \int_M f\omega$ for any smooth m form. Here f can be any continuous real-valued function of compact support. This notion of integration behaves in a predictable way under diffeomorphism. When ω satisfies a positivity condition relative to the orientation, the integration defines a measure on M. A smooth map $M \to N$ with $\dim M < \dim N$ carries M to a set of measure zero.

For a Lie group G, a left Haar measure is a nonzero Borel measure invariant under left translations. Such a measure results from integration of ω if $M = G$ and if the form ω is positive and left invariant. A left Haar measure is unique up to a multiplicative constant. Left and right Haar measures are related by the modular function, which is given in terms of the adjoint representation of G on its Lie algebra. A group is unimodular if its Haar measure is two-sided invariant. Unimodular Lie groups include those that are abelian or compact or semisimple or reductive or nilpotent.

When a Lie group G has the property that almost every element is a product of elements of two closed subgroups S and T with compact intersection, then the left Haar measures on G, S, and T are related. As a consequence, Haar measure on a reductive Lie group has a decomposition that mirrors the Iwasawa decomposition, and also Haar measure satisfies various relationships with the Haar measures of parabolic subgroups. These integration formulas lead to a theorem of Helgason that characterizes and parametrizes irreducible finite-dimensional representations of G with a nonzero K fixed vector.

The Weyl Integration Formula tells how to integrate over a compact connected Lie group by first integrating over conjugacy classes. It is a starting point for an analytic treatment of parts of representation theory for such groups. Harish-Chandra generalized the Weyl Integration Formula to reductive Lie groups that are not necessarily compact. The formula relies on properties of Cartan subgroups proved in Chapter VII.

1. Differential Forms and Measure Zero

Let M be an m-dimensional manifold, understood to be smooth and to have a countable base for its topology; M need not be connected. We say that M is **oriented** if an atlas of compatible charts $(U_\alpha, \varphi_\alpha)$ is given with the property that the m-by-m derivative matrices of all coordinate changes

$$(8.1)\qquad \varphi_\beta \circ \varphi_\alpha^{-1} : \varphi_\alpha(U_\alpha \cap U_\beta) \to \varphi_\beta(U_\alpha \cap U_\beta)$$

have everywhere positive determinant. When M is oriented, a compatible chart (U, φ) is said to be **positive** relative to $(U_\alpha, \varphi_\alpha)$ if the derivative matrix of $\varphi \circ \varphi_\alpha^{-1}$ has everywhere positive determinant for all α. We always have the option of adjoining to the given atlas of charts for an oriented M any or all other compatible charts (U, φ) that are positive relative to all $(U_\alpha, \varphi_\alpha)$, and M will still be oriented.

On an oriented M as above, there is a well defined notion of integration involving smooth m forms, which is discussed in Chapter V of Chevalley [1946], Chapter X of Helgason [1962], and elsewhere. In this section we shall review the definition and properties, and then we shall apply the theory in later sections in the context of Lie groups.

We shall make extensive use of **pullbacks** of differential forms. If $\Phi : M \to N$ is smooth and if ω is a smooth k form on N, then $\Phi^*\omega$ is the smooth k form on M given by

$$(8.2)\qquad (\Phi^*\omega)_p(\xi_1, \ldots, \xi_k) = \omega_{\Phi(p)}(d\Phi_p(\xi_1), \ldots, d\Phi_p(\xi_k))$$

for p in M and $\xi_1, \ldots, \xi_k$ in the tangent space $T_p(M)$; here $d\Phi_p$ is the differential of Φ at p. In case M and N are open subsets of $\mathbb{R}^m$ and ω is the smooth m form $F(y_1, \ldots, y_m)\, dy_1 \wedge \cdots \wedge dy_m$ on N, the formula for $\Phi^*\omega$ on M is

$$(8.3)\qquad \Phi^*\omega = (F \circ \Phi)(x_1, \ldots, x_m) \det(\Phi'(x_1, \ldots, x_m))\, dx_1 \wedge \cdots \wedge dx_m,$$

where Φ has m entries $y_1(x_1, \ldots, x_m), \ldots, y_m(x_1, \ldots, x_m)$ and where Φ' denotes the derivative matrix $\left(\dfrac{\partial y_i}{\partial x_j}\right)$.

Let ω be a smooth m form on M. The theory of integration provides a definition of $\int_M f\omega$ for all f in the space $C_{\text{com}}(M)$ of continuous functions of compact support on M. Namely we first assume that f is compactly supported in a coordinate neighborhood U_α. The local expression for ω in $\varphi_\alpha(U_\alpha)$ is

$$(8.4)\qquad (\varphi_\alpha^{-1})^*\omega = F_\alpha(x_1, \ldots, x_m)\, dx_1 \wedge \cdots \wedge dx_m$$

with $F_\alpha : \varphi_\alpha(U_\alpha) \to \mathbb{R}$ smooth. Since $f \circ \varphi_\alpha^{-1}$ is compactly supported in $\varphi_\alpha(U_\alpha)$, it makes sense to define

$$(8.5a)\qquad \int_M f\omega = \int_{\varphi_\alpha(U_\alpha)} (f \circ \varphi_\alpha^{-1})(x_1, \ldots, x_m) F_\alpha(x_1, \ldots, x_m)\, dx_1 \cdots dx_m.$$

If f is compactly supported in an intersection $U_\alpha \cap U_\beta$, then the integral is given also by

$$(8.5\text{b}) \qquad \int_M f\omega = \int_{\varphi_\beta(U_\beta)} (f \circ \varphi_\beta^{-1})(y_1, \ldots, y_m) F_\beta(y_1, \ldots, y_m)\, dy_1 \cdots dy_m.$$

To see that the right sides of (8.5) are equal, we use the change of variables formula for multiple integrals. The change of variables $y = \varphi_\beta \circ \varphi_\alpha^{-1}(x)$ in (8.1) expresses $y_1, \ldots, y_m$ as functions of $x_1, \ldots, x_m$, and therefore (8.5b) is

$$\begin{aligned} &= \int_{\varphi_\beta(U_\alpha \cap U_\beta)} (f \circ \varphi_\beta^{-1})(y_1, \ldots, y_m) F_\beta(y_1, \ldots, y_m)\, dy_1 \cdots dy_m \\ &= \int_{\varphi_\alpha(U_\alpha \cap U_\beta)} f \circ \varphi_\beta^{-1} \circ \varphi_\beta \circ \varphi_\alpha^{-1}(x_1, \ldots, x_m) \\ &\qquad\qquad \times F_\beta \circ \varphi_\beta \circ \varphi_\alpha^{-1}(x_1, \ldots, x_m) |\det(\varphi_\beta \circ \varphi_\alpha^{-1})'|\, dx_1 \cdots dx_m. \end{aligned}$$

The right side here will be equal to the right side of (8.5a) if it is shown that

$$(8.6) \qquad F_\alpha \overset{?}{=} (F_\beta \circ \varphi_\beta \circ \varphi_\alpha^{-1}) |\det(\varphi_\beta \circ \varphi_\alpha^{-1})'|.$$

Now

$$\begin{aligned} F_\alpha\, dx_1 \wedge \cdots \wedge dx_m &= (\varphi_\alpha^{-1})^* \omega && \text{from (8.4)} \\ &= (\varphi_\beta \circ \varphi_\alpha^{-1})^* (\varphi_\beta^{-1})^* \omega \\ &= (\varphi_\beta \circ \varphi_\alpha^{-1})^* (F_\beta\, dy_1 \wedge \cdots dy_m) && \text{from (8.4)} \\ &= (F_\beta \circ \varphi_\beta \circ \varphi_\alpha^{-1}) \det(\varphi_\beta \circ \varphi_\alpha^{-1})'\, dx_1 \wedge \cdots \wedge dx_m && \text{by (8.3).} \end{aligned}$$

Thus

$$(8.7\text{a}) \qquad F_\alpha = (F_\beta \circ \varphi_\beta \circ \varphi_\alpha^{-1}) \det(\varphi_\beta \circ \varphi_\alpha^{-1})'.$$

Since $\det(\varphi_\beta \circ \varphi_\alpha^{-1})'$ is everywhere positive, (8.6) follows from (8.7a). Therefore $\int_M f\omega$ is well defined if f is compactly supported in $U_\alpha \cap U_\beta$.

For future reference we rewrite (8.7a) in terms of coordinates as

$$(8.7\text{b}) \qquad F_\beta(y_1, \ldots, y_m) = F_\alpha(x_1, \ldots, x_m) \det\left(\frac{\partial y_i}{\partial x_j}\right)^{-1}.$$

To define $\int_M f\omega$ for general f in $C_{\text{com}}(M)$, we make use of a smooth partition of unity $\{\psi_\alpha\}$ such that ψ_α is compactly supported in U_α and only finitely many ψ_α are nonvanishing on each compact set. Then $f = \sum \psi_\alpha f$ is actually a finite sum, and we can define

$$\int_M f\omega = \sum \int_M (\psi_\alpha f)\omega. \tag{8.8}$$

Using the consistency result proved above by means of (8.6), one shows that this definition is unchanged if the partition of unity is changed, and then $\int_M f_\omega$ is well defined. (See either of the above references.)

When ω is fixed, it is apparent from (8.5a) and (8.8) that the map $f \mapsto \int_M f\omega$ is a linear functional on $C_{\text{com}}(M)$. We say that ω is **positive** relative to the given atlas if each local expression (8.4) has $F_\alpha(x_1, \ldots, x_m)$ everywhere positive on $\varphi_\alpha(U_\alpha)$. In this case the linear functional $f \mapsto \int_M f\omega$ is positive in the sense that $f \geq 0$ implies $\int_M f\omega \geq 0$. By the Riesz Representation Theorem there exists a Borel measure $d\mu_\omega$ on M such that $\int_M f\omega = \int_M f(x)\, du_\omega(x)$ for all $f \in C_{\text{com}}(M)$. The first two propositions tell how to create and recognize positive ω's.

Proposition 8.9. If an m-dimensional manifold M admits a nowhere-vanishing m form ω, then M can be oriented so that ω is positive.

PROOF. Let $\{(U_\alpha, \varphi_\alpha)\}$ be an atlas for M. The components of each U_α are open and cover U_α. Thus there is no loss of generality in assuming that each coordinate neighborhood U_α is connected. For each U_α, let F_α be the function in (8.4) in the local expression for ω in $\varphi_\alpha(U_\alpha)$. Since ω is nowhere vanishing and U_α is connected, F_α has constant sign. If the sign is negative, we redefine φ_α by following it with the map $(x_1, x_2, \ldots, x_m) \mapsto (-x_1, x_2, \ldots, x_m)$, and then F_α is positive. In this way we can arrange that all F_α are positive on their domains. Referring to (8.7b), we see that each function $\det\left(\frac{\partial y_i}{\partial x_j}\right)$ is positive on its domain. Hence M is oriented. Since the F_α are all positive, ω is positive relative to this orientation.

Proposition 8.10. If a connected manifold M is oriented and if ω is a nowhere-vanishing smooth m form on M, then either ω is positive or $-\omega$ is positive.

PROOF. At each point p of M, all the functions F_α representing ω locally as in (8.4) have $F_\alpha(\varphi_\alpha(p))$ nonzero of the same sign because of (8.7b), the nowhere-vanishing of ω, and the fact that M is oriented. Let S be the set where this common sign is positive. Possibly replacing ω

by $-\omega$, we may assume that S is nonempty. We show that S is open and closed. Let p be in S and let U_α be a coordinate neighborhood containing p. Then $F_\alpha(\varphi_\alpha(p)) > 0$ since p is in S, and hence $F_\alpha \circ \varphi_\alpha$ is positive in a neighborhood of p. Hence S is open. Let $\{p_n\}$ be a sequence in S converging to p in M, and let U_α be a coordinate neighborhood containing p. Then $F_\alpha(\varphi_\alpha(p_n)) > 0$ and $F_\alpha(\varphi_\alpha(p)) \neq 0$. Since $\lim F_\alpha(\varphi_\alpha(p_n)) = F_\alpha(\varphi_\alpha(p))$, $F_\alpha(\varphi_\alpha(p))$ is > 0. Therefore p is in S, and S is closed. Since M is connected and S is nonempty open closed, $S = M$.

The above theory allows us to use nowhere-vanishing smooth m forms to define measures on manifolds. But we can define sets of measure zero without m forms and orientations. Let $\{(U_\alpha, \varphi_\alpha)\}$ be an atlas for the m-dimensional manifold M. We say that a subset S of M has **measure zero** if $\varphi_\alpha(S \cap U_\alpha)$ has m-dimensional Lebesgue measure 0 for all α.

Suppose that M is oriented and ω is a positive m form. If $d\mu_\omega$ is the associated measure and if ω has local expressions as in (8.4), then (8.5a) shows that

$$d\mu_\omega(S \cap U_\alpha) = \int_{\varphi_\alpha(S\cap U_\alpha)} F_\alpha(x_1, \ldots, x_m)\, dx_1 \cdots dx_m. \tag{8.11}$$

If S has measure zero in the sense of the previous paragraph, then the right side is 0 and hence $d\mu_\omega(S \cap U_\alpha) = 0$. Since a countable collection of U_α's suffices to cover M, $d\mu_\omega(S) = 0$. Thus a set a measure zero as in the previous paragraph has $d\mu_\omega(S) = 0$.

Conversely if ω is a nowhere-vanishing positive m form, $d\mu_\omega(S) = 0$ implies that S has measure zero as above. In fact, the left side of (8.11) is 0, and the integrand on the right side is > 0 everywhere. Therefore $\varphi_\alpha(S \cap U_\alpha)$ has Lebesgue measure 0.

Let $\Phi : M \to N$ be a smooth map between m-dimensional manifolds. A **critical point** p of Φ is a point where $d\Phi_p$ has rank $< m$. In this case, $\Phi(p)$ is called a **critical value**.

Theorem 8.12 (Sard's Theorem). If $\Phi : M \to N$ is a smooth map between m-dimensional manifolds, then the set of critical values of Φ has measure zero in N.

PROOF. About each point of M, we can choose a compatible chart (U, φ) so that $\Phi(U)$ is contained in a coordinate neighborhood of N. Countably many of these charts in M cover M, and it is enough to consider one of them. We may then compose with the coordinate mappings to see that it is enough to treat the following situation: Φ is a smooth map defined on a neighborhood of $C = \{x \in \mathbb{R}^m \mid 0 \leq x_i \leq 1 \text{ for } 1 \leq i \leq m\}$ with values in $\mathbb{R}^m$, and we are to prove that Φ of the critical points in C has Lebesgue measure 0 in $\mathbb{R}^m$.

For points $x = (x_1, \ldots, x_m)$ and $x' = (x_1', \ldots, x_m')$ in $\mathbb{R}^m$, the Mean Value Theorem gives

$$\Phi_i(x') - \Phi_i(x) = \sum_{j=1}^{m} \frac{\partial \Phi_i}{\partial x_j}(z_i)(x_j' - x_j), \tag{8.13}$$

where z_i is a point on the line segment from x to x'. Since the $\dfrac{\partial \Phi_i}{\partial x_j}$ are bounded on C, we see as a consequence that

$$\|\Phi(x') - \Phi(x)\| \le a\|x' - x\| \tag{8.14}$$

with a independent of x and x'. Let $L_x(x') = (L_{x,1}(x'), \ldots, L_{x,m}(x'))$ be the best first-order approximation to Φ about x, namely

$$L_{x,i}(x') = \Phi_i(x) + \sum_{j=1}^{m} \frac{\partial \Phi_i}{\partial x_j}(x)(x_j' - x_j). \tag{8.15}$$

Subtracting (8.15) from (8.13), we obtain

$$\Phi_i(x') - L_{x,i}(x') = \sum_{j=1}^{m} \left(\frac{\partial \Phi_i}{\partial x_j}(z_i) - \frac{\partial \Phi_i}{\partial x_j}(x) \right) (x_j' - x_j).$$

Since $\dfrac{\partial \Phi_i}{\partial x_j}$ is smooth and $\|z_i - x\| \le \|x' - x\|$, we deduce that

$$\|\Phi(x') - L_x(x')\| \le b\|x' - x\|^2 \tag{8.16}$$

with b independent of x and x'.

If x is a critical point, let us bound the image of the set of x' with $\|x' - x\| \le c$. The determinant of the linear part of L_x is 0, and hence L_x has image in a hyperplane. By (8.16), $\Phi(x')$ has distance $\le bc^2$ from this hyperplane. In each of the $m - 1$ perpendicular directions, (8.14) shows that $\Phi(x')$ and $\Phi(x)$ are at distance $\le ac$ from each other. Thus $\Phi(x')$ is contained in a rectangular solid about $\Phi(x)$ of volume $2^m(ac)^{m-1}(bc^2) = 2^m a^{m-1} b c^{m+1}$.

We subdivide C into N^m smaller cubes of side $1/N$. If one of these smaller cubes contains a critical point x, then any point x' in the smaller cube has $\|x' - x\| \le \sqrt{m}/N$. By the result of the previous paragraph, Φ of the cube is contained in a solid of volume $2^m a^{m-1} b(\sqrt{m}/N)^{m+1}$. The union of these solids, taken over all small cubes containing a critical point, contains the critical values. Since there are at most N^m cubes, the outer measure of the set of critical values is $\le 2^m a^{m-1} b m^{\frac{1}{2}(m+1)} N^{-1}$. This estimate is valid for all N, and hence the set of critical values has Lebesgue measure 0.

Corollary 8.17. If $\Phi : M \to N$ is a smooth map between manifolds with $\dim M < \dim N$, then the image of Φ has measure zero in N.

PROOF. Let $\dim M = k < m = \dim N$. Without loss of generality we may assume that $M \subseteq \mathbb{R}^k$. Sard's Theorem (Theorem 8.12) applies to the composition of the projection $\mathbb{R}^m \to \mathbb{R}^k$ followed by Φ. Every point of the domain is a critical point, and hence every point of the image is a critical value. The result follows.

We define a **lower-dimensional set** in N to be any set contained in the countable union of smooth images of manifolds M with $\dim M < \dim N$. It follows from Corollary 8.17 that

(8.18) any lower-dimensional set in N has measure zero.

Let M and N be oriented m-dimensional manifolds, and let $\Phi : M \to N$ be a diffeomorphism. We say that Φ is **orientation preserving** if, for every chart $(U_\alpha, \varphi_\alpha)$ in the atlas for M, the chart $(\Phi(U_\alpha), \varphi_\alpha \circ \Phi^{-1})$ is positive relative to the atlas for N. In this case the atlas of charts for N can be taken to be $\{(\Phi(U_\alpha), \varphi_\alpha \circ \Phi^{-1})\}$. Then the change of variables formula for multiple integrals may be expressed using pullbacks as in the following proposition.

Proposition 8.19. Let M and N be oriented m-dimensional manifolds, and let $\Phi : M \to N$ be an orientation-preserving diffeomorphism. If ω is a smooth m form on N, then

$$\int_N f\omega = \int_M (f \circ \Phi)\Phi^*\omega$$

for every $f \in C_{\text{com}}(N)$.

PROOF. Let the atlas for M be $\{(U_\alpha, \varphi_\alpha)\}$, and take the atlas for N to be $\{(\Phi(U_\alpha), \varphi_\alpha \circ \Phi^{-1})\}$. It is enough to prove the result for f compactly supported in a particular $\Phi(U_\alpha)$. For such f, (8.5) gives

(8.20a)

$$\int_N f\omega = \int_{\varphi_\alpha \circ \Phi^{-1}(\Phi(U_\alpha))} f \circ \Phi \circ \varphi_\alpha^{-1}(x_1, \dots, x_m) F_\alpha(x_1, \dots, x_m)\, dx_1 \cdots dx_m,$$

where F_α is the function with

(8.20b) $$((\varphi_\alpha \circ \Phi^{-1})^{-1})^*\omega = F_\alpha(x_1, \dots, x_m)\, dx_1 \wedge \cdots \wedge dx_m.$$

The function $f \circ \Phi$ is compactly supported in U_α, and (8.5) gives also

(8.20c)
$$\int_M (f \circ \Phi)\Phi^*\omega = \int_{\varphi_\alpha(U_\alpha)} f \circ \Phi \circ \varphi_\alpha^{-1}(x_1, \ldots, x_m) F_\alpha(x_1, \ldots, x_m)\, dx_1 \cdots dx_m$$

since

$$(\varphi_\alpha^{-1})^*\Phi^*\omega = ((\varphi_\alpha \circ \Phi^{-1})^{-1})^*\omega = F_\alpha(x_1, \ldots, x_m)\, dx_1 \wedge \cdots \wedge dx_m$$

by (8.20b). The right sides of (8.20a) and (8.20c) are equal, and hence so are the left sides.

2. Haar Measure for Lie Groups

Let G be a Lie group, and let $\mathfrak{g}$ be its Lie algebra. For $g \in G$, let $L_g : G \to G$ and $R_g : G \to G$ be the left and right translations $L_g(x) = gx$ and $R_g(x) = xg$. A smooth k form ω on G is **left invariant** if $L_g^*\omega$ for all $g \in G$, **right invariant** if $R_g^*\omega = \omega$ for all $g \in G$.

Regarding $\mathfrak{g}$ as the tangent space at 1 of G, let $X_1, \ldots, X_m$ be a basis of $\mathfrak{g}$, and let $\tilde{X}_1, \ldots, \tilde{X}_m$ be the corresponding left-invariant vector fields on G. We can define smooth 1 forms $\omega_1, \ldots, \omega_m$ on G by the condition that $(\omega_i)_p((\tilde{X}_j)_p) = \delta_{ij}$ for all p. Then $\omega_1, \ldots, \omega_m$ are left invariant, and at each point of G they form a basis of the dual of the tangent space at that point. The differential form $\omega = \omega_1 \wedge \cdots \wedge \omega_m$ is therefore a smooth m form that is nowhere vanishing on G. Since pullback commutes with $\wedge$, ω is left invariant. Using Proposition 8.9, we can orient G so that ω is positive. This proves part of the following theorem.

Theorem 8.21. If G is a Lie group of dimension m, then G admits a nowhere-vanishing left-invariant smooth m form ω. Then G can be oriented so that ω is positive, and ω defines a nonzero Borel measure $d\mu_l$ on G that is left invariant in the sense that $d\mu_l(L(g)E) = d\mu_l(E)$ for all $g \in G$ and every Borel set E in G.

PROOF. We have seen that ω exists and that G may be oriented so that ω is positive. Let $d\mu_l$ be the associated measure, so that $\int_G f\omega = \int_G f(x)\, d\mu_l(x)$ for all $f \in C_{\text{com}}(G)$. From Proposition 8.19 and the equality $L_g^*\omega = \omega$, we have

(8.22)
$$\int_G f(gx)\, d\mu_l(x) = \int_G f(x)\, d\mu_l(x)$$

for all $f \in C_{\mathrm{com}}(G)$. If K is a compact set in G, we can apply (8.22) to all f that are $\geq$ the characteristic function of K. Taking the infimum shows that $d\mu_l(L(g^{-1})K) = d\mu_l(K)$. Since G has a countable base, the measure $d\mu_l$ is automatically regular, and hence $d\mu_l(l(g^{-1})E) = d\mu_l(E)$ for all Borel sets E.

A nonzero Borel measure on G invariant under left translation is called a **left Haar measure** on G. Theorem 8.21 thus says that a left Haar measure exists.

In the construction of the left-invariant m form ω before Theorem 8.21, a different basis of G would have produced a multiple of ω, hence a multiple of the left Haar measure in Theorem 8.21. If the second basis is $Y_1, \ldots, Y_m$ and if $Y_j = \sum_{i=1}^m a_{ij} X_i$, then the multiple is $\det(a_{ij})^{-1}$. When the determinant is positive, we are led to orient G in the same way, otherwise oppositely. The new left Haar measure is $|\det(a_{ij})|^{-1}$ times the old. The next result strengthens this assertion of uniqueness of Haar measure.

Theorem 8.23. If G is a Lie group, then any two left Haar measures on G are proportional.

PROOF. Let $d\mu_1$ and $d\mu_2$ be left Haar measures. Then the sum $d\mu = d\mu_1 + d\mu_2$ is a left Haar measure, and $d\mu(E) = 0$ implies $d\mu_1(E) = 0$. By the Radon-Nikodym Theorem there exists a Borel function $h_1 \geq 0$ such that $d\mu_1 = h_1\, d\mu$. Fix g in G. By the left invariance of $d\mu_1$ and $d\mu$, we have

$$\begin{aligned}\int_G f(x)h_1(g^{-1}x)\,d\mu(x) &= \int_G f(gx)h_1(x)\,d\mu(x) = \int_G f(gx)\,d\mu_1(x)\\ &= \int_G f(x)\,d\mu_1(x) = \int_G f(x)h_1(x)\,d\mu(x)\end{aligned}$$

for every Borel function $f \geq 0$. Therefore the measures $h_1(g^{-1}x)\,d\mu(x)$ and $h_1(x)\,d\mu(x)$ are equal, and $h_1(g^{-1}x) = h_1(x)$ for almost every $x \in G$ (with respect to $d\mu$). We can regard $h_1(g^{-1}x)$ and $h_1(x)$ as functions of $(g, x) \in G \times G$, and these are Borel functions since the group operations are continuous. For each g, they are equal for almost every x. By Fubini's Theorem they are equal for almost every pair (g, x) (with respect to the product measure), and then for almost every x they are equal for almost every g. Pick such an x, say x_0. Then it follows that $h_1(x) = h_1(x_0)$ for almost every x. Thus $d\mu_1 = h_1(x_0)\,d\mu$. So $d\mu_1$ is a multiple of $d\mu$, and so is $d\mu_2$.

A **right Haar measure** on G is a nonzero Borel measure invariant under right translations. Such a measure may be constructed similarly by starting from right-invariant 1 forms and creating a nonzero right-invariant m form. As is true for left Haar measures, any two right Haar measures are proportional. To simplify the notation, we shall denote particular left and right Haar measures on G by $d_l x$ and $d_r x$, respectively.

An important property of left and right Haar measures is that

(8.24) any nonempty open set has nonzero Haar measure.

In fact, in the case of a left Haar measure if any compact set is given, finitely many left translates of the given open set together cover the compact set. If the open set had 0 measure, so would its left translates and so would every compact set. Then the measure would be identically 0 by regularity.

Another important property is that

(8.25) any lower-dimensional set in G has 0 Haar measure.

In fact, Theorems 8.21 and 8.23 show that left and right Haar measures are given by nowhere-vanishing differential forms. The sets of measure 0 relative to Haar measure are therefore the same as the sets of measure zero in the sense of Sard's Theorem, and (8.25) is a special case of (8.18).

Since left translations on G commute with right translations, $d_l(\cdot\, t)$ is a left Haar measure for any $t \in G$. Left Haar measures are proportional, and we therefore define the **modular function** $\Delta : G \to \mathbb{R}^+$ of G by

$$d_l(\cdot\, t) = \Delta(t)^{-1} d_l(\cdot). \tag{8.26}$$

Proposition 8.27. If G is a Lie group, then the modular function for G is given by $\Delta(t) = |\det \mathrm{Ad}(t)|$.

PROOF. If X is in $\mathfrak{g}$ and $\tilde{X}$ is the corresponding left-invariant vector field, then we can use Proposition 1.88 to make the computation

$$\begin{aligned}(dR_{t^{-1}})_p(\tilde{X}_p)h &= \tilde{X}_p(h \circ R_{t^{-1}}) = \frac{d}{dr} h(p(\exp rX)t^{-1})|_{r=0} \\ &= \frac{d}{dr} h(pt^{-1} \exp r\mathrm{Ad}(t)X)|_{r=0} = (\mathrm{Ad}(t)X)^{\sim} h(pt^{-1}),\end{aligned}$$

and the conclusion is that

$$(dR_{t^{-1}})_p(\tilde{X}_p) = (\mathrm{Ad}(t)X)^{\sim}_{pt^{-1}}. \tag{8.28}$$

Therefore the left-invariant m form ω has

$$\begin{aligned}(R^*_{t^{-1}}\omega)_p((\tilde{X}_1)_p, \dots, (\tilde{X}_m)_p) &= \omega_{pt^{-1}}((dR_{t^{-1}})_p(\tilde{X}_1)_p, \dots, (dR_{t^{-1}})_p(\tilde{X}_m)_p) \\ &= \omega_{pt^{-1}}((\mathrm{Ad}(t)X_1)\tilde{}_{pt^{-1}}, \dots, (\mathrm{Ad}(t)X_m)\tilde{}_{pt^{-1}}) \qquad \text{by (8.28)} \\ &= (\det \mathrm{Ad}(t))\omega_{pt^{-1}}((\tilde{X}_1)_{pt^{-1}}, \dots, (\tilde{X}_m)_{pt^{-1}}),\end{aligned}$$

and we obtain

$$R^*_{t^{-1}}\omega = (\det \mathrm{Ad}(t))\omega. \tag{8.29}$$

The assumption is that ω is positive, and therefore $R^*_{t^{-1}}\omega$ or $-R^*_{t^{-1}}\omega$ is positive according as the sign of $\det \mathrm{Ad}(t)$. When $\det \mathrm{Ad}(t)$ is positive, (8.29) and Proposition 8.19 give

$$\begin{aligned}(\det \mathrm{Ad}(t)) \int_G f(x)\, d_l x = (\det \mathrm{Ad}(t)) \int_G f\omega &= \int_G f\, R^*_{t^{-1}}\omega \\ &= \int_G (f \circ R_t)\omega = \int_G f(xt)\, d_l x \\ &= \int_G f(x)\, d_l(xt^{-1}) = \Delta(t) \int_G f(x)\, d_l x,\end{aligned}$$

and thus $\det \mathrm{Ad}(t) = \Delta(t)$. When $\det \mathrm{Ad}(t)$ is negative, every step of this computation is valid except for the first equality of the second line. Since $-R^*_{t^{-1}}\omega$ is positive, Proposition 8.19 requires a minus sign in its formula in order to apply to $\Phi = R_{t^{-1}}$. Thus $-\det \mathrm{Ad}(t) = \Delta(t)$. For all t, we therefore have $\Delta(t) = |\det \mathrm{Ad}(t)|$.

Corollary 8.30. The modular function Δ for G has the properties that

(a) $\Delta : G \to \mathbb{R}^+$ is a smooth homomorphism
(b) $\Delta(t) = 1$ for t in any compact subgroup of G and in any semisimple analytic subgroup of G
(c) $d_l(x^{-1})$ and $\Delta(x)\, d_l x$ are right Haar measures and are equal
(d) $d_r(x^{-1})$ and $\Delta(x)^{-1}\, d_r x$ are left Haar measures and are equal
(e) $d_r(t\,\cdot) = \Delta(t)\, d_r(\cdot)$ for any right Haar measure on G.

PROOF. Conclusion (a) is immediate from Proposition 8.27. The image under Δ of any compact subgroup of G is a compact subgroup of $\mathbb{R}^+$ and hence is $\{1\}$. This proves the first half of (b), and the second half follows from Lemma 4.28.

In (c) put $d\mu(x) = \Delta(x)\,d_l x$. This is a Borel measure since Δ is continuous (by (a)). Since Δ is a homomorphism, (8.26) gives

$$\begin{aligned}\int_G f(xt)\,d\mu(x) &= \int_G f(xt)\Delta(x)\,d_l x = \int_G f(x)\Delta(xt^{-1})\,d_l(xt^{-1}) \\ &= \int_G f(x)\Delta(x)\Delta(t^{-1})\Delta(t)\,d_l x \\ &= \int_G f(x)\Delta(x)\,d_l x = \int_G f(x)\,d\mu(x).\end{aligned}$$

Hence $d\mu(x)$ is a right Haar measure. It is clear that $d_l(x^{-1})$ is a right Haar measure, and thus Theorem 8.23 for right Haar measures implies that $d_l(x^{-1}) = c\Delta(x)\,d_l x$ for some constant $c > 0$. Changing x to x^{-1} in this formula, we obtain

$$d_l x = c\Delta(x^{-1})\,d_l(x^{-1}) = c^2\Delta(x^{-1})\Delta(x)\,d_l x = c^2\,d_l x.$$

Hence $c = 1$, and (c) is proved.

For (d) and (e) there is no loss of generality in assuming that $d_r x = d_l(x^{-1}) = \Delta(x)\,d_l x$, in view of (c). Conclusion (d) is immediate from this identity if we replace x by x^{-1}. For (e) we have

$$\begin{aligned}\int_G f(x)\,d_r(tx) &= \int_G f(t^{-1}x)\,d_r x = \int_G f(t^{-1}x)\Delta(x)\,d_l x \\ &= \int_G f(x)\Delta(tx)\,d_l x \\ &= \Delta(t)\int_G f(x)\Delta(x)\,d_l x = \Delta(t)\int_G f(x)\,d_r x,\end{aligned}$$

and we conclude that $d_r(t\,\cdot) = \Delta(t)\,d_r(\cdot)$.

The Lie group G is said to be **unimodular** if every left Haar measure is a right Haar measure (and vice versa). In this case we can speak of **Haar measure** on G. In view of (8.26), G is unimodular if and only if $\Delta(t) = 1$ for all $t \in G$.

Corollary 8.31. The following kinds of Lie groups are always unimodular:

(a) abelian Lie groups
(b) compact Lie groups
(c) semisimple Lie groups
(d) reductive Lie groups
(e) nilpotent Lie groups.

PROOF. Conclusion (a) is trivial, and (b) and (c) follow from Corollary 8.30b. For (d) let (G, K, θ, B) be reductive. By Proposition 7.27, $G \cong {}^0G \times Z_{vec}$. A left Haar measure for G may be obtained as the product of the left Haar measures of the factors, and (a) shows that Z_{vec} is unimodular. Hence it is enough to consider ${}^0G$, which is reductive by Proposition 7.27c. The modular function for ${}^0G$ must be 1 on K by Corollary 8.30b, and K meets every component of ${}^0G$. Thus it is enough to prove that ${}^0G_0$ is unimodular. This group is generated by its center and its semisimple part. The center is compact by Proposition 7.27, and the modular function must be 1 there, by Corollary 8.30b. Again by Corollary 8.30b, the modular function must be 1 on the semisimple part. Then (d) follows.

For (e) we appeal to Proposition 8.27. It is enough to prove that $\det \operatorname{Ad}(x) = 1$ for all x in G. By Theorem 1.104 the exponential map carries the Lie algebra $\mathfrak{g}$ onto G. If $x = \exp X$, then $\det \operatorname{Ad}(x) = \det e^{\operatorname{ad} X} = e^{\operatorname{Tr} \operatorname{ad} X}$. Since $\mathfrak{g}$ is nilpotent, (1.31) shows that $\operatorname{ad} X$ is a nilpotent linear transformation. Therefore 0 is the only generalized eigenvalue of $\operatorname{ad} X$, and $\operatorname{Tr} \operatorname{ad} X = 0$. This proves (e).

3. Decompositions of Haar Measure

In this section we let G be a Lie group, and we let $d_l x$ and $d_r x$ be left and right Haar measures for it.

Theorem 8.32. Let G be a Lie group, and let S and T be closed subgroups such that $S \cap T$ is compact, multiplication $S \times T \to G$ is an open map, and the set of products ST exhausts G except possibly for a set of Haar measure 0. Let Δ_T and Δ_G denote the modular functions of T and G. Then the left Haar measures on G, S, and T can be normalized so that

$$\int_G f(x)\, d_l x = \int_{S \times T} f(st) \frac{\Delta_T(t)}{\Delta_G(t)}\, d_l s\, d_l t$$

for all Borel functions $f \geq 0$ on G.

PROOF. Let $\Omega \subseteq G$ be the set of products ST, and let $K = S \cap T$. The group $S \times T$ acts continuously on Ω by $(s, t)\omega = s\omega t^{-1}$, and the isotropy subgroup at 1 is $\operatorname{diag} K$. Thus the map $(s, t) \mapsto st^{-1}$ descends to a map $(S \times T)/\operatorname{diag} K \to \Omega$. This map is a homeomorphism since multiplication $S \times T \to G$ is an open map.

Hence any Borel measure on Ω can be reinterpreted as a Borel measure on $(S \times T)/\operatorname{diag} K$. We apply this observation to the restriction of a left

Haar measure $d_l x$ for G from G to Ω, obtaining a Borel measure $d\mu$ on $(S \times T)/\operatorname{diag} K$. On Ω, we have

$$d_l(L_{s_0} R_{t_0^{-1}} x) = \Delta_G(t_0)\, d_l x$$

by (8.26), and the action unwinds to

$$d\mu(L_{(s_0,t_0)} x) = \Delta_G(t_0)\, d\mu(x) \tag{8.33}$$

on $(S \times T)/\operatorname{diag} K$. Define a measure $d\tilde{\mu}(s,t)$ on $S \times T$ by

$$\int_{S\times T} f(s,t)\, d\tilde{\mu}(s,t) = \int_{(S\times T)/\operatorname{diag} K} \left[\int_K f(sk,tk)\, dk\right] d\mu((s,t)K),$$

where dk is a Haar measure on K normalized to have total mass 1. From (8.33) it follows that

$$d\tilde{\mu}(s_0 s, t_0 t) = \Delta_G(t_0)\, d\tilde{\mu}(s,t).$$

The same proof as for Theorem 8.23 shows that any two Borel measures on $S \times T$ with this property are proportional, and $\Delta_G(t)\, d_l s\, d_l t$ is such a measure. Therefore

$$d\tilde{\mu}(s,t) = \Delta_G(t)\, d_l s\, d_l t$$

for a suitable normalization of $d_l s\, d_l t$.

The resulting formula is

$$\int_\Omega f(x)\, d_l x = \int_{S\times T} f(st^{-1}) \Delta_G(t)\, d_l s\, d_l t$$

for all Borel functions $f \geq 0$ on Ω. On the right side the change of variables $t \mapsto t^{-1}$ makes the right side become

$$\int_{S\times T} f(st) \Delta_G(t)^{-1}\, d_l s\, \Delta_T(t)\, d_l t,$$

according to Corollary 8.30c, and we can replace Ω by G on the left side since the complement of Ω in G has measure 0. This completes the proof.

If H is a closed subgroup of G, then we can ask whether G/H has a nonzero G invariant Borel measure. Theorem 8.36 below will give a necessary and sufficient condition for this existence, but we need some preparation. Fix a left Haar measure $d_l h$ for H. If f is in $C_{\text{com}}(G)$, define

$$f^{\#}(g) = \int_{G/H} f(gh)\, d_l h. \tag{8.34a}$$

This function is invariant under right translation by H, and we can define

$$f^{\#\#}(gH) = f^{\#}(g). \tag{8.34b}$$

The function $f^{\#\#}$ has compact support on G/H.

Lemma 8.35. The map $f \mapsto f^{\#\#}$ carries $C_{\text{com}}(G)$ onto $C_{\text{com}}(G/H)$, and a nonnegative member of $C_{\text{com}}(G/H)$ has a nonnegative preimage in $C_{\text{com}}(G)$.

PROOF. Let $\pi : G \to G/H$ be the quotient map. Let $F \in C_{\text{com}}(G/H)$ be given, and let K be a compact set in G/H with $F = 0$ off K. We first produce a compact set $\tilde{K}$ in G with $\pi(\tilde{K}) = K$. For each coset in K, select an inverse image x and let N_x be a compact neighborhood of x in G. Since π is open, π of the interior of N_x is open. These open sets cover K, and a finite number of them suffices. Then we can take $\tilde{K}$ to be the intersection of $\pi^{-1}(K)$ with the union of the finitely many N_x's.

Next let K_H be a compact neighborhood of 1 in H. By (8.24) the left Haar measure on H is positive on K_H. Let $\tilde{K}'$ be the compact set $\tilde{K}' = \tilde{K}K_H$, so that $\pi(\tilde{K}') = \pi(\tilde{K}) = K$. Choose $f_1 \in C_{\text{com}}(G)$ with $f_1 \geq 0$ everywhere and with $f_1 = 1$ on $\tilde{K}'$. If g is in $\tilde{K}'$, then $\int_H f_1(gh)\,d_lh$ is $\geq$ the H measure of K_H, and hence $f_1^{\#\#}$ is > 0 on K. Define

$$f(g) = \begin{cases} f_1(g)\dfrac{F(\pi(g))}{f_1^{\#\#}(\pi(g))} & \text{if } \pi(g) \in K \\ 0 & \text{otherwise.} \end{cases}$$

Then $f^{\#\#}$ is F on K and is 0 off K, so that $f^{\#\#} = F$ everywhere.

Certainly f has compact support. To see that f is continuous, it suffices to check that the two formulas for $f(g)$ fit together continuously at points g of $\pi^{-1}(K)$. It is enough to check points where $f(g) \neq 0$. Say $g_n \to g$. We must have $F(\pi(g)) \neq 0$. Since F is continuous, $F(\pi(g_n)) \neq 0$ eventually. Thus for all n sufficiently large, $f(g_n)$ is given by the first of the two formulas. Thus f is continuous.

Theorem 8.36. Let G be a Lie group, let H be a closed subgroup, and let Δ_G and Δ_H be the respective modular functions. Then a necessary and sufficient condition for G/H to have a nonzero G invariant Borel measure is that the restriction to H of Δ_G equal Δ_H. In this case such a measure $d\mu(gH)$ is unique up to a scalar, and it can be normalized so that

$$\int_G f(g)\,d_lg = \int_{G/H}\Big[\int_H f(gh)\,d_lh\Big]\,d\mu(gH) \tag{8.37}$$

for all $f \in C_{\text{com}}(G)$.

PROOF. Let $d\mu(gH)$ be such a measure. In the notation of (8.34), we can define a measure $d\tilde{\mu}(g)$ on G by

$$\int_G f(g)\, d\tilde{\mu}(g) = \int_{G/H} f^{\#\#}(gH)\, d\mu(gH).$$

Since $f \mapsto f^{\#\#}$ commutes with left translation by G, $d\tilde{\mu}$ is a left Haar measure on G. By Theorem 8.23, $d\tilde{\mu}$ is unique up to a scalar; hence $d\mu(gH)$ is unique up to a scalar.

Under the assumption that G/H has a nonzero invariant Borel measure, we have just seen in essence that we can normalize the measure so that (8.37) holds. If we replace f in (8.37) by $f(\cdot\, h_0)$, then the left side is multiplied by $\Delta_G(h_0)$, and the right side is multiplied by $\Delta_H(h_0)$. Hence $\Delta_G|_H = \Delta_H$ is necessary for existence.

Let us prove that this condition is sufficient for existence. Given $h \in C_{\text{com}}(G/H)$, we can choose $f \in C_{\text{com}}(G)$ by Lemma 8.35 so that $f^{\#\#} = h$. Then we define $L(h) = \int_G f(g)\, d_l g$. If L is well defined, then it is linear, Lemma 8.35 shows that it is positive, and L certainly is the same on a function as on its G translates. Therefore L defines a G invariant Borel measure $d\mu(gH)$ on G/H such that (8.37) holds.

Thus all we need to do is see that L is well defined if $\Delta_G|_H = \Delta_H$. We are thus to prove that if $f \in C_{\text{com}}(G)$ has $f^{\#} = 0$, then $\int_G f(g)\, d_l g = 0$. Let ψ be in $C_{\text{com}}(G)$. Then we have

$$\begin{aligned}
0 &= \int_G \psi(g) f^{\#}(g)\, d_l g = 0 \\
&= \int_G \Big[\int_H \psi(g) f(gh)\, d_l h\Big]\, d_l g \\
&= \int_H \Big[\int_G \psi(g) f(gh)\, d_l g\Big]\, d_l h \\
&= \int_H \Big[\int_G \psi(gh^{-1}) f(g)\, d_l g\Big]\, \Delta_G(h)\, d_l h && \text{by (8.26)} \\
&= \int_G f(g) \Big[\int_H \psi(gh^{-1}) \Delta_G(h)\, d_l h\Big]\, d_l g \\
&= \int_G f(g) \Big[\int_H \psi(gh) \Delta_G(h)^{-1} \Delta_H(h)\, d_l h\Big]\, d_l g && \text{by Corollary 8.30c} \\
&= \int_G f(g) \psi^{\#}(g)\, d_l g && \text{since } \Delta_G|_H = \Delta_H.
\end{aligned}$$

By Lemma 8.35 we can choose $\psi \in C_{\text{com}}(G)$ so that $\psi^{\#\#} = 1$ on the projection to G/H of the support of f. Then the right side is $\int_G f(g)\, d_l g$, and the conclusion is that this is 0. Thus L is well defined, and existence is proved.

4. Application to Reductive Lie Groups

Let (G, K, θ, B) be a reductive Lie group. We shall use the notation of Chapter VII, but we drop the subscripts 0 from real Lie algebras since we shall have relatively few occurrences of their complexifications. Thus, for example, the Cartan decomposition of the Lie algebra of G will be written $\mathfrak{g} = \mathfrak{k} \oplus \mathfrak{p}$.

In this section we use Theorem 8.32 and Proposition 8.27 to give decompositions of Haar measures that mirror group decompositions in Chapter VII. The group G itself is unimodular by Corollary 8.31d, and we write dx for a two-sided Haar measure. We shall be interested in parabolic subgroups MAN, and we need to compute the corresponding modular function that is given by Proposition 8.27 as

$$\Delta_{MAN}(man) = |\det \mathrm{Ad}_{\mathfrak{m}+\mathfrak{a}+\mathfrak{n}}(man)|.$$

For the element m, $|\det \mathrm{Ad}_{\mathfrak{m}+\mathfrak{a}+\mathfrak{n}}(m)| = 1$ by Corollary 8.30b. The element a acts as 1 on $\mathfrak{m}$ and $\mathfrak{a}$, and hence $\det \mathrm{Ad}_{\mathfrak{m}+\mathfrak{a}+\mathfrak{n}}(a) = \det \mathrm{Ad}_{\mathfrak{n}}(a)$. On an $\mathfrak{a}$ root space $\mathfrak{g}_\lambda$, a acts by $e^{\lambda \log a}$, and thus $\det \mathrm{Ad}_{\mathfrak{n}}(a) = e^{2\rho_A \log a}$, where $2\rho_A$ is the sum of all the positive $\mathfrak{a}$ roots with multiplicities counted. Finally $\det \mathrm{Ad}_{\mathfrak{m}+\mathfrak{a}+\mathfrak{n}}(n) = 1$ for the same reasons as in the proof of Corollary 8.31e. Therefore

$$\Delta_{MAN}(man) = |\det \mathrm{Ad}_{\mathfrak{m}+\mathfrak{a}+\mathfrak{n}}(man)| = e^{2\rho_A \log a}. \tag{8.38}$$

We can then apply Theorem 8.32 and Corollary 8.31 to obtain

$$d_l(man) = \frac{\Delta_N(n)}{\Delta_{MAN}(n)} d_l(ma)\, d_l n = dm\, da\, dn. \tag{8.39a}$$

By (8.38) and Corollary 8.30c,

$$d_r(man) = e^{2\rho_A \log a}\, dm\, da\, dn. \tag{8.39b}$$

Similarly for the subgroup AN of MAN, we have

$$\Delta_{AN}(an) = e^{2\rho_A \log a} \tag{8.40}$$

and

$$\begin{aligned} d_l(an) &= da\, dn \\ d_r(an) &= e^{2\rho_A \log a}\, da\, dn. \end{aligned} \tag{8.41}$$

Now we shall apply Theorem 8.32 to G itself. Combining Corollary 8.30c with the fact that G is unimodular, we can write

$$dx = d_l s\, d_r t \tag{8.42}$$

whenever the hypotheses in the theorem for S and T are satisfied.

Proposition 8.43. If $G = KA_\mathfrak{p}N_\mathfrak{p}$ is an Iwasawa decomposition of the reductive Lie group G, then the Haar measures of G, $A_\mathfrak{p}N_\mathfrak{p}$, $A_\mathfrak{p}$, and $N_\mathfrak{p}$ can be normalized so that

$$dx = dk\, d_r(an) = e^{2\rho_{A_\mathfrak{p}} \log a}\, dk\, da\, dn.$$

If the Iwasawa decomposition is written instead as $G = A_\mathfrak{p}N_\mathfrak{p}K$, then the decomposition of measures is

$$dx = d_l(an)\, dk = da\, dn\, dk.$$

PROOF. If G is written as $G = KA_\mathfrak{p}N_\mathfrak{p}$, then we use $S = K$ and $T = A_\mathfrak{p}N_\mathfrak{p}$ in Theorem 8.32. The hypotheses are satisfied since Proposition 7.31 shows that $S \times T \to G$ is a diffeomorphism. The second equality follows from (8.41). The argument when $G = A_\mathfrak{p}N_\mathfrak{p}K$ is similar.

Proposition 8.44. If G is a reductive Lie group and MAN is a parabolic subgroup, so that $G = KMAN$, then the Haar measures of G, MAN, M, A, and N can be normalized so that

$$dx = dk\, d_r(man) = e^{2\rho_A \log a} dk\, dm\, da\, dn.$$

PROOF. We use $S = K$ and $T = MAN$ in Theorem 8.32. Here $S \cap T = K \cap M$ is compact, and we know that $G = KMAN$. Since $A_\mathfrak{p}N_\mathfrak{p} \subseteq MAN$ and $K \times A_\mathfrak{p}N_\mathfrak{p} \to G$ is open, $K \times MAN \to G$ is open. Then Theorem 8.32 gives the first equality, and the second equality follows from (8.39b).

Proposition 8.45. If MAN is a parabolic subgroup of the reductive Lie group G, then N^-MAN is open in G and its complement is a lower-dimensional set, hence a set of measure 0. The Haar measures of G, MAN, N^-, M, A, and N can be normalized so that

$$dx = d\bar{n}\, d_r(man) = e^{2\rho_A \log a} d\bar{n}\, dm\, da\, dn \qquad (\bar{n} \in N^-).$$

PROOF. We use $S = N^-$ and $T = MAN$ in Theorem 8.32. Here $S \cap T = \{1\}$ by Lemma 7.64, and $S \times T \to G$ is everywhere regular (hence open) by Lemma 6.44. We need to see that the complement of N^-MAN is lower dimensional and has measure 0. Let $M_\mathfrak{p}A_\mathfrak{p}N_\mathfrak{p} \subseteq MAN$ be a minimal parabolic subgroup. In the Bruhat decomposition of G as in Theorem 7.40, a double coset of $M_\mathfrak{p}A_\mathfrak{p}N_\mathfrak{p}$ is of the form

$$M_\mathfrak{p}A_\mathfrak{p}N_\mathfrak{p}wM_\mathfrak{p}A_\mathfrak{p}N_\mathfrak{p} = N_\mathfrak{p}wM_\mathfrak{p}A_\mathfrak{p}N_\mathfrak{p} = w(w^{-1}N_\mathfrak{p}w)M_\mathfrak{p}A_\mathfrak{p}N_\mathfrak{p},$$

where w is a representative in $N_K(\mathfrak{a}_\mathfrak{p})$ of a member of $N_K(\mathfrak{a}_\mathfrak{p})/M_\mathfrak{p}$. The double coset is thus a translate of $(w^{-1}N_\mathfrak{p}w)M_\mathfrak{p}A_\mathfrak{p}N_\mathfrak{p}$. To compute the dimension of this set, we observe that

$$\dim \operatorname{Ad}(w)^{-1}\mathfrak{n}_\mathfrak{p} + \dim(\mathfrak{m}_\mathfrak{p} \oplus \mathfrak{a}_\mathfrak{p} \oplus \mathfrak{n}_\mathfrak{p}) = \dim \mathfrak{g}.$$

Now $\operatorname{Ad}(w)^{-1}\mathfrak{n}_\mathfrak{p}$ has 0 intersection with $\mathfrak{m}_\mathfrak{p} \oplus \mathfrak{a}_\mathfrak{p} \oplus \mathfrak{n}_\mathfrak{p}$ if and only if $\operatorname{Ad}(w)^{-1}\mathfrak{n}_\mathfrak{p} = \theta\mathfrak{n}_\mathfrak{p}$, which happens for exactly one coset $wM_\mathfrak{p}$ by Proposition 7.32 and Theorem 2.63. This case corresponds to the open set $N_\mathfrak{p}^- M_\mathfrak{p}A_\mathfrak{p}N_\mathfrak{p}$. In the other cases, there is a closed positive-dimensional subgroup R_w of $w^{-1}N_\mathfrak{p}w$ such that the smooth map

$$w^{-1}N_\mathfrak{p}w \times M_\mathfrak{p}A_\mathfrak{p}N_\mathfrak{p} \to (w^{-1}N_\mathfrak{p}w)M_\mathfrak{p}A_\mathfrak{p}N_\mathfrak{p}$$

given by $(x, y) \mapsto xy^{-1}$ factors to a smooth map

$$(w^{-1}N_\mathfrak{p}w \times M_\mathfrak{p}A_\mathfrak{p}N_\mathfrak{p})/\operatorname{diag} R_w \to (w^{-1}N_\mathfrak{p}w)M_\mathfrak{p}A_\mathfrak{p}N_\mathfrak{p}.$$

Hence in these cases $(w^{-1}N_\mathfrak{p}w)M_\mathfrak{p}A_\mathfrak{p}N_\mathfrak{p}$ is the smooth image of a manifold of dimension $< \dim G$ and is lower dimensional in G.

This proves for $M_\mathfrak{p}A_\mathfrak{p}N_\mathfrak{p}$ that $N_\mathfrak{p}^- M_\mathfrak{p}A_\mathfrak{p}N_\mathfrak{p}$ is open with complement of lower dimension. By (8.25) the complement is of Haar measure 0. Now let us consider N^-MAN. Since $M_\mathfrak{p}A_\mathfrak{p}N_\mathfrak{p} \subseteq MAN$, we have

$$\begin{aligned} N_\mathfrak{p}^- M_\mathfrak{p}A_\mathfrak{p}N_\mathfrak{p} &= (M_\mathfrak{p}A_\mathfrak{p}N_\mathfrak{p}^-)M_\mathfrak{p}A_\mathfrak{p}N_\mathfrak{p} \\ &\subseteq (MAN^-)MAN = N^-MAN. \end{aligned}$$

Thus the open set N^-MAN has complement of lower dimension and hence of Haar measure 0.

Theorem 8.32 is therefore applicable, and we obtain $dx = d\bar{n}\, d_r(man)$. The equality $d\bar{n}\, d_r(man) = e^{2\rho_A \log a}\, d\bar{n}\, dm\, da\, dn$ follows from (8.39b).

Proposition 8.46. Let MAN be a parabolic subgroup of the reductive Lie group G, and let ρ_A be as in (8.38). For g in G, decompose g according to $G = KMAN$ as

$$g = \kappa(g)\mu(g)\exp H(g)\, n.$$

Then Haar measures, when suitably normalized, satisfy

$$\int_K f(k)\, dk = \int_{N^-} f(\kappa(\bar{n}))e^{-2\rho_A H(\bar{n})}\, d\bar{n}$$

for all continuous functions on K that are right invariant under $K \cap M$.

REMARK. The expressions $\kappa(g)$ and $\mu(g)$ are not uniquely defined, but $H(g)$ is uniquely defined, as a consequence of the Iwasawa decomposition, and $f(\kappa(\bar{n}))$ will be seen to be well defined because of the assumed right invariance under $K \cap M$.

PROOF. Given f continuous on K and right invariant under $K \cap M$, extend f to a function F on G by

$$F(kman) = e^{-2\rho_A \log a} f(k). \tag{8.47}$$

The right invariance of f under $K \cap M$ makes F well defined since $K \cap MAN = K \cap M$. Fix $\varphi \geq 0$ in $C_{\text{com}}(MAN)$ with

$$\int_{MAN} \varphi(man)\, d_l(man) = 1;$$

by averaging over $K \cap M$, we may assume that φ is left invariant under $K \cap M$. Extend φ to G by the definition $\varphi(kman) = \varphi(man)$; the left invariance of φ under $K \cap M$ makes φ well defined. Then

$$\int_{MAN} \varphi(xman)\, d_l(man) = 1 \qquad \text{for all } x \in G.$$

The left side of the formula in the conclusion is

$$\begin{aligned}
&\int_K f(k)\, dk \\
&\quad = \int_K f(k) \Big[\int_{MAN} \varphi(kman)\, d_l(man) \Big] dk \\
&\quad = \int_{K \times MAN} f(k) \varphi(kman) e^{-2\rho_A \log a}\, dk\, d_r(man) && \text{by (8.39)} \\
&\quad = \int_{K \times MAN} F(kman) \varphi(kman)\, dk\, d_r(man) && \text{by (8.47)} \\
&\quad = \int_G F(x) \varphi(x)\, dx && \text{by Proposition 8.44,}
\end{aligned}$$

while the right side of the formula is

$$\begin{aligned}
&\int_{N^-} f(\kappa(\bar{n})) e^{-2\rho_A H(\bar{n})}\, d\bar{n} \\
&\quad = \int_{N^-} F(\bar{n}) \Big[\int_{MAN} \varphi(\bar{n}man)\, d_l(man) \Big] d\bar{n} && \text{by (8.47)}
\end{aligned}$$

$$= \int_{N^- \times MAN} F(\bar{n}) e^{-2\rho_A \log a} \varphi(\bar{n}man)\, d\bar{n}\, d_r(man) \qquad \text{by (8.39)}$$

$$= \int_{N^- \times MAN} F(\bar{n}man) \varphi(\bar{n}man)\, d\bar{n}\, d_r(man) \qquad \text{by (8.47)}$$

$$= \int_G F(x)\varphi(x)\, dx \qquad \text{by Proposition 8.45.}$$

The proposition follows.

For an illustration of the use of Proposition 8.46, we shall prove a theorem of Helgason that has important applications in the harmonic analysis of G/K. We suppose that the reductive group G is semisimple and has a complexification $G^{\mathbb{C}}$. We fix an Iwasawa decomposition $G = KA_{\mathfrak{p}}N_{\mathfrak{p}}$. Let $\mathfrak{t}_{\mathfrak{p}}$ be a maximal abelian subspace of $\mathfrak{m}_{\mathfrak{p}}$, so that $\mathfrak{t}_{\mathfrak{p}} \oplus \mathfrak{a}_{\mathfrak{p}}$ is a maximally noncompact θ stable Cartan subalgebra of $\mathfrak{g}$. Representations of G yield representations of $\mathfrak{g}$, hence complex-linear representations of $\mathfrak{g}^{\mathbb{C}}$. Then the theory of Chapter V is applicable, and we use the complexification of $\mathfrak{t}_{\mathfrak{p}} \oplus \mathfrak{a}_{\mathfrak{p}}$ as Cartan subalgebra for that purpose. Let Δ and Σ be the sets of roots and restricted roots, respectively, and let Σ^+ be the set of positive restricted roots relative to $\mathfrak{n}_{\mathfrak{p}}$.

Roots and weights are real on $i\mathfrak{t}_{\mathfrak{p}} \oplus \mathfrak{a}_{\mathfrak{p}}$, and we introduce an ordering such that the nonzero restriction to $\mathfrak{a}_{\mathfrak{p}}$ of a member of Δ^+ is a member of Σ^+. By a **restricted weight** of a finite-dimensional representation, we mean the restriction to $\mathfrak{a}_{\mathfrak{p}}$ of a weight. We introduce in an obvious fashion the notions of **restricted-weight spaces** and **restricted-weight vectors**. Because of our choice of ordering, the restriction to $\mathfrak{a}_{\mathfrak{p}}$ of the highest weight of a finite-dimensional representation is the highest restricted weight.

Lemma 8.48. Let the reductive Lie group G be semisimple. If π is an irreducible complex-linear representation of $\mathfrak{g}^{\mathbb{C}}$, then $\mathfrak{m}_{\mathfrak{p}}$ acts in each restricted weight space of π, and the action by $\mathfrak{m}_{\mathfrak{p}}$ is irreducible in the highest restricted-weight space.

PROOF. The first conclusion follows at once since $\mathfrak{m}_{\mathfrak{p}}$ commutes with $\mathfrak{a}_{\mathfrak{p}}$. Let $v \neq 0$ be a highest restricted-weight vector, say with weight ν. Let V be the space for π, and let V_ν be the restricted-weight space corresponding to ν. We write $\mathfrak{g} = \theta\eta_{\mathfrak{p}} \oplus \mathfrak{m}_{\mathfrak{p}} \oplus \mathfrak{a}_{\mathfrak{p}} \oplus \mathfrak{n}_{\mathfrak{p}}$, express members of $U(\mathfrak{g}^{\mathbb{C}})$ in the corresponding basis given by the Poincaré-Birkhoff-Witt Theorem, and apply an element to v. Since $\mathfrak{n}_{\mathfrak{p}}$ pushes restricted weights up and $\mathfrak{a}_{\mathfrak{p}}$ acts by scalars in V_ν and $\theta\mathfrak{n}_{\mathfrak{p}}$ pushes weights down, we see from the irreducibility of π on V that $U(\mathfrak{m}_{\mathfrak{p}}^{\mathbb{C}})v = V_\nu$. Since v is an arbitrary nonzero member of V_ν, $\mathfrak{m}_{\mathfrak{p}}$ acts irreducibly on V_ν.

Theorem 8.49 (Helgason). Let the reductive Lie group G be semisimple and have a complexification $G^{\mathbb{C}}$. For an irreducible finite-dimensional representation π of G, the following statements are equivalent:

(a) π has a nonzero K fixed vector
(b) $M_{\mathfrak{p}}$ acts by the 1-dimensional trivial representation in the highest restricted-weight space of π
(c) the highest weight $\tilde{\nu}$ of π vanishes on $\mathfrak{t}_{\mathfrak{p}}$, and the restriction ν of $\tilde{\nu}$ to $\mathfrak{a}_{\mathfrak{p}}$ is such that $\langle \nu, \beta \rangle / |\beta|^2$ is an integer for every restricted root β.

Conversely any dominant $\nu \in \mathfrak{a}_{\mathfrak{p}}^*$ such that $\langle \nu, \beta \rangle / |\beta|^2$ is an integer for every restricted root β is the highest restricted weight of some irreducible finite-dimensional π with a nonzero K fixed vector.

PROOF. For the proofs that (a) through (c) are equivalent, there is no loss in generality in assuming that $G^{\mathbb{C}}$ is simply connected, as we may otherwise take a simply connected cover of $G^{\mathbb{C}}$ and replace G by the analytic subgroup of this cover with Lie algebra $\mathfrak{g}$. With $G^{\mathbb{C}}$ simply connected, the representation π of G yields a representation of $\mathfrak{g} = \mathfrak{k} \oplus \mathfrak{p}$, then of $\mathfrak{g}^{\mathbb{C}}$, and then of the compact form $\mathfrak{u} = \mathfrak{k} \oplus i\mathfrak{p}$. Since $G^{\mathbb{C}}$ is simply connected, so is the analytic subgroup U with Lie algebra $\mathfrak{u}$ (Theorem 6.31). The representation π therefore lifts from $\mathfrak{u}$ to U. By Proposition 4.6 we can introduce a Hermitian inner product on the representation space so that U acts by unitary operators. Then it folows that K acts by unitary operators and $i\mathfrak{t}_{\mathfrak{p}} \oplus \mathfrak{a}_{\mathfrak{p}}$ acts by Hermitian operators. In particular, distinct weight spaces are orthogonal, and so are distinct restricted-weight spaces.

(a) $\Rightarrow$ (b). Let ϕ_ν be a nonzero highest restricted-weight vector, and let ϕ_K be a nonzero K fixed vector. Since $\mathfrak{n}_{\mathfrak{p}}$ pushes restricted weights up and since the exponential map carries $\mathfrak{n}_{\mathfrak{p}}$ onto $N_{\mathfrak{p}}$ (Theorem 1.104), $\pi(n)\phi_\nu = \phi_\nu$ for $n \in N_{\mathfrak{p}}$. Therefore

$$(\pi(kan)\phi_\nu, \phi_K) = (\pi(a)\phi_\nu, \pi(k)^{-1}\phi_K) = e^{\nu \log a}(\phi_\nu, \phi_K).$$

By the irreducibility of π and the fact that $G = KA_{\mathfrak{p}}N_{\mathfrak{p}}$, the left side cannot be identically 0, and hence (ϕ_ν, ϕ_K) on the right side is nonzero. The inner product with ϕ_K is then an everywhere-nonzero linear functional on the highest restricted-weight space, and the highest restricted-weight space must be 1-dimensional. If ϕ_ν is a nonzero vector of norm 1 in this space, then $(\phi_K, \phi_\nu)\phi_\nu$ is the orthogonal projection of ϕ_K into this space. Since $M_{\mathfrak{p}}$ commutes with $\mathfrak{a}_{\mathfrak{p}}$, the action by $M_{\mathfrak{p}}$ commutes with this projection. But $M_{\mathfrak{p}}$ acts trivially on ϕ_K since $M_{\mathfrak{p}} \subseteq K$, and therefore $M_{\mathfrak{p}}$ acts trivially on ϕ_ν.

(b) $\Rightarrow$ (a). Let $v \neq 0$ be in the highest restricted-weight space, with restricted weight ν. Then $\int_K \pi(k)\,dk$ is obviously fixed by K, and the problem is to see that it is not 0. Since v is assumed to be fixed by $M_\mathfrak{p}$, $k \mapsto \pi(k)v$ is a function on K right invariant under $M_\mathfrak{p}$. By Proposition 8.46,

$$\int_K \pi(k)v\,dk = \int_{N_\mathfrak{p}^-} \pi(\kappa(\bar{n}))ve^{-2\rho_{A_\mathfrak{p}}H(\bar{n})}\,d\bar{n} = \int_{N_\mathfrak{p}^-} \pi(\bar{n})ve^{(-\nu-2\rho_{A_\mathfrak{p}})H(\bar{n})}\,d\bar{n}.$$

Here $e^{(-\nu-2\rho_{A_\mathfrak{p}})H(\bar{n})}$ is everywhere positive since ν is real, and $(\pi(\bar{n})v, v) = |v|^2$ since the exponential map carries $\theta\mathfrak{n}_\mathfrak{p}$ onto $N_\mathfrak{p}^-$, $\theta\mathfrak{n}_\mathfrak{p}$ lowers restricted weights, and the different restricted-weight spaces are orthogonal. Therefore $\left(\int_K \pi(k)v\,dk,\, v\right)$ is positive, and $\int_K \pi(k)v\,dk$ is not 0.

(b) $\Rightarrow$ (c). Since $(M_\mathfrak{p})_0$ acts trivially, it follows immediately that $\tilde{\nu}$ vanishes on $\mathfrak{t}_\mathfrak{p}$. For each restricted root β, define $\gamma_\beta = \exp 2\pi i|\beta|^{-2}H_\beta$ as in (7.57). This element is in $M_\mathfrak{p}$ by (7.58). Since $G^\mathbb{C}$ is simply connected, π extends to a holomorphic representation of $G^\mathbb{C}$. Then we can compute $\pi(\gamma_\beta)$ on a vector v of restricted weight ν as

$$\pi(\gamma_\beta)v = \pi(\exp(2\pi i|\beta|^{-2}\operatorname{ad} H_\beta))v = e^{2\pi i\langle\nu,\beta\rangle/|\beta|^2}v. \tag{8.50}$$

Since the left side equals v by (b), $\langle\nu,\beta\rangle/|\beta|^2$ must be an integer.

(c) $\Rightarrow$ (b). The action of $(M_\mathfrak{p})_0$ on the highest restricted-weight space is irreducible by Lemma 8.48. Since $\tilde{\nu}$ vanishes on $\mathfrak{t}_\mathfrak{p}$, the highest weight of this representation of $(M_\mathfrak{p})_0$ is 0. Thus $(M_\mathfrak{p})_0$ acts trivially, and the space is 1-dimensional. The calculation (8.50), in the presence of (c), shows that each γ_β acts trivially. Since the γ_β that come from real roots generate F (by Theorem 7.55) and since $M_\mathfrak{p} = (F)(M_\mathfrak{p})_0$ (by Corollary 7.52), $M_\mathfrak{p}$ acts trivially.

We are left with the converse statement. Suppose $\nu \in \mathfrak{a}_\mathfrak{p}^*$ is such that $\langle\nu,\beta\rangle/|\beta|^2$ is an integer ≥ 0 for all $\beta \in \Sigma^+$. Define $\tilde{\nu}$ to be ν on $\mathfrak{a}_\mathfrak{p}$ and 0 on $\mathfrak{t}_\mathfrak{p}$. We are to prove that $\tilde{\nu}$ is the highest weight of an irreducible finite-dimensional representation of G with a K fixed vector. The form $\tilde{\nu}$ is dominant. If it is algebraically integral, then Theorem 5.5 gives us a complex-linear representation π of $\mathfrak{g}^\mathbb{C}$ with highest weight $\tilde{\nu}$. Some finite covering group $\tilde{G}$ of G will have a simply connected complexification, and then π lifts to $\tilde{G}$. By the implication (c) $\Rightarrow$ (a), π has a nonzero $\tilde{K}$ fixed vector. Since the kernel of $\tilde{G} \to G$ is in $\tilde{K}$ and since such elements must then act trivially, π descends to a representation of G with a nonzero K fixed vector. In other words, it is enough to prove that $\tilde{\nu}$ is algebraically integral.

Let α be a root, and let β be its restriction to $\mathfrak{a}_\mathfrak{p}$. Since $\langle\tilde{\nu},\alpha\rangle = \langle\nu,\beta\rangle$, we may assume that $\beta \neq 0$. Let $|\alpha|^2 = C|\beta|^2$. Then

$$\frac{2\langle\tilde{\nu},\alpha\rangle}{|\alpha|^2} = \frac{2\langle\nu,\beta\rangle}{C|\beta|^2},$$

and it is enough to show that either

$$\text{(8.51a)} \qquad 2/C \text{ is an integer}$$

or

$$\text{(8.51b)} \qquad |2/C| = \tfrac{1}{2} \text{ and } \langle\nu,\beta\rangle/|\beta|^2 \text{ is even.}$$

Write $\alpha = \beta + \varepsilon$ with $\varepsilon \in i\mathfrak{t}_\mathfrak{p}^*$. Then $\theta\alpha$ is the root $\theta\alpha = -\beta + \varepsilon$. Thus $-\theta\alpha = \beta - \varepsilon$ is a root with the same length as α.

If α and $-\theta\alpha$ are multiples of one another, then $\varepsilon = 0$ and $C = 1$, so that $2/C$ is an integer. If α and $-\theta\alpha$ are not multiples of one another, then the Schwarz inequality gives

$$\text{(8.52)} \qquad \begin{aligned}(-1 \text{ or } 0 \text{ or } +1) &= \frac{2\langle\alpha,-\theta\alpha\rangle}{|\alpha|^2} = \frac{2\langle\beta+\varepsilon,\beta-\varepsilon\rangle}{|\alpha|^2} \\ &= \frac{2(|\beta|^2-|\varepsilon|^2)}{|\alpha|^2} = \frac{2(2|\beta|^2-|\alpha|^2)}{|\alpha|^2} = \frac{4}{C}-2.\end{aligned}$$

If the left side of (8.52) is -1, then $2/C = \frac{1}{2}$. Since the left side of (8.52) is -1, $\alpha - \theta\alpha = 2\beta$ is a root, hence also a restricted root. By assumption, $\langle\nu,2\beta\rangle/|2\beta|^2$ is an integer; hence $\langle\nu,\beta\rangle/|\beta|^2$ is even. Thus (8.51b) holds. If the left side of (8.52) is 0, then $2/C = 1$ and (8.51a) holds.

To complete the proof, we show that the left side of (8.52) cannot be $+1$. If it is $+1$, then $\alpha - (-\theta\alpha) = 2\varepsilon$ is a root vanishing on $\mathfrak{a}_\mathfrak{p}$, and hence any root vector for it is in $\mathfrak{m}_\mathfrak{p}^\mathbb{C} \subseteq \mathfrak{k}^\mathbb{C}$. However this root is also equal to $\alpha + \theta\alpha$, and $[X_\alpha, \theta X_\alpha]$ must be a root vector. Since $\theta[X_\alpha,\theta X_\alpha] = -[X_\alpha,\theta X_\alpha]$, $[X_\alpha,\theta X_\alpha]$ is in $\mathfrak{p}^\mathbb{C}$. Thus the root vector is in $\mathfrak{k}^\mathbb{C} \cap \mathfrak{p}^\mathbb{C} = 0$, and we have a contradiction.

5. Weyl Integration Formula

The original Weyl Integration Formula tells how to integrate over a compact connected Lie group by first integrating over each conjugacy class and then integrating over the set of conjugacy classes. Let G be a compact connected Lie group, let T be a maximal torus, and let $\mathfrak{g}_0$

and $\mathfrak{t}_0$ be the respective Lie algebras. Let $m = \dim G$ and $l = \dim T$. As in §VII.8, an element g of G is **regular** if the eigenspace of $\text{Ad}(g)$ for eigenvalue 1 has dimension l. Let G' and T' be the sets of regular elements in G and T; these are open subsets of G and T, respectively.

Theorem 4.36 implies that the smooth map $G \times T \to G$ given by $\psi(g, t) = gtg^{-1}$ is onto G. Fix $g \in G$ and $t \in T$. If we identify tangent spaces at g, t, and gtg^{-1} with $\mathfrak{g}_0$, $\mathfrak{t}_0$, and $\mathfrak{g}_0$ by left translation, then (4.45) computes the differential of ψ at (g, t) as

$$\psi(X, H) = \text{Ad}(g)((\text{Ad}(t^{-1}) - 1)X + H) \qquad \text{for } X \in \mathfrak{g}_0,\ H \in \mathfrak{t}_0.$$

The map ψ descends to $G/T \times T \to G$, and we call the descended map ψ also. We may identify the tangent space of G/T with an orthogonal complement $\mathfrak{t}_0^\perp$ to $\mathfrak{t}_0$ in $\mathfrak{g}_0$ (relative to an invariant inner product). The space $\mathfrak{t}_0^\perp$ is invariant under $\text{Ad}(t^{-1}) - 1$, and we can write

$$d\psi(X, H) = \text{Ad}(g)((\text{Ad}(t^{-1}) - 1)X + H) \qquad \text{for } X \in \mathfrak{t}_0^\perp,\ H \in \mathfrak{t}_0.$$

Now $d\psi$ at (g, t) is essentially a map of $\mathfrak{g}_0$ to itself, with matrix

$$\begin{array}{cc} & \begin{array}{cc} \mathfrak{t}_0 & \quad\mathfrak{t}_0^\perp \end{array} \\ (d\psi)_{(g,t)} = \text{Ad}(g) & \begin{pmatrix} 1 & 0 \\ 0 & \text{Ad}(t^{-1}) - 1 \end{pmatrix}. \end{array}$$

Since $\det \text{Ad}(g) = 1$ by compactness and connectedness of G,

$$\det(d\psi)_{(g,t)} = \det((\text{Ad}(t^{-1}) - 1)|_{\mathfrak{t}_0^\perp}). \tag{8.53}$$

We can think of building a left-invariant $(m - l)$ form on G/T from the duals of the X's in $\mathfrak{t}_0^\perp$ and a left-invariant l form on T from the duals of the H's in $\mathfrak{t}_0$. We may think of a left-invariant m form on G as the wedge of these forms. Referring to Proposition 8.19 and (8.7b) and taking (8.53) into account, we at first expect an integral formula

(8.54a)
$$\int_G f(x)\,dx \overset{?}{=} \int_T \Big[\int_{G/T} f(gtg^{-1})\,d(gT)\Big] \,|\det(\text{Ad}(t^{-1}) - 1)|_{\mathfrak{t}_0^\perp}|\,dt$$

if the measures are normalized so that

$$\int_G f(x)\,dx = \int_{G/T} \Big[\int_T f(xt)\,dt\Big]\,d(xT). \tag{8.54b}$$

But Proposition 8.19 fails to be applicable in two ways. One is that the onto map $\psi : G/T \times T \to G$ has differential of determinant 0 at some points, and the other is that ψ is not one-one even if we exclude points of the domain where the differential has determinant 0.

From (8.53) we can exclude the points where the differential has determinant 0 if we restrict ψ to a map $\psi : G/T \times T' \to G'$. To understand T', consider $\mathrm{Ad}(t^{-1})-1$ as a linear map of the complexification $\mathfrak{g}$ to itself. If $\Delta = \Delta(\mathfrak{g}, \mathfrak{t})$ is the set of roots, then $\mathrm{Ad}(t^{-1}) - 1$ is diagonable with eigenvalues 0 with multiplicity l and also $\xi_\alpha(t^{-1}) - 1$ with multiplicity 1 each. Hence $\left|\det(\mathrm{Ad}(t^{-1}) - 1)|_{\mathfrak{t}_0^\perp}\right| = \left|\prod_{\alpha\in\Delta} (\xi_\alpha(t^{-1}) - 1)\right|$. If we fix a positive system Δ^+ and recognize that $\xi_\alpha(t^{-1}) = \overline{\xi_{-\alpha}(t^{-1})}$, then we see that

$$\left|\det(\mathrm{Ad}(t^{-1}) - 1)|_{\mathfrak{t}_0^\perp}\right| = \prod_{\alpha\in\Delta^+} |\xi_\alpha(t^{-1}) - 1|^2. \tag{8.55}$$

Putting $t = \exp iH$ with $iH \in \mathfrak{t}_0$, we have $\xi_\alpha(t^{-1}) = e^{-i\alpha(H)}$. Thus the set in the torus where (8.55) is 0 is a countable union of lower-dimensional sets and is a lower-dimensional set. By (8.25) the singular set in T has dt measure 0. The singular set in G is the smooth image of the product of G/T and the singular set in T, hence is lower dimensional and is of measure 0 for $d\mu(gT)$. Therefore we may disregard the singular set and consider ψ as a map $G/T \times T' \to G'$.

The map $\psi : G/T \times T' \to G'$ is not, however, one-one. If w is in $N_G(\mathfrak{t}_0)$, then

$$\psi(gwT, w^{-1}tw) = \psi(gT, t). \tag{8.56}$$

Since $gwT \neq gT$ when w is not in $Z_G(\mathfrak{t}_0) = T$, each member of G' has at least $|W(G, T)|$ preimages. On the other hand, if $\psi(gT, s) = \psi(hT, t)$, then Proposition 4.53 shows that s and t are conjugate via $N_G(\mathfrak{t}_0)$. Say $s = w^{-1}tw$. Then (8.56) gives

$$\psi(hT, t) = \psi(gT, w^{-1}tw) = \psi(gw^{-1}T, t).$$

So $hth^{-1} = gw^{-1}t(gw^{-1})^{-1}$ and $wg^{-1}h$ centralizes t. Since t is regular and G has a complexification, Corollary 7.106 shows that $wg^{-1}h$ is in $N_G(\mathfrak{t}_0)$, say $wg^{-1}h = w'$. Then $h = gw^{-1}w'$, and the new feature beyond (8.56) is that

$$\psi(hT, t) = \psi(hw'^{-1}T, t) \qquad \text{if } w'^{-1}tw' = t. \tag{8.57}$$

Here is an example of how (8.57) can happen.

EXAMPLE. Let $G = SO(3)$ and $T = \begin{pmatrix} \cos\theta & \sin\theta & 0 \\ -\sin\theta & \cos\theta & 0 \\ 0 & 0 & 1 \end{pmatrix}$. Here T' corresponds to $\theta \neq 2\pi n$. If θ is not a multiple of π, then the corresponding element t of T' is not centralized by $w' = \begin{pmatrix} 0 & 1 & 0 \\ -1 & 0 & 0 \\ 0 & 0 & 1 \end{pmatrix}$. Thus for θ not a multiple of π, t has two preimages under ψ. But for $\theta = \pi$, the element t is regular but is centralized by w'. Thus t has more than two preimages under ψ:

$$(1, t), \quad (w', t), \quad (\operatorname{diag}(-1, 1, -1), t).$$

Lemma 8.58. The set of elements $t \in T'$ where $wtw^{-1} = t$ for some member w of $N_G(\mathfrak{t}_0)$ not in T is a relatively closed lower-dimensional set of dt measure 0.

PROOF. It is clear that the set in question is relatively closed, and (8.25) shows that it is enough to exhibit it as lower dimensional. Write such an element t as $\exp iH$ with $iH \in \mathfrak{t}_0$. Since $wtw^{-1} = t$, we have $\exp i\operatorname{Ad}(w)H = \exp iH$. Applying ξ_α for $\alpha \in \Delta$, we obtain $e^{i\alpha(\operatorname{Ad}(w)H)} = e^{i\alpha(H)}$. Thus $\alpha(\operatorname{Ad}(w)H) = \alpha(H) + 2\pi n$ for some n. If w is nontrivial, then the set of H's satisfying this equation for a single α and n is lower dimensional. Hence the set in question is lower dimensional.

Let T'' be the complement in T' of the exceptional set in Lemma 8.58. Put $G'' = \psi(G/T \times T'')$. Since T'' is open and ψ is everywhere regular on $G/T \times T'$, G'' is open. Lemma 8.58 and (8.25) show that the complement of T'' in T and the complement of G'' in G have measure 0. Thus in establishing an integration formula, we may consider ψ as a map $G/T \times T'' \to G''$. By restricting from T' to T'', we have eliminated the phenomenon (8.57). Therefore each member of G'' has exactly $|W(G,T)|$ preimages under ψ, given as in (8.56).

Now we return to Proposition 8.19. Instead of assuming that $\Phi : M \to N$ is an orientation-preserving diffeomorphism, we assume that Φ is an everywhere regular n-to-1 map of M onto N with $\dim M = \dim N$. Then the proof of Proposition 8.19 applies with easy modifications to give

$$n \int_N f\omega = \int_M (f \circ \Phi)\Phi^*\omega. \tag{8.59}$$

Therefore we have the following result in place of (8.54).

Theorem 8.60 (Weyl Integration Formula). Let T be a maximal torus of the compact connected Lie group G, and let invariant measures on G, T, and G/T be normalized so that

$$\int_G f(x)\,dx = \int_{G/T}\Big[\int_T f(xt)\,dt\Big]\,d(xT)$$

for all continuous f on G. Then every Borel function $F \geq 0$ on G satisfies

$$\int_G F(x)\,dx = \frac{1}{|W(G,T)|}\int_T\Big[\int_{G/T} F(gtg^{-1})\,d(gT)\Big]\,|D(t)|^2\,dt,$$

where
$$|D(t)|^2 = \prod_{\alpha\in\Delta^+} |1-\xi_\alpha(t^{-1})|^2.$$

The integration formula in Theorem 8.60 is a starting point for an analytic treatment of parts of representation theory for compact connected Lie groups. For a given such group for which δ is analytically integral, let us sketch how the theorem leads simultaneously to a construction of an irreducible representation with given dominant analytically integral highest weight and to a proof of the Weyl Character Formula.

Define

$$D(t) = \xi_\delta(t)\prod_{\alpha\in\Delta^+}(1-\xi_{-\alpha}(t)), \tag{8.61}$$

so that Theorem 8.60 for any Borel function f constant on conjugacy classes and either nonnegative or integrable reduces to

$$\int_G f(x)\,dx = \frac{1}{|W(G,T)|}\int_T f(t)|D(t)|^2\,dt \tag{8.62}$$

if we take dx, dt, and $d(gT)$ to have total mass one. For $\lambda\in\mathfrak{t}^*$ dominant and analytically integral, define

$$\chi_\lambda(t) = \frac{\sum_{s\in W(G,T)}\varepsilon(s)\xi_{s(\lambda+\delta)}(t)}{D(t)}.$$

Then χ_λ is invariant under $W(G,T)$, and Proposition 4.53 shows that $\chi_\lambda(t)$ extends to a function χ_λ on G constant on conjugacy classes. Applying (8.62) with $f=|\chi_\lambda|^2$, we see that

$$\int_G |\chi_\lambda|^2\,dx = 1. \tag{8.63a}$$

Applying (8.62) with $f = \chi_\lambda \overline{\chi_{\lambda'}}$, we see that

$$\int_G \chi_\lambda(x)\overline{\chi_{\lambda'}(x)}\,dx = 0 \qquad \text{if } \lambda \neq \lambda'. \tag{8.63b}$$

Let χ be the character of an irreducible finite-dimensional representation of G. On T, $\chi(t)$ must be of the form $\sum_\mu \xi_\mu(t)$, where the μ's are the weights repeated according to their multiplicities. Also $\chi(t)$ is even under $W(G,T)$. Then $D(t)\chi(t)$ is odd under $W(G,T)$ and is of the form $\sum_\nu n_\nu \xi_\nu(t)$ with each n_ν in $\mathbb{Z}$. Focusing on the dominant ν's and seeing that the ν's orthogonal to a root must drop out, we find that $\chi(t) = \sum_\lambda a_\lambda \chi_\lambda(t)$ with $a_\lambda \in \mathbb{Z}$. By (8.63),

$$\int_G |\chi(x)|^2\,dx = \sum_\lambda |a_\lambda|^2.$$

For an irreducible character Corollary 4.16 shows that the left side is 1. So one a_λ is ± 1 and the others are 0. Since $\chi(t)$ is of the form $\sum_\mu \xi_\mu(t)$, we readily find that $a_\lambda = +1$ for some λ. Hence every irreducible character is of the form $\chi = \chi_\lambda$ for some λ. This proves the Weyl Character Formula. Using the Peter-Weyl Theorem (Theorem 4.20), we readily see that no L^2 function on G that is constant on conjugacy classes can be orthogonal to all irreducible characters. Then it follows from (8.63b) that every χ_λ is an irreducible character. This proves the existence of an irreducible representation corresponding to a given dominant analytically integral form as highest weight.

For reductive Lie groups that are not necessarily compact, there is a formula analogous to Theorem 8.60. This formula is a starting point for the analytic treatment of representation theory on such groups. We state the result as Theorem 8.64 but omit the proof. The proof makes use of Theorem 7.108 and of other variants of results that we applied in the compact case.

Theorem 8.64 (Harish-Chandra). Let G be a reductive Lie group, let $(\mathfrak{h}_1)_0, \ldots, (\mathfrak{h}_r)_0$ be a maximal set of nonconjugate θ stable Cartan subalgebras of $\mathfrak{g}_0$, and let $H_1, \ldots, H_r$ be the corresponding Cartan subgroups. Let the invariant measures on each H_j and G/H_j be normalized so that

$$\int_G f(x)\,dx = \int_{G/H_j} \Big[\int_{H_j} f(gh)\,dh\Big]\,d(gH_j) \qquad \text{for all } f \in C_{\text{com}}(G).$$

Then every Borel function $F \geq 0$ on G satisfies

$$\int_G F(x)\,dx = \sum_{j=1}^{r} \frac{1}{|W(G, H_j)|} \int_{H_j} \Big[\int_{G/H_j} F(ghg^{-1})\,d(gH_j) \Big] |D_{H_j}(h)|^2\,dh,$$

where
$$|D_{H_j}(h)|^2 = \prod_{\alpha \in \Delta(\mathfrak{g}, \mathfrak{h}_j)} |1 - \xi_\alpha(h^{-1})|.$$

6. Problems

1. Prove that if M is an oriented m-dimensional manifold, then M admits a nowhere-vanishing smooth m form.
2. Prove that the zero locus of a nonzero real analytic function on a cube in $\mathbb{R}^n$ has Lebesgue measure 0.
3. Let G be the group of all real matrices $\begin{pmatrix} a & b \\ 0 & 1 \end{pmatrix}$ with $a > 0$. Show that $a^{-2}\,da\,db$ is a left Haar measure and that $a^{-1}\,da\,db$ is a right Haar measure.
4. Let G be a noncompact semisimple Lie group with finite center, and let $M_\mathfrak{p} A_\mathfrak{p} N_\mathfrak{p}$ be a minimal parabolic subgroup. Prove that $G/M_\mathfrak{p} A_\mathfrak{p} N_\mathfrak{p}$ has no nonzero G invariant Borel measure.
5. Prove that the complement of the set of regular points in a reductive Lie group G is a closed set of Haar measure 0.

Problems 6–8 concern Haar measure on $GL(n, \mathbb{R})$.

6. Why is Haar measure on $GL(n, \mathbb{R})$ two-sided invariant?
7. Regard $\mathfrak{gl}(n, \mathbb{R})$ as an n^2-dimensional vector space over $\mathbb{R}$. For each $x \in GL(n, \mathbb{R})$, let L_x denote left multiplication by x. Prove that $\det L_x = (\det x)^n$.
8. Let E_{ij} be the matrix that is 1 in the $(i, j)^{\text{th}}$ place and is 0 elsewhere. Regard $\{E_{ij}\}$ as the standard basis of $\mathfrak{gl}(n, \mathbb{R})$, and introduce Lebesgue measure accordingly.
   (a) Why is the set of $x \in \mathfrak{gl}(n, \mathbb{R})$ with $\det x = 0$ a set of Lebesgue measure 0?
   (b) Deduce from Problem 7 that $|\det y|^{-n}\,dy$ is a Haar measure for $GL(n, \mathbb{R})$.

Problems 9–12 concern the function $e^{\nu H_{\mathfrak{p}}(x)}$ for a semisimple Lie group G with a complexification $G^{\mathbb{C}}$. Here it is assumed that $G = KA_{\mathfrak{p}}N_{\mathfrak{p}}$ is an Iwasawa decomposition of G and that elements decompose as $x = \kappa(g)\exp H_{\mathfrak{p}}(x)\, n$. Let $\mathfrak{a}_{\mathfrak{p}}$ be the Lie algebra of $A_{\mathfrak{p}}$, and let ν be in $\mathfrak{a}_{\mathfrak{p}}^*$.

9. Let π be an irreducible finite-dimensional representation of G on V, and introduce a Hermitian inner product in V as in the proof of Theorem 8.49. If π has highest restricted weight ν and if v is in the restricted-weight space for ν, prove that $\|\pi(x)v\|^2 = e^{2\nu H_{\mathfrak{p}}(x)}\|v\|^2$.

10. In $G = SL(3,\mathbb{R})$, let $K = SO(3)$ and let $M_{\mathfrak{p}}A_{\mathfrak{p}}N_{\mathfrak{p}}$ be upper-triangular. Introduce parameters for $N_{\mathfrak{p}}^-$ by writing $N_{\mathfrak{p}}^- = \left\{\bar{n} = \begin{pmatrix} 1 & 0 & 0 \\ x & 1 & 0 \\ z & y & 1 \end{pmatrix}\right\}$. Let $f_1 - f_2$, $f_2 - f_3$, and $f_1 - f_3$ be the positive restricted roots as usual, and let $\rho_{\mathfrak{p}}$ denote half their sum (namely $f_1 - f_3$).
    (a) Show that $e^{2f_1 H_{\mathfrak{p}}(\bar{n})} = 1+x^2+z^2$ and $e^{2(f_1+f_2)H_{\mathfrak{p}}(\bar{n})} = 1+y^2+(z-xy)^2$ for $\bar{n} \in N_{\mathfrak{p}}^-$.
    (b) Deduce that $e^{2\rho_{\mathfrak{p}} H_{\mathfrak{p}}(\bar{n})} = (1+x^2+z^2)(1+y^2+(z-xy)^2)$ for $\bar{n} \in N_{\mathfrak{p}}^-$.

11. In $G = SO(n,1)_0$, let $K = SO(n) \times \{1\}$ and $\mathfrak{a}_{\mathfrak{p}} = \mathbb{R}(E_{1,n+1} + E_{n+1,1})$, with E_{ij} as in Problem 8. If $\lambda(E_{1,n+1} + E_{n+1,1}) > 0$, say that $\lambda \in \mathfrak{a}_{\mathfrak{p}}^*$ is positive, and obtain $G = KA_{\mathfrak{p}}N_{\mathfrak{p}}$ accordingly.
    (a) Using the standard representation of $SO(n,1)_0$, compute $e^{2\lambda H_{\mathfrak{p}}(x)}$ for a suitable λ and all $x \in G$.
    (b) Deduce a formula for $e^{2\rho_{\mathfrak{p}} H_{\mathfrak{p}}(x)}$ from the result of (a). Here $\rho_{\mathfrak{p}}$ is half the sum of the positive restricted roots repeated according to their multiplicities.

12. In $G = SU(n,1)$, let $K = S(U(n) \times U(1))$, and let $\mathfrak{a}_{\mathfrak{p}}$ and positivity be as in Problem 11. Repeat the two parts of Problem 11 for this group.

APPENDIX A

Tensors, Filtrations, and Gradings

Abstract. If E is a vector space, the tensor algebra $T(E)$ of E is the direct sum over $n \geq 0$ of the n-fold tensor product of E with itself. This is an associative algebra with the following universal mapping property: Any linear mapping of E into an associative algebra A with identity extends to an algebra homomorphism of $T(E)$ into A carrying 1 into 1.

The symmetric algebra $S(E)$ is a quotient of E with the following universal mapping property: Any linear mapping of E into a commutative associative algebra A with identity extends to an algebra homomorphism of $S(E)$ into A carrying 1 into 1. The symmetric algebra is commutative.

Similarly the exterior algebra $\bigwedge(E)$ is a quotient of E with this universal mapping property: Any linear mapping l of E into an associative algebra A with identity such that $l(v)^2 = 0$ for all $v \in E$ extends to an algebra homomorphism of $\bigwedge(E)$ into A carrying 1 into 1.

The tensor algebra, the symmetric algebra, and the exterior algebra are all examples of graded associative algebras. A more general notion than a graded algebra is that of a filtered algebra. A filtered associative algebra has an associated graded algebra. The notions of gradings and filtrations make sense in the context of vector spaces, and a linear map between filtered vector spaces that respects the filtration induces an associated graded map between the associated graded vector spaces. If the associated graded map is an isomorphism, then the original map is an isomorphism.

1. Tensor Algebra

Just as polynomial rings are often used in the construction of more general commutative rings, so tensor algebras are often used in the construction of more general rings that may not be commutative. In this section we construct the tensor algebra of a vector space as a direct sum of iterated tensor products of the vector space with itself, and we establish its properties. We shall proceed with care, in order to provide a complete proof of the associativity of the multiplication.

Fix a field $\mathbb{k}$. Let E and F be vector spaces over the field $\mathbb{k}$. A **tensor product** V of E and F is a pair (V, ι) consisting of a vector space V over $\mathbb{k}$ together with a bilinear map $\iota : E \times F \to V$, with the following universal mapping property: Whenever b is a bilinear mapping of $E \times F$ into a vector space U over $\Bbbk$, then there exists a unique linear mapping B of V into U such that the diagram

(A.1)

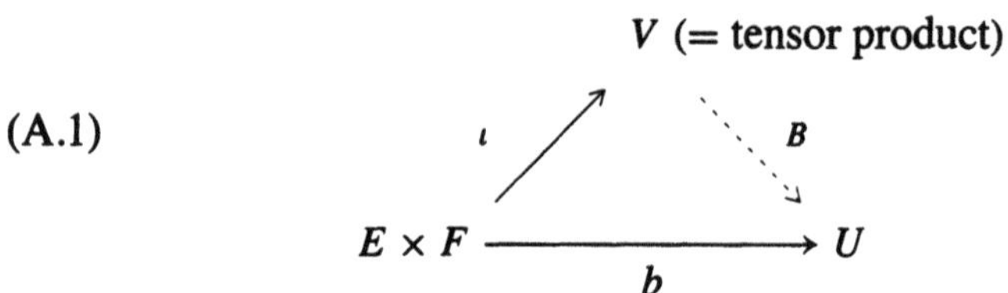

commutes. We call B the **linear extension** of b to the tensor product.

It is well known that a tensor product of E and F exists and is unique up to canonical isomorphism, and we shall not repeat the proof. One feature of the proof is that it gives an explicit construction of a vector space that has the required property.

A tensor product of E and F is denoted $E \otimes_{\Bbbk} F$, and the associated bilinear map ι is written $(e, f) \mapsto e \otimes f$. The elements $e \otimes f$ generate $E \otimes_{\Bbbk} F$, as a consequence of a second feature of the proof of existence of a tensor product.

There is a canonical isomorphism

$$E \otimes_{\Bbbk} F \cong F \otimes_{\Bbbk} E \tag{A.2}$$

given by taking the linear extension of $(e, f) \mapsto f \otimes e$ as the map from left to right. The linear extension of $(f, e) \mapsto e \otimes f$ gives a two-sided inverse.

Another canonical isomorphism of interest is

$$E \otimes_{\Bbbk} \Bbbk \cong E. \tag{A.3}$$

Here the map from left to right is the linear extension of $(e, c) \mapsto ce$, while the map from right to left is $e \mapsto e \otimes 1$. In view of (A.2) we have $\Bbbk \otimes_{\Bbbk} E \cong E$ also.

Tensor product distributes over direct sums, even infinite direct sums:

$$E \otimes_{\Bbbk} \Big(\bigoplus_{\alpha} F_{\alpha}\Big) \cong \bigoplus_{\alpha} (E \otimes_{\Bbbk} F_{\alpha}). \tag{A.4}$$

The map from left to right is the linear extension of the bilinear map $(e, \sum f_{\alpha}) \mapsto \sum (e \otimes f_{\alpha})$. To define the inverse, we have only to define it on each $E \otimes_{\Bbbk} F_{\alpha}$, where it is the linear extension of $(e, f_{\alpha}) \mapsto e \otimes (i_{\alpha}(f_{\alpha}))$; here $i_{\alpha} : F_{\alpha} \to \bigoplus F_{\beta}$ is the injection corresponding to α. It follows from

(A.3) and (A.4) that if $\{x_i\}$ is a basis of E and $\{y_j\}$ is a basis of F, then $\{x_i \otimes y_j\}$ is a basis of $E \otimes_{\mathbb{k}} F$. Consequently

$$\dim(E \otimes_{\mathbb{k}} F) = (\dim E)(\dim F). \tag{A.5}$$

Let $\operatorname{Hom}_{\mathbb{k}}(E, F)$ be the vector space of $\mathbb{k}$ linear maps from E into F. One special case is $E = \mathbb{k}$, and we have

$$\operatorname{Hom}_{\mathbb{k}}(\mathbb{k}, F) \cong F. \tag{A.6}$$

The map from left to right sends φ into $\varphi(1)$, while the map from right to left sends f into φ with $\varphi(c) = cf$. Another special case of interest occurs when $F = \mathbb{k}$. Then $\operatorname{Hom}(E, \mathbb{k}) = E^*$ is just the vector space **dual** of E.

We can use $\otimes_{\mathbb{k}}$ to construct new linear mappings. Let E_1, F_1, E_2 and F_2 be vector spaces, Suppose that L_1 is in $\operatorname{Hom}_{\mathbb{k}}(E_1, F_1)$ and L_2 is in $\operatorname{Hom}_{\mathbb{k}}(E_2, F_2)$. Then we can define

$$L_1 \otimes L_2 \quad \text{in} \quad \operatorname{Hom}_{\mathbb{k}}(E_1 \otimes_{\mathbb{k}} E_2,\ F_1 \otimes_{\mathbb{k}} F_2) \tag{A.7}$$

as follows: The map $(e_1, e_2) \mapsto L_1(e_1) \otimes L_2(e_2)$ is bilinear from $E_1 \times E_2$ into $F_1 \otimes_{\mathbb{k}} F_2$, and we let $L_1 \otimes L_2$ be its linear extension to $E_1 \otimes_{\mathbb{k}} E_2$. The uniqueness in the universal mapping property allows us to conclude that

$$(L_1 \otimes L_2)(M_1 \otimes M_2) = L_1 M_1 \otimes L_2 M_2 \tag{A.8}$$

when the domains and ranges match in the obvious way.

Let A, B, and C be vector spaces over $\mathbb{k}$. A **triple tensor product** $V = A \otimes_{\mathbb{k}} B \otimes_{\mathbb{k}} C$ is a vector space over $\mathbb{k}$ with a trilinear map $\iota : A \times B \times C \to V$ having the following universal mapping property: Whenever t is a trilinear mapping of $A \times B \times C$ into a vector space U over $\mathbb{k}$, then there exists a linear linear mapping T of V into U such that the diagram

(A.9)

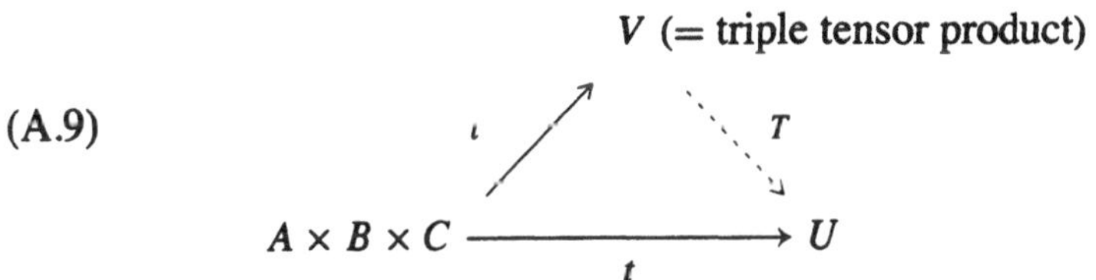

commutes. It is clear that there is at most one triple tensor product up to canonical isomorphism, and one can give an explicit construction just as for ordinary tensor products $E \otimes_{\mathbb{k}} F$. We shall use triple tensor products to establish an associativity formula for ordinary tensor products.

Proposition A.10.

(a) $(A \otimes_{\mathbb{k}} B) \otimes_{\mathbb{k}} C$ and $A \otimes_{\mathbb{k}} (B \otimes_{\mathbb{k}} C)$ are triple tensor products.

(b) There exists a unique isomorphism Φ from left to right in

$$(A \otimes_{\mathbb{k}} B) \otimes_{\mathbb{k}} C \cong A \otimes_{\mathbb{k}} (B \otimes_{\mathbb{k}} C) \tag{A.11}$$

such that $\Phi((a \otimes b) \otimes c) = a \otimes (b \otimes c)$ for all $a \in A$, $b \in B$, and $c \in C$.

PROOF.

(a) Consider $(A \otimes_{\mathbb{k}} B) \otimes_{\mathbb{k}} C$. Let $t : A \times B \times C \to U$ be trilinear. For $c \in C$, define $t_c : A \times B \to U$ by $t_c(a, b) = t(a, b, c)$. Then t_c is bilinear and hence extends to a linear $T_c : A \otimes_{\mathbb{k}} B \to U$. Since t is trilinear, $t_{c_1+c_2} = t_{c_1} + t_{c_2}$ and $t_{xc} = xt_c$ for scalar x; thus uniqueness of the linear extension forces $T_{c_1+c_2} = T_{c_1} + T_{c_2}$ and $T_{xc} = xT_c$. Consequently

$$t' : (A \otimes_{\mathbb{k}} B) \times C \to U$$

given by $t'(d, c) = T_c(d)$ is bilinear and hence extends to a linear $T : (A \otimes_{\mathbb{k}} B) \otimes_{\mathbb{k}} C \to U$. This T proves existence of the linear extension of the given t. Uniqueness is trivial, since the elements $(a \otimes b) \otimes c$ generate $(A \otimes_{\mathbb{k}} B) \otimes_{\mathbb{k}} C$. So $(A \otimes_{\mathbb{k}} B) \otimes_{\mathbb{k}} C$ is a triple tensor product. In a similar fashion, $A \otimes_{\mathbb{k}} (B \otimes_{\mathbb{k}} C)$ is a triple tensor product.

(b) In (A.9) take $V = (A \otimes_{\mathbb{k}} B) \otimes_{\mathbb{k}} C$, $U = A \otimes_{\mathbb{k}} (B \otimes_{\mathbb{k}} C)$, and $t(a, b, c) = a \otimes (b \otimes c)$. We have just seen in (a) that V is a triple tensor product with $\iota(a, b, c) = (a \otimes b) \otimes c$. Thus there exists a linear $T : V \to U$ with $T\iota(a, b, c) = t(a, b, c)$. This equation means that $T((a \otimes b) \otimes c) = a \otimes (b \otimes c)$. Interchanging the roles of $(A \otimes_{\mathbb{k}} B) \otimes_{\mathbb{k}} C$ and $A \otimes_{\mathbb{k}} (B \otimes_{\mathbb{k}} C)$, we obtain a two-sided inverse for T. Thus T will serve as Φ in (b), and existence is proved. Uniqueness is trivial, since the elements $(a \otimes b) \otimes c$ generate $(A \otimes_{\mathbb{k}} B) \otimes_{\mathbb{k}} C$.

When this proposition is used, it is often necessary to know that the isomorphism Φ is compatible with maps $A \to A'$, $B \to B'$, and $C \to C'$. This property is called **naturality** in the variables A, B, and C, and we make it precise in the next proposition.

Proposition A.12. Let A, B, C, A', B', and C' be vector spaces over $\mathbb{k}$, and let $L_A : A \to A'$, $L_B : B \to B'$, and $L_C : C \to C'$ be linear maps. Then the isomorphism Φ of Proposition A.10b is natural in the sense that the diagram

$$\begin{array}{ccc} (A \otimes_{\mathbb{k}} B) \otimes_{\mathbb{k}} C & \xrightarrow{\Phi} & A \otimes_{\mathbb{k}} (B \otimes_{\mathbb{k}} C) \\ \Big\downarrow{\scriptstyle (L_A \otimes L_B) \otimes L_C} & & \Big\downarrow{\scriptstyle L_A \otimes (L_B \otimes L_C)} \\ (A' \otimes_{\mathbb{k}} B') \otimes_{\mathbb{k}} C' & \xrightarrow{\Phi} & A' \otimes_{\mathbb{k}} (B' \otimes_{\mathbb{k}} C') \end{array}$$

commutes.

PROOF. We have

$$\begin{aligned}((L_A \otimes (L_B \otimes L_C)) \circ \Phi)((a \otimes b) \otimes c) \\ &= (L_A \otimes (L_B \otimes L_C))(a \otimes (b \otimes c)) \\ &= L_A a \otimes (L_B \otimes L_C)(b \otimes c) \\ &= \Phi((L_A a \otimes L_B b) \otimes L_C c) \\ &= \Phi((L_A \otimes L_B)(a \otimes b) \otimes L_C c) \\ &= (\Phi \circ ((L_A \otimes L_B) \otimes L_C))((a \otimes b) \otimes c),\end{aligned}$$

and the proposition follows.

There is no difficulty in generalizing matters to n-fold tensor products by induction. An **n-fold tensor product** is to be universal for n-multilinear maps. It is clearly unique up to canonical isomorphism. A direct construction is possible. Another such tensor product is the $(n-1)$-fold tensor product of the first $n-1$ spaces, tensored with the nth space. Proposition A.10b allows us to regroup parentheses (inductively) in any fashion we choose, and iterated application of Proposition A.12 shows that we get a well defined notion of the tensor product of n linear maps.

Fix a vector space E over $\mathbb{k}$, and let $T^n(E)$ be the n-fold tensor product of E with itself. In the case $n = 0$, we let $T^0(E)$ be the field $\mathbb{k}$. Define, initially as a vector space, $T(E)$ to be the direct sum

$$T(E) = \bigoplus_{n=0}^{\infty} T^n(E) \qquad \text{(A.13)}$$

The elements that lie in one or another $T^n(E)$ are called **homogeneous**. We define a bilinear multiplication on homogeneous elements

$$T^m(E) \times T^n(E) \to T^{m+n}(E)$$

to be the restriction of the above canonical isomorphism

$$T^m(E) \otimes_{\mathbb{k}} T^n(E) \to T^{m+n}(E).$$

This multiplication is associative because the restriction of the isomorphism

$$T^l(E) \otimes_{\mathbb{k}} (T^m(E) \otimes_{\mathbb{k}} T^n(E)) \to (T^l(E) \otimes_{\mathbb{k}} T^m(E)) \otimes_{\mathbb{k}} T^n(E)$$

to $T^l(E) \times (T^m(E) \times T^n(E))$ factors through the map

$$T^l(E) \times (T^m(E) \times T^n(E)) \to (T^l(E) \times T^m(E)) \times T^n(E)$$

given by $(r, (s, t)) \mapsto ((r, s), t)$. Thus $T(E)$ becomes an associative algebra with identity and is known as the **tensor algebra** of E. The algebra $T(E)$ has the following universal mapping property.

Proposition A.14. $T(E)$ has the following universal mapping property: Let ι be the map that embeds E as $T^1(E) \subseteq T(E)$. If $l : E \to A$ is any linear map of E into an associative algebra with identity, then there exists a unique associative algebra homomorphism $L : T(E) \to A$ with $L(1) = 1$ such that the diagram

(A.15)

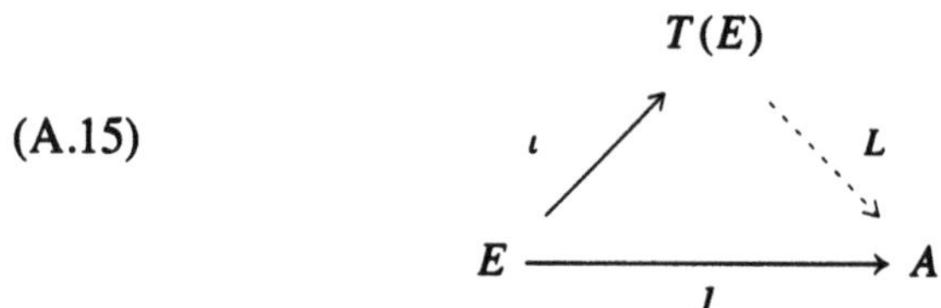

commutes.

PROOF. Uniqueness is clear, since E and 1 generate $T(E)$ as an algebra. For existence we define $L^{(n)}$ on $T^n(E)$ to be the linear extension of the n-multilinear map

$$(v_1, v_2, \ldots, v_n) \mapsto l(v_1)l(v_2)\cdots l(v_n),$$

and we let $L = \bigoplus L^{(n)}$ in obvious notation. Let $u_1 \otimes \cdots \otimes u_m$ be in $T^m(E)$ and $v_1 \otimes \cdots \otimes v_n$ be in $T^n(E)$. Then we have

$$L^{(m)}(u_1 \otimes \cdots \otimes u_m) = l(u_1)\cdots l(u_m)$$

$$L^{(n)}(v_1 \otimes \cdots \otimes v_n) = l(v_1)\cdots l(v_n)$$

$$L^{(m+n)}(u_1 \otimes \cdots \otimes u_m \otimes v_1 \otimes \cdots \otimes v_n) = l(u_1)\cdots l(u_m)l(v_1)\cdots l(v_n).$$

Hence

$$L^{(m)}(u_1 \otimes \cdots \otimes u_m)L^{(n)}(v_1 \otimes \cdots \otimes v_n) = L^{(m+n)}(u_1 \otimes \cdots \otimes u_m \otimes v_1 \otimes \cdots \otimes v_n).$$

Taking linear combinations, we see that L is a homomorphism.

2. Symmetric Algebra

We continue to allow $\Bbbk$ to be an arbitrary field. Let E be a vector space over $\Bbbk$, and let $T(E)$ be the tensor algebra. We begin by defining the symmetric algebra $S(E)$. The elements of $S(E)$ are to be all the symmetric tensors, and so we want to force $u \otimes v = v \otimes u$. Thus we define the **symmetric algebra** by

(A.16a) $$S(E) = T(E)/I,$$

where

$$I = \begin{pmatrix} \text{two-sided ideal generated by all} \\ u \otimes v - v \otimes u \text{ with } u \text{ and } v \\ \text{in } T^1(E) \end{pmatrix}. \tag{A.16b}$$

Then $S(E)$ is an associative algebra with identity.

Since the generators of I are homogeneous elements (all in $T^2(E)$), it is clear that the ideal I satisfies

$$I = \bigoplus_{n=0}^{\infty} (I \cap T^n(E)).$$

An ideal with this property is said to be **homogeneous**. Since I is homogeneous,

$$S(E) = \bigoplus_{n=0}^{\infty} T^n(E)/(I \cap T^n(E)).$$

We write $S^n(E)$ for the nth summand on the right side, so that

$$S(E) = \bigoplus_{n=0}^{\infty} S^n(E). \tag{A.17}$$

Since $I \cap T^1(E) = 0$, the map of E into first-order elements $S^1(E)$ is one-one onto. The product operation in $S(E)$ is written without a product sign, the image in $S^n(E)$ of $v_1 \otimes \cdots \otimes v_n$ in $T^n(E)$ being denoted $v_1 \cdots v_n$. If a is in $S^m(E)$ and b is in $S^n(E)$, then ab is in $S^{m+n}(E)$. Moreover $S^n(E)$ is generated by elements $v_1 \cdots v_n$ with all v_j in $S^1(E) \cong E$, since $T^n(E)$ is generated by corresponding elements $v_1 \otimes \cdots \otimes v_n$. The defining relations for $S(E)$ make $v_i v_j = v_j v_i$ for v_i and v_j in $S^1(E)$, and it follows that $S(E)$ is commutative.

Proposition A.18.

(a) $S^n(E)$ has the following universal mapping property: Let ι be the map $\iota(v_1, \ldots, v_n) = v_1 \cdots v_n$ of $E \times \cdots \times E$ into $S^n(E)$. If l is any symmetric n-multilinear map of $E \times \cdots \times E$ into a vector space U, then there exists a unique linear map $L : S^n(E) \to U$ such that the diagram

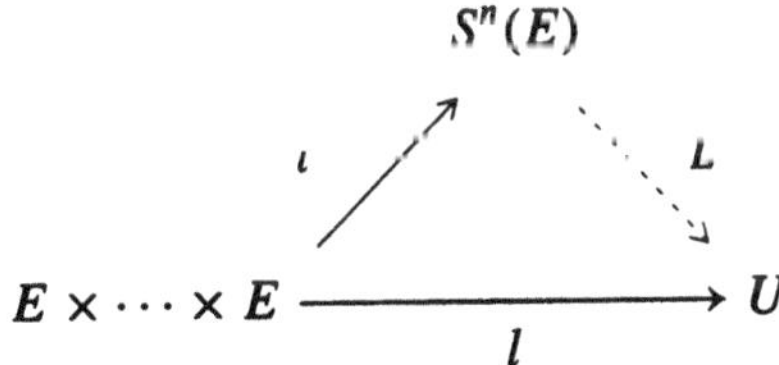

commutes.

(b) $S(E)$ has the following universal mapping property: Let ι be the map that embeds E as $S^1(E) \subseteq S(E)$. If l is any linear map of E into a commutative associative algebra A with identity, then there exists a unique algebra homomorphism $L : S(E) \to A$ with $L(1) = 1$ such that the diagram

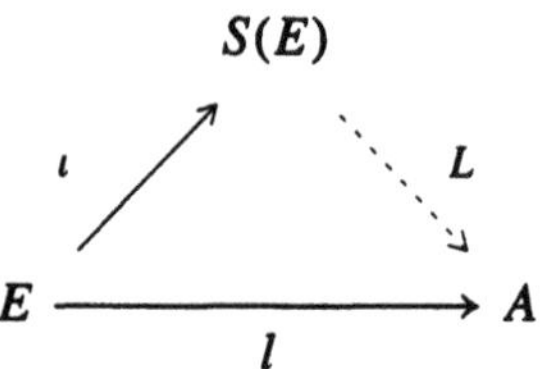

commutes.

PROOF. In both cases uniqueness is trivial. For existence we use the universal mapping properties of $T^n(E)$ and $T(E)$ to produce $\tilde{L}$ on $T^n(E)$ or $T(E)$. If we can show that $\tilde{L}$ annihilates the appropriate subspace so as to descend to $S^n(E)$ or $S(E)$, then the resulting map can be taken as L, and we are done. For (a) we have $\tilde{L} : T^n(E) \to U$, and we are to show that $\tilde{L}(T^n(E) \cap I) = 0$, where I is generated by all $u \otimes v - v \otimes u$ with u and v in $T^1(E)$. A member of $T^n(E) \cap I$ is thus of the form $\sum a_i \otimes (u_i \otimes v_i - v_i \otimes u_i) \otimes b_i$ with each term in $T^n(E)$. Each term here is a sum of pure tensors

(A.19) $x_1 \otimes \cdots \otimes x_r \otimes u_i \otimes v_i \otimes y_1 \otimes \cdots \otimes y_s - x_1 \otimes \cdots \otimes x_r \otimes v_i \otimes u_i \otimes y_1 \otimes \cdots \otimes y_s$

with $r + 2 + s = n$. Since l by assumption takes equal values on

$$x_1 \times \cdots \times x_r \times u_i \times v_i \times y_1 \times \cdots \times y_s$$

and

$$x_1 \times \cdots \times x_r \times v_i \times u_i \times y_1 \times \cdots \times y_s,$$

$\tilde{L}$ vanishes on (A.19), and it follows that $\tilde{L}(T^n(E) \cap I) = 0$.

For (b) we are to show that $\tilde{L} : T(E) \to A$ vanishes on I. Since $\ker \tilde{L}$ is an ideal, it is enough to check that $\tilde{L}$ vanishes on the generators of I. But $\tilde{L}(u \otimes v - v \otimes u) = l(u)l(v) - l(v)l(u) = 0$ by the commutativity of A, and thus $L(I) = 0$.

Corollary A.20. If E and F are vector spaces over $\Bbbk$, then $\mathrm{Hom}_{\Bbbk}(S^n(E), F)$ is canonically isomorphic (via restriction to pure tensors) to the vector space of F valued symmetric n-multilinear functions on $E \times \cdots \times E$.

PROOF. Restriction is linear and one-one. It is onto by Proposition A.18a.

Next we shall identify a basis for $S^n(E)$ as a vector space. The union of such bases as n varies will then be a basis of $S(E)$. Let $\{u_i\}_{i\in A}$ be a basis of E. A **simple ordering** on the index set A is a partial ordering in which every pair of elements is comparable.

Proposition A.21. Let E be a vector space over $\Bbbk$, let $\{u_i\}_{i\in A}$ be a basis of E, and suppose that a simple ordering has been imposed on the index set A. Then the set of all monomials $u_{i_1}^{j_1}\cdots u_{i_k}^{j_k}$ with $i_1 < \cdots < i_k$ and $\sum_m j_m = n$ is a basis of $S^n(E)$.

REMARK. In particular if E is finite-dimensional with ordered basis $u_1, \dots, u_N$, then the monomials $u_1^{j_1}\cdots u_N^{j_N}$ of total degree n form a basis of $S^n(E)$.

PROOF. Since $S(E)$ is commutative and since monomials span $T^n(E)$, the indicated set spans $S^n(E)$. Let us see independence. The map $\sum c_i u_i \mapsto \sum c_i X_i$ of E into the polynomial algebra $\Bbbk[\{X_i\}_{i\in A}]$ is linear into a commutative algebra with identity. Its extension via Proposition A.18b maps our spanning set for $S^n(E)$ to distinct monomials in $\Bbbk[\{X_i\}_{i\in A}]$, which are necessarily linearly independent. Hence our spanning set is a basis.

The proof of Proposition A.21 may suggest that $S(E)$ is just polynomials in disguise, but this suggestion is misleading, even if E is finite-dimensional. The isomorphism with $\Bbbk[\{X_i\}_{i\in A}]$ in the proof depended on choosing a basis of E. The canonical isomorphism is between $S(E^*)$ and polynomials on E. Part (b) of the corollary below goes in the direction of establishing such an isomorphism.

Corollary A.22. Let E be a finite-dimensional vector space over $\Bbbk$ of dimension N. Then

(a) $\dim S^n(E) = \binom{n+N-1}{N-1}$ for $0 \le n < \infty$.

(b) $S^n(E^*)$ is canonically isomorphic to $S^n(E)^*$ by

$$(f_1\cdots f_n)(w_1,\dots,w_n) = \sum_{\tau\in\mathfrak{S}_n}\prod_{j=1}^{n} f_j(w_{\tau(j)}),$$

where $\mathfrak{S}_n$ is the symmetric group on n letters.

PROOF.

(a) A basis has been described in Proposition A.21. To see its cardinality, we recognize that picking out $N-1$ objects from $n+N-1$ to label as dividers is a way of assigning exponents to the u_j's in an ordered basis; thus the cardinality of the indicated basis is $\binom{n+N-1}{N-1}$.

(b) Let $f_1, \ldots, f_n$ be in E^*, and define

$$l_{f_1,\ldots,f_n}(w_1, \ldots, w_n) = \sum_{\tau \in \mathfrak{S}_n} \prod_{j=1}^{n} f_j(w_{\tau(j)}).$$

Then $l_{f_1,\ldots,f_n}$ is symmetric n-multilinear from $E \times \cdots \times E$ into $\Bbbk$ and extends by Proposition A.18a to a linear $L_{f_1,\ldots,f_n} : S^n(E) \to \Bbbk$. Thus $l(f_1, \ldots, f_n) = L_{f_1,\ldots,f_n}$ defines a symmetric n-multilinear map of $E^* \times \cdots \times E^*$ into $S^n(E^*)$. Its linear extension L maps $S^n(E^*)$ into $S^n(E)^*$.

To complete the proof, we shall show that L carries basis to basis. Let $u_1, \ldots, u_N$ be an ordered basis of E, and let $u_1^*, \ldots, u_N^*$ be the dual basis. Part (a) shows that the elements $(u_1^*)^{j_1} \cdots (u_N^*)^{j_N}$ with $\sum_m j_m = n$ form a basis of $S^n(E^*)$ and that the elements $(u_1)^{k_1} \cdots (u_N)^{k_N}$ with $\sum_m k_m = n$ form a basis of $S^n(E)$. We show that L of the basis of $S^n(E^*)$ is the dual basis of the basis of $S^n(E)$, except for nonzero scalar factors. Thus let $f_1, \ldots, f_{j_1}$ all be u_1^*, let $f_{j_1+1}, \ldots, f_{j_1+j_2}$ all be u_2^*, and so on. Similarly let $w_1, \ldots, w_{k_1}$ all be u_1, let $w_{k_1+1}, \ldots, w_{k_1+k_2}$ all be u_2, and so on. Then

$$\begin{aligned} L((u_1^*)^{j_1} \cdots (u_N^*)^{j_N})((u_1)^{k_1} \cdots (u_N)^{k_N}) &= L(f_1 \cdots f_n)(w_1 \cdots w_n) \\ &= l(f_1, \ldots, f_n)(w_1 \cdots w_n) \\ &= \sum_{\tau \in \mathfrak{S}_n} \prod_{i=1}^{n} f_i(w_{\tau(i)}). \end{aligned}$$

For given τ, the product on the right side is 0 unless, for each index i, an inequality $j_{m-1} + 1 \leq i \leq j_m$ implies that $k_{m-1} + 1 \leq \tau(i) \leq k_m$. In this case the product is 1; so the right side counts the number of such τ's. For given τ, getting product nonzero forces $k_m = j_m$ for all m. And when $k_m = j_m$ for all m, the choice $\tau = 1$ does lead to product 1. Hence the members of L of the basis are nonzero multiples of the members of the dual basis, as asserted.

Now let us suppose that $\Bbbk$ has characteristic 0. We define an n-multilinear function from $E \times \cdots \times E$ into $T^n(E)$ by

$$(v_1, \ldots, v_n) \mapsto \frac{1}{n!} \sum_{\tau \in \mathfrak{S}_n} v_{\tau(1)} \otimes \cdots \otimes v_{\tau(n)},$$

and let $\sigma : T^n(E) \to T^n(E)$ be its linear extension. We call σ the **symmetrizer** operator. The image of σ is denoted $\tilde{S}^n(E)$, and the members of this subspace are called **symmetrized** tensors.

Corollary A.23. Let $\Bbbk$ have characteristic 0, and let E be a vector space over $\Bbbk$. Then the symmetrizer operator σ satisfies $\sigma^2 = \sigma$. The kernel of σ is exactly $T^n(E) \cap I$, and therefore

$$T^n(E) = \tilde{S}^n(E) \oplus (T^n(E) \cap I).$$

REMARK. In view of this corollary, the quotient map $T^n(E) \to S^n(E)$ carries $\tilde{S}^n(E)$ one-one onto $S(E)$. Thus $\tilde{S}^n(E)$ can be viewed as a copy of $S(E)$ embedded as a direct summand of $T^n(E)$.

PROOF. We have

$$\begin{aligned}
\sigma^2(v_1 \otimes \cdots \otimes v_n) &= \frac{1}{(n!)^2} \sum_{\rho,\tau \in \mathfrak{S}_n} v_{\rho\tau(1)} \otimes \cdots \otimes v_{\rho\tau(n)} \\
&= \frac{1}{(n!)^2} \sum_{\rho \in \mathfrak{S}_n} \sum_{\substack{\omega \in \mathfrak{S}_n, \\ (\omega = \rho\tau)}} v_{\omega(1)} \otimes \cdots \otimes v_{\omega(n)} \\
&= \frac{1}{n!} \sum_{\rho \in \mathfrak{S}_n} \sigma(v_1 \otimes \cdots \otimes v_n) \\
&= \sigma(v_1 \otimes \cdots \otimes v_n).
\end{aligned}$$

Hence $\sigma^2 = \sigma$. Consequently $T^n(E)$ is the direct sum of image σ and $\ker \sigma$. We thus are left with identifying $\ker \sigma$ as $T^n(E) \cap I$.

The subspace $T^n(E) \cap I$ is spanned by elements

$$x_1 \otimes \cdots \otimes x_r \otimes u \otimes v \otimes y_1 \otimes \cdots \otimes y_s - x_1 \otimes \cdots \otimes x_r \otimes v \otimes u \otimes y_1 \otimes \cdots \otimes y_s$$

with $r+2+s = n$, and it is clear that σ vanishes on such elements. Hence $T^n(E) \cap I \subseteq \ker \sigma$. Suppose that the inclusion is strict, say with t in $\ker \sigma$ but t not in $T^n(E) \cap I$. Let q be the quotient map $T^n(E) \to S^n(E)$. The kernel of q is $T^n(E) \cap I$, and thus $q(t) \neq 0$. From Proposition A.21 it is clear that q carries $\tilde{S}^n(E) = \text{image}\,\sigma$ onto $S^n(E)$. Thus choose $t' \in \tilde{S}^n(E)$ with $q(t') = q(t)$. Then $t' - t$ is in $\ker q = T^n(E) \cap I \subseteq \ker \sigma$. Since $\sigma(t) = 0$, we see that $\sigma(t') = 0$. Consequently t' is in $\ker \sigma \cap \text{image}\,\sigma = 0$, and we obtain $t' = 0$ and $q(t) = q(t') = 0$, contradiction.

3. Exterior Algebra

We turn to a discussion of the exterior algebra. Let $\Bbbk$ be an arbitrary field, and let E be a vector space over $\Bbbk$. The construction, results, and proofs for the exterior algebra $\bigwedge(E)$ are similar to those for the symmetric algebra $S(E)$. The elements of $\bigwedge(E)$ are to be all the alternating tensors (= skew-symmetric if $\Bbbk$ has characteristic $\neq 2$), and so we want to force $v \otimes v = 0$. Thus we define the **exterior algebra** by

$$\bigwedge(E) = T(E)/I', \tag{A.24a}$$

where

$$I' = \begin{pmatrix} \text{two-sided ideal generated by all} \\ v \otimes v \text{ with } v \text{ in } T^1(E) \end{pmatrix}. \tag{A.24b}$$

Then $\bigwedge(E)$ is an associative algebra with identity.

It is clear that I' is homogeneous: $I' = \bigoplus_{n=0}^{\infty} (I' \cap T^n(E))$. Thus we can write

$$\bigwedge(E) = \bigoplus_{n=0}^{\infty} T^n(E)/(I' \cap T^n(E)).$$

We write $\bigwedge^n(E)$ for the nth summand on the right side, so that

$$\bigwedge(E) = \bigoplus_{n=0}^{\infty} \bigwedge^n(E). \tag{A.25}$$

Since $I' \cap T^1(E) = 0$, the map of E into first-order elements $\bigwedge^1(E)$ is one-one onto. The product operation in $\bigwedge(E)$ is denoted $\wedge$ rather than $\otimes$, the image in $\bigwedge^n(E)$ of $v_1 \otimes \cdots v_n$ in $T^n(E)$ being denoted $v_1 \wedge \cdots \wedge v_n$. If a is in $\bigwedge^m(E)$ and b is in $\bigwedge^n(E)$, then $a \wedge b$ is in $\bigwedge^{m+n}(E)$. Moreover $\bigwedge^n(E)$ is generated by elements $v_1 \wedge \cdots \wedge v_n$ with all v_j in $\bigwedge^1(E) \cong E$, since $T^n(E)$ is generated by corresponding elements $v_1 \otimes \cdots \otimes v_n$. The defining relations for $\bigwedge(E)$ make $v_i \wedge v_j = -v_j \wedge v_i$ for v_i and v_j in $\bigwedge^1(E)$, and it follows that

$$a \wedge b = (-1)^{mn} b \wedge a \qquad \text{if } a \in \textstyle\bigwedge^m(E) \text{ and } b \in \bigwedge^n(E). \tag{A.26}$$

Proposition A.27.

(a) $\bigwedge^n(E)$ has the following universal mapping property: Let ι be the map $\iota(v_1, \ldots, v_n) = v_1 \wedge \cdots \wedge v_n$ of $E \times \cdots \times E$ into $\bigwedge^n(E)$. If l is any alternating n-multilinear map of $E \times \cdots \times E$ into a vector space U, then there exists a unique linear map $L : \bigwedge^n(E) \to U$ such that the diagram

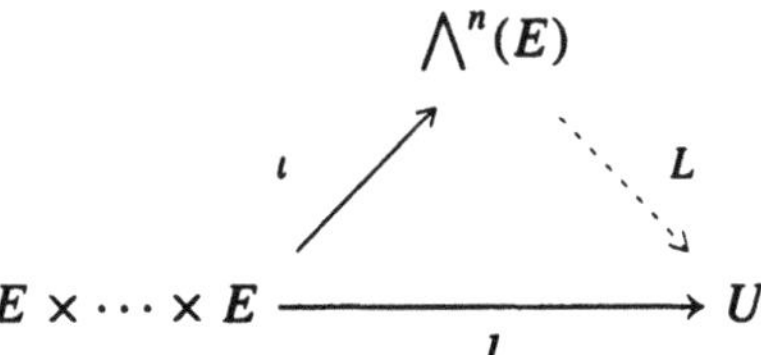

commutes.

(b) $\bigwedge(E)$ has the following universal mapping property: Let ι be the map that embeds E as $\bigwedge^1(E) \subseteq \bigwedge(E)$. If l is any linear map of E into an associative algebra A with identity such that $l(v)^2 = 0$ for all $v \in E$, then there exists a unique algebra homomorphism $L : \bigwedge(E) \to A$ with $L(1) = 1$ such that the diagram

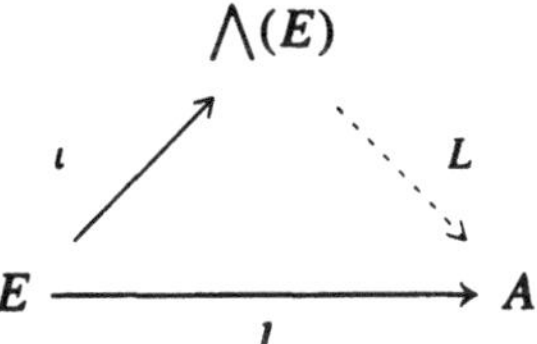

commutes.

PROOF. The proof is completely analogous to the proof of Proposition A.18.

Corollary A.28. If E and F are vector spaces over $\mathbb{k}$, then $\operatorname{Hom}_{\mathbb{k}}(\bigwedge^n(E), F)$ is canonically isomorphic (via restriction to pure tensors) to the vector space of F valued alternating n multilinear functions on $E \times \cdots \times E$.

PROOF. Restriction is linear and one-one. It is onto by Proposition A.27a.

Next we shall identify a basis for $\bigwedge^n(E)$ as a vector space. The union of such bases as n varies will then be a basis of $\bigwedge(E)$.

Proposition A.29. Let E be a vector space over $\mathbb{k}$, let $\{u_i\}_{i \in A}$ be a basis of E, and suppose that a simple ordering has been imposed on the index set A. Then the set of all monomials $u_{i_1} \wedge \cdots \wedge u_{i_n}$ with $i_1 < \cdots < i_n$ is a basis of $\bigwedge^n(E)$.

PROOF. Since multiplication in $\bigwedge(E)$ satisfies (A.26) and since monomials span $T^n(E)$, the indicated set spans $\bigwedge^n(E)$. Let us see independence. For $i \in A$, let u_i^* be the member of E^* with $u_i^*(u_j)$ equal to 1 for $j = i$ and equal to 0 for $j \neq i$. Fix $r_1 < \cdots < r_n$, and define

$$l(w_1, \dots, w_n) = \det\{u_{r_i}^*(w_j)\} \qquad \text{for } w_1, \dots, w_n \text{ in } E.$$

Then l is alternating n-multilinear from $E \times \cdots \times E$ into $\Bbbk$ and extends by Proposition A.27a to $L : \bigwedge^n(E) \to \Bbbk$. If $k_1 < \cdots < k_n$, then

$$L(u_{k_1} \wedge \cdots \wedge u_{k_n}) = l(u_{k_1}, \dots, u_{k_n}) = \det\{u_{r_i}^*(u_{k_j})\},$$

and the right side is 0 unless $r_1 = k_1, \dots, r_n = k_n$, in which case it is 1. This proves that the $u_{r_1} \wedge \cdots \wedge u_{r_n}$ are linearly independent in $\bigwedge^n(E)$.

Corollary A.30. Let E be a finite-dimensional vector space over $\Bbbk$ of dimension N. Then

(a) $\dim \bigwedge^n(E) = \binom{N}{n}$ for $0 \le n \le N$ and $= 0$ for $n > N$.

(b) $\bigwedge^n(E^*)$ is canonically isomorphic to $\bigwedge^n(E)^*$ by

$$(f_1 \wedge \cdots \wedge f_n)(w_1, \dots, w_n) = \det\{f_i(w_j)\}.$$

PROOF. Part (a) is an immediate consequence of Proposition A.29, and (b) is proved in the same way as Corollary A.22b, using Proposition A.27a as a tool.

Now let us suppose that $\Bbbk$ has characteristic 0. We define an n-multilinear function from $E \times \cdots \times E$ into $T^n(E)$ by

$$(v_1, \dots, v_n) \mapsto \frac{1}{n!} \sum_{\tau \in \mathfrak{S}_n} (\operatorname{sgn} \tau) v_{\tau(1)} \otimes \cdots \otimes v_{\tau(n)},$$

and let $\sigma' : T^n(E) \to T^n(E)$ be its linear extension. We call σ' the **antisymmetrizer** operator. The image of σ' is denoted $\widetilde{\bigwedge}^n(E)$, and the members of this subspace are called **antisymmetrized** tensors.

Corollary A.31. Let $\Bbbk$ have characteristic 0, and let E be a vector space over $\Bbbk$. Then the antisymmetrizer operator σ' satisfies $\sigma'^2 = \sigma'$. The kernel of σ' is exactly $T^n(E) \cap I'$, and therefore

$$T^n(E) = \widetilde{\bigwedge}^n(E) \oplus (T^n(E) \cap I').$$

REMARK. In view of this corollary, the quotient map $T^n(E) \to \bigwedge^n(E)$ carries $\widetilde{\bigwedge}^n(E)$ one-one onto $\bigwedge(E)$. Thus $\widetilde{\bigwedge}^n(E)$ can be viewed as a copy of $\bigwedge(E)$ embedded as a direct summand of $T^n(E)$.

PROOF. We have

$$\begin{aligned}\sigma'^2(v_1 \otimes \cdots \otimes v_n) &= \frac{1}{(n!)^2} \sum_{\rho,\tau \in \mathfrak{S}_n} (\operatorname{sgn} \rho\tau) v_{\rho\tau(1)} \otimes \cdots \otimes v_{\rho\tau(n)} \\ &= \frac{1}{(n!)^2} \sum_{\rho \in \mathfrak{S}_n} \sum_{\substack{\omega \in \mathfrak{S}_n, \\ (\omega = \rho\tau)}} (\operatorname{sgn} \omega) v_{\omega(1)} \otimes \cdots \otimes v_{\omega(n)} \\ &= \frac{1}{n!} \sum_{\rho \in \mathfrak{S}_n} \sigma'(v_1 \otimes \cdots \otimes v_n) \\ &= \sigma'(v_1 \otimes \cdots \otimes v_n).\end{aligned}$$

Hence $\sigma'^2 = \sigma'$. Consequently $T^n(E)$ is the direct sum of image σ' and $\ker \sigma'$. We thus are left with identifying $\ker \sigma'$ as $T^n(E) \cap I'$.

The subspace $T^n(E) \cap I'$ is spanned by elements

$$x_1 \otimes \cdots \otimes x_r \otimes v \otimes v \otimes y_1 \otimes \cdots \otimes y_s$$

with $r + 2 + s = n$, and it is clear that σ' vanishes on such elements. Hence $T^n(E) \cap I' \subseteq \ker \sigma'$. Suppose that the inclusion is strict, say with t in $\ker \sigma'$ but t not in $T^n(E) \cap I'$. Let q be the quotient map $T^n(E) \to \bigwedge^n(E)$. The kernel of q is $T^n(E) \cap I'$, and thus $q(t) \neq 0$. From Proposition A.29 it is clear that q carries $\tilde{\bigwedge}^n(E) = \operatorname{image} \sigma'$ onto $\bigwedge^n(E)$. Thus choose $t' \in \tilde{\bigwedge}^n(E)$ with $q(t') = q(t)$. Then $t' - t$ is in $\ker q = T^n(E) \cap I' \subseteq \ker \sigma'$. Since $\sigma'(t) = 0$, we see that $\sigma'(t') = 0$. Consequently t' is in $\ker \sigma' \cap \operatorname{image} \sigma' = 0$, and we obtain $t' = 0$ and $q(t) = q(t') = 0$, contradiction.

4. Filtrations and Gradings

Let $\Bbbk$ be any field. A vector space V over $\Bbbk$ will be said to be **filtered** if there is a specified increasing sequence of subspaces

$$V_0 \subseteq V_1 \subseteq V_2 \subseteq \cdots \tag{A.32}$$

with union V. In this case we put $V_{-1} = 0$ by convention. We shall say that V is **graded** if there is a specified sequence of subspaces $V^0, V^1, V^2, \ldots$ such that

$$V = \bigoplus_{n=0}^{\infty} V^n. \tag{A.33}$$

When V is graded, there is a natural filtration of V given by

$$V_n = \bigoplus_{k=0}^{n} V^k. \tag{A.34}$$

When E is a vector space, the tensor algebra $V = T(E)$ is graded as a vector space, and the same thing is true of the symmetric algebra $S(E)$ and the exterior algebra $\bigwedge(E)$. In each case the nth subspace of the grading consists of the subspace of tensors that are homogeneous of degree n.

When V is a filtered vector space as in (A.32), the associated graded vector space is

$$\operatorname{gr} V = \bigoplus_{n=0}^{\infty} V_n/V_{n-1}. \tag{A.35}$$

In the case that V is graded and its filtration is the natural one given in (A.34), $\operatorname{gr} V$ recovers the given grading on V, i.e., $\operatorname{gr} V$ is canonically isomorphic with V in a way that preserves the grading.

Let V and V' be two filtered vector spaces, and let φ be a linear map between them such that $\varphi(V_n) \subseteq V_n'$ for all n. Since the restriction of φ to V_n carries V_{n-1} into V_{n-1}', this restriction induces a linear map $\operatorname{gr}^n \varphi : (V_n/V_{n-1}) \to (V_n'/V_{n-1}')$. The direct sum of these linear maps is then a linear map

$$\operatorname{gr} \varphi : \operatorname{gr} V \to \operatorname{gr} V' \tag{A.36}$$

called the **associated graded map** for φ.

Proposition A.37. Let V and V' be two filtered vector spaces, and let φ be a linear map between them such that $\varphi(V_n) \subseteq V_n'$ for all n. If $\operatorname{gr} \varphi$ is an isomorphism, then φ is an isomorphism.

PROOF. It is enough to prove that $\varphi|_{V_n} : V_n \to V_n'$ is an isomorphism for every n. We establish this property by induction on n, the trivial case for the induction being $n = -1$. Suppose that

$$\varphi|_{V_{n-1}} : V_{n-1} \to V_{n-1}' \quad \text{is an isomorphism.} \tag{A.38}$$

By assumption

$$\operatorname{gr}^n \varphi : (V_n/V_{n-1}) \to (V_n'/V_{n-1}') \quad \text{is an isomorphism.} \tag{A.39}$$

If v is in $\ker(\varphi|_{V_n})$, then $(\operatorname{gr}^n \varphi)(v + V_{n-1}) = 0 + V_{n-1}'$, and (A.39) shows that v is in V_{n-1}. By (A.38), $v = 0$. Thus $\varphi|_{V_n}$ is one-one. Next suppose that v' is in V_n'. By (A.39) there exists v_n in V_n such that $(\operatorname{gr}^n \varphi)(v_n + V_{n-1}) = v' + V_{n-1}'$. Write $\varphi(v_n) = v' + v_{n-1}'$ with v_{n-1}' in V_{n-1}'. By (A.38) there exists v_{n-1} in V_{n-1} with $\varphi(v_{n-1}) = v_{n-1}'$. Then $\varphi(v_n - v_{n-1}) = v'$, and thus $\varphi|_{V_n}$ is onto. This completes the induction.

Now let A be an associative algebra over $\Bbbk$ with identity. If A has a filtration $A_0, A_1, \ldots$ of vector subspaces with $1 \in A_0$ such that $A_m A_n \subseteq A_{m+n}$ for all m and n, then we say that A is a **filtered associative algebra**. Similarly if A is graded as $A = \bigoplus_{n=0}^{\infty} A^n$ in such a way that $A^m A^n \subseteq A^{m+n}$ for all m and n, then we say that A is a **graded associative algebra**.

Proposition A.40. If A is a filtered associative algebra with identity, then the graded vector space $\operatorname{gr} A$ acquires a multiplication in a natural way making it into a graded associative algebra with identity.

PROOF. We define a product

$$(A_m/A_{m-1}) \times (A_n/A_{n-1}) \to A_{m+n}/A_{m+n-1}$$

by
$$(a_m + A_{m-1})(a_n + A_{n-1}) = a_m a_n + A_{m+n-1}.$$

This is well defined since $a_m A_{n-1}$, $A_{m-1}a_n$, and $A_{m-1}A_{n-1}$ are all contained in A_{m+n-1}. It is clear that this multiplication is distributive and associative as far as it is defined. We extend the definition of multiplication to all of $\operatorname{gr} A$ by taking sums of products of homogeneous elements, and the result is an associative algebra. The identity is the element $1 + A_{-1}$ of A_0/A_{-1}.

APPENDIX B

Lie's Third Theorem

Abstract. A finite-dimensional real Lie algebra is the semidirect product of a semisimple subalgebra and the solvable radical, according to the Levi decomposition. As a consequence of this theorem and the correspondence between semidirect products of Lie algebras and semidirect products of simply connected analytic groups, every finite-dimensional real Lie algebra is the Lie algebra of an analytic group.

1. Levi Decomposition

Chapter I omits two important theorems about general finite-dimensional Lie algebras over $\mathbb{R}$ that need to be mentioned, and those results appear in this appendix. They were omitted from Chapter I because they use a result about semisimple Lie algebras that was not proved until Chapter V.

Lemma B.1. Let φ be an $\mathbb{R}$ linear representation of the real semisimple Lie algebra $\mathfrak{g}$ on a finite-dimensional real vector space V. Then V is completely reducible in the sense that there exist invariant subspaces $U_1, \ldots, U_r$ of V such that $V = U_1 \oplus \cdots \oplus U_r$ and such that the restriction of the representation to each U_i is irreducible.

PROOF. It is enough to prove that any invariant subspace U of V has an invariant complement W. By Theorem 5.29, there exists an invariant complex subspace W' of $V^{\mathbb{C}}$ such that $V^{\mathbb{C}} = U^{\mathbb{C}} \oplus W'$. Let P be the $\mathbb{R}$ linear projection of $V^{\mathbb{C}}$ on V along iV, and put

$$W = P(W' \cap (V \oplus iU)).$$

Since P commutes with $\varphi(\mathfrak{g})$, we see that $\varphi(\mathfrak{g})(W) \subseteq W$. To complete the proof, we show that $V = U \oplus W$.

Let a be in $U \cap W$. Then $a + ib$ is in $W' \cap (V \oplus iU)$ for some $b \in V$. The element b must be in U, and we know that a is in U. Hence $a + ib$ is in $U^{\mathbb{C}}$. But then $a + ib$ is in $U^{\mathbb{C}} \cap W' = 0$, and $a = 0$. Hence $U \cap W = 0$.

Next let $v \in V$ be given. Since $V^{\mathbb{C}} = U^{\mathbb{C}} + W'$, we can write $v = (a+ib)+(x+iy)$ with $a \in U, b \in U$, and $x+iy \in W'$. Since v is in V, $y=-b$. Therefore $x+iy$ is in $V \oplus iU$, as well as W'. Since $P(x+iy)=x$, x is in W. Then $v = a+x$ with $a \in U$ and $x \in W$, and $V = U+W$.

Theorem B.2 (Levi decomposition). If $\mathfrak{g}$ is a finite-dimensional Lie algebra over $\mathbb{R}$, then there exists a semisimple subalgebra $\mathfrak{s}$ of $\mathfrak{g}$ such that $\mathfrak{g}$ is the semidirect product $\mathfrak{g} = \mathfrak{s} \oplus_\pi (\operatorname{rad} \mathfrak{g})$ for a suitable homomorphism $\pi : \mathfrak{s} \to \operatorname{Der}_{\mathbb{R}}(\operatorname{rad} \mathfrak{g})$.

PROOF. Let $\mathfrak{r} = \operatorname{rad} \mathfrak{g}$. We begin with two preliminary reductions. The first reduction will enable us to assume that there is no nonzero ideal $\mathfrak{a}$ of $\mathfrak{g}$ properly contained in $\mathfrak{r}$. In fact, an argument by induction on the dimension would handle such a situation: Proposition 1.11 shows that the radical of $\mathfrak{g}/\mathfrak{a}$ is $\mathfrak{r}/\mathfrak{a}$. Hence induction gives $\mathfrak{g}/\mathfrak{a} = \mathfrak{s}/\mathfrak{a} \oplus \mathfrak{r}/\mathfrak{a}$ with $\mathfrak{s}/\mathfrak{a}$ semisimple. Since $\mathfrak{s}/\mathfrak{a}$ is semisimple, $\mathfrak{a} = \operatorname{rad} \mathfrak{s}$. Then induction gives $\mathfrak{s} = \mathfrak{s}' \oplus \mathfrak{a}$ with $\mathfrak{s}'$ semisimple. Consequently $\mathfrak{g} = \mathfrak{s}' \oplus \mathfrak{r}$, and $\mathfrak{s}'$ is the required complementary subalgebra.

As a consequence, $\mathfrak{r}$ is abelian. In fact, otherwise Proposition 1.7 shows that $[\mathfrak{r}, \mathfrak{r}]$ is an ideal in $\mathfrak{g}$, necessarily nonzero and properly contained in $\mathfrak{r}$. So the first reduction eliminates this case.

The second reduction will enable us to assume that $[\mathfrak{g}, \mathfrak{r}] = \mathfrak{r}$. In fact, $[\mathfrak{g}, \mathfrak{r}]$ is an ideal of $\mathfrak{g}$ contained in $\mathfrak{r}$. The first reduction shows that we may assume it is 0 or $\mathfrak{r}$. If $[\mathfrak{g}, \mathfrak{r}] = 0$, then the real representation ad of $\mathfrak{g}$ on $\mathfrak{g}$ descends to a real representation of $\mathfrak{g}/\mathfrak{r}$ on $\mathfrak{g}$. Since $\mathfrak{g}/\mathfrak{r}$ is semisimple, Lemma B.1 shows that the action is completely reducible. Thus $\mathfrak{r}$, which is an invariant subspace in $\mathfrak{g}$, has an invariant complement, and we may take this complement as $\mathfrak{s}$.

As a consequence,

$$\mathfrak{r} \cap Z_{\mathfrak{g}} = 0. \tag{B.3}$$

In fact $\mathfrak{r} \cap Z_{\mathfrak{g}}$ is an ideal of $\mathfrak{g}$. It is properly contained in $\mathfrak{r}$ since $\mathfrak{r} \cap Z_{\mathfrak{g}} = \mathfrak{r}$ implies that $[\mathfrak{g}, \mathfrak{r}] = 0$, in contradiction with the second reduction. Therefore the first reduction implies (B.3).

With the reductions in place, we imitate some of the proof of Theorem 5.29. That is, we put

$$V = \{\gamma \in \operatorname{End} \mathfrak{g} \mid \gamma(\mathfrak{g}) \subseteq \mathfrak{r} \text{ and } \gamma|_{\mathfrak{r}} \text{ is scalar}\}$$

and define a representation σ of $\mathfrak{g}$ on $\operatorname{End} \mathfrak{g}$ by

$$\sigma(X)\gamma = (\operatorname{ad} X)\gamma - \gamma(\operatorname{ad} X) \qquad \text{for } \gamma \in \operatorname{End} \mathfrak{g} \text{ and } X \in \mathfrak{g}.$$

The subspace V is an invariant subspace under σ, and

$$U = \{\gamma \in V \mid \gamma = 0 \text{ on } \mathfrak{r}\}$$

is an invariant subspace of codimension 1 in V such that $\sigma(X)(V) \subseteq U$ for $X \in \mathfrak{g}$. Let

$$T = \{\operatorname{ad} Y \mid Y \in \mathfrak{r}\}.$$

This is a subspace of U since $\mathfrak{r}$ is an abelian Lie subalgebra. If X is in $\mathfrak{g}$ and $\gamma = \operatorname{ad} Y$ is in T, then $\sigma(X)\gamma = \operatorname{ad}[X, Y]$ with $[X, Y] \in \mathfrak{r}$. Hence T is an invariant subspace under σ.

From $V \supseteq U \supseteq T$, we can form the quotient representations V/T and V/U. The natural map of V/T onto V/U respects the $\mathfrak{g}$ actions, and the $\mathfrak{g}$ action of V/U is 0 since $\sigma(X)(V) \subseteq U$ for $X \in \mathfrak{g}$. If X is in $\mathfrak{r}$ and γ is in V, then

$$\sigma(X)\gamma = (\operatorname{ad} X)\gamma - \gamma(\operatorname{ad} X) = -\gamma(\operatorname{ad} X)$$

since image $\gamma \subseteq \mathfrak{r}$ and $\mathfrak{r}$ is abelian. Since γ is a scalar $\lambda(\gamma)$ on $\mathfrak{r}$, we can rewrite this formula as

$$\sigma(X)\gamma = \operatorname{ad}(-\lambda(\gamma)X). \tag{B.4}$$

Equation (B.4) exhibits $\sigma(X)\gamma$ as in T. Thus $\sigma|_{\mathfrak{r}}$ maps V into T, and σ descends to representations of $\mathfrak{g}/\mathfrak{r}$ on V/T and V/U. The natural map of V/T onto V/U respects these $\mathfrak{g}/\mathfrak{r}$ actions.

Since $\dim V/U = 1$, the kernel of $V/T \to V/U$ is a $\mathfrak{g}/\mathfrak{r}$ invariant subspace of V/T of codimension 1, necessarily of the form W/T with $W \subseteq V$. Since $\mathfrak{g}/\mathfrak{r}$ is semisimple, Lemma B.1 allows us to write

$$V/T = W/T \oplus (\mathbb{R}\gamma_0 + T)/T \tag{B.5}$$

for a 1-dimensional invariant subspace $(\mathbb{R}\gamma_0 + T)/T$. The directness of this sum means that γ_0 is not in U. So γ_0 is not 0 on $\mathfrak{r}$. Normalizing, we may assume that γ_0 acts by the scalar -1 on $\mathfrak{r}$. In view of (B.4), we have

$$\sigma(X)\gamma_0 = \operatorname{ad} X \qquad \text{for } X \in \mathfrak{r}. \tag{B.6}$$

Since $(\mathbb{R}\gamma_0 + T)/T$ is invariant in (B.5), we have $\sigma(X)\gamma_0 \in T$ for each $X \in \mathfrak{g}$. Thus we can write $\sigma(X)\gamma_0 = \operatorname{ad}\varphi(X)$ for some $\varphi(X) \in \mathfrak{r}$. The element $\varphi(X)$ is unique by (B.3), and therefore φ is a linear function $\varphi : \mathfrak{g} \to \mathfrak{r}$. By (B.6), φ is a projection. If we put $\mathfrak{s} = \ker\varphi$, then we have $\mathfrak{g} = \mathfrak{s} \oplus \mathfrak{r}$ as vector spaces, and we have only to show that $\mathfrak{s}$ is a Lie subalgebra. The subspace $\mathfrak{s} = \ker\varphi$ is the set of all X such that $\sigma(X)\gamma_0 = 0$. This is the set of all X such that $(\operatorname{ad} X)\gamma_0 = \gamma_0(\operatorname{ad} X)$. Actually if γ is any element of $\operatorname{End}\mathfrak{g}$, then the set of $X \in \mathfrak{g}$ such that $(\operatorname{ad} X)\gamma = \gamma(\operatorname{ad} X)$ is always a Lie subalgebra. Hence $\mathfrak{s}$ is a Lie subalgebra, and the proof is complete.

2. Lie's Third Theorem

Lie's Third Theorem, which Lie proved as a result about vector fields and local Lie groups, has come to refer to the following improved theorem due to Cartan.

Theorem B.7. Every finite-dimensional Lie algebra over $\mathbb{R}$ is isomorphic to the Lie algebra of an analytic group.

PROOF. Let $\mathfrak{g}$ be given, and write $\mathfrak{g} = \mathfrak{s} \oplus_\pi \mathfrak{r}$ as in Theorem B.2, with $\mathfrak{s}$ semisimple and $\mathfrak{r}$ solvable. Corollary 1.103 shows that there is a simply connected Lie group R with Lie algebra isomorphic to $\mathfrak{r}$. The group $\operatorname{Int}\mathfrak{s}$ is an analytic group with Lie algebra $\operatorname{ad}\mathfrak{s}$ isomorphic to $\mathfrak{s}$ since $\mathfrak{s}$ has center 0. Let S be the universal covering group of $\operatorname{Int}\mathfrak{s}$. By Theorem 1.102 there exists a unique action τ of S on R by automorphisms such that $d\bar{\tau} = \pi$, and $G = S \times_\tau R$ is a simply connected analytic group with Lie algebra isomorphic to $\mathfrak{g} = \mathfrak{s} \oplus_\pi \mathfrak{r}$.

APPENDIX C

Data for Simple Lie Algebras

Abstract. This appendix contains information about irreducible root systems, simple Lie algebras over $\mathbb{C}$ and $\mathbb{R}$, and Lie groups whose Lie algebras are simple, noncompact, and noncomplex. The first two sections deal with the root systems themselves and the corresponding complex simple Lie algebras. The last two sections deal with the simple real Lie algebras that are noncompact and noncomplex and with their corresponding Lie groups.

1. Classical Irreducible Reduced Root Systems

This section collects information about the classical irreducible reduced root systems, those of types A_n for $n \geq 1$, B_n for $n \geq 2$, C_n for $n \geq 3$, and D_n for $n \geq 4$.

The first three items describe the underlying vector space V, the root system Δ as a subset of V, and the usual complex semisimple Lie algebra $\mathfrak{g}$ associated with Δ. All this information appears also in (2.43). In each case the root system is a subspace of some $\mathbb{R}^k = \left\{ \sum_{i=1}^k a_i e_i \right\}$. Here $\{e_i\}$ is the standard orthonormal basis, and the a_i's are real.

The next four items give the number $|\Delta|$ of roots, the dimension $\dim \mathfrak{g}$ of the Lie algebra $\mathfrak{g}$, the order $|W|$ of the Weyl group of Δ, and the determinant $\det(A_{ij})$ of the Cartan matrix. All this information appears also in Problems 15 and 28 for Chapter II.

The next two items give the customary choice of positive system Δ^+ and the associated set Π of simple roots. This information appears also in (2.50), and the corresponding Dynkin diagrams appear in Figure 2.3 and again in Figure 2.4.

The last three items give, relative to the listed positive system Δ^+, the fundamental weights $\varpi_1, \ldots \varpi_n$, the largest root, and the half sum δ of the positive roots. The fundamental weights ϖ_j are defined by the condition $2\langle \varpi_j, \alpha_i \rangle / |\alpha_i|^2 = \delta_{ij}$ if Π is regarded as the ordered set $\{\alpha_1, \ldots, \alpha_n\}$. Their significance is explained in Problems 36–41 for Chapter V. The ϖ_i are expressed as members of V. The largest root is listed in two formats: (a) as a tuple like $(11 \cdots 1)$ that indicates the expansion in terms of the simple roots $\alpha_1, \ldots, \alpha_n$ and (b) as a member of V.

A_n

$V = \{v \in \mathbb{R}^{n+1} \mid \langle v, e_1 + \cdots + e_{n+1} \rangle = 0\}$
$\Delta = \{e_i - e_j \mid i \neq j\}$
$\mathfrak{g} = \mathfrak{sl}(n+1, \mathbb{C})$

$|\Delta| = n(n+1)$
$\dim \mathfrak{g} = n(n+2)$
$|W| = (n+1)!$
$\det(A_{ij}) = n+1$

$\Delta^+ = \{e_i - e_j \mid i < j\}$
$\Pi = \{e_1 - e_2,\ e_2 - e_3, \ldots, e_n - e_{n+1}\}$

Fundamental weights:

$\varpi_i = e_1 + \cdots + e_i$ projected to V
$\quad = e_1 + \cdots + e_i - \frac{i}{n+1}(e_1 + \cdots + e_{n+1})$

Largest root $= (11 \cdots 1) = e_1 - e_{n+1}$
$\delta = (\frac{n}{2})e_1 + (\frac{n-2}{2})e_2 + \cdots + (-\frac{n}{2})e_{n+1}$

B_n

$V = \mathbb{R}^n$
$\Delta = \{\pm e_i \pm e_j \mid i < j\} \cup \{\pm e_i\}$
$\mathfrak{g} = \mathfrak{so}(2n+1, \mathbb{C})$

$|\Delta| = 2n^2$
$\dim \mathfrak{g} = n(2n+1)$
$|W| = n!2^n$
$\det(A_{ij}) = 2$

$\Delta^+ = \{e_i \pm e_j \mid i < j\} \cup \{e_i\}$
$\Pi = \{e_1 - e_2,\ e_2 - e_3, \ldots, e_{n-1} - e_n,\ e_n\}$

Fundamental weights:

$\varpi_i = e_1 + \cdots + e_i$ for $i < n$
$\varpi_n = \frac{1}{2}(e_1 + \cdots + e_n)$

Largest root $= (122 \cdots 2) = e_1 + e_2$
$\delta = (n - \frac{1}{2})e_1 + (n - \frac{3}{2})e_2 + \cdots + \frac{1}{2}e_n$

C_n

$V = \mathbb{R}^n$
$\Delta = \{\pm e_i \pm e_j \mid i < j\} \cup \{\pm 2e_i\}$
$\mathfrak{g} = \mathfrak{sp}(n, \mathbb{C})$

$|\Delta| = 2n^2$
$\dim \mathfrak{g} = n(2n+1)$
$|W| = n!2^n$
$\det(A_{ij}) = 2$

$\Delta^+ = \{e_i \pm e_j \mid i < j\} \cup \{2e_i\}$
$\Pi = \{e_1 - e_2,\ e_2 - e_3, \ldots, e_{n-1} - e_n,\ 2e_n\}$

Fundamental weights:
$\varpi_i = e_1 + \cdots + e_i$
Largest root $= (22 \cdots 21) = 2e_1$
$\delta = ne_1 + (n-1)e_2 + \cdots + 1e_n$

D_n

$V = \mathbb{R}^n$
$\Delta = \{\pm e_i \pm e_j \mid i < j\}$
$\mathfrak{g} = \mathfrak{so}(2n, \mathbb{C})$

$|\Delta| = 2n(n-1)$
$\dim \mathfrak{g} = n(2n-1)$
$|W| = n!2^{n-1}$
$\det(A_{ij}) = 4$

$\Delta^+ = \{e_i \pm e_j \mid i < j\}$
$\Pi = \{e_1 - e_2,\ e_2 - e_3, \ldots, e_{n-1} - e_n,\ e_{n-1} + e_n\}$

Fundamental weights:
$\varpi_i = e_1 + \cdots + e_i$ for $i \leq n-2$
$\varpi_{n-1} = \frac{1}{2}(e_1 + \cdots + e_{n-1} - e_n)$
$\varpi_n = \frac{1}{2}(e_1 + \cdots + e_{n-1} + e_n)$
Largest root $= (122 \cdots 211) = e_1 + e_2$
$\delta = (n-1)e_1 + (n-2)e_2 + \cdots + 1e_{n-1}$

---

2. Exceptional Irreducible Reduced Root Systems

This section collects information about the exceptional irreducible reduced root systems, those of types E_6, E_7, E_8, F_4, and G_2.

The first two items describe the underlying vector space V and the root system Δ as a subset of V. All this information appears also in Proposition 2.87 and in the last diagram of Figure 2.2. In each case the root system is a subspace of some $\mathbb{R}^k = \left\{\sum_{i=1}^k a_i e_i\right\}$. Here $\{e_i\}$ is the standard orthonormal basis, and the a_i's are real.

The next four items give the number $|\Delta|$ of roots, the dimension $\dim \mathfrak{g}$ of a Lie algebra $\mathfrak{g}$ with Δ as root system, the order $|W|$ of the Weyl group of Δ, and the determinant $\det(A_{ij})$ of the Cartan matrix. All this information appears also in Problems 16 and 29–34 for Chapter II.

The next three items give the customary choice of positive system Δ^+, the associated set Π of simple roots, and the numbering of the simple roots in the Dynkin diagram. This information about Π appears also in (2.85b) and (2.86b), and the corresponding Dynkin diagrams appear in Figure 2.4.

The last three items give, relative to the listed positive system Δ^+, the fundamental weights $\varpi_1, \dots \varpi_n$, the positive roots with a coefficient ≥ 2, and the half sum δ of the positive roots. The fundamental weights ϖ_j are defined by the condition $2\langle \varpi_j, \alpha_i\rangle / |\alpha_i|^2 = \delta_{ij}$ if Π is regarded as the ordered set $\{\alpha_1, \dots, \alpha_n\}$. Their significance is explained in Problems 36–41 for Chapter V. Let the fundamental weights be expressed in terms of the simple roots as $\varpi_j = \sum_i C_{ij}\alpha_i$. Taking the inner product of both sides with $2|\alpha_k|^{-2}\alpha_k$, we see that the matrix (C_{ij}) is the inverse of the Cartan matrix (A_{ij}). Alternatively taking the inner product of both sides with ϖ_i, we see that C_{ij} is a positive multiple of $\langle \varpi_j, \varpi_i\rangle$, which is > 0 by Lemma 6.97. The forms ω_i that appear in §VI.10 are related to the fundamental weights ω_i by $\varpi_i = \frac{1}{2}|\alpha_i|^2\omega_i$. The positive roots with a coefficient ≥ 2 are listed in a format that indicates the expansion in terms of the simple roots $\alpha_1, \dots, \alpha_n$. The last root in the list is the largest root.

The displays of the last two sets of items have been merged in the case of G_2.

***E*₆**

$V = \{v \in \mathbb{R}^8 \mid \langle v, e_6 - e_7\rangle = \langle v, e_7 + e_8\rangle = 0)\}$
$\Delta = \{\pm e_i \pm e_j \mid i < j \leq 5\} \cup \left\{\frac{1}{2}\sum_{i=1}^{8}(-1)^{n(i)}e_i \in V \mid \sum_{i=1}^{8}(-1)^{n(i)} \text{ even}\right\}$

$|\Delta| = 72$
$\dim \mathfrak{g} = 78$
$|W| = 2^7 \cdot 3^4 \cdot 5$
$\det(A_{ij}) = 3$

$\Delta^+ = \{e_i \pm e_j \mid i > j\}$
$\qquad \cup \left\{\frac{1}{2}(e_8 - e_7 - e_6 + \sum_{i=1}^{5}(-1)^{n(i)}e_i \mid \sum_{i=1}^{5}(-1)^{n(i)} \text{ even}\right\}$
$\Pi = \{\alpha_1, \alpha_2, \alpha_3, \alpha_4, \alpha_5, \alpha_6\}$
$\quad = \{\frac{1}{2}(e_8 - e_7 - e_6 - e_5 - e_4 - e_3 - e_2 + e_1),$
$\qquad e_2 + e_1, e_2 - e_1, e_3 - e_2, e_4 - e_3, e_5 - e_4\}$

Numbering of simple roots in Dynkin diagram $= \begin{pmatrix} 2 \\ 65431 \end{pmatrix}$

Fundamental weights in terms of simple roots:

$\varpi_1 = \frac{1}{3}(4\alpha_1 + 3\alpha_2 + 5\alpha_3 + 6\alpha_4 + 4\alpha_5 + 2\alpha_6)$
$\varpi_2 = 1\alpha_1 + 2\alpha_2 + 2\alpha_3 + 3\alpha_4 + 2\alpha_5 + 1\alpha_6$
$\varpi_3 = \frac{1}{3}(5\alpha_1 + 6\alpha_2 + 10\alpha_3 + 12\alpha_4 + 8\alpha_5 + 4\alpha_6)$
$\varpi_4 = 2\alpha_1 + 3\alpha_2 + 4\alpha_3 + 6\alpha_4 + 4\alpha_5 + 2\alpha_6$
$\varpi_5 = \frac{1}{3}(4\alpha_1 + 6\alpha_2 + 8\alpha_3 + 12\alpha_4 + 10\alpha_5 + 5\alpha_6)$
$\varpi_6 = \frac{1}{3}(2\alpha_1 + 3\alpha_2 + 4\alpha_3 + 6\alpha_4 + 5\alpha_5 + 4\alpha_6)$

Positive roots having a coefficient ≥ 2:

$\begin{pmatrix} 1 \\ 01210 \end{pmatrix}, \begin{pmatrix} 1 \\ 11210 \end{pmatrix}, \begin{pmatrix} 1 \\ 01211 \end{pmatrix}, \begin{pmatrix} 1 \\ 12210 \end{pmatrix}, \begin{pmatrix} 1 \\ 11211 \end{pmatrix}, \begin{pmatrix} 1 \\ 01221 \end{pmatrix},$
$\begin{pmatrix} 1 \\ 12211 \end{pmatrix}, \begin{pmatrix} 1 \\ 11221 \end{pmatrix}, \begin{pmatrix} 1 \\ 12221 \end{pmatrix}, \begin{pmatrix} 1 \\ 12321 \end{pmatrix}, \begin{pmatrix} 2 \\ 12321 \end{pmatrix}$

$\delta = e_2 + 2e_3 + 3e_4 + 4e_5 - 4e_6 - 4e_7 + 4e_8$

E_7

$V = \{v \in \mathbb{R}^8 \mid \langle v, e_7 + e_8 \rangle = 0\}$

$\Delta = \{\pm e_i \pm e_j \mid i < j \leq 6\} \cup \{\pm(e_7 - e_8)\}$
$\quad \cup \left\{ \frac{1}{2} \sum_{i=1}^{8} (-1)^{n(i)} e_i \in V \;\middle|\; \sum_{i=1}^{8} (-1)^{n(i)} \text{ even} \right\}$

$|\Delta| = 126$
$\dim \mathfrak{g} = 133$
$|W| = 2^{10} \cdot 3^4 \cdot 5 \cdot 7$
$\det(A_{ij}) = 2$

$\Delta^+ = \{e_i \pm e_j \mid i > j\} \cup \{e_8 - e_7\}$
$\quad \cup \left\{ \frac{1}{2}(e_8 - e_7 + \sum_{i=1}^{6} (-1)^{n(i)} e_i \;\middle|\; \sum_{i=1}^{6} (-1)^{n(i)} \text{ odd} \right\}$

$\Pi = \{\alpha_1, \alpha_2, \alpha_3, \alpha_4, \alpha_5, \alpha_6, \alpha_7\}$
$\quad = \{\frac{1}{2}(e_8 - e_7 - e_6 - e_5 - e_4 - e_3 - e_2 + e_1,$
$\qquad e_2 + e_1, e_2 - e_1, e_3 - e_2, e_4 - e_3, e_5 - e_4, e_6 - e_5\}$

Numbering of simple roots in Dynkin diagram $= \binom{2}{765431}$

Fundamental weights in terms of simple roots:

$\varpi_1 = 2\alpha_1 + 2\alpha_2 + 3\alpha_3 + 4\alpha_4 + 3\alpha_5 + 2\alpha_6 + 1\alpha_7$
$\varpi_2 = \frac{1}{2}(4\alpha_1 + 7\alpha_2 + 8\alpha_3 + 12\alpha_4 + 9\alpha_5 + 6\alpha_6 + 3\alpha_7)$
$\varpi_3 = 3\alpha_1 + 4\alpha_2 + 6\alpha_3 + 8\alpha_4 + 6\alpha_5 + 4\alpha_6 + 2\alpha_7$
$\varpi_4 = 4\alpha_1 + 6\alpha_2 + 8\alpha_3 + 12\alpha_4 + 9\alpha_5 + 6\alpha_6 + 3\alpha_7$
$\varpi_5 = \frac{1}{2}(6\alpha_1 + 9\alpha_2 + 12\alpha_3 + 18\alpha_4 + 15\alpha_5 + 10\alpha_6 + 5\alpha_7)$
$\varpi_6 = 2\alpha_1 + 3\alpha_2 + 4\alpha_3 + 6\alpha_4 + 5\alpha_5 + 4\alpha_6 + 2\alpha_7$
$\varpi_7 = \frac{1}{2}(2\alpha_1 + 3\alpha_2 + 4\alpha_3 + 6\alpha_4 + 5\alpha_5 + 4\alpha_6 + 3\alpha_7)$

Positive roots having a coefficient ≥ 2 and involving α_7:

$\binom{1}{111210}, \binom{1}{111211}, \binom{1}{112210}, \binom{1}{111221}, \binom{1}{112211},$
$\binom{1}{122210}, \binom{1}{112221}, \binom{1}{122211}, \binom{1}{122221}, \binom{1}{112321},$
$\binom{1}{122321}, \binom{2}{112321}, \binom{1}{123321}, \binom{2}{122321}, \binom{2}{123321},$
$\binom{2}{123421}, \binom{2}{123431}, \binom{2}{123432}$

$\delta = \frac{1}{2}(2e_2 + 4e_3 + 6e_4 + 8e_5 + 10e_6 - 17e_7 + 17e_8)$

E_8

$V = \mathbb{R}^8$

$\Delta = \{\pm e_i \pm e_j \mid i < j\}$
$\quad \cup \left\{\frac{1}{2}\sum_{i=1}^{8}(-1)^{n(i)}e_i \mid \sum_{i=1}^{8}(-1)^{n(i)} \text{ even}\right\}$

$|\Delta| = 240$
$\dim \mathfrak{g} = 248$
$|W| = 2^{14} \cdot 3^5 \cdot 5^2 \cdot 7$
$\det(A_{ij}) = 1$

$\Delta^+ = \{e_i \pm e_j \mid i > j\}$
$\quad \cup \left\{\frac{1}{2}(e_8 + \sum_{i=1}^{7}(-1)^{n(i)}e_i \mid \sum_{i=1}^{7}(-1)^{n(i)} \text{ even}\right\}$

$\Pi = \{\alpha_1, \alpha_2, \alpha_3, \alpha_4, \alpha_5, \alpha_6, \alpha_7, \alpha_8\}$
$\quad = \{\frac{1}{2}(e_8 - e_7 - e_6 - e_5 - e_4 - e_3 - e_2 + e_1,$
$\quad e_2 + e_1,\ e_2 - e_1,\ e_3 - e_2,\ e_4 - e_3,\ e_5 - e_4,\ e_6 - e_5,\ e_7 - e_6\}$

Numbering of simple roots in Dynkin diagram $= \begin{pmatrix} \ \ \ \ 2 \\ 8765431 \end{pmatrix}$

Fundamental weights in terms of simple roots:

$$
\begin{aligned}
\varpi_1 &= 4\alpha_1 + 5\alpha_2 + 7\alpha_3 + 10\alpha_4 + 8\alpha_5 + 6\alpha_6 + 4\alpha_7 + 2\alpha_8 \\
\varpi_2 &= 5\alpha_1 + 8\alpha_2 + 10\alpha_3 + 15\alpha_4 + 12\alpha_5 + 9\alpha_6 + 6\alpha_7 + 3\alpha_8 \\
\varpi_3 &= 7\alpha_1 + 10\alpha_2 + 14\alpha_3 + 20\alpha_4 + 16\alpha_5 + 12\alpha_6 + 8\alpha_7 + 4\alpha_8 \\
\varpi_4 &= 10\alpha_1 + 15\alpha_2 + 20\alpha_3 + 30\alpha_4 + 24\alpha_5 + 18\alpha_6 + 12\alpha_7 + 6\alpha_8 \\
\varpi_5 &= 8\alpha_1 + 12\alpha_2 + 16\alpha_3 + 24\alpha_4 + 20\alpha_5 + 15\alpha_6 + 10\alpha_7 + 5\alpha_8 \\
\varpi_6 &= 6\alpha_1 + 9\alpha_2 + 12\alpha_3 + 18\alpha_4 + 15\alpha_5 + 12\alpha_6 + 8\alpha_7 + 4\alpha_8 \\
\varpi_7 &= 4\alpha_1 + 6\alpha_2 + 8\alpha_3 + 12\alpha_4 + 10\alpha_5 + 8\alpha_6 + 6\alpha_7 + 3\alpha_8 \\
\varpi_8 &= 2\alpha_1 + 3\alpha_2 + 4\alpha_3 + 6\alpha_4 + 5\alpha_5 + 4\alpha_6 + 3\alpha_7 + 2\alpha_8
\end{aligned}
$$

Positive roots having a coefficient ≥ 2 and involving α_8:

$$
\begin{pmatrix} \ \ \ \ 1 \\ 1111210 \end{pmatrix}, \begin{pmatrix} \ \ \ \ 1 \\ 1112210 \end{pmatrix}, \begin{pmatrix} \ \ \ \ 1 \\ 1111211 \end{pmatrix}, \begin{pmatrix} \ \ \ \ 1 \\ 1122210 \end{pmatrix}, \begin{pmatrix} \ \ \ \ 1 \\ 1111221 \end{pmatrix},
$$

$$
\begin{pmatrix} \ \ \ \ 1 \\ 1112211 \end{pmatrix}, \begin{pmatrix} \ \ \ \ 1 \\ 1112221 \end{pmatrix}, \begin{pmatrix} \ \ \ \ 1 \\ 1122211 \end{pmatrix}, \begin{pmatrix} \ \ \ \ 1 \\ 1222210 \end{pmatrix}, \begin{pmatrix} \ \ \ \ 1 \\ 1112321 \end{pmatrix},
$$

$$
\begin{pmatrix} \ \ \ \ 1 \\ 1122221 \end{pmatrix}, \begin{pmatrix} \ \ \ \ 1 \\ 1222211 \end{pmatrix}, \begin{pmatrix} \ \ \ \ 2 \\ 1112321 \end{pmatrix}, \begin{pmatrix} \ \ \ \ 1 \\ 1122321 \end{pmatrix}, \begin{pmatrix} \ \ \ \ 1 \\ 1222221 \end{pmatrix},
$$

$$
\begin{pmatrix} \ \ \ \ 2 \\ 1122321 \end{pmatrix}, \begin{pmatrix} \ \ \ \ 1 \\ 1123321 \end{pmatrix}, \begin{pmatrix} \ \ \ \ 1 \\ 1222321 \end{pmatrix}, \begin{pmatrix} \ \ \ \ 2 \\ 1123321 \end{pmatrix}, \begin{pmatrix} \ \ \ \ 2 \\ 1222321 \end{pmatrix},
$$

$$
\begin{pmatrix} \ \ \ \ 1 \\ 1223321 \end{pmatrix}, \begin{pmatrix} \ \ \ \ 2 \\ 1123421 \end{pmatrix}, \begin{pmatrix} \ \ \ \ 2 \\ 1223321 \end{pmatrix}, \begin{pmatrix} \ \ \ \ 1 \\ 1233321 \end{pmatrix}, \begin{pmatrix} \ \ \ \ 2 \\ 1123431 \end{pmatrix},
$$

$$\begin{pmatrix} 2 \\ 1223421 \end{pmatrix}, \begin{pmatrix} 2 \\ 1233321 \end{pmatrix}, \begin{pmatrix} 2 \\ 1123432 \end{pmatrix}, \begin{pmatrix} 2 \\ 1223431 \end{pmatrix}, \begin{pmatrix} 2 \\ 1233421 \end{pmatrix},$$
$$\begin{pmatrix} 2 \\ 1223432 \end{pmatrix}, \begin{pmatrix} 2 \\ 1233431 \end{pmatrix}, \begin{pmatrix} 2 \\ 1234421 \end{pmatrix}, \begin{pmatrix} 2 \\ 1233432 \end{pmatrix}, \begin{pmatrix} 2 \\ 1234431 \end{pmatrix},$$
$$\begin{pmatrix} 2 \\ 1234531 \end{pmatrix}, \begin{pmatrix} 2 \\ 1234432 \end{pmatrix}, \begin{pmatrix} 3 \\ 1234531 \end{pmatrix}, \begin{pmatrix} 2 \\ 1234532 \end{pmatrix}, \begin{pmatrix} 3 \\ 1234532 \end{pmatrix},$$
$$\begin{pmatrix} 2 \\ 1234542 \end{pmatrix}, \begin{pmatrix} 3 \\ 1234542 \end{pmatrix}, \begin{pmatrix} 3 \\ 1234642 \end{pmatrix}, \begin{pmatrix} 3 \\ 1235642 \end{pmatrix}, \begin{pmatrix} 3 \\ 1245642 \end{pmatrix},$$
$$\begin{pmatrix} 3 \\ 1345642 \end{pmatrix}, \begin{pmatrix} 3 \\ 2345642 \end{pmatrix}$$

$\delta = e_2 + 2e_3 + 3e_4 + 4e_5 + 5e_6 + 6e_7 + 23e_8$

***F*₄**

$V = \mathbb{R}^4$
$\Delta = \{\pm e_i \pm e_j \mid i < j\} \cup \{\pm e_i\} \cup \{\frac{1}{2}(\pm e_1 \pm e_2 \pm e_3 \pm e_4)\}$

$|\Delta| = 48$
$\dim \mathfrak{g} = 52$
$|W| = 2^7 \cdot 3^2$
$\det(A_{ij}) = 1$

$\Delta^+ = \{e_i \pm e_j \mid i < j\} \cup \{e_i\} \cup \{\frac{1}{2}(e_1 \pm e_2 \pm e_3 \pm e_4\}$
$\Pi = \{\alpha_1, \alpha_2, \alpha_3, \alpha_4\}$
$\quad = \{\frac{1}{2}(e_1 - e_2 - e_3 - e_4), e_4, e_3 - e_4, e_2 - e_3\}$
Numbering of simple roots in Dynkin diagram = (1234)

Fundamental weights in terms of simple roots:
$\varpi_1 = 2\alpha_1 + 3\alpha_2 + 2\alpha_3 + 1\alpha_4$
$\varpi_2 = 3\alpha_1 + 6\alpha_2 + 4\alpha_3 + 2\alpha_4$
$\varpi_3 = 4\alpha_1 + 8\alpha_2 + 6\alpha_3 + 3\alpha_4$
$\varpi_4 = 2\alpha_1 + 4\alpha_2 + 3\alpha_3 + 2\alpha_4$

Positive roots having a coefficient ≥ 2:
(0210), (0211), (1210), (0221), (1211), (2210), (1221),
(2211), (1321), (2221), (2321), (2421), (2431), (2432)

$\delta = 11e_1 + 5e_2 + 3e_3 + e_4$

G_2

$V = \{v \in \mathbb{R}^3 \mid \langle v, e_1 + e_2 + e_3 \rangle = 0\}$
$\Delta = \{\pm(e_1 - e_2),\ \pm(e_2 - e_3),\ \pm(e_2 - e_3)\}$
$\quad \cup \{\pm(2e_1 - e_2 - e_3),\ \pm(2e_2 - e_1 - e_3),\ \pm(2e_3 - e_1 - e_2)\}$

$|\Delta| = 12$
$\dim \mathfrak{g} = 14$
$|W| = 2^2 \cdot 3$
$\det(A_{ij}) = 1$

$\Pi = \{\alpha_1, \alpha_2\}$
$\quad = \{e_1 - e_2,\ -2e_1 + e_2 + e_3\}$
Numbering of simple roots in Dynkin diagram = (12)
$\Delta^+ = \{(10), (01), (11), (21), (31), (32)\}$
Fundamental weights in terms of simple roots:
$\quad \varpi_1 = 2\alpha_1 + 1\alpha_2$
$\quad \varpi_2 = 3\alpha_1 + 2\alpha_2$
$\delta = 5\alpha_1 + 3\alpha_2$

---

3. Classical Noncompact Simple Real Lie Algebras

This section shows for the classical noncompact noncomplex simple real Lie algebras how the methods of §§VI.10–11 reveal the structure of each of these examples.

The first three items, following the name of a Lie algebra $\mathfrak{g}_0$, describe a standard Vogan diagram of $\mathfrak{g}_0$, the fixed subalgebra $\mathfrak{k}_0$ of a Cartan involution, and the simple roots for $\mathfrak{k}_0$. In §VI.10 each $\mathfrak{g}_0$ has at most two standard Vogan diagrams, and one of them is selected and described here. References to roots use the notation of §1 of this appendix. If the Dynkin diagram has a double line or a triple point, then the double line or triple point is regarded as near the right end.

The simple roots of $\mathfrak{t}_0$ are obtained as follows. When the automorphism in the Vogan diagram is nontrivial, the remarks before Lemma 7.127 show that $\mathfrak{k}_0$ is semisimple. The simple roots for $\mathfrak{k}_0$ then include the compact imaginary simple roots and the average of the members of each 2-element orbit of simple roots. If the Vogan diagram has no painted imaginary root, there is no other simple root for $\mathfrak{k}_0$. Otherwise there is one other simple root for $\mathfrak{k}_0$, obtained by taking a minimal complex root containing the painted imaginary root in its expansion and averaging it over its 2-element orbit under the automorphism. When the automorphism is trivial, the remarks near the end of §VII.9 show that either $\dim \mathfrak{c}_0 = 1$, in which case the simple roots for $\mathfrak{k}_0$ are the compact simple roots for $\mathfrak{g}_0$, or else $\dim \mathfrak{c}_0 = 0$, in which case the simple roots for $\mathfrak{k}_0$ are the compact imaginary simple roots for $\mathfrak{g}_0$ and one other compact imaginary root. In the latter case this other compact imaginary root is the unique smallest root containing the noncompact simple root twice in its expansion.

The next two items give the real rank and a list of roots to use in Cayley transforms to pass from a maximally compact Cartan subalgebra to a maximally noncompact Cartan subalgebra. The list of roots is obtained by an algorithm described in §VI.11. In every case the members of the list are strongly orthogonal noncompact imaginary roots. When the automorphism in the Vogan diagram is nontrivial, the noncompactness of the roots is not necessarily obvious but may be verified with the aid of Proposition 6.104. The real rank of $\mathfrak{g}_0$ is the sum of the number of 2-element orbits among the simple roots, plus the number of roots in the Cayley transform list. The Cayley transform list is empty if and only if $\mathfrak{g}_0$ has just one conjugacy class of Cartan subalgebras.

The next three items identify the system of restricted roots, the real-rank-one subalgebras associated to each restricted root, and the subalgebra $\mathfrak{m}_{\mathfrak{p},0}$. Let $\mathfrak{h}_0 = \mathfrak{t}_0 \oplus \mathfrak{a}_0$ be the given maximally compact Cartan subalgebra. The information in the previous items has made it possible to identify the Cayley transform of $\mathfrak{a}_{\mathfrak{p},0}$ as a subspace of $i\mathfrak{t}_0 \oplus \mathfrak{a}_0$. The restriction of the roots to this subspace therefore identifies the restricted roots. By (6.109) the multiplicities of the restricted roots determine the real-rank-one subalgebras associated by §VII.6 to each restricted root λ for which $\frac{1}{2}\lambda$ is not a restricted root. The computation of these subalgebras is simplified by the fact that any two restricted roots of the same length are conjugate by the Weyl group of the restricted roots; the associated subalgebras are then conjugate. The roots of $\mathfrak{g}_0$ orthogonal to all roots in the Cayley transform list and to the -1 eigenspace of the automorphism are the roots of $\mathfrak{m}_{\mathfrak{p},0}$; such roots therefore determine the semisimple part of $\mathfrak{m}_{\mathfrak{p},0}$. The dimension of the center of $\mathfrak{m}_{\mathfrak{p},0}$ can then be deduced by comparing rank $\mathfrak{g}_0$, dim $\mathfrak{a}_{\mathfrak{p},0}$, and rank $[\mathfrak{m}_{\mathfrak{p},0}, \mathfrak{m}_{\mathfrak{p},0}]$.

The next three items refer to the customary analytic group G with Lie algebra $\mathfrak{g}_0$. The group G is listed together with the customary maximal compact subgroup K and the number of components of $M_{\mathfrak{p}}$. For results about the structure of $M_{\mathfrak{p}}$, see §§VII.5–6.

An item of "special features" notes if G/K is Hermitian or if $\mathfrak{g}_0$ is a split real form or if $\mathfrak{g}_0$ has just one conjugacy class of Cartan subalgebras. Finally an item of "further information" points to some places in the book where this $\mathfrak{g}_0$ or G has been discussed as an example.

Vogan diagrams of the Lie algebras in this section are indicated in Figure 6.1. A table of real ranks and restricted-root systems appears as (6.107).

$\mathfrak{sl}(n, \mathbb{R}), n$ odd ≥ 3

Vogan diagram:
  A_{n-1}, nontrivial automorphism,
  no imaginary simple roots
$\mathfrak{k}_0 = \mathfrak{so}(n)$
Simple roots for $\mathfrak{k}_0$:
  $\frac{1}{2}(e_{\frac{1}{2}(n-1)} - e_{\frac{1}{2}(n+3)})$ and
  all $\frac{1}{2}(e_i - e_{i+1} + e_{n-i} - e_{n+1-i})$ for $1 \leq i \leq \frac{1}{2}(n-3)$

Real rank $= n - 1$
Cayley transform list:
  all $e_i - e_{n+1-i}$ for $1 \leq i \leq \frac{1}{2}(n-1)$

$\Sigma = A_{n-1}$
Real-rank-one subalgebras:
  $\mathfrak{sl}(2, \mathbb{R})$ for all restricted roots
$\mathfrak{m}_{\mathfrak{p},0} = 0$

$G = SL(n, \mathbb{R})$
$K = SO(n)$
$|M_{\mathfrak{p}}| = 2^{n-1}$

Special feature:
  $\mathfrak{g}_0$ is a split real form

Further information:
  For $\mathfrak{h}_0$, compare with Example 2 in §VI.8.
  For $M_{\mathfrak{p}}$, see Example 1 in §VI.5.

$\mathfrak{sl}(n, \mathbb{R})$, n even ≥ 2

Vogan diagram:
A_{n-1}, nontrivial automorphism (if $n > 2$),
unique imaginary simple root $e_{\frac{1}{2}n} - e_{\frac{1}{2}n+1}$ painted
$\mathfrak{k}_0 = \mathfrak{so}(n)$
Simple roots for $\mathfrak{k}_0$:
$\frac{1}{2}(e_{\frac{1}{2}(n-2)} + e_{\frac{1}{2}n} - e_{\frac{1}{2}(n+2)} - e_{\frac{1}{2}(n+4)})$ and
all $\frac{1}{2}(e_i - e_{i+1} + e_{n-i} - e_{n+1-i})$ for $1 \leq i \leq \frac{1}{2}(n-2)$

Real rank $= n - 1$
Cayley transform list:
all $e_i - e_{n+1-i}$ for $1 \leq i \leq \frac{1}{2}n$

$\Sigma = A_{n-1}$
Real-rank-one subalgebras:
$\mathfrak{sl}(2, \mathbb{R})$ for all restricted roots
$m_{\mathfrak{p},0} = 0$

$G = SL(n, \mathbb{R})$
$K = SO(n)$
$|M_{\mathfrak{p}}| = 2^{n-1}$

Special features:
G/K is Hermitian when $n = 2$,
$\mathfrak{g}_0$ is a split real form for all n

Further information:
For $\mathfrak{h}_0$, see Example 2 in §VI.8.
For $M_{\mathfrak{p}}$, see Example 1 in §VI.5.
For Cartan subalgebras see Problem 13 for Chapter VI.

$\mathfrak{sl}(n, \mathbb{H}), n \geq 2$

Vogan diagram:

A_{2n-1}, nontrivial automorphism,

unique imaginary simple root $e_n - e_{n+1}$ unpainted

$\mathfrak{k}_0 = \mathfrak{sp}(n)$

Simple roots for $\mathfrak{k}_0$:

$e_n - e_{n+1}$ and

all $\frac{1}{2}(e_i - e_{i+1} + e_{2n-i} - e_{2n+1-i})$ for $1 \leq i \leq n-1$

Real rank $= n - 1$

Cayley transform list: empty

$\Sigma = A_{n-1}$

Real-rank-one subalgebras:

$\mathfrak{so}(5, 1)$ for all restricted roots

$\mathfrak{m}_{\mathfrak{p},0} \cong \mathfrak{su}(2)^n$,

simple roots equal to

all $e_i - e_{2n+1-i}$ for $1 \leq i \leq n$

$G = SL(n, \mathbb{H})$

$K = Sp(n)$

$|M_{\mathfrak{p}}/(M_{\mathfrak{p}})_0| = 1$

Special feature:

$\mathfrak{g}_0$ has one conjugacy class of Cartan subalgebras

Further information:

For $M_{\mathfrak{p}}$, see Example 1 in §VI.5.

$\mathfrak{su}(p, q)$, $1 \leq p \leq q$

Vogan diagram:
  A_{p+q-1}, trivial automorphism,
  p^{th} simple root $e_p - e_{p+1}$ painted
$\mathfrak{k}_0 = \mathfrak{s}(\mathfrak{u}(p) \oplus \mathfrak{u}(q))$
Simple roots for $\mathfrak{k}_0$: compact simple roots only

Real rank $= p$
Cayley transform list:
  all $e_i - e_{2p+1-i}$ for $1 \leq i \leq p$

$$\Sigma = \begin{Bmatrix} (BC)_p \text{ if } p < q \\ C_p \quad \text{if } p = q \end{Bmatrix}$$

Real-rank-one subalgebras:
  $\mathfrak{sl}(2, \mathbb{C})$ for all restricted roots $\pm f_i \pm f_j$
  $\mathfrak{su}(q - p + 1, 1)$ for all $\pm\{f_i, 2f_i\}$

$$\mathfrak{m}_{\mathfrak{p},0} = \begin{Bmatrix} \mathbb{R}^p \oplus \mathfrak{su}(q-p) \text{ if } p < q \\ \mathbb{R}^{p-1} \quad\quad \text{if } p = q \end{Bmatrix},$$

  simple roots by Cayley transform from
  all $e_{2p+i} - e_{2p+i+1}$ for $1 \leq i \leq q - p - 1$

$G = SU(p, q)$
$K = S(U(p) \times U(q))$
$$|M_\mathfrak{p}/(M_\mathfrak{p})_0| = \begin{Bmatrix} 1 \text{ if } p < q \\ 2 \text{ if } p = q \end{Bmatrix}$$

Special feature:
  G/K is Hermitian

Further information:
  For $\mathfrak{h}_0$, see Example 1 in §VI.8.
  For $M_\mathfrak{p}$, see Example 2 in §VI.5.
  For the Hermitian structure see the example in §VII.9.
  For restricted roots see the example with (6.106) and see also Problem 37 for Chapter VII.

$\mathfrak{so}(2p, 2q+1), 1 \le p \le q$

Vogan diagram:
B_{p+q}, trivial automorphism,
p^{th} simple root $e_p - e_{p+1}$ painted
$\mathfrak{k}_0 = \mathfrak{so}(2p) \oplus \mathfrak{so}(2q+1)$
Simple roots for $\mathfrak{k}_0$: compact simple roots and
$$\begin{Bmatrix} e_{p-1} + e_p \text{ when } p > 1 \\ \text{no other when } p = 1 \end{Bmatrix}$$

Real rank $= 2p$
Cayley transform list:
all $e_i \pm e_{2p+1-i}$ for $1 \le i \le p$

$\Sigma = B_{2p}$
Real-rank-one subalgebras:
$\mathfrak{sl}(2, \mathbb{R})$ for all long restricted roots
$\mathfrak{so}(2q - 2p + 2, 1)$ for all short restricted roots
$\mathfrak{m}_{\mathfrak{p},0} = \mathfrak{so}(2q - 2p + 1)$,
simple roots when $p < q$ by Cayley transform from
e_{p+q} and all $e_{2p+i} - e_{2p+i+1}$ for $1 \le i \le q - p - 1$

$G = SO(2p, 2q+1)_0$
$K = SO(p) \times SO(2q+1)$
$|M_{\mathfrak{p}}/(M_{\mathfrak{p}})_0| = 2^{2p-1}$

Special features:
G/K is Hermitian when $p = 1$,
$\mathfrak{g}_0$ is a split real form when $p = q$

Further information:
For $M_{\mathfrak{p}}$, see Example 3 in §VI.5.

$\mathfrak{so}(2p, 2q+1)$, $p > q \geq 0$

Vogan diagram:
B_{p+q}, trivial automorphism,
p^{th} simple root $e_p - e_{p+1}$ painted
$\mathfrak{k}_0 = \mathfrak{so}(2p) \oplus \mathfrak{so}(2q+1)$
Simple roots for $\mathfrak{k}_0$: compact simple roots and
$\begin{Bmatrix} e_{p-1} + e_p \text{ when } p > 1 \\ \text{no other when } p = 1 \text{ and } q = 0 \end{Bmatrix}$

Real rank $= 2q + 1$
Cayley transform list:
e_{p-q} and all $e_i \pm e_{2p+1-i}$ for $p - q + 1 \leq i \leq p$

$\Sigma = B_{2q+1}$
Real-rank-one subalgebras:
$\mathfrak{sl}(2, \mathbb{R})$ for all long restricted roots
$\mathfrak{so}(2p - 2q, 1)$ for all short restricted roots
$\mathfrak{m}_{\mathfrak{p},0} = \mathfrak{so}(2p - 2q - 1)$,
simple roots when $p > q + 1$ by Cayley transform from
e_{p-q-1} and all $e_i - e_{i+1}$ for $1 \leq i \leq p - q - 2$

$G = SO(2p, 2q+1)_0$
$K = SO(p) \times SO(2q+1)$
$|M_{\mathfrak{p}}/(M_{\mathfrak{p}})_0| = 2^{2q}$

Special feature:
G/K is Hermitian when $p = 1$ and $q = 0$,
$\mathfrak{g}_0$ is a split real form when $p = q + 1$

Further information:
For $M_{\mathfrak{p}}$, see Example 3 in §VI.5.

$\mathfrak{sp}(p, q)$, $1 \le p \le q$

Vogan diagram:
    C_{p+q}, trivial automorphism,
    p^{th} simple root $e_p - e_{p+1}$ painted
$\mathfrak{k}_0 = \mathfrak{sp}(p) \times \mathfrak{sp}(q)$
Simple roots for $\mathfrak{k}_0$: compact simple roots and
    $2e_p$

Real rank $= p$
Cayley transform list:
    all $e_i - e_{2p+1-i}$ for $1 \le i \le p$

$$\Sigma = \begin{Bmatrix} (BC)_p \text{ if } p < q \\ C_p \quad \text{if } p = q \end{Bmatrix}$$

Real-rank-one subalgebras:
    $\mathfrak{so}(5, 1)$ for all restricted roots $\pm f_i \pm f_j$
    $\mathfrak{sp}(q - p + 1, 1)$ for all $\pm\{f_i, 2f_i\}$
$\mathfrak{m}_{\mathfrak{p},0} = \mathfrak{su}(2)^p \oplus \mathfrak{sp}(q - p)$,
    simple roots by Cayley transform from
    all $e_i + e_{2p+1-i}$ for $1 \le i \le p$,
    all $e_{2p+i} - e_{2p+i+1}$ for $1 \le i \le q - p - 1$,
    and also $2e_{p+q}$ if $p < q$

$G = Sp(p, q)$
$K = Sp(p) \times Sp(q)$
$|M_{\mathfrak{p}}/(M_{\mathfrak{p}})_0| = 1$

Further information:
    $M_{\mathfrak{p}}$ is connected by Corollary 7.69 and Theorem 7.55.

$\mathfrak{sp}(n, \mathbb{R}), n \geq 1$

Vogan diagram:
 C_n, trivial automorphism,
 n^{th} simple root $2e_n$ painted
$\mathfrak{k}_0 = \mathfrak{u}(n)$
Simple roots for $\mathfrak{k}_0$: compact simple roots only

Real rank $= n$
Cayley transform list:
 all $2e_i$, $1 \leq i \leq n$

$\Sigma = C_n$
Real-rank-one subalgebras:
 $\mathfrak{sl}(2, \mathbb{R})$ for all restricted roots
$m_{\mathfrak{p},0} = 0$

$G = Sp(n, \mathbb{R})$
$K = U(n)$
$|M_{\mathfrak{p}}| = 2^n$

Special features:
 G/K is Hermitian,
 $\mathfrak{g}_0$ is a split real form

Further information:
 For isomorphisms see Problem 15 for Chapter VI and Problem 30 for Chapter VII.
 For the Hermitian structure see Problems 31–33 for Chapter VII.
 For restricted roots see Problem 38 for Chapter VII.

$\mathfrak{so}(2p+1, 2q+1), 0 \leq p \leq q$ but not $\mathfrak{so}(1,1)$ or $\mathfrak{so}(1,3)$

Vogan diagram:
D_{p+q+1}, nontrivial automorphism,
p^{th} simple root $e_p - e_{p+1}$ painted, none if $p = 0$
$\mathfrak{k}_0 = \mathfrak{so}(2p+1) \oplus \mathfrak{so}(2q+1)$
Simple roots for $\mathfrak{k}_0$:
e_p (if $p > 0$), e_{p+q}, and
all $e_i - e_{i+1}$ with $1 \leq i \leq p-1$ or $p+1 \leq i \leq p+q-1$

Real rank $= 2p+1$
Cayley transform list:
all $e_i \pm e_{2p+1-i}$, $1 \leq i \leq p$

$$\Sigma = \begin{Bmatrix} B_p \text{ if } p < q \\ D_p \text{ if } p = q \end{Bmatrix}$$

Real-rank-one subalgebras:
$\mathfrak{sl}(2, \mathbb{R})$ for all long restricted roots
$\mathfrak{so}(2q-2p+1, 1)$ for all short restricted roots when $p < q$
$\mathfrak{m}_{\mathfrak{p},0} = \mathfrak{so}(2q-2p)$,
simple roots when $p < q-1$ by Cayley transform from
$e_{p+q-1} + e_{p+q}$ and all $e_{2p+i} - e_{2p+i+1}$ for $1 \leq i \leq q-p-1$

$G = SO(2p+1, 2q+1)_0$
$K = SO(2p+1) \times SO(2q+1)$
$|M_{\mathfrak{p}}/(M_{\mathfrak{p}})_0| = 2^{2p}$

Special feature:
$\mathfrak{g}_0$ has one conjugacy class of Cartan subalgebras when $p = 0$

Further information:
For $p = 0$ and $q = 1$, $\mathfrak{so}(1,3) \cong \mathfrak{sl}(2, \mathbb{C})$ is complex.
For $M_{\mathfrak{p}}$, see Example 3 in §VII.5.

$\mathfrak{so}(2p, 2q)$, $1 \le p \le q$ but not $\mathfrak{so}(2, 2)$

Vogan diagram:
D_{p+q}, trivial automorphism,
p^{th} simple root $e_p - e_{p+1}$ painted
$\mathfrak{k}_0 = \mathfrak{so}(2p) \oplus \mathfrak{so}(2q)$
Simple roots for $\mathfrak{k}_0$: compact simple roots and
$\left\{ \begin{array}{l} e_{p-1} + e_p \text{ when } p > 1 \\ \text{no other when } p = 1 \end{array} \right\}$

Real rank $= 2p$
Cayley transform list:
all $e_i \pm e_{2p+1-i}$, $1 \le i \le p$

$\Sigma = \left\{ \begin{array}{l} B_p \text{ if } p < q \\ D_p \text{ if } p = q \end{array} \right\}$
Real-rank-one subalgebras:
$\mathfrak{sl}(2, \mathbb{R})$ for all long restricted roots
$\mathfrak{so}(2q - 2p + 1, 1)$ for all short restricted roots when $p < q$
$\mathfrak{m}_{\mathfrak{p},0} = \mathfrak{so}(2q - 2p)$,
simple roots when $p < q - 1$ by Cayley transform from
$e_{p+q-1} + e_{p+q}$ and all $e_{2p+i} - e_{2p+i+1}$ for $1 \le i \le q - p - 1$

$G = SO(2p, 2q)_0$
$K = SO(2p) \times SO(2q)$
$|M_{\mathfrak{p}}/(M_{\mathfrak{p}})_0| = 2^{2p-1}$

Special features:
G/K is Hermitian when $p = 1$,
$\mathfrak{g}_0$ is a split real form when $p = q$

Further information:
For $p = q = 2$, $\mathfrak{so}(2, 2) \cong \mathfrak{sl}(2, \mathbb{R}) \oplus \mathfrak{sl}(2, \mathbb{R})$ is not simple.
For $M_{\mathfrak{p}}$, see Example 3 in §VI.5.

$\mathfrak{so}^*(2n), n \geq 3$

Vogan diagram:

D_n, trivial automorphism,

n^{th} simple root $e_{n-1} + e_n$ painted

$\mathfrak{k}_0 = \mathfrak{u}(n)$

Simple roots for $\mathfrak{k}_0$: compact simple roots only

Real rank $= [n/2]$

Cayley transform list:

all $e_{n-2i+1} + e_{n-2i+2}$, $1 \leq i \leq [n/2]$

$$\Sigma = \begin{Bmatrix} (BC)_{\frac{1}{2}(n-1)} & \text{if } n \text{ odd} \\ C_{\frac{1}{2}n} & \text{if } n \text{ even} \end{Bmatrix}$$

Real-rank-one subalgebras:

$$\begin{Bmatrix} \mathfrak{sl}(2,\mathbb{R}) \text{ for all } \pm 2f_i & \text{if } n \text{ even} \\ \mathfrak{su}(3,1) \text{ for all } \pm\{f_i, 2f_i\} & \text{if } n \text{ odd} \end{Bmatrix}$$

$\mathfrak{so}(5,1)$ for all $\pm f_i \pm f_j$

$$\mathfrak{m}_{\mathfrak{p},0} = \begin{Bmatrix} \mathfrak{su}(2)^{\frac{1}{2}n} & \text{if } n \text{ even} \\ \mathfrak{su}(2)^{\frac{1}{2}(n-1)} \oplus \mathbb{R} & \text{if } n \text{ odd} \end{Bmatrix},$$

simple roots by Cayley transform from

all $e_{n-2i+1} - e_{n-2i+2}$ for $1 \leq i \leq [n/2]$

$G = SO^*(2n)$

$K = U(n)$

$$|M_{\mathfrak{p}}/(M_{\mathfrak{p}})_0| = \begin{Bmatrix} 2 \text{ if } n \text{ even} \\ 1 \text{ if } n \text{ odd} \end{Bmatrix}$$

Special feature:

G/K is Hermitian

Further information:

When $n = 2$, $\mathfrak{so}^*(4) \cong \mathfrak{sl}(2,\mathbb{R}) \oplus \mathfrak{su}(2)$ is not simple.

For $\mathfrak{h}_0$ and explicit root structure, see Problem 6 for Chapter VI.

$M_{\mathfrak{p}}$ is connected when n is odd by Corollary 7.69 and Theorem 7.55.

For Hermitian structure see Problems 34–36 for Chapter VII.

For restricted roots see Problem 39 for Chapter VII.

---

4. Exceptional Noncompact Simple Real Lie Algebras

This section exhibits the structure of the exceptional noncompact noncomplex simple real Lie algebras by using the methods of §§VI.10–11.

The format is rather similar to that in the previous section. The first three items following the name of a Lie algebra $\mathfrak{g}_0$ (given as in the listing of Cartan [1927a] and Helgason [1978]) describe the standard Vogan diagram of $\mathfrak{g}_0$, the fixed subalgebra $\mathfrak{k}_0$ of a Cartan involution, and the simple roots of $\mathfrak{k}_0$. In the cases of F_4 and G_2, the left root in a Dynkin diagram is short. Techniques for obtaining the roots of $\mathfrak{k}_0$ are described in §3, and references to explicit roots use the notation of §2. As in §3, when the automorphism in the Vogan diagram is trivial and $\dim \mathfrak{c}_0=0$, there is one simple root of $\mathfrak{k}_0$ that is not simple for $\mathfrak{g}_0$. This root is the unique smallest root containing the noncompact simple root twice in its expansion. It may be found by referring to the appropriate table in §2 of "positive roots having a coefficient ≥ 2."

One difference in format in this section, by comparison with §3, is that roots are displayed in two ways. The first way gives the expansion in terms of simple roots, using notation introduced in §2. The second way is in terms of the underlying space V of the root system.

The next two items give the real rank and a list of roots to use in Cayley transforms to obtain a maximally noncompact Cartan subalgebra. The three items after that identify the system of restricted roots, the real-rank-one subalgebras associated to each restricted root, and the subalgebra $\mathfrak{m}_{\mathfrak{p},0}$. The techniques are unchanged from §3.

The final item is the mention of any special feature. A notation appears if G/K is Hermitian or if $\mathfrak{g}_0$ is a split real form or if $\mathfrak{g}_0$ has just one conjugacy class of Cartan subalgebras. For each complex simple Lie algebra, there is a unique real form such that G/K has a kind of quaternion structure (see the Notes). Except in type A_n, $\mathfrak{k}_0$ has a summand $\mathfrak{su}(2)$ for this case. Under the item of special features, a notation appears if G/K is quaternionic.

Vogan diagrams of the Lie algebras in this section appear also in Figures 6.2 and 6.3, and $\mathfrak{k}_0$ for each diagram is indicated in those figures. A table of real ranks and restricted-root systems appears as (6.108).

E I

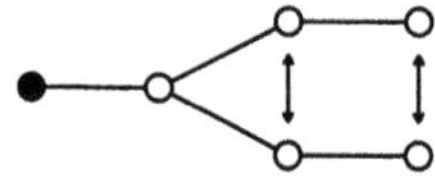

$\mathfrak{k}_0 = \mathfrak{sp}(4)$

Simple roots for $\mathfrak{k}_0$:

1) $\begin{pmatrix} 0 \\ 00100 \end{pmatrix}, \begin{pmatrix} 0 \\ 0\frac{1}{2}0\frac{1}{2}0 \end{pmatrix}, \begin{pmatrix} 0 \\ \frac{1}{2}000\frac{1}{2} \end{pmatrix}, \begin{pmatrix} 1 \\ 0\frac{1}{2}1\frac{1}{2}0 \end{pmatrix}$

2) $e_3 - e_2, \frac{1}{2}(e_4 - e_3 + e_2 - e_1), \frac{1}{4}(e_8 - e_7 - e_6 + e_5 - 3e_4 - e_3 - e_2 + e_1),$
$\frac{1}{2}(e_4 + e_3 + e_2 + e_1)$

Real rank $= 6$

Cayley transform lists:

1) $\begin{pmatrix} 1 \\ 00000 \end{pmatrix}, \begin{pmatrix} 1 \\ 01210 \end{pmatrix}, \begin{pmatrix} 1 \\ 11211 \end{pmatrix}, \begin{pmatrix} 1 \\ 12221 \end{pmatrix}$

2) $e_2 + e_1, e_4 + e_3, \frac{1}{2}(e_8 - e_7 - e_6 + e_5 - e_4 + e_3 - e_2 + e_1),$
$\frac{1}{2}(e_8 - e_7 - e_6 + e_5 + e_4 - e_3 + e_2 - e_1)$

$\Sigma = E_6$

Real-rank-one subalgebras:

$\mathfrak{sl}(2, \mathbb{R})$ for all restricted roots

$\mathfrak{m}_{\mathfrak{p},0} = 0$

Special feature:

$\mathfrak{g}_0$ is a split real form

E II

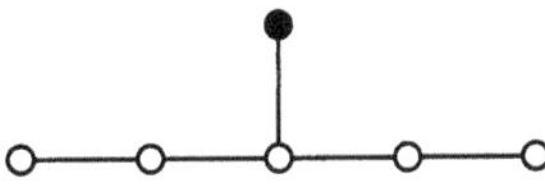

$\mathfrak{k}_0 = \mathfrak{su}(6) \oplus \mathfrak{su}(2)$
Simple roots for $\mathfrak{k}_0$: compact simple roots and

$$\begin{pmatrix} 2 \\ 12321 \end{pmatrix} = \tfrac{1}{2}(e_8 - e_7 - e_6 + e_5 + e_4 + e_3 + e_2 + e_1)$$

Real rank $= 4$
Cayley transform lists:

1) $\begin{pmatrix} 1 \\ 00000 \end{pmatrix}, \begin{pmatrix} 1 \\ 01210 \end{pmatrix}, \begin{pmatrix} 1 \\ 11211 \end{pmatrix}, \begin{pmatrix} 1 \\ 12221 \end{pmatrix}$
2) $e_2 + e_1, e_4 + e_3, \tfrac{1}{2}(e_8 - e_7 - e_6 + e_5 - e_4 + e_3 - e_2 + e_1),$
$\tfrac{1}{2}(e_8 - e_7 - e_6 + e_5 + e_4 - e_3 + e_2 - e_1)$

$\Sigma = F_4$
Real-rank-one subalgebras:
$\mathfrak{sl}(2, \mathbb{R})$ for all long restricted roots
$\mathfrak{sl}(2, \mathbb{C})$ for all short restricted roots
$\mathfrak{m}_{\mathfrak{p},0} = \mathbb{R}^2$

Special feature:
G/K is of quaternion type

E III

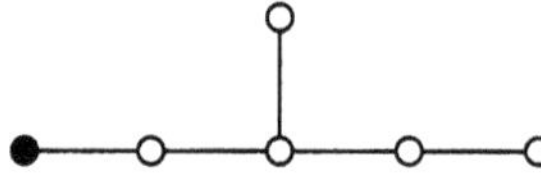

$\mathfrak{k}_0 = \mathfrak{so}(10) \oplus \mathbb{R}$
Simple roots for $\mathfrak{k}_0$: compact simple roots only

Real rank = 2
Cayley transform lists:

1) $\begin{pmatrix} 0 \\ 10000 \end{pmatrix}, \begin{pmatrix} 1 \\ 12210 \end{pmatrix}$
2) $e_5 - e_4,\ e_5 + e_4$

$\Sigma = (BC)_2$
Real-rank-one subalgebras:

$\mathfrak{so}(7,1)$ for restricted roots $\pm f_1 \pm f_2$
$\mathfrak{su}(5,1)$ for $\pm\{f_i,\ 2f_i\}$

$\mathfrak{m}_{\mathfrak{p},0} = \mathfrak{su}(4) \oplus \mathbb{R}$, simple roots by Cayley transform from

1) $\begin{pmatrix} 1 \\ 00000 \end{pmatrix}, \begin{pmatrix} 0 \\ 00010 \end{pmatrix}, \begin{pmatrix} 0 \\ 00100 \end{pmatrix}$
2) $e_2 + e_1,\ e_2 - e_1,\ e_3 - e_2$

Special feature:

G/K is Hermitian

E IV

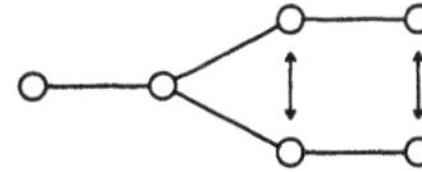

$\mathfrak{k}_0 = \mathfrak{f}_4$
Simple roots for $\mathfrak{k}_0$:

1) $\begin{pmatrix} 0 \\ 00100 \end{pmatrix}, \begin{pmatrix} 0 \\ 0\frac{1}{2}0\frac{1}{2}0 \end{pmatrix}, \begin{pmatrix} 0 \\ \frac{1}{2}000\frac{1}{2} \end{pmatrix}, \begin{pmatrix} 1 \\ 00000 \end{pmatrix}$
2) $e_3 - e_2, \frac{1}{2}(e_4 - e_3 + e_2 - e_1), \frac{1}{4}(e_8 - e_7 - e_6 + e_5 - 3e_4 - e_3 - e_2 + e_1),$
$e_2 + e_1$

Real rank $= 2$
Cayley transform list: empty

$\Sigma = A_2$
Real-rank-one subalgebras:
$\mathfrak{so}(9, 1)$ for all restricted roots
$\mathfrak{m}_{\mathfrak{p},0} = \mathfrak{so}(8)$, simple roots by Cayley transform from

1) $\begin{pmatrix} 1 \\ 00000 \end{pmatrix}, \begin{pmatrix} 0 \\ 00100 \end{pmatrix}, \begin{pmatrix} 0 \\ 01110 \end{pmatrix}, \begin{pmatrix} 0 \\ 11111 \end{pmatrix}$
2) $e_2 + e_1,\ e_3 - e_2,\ e_4 - e_1,\ \frac{1}{2}(e_8 - e_7 - e_6 + e_5 - e_4 - e_3 - e_2 - e_1)$

Special feature:
$\mathfrak{g}_0$ has one conjugacy class of Cartan subalgebras

---

E V

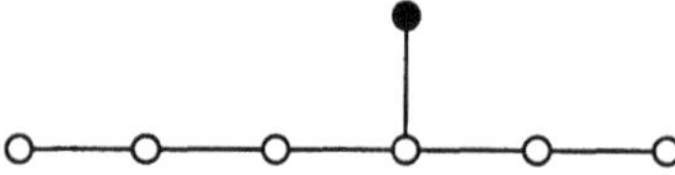

$\mathfrak{k}_0 = \mathfrak{su}(8)$

Simple roots for $\mathfrak{k}_0$: compact simple roots and

$$\begin{pmatrix} 2 \\ 012321 \end{pmatrix} = \tfrac{1}{2}(e_8 - e_7 - e_6 + e_5 + e_4 + e_3 + e_2 + e_1)$$

Real rank = 7

Cayley transform lists:

1) $\begin{pmatrix} 1 \\ 000000 \end{pmatrix}$, $\begin{pmatrix} 1 \\ 001210 \end{pmatrix}$, $\begin{pmatrix} 1 \\ 122210 \end{pmatrix}$, $\begin{pmatrix} 1 \\ 011211 \end{pmatrix}$, $\begin{pmatrix} 1 \\ 012221 \end{pmatrix}$, $\begin{pmatrix} 1 \\ 111221 \end{pmatrix}$, $\begin{pmatrix} 1 \\ 112211 \end{pmatrix}$

2) $e_2 + e_1$, $e_4 + e_3$, $e_6 + e_5$, $\tfrac{1}{2}(e_8 - e_7 - e_6 + e_5 - e_4 + e_3 - e_2 + e_1)$, $\tfrac{1}{2}(e_8 - e_7 - e_6 + e_5 + e_4 - e_3 + e_2 - e_1)$, $\tfrac{1}{2}(e_8 - e_7 + e_6 - e_5 - e_4 + e_3 + e_2 - e_1)$, $\tfrac{1}{2}(e_8 - e_7 + e_6 - e_5 + e_4 - e_3 - e_2 + e_1)$

$\Sigma = E_7$

Real-rank-one subalgebras:

$\mathfrak{sl}(2, \mathbb{R})$ for all restricted roots

$\mathfrak{m}_{\mathfrak{p},0} = 0$

Special feature:

$\mathfrak{g}_0$ is a split real form

E VI

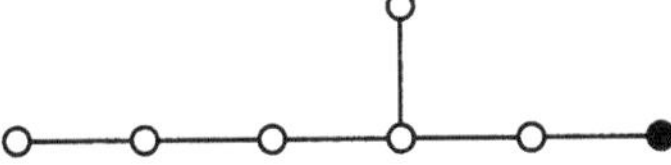

$\mathfrak{k}_0 = \mathfrak{so}(12) \oplus \mathfrak{su}(2)$
Simple roots for $\mathfrak{k}_0$: compact simple roots and

$$\begin{pmatrix} \;\;\;2 \\ 123432 \end{pmatrix} = e_8 - e_7$$

Real rank = 4
Cayley transform lists:

1) $\begin{pmatrix} \;\;\;0 \\ 000001 \end{pmatrix}, \begin{pmatrix} \;\;\;1 \\ 001221 \end{pmatrix}, \begin{pmatrix} \;\;\;1 \\ 122221 \end{pmatrix}, \begin{pmatrix} \;\;\;2 \\ 123421 \end{pmatrix}$

2) $\frac{1}{2}(e_8-e_7-e_6-e_5-e_4-e_3-e_2+e_1), \frac{1}{2}(e_8-e_7-e_6-e_5+e_4+e_3+e_2-e_1),$
$\frac{1}{2}(e_8-e_7+e_6+e_5-e_4-e_3+e_2-e_1), \frac{1}{2}(e_8-e_7+e_6+e_5+e_4+e_3-e_2+e_1)$

$\Sigma = F_4$
Real-rank-one subalgebras:
 $\mathfrak{sl}(2,\mathbb{R})$ for all long restricted roots
 $\mathfrak{so}(5,1)$ for all short restricted roots
$\mathfrak{m}_{\mathfrak{p},0} = \mathfrak{su}(2) \oplus \mathfrak{su}(2) \oplus \mathfrak{su}(2)$, simple roots by Cayley transform from

1) $\begin{pmatrix} \;\;\;1 \\ 000000 \end{pmatrix}, \begin{pmatrix} \;\;\;0 \\ 001000 \end{pmatrix}, \begin{pmatrix} \;\;\;0 \\ 100000 \end{pmatrix}$

2) $e_2 + e_1,\ e_4 - e_3,\ e_6 - e_5$

Special feature:
 G/K is of quaternion type

E VII

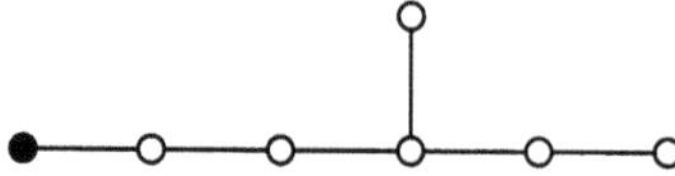

$\mathfrak{k}_0 = \mathfrak{e}_6 \oplus \mathbb{R}$
Simple roots for $\mathfrak{k}_0$: compact simple roots only

Real rank = 3
Cayley transform lists:

1) $\begin{pmatrix} 0 \\ 100000 \end{pmatrix}, \begin{pmatrix} 1 \\ 122210 \end{pmatrix}, \begin{pmatrix} 2 \\ 123432 \end{pmatrix}$
2) $e_6 - e_5,\ e_6 + e_5,\ e_8 - e_7$

$\Sigma = C_3$
Real-rank-one subalgebras:

$\mathfrak{sl}(2, \mathbb{R})$ for all long restricted roots
$\mathfrak{so}(9, 1)$ for all short restricted roots

$\mathfrak{m}_{\mathfrak{p},0} = \mathfrak{so}(8)$, simple roots by Cayley transform from

1) $\begin{pmatrix} 1 \\ 000000 \end{pmatrix}, \begin{pmatrix} 0 \\ 000010 \end{pmatrix}, \begin{pmatrix} 0 \\ 000100 \end{pmatrix}, \begin{pmatrix} 0 \\ 001000 \end{pmatrix}$
2) $e_2 + e_1,\ e_2 - e_1,\ e_3 - e_2,\ e_4 - e_3$

Special feature:

G/K is Hermitian

E VIII

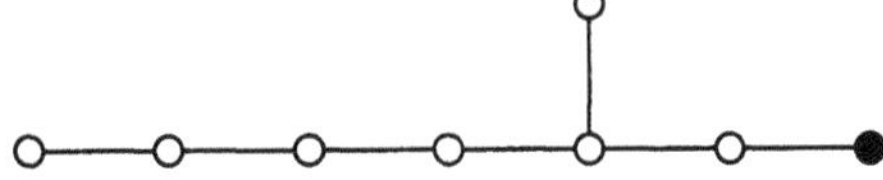

$\mathfrak{k}_0 = \mathfrak{so}(16)$
Simple roots for $\mathfrak{k}_0$: compact simple roots and

$$\begin{pmatrix} 2 \\ 0123432 \end{pmatrix} = e_8 - e_7$$

Real rank = 8
Cayley transform lists:

1) $\begin{pmatrix} 0 \\ 0000001 \end{pmatrix}, \begin{pmatrix} 1 \\ 0001221 \end{pmatrix}, \begin{pmatrix} 1 \\ 0122221 \end{pmatrix}, \begin{pmatrix} 2 \\ 0123421 \end{pmatrix},$
$\begin{pmatrix} 1 \\ 1122321 \end{pmatrix}, \begin{pmatrix} 1 \\ 1223321 \end{pmatrix}, \begin{pmatrix} 2 \\ 1222321 \end{pmatrix}, \begin{pmatrix} 2 \\ 1123321 \end{pmatrix}$

2) $\frac{1}{2}(e_8-e_7-e_6-e_5-e_4-e_3-e_2+e_1), \frac{1}{2}(e_8-e_7-e_6-e_5+e_4+e_3+e_2-e_1),$
$\frac{1}{2}(e_8-e_7+e_6+e_5-e_4-e_3+e_2-e_1), \frac{1}{2}(e_8-e_7+e_6+e_5+e_4+e_3-e_2+e_1),$
$\frac{1}{2}(e_8+e_7-e_6+e_5-e_4+e_3-e_2-e_1), \frac{1}{2}(e_8+e_7+e_6-e_5+e_4-e_3-e_2-e_1),$
$\frac{1}{2}(e_8+e_7+e_6-e_5-e_4+e_3+e_2+e_1), \frac{1}{2}(e_8+e_7-e_6+e_5+e_4-e_3+e_2+e_1)$

$\Sigma = E_8$
Real-rank-one subalgebras:
$\mathfrak{sl}(2, \mathbb{R})$ for all restricted roots
$\mathfrak{m}_{\mathfrak{p},0} = 0$

Special feature:
$\mathfrak{g}_0$ is a split real form

E IX

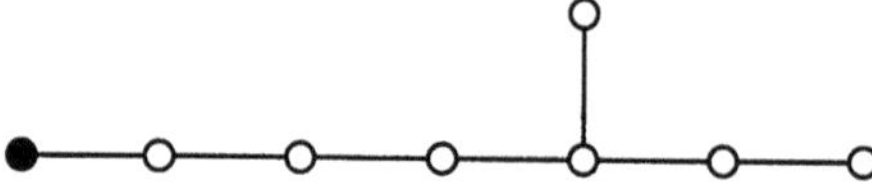

$\mathfrak{k}_0 = \mathfrak{e}_7 \oplus \mathfrak{su}(2)$

Simple roots for $\mathfrak{k}_0$: compact simple roots and

$$\begin{pmatrix} 3 \\ 2345642 \end{pmatrix} = e_8 + e_7$$

Real rank = 4

Cayley transform lists:

1) $\begin{pmatrix} 0 \\ 1000000 \end{pmatrix}, \begin{pmatrix} 1 \\ 1222210 \end{pmatrix}, \begin{pmatrix} 2 \\ 1223432 \end{pmatrix}, \begin{pmatrix} 3 \\ 1245642 \end{pmatrix}$

2) $e_7 - e_6,\ e_7 + e_6,\ e_8 - e_5,\ e_8 + e_5$

$\Sigma = F_4$

Real-rank-one subalgebras:

$\mathfrak{sl}(2, \mathbb{R})$ for all long restricted roots

$\mathfrak{so}(9, 1)$ for all short restricted roots

$\mathfrak{m}_{\mathfrak{p},0} = \mathfrak{so}(8)$, simple roots by Cayley transform from

1) $\begin{pmatrix} 1 \\ 0000000 \end{pmatrix}, \begin{pmatrix} 0 \\ 0000010 \end{pmatrix}, \begin{pmatrix} 0 \\ 0000100 \end{pmatrix}, \begin{pmatrix} 0 \\ 0001000 \end{pmatrix}$

2) $e_2 + e_1,\ e_2 - e_1,\ e_3 - e_2,\ e_4 - e_3$

Special feature:

G/K is of quaternion type

F I

$\mathfrak{k}_0 = \mathfrak{sp}(3) \oplus \mathfrak{su}(2)$
Simple roots for $\mathfrak{k}_0$: compact simple roots and
 $(2432) = e_1 + e_2$, with (1000) short

Real rank = 4
Cayley transform lists:
 1) (0001), (0221), (2221), (2421)
 2) $e_2 - e_3$, $e_2 + e_3$, $e_1 - e_4$, $e_1 + e_4$

$\Sigma = F_4$
Real-rank-one subalgebras:
 $\mathfrak{sl}(2, \mathbb{R})$ for all restricted roots
$\mathfrak{m}_{\mathfrak{p},0} = 0$

Special feature:
 $\mathfrak{g}_0$ is a split real form

F II

●—○⇒○—○

$\mathfrak{k}_0 = \mathfrak{so}(9)$

Simple roots for $\mathfrak{k}_0$: compact simple roots and

$(2210) = e_1 - e_2$, with (1000) short

Real rank = 1

Cayley transform lists:

1) (1000)
2) $\frac{1}{2}(e_1 - e_2 - e_3 - e_4)$

$\Sigma = (BC)_1$

Real-rank-one subalgebra:

F II

$\mathfrak{m}_{\mathfrak{p},0} = \mathfrak{so}(7)$, simple roots by Cayley transform from

1) (0010), (0001), (1210)
2) $e_3 - e_4$, $e_2 - e_3$, $\frac{1}{2}(e_1 - e_2 + e_3 + e_4)$.

G

$\mathfrak{k}_0 = \mathfrak{su}(2) \oplus \mathfrak{su}(2)$
Simple roots for $\mathfrak{k}_0$: compact simple roots and
(32), with (10) short

Real rank = 2
Cayley transform list:
(01), (21)

$\Sigma = G_2$
Real-rank-one subalgebras:
$\mathfrak{sl}(2, \mathbb{R})$ for all restricted roots
$\mathfrak{m}_{\mathfrak{p},0} = 0$

Special feature:
$\mathfrak{g}_0$ is a split real form,
G/K is of quaternion type

HINTS FOR SOLUTIONS OF PROBLEMS

Chapter I

1. In Example 12a, $[\mathfrak{g}, \mathfrak{g}]$ has $a = b = 0$, and $[\mathfrak{g}, [\mathfrak{g}, \mathfrak{g}]] = 0$. In Example 12b an elementary sequence has $\mathfrak{a}_1$ with $t = 0$ and $\mathfrak{a}_2$ with $t = x = 0$.

2. For (c), the span of X and Y is characterized as $[\mathfrak{g}, \mathfrak{g}]$. The given Z is an element not in $[\mathfrak{g}, \mathfrak{g}]$, and $\operatorname{ad} Z$ has eigenvalues 0 (from Z), α (from X), and 1 (from Y). If Z is replaced by $Z + sX + tY$, then the eigenvalues are unchanged. If we multiply by $c \in \mathbb{R}$, the eigenvalues are multiplied by c. Hence α is characterized as follows: Let Z be any vector not in $[\mathfrak{g}, \mathfrak{g}]$. Then α is the ratio of the larger nonzero absolute value of an eigenvalue to the smaller one.

4. The complexifications in this sense are both $\mathfrak{sl}(2, \mathbb{C})$.

8. Compute $B(X, Y)$ and $C(X, Y)$ for $X = Y = \operatorname{diag}(1, -1, 0, \dots, 0)$ and find that $B = 2nC$.

9. Abbreviate the displayed matrix as (θ, x, y). For (a) we have

$$[(1, 0, 0), (\theta, x, y)] = (0, y, -x).$$

Hence if $\mathbb{C}(\theta, x, y)$ is an ideal, $x = y = 0$. But then $\theta = 0$ also from the same bracket formula with $x = 1, y = 0$. For (b), $\operatorname{ad}(1, 0, 0)$ has eigenvalues 0 and $\pm i$.

10. Complexify and apply Lie's Theorem to $\operatorname{ad} \mathfrak{g}$.

11. Lie's Theorem shows that $\operatorname{ad} \mathfrak{g}$ can be taken simultaneously upper triangular, and (1.31) shows that the diagonal entries are then 0.

12. The Killing form B, being nondegenerate, gives a vector space isomorphism $b : \mathfrak{g} \to \mathfrak{g}^*$, while C gives a linear map $c : \mathfrak{g} \to \mathfrak{g}^*$. Then $b^{-1}c : \mathfrak{g} \to \mathfrak{g}$ is a linear map that commutes with $\operatorname{ad} \mathfrak{g}$. Since $\mathfrak{g}$ is simple, $\operatorname{ad} : \mathfrak{g} \to \operatorname{End} \mathfrak{g}$ is an irreducible representation. As in Lemma 1.66, $b^{-1}c$ must then be scalar.

13. For $\mathfrak{sl}(2, \mathbb{R})$, there is a 2-dimensional subalgebra, while for $\mathfrak{su}(2)$, there is not.

14. For (b) use the remarks at the end of §2. For (a), $SU(2)$ is topologically a 3-sphere and is therefore simply connected. Having a simply connected domain is a sufficient condition for a homomorphism of Lie algebras to lift to a homomorphism of analytic groups.

16. No if $n > 1$. $SU(n)$ and Z have finite nontrivial intersection.

17. The linear map $\varphi\begin{pmatrix}1 & 0\\ 0 & -1\end{pmatrix}$ acts on $P\begin{pmatrix}z_1\\ z_2\end{pmatrix} = z_2^n$ with eigenvalue n. Since φ is a direct sum of irreducibles and n is an eigenvalue of $\varphi\begin{pmatrix}1 & 0\\ 0 & -1\end{pmatrix}$, it follows that an irreducible of some dimension $n+2k$ with $k \geq 0$ occurs in V_n. Dimensionality forces $k = 0$ and gives the result.

19. We use Problem 18. Direct computation shows that $[\mathfrak{g}, \mathfrak{g}]$ is contained in the subspace with $\theta = 0$. One still has to show that equality holds. For this purpose one is allowed to pick particular matrices to bracket and show that the span of such brackets is 3-dimensional.

20. The starting point is Theorem 1.104. For details of how to apply this theorem, see Lemma 1.1.4.1 of Warner [1972a].

21. $n; 2 - (n \bmod 2)$ if $n \geq 3$; 2; n; $2 - (n \bmod 2)$ if $n \geq 3$; 2.

22. Let M be the diagonal matrix with n diagonal entries of i and then n diagonal entries of 1. An isomorphism $G \to SO(2n, \mathbb{C})$ is $x \mapsto y$, where $y = MxM^{-1}$.

24. $G = \{\mathrm{diag}(e^k, e^{-k}\}_{k=-\infty}^{\infty}$.

27. A suitable linear combination L of the two given linear mappings lowers the degree of $P(s)$ in $e^{-\pi s^2}P(s)$ by exactly one. Take a nonzero $e^{-\pi s^2}P(s)$ in an invariant subspace U and apply $L^{\deg P}$ to $e^{-\pi s^2}P(s)$ to see that $e^{-\pi s^2}$ is in U. Apply powers of "multiplication by $-i\hbar s$" to this to see that all of V is contained in U.

28. Let Z be nonzero in $[\mathfrak{g}, \mathfrak{g}]$. Extend to a basis $\{X, Y, Z\}$. If $[\mathfrak{g}, Z] = 0$, then $[\mathfrak{g}, \mathfrak{g}] = \mathbb{R}[X, Y]$ and hence $[X, Y] = cZ$ with $c \neq 0$; in this case we can easily set up an isomorphism with the Heisenberg algebra. Otherwise $[\mathfrak{g}, Z] = \mathbb{R}Z$. Since $[X, Z]$ and $[Y, Z]$ are multiples of Z, some nonzero linear combination of X and Y brackets Z to 0. Thus we can find a basis $\{X', Y', Z\}$ with $[X', Z] = 0$, $[Y', Z] = Z$, $[X', Y'] = cZ$. Then $\{X' + cZ, Y', Z\}$ has $[X' + cZ, Z] = 0$, $[Y', Z] = Z$, $[X' + cZ, Y'] = 0$. Then $\mathfrak{g} = \mathbb{R}(X' + cZ) \oplus \mathrm{span}\{Y', Z\}$ as required.

29. A 2-dimensional nilpotent Lie algebra is abelian; hence $[\mathfrak{g}, \mathfrak{g}]$ is abelian. The matrix $\begin{pmatrix}\alpha & \beta\\ \gamma & \delta\end{pmatrix}$ is nonsingular since otherwise $\dim[\mathfrak{g}, \mathfrak{g}] < 2$.

30. Classify by $\dim[\mathfrak{g}, \mathfrak{g}]$. If this is 3, $\mathfrak{g}$ is simple by the remarks at the end of §2. If it is 2 or 1, $\mathfrak{g}$ is analyzed by Problem 29 or Problem 28. If it is 0, $\mathfrak{g}$ is abelian.

31. If $X = [Y, Z]$, then $\mathrm{ad}\, X = \mathrm{ad}\, Y\, \mathrm{ad}\, Z - \mathrm{ad}\, Z\, \mathrm{ad}\, Y$, and the trace is 0.

32. One of the eigenvalues of $\mathrm{ad}\, X_0$ is 0, and the sum of the eigenvalues is 0 by Problem 31. Hence the eigenvalues are $0, \lambda, -\lambda$ with $\lambda \in \mathbb{C}$. The number λ cannot be 0 since $\mathrm{ad}\, X_0$ is by assumption not nilpotent. Since the characteristic polynomial of $\mathrm{ad}\, X_0$ is real, λ is real or purely imaginary. If λ is real, then the sum of the eigenspaces in $\mathfrak{g}$ for λ and $-\lambda$ is the required complement. If λ

is purely imaginary, then the intersection of $\mathfrak{g}$ with the sum of the λ and $-\lambda$ eigenspaces in $\mathfrak{g}^{\mathbb{C}}$ is the required complement.

33. Let λ be as in Problem 32. If λ is real, then scale X_0 to X to make $\lambda = 2$ and show that ad X has the first form. If λ is purely imaginary, scale X_0 to X to make $\lambda = i$ and show that ad X has the second form.

34. The Jacobi identity gives $(\operatorname{ad} X)[Y, Z] = 0$. So $[Y, Z] = aX$, and a cannot be 0 since $[\mathfrak{g}, \mathfrak{g}] = \mathfrak{g}$. Scale one of Y and Z to make $a = 1$, and then compare with (1.6) and (1.3) in the two cases.

35. In Problem 32 we still obtain $\lambda \in \mathbb{C}$ with $\lambda \neq 0$. Since the field is $\mathbb{C}$, we obtain eigenvectors for ad X_0 with eigenvalues λ and $-\lambda$, respectively. Then we can proceed as in the first case of Problem 33.

Chapter II

1. $(2n+2)^{-1}, (2n-1)^{-1}, (2n+2)^{-1}, (2n-2)^{-1}$.

2. One of the ideals is the complex span of

$$\begin{pmatrix} 0 & 0 & 1 & i \\ 0 & 0 & -i & 1 \\ -1 & i & 0 & 0 \\ -i & -1 & 0 & 0 \end{pmatrix}, \quad \text{its conjugate, and} \quad \begin{pmatrix} 0 & 1 & 0 & 0 \\ -1 & 0 & 0 & 0 \\ 0 & 0 & 0 & -1 \\ 0 & 0 & 1 & 0 \end{pmatrix}.$$

3. For (c), $\mathfrak{g} = \mathbb{C}X \oplus \mathbb{C}Y$ with $\mathbb{C}Y$ the weight space for the linear functional $cX \to c$.

4. For (a), take $\mathfrak{g} = \mathfrak{sp}(2, \mathbb{C})$ and $\Delta' = \{\pm e_1 \pm e_2\}$.

6. Propositions 2.17c and 2.17e show that $\dim \mathfrak{g} \geq 3 \dim \mathfrak{h}$ and that $\dim \mathfrak{g} \equiv \dim \mathfrak{h} \bmod 2$. Thus $\dim \mathfrak{h} \geq 3$ implies $\dim \mathfrak{g} \geq 9$. If $\dim \mathfrak{h} = 2$, then $\dim \mathfrak{g}$ is even and is ≥ 6. Hence $\dim \mathfrak{g} = 4, 5$, or 7 implies $\dim \mathfrak{h} = 1$. Meanwhile Propositions 2.21 and 2.29 show that $\dim \mathfrak{h} = 1$ implies $\dim \mathfrak{g} = 3$. Hence $\dim \mathfrak{g} = 4, 5$, and 7 cannot occur.

7. From $|\alpha|^2 > 0$, we get $\langle \alpha, \alpha_i \rangle > 0$ for some simple α_i. Use this i as i_k, repeat with $\alpha - \alpha_{i_k}$, and iterate.

8. Proposition 2.48e for the first conclusion. For the second conclusion use the positive roots in (2.50) and take $\beta_0 = e_1$.

10. In (a) any two roots at an angle 150° will do.

11. For (a) if the two roots are α and β, then $s_\alpha s_\beta(\alpha) = \beta$. For (b) combine (a) with Proposition 2.62, using a little extra argument in the nonreduced case.

13. By induction Chevalley's Lemma identifies the subgroup of the Weyl group fixing a given vector subspace as generated by its root reflections. For (a) use this extended result for the $+1$ eigenspace, inducting on the dimension of the -1 eigenspace. Then (b) is a special case.

14. Choose w with $l(w)$ as small as possible so that $w\lambda$ and λ are both dominant but $w\lambda \neq \lambda$. Write $w = vs_\alpha$ with α simple and $l(v) = l(w) - 1$. Then $v\alpha > 0$ by Lemma 2.71. So $\langle w\lambda, v\alpha\rangle \geq 0$, and we get $\langle \lambda, \alpha\rangle \leq 0$. Since λ is dominant, $\langle \lambda, \alpha\rangle = 0$. Then $\lambda \neq w\lambda = v\lambda$, in contradiction with the minimal choice of w.

17. For (b) if $\alpha = \sum c_i\alpha_i$ with all $c_i \geq 0$, then $\alpha^\vee = \sum d_i\alpha_i^\vee$ with $d_i = c_i|\alpha|^{-2}|\alpha_i|^2 \geq 0$.

18. For (a) use Theorem 2.63. For (b) the indicated Dynkin diagrams admit no nontrivial automorphisms. For (c) and (d) use the explicit descriptions of the Weyl groups in Example 1 of §6.

19. $(BC)_1 \oplus A_1$ and $(BC)_1 \oplus (BC)_1$ are missing.

20. Here $\mathfrak{g}' = \mathfrak{g}$, $\mathfrak{h}' = \mathfrak{h}$, and $\varphi = w$. Fix Π, and choose nonzero root vectors E_α for $\alpha \in \Pi$. For each $\alpha \in \Pi$, choose any nonzero root vector $E_{w\alpha}$, and require that E_α map to $E_{w\alpha}$.

22. For (a) Lemma 2.71 allows us to see that $\operatorname{sgn} w = -1$ if w is a product of an odd number of simple reflections in any fashion and $\operatorname{sgn} w = +1$ if w is a product of an even number of simple reflections in any fashion. The homomorphism property follows. In (b), $\operatorname{sgn} w$ and $\det w$ are multiplicative and agree on simple reflections. Part (c) follows from (b).

23. We have

$$\begin{aligned}
l(w_1w_2) &= \#\{\alpha > 0 \mid w_1w_2\alpha < 0\} \\
&= \#\{\alpha > 0 \mid w_2\alpha > 0,\, w_1w_2\alpha < 0\} \\
&\quad + \#\{\alpha > 0 \mid w_2\alpha < 0,\, w_1w_2\alpha < 0\} \\
&= \#\{\alpha \mid \alpha > 0,\, w_2\alpha > 0,\, w_1w_2\alpha < 0\} \\
&\quad + l(w_2) - \#\{\alpha \mid \alpha > 0,\, w_2\alpha < 0,\, w_1w_2\alpha > 0\} \\
&= \#\{\beta \mid w_2^{-1}\beta > 0,\, \beta > 0,\, w_1\beta < 0\} \\
&\quad + l(w_2) - \#\{\gamma \mid w_2^{-1}\gamma < 0,\, \gamma > 0,\, w_1\gamma < 0\} \\
&= l(w_1) - \#\{\beta \mid w_2^{-1}\beta < 0,\, \beta > 0,\, w_1\beta < 0\} \\
&\quad + l(w_2) - \#\{\gamma \mid w_2^{-1}\gamma < 0,\, \gamma > 0,\, w_1\gamma < 0\}
\end{aligned}$$

with $\beta = w_2\alpha$ and $\gamma = -w_2\alpha$.

24. By Problem 23,

$$l(ws_\alpha) = l(w) + l(s_\alpha) - 2\#\{\beta > 0 \mid w\beta < 0 \text{ and } s_\alpha\beta < 0\}.$$

For the first conclusion we thus are to prove that $w\alpha < 0$ implies

$$(*) \qquad l(s_\alpha) < 2\#\{\beta > 0 \mid w\beta < 0 \text{ and } s_\alpha\beta < 0\}.$$

Here the left side is $\#\{\gamma > 0 \mid s_\alpha\gamma < 0\}$. Except for α, such γ's come in pairs, γ and $-s_\alpha\gamma$. From each pair at least one of γ and $-s_\alpha\gamma$ is a β for the right side of $(*)$ because $\gamma - s_\alpha\gamma = \frac{2\langle\gamma,\alpha\rangle}{|\alpha|^2}\alpha$ is > 0 and $w\alpha$ is < 0. So each pair, γ and $-s_\alpha\gamma$, makes a contribution to $(*)$ for which the left side is $\leq$ the right side. The root α contributes 1 to the left side of $(*)$ and 2 to the right side. So the inequality $(*)$ is strict.

25. Use expansion in cofactors about the first column.

26. The Dynkin diagram should consist consecutively of vertex, single edge, vertex, single edge, and then the rest of the diagram.

28. In handling C_n and D_n, take into account that $C_2 \cong B_2$ and $D_3 \cong A_3$.

31. In (a) the long roots are already as in (2.43); no isomorphism is involved. In (b) each member of W_F preserves length when operating on roots. In (c) the two indicated reflections correspond to two distinct transpositions of the three outer roots of the Dynkin diagram of D_4, and together they generate the symmetric group on three letters. This group is the full group of automorphisms of the Dynkin diagram of D_4. For (d) the order of W_D is given in Problem 15.

32. For (a) use Problem 11b. Let α be the root in (a). In (b) there are five simple roots orthogonal to α, and all the roots orthogonal to α then have to be in the space spanned by these simple roots. For (c) apply Chevalley's Lemma to $-s_\alpha$. For (d) use Chevalley's Lemma directly. For (e) the number of roots for E_6 is given in Problem 16, and the order of the Weyl group fixing α is given in Problem 15, by (d).

33. Same idea as for Problem 32.

34. Same idea as for Problem 32 once the result of Problem 33d is taken into account.

35. Multiply $X^t I_{3,3} + I_{3,3}X = 0$ through on the left and right by S^{-1}.

36. Use the basis in the order

$$\begin{gathered}(e_1 \wedge e_4) + (e_2 \wedge e_3), \quad (e_1 \wedge e_2) + (e_3 \wedge e_4), \quad (e_1 \wedge e_3) - (e_2 \wedge e_4), \\ (e_1 \wedge e_4) - (e_2 \wedge e_3), \quad (e_1 \wedge e_2) - (e_3 \wedge e_4), \quad (e_1 \wedge e_3) + (e_2 \wedge e_4).\end{gathered}$$

Then the matrix of M is of the form $\begin{pmatrix} a & b \\ c & d \end{pmatrix}$ with a, b, c, d each 3-by-3, with a and d skew symmetric, and with $c = b^t$. This is the condition that M be in $\mathfrak{g}$.

37. Since $\mathfrak{sl}(4, \mathbb{C})$ is simple and the kernel is an ideal, it suffices to find one element that does not act as 0, and a diagonal element will serve this purpose. Then the homomorphism is one-one. A count of dimensions shows it is onto.

38. The condition for a 4-by-4 matrix $\begin{pmatrix} a & b \\ c & d \end{pmatrix}$ with a, b, c, d to be in $\mathfrak{sp}(2, \mathbb{C})$ is that $d = -a^t$ and $c = b^t$. Putting this condition into place in Problem 36 as solved above, we find that the last row and column of the image matrix are always 0.

39. The homomorphism is one-one by Problem 37, and a count of dimensions shows it is onto.

40. The projected system consists of the six vectors obtained by permuting the indices of $\pm\frac{1}{3}H_{e_1+e_2-2e_3}$, together with the six vectors $H_{e_i-e_j}$ for $i \neq j$.

41. The centralizer is the direct sum of the Cartan subalgebra and the six 1-dimensional spaces $\mathbb{C}H_{e_i-e_j}$.

42. Showing closure under brackets involves several cases and makes systematic use of Corollary 2.37. Under the action of the complementary space to $H_{e_1+e_2+e_3}$ in the Cartan subalgebra, the roots are those in problem 40 and form a system of type G_2.

Chapter III

1. For (a) the argument is essentially the same as the proof of Lemma 1.65. Part (b) is trivial.

2. The finite-dimensional subspaces $U_n(\mathfrak{g})$ are invariant.

3. Use Proposition 3.16 and the fact that $S(\mathfrak{g})$ has no zero divisors.

4. For (a), $\mathfrak{F}$ is 1-dimensional abelian. For (b) let V have basis $\{X, Y\}$. Then the element $[X, [\dots, [X, Y]]]$ is in $\mathfrak{F}$ and is in $T^{n+1}(V)$ if there are n factors X. When expanded out, this element contains the term $X \otimes \cdots \otimes X \otimes Y$ only once, the other terms being independent of this term. Hence the element is not 0.

5. For (a) a basis is $X_1, X_2, X_3, [X_1, X_2], [X_2, X_3], [X_3, X_1]$. Any triple bracket is 0, and hence $\mathfrak{g}$ is nilpotent. The bracket of X_1 and X_2 is not zero, and hence $\mathfrak{g}$ is not abelian. In (b) one writes down the 6-by-6 symmetric matrix that incorporates the given values for B and checks that it is nonsingular. This proves nondegeneracy. For invariance it is enough to check behavior on the basis, and expressions $B(X_i, [X_j, X_k])$ are the only ones that need to be checked.

6. Let $\mathfrak{F}$ be a free Lie algebra on n elements $X_1, \dots, X_n$, and let $\mathfrak{N}$ be the two-sided ideal generated by all $[X_i, [X_j, X_k]]$. Then $\mathfrak{F}/\mathfrak{N}$ is two-step nilpotent and has the required universal property. The elements of $\mathfrak{F} \cap T_2(\mathrm{span}\{X_i\}_{i=1}^n)$ map onto $\mathfrak{F}/\mathfrak{N}$, and finite-dimensionality follows.

7. See the comparable construction for Lie algebras in §I.3.

8. This is an application of Proposition 3.3.

10. Use Proposition 3.3.

11. See Knapp [1986], proof of Theorem 3.6.

12. See Knapp [1986], proof of Lemma 3.5.

Chapter IV

1. For (a), $\Phi(t_\theta)(z_1^k z_2^{N-k}) = (e^{-i\theta}z_1)^k(e^{i\theta}z_2)^{N-k} = e^{i(N-2k)\theta}z_1^k z_2^{N-k}$. For (c),

$$\chi(t_\theta) = \sum_{k=0}^{N} e^{i(N-2k)\theta} = \frac{e^{i(N+1)\theta} - e^{-i(N+1)\theta}}{e^{i\theta} - e^{-i\theta}}.$$

For (d) write χ_M as a sum and χ_N as a quotient, and then multiply and sort out.

2. If $x \in G$ is given, choose X in the Lie algebra with $\exp X = x$. By Theorem 4.34 there is an element $g \in G$ with $\mathrm{Ad}(g)X$ in the Lie algebra of the given torus. Then $gxg^{-1} = \exp \mathrm{Ad}(g)X$ is in the given torus.

3. See Knapp [1986], 86–87.

4. Use $\mathrm{diag}(-1, -1, 1)$.

5. Matrices with one nonzero element in each row and column.

6. Let $\tilde{G}$ be a nontrivial finite cover of G. Then Proposition 4.67 shows that there are analytically integral forms for $\tilde{G}$ that are not algebraically integral, in contradiction with Proposition 4.59.

7. $\dim V$.

8. It is enough to check that $\mathfrak{so}(n)$ acts in a skew-symmetric fashion, and this reduces to checking what happens with a Lie algebra element that is $\begin{pmatrix} 0 & 1 \\ -1 & 0 \end{pmatrix}$ in the upper left 2-by-2 block and is 0 elsewhere.

10. Let v_{N-2} be in V_{N-2} and v_N be in V_N. Then

$$\langle |x|^2 v_{N-2}, v_N \rangle = \langle v_{N-2}, \partial(|x|^2)v_N \rangle.$$

If v_N is harmonic, then the right side vanishes and we see from the left side that $|x|^2 V_{N-2}$ is orthogonal to H_N. In the reverse direction if v_N is orthogonal to $|x|^2 V_{N-2}$, then the left side vanishes and we see from the right side that $\partial(|x|^2)v_N$ is orthogonal to V_{N-2} and must be 0.

11. The dimension of the image of Δ in V_{N-2} must equal the dimension of the orthogonal complement to the kernel in the domain.

12. Induction.

13. Use $\dim H_N = \dim V_N - \dim V_{N-2}$. The number of monomials in V_N is the number of ways of choosing $n - 1$ dividers from among $N + n - 1$ contributions of 1 to an exponent, thus is $\binom{N+n-1}{n-1}$.

15. We have $\bigoplus_{p+q=N} V_{p,q} = V_N$ and $\bigoplus_{p+q=N} V_{p-1,q-1} = V_{N-2}$. Certainly $\Delta(V_{p,q}) \subseteq V_{p-1,q-1}$. If the inclusion is proper for one pair (p, q), then Δ cannot map V_N onto V_{N-2}.

16. Use $\dim V_{p,q} = (\dim V_p)(\dim V_q)$ and the computation for Problem 13.

17. For (c) let $\lambda = \sum c_j e_j$. If $c_j = a_j + \frac{k}{n}$ with $a_j \in \mathbb{Z}$ for all j, then $\xi_\lambda(\mathrm{diag}(e^{i\theta_1}, \ldots, e^{i\theta_n})) = \exp i(\sum a_j \theta_j)$. For (d), the quotient is canonically isomorphic to the set of all k/n with k taken modulo n.

18. For (c), use Proposition 4.68.

19. For (c), $\xi_{e_k}(\mathrm{diag}(e^{i\theta_1}, \ldots, e^{i\theta_n}, e^{-i\theta_1}, \ldots, e^{-i\theta_n}) = e^{i\theta_k}$.

20. For (d) the group is cyclic for $SO(2n)$ with n odd, and it is the direct sum of two groups of order 2 for $SO(2n)$ with n even. In fact, two distinct nontrivial coset representatives are e_1 and $\frac{1}{2}(e_1 + \cdots + e_n)$. The first one has order 2 as a coset, while the second one has order 2 as a coset if n is even but order 4 if n is odd.

Chapter V

1. For (b) apply w_0 to (d) or (e) in Theorem 5.5.

2. For (a) Problem 10 in Chapter IV gives $P_N = P_{N-2} \oplus |x|^2 H_N$. Once one has shown that Ne_1 is the highest weight of P_n, then $(N-2)e_1$ must be the highest weight of P_{N-2}, and Ne_1 must be the highest weight of H_N. For (b) the result of Problem 13 in Chapter IV is $\binom{N+n-1}{n-1} - \binom{N+n-3}{n-1} = \frac{(N+n-3)!(2N+n-2)}{N!(n-2)!}$. When $n = 2m+1$ is odd, use $\delta = (m-\frac{1}{2}, m-\frac{3}{2}, \ldots, \frac{1}{2})$ and $Ne_1 + \delta = (N+m-\frac{1}{2}, m-\frac{3}{2}, \ldots, \frac{1}{2})$ in the Weyl Dimension Formula to obtain

$$\Big(\frac{N+m-\frac{1}{2}}{m-\frac{1}{2}}\Big)\Big(\prod_{j=2}^{m} \frac{N+j-1}{j-1}\Big)\Big(\prod_{j=2}^{m} \frac{N+2m-j}{2m-j}\Big),$$

which reduces to the same result.

3. For (a) argue as in Problem 2a. For (b) the result of Problem 16 in Chapter IV is $\frac{(p+n-2)!(q+n-2)!(p+q+n-1)}{p!q!(n-1)!(n-2)!}$. Use $\langle \delta, e_i - e_j \rangle = j - i$ in the Weyl Dimension Formula to obtain

$$\Big(\frac{p+q+n-1}{n-1}\Big)\Big(\prod_{j=2}^{n-1} \frac{q+j-1}{j-1}\Big)\Big(\prod_{i=2}^{n-1} \frac{p+n-i}{n-i}\Big),$$

which reduces to the same result.

4. Abbreviate $H_{e_1-e_2}$ as H_{12}, etc. A nonzero homogeneous element of degree 3 is $(H_{12} + H_{13})(H_{21} + H_{23})(H_{31} + H_{32})$.

5. Let $t = \operatorname{diag}(t_1, \ldots, t_n)$ with $\prod t_i = 1$. Then W is the symmetric group on n letters, and $\varepsilon(w)$ is the sign of the permutation. The right side of the formula in Corollary 5.76, evaluated at t, is the determinant of the Vandermonde matrix with $(i, j)^{\text{th}}$ entry $(t_{n+1-j})^{i-1}$. The left side, evaluated at t, is the value of the determinant, namely $\prod_{i<j}(t_i - t_j)$.

6. Use the Kostant Multiplicity Formula for occurrence of the weight λ in the trivial representation.

7. The Weyl Dimension Formula gives

$$\frac{\prod_{i<j}\left\langle \sum_{k=1}^{l} e_k + \delta, e_i - e_j \right\rangle}{\prod_{i<j}\langle \delta, e_i - e_j \rangle} = \prod_{i\le l<j}\left(\frac{j-i+1}{j-i}\right) = \binom{n}{l}.$$

8. Here $\delta = (n - \frac{1}{2}, n - \frac{3}{2}, \ldots, \frac{1}{2})$, and the Weyl Dimension Formula gives us nontrivial factors for the $e_i - e_j$ with $i \le l < j$, the e_i with $i \le l$, the $e_i + e_j$ with $i < j \le l$, and the $e_i + e_j$ with $i \le l < j$, namely

$$\begin{aligned}\Big(\prod_{i\le l<j} \tfrac{j-i+1}{j-i}\Big)\Big(\prod_{i\le l} \tfrac{n+\frac{3}{2}-i}{n+\frac{1}{2}-i}\Big)\Big(\prod_{i<j\le l} \tfrac{2n+3-i-j}{2n+1-i-j}\Big)\Big(\prod_{i\le l<j} \tfrac{2n+2-i-j}{2n+1-i-j}\Big)& \\ = \binom{n}{l}\frac{n+\frac{1}{2}}{n+\frac{1}{2}-l}\Big(\prod_{i\le l}\tfrac{2n+1-i-l}{n+1-i}\Big)\Big(\prod_{i<l}\tfrac{(2+2n-2i)(1+2n-2i)}{(2+2n-i-l)(1+2n-i-l)}\Big)&,\end{aligned}$$

and this reduces to $\binom{2n+1}{l}$.

9. Similar to Problem 8 but without factors from the e_l's.

10. The dimension of $\bigwedge^n \mathbb{C}^{2n}$ is $\binom{2n}{n}$, and the dimensions of the indicated irreducible representations are seen to be each $\frac{1}{2}\binom{2n}{n}$. Each weight of $\bigwedge^n \mathbb{C}^{2n}$ is of the form $\sum_{j=1}^{n} a_j e_j$ with $a_j = 1, 0,$ or -1 and with $\sum a_j$ of the same parity as n. Here $\sum_{j=1}^{n} e_j$ has multiplicity one and corresponds to one of the irreducible constituents. The next highest weight is $\left(\sum_{j=1}^{n-1} e_j\right) - e_n$, which is not a weight of this irreducible constituent by Theorem 5.5d. Hence it leads to a second irreducible constituent. These two constituents account for the full dimension of $\bigwedge^n \mathbb{C}^{2n}$.

11. If not, then the action of $U(\mathfrak{n}^-)$ in $V(\lambda)$ would not be one-one, in contradiction with Proposition 5.14b.

12. By Proposition 5.11c, $\mu - \delta = \lambda - \delta - q^+$ with q^+ in Q^+. Also μ is in $W\lambda$ by Theorem 5.62 and Example 2 at the end of §5.

13. M must have at least one highest weight vector, and irreducibility implies that that vector must generate. By Proposition 5.14c, M is isomorphic to a quotient of some $V(\mu)$. By Proposition 5.15, M is isomorphic to $L(\mu)$.

16. Write the λ'' highest weight vector as $v = \sum_{\mu+\mu'=\lambda''}(v_\mu \otimes v_{\mu'})$, allowing more than one term per choice of μ and taking the $v_{\mu'}$'s to be linearly independent. Choose $\mu = \mu_0$ as large as possible so that there is a nonzero term $v_\mu \otimes v_{\mu'}$. Apply root vectors for positive roots and see that v_μ is highest for φ_λ.

17. Changing notation, suppose that the weights of $\varphi_{\lambda'}$ have multiplicity one. Let $\varphi_{\lambda''}$ occur more than once. By Problem 16 write $\lambda'' = \lambda + \mu'$ for a weight μ' of $\varphi_{\lambda'}$. The solution to Problem 16 shows that a highest weight vector for each occurrence of $\varphi_{\lambda''}$ contains a term equal to a nonzero multiple of $v_\lambda \otimes v_{\mu'}$. A suitable linear combination of these vectors does not contain such a term, in contradiction with Problem 16.

18. By Chevalley's Lemma, $\langle \lambda, \alpha \rangle = 0$ for some root α. Rewrite the sum as an iterated sum, the inner sum over $\{1, s_\alpha\}$ and the outer sum over cosets of this subgroup.

19. Putting $\mu'' = w\lambda''$ and using that $m_\lambda(w\lambda'') = m_\lambda(\lambda'')$, we have

$$\begin{aligned}
\chi_\lambda \chi_{\lambda'} &= d^{-1} \sum_{w \in W} \sum_{\mu''=\text{weight of } \varphi_\lambda} m_\lambda(\mu'')\varepsilon(w)\xi_{\mu''}\xi_{w(\lambda'+\delta)} \\
&= d^{-1} \sum_{w \in W} \sum_{\lambda''=\text{weight of } \varphi_\lambda} m_\lambda(\lambda'')\varepsilon(w)\xi_{w(\lambda''+\lambda'+\delta)} \\
&= d^{-1} \sum_{\lambda''=\text{weight of } \varphi_\lambda} m_\lambda(\lambda'') \sum_{w \in W} \varepsilon(w)\xi_{w(\lambda''+\lambda'+\delta)^\vee} \\
&= d^{-1} \sum_{\lambda''=\text{weight of } \varphi_\lambda} m_\lambda(\lambda'')\chi_{(\lambda''+\lambda'+\delta)^\vee-\delta}.
\end{aligned}$$

20. The lowest weight $-\mu$ has $m_\lambda(-\mu) = 1$ by Theorem 5.5e. If $\lambda' - \mu$ is dominant, then $\operatorname{sgn}(-\mu + \lambda' + \delta) = 1$. So $\lambda'' = -\mu$ contributes $+1$ to the coefficient of $\chi_{\lambda'-\mu}$. Suppose some other λ'' contributes. Then $(\lambda''+\lambda'+\delta)^\vee - \delta = \lambda' - \mu$. So $(\lambda'' + \lambda' + \delta)^\vee = \lambda' - \mu + \delta$, $\lambda'' + \lambda' + \delta = s(\lambda' - \mu + \delta) = \lambda' - \mu + \delta - \sum_{\alpha>0} n_\alpha \alpha$, and $\lambda'' = -\mu - \sum_{\alpha>0} n_\alpha \alpha$. This says that λ'' is lower than the lowest weight unless $\lambda'' = -\mu$.

22. Write $(\lambda' + \delta + \lambda'')^\vee = \lambda' + w\lambda + \delta$, $\lambda' + \delta + \lambda'' = s(\lambda' + w\lambda + \delta)$. Subtract $\lambda' + \delta$ from both sides and compute the length squared, taking into account that $\lambda' + \delta$ is strictly dominant and $\lambda' + w\lambda + \delta$ is dominant:

$$\begin{aligned}
|\lambda''|^2 &= |s(\lambda' + w\lambda + \delta) - (\lambda' + \delta)|^2 \\
&= |\lambda' + w\lambda + \delta|^2 - 2\langle s(\lambda' + w\lambda + \delta), \lambda' + \delta\rangle + |\lambda' + \delta|^2 \\
&\geq |\lambda' + w\lambda + \delta|^2 - 2\langle \lambda' + w\lambda + \delta, \lambda' + \delta\rangle + |\lambda' + \delta|^2 \\
&= |(\lambda' + w\lambda + \delta) - (\lambda' + \delta)|^2 \\
&= |\lambda|^2.
\end{aligned}$$

Equality holds, and this forces $s(\lambda'+w\lambda+\delta) = \lambda'+w\lambda+\delta$. Hence $\lambda'+\delta+\lambda'' = \lambda'+w\lambda+\delta$, and $\lambda'' = w\lambda$.

23. Use Corollary 4.16.

24. To multiply two basis vectors, one deletes the common pairs of u_i's, inserts a factor of -1 for each such pair, puts the remaining u_i's in order, and inserts the sign of the permutation used to put the u_i's in order. In the same way it is possible to give a description of how to multiply three basis vectors, and associativity comes down to knowing that the sign function is multiplicative on the permutation group.

25. The bracket $[u_i u_j, u_{i'} u_{j'}]$ is 0 if all indices are different, is $2u_i u_{j'}$ if $j = i'$ and $i \neq j'$, and so on.

28. $c(u_{2m+1})z_S = \pm z_{S'}$ and $c(u_{2m+1})z_{S'} = \pm z_S$.

29. This follows from Problems 27 and 28.

30. The parity of the number of elements of S changes under each $c(z_j)$ or $c(\bar{z}_j)$, hence under each $c(u_{2j-1})$ or $c(u_{2j})$. Hence $c(\mathfrak{q}^{\mathbb{C}})$ leaves $\mathcal{S}^+$ and $\mathcal{S}^-$ invariant.

31. Argue as in Problem 30, taking the result of Problem 28 into account.

34. The computation in (b) is similar to the one for Problem 10.

35. The computation in (b) is similar to the one in Problem 8 when $l = n$.

36. For (a), $2\langle \sum_{k=1}^{l} e_k, e_i - e_{i+1}\rangle/|e_i - e_{i+1}|^2$ is 1 if $i = l$, 0 if $i \neq l$. For (b) Problem 7 shows that the alternating-tensor representation in $\bigwedge^l \mathbb{C}^n$ is irreducible with highest weight $\sum_{k=1}^{l} e_k$.

39. For the αth factor of the Weyl Dimension Formula, we have

$$\frac{\langle \lambda+\delta, \alpha\rangle}{\langle \delta, \alpha\rangle} = \frac{\langle \lambda'+\delta, \alpha\rangle}{\langle \delta, \alpha\rangle} + \frac{\langle \lambda-\lambda', \alpha\rangle}{\langle \delta, \alpha\rangle}.$$

The right side is $\geq$ the first term on the right for every α, and there is some α for which the inequality is strict.

40. It follows from Problem 39 by induction that if $\lambda = \sum n_i \varpi_i$ and $M = \sum n_i$, then the dimension of the irreducible representation with highest weight λ is $\geq M+1$.

41. For (a) Problem 42 exhibits a complex simple Lie algebra of type G_2 inside a complex simple Lie algebra of type B_3. The latter can be taken to be $\mathfrak{so}(7,\mathbb{C})$, for which the standard representation has dimension 7. The Cartan subalgebra of G_2 does not act trivially in this representation, and hence the representation is not 0. For (c) the dimension of the fundamental representation attached to α_2 is 7. For (d) let φ_λ be irreducible with highest weight $n_1\varpi_1 + n_2\varpi_2$. Problem 39 shows that $\dim \varphi_\lambda \geq 14$ if $n_1 \geq 1$. Also $\dim \varphi_\lambda \geq 7$ if $n_2 \geq 1$ with equality only if $(n_1, n_2) = (0, 1)$. Hence a nonzero

irreducible representation must have dimension ≥ 7. Since the representation in (a) is completely reducible (Theorem 5.29) and nonzero, one of its irreducible constituents has dimension ≥ 7. This irreducible constituent must exhaust the representation.

Chapter VI

1. By Theorem 2.15 we can assume that the two split real forms $\mathfrak{g}_0$ and $\mathfrak{g}_0'$ have a common Cartan subalgebra $\mathfrak{h}_0$. Fix a positive system of roots. For each simple root α, choose root vectors $X_\alpha \in \mathfrak{g}_0$ and $X_\alpha' \in \mathfrak{g}_0'$. Using the Isomorphism Theorem, construct an isomorphism $\varphi : \mathfrak{g} \to \mathfrak{g}$ that is the identity on $\mathfrak{h}_0$ and carries X_α to X_α' for each simple root α.

2. Let G be semisimple with Lie algebra $\mathfrak{g}_0$, and let K correspond to $\mathfrak{k}_0$. Then $\mathrm{Ad}_{\mathfrak{g}_0}(K)$ is compact with Lie algebra $\mathrm{ad}_{\mathfrak{g}_0}(\mathfrak{k}_0)$, and hence $\mathfrak{k}_0$ is compactly embedded. If $\mathfrak{k}_0 \subseteq \mathfrak{k}_1$ with $\mathfrak{k}_1$ compactly embedded, let K_1 correspond to $\mathfrak{k}_1$. Since $\mathrm{Int}\,\mathfrak{g}_0 \cong \mathrm{Ad}(G)$, the analytic subgroup of $\mathrm{Int}\,\mathfrak{g}_0$ with Lie algebra $\mathrm{ad}_{\mathfrak{g}_0}(\mathfrak{k}_1)$, which is compact by assumption, is $\mathrm{Ad}_{\mathfrak{g}_0}(K_1)$. Apply Theorem 6.31g to the group $\mathrm{Ad}(G)$.

3. Write $\exp \frac{1}{2}Y \exp X = k \exp X'$ with $k \in K$ and $X' \in \mathfrak{p}_0$. Apply θ to this identity, take the inverse, and multiply.

4. Write $g^{1/2} = kan$ with $k \in K, a \in A, n \in N$. Apply θ to this identity, take the inverse, and multiply.

7. For (b) let $\alpha_1, \ldots, \alpha_l$ be the simple roots of Δ^+, with $\alpha_{s+1}, \ldots, \alpha_l$ spanning V. For $i \leq s$, the root $-\theta\alpha_i$ has the same restriction to $\mathfrak{a}$ as α_i. In particular it is positive. Write $-\theta\alpha_i = \sum_{j=1}^{l} n_{ij}\alpha_j$ with n_{ij} an integer ≥ 0. Then $-\theta\alpha_i$ is in $\sum_{j=1}^{s} n_{ij}\alpha_j + V$. Application of $-\theta$ shows that α_i is in $\sum_{j=1}^{s}\sum_{k=1}^{s} n_{ij}n_{jk}\alpha_k + V$. Hence $(n_{ij})^2 = (\delta_{ij})$. Given i, choose i' with $n_{ii'}n_{i'i} = 1$. Then $n_{ii'}n_{i'k} = 0$ for $k \neq i$ says $n_{i'k} = 0$ for $k \neq i'$. It follows that (n_{ij}) is the matrix of a permutation of order 2. For (c) the definition makes $-\theta\alpha_i(H) = \alpha_i(H)$ for $1 \leq i \leq l$. So $\alpha_i(-\theta H) = \alpha_i(H)$ for all i, and $\theta H = -H$. Result (d) is immediate from (c). For (e) suppose that $\alpha_i|_{\mathfrak{a}_0} = \beta' + \beta''$ with β' and β'' in Σ^+. By (a), $\beta' = \alpha_{i'}|_{\mathfrak{a}_0}$ and $\beta'' = \alpha_{i''}|_{\mathfrak{a}_0}$ with $\alpha_{i'}$ and $\alpha_{i''}$ simple. Then the set of restrictions from the orbits is dependent, in contradiction with (d).

8. $\tilde{K} \cong SU(2)$, and $\tilde{M}$ is a subgroup of $\tilde{K}$ of order 8 with M as a quotient. Since $\tilde{M} \subseteq SU(2)$, $\tilde{M}$ has a unique element of order 2. Considering the five abstract groups of order 8, we see that $\{\pm 1, \pm i, \pm j, \pm k\}$ is the only possibility.

9. Theorem 6.74 reduces this to showing that a subdiagram D of D' closed under the automorphism yields a subalgebra of $\mathfrak{g}_0'$. There is no loss of generality in assuming that $\mathfrak{g}_0'$ is set up as in the proof of Theorem 6.88, with corresponding

compact real form $\mathfrak{u}_0'$ as in (6.89). Let $\mathfrak{u}_0$ be the sum as in (6.89) but taken just over roots α for D. Since D is closed under the automorphism, $\theta'\mathfrak{u}_0 = \mathfrak{u}_0$. Thus $\mathfrak{u}_0 = (\mathfrak{u}_0 \cap \mathfrak{k}') \oplus (\mathfrak{u}_0 \cap \mathfrak{p}')$, and we can take $\mathfrak{g}_0 = (\mathfrak{u}_0 \cap \mathfrak{k}') \oplus i(\mathfrak{u}_0 \cap \mathfrak{p}')$.

10. For (a) let $\mathfrak{h}_0$ be the Cartan subalgebra, and let $\alpha_1, \ldots, \alpha_l$ be the simple roots. Define $H \in i\mathfrak{h}_0$ by $\alpha_j(H) = +1$ if α_j is compact, -1 if α_j is noncompact. Put $k = \exp \pi i H \in K$. Then $\mathrm{Ad}(k)X_\alpha = e^{\pi i \alpha(H)}X_\alpha$ for α simple. By the uniqueness in the Isomorphism Theorem, $\mathrm{Ad}(k) = \theta$. For (b), $\mathrm{Ad}(k)$ is -1 on $\mathfrak{p}_0$, hence on $\mathfrak{a}_0$. So $W(G, A)$ contains -1. By Theorem 6.57, -1 is in $W(\Sigma)$.

11. Let $\mathfrak{h}_0 = \mathfrak{t}_0 \oplus \mathfrak{a}_0$ be a maximally compact Cartan subalgebra, choose Δ^+ taking $i\mathfrak{t}_0$ before $\mathfrak{a}_0$, and suppose k exists. Since $\mathrm{Ad}(k)$ fixes $\mathfrak{t}_0$, k is in the analytic subgroup T corresponding to $\mathfrak{t}_0$ (Corollary 4.51). Let U be the adjoint group of the compact real form $\mathfrak{k}_0 \oplus i\mathfrak{p}_0$, and let S be the maximal torus with Lie algebra $\mathfrak{t}_0 \oplus i\mathfrak{a}_0$. Then $\mathrm{Ad}(k)$ is in S. By Theorem 4.54, $\mathrm{Ad}(k)$ is in $W(\Delta)$. But $\theta\Delta^+ = \Delta^+$. So Theorem 2.63 says that $\theta = 1$ on $\mathfrak{h}_0$, in contradiction with $\mathfrak{a}_0 \neq 0$.

12. If E_α is a root vector for α, then $\theta[E_\alpha, \theta E_\alpha] = -[E_\alpha, \theta E_\alpha]$. So if $\alpha + \theta\alpha$ is a root (necessarily imaginary), it is noncompact. This contradicts Proposition 6.70.

14. Let the third subalgebra be $\{H(t, \theta)\}$. Let $g \in GL(4, \mathbb{R})$ conjugate the third subalgebra so that $g\{H(t, 0)\}g^{-1}$ is contained in $\mathfrak{h}_0$ and $g\{H(0, \theta)\}g^{-1}$ is contained in $\mathfrak{h}_0'$. Then $gH(t, 0)g^{-1}$ must be diagonal with diagonal entries $(t, t, -t, -t)$ or $(-t, t, t, -t)$ or $(t, -t, -t, t)$ or $(-t, -t, t, t)$. Since $gH(0, \theta)g^{-1}$ has to commute with one of these for all t, it has to occur in blocks, using entries $(12, 21)$ and $(34, 43)$, or $(14, 41)$ and $(23, 32)$. However, $\mathfrak{h}_0'$ uses entries (13) (31), (24), (42), which are different.

15. The map carries $\mathfrak{sp}(n, \mathbb{C})$ to itself. Thus in (b) the members Y of the image satisfy $JY + Y^*J = 0$ and $JY + Y^tJ = 0$. Hence Y is real.

16. The computation in (a) should be compared with the example at the beginning of §6. In (b) the exponential map commutes with matrix conjugation. Hence it is enough to find which matrices of (a) are in the image. To do so, one first checks that any matrix (like the X that exponentiates to it) that commutes with $\begin{pmatrix} a & 0 \\ 0 & a^{-1} \end{pmatrix}$ for some a with $a \neq a^{-1}$ is itself diagonal. Similarly, any matrix that commutes with $\pm\begin{pmatrix} 1 & t \\ 0 & 1 \end{pmatrix}$ for some $t \neq 0$ is upper triangular.

17. Two copies of the Dynkin diagram of $\mathfrak{g}$ with an arrow between each vertex of one diagram and the corresponding vertex of the other diagram.

18. A simple root β in the Vogan diagram gets replaced by $s_\alpha\beta = \beta - 2\langle\beta, \alpha\rangle/|\alpha|^2$. Since θ fixes α, expansion of $s_\alpha\beta$ in terms of α and β shows that θ commutes with s_α. Hence the automorphism of the Dynkin diagram of $s_\alpha\Delta^+$ has the same effect as it does on the Dynkin diagram of Δ^+. For the imaginary roots the key fact for the painting is (6.99). Suppose α is compact.

Since $s_\alpha\beta$ is the sum of β and a multiple of α, β and $s_\alpha\beta$ get the same painting. If α is noncompact, $s_\alpha\beta = \beta - 2\langle\beta,\alpha\rangle/|\alpha|^2$. For $\beta = \alpha$, α gets replaced by $-\alpha$, for no change in painting. For β orthogonal to α, β is left unchanged. This proves (a) and (b). For β adjacent to α, the painting of β gets reversed unless $2\langle\beta,\alpha\rangle/|\alpha|^2 = -2$, in which case it is unchanged. This proves (c). The algorithm for (d) is repeatedly to let s_α be the second painted simple root from the left and to apply s_α.

19. In (a) the root is the sum of $\frac{1}{2}(e_1 - e_2 - e_3 - e_4)$, $e_3 - e_4$, e_4 and e_4. In (b) the roots are orthogonal but not strongly orthogonal. If a Cayley transform is performed relative to one of the two roots, the other root becomes compact by Proposition 6.72b.

20. $e_2 - e_3, e_2 + e_3, e_1 - e_4, e_1 + e_4$.

21. In the notation of (2.86), the algorithm gives $e_6 - e_5, e_6 + e_5, e_8 - e_7$ as the strongly orthogonal sequence of noncompact roots.

22. $B([\mathfrak{p}_0', \mathfrak{p}_0'^{\perp}], \mathfrak{k}_0) = B(\mathfrak{p}_0', [\mathfrak{p}_0'^{\perp}, \mathfrak{k}_0]) \subseteq B(\mathfrak{p}_0', \mathfrak{p}_0'^{\perp}) = 0$. Since $[\mathfrak{p}_0', \mathfrak{p}_0'^{\perp}] \subseteq \mathfrak{k}_0$ and since $B|_{\mathfrak{k}_0\times\mathfrak{k}_0}$ is negative definite, $[\mathfrak{p}_0', \mathfrak{p}_0'^{\perp}] = 0$.

23. Invariance under $\operatorname{ad}\mathfrak{k}_0$ follows from the Jacobi identity. Since $B|_{\mathfrak{p}_0\times\mathfrak{p}_0}$ is positive definite, $\mathfrak{p}_0 = \mathfrak{p}_0' \oplus \mathfrak{p}_0'^{\perp}$. By Problem 22, $[\mathfrak{p}_0', \mathfrak{p}_0] = [\mathfrak{p}_0', \mathfrak{p}_0']$. Then $[\mathfrak{p}_0, \{[\mathfrak{p}_0', \mathfrak{p}_0] \oplus \mathfrak{p}_0'\}] \subseteq [\mathfrak{p}_0, [\mathfrak{p}_0', \mathfrak{p}_0']] + [\mathfrak{p}_0, \mathfrak{p}_0']$. The first term, by the Jacobi identity, is $\subseteq [\mathfrak{p}_0', [\mathfrak{p}_0, \mathfrak{p}_0']] \subseteq [\mathfrak{p}_0', \mathfrak{k}_0] \subseteq \mathfrak{p}_0'$.

24. In (a) Problem 23 says that $[\mathfrak{p}_0, \mathfrak{p}_0] \oplus \mathfrak{p}_0$ is an ideal in $\mathfrak{g}_0$. Since $\mathfrak{g}_0$ is simple and $\mathfrak{p}_0 \neq 0$, this ideal is $\mathfrak{g}_0$. Hence $[\mathfrak{p}_0, \mathfrak{p}_0] = \mathfrak{k}_0$. In (b) a larger subalgebra has to be of the form $\mathfrak{k}_0 \oplus \mathfrak{p}_0'$, where $\mathfrak{p}_0'$ is an $\operatorname{ad}\mathfrak{k}_0$ invariant subspace of $\mathfrak{p}_0$. If $\mathfrak{p}_0' \neq 0$, Problem 23 forces $[\mathfrak{p}_0', \mathfrak{p}_0] \oplus \mathfrak{p}_0' = \mathfrak{g}_0$, and then $\mathfrak{p}_0'$ has to be $\mathfrak{p}_0$.

25. All the necessary Vogan diagrams are the special ones from Theorem 6.96. So this problem is a routine computation.

26. These are the restrictions to real matrices, and the images are the sets of real matrices in $\mathfrak{so}(3,3)^{\mathbb{C}}$ and $\mathfrak{so}(3,2)^{\mathbb{C}}$, respectively.

27. The point here is that the domain groups are not simply connected. But the analytic subgroups of matrices corresponding to the complexified Lie algebras are simply connected. The kernel in each case has order 2.

Chapter VII

1. When $n > 1$, the element $g = \operatorname{diag}(1, \dots, 1, -1)$ yields a nontrivial automorphism of the Dynkin diagram. Then Theorem 7.8 implies that $\operatorname{Ad}(g)$ is not in $\operatorname{Int}\mathfrak{g}$.

2. In (a) the main step is to show that K is compact. The subgroup of $\tilde{G}$ where Θ is 1 is $\mathbb{R}^2$, and K is of the form $\mathbb{R}^2/D$, which is compact. In (b), G_{ss} is

$(\widetilde{SL}(2,\mathbb{R}) \times \{0\})/(D \cap (\widetilde{SL}(2,\mathbb{R}) \times \{0\}))$, and the intersection on the bottom is trivial. Thus G_{ss} has infinite center. If G_{ss} were closed in G, K_{ss} would be closed in G, hence in K. Then K_{ss} would be compact, contradiction.

5. Let MAN be block upper-triangular with respective blocks of sizes 2 and 1. Then M is isomorphic to the group of 2-by-2 real matrices of determinant ± 1 and has a compact Cartan subalgebra. The group M is disconnected, and its center $Z_M = \{\pm 1\}$ is contained in M_0. Therefore $M \neq M_0 Z_M$.

6. Refer to the diagram of the root system G_2 in Figure 2.2. Take this to be the diagram of the restricted roots. Arrange for $\mathfrak{a}_0$ to correspond to the vertical axis and for $\mathfrak{t}_0$ to correspond to the horizontal axis. The nonzero projections of the roots on the $\mathfrak{a}_0$ axis are of the required form.

7. In (b) one MA is $\cong GL^+(2,\mathbb{R}) \times \mathbb{Z}/2\mathbb{Z}$ (the plus referring to positive determinant), and the other is $\cong GL(2,\mathbb{R})$. If the two Cartan subalgebras were conjugate, the two MA's would be conjugate.

8. It is easier to work with $SO(2,n)_0$. For (a), conjugate the Lie algebra by $\operatorname{diag}(i, i, 1, \dots, 1)$. In (b), $\mathfrak{c}_0$ comes from the upper left 2-by-2 block. For (c) the Cartan subalgebra $\mathfrak{h}$ given in §II.1 is fixed by the conjugation in (a) and intersects with $\mathfrak{g}_0$ in a compact Cartan subalgebra of $\mathfrak{g}_0$. The noncompact roots are those that involve $\pm e_1$, and all others are compact. For (d) the usual ordering makes $e_1 \pm e_j$ and e_1 larger than all compact roots; hence it is good.

9. It is one-one since $N_K(\mathfrak{a}_0) \cap Z_G(\mathfrak{a}_0) = Z_K(\mathfrak{a}_0)$. To see that it is onto, let $g \in N_G(\mathfrak{a}_0)$ be given, and write $g = k \exp X$. By Lemma 7.22, k and X normalize $\mathfrak{a}_0$. Then X centralizes $\mathfrak{a}_0$. Hence g can be adjusted by the member $\exp X$ of $Z_G(\mathfrak{a}_0)$ so as to be in $N_K(\mathfrak{a}_0)$.

10. Imitate the proof of Proposition 7.85.

11. The given ordering on roots is compatible with an ordering on restricted roots. Any real or complex root whose restriction to $\mathfrak{a}_0$ is positive contributes to both $\mathfrak{b}$ and $\bar{\mathfrak{b}}$. Any imaginary root contributes either to $\mathfrak{b}$ or to $\bar{\mathfrak{b}}$. Therefore $\mathfrak{m} \oplus \mathfrak{a} \oplus \mathfrak{n} = \mathfrak{b} + \bar{\mathfrak{b}}$.

12. For (a) when α is real, form the associated Lie subalgebra $\mathfrak{sl}(2,\mathbb{R})$ and argue as in Proposition 6.52c. When α is compact imaginary, reduce matters to $SU(2)$. For (b), fix a positive system $\Delta^+(\mathfrak{k},\mathfrak{h})$ of compact roots. If s_α is in $W(G,H)$, choose $w \in W(\Delta(\mathfrak{k},\mathfrak{h}))$ with $w s_\alpha \Delta^+(\mathfrak{k},\mathfrak{h}) = \Delta^+(\mathfrak{k},\mathfrak{h})$. Let $\tilde{w}$ and $\tilde{s}_\alpha$ be representatives. By Theorem 7.8, $\operatorname{Ad}(\tilde{w}\tilde{s}_\alpha) = 1$ on $\mathfrak{h}$. Hence s_α is in $W(\Delta(\mathfrak{k},\mathfrak{h}))$. By Chevalley's Lemma some multiple of α is in $\Delta(\mathfrak{k},\mathfrak{h})$, contradiction. For (c) use the group of 2-by-2 real matrices of determinant ± 1.

13. Parts (a) and (c) are trivial. In (b) put $M = {}^0Z_G(\mathfrak{a}_0)$. If k is in $N_K(\mathfrak{a}_0)$, then $\operatorname{Ad}(k)$ carries $\mathfrak{t}_0$ to a compact Cartan subalgebra of $\mathfrak{m}_0$ and can be carried back to $\mathfrak{t}_0$ by Ad of a member of $K \cap M$, essentially by Proposition 6.61.

14. Otherwise $N_{\mathfrak{g}_0}(\mathfrak{k}_0)$ would contain a nonzero member X of $\mathfrak{p}_0$. Then $\operatorname{ad} X$ carries $\mathfrak{k}_0$ to $\mathfrak{k}_0$ because X is in the normalizer, and $\operatorname{ad} X$ carries $\mathfrak{k}_0$ to $\mathfrak{p}_0$

since X is in $\mathfrak{p}_0$. So ad X is 0 on $\mathfrak{k}_0$. It follows that $(\text{ad } X)^2$ is 0 on $\mathfrak{g}_0$. If B is the Killing form, then $B(X, X) = 0$. Since B is positive definite on $\mathfrak{p}_0$, $X = 0$.

15. Using Corollary 7.6, we can set G up as a closed linear group of matrices closed under conjugate transpose. Then Example 4 of reductive Lie groups will show that $N_{G^{\mathbb{C}}}(\mathfrak{g}_0)$ is reductive.

16. Without loss of generality, $G^{\mathbb{C}}$ is simply connected, so that θ extends to $\mathfrak{g}$ and lifts to Θ on $G^{\mathbb{C}}$. The closure of $\exp i\mathfrak{a}_0$ is a torus, and it is contained in the maximal torus $\exp(\mathfrak{t}_0 \oplus i\mathfrak{a}_0)$. If $\exp i\mathfrak{a}_0$ is not closed, then there is some nonzero $X \in \mathfrak{t}_0$ such that $\exp rX$ is in the closure for all real r. Every element x of $\exp i\mathfrak{a}_0$ has the property that $\Theta x = x^{-1}$. If $\exp rX$ has this property for all r, then $\theta X = -X$. Since X is in $\mathfrak{t}_0$, $\theta X = X$. Hence $X = 0$.

17. $G = SL(2, \mathbb{C})$ contains elements $\gamma_\beta \neq 1$ as in (7.57), but K_{split} is trivial.

18. Let T be a maximal torus of K with Lie algebra $\mathfrak{t}_0$. Let U be the analytic subgroup of $G^{\mathbb{C}}$ with Lie algebra $\mathfrak{k}_0 \oplus i\mathfrak{p}_0$. The analytic subgroup H of $G^{\mathbb{C}}$ with Lie algebra $(\mathfrak{t}_0)^{\mathbb{C}}$ is a Cartan subgroup of $G^{\mathbb{C}}$ and is of the form $H = TA$ for a Euclidean group A. The center of $G^{\mathbb{C}}$ lies in $U \cap H = T$ and hence lies in G.

19. Let $G \subseteq G_1^{\mathbb{C}}$ and $G \subseteq G_2^{\mathbb{C}}$. Define $\tilde{G}^{\mathbb{C}}$ to be a simply connected cover of $G_1^{\mathbb{C}}$, and let $\tilde{G}^{\mathbb{C}} \to G_1^{\mathbb{C}}$ be the covering map. Let $\tilde{G}$ be the analytic subgroup of $\tilde{G}^{\mathbb{C}}$ with Lie algebra $\mathfrak{g}_0$. The isomorphism between the Lie algebras of $G_1^{\mathbb{C}}$ and $G_2^{\mathbb{C}}$ induced by the identity map of G yields a holomorphic homomorphism $\tilde{G}^{\mathbb{C}} \to G_2^{\mathbb{C}}$, and the main step is to show that this map descends to $G_1^{\mathbb{C}}$. By Problem 18 the kernel of the holomorphic covering map $\tilde{G}^{\mathbb{C}} \to G_1^{\mathbb{C}}$ and the constructed map $\tilde{G}^{\mathbb{C}} \to G_2^{\mathbb{C}}$ are both equal to the kernel of $\tilde{G} \to G$, hence are equal to each other. Therefore $\tilde{G}^{\mathbb{C}} \to G_2^{\mathbb{C}}$ descends to a one-one holomorphic homomorphism $G_1^{\mathbb{C}} \to G_2^{\mathbb{C}}$. Reversing the roles of $G_1^{\mathbb{C}}$ and $G_2^{\mathbb{C}}$ shows that this is an isomorphism.

20. G is isomorphic to the group $\text{Ad}(G)$ of 8-by-8 matrices, but $SL(3, \mathbb{C})$ is not isomorphic to $\text{Ad}(SL(3, \mathbb{C}))$.

21. The multiplication-by-i mapping $J : \mathfrak{p}_0 \to \mathfrak{p}_0$ has to come from $\mathfrak{c}_0$ by Theorem 7.117, and $\mathfrak{g}_0$ simple implies that $\dim \mathfrak{c}_0 = 1$. Since $J^2 = -1$, the only possibilities for J are some operator and its negative.

24. Since G/K is not Hermitian, there exist noncompact roots. Problem 23 shows that the lattices are distinct. By Theorem 6.96 we may assume that the simple roots are $\alpha_1, \ldots, \alpha_l$ with exactly one noncompact, say α_l. Since G/K is not Hermitian, some expression $2\alpha_l + \sum_{j=1}^{l-1} n_j\alpha_j$ is a root, necessarily compact. Then the lattice generated by the compact roots has $\mathbb{Z}$ basis $\alpha_1, \ldots, \alpha_{l-1}, 2\alpha_l$, while the lattice generated by all the roots has $\mathbb{Z}$ basis $\alpha_1, \ldots, \alpha_l$. Thus the index is 2.

25. This is a special case of (6.103).

26. If $\mathfrak{s}$ is the subalgebra, then the fact that $\mathfrak{t} \subseteq \mathfrak{s}$ means that

$$\mathfrak{s} = \mathfrak{t} \oplus \bigoplus_{\alpha} (\mathfrak{s} \cap \mathfrak{g}_\alpha).$$

Then $\mathfrak{s} = \mathfrak{k} \oplus \bigoplus_{\alpha \in E} \mathfrak{g}_\alpha$ since each root space has dimension 1.

27. This follows from the fact that $\overline{\mathfrak{g}_\alpha} = \mathfrak{g}_{-\alpha}$.

28. If V is an invariant subspace of $\mathfrak{p}$, then $\mathfrak{k} \oplus V$ is a Lie subalgebra of $\mathfrak{g}$, hence is of the form in Problem 26 for some $E \subseteq \Delta_n$. By Problem 24b in Chapter VI, any proper nonempty E satisfying the conditions in Problem 27 must have $E \cup (-E) = \Delta_n$ and $E \cap (-E) = \emptyset$. Since $\mathfrak{p}$ is completely reducible, it follows that the only nontrivial splitting of $\mathfrak{p}$ can involve some E and its complement. Hence there are at most two irreducible pieces.

29. Let $\mathfrak{p}_1$ be one of the two irreducible pieces, and let it correspond to E as above. Let $\mathfrak{p}_2$ be the other irreducible piece. If α_1 and α_2 are in E and $\alpha_1 + \alpha_2$ is a root, then a nonzero root vector for $-(\alpha_1 + \alpha_2)$ carries a nonzero root vector for α_1 to a nonzero root vector for $-\alpha_2$, hence a nonzero member of $\mathfrak{p}_1$ to a nonzero member of $\mathfrak{p}_2$, contradiction. Thus a sum of two members of E cannot be a root. Consequently $\langle \alpha, \beta \rangle \geq 0$ for all $\alpha, \beta \in E$. Let $\sigma = \sum_{\alpha \in E} \alpha$. Then it follows that $\langle \sigma, \alpha \rangle > 0$ for all $\alpha \in E$. Proposition 5.99 implies that σ is orthogonal to all compact roots. Hence iH_σ is in $\mathfrak{c}_0$. If we determine an ordering by using H_σ first, then $\mathfrak{p}_1 = \mathfrak{p}^+$ and $\mathfrak{p}_2 = \mathfrak{p}^-$.

30. Problem 15b of Chapter VI gives a one-one map on matrices that exhibits the Lie algebras of the two groups as isomorphic. The group $Sp(n, \mathbb{R})$ is connected by Proposition 1.124, and it is enough to prove that $SU(n, n) \cap Sp(n, \mathbb{C})$ is connected. For this connectivity it is enough by Proposition 1.122 to prove that $U(2n) \cap SU(n, n) \cap Sp(n, \mathbb{C})$ is connected, i.e., that the unitary matrices $\begin{pmatrix} u_1 & 0 \\ 0 & u_2 \end{pmatrix}$ in $Sp(n, \mathbb{C})$ are exactly those with $u_2 = \bar{u}_1$. This is an easy computation from the definition of $Sp(n, \mathbb{C})$.

31. The example in §9 shows that $SU(n, n)$ preserves the condition that $1_n - Z^*Z$ is positive definite. Let us check that the preservation of the condition $Z = Z^t$ depends only on $Sp(n, \mathbb{C})$. The conditions for $\begin{pmatrix} A & B \\ C & D \end{pmatrix}$ to be in $Sp(n, \mathbb{C})$ are that $A^tC = C^tA$, $B^tD = D^tB$, and $A^tD - C^tB = 1$. These conditions imply that $(ZC^t + D^t)(AZ + B) = (ZA^t + B^t)(CZ + D)$ when $Z = Z^t$, and it follows that $(AZ + B)(CZ + D)^{-1}$ is symmetric when $(CZ + D)^{-1}$ is defined.

33. $e_1 \geq e_2 \geq \cdots \geq e_n$.

37. For (b) the question concerns the projection of a root $e_r - e_s$ on the linear span of the γ_j. The projection of $e_r - e_s$ can involve only those γ_j's containing $\pm e_r$ or $\pm e_s$. Hence there are at most two. The projection of $\pm(e_i - e_{n+m+1-i})$ is $\pm\gamma_i$ if $i \leq m$, and the projection of $e_i - e_j$ is $\frac{1}{2}(\gamma_i - \gamma_j)$ if i and j are $\leq m$.

Applying root reflections, we must get all $\frac{1}{2}(\pm\gamma_i \pm \gamma_j)$. If $m = n$, all e_r's contribute to the γ_i's, and we get no other restricted roots. If $m < n$, then e_n does not contribute to the γ_i's. If r is any index such that e_r does not contribute, then $\pm(e_i - e_r)$ has projection $\pm\frac{1}{2}\gamma_i$.

40. The roots $\alpha + \beta$ and γ are positive noncompact, and their sum cannot be a root in a good ordering since $[\mathfrak{p}^+, \mathfrak{p}^+] = 0$.

41. For (a), $2\langle\gamma, \gamma_i\rangle/|\gamma_i|^2 = 2c_i$ is an integer ≤ 3 in absolute value. For (b) if $c_i = -\frac{3}{2}$, then γ, $\gamma + \gamma_i$, $\gamma + 2\gamma_i$, $\gamma + 3\gamma_i$ are roots. Either γ or $\gamma + \gamma_i$ is compact, and then Problem 40 applies either to the first three roots or to the last three. For (c) let $c_i = \pm 1$ and $c_j \neq 0$. Applying root reflections suitably, we obtain a root γ with $c_i = -1$ and $c_j < 0$. Then we can argue as in (b) for the sequence γ, $\gamma + \gamma_i$, $\gamma + 2\gamma_i$, $\gamma + 2\gamma_i + \gamma_j$. In (d) if $c_i \neq 0, c_j \neq 0$, and $c_k \neq 0$, we may assume γ has $c_i < 0, c_j < 0, c_k < 0$. Then we argue similarly with γ, $\gamma + \gamma_i$, $\gamma + \gamma_i + \gamma_j$, $\gamma + \gamma_i + \gamma_j + \gamma_k$. In (e) the restricted roots are all possibilities for $\sum_{i=1}^s c_i\gamma_i$, and parts (a) through (d) have limited these to $\pm\gamma_i$, $\frac{1}{2}(\pm\gamma_i \pm \gamma_j)$, $\pm\frac{1}{2}\gamma_i$. For (f) the $\pm\gamma_i$ are restricted roots, and the system is irreducible. If some $\pm\frac{1}{2}\gamma_i$ is a restricted root, then the system is $(BC)_s$ by Proposition 2.92. Otherwise the system is an irreducible subsystem of rank s within all $\pm\gamma_i$ and $\frac{1}{2}(\pm\gamma_i \pm \gamma_j)$, and it must be C_s.

42. From $uGu^{-1} = G'$, we have $uGB = G'uB$. Also $GB = \Omega K^{\mathbb{C}}P^-$ implies $uGB = u\Omega K^{\mathbb{C}}P^-$. Here $u\begin{pmatrix}1 & z\\ 0 & 1\end{pmatrix} = \frac{1}{\sqrt{2}}\begin{pmatrix}1 & z+i\\ i & iz+1\end{pmatrix}$ has P^+ component $\begin{pmatrix}1 & \frac{z+i}{iz+1}\\ 0 & 1\end{pmatrix}$, and hence $uGB = \Omega'' K^{\mathbb{C}}P^-$, where Ω'' consists of all $\begin{pmatrix}1 & w\\ 0 & 1\end{pmatrix}$ with $w = \dfrac{z+i}{iz+1}$ and $|z| < 1$. Then Ω'' is just Ω' (the mapping from Ω to Ω' being the classical **Cayley transform**). The action of G' on Ω' is by $g(\omega') = (P^+$ component of $g\omega')$, and this is given by linear fractional transformations by the same computation as for the action of G on Ω.

43. For $\dfrac{1}{\sqrt{2}}\begin{pmatrix}1 & i\\ i & 1\end{pmatrix}$, the decomposition into $P^+K^{\mathbb{C}}P^-$ is

$$\begin{pmatrix}1 & i\\ 0 & 1\end{pmatrix}\begin{pmatrix}\sqrt{2} & 0\\ 0 & 1/\sqrt{2}\end{pmatrix}\begin{pmatrix}1 & 0\\ i & 0\end{pmatrix}.$$

The element u_j is the Cayley transform $\mathbf{c}_{\gamma_j}$ defined as in (6.65a), with root vectors normalized so that $[E_{\gamma_j}, \overline{E_{\gamma_j}}] = 2|\gamma_j|^{-2}H_{\gamma_j} = H'_{\gamma_j}$. More precisely we are to think of $E_{\gamma_j} \leftrightarrow \begin{pmatrix}0 & -i\\ 0 & 0\end{pmatrix}$ and $\overline{E_{\gamma_j}} \leftrightarrow \begin{pmatrix}0 & 0\\ i & 0\end{pmatrix}$, so that $u_j = \exp\frac{\pi}{4}(\overline{E_{\gamma_j}} - E_{\gamma_j}) \leftrightarrow \dfrac{1}{\sqrt{2}}\begin{pmatrix}1 & i\\ i & 1\end{pmatrix}$. Then the decomposition for (a) is

$$u_j = \exp(-E_{\gamma_j})\exp((\tfrac{1}{2}\log 2)H'_{\gamma_j})\exp(\overline{E_{\gamma_j}}).$$

In (b) the factor u_j of u affects only the jth factor of $\exp\left(\sum c_j(E_{\gamma_j} + \overline{E_{\gamma_j}})\right)$ when it is expanded out, and the result of applying $\mathrm{Ad}(u)$ is therefore $\exp\sum c_j \mathrm{Ad}(u_j)(E_{\gamma_j} + \overline{E_{\gamma_j}}) = \exp\sum(-c_j H'_{\gamma_j})$ by a computation in $\mathfrak{sl}(2,\mathbb{C})$. In (c) define a restricted root β to be positive if $\beta(E_{\gamma_j} + \overline{E_{\gamma_j}}) < 0$ for the first j having $\beta(E_{\gamma_j} + \overline{E_{\gamma_j}}) \neq 0$. If X is a restricted-root vector for such a β and j is the distinguished index, then $[E_{\gamma_i} + \overline{E_{\gamma_i}}, X] = -c_i X$ for all i, with $c_1 = \cdots = c_{j-1} = 0$ and $c_j > 0$. Then

$$[H'_{\gamma_i}, \mathrm{Ad}(u)X] = -[\mathrm{Ad}(u)(E_{\gamma_i} + \overline{E_{\gamma_i}}), \mathrm{Ad}(u)X] = c_i \mathrm{Ad}(u)X.$$

So $\mathrm{Ad}(u)X$ is a sum of root vectors for roots $\tilde{\beta}$ such that $\tilde{\beta}(H'_{\gamma_i}) = c_i$. If $\tilde{\beta}$ is negative and noncompact, then $\langle\tilde{\beta}, \gamma_i\rangle$ is < 0 when it is $\neq 0$ for the first time. But $\langle\tilde{\beta}, \gamma_j\rangle = c_j > 0$. Hence $\tilde{\beta}$ is compact or positive noncompact. Then (c) follows, and (d) is a consequence of $uGB = (uN_{\mathfrak{p}}u^{-1})(uA_{\mathfrak{p}}u^{-1})uKB \subseteq P^+K^{\mathbb{C}} \cdot K^{\mathbb{C}} \cdot P^+K^{\mathbb{C}}P^- \cdot KB \subseteq P^+K^{\mathbb{C}}P^-$.

44. This follows from Problem 43 and the style of argument used in the proof of Theorem 7.129.

Chapter VIII

1. Let $\{\psi_\alpha\}$ be a smooth partition of unity as in (8.8). Define a smooth m form ω_α on U_α by $\omega_\alpha = \varphi_\alpha^*(dx_1 \wedge \cdots \wedge dx_m)$. Then $\omega = \sum_\alpha \psi_\alpha \omega_\alpha$ is a smooth m form on M. Since M is oriented, the local coefficient (8.4) of each ω_α is ≥ 0 in each coordinate neighborhood. Hence the sum defining ω involves no cancellation in local coordinates and is everywhere positive.

2. It is assumed that F is real analytic on a neighborhood of a cube, say with sides $0 \leq x_j \leq 1$. The set of a with $0 < a < 1$ such that $F(a, x_2, \ldots, x_n)$ is identically 0 is finite since otherwise there would be an accumulation point and a power series expansion about the limiting point would show that F vanishes on an open set. This fact may be combined with Fubini's Theorem and induction to give a proof.

3. We have $\begin{pmatrix} a_0 & b_0 \\ 0 & 1 \end{pmatrix}\begin{pmatrix} a & b \\ 0 & 1 \end{pmatrix} = \begin{pmatrix} a_0 a & a_0 b + b_0 \\ 0 & 1 \end{pmatrix}$. Thus left translation carries $da\,db$ to $d(a_0 a)\,d(a_0 b) = a_0^2\,da\,db$, and it carries $a^{-2}\,da\,db$ to $(a_0 a)^{-2} a_0^2\,da\,db = a^{-2}\,da\,db$. So $a^{-2}\,da\,db$ is a left Haar measure. The computation for a right Haar measure is similar.

4. G is unimodular by Corollary 8.31, and $M_{\mathfrak{p}}A_{\mathfrak{p}}N_{\mathfrak{p}}$ is not (by (8.38)). Apply Theorem 8.36.

5. Use Problem 2.

6. $GL(n, \mathbb{R})$ is reductive.

7. With E_{ij} as in Problem 8, use

$$E_{11}, E_{21}, \ldots, E_{n1}, E_{12}, \ldots, E_{n2}, \ldots, E_{1n}, \ldots, E_{nn}$$

as a basis. Then L_x is linear, and its expression in this basis is block diagonal with each block a copy of x. Hence $\det L_x = (\det x)^n$.

8. Part (a) uses Problem 2. For (b) we use Problem 7 and the change-of-variables formula for multiple integrals to write

$$\begin{aligned}\int_{GL(n,\mathbb{R})} f(y)\,dy &= \int_{GL(n,\mathbb{R})} f(L_x y)|\det L_x|\,dy \\ &= \int_{GL(n,\mathbb{R})} f(xy)|\det x|^n\,dy = \int_{GL(n,\mathbb{R})} f(y)|\det x|^n\,d(x^{-1}y),\end{aligned}$$

where dy denotes Lebesgue measure restricted to the open set $GL(n,\mathbb{R})$. This shows that $|\det x|^n\,d(x^{-1}y) = dy$, and it follows that $|\det y|^{-n}\,dy$ is left invariant.

9. Write $x = kan$. Then $\pi(n)v = v$, and $\pi(a)v = e^{\nu \log a}v$. Hence $\|\pi(x)v\|^2 = \|e^{\nu \log a}\pi(k)v\|^2 = e^{2\nu \log a}\|v\|^2$.

10. Part (a) uses Problem 9, first with the standard representation (with $v = \begin{pmatrix}1\\0\\0\end{pmatrix}$) and then with $\bigwedge^2$ of the standard representation (with $v = \begin{pmatrix}1\\0\\0\end{pmatrix} \wedge \begin{pmatrix}0\\1\\0\end{pmatrix}$). For (b), $(2f_1) + 2(f_1 + f_2) = 4f_1 + 2f_2 = 2f_1 - 2f_3 = 2\rho_{\mathfrak{p}}$.

11. For (a) use the standard representation with $v = \begin{pmatrix}1\\0\\ \vdots\\0\\1\end{pmatrix}$. The highest restricted weight λ is 1 on $E_{1,n+1} + E_{n+1,1}$. Then

$$\pi(x)v = \begin{pmatrix} x_{11} + x_{1,n+1} \\ \vdots \\ x_{n+1,1} + x_{n+1,n+1} \end{pmatrix},$$

so that $\|\pi(x)v\|^2 = \sum_{j=1}^{n+1}(x_{j1} + x_{j,n+1})^2$ and

$$e^{2\lambda H_{\mathfrak{p}}(x)} = \tfrac{1}{2}\sum_{j=1}^{n+1}(x_{j1} + x_{j,n+1})^2.$$

In (b) the unique positive restricted root α is 2 on $E_{1,n+1} + E_{n+1,1}$, and $\rho_{\mathfrak{p}} = \frac{1}{2}(n-1)\alpha$. Hence $e^{2\rho_{\mathfrak{p}} H_{\mathfrak{p}}(x)} = (e^{2\lambda H_{\mathfrak{p}}(x)})^{n-1}$.

NOTES

Background

The theory of Lie groups, as it came to be known in the 20th century, was begun single-handedly by Sophus Lie in 1873. Lie developed the theory over a period of many years, and then he gave a systematic exposition as part of a three-volume work written jointly with the younger F. Engel (Lie-Engel [1888–90–93]). A detailed summary of this early theory, with extensive references, appears in Bourbaki [1972], 286–308.

Lie worked with families of (not necessarily linear) transformations of n complex variables given by holomorphic functions

$$x_i' = f_i(x_1, \ldots, x_n, a_1, \ldots, a_r), \qquad 1 \le i \le n,$$

the family given by the complex parameters $a_1, \ldots, a_r$. Later $a_1, \ldots, a_r$ were allowed to be real. It was assumed that the transformation corresponding to some set $a_1^0, \ldots, a_r^0$ of parameters reduced to the identity and that, roughly speaking, the family was effective and was closed under composition. The result was a "transformation group, finite and continuous." For more detail about the composition law, see Cartan [1894], 13–14, and for the definition of "effective," see Bourbaki [1972], 290.

Such a transformation group was not literally closed under composition, the functions f_i not being globally defined. Thus it had a local nature, and Lie and Engel assumed that it was local when necessary. On the other hand, a transformation group in the sense of Lie is not quite what is now meant by a "local Lie group," because the space variables x_i and the group variables a_j were inseparable, at least at first. In any event, to a "finite and continuous" transformation group, Lie associated a family of "infinitely small transformations" or "infinitesimal transformations," which carried the information now associated with the Lie algebra. In terms of a Taylor development through order 1, namely,

$$f_i(x_1, \ldots, x_n, a_1^0 + z_1, \ldots, a_r^0 + z_r) = x_i + \sum_{k=1}^{r} z_k X_{ki}(x_1, \ldots, x_n) + \cdots,$$

the infinitesimal transformations were given by

$$dx_i = \Big(\sum_{k=1}^{r} z_k X_{ki}(x_1, \ldots, x_n)\Big)\, dt, \qquad 1 \le i \le n.$$

The main results of Lie-Engel [1888–90–93] for current purposes were three theorems of Lie for passing back and forth between "finite continuous" transformation groups and their families of infinitesimal transformations, each theorem consisting of a statement and its converse. For precise statements, see Bourbaki [1972], 294–296.

Lie did observe that the (local) group of transformations of $\mathbb{C}^n$ yielded a new transformation group whose space variables were the parameters, and he called this the "parameter group." There were in effect two parameter groups, one given by the group of left translations of the parameters and one given by the group of right translations of the parameters. Lie showed that two transformation groups have isomorphic parameter groups if and only if their families of infinitesimal transformations are isomorphic.

Although Lie did have occasion to work with particular global groups (such as the complex classical groups), he did not raise the overall question of what constitutes a global group. He was able to study his particular global groups as transformation groups with their standard linear actions.

The idea of treating global groups systematically did not arise until Weyl, inspired by work of I. Schur [1924] that extended representation theory from finite groups to the orthogonal and unitary groups, began his study Weyl [1924] and [1925–26] of compact connected groups. Schreier [1926–27] defined topological groups and proved the existence of universal covering groups of global Lie groups, and Cartan [1930] underlined the importance of global groups by proving a global version of Lie's third theorem—that every finite-dimensional real Lie algebra is the Lie algebra of a Lie group. Pontrjagin [1939] gave a systematic exposition of topological groups, carefully distinguishing local and global results for Lie groups and proving global results where he could. Finally Chevalley [1946] provided the first systematic treatment of a global theory, introducing analytic subgroups and establishing a one-one correspondence between Lie subalgebras and analytic subgroups.

The term "Lie group" came into widespread use in the early 1930s, and the term "Lie algebra" appeared shortly thereafter. In retrospect much early work in Lie theory was on Lie algebras because of Lie's three theorems that in effect reduced properties of local Lie groups to properties of Lie algebras.

Chapter I

The beginning properties of finite-dimensional Lie algebras in Chapter I are all due to Lie (see Lie-Engel [1888–90–93]). Lie classified the complex Lie algebras of dimension ≤ 4, introduced solvable Lie algebras (calling them "integrable"), proved Proposition 1.23, and proved Lie's Theorem (Theorem 1.25 as Satz 2 on p. 678 of Vol. III and Corollary 1.29 as Satz 9 on p. 681 of Vol. III). Lie defined simple Lie algebras and showed that the complex classical

Lie algebras $\mathfrak{sl}(n, \mathbb{C})$, $\mathfrak{so}(n, \mathbb{C})$, and $\mathfrak{sp}(n, \mathbb{C})$ are simple for the appropriate values of n.

The original form of Engel's Theorem is that $\mathfrak{g}$ is solvable if ad X is nilpotent for all $X \in \mathfrak{g}$. Application of Lie's Theorem yields Corollary 1.38. The result of Engel's Theorem came out of an incomplete discusion in Killing [1888–89–90]. Engel had his student Umlauf in his thesis (Umlauf [1891]) make a number of Killing's results rigorous, and this was one of them. Cartan had access to Umlauf's thesis but gave Engel principal credit for the theorem (see Cartan [1894], 46), and "Engel's Theorem" has come to be the accepted name.

Killing [1888–89–90] proved the existence of the radical and defined a Lie algebra to be semisimple if it has radical 0. Theorem 1.51, relating semisimplicity to simplicity, is due to Killing.

The Killing form defined in (1.18) came later (see Weyl [1925–26], Kap. III, §3). Killing used no bilinear form of this kind, and Cartan [1894] used a variant. If $\dim \mathfrak{g} = n$, Cartan defined $\psi_2(X)$ to be the coefficient of λ^{n-2} in the characteristic polynomial $\det(\lambda 1 - \operatorname{ad} X)$. Then $\psi_2(X)$ is a quadratic form in X given by $\psi_2(X) = \frac{1}{2}\big((\operatorname{Tr}(\operatorname{ad} X))^2 - \operatorname{Tr}((\operatorname{ad} X)^2)\big)$. The form $\psi_2(X)$ reduces to a multiple of the Killing form if $\operatorname{Tr}(\operatorname{ad} X) = 0$ for all X, as is true when $[\mathfrak{g}, \mathfrak{g}] = \mathfrak{g}$. Cartan [1894] established the criteria for solvability and semisimplicity (Proposition 1.43 and Theorem 1.42), but the criteria are stated in terms of ψ_2 rather than the Killing form.

Most of the results on Lie algebras in §§1–7 are valid whenever the underlying field has characteristic 0; occasionally (as in Lie's Theorem) it is necessary to assume also that the field is algebraically closed or at least that some eigenvalues lie in the field. Some proofs are easier when the underlying field is a subfield of $\mathbb{C}$, and the goal for this book of working with Lie groups has led us to give only the easier proofs in such cases. An example occurs with Cartan's Criterion for Solvability. The part of the proof where it is easier to handle subfields of $\mathbb{C}$ rather than general fields of characteristic 0 is that $[\mathfrak{g}, \mathfrak{g}] \subseteq \operatorname{rad} B$ implies $\mathfrak{g}$ solvable. One general proof of this assertion regards the base field as a vector space over $\mathbb{Q}$ and works with $\mathbb{Q}$ linear functions on the base field in a complicated way; see Varadarajan [1974] for this proof. Another general proof, which was pointed out to the author by R. Scott Fowler, uses the theory of real closed fields to generalize the argument that we have given.

The simple Lie algebras over $\mathbb{R}$ were classified in Cartan [1914], and Cartan must accordingly be given credit for the discovery of any of the classical simple Lie algebras that were not known from geometry. The irreducible finite-dimensional complex linear representations of $\mathfrak{sl}(2, \mathbb{C})$ as in Theorem 1.63 are implicit in Cartan [1894] and explicit in Cartan [1913]. Complete reducibility of finite-dimensional complex linear representations of $\mathfrak{sl}(2, \mathbb{C})$ (Theorem 1.64) was proved by E. Study, according to Lie-Engel [1888–90–93]. The expression $\frac{1}{2}h^2 + h + 2fe$ that appears in Lemma 1.65 is called the "Casimir operator" for $\mathfrak{sl}(2, \mathbb{C})$; see the Notes for Chapter V.

The name "Schur's Lemma" is attached to many results like Lemma 1.66. Burnside [1904] proved in the language of matrix representations that a linear map carrying one irreducible representation space for a finite group to another and commuting with the group is 0 or is nonsingular. I. Schur [1905] proved in the same language that if the linear map carries one irreducible representation space for a finite group to itself and commutes with the group, then the map is scalar.

The origins of the theory of Lie groups in §10 have been discussed above. The text weaves together the general theory from Chevalley [1946] with its concrete interpretation for Lie groups of matrices as given in Knapp [1988], Chapter I. The exponential mapping was already part of the work in Lie-Engel [1888–90–93], and Lie understood the exponential's behavior through quadratic terms (in a form equivalent with Lemma 1.92). Higher-order terms are related to the Campbell-Hausdorff formula and are discussed in the Notes for Appendix B. The adjoint representation is due to Lie.

The treatment in §10 uses C^∞ functions rather than real analytic functions, and it is remarked that the Lie groups under discussion are the same in the two cases. This fact had already been noticed by Lie. F. Schur [1893] gave a proof essentially that a C^2 Lie group could be made into a real analytic group, and Hilbert in 1900 raised the question whether Lie's transformation groups might be approached without the assumption of differentiability (**Hilbert's fifth problem**). An affirmative answer to the question whether every locally Euclidean group is Lie was provided by Gleason, Montgomery, and Zippin, and an exposition appears in Montgomery-Zippin [1955]. See Yang [1976] for a discussion of progress on the full question of Hilbert's.

At a certain stage in Lie theory, analyticity plays a vital role, but not before the end of this book. In infinite-dimensional representation theory real analyticity is crucial. The group $\mathbb{R} = \{r \in \mathbb{R}\}$ acts continuously by unitary transformations on $L^2(\mathbb{R})$ when it acts by translations, and this action is reflected on the Lie algebra level on smooth functions, the members of the Lie algebra acting by multiples of d/dr. The subspace of smooth functions with support in the unit interval is carried to itself by the Lie algebra of differentiations, but not even the closure of this subspace is carried to itself by the group of translations. Harish-Chandra [1953] showed how to avoid this pathology in all situations by using real analyticity.

The results of §12 are implicit in Cartan [1930], which proves the existence of a Lie group corresponding to each real Lie algebra. Cartan [1927b] lists the classical groups of §14, and the geometric methods of that paper yield the polar decomposition of Proposition 1.122 for those groups. The actual method of proof that we have used for Proposition 1.122 is taken from Mostow [1949].

A number of books treat elementary Lie theory. In addition to Chevalley [1946], the list includes Adams [1969], Bourbaki [1960] and [1972], Cohn [1957], Freudenthal and de Vries [1969], Helgason [1962] and [1978],

Hochschild [1965], Pontrjagin [1939], Serre [1965], Spivak [1970], Varadarajan [1974], and F. Warner [1971]. See Séminaire "Sophus Lie" [1955] for a treatment using Chevalley [1946] as a prerequisite. The books Dixmier [1974], Humphreys [1972], and Jacobson [1962] treat Lie algebras.

Chapter II

Although the four families of classical complex simple Lie algebras in §1 were known to Lie, the general theory and classification of complex simple Lie algebras are largely due to Killing [1888–89–90] and Cartan [1894]. Many of the results in §§1–5 and §7 were announced by Killing, but Killing's proofs were often incomplete or incorrect, and sometimes proofs were absent altogether. Umlauf [1891] in his thesis under the direction of Engel undertook to give rigorous proofs of some of Killing's work. Cartan had access to Umlauf's thesis, and Cartan [1894] gives a rigorous treatment of the classification of complex simple Lie algebras. Cartan [1894] repeatedly gives page references to both Killing's work and Umlauf's work, but Cartan's thesis gives principal credit to Engel for Umlauf's work. Cartan was generous to Killing both in 1894 and later for the contributions Killing had made, but others were less kind, dismissing Killing's work completely because of its gaps and errors.

The characteristic polynomial $\det(\lambda 1 - \operatorname{ad} X)$ had already been considered by Lie, and Killing [1888–89–90] investigated its roots systematically. Umlauf [1891] was able to take the crucial step of dropping all special assumptions about multiplicities of the roots. Umlauf's work contains a proof of the existence of Cartan subalgebras in the style of Theorem 2.9′: $\mathfrak{g}_{0,X}$ is a Cartan subalgebra if the lowest-order nonzero term of the characteristic polynomial is nonzero on X. Elementary properties of roots and root strings were established by Umlauf without the assumption of semisimplicity, and Cartan [1894] reproduces all this work. Then Cartan [1894] brings in the assumption of semisimplicity and makes use of Cartan's Criterion (Theorem 1.42). Cartan [1894] defines Weyl group reflections (§IV.6) and uses "fundamental roots" rather than simple roots. An $\mathbb{R}$ basis of roots is **fundamental** if when the reflections in these roots are applied to the basis and iterated, all roots are obtained.

Killing's main result had been a classification of the complex simple Lie algebras. Killing [1888–89–90] correctly limited the possible exceptional algebras to ones in dimensions 14, 52, 78, 133, 248. He found two possibilities in dimension 52, and he did not address the question of existence. Engel [1893] constructed what we now call G_2.

Cartan [1894] redid the classification, pointing out (p. 94) a simple isomorphism between Killing's two 52-dimensional exceptional cases. Effectively Cartan also showed that the passage to roots is one-one in the semisimple case, and he proved existence. Since Cartan's definition of what we now call a Cartan subalgebra for a given $\mathfrak{g}$ involved regular elements, Cartan knew that all such subalgebras had a common dimension, namely the number of low-order 0 terms in the characteristic polynomial. Thus to show that the passage to roots is one-one, Cartan had only to investigate cases of equal rank and dimension, showing how the Lie algebras can be distinguished. This he did case by case. He proved existence case-by-case as well, giving multiplication tables for the root vectors. He omitted the details of these computations on the grounds of their length.

The proof of the classification was simplified over a period of time. Simple roots do not appear in Cartan [1894] and [1913]. Weyl [1925–26], Kap. IV, §5, introduces lexicographic orderings and positive roots as a tool in working with roots. Van der Waerden [1933] simplified the proof of classification, and then Dynkin [1946] and [1947] used the diagrams bearing his name and simplified the proof still further. Dynkin diagrams are instances of Coxeter graphs (Coxeter [1934]), and Witt [1941] makes use of these graphs in the context of complex semisimple Lie algebras. The second Dynkin paper, Dynkin [1947], acknowledges this work of Coxeter and Witt. For a fuller discussion of Coxeter graphs, see Bourbaki [1968], Ch. IV, and Humphreys [1990]. The proof of classification given here is now standard except for minor variations; see Jacobson [1962] and Humphreys [1972], for example.

Abstract root systems occur implicitly in Witt [1941] and explicitly in Bourbaki [1968], and the Weyl group makes appearances as a group in Weyl [1925–26] and Cartan [1925b]. Chevalley's Lemma (Proposition 2.72) appears without proof in a setting in Harish-Chandra [1958] where it is combined with Theorem 6.57, and it is attributed to Chevalley.

Although Cartan had proved what amounts to the Existence Theorem (Theorem 2.111), he had done so case by case. Witt [1941], Satz 15, gave what amounted to a general argument, provided one knew existence in rank ≤ 4. Chevalley [1948a] and [1948b] and Harish-Chandra [1951] gave the first completely general arguments, starting from a free Lie algebra and factoring out a certain ideal. See Jacobson [1962] for an exposition. Serre [1966] improved the argument by redefining the ideal more concretely; the Serre relations of Proposition 2.95 are generators of this ideal. Serre's argument is reproduced in Humphreys [1972]; we have given the same argument here but in a different order. See Helgason [1978], §X.4, for this kind of argument in a more general context.

The uniqueness aspect of the Isomorphism Theorem is in Cartan [1894], and the existence aspect is in Weyl [1925–26] and van der Waerden [1933]. The argument here is built around the Serre relations.

The result of Problem 7 is from Cartan [1894] and is used over and over in the theory. The result in Problem 11 appears in Kostant [1955]. The length function of Problems 21–24 is in Bourbaki [1968]. The realization of G_2 in Problem 40 is the one that Bourbaki [1968] gives and that we repeat in §2 of Appendix C. The connection with C_3 was pointed out to the author by J.-S. Huang; a connection with D_4 was known much earlier.

The results about complex semisimple Lie algebras are essentially unchanged if one replaces $\mathbb{C}$ by an arbitrary algebraically closed field of characteristic 0, but a little algebraic geometry needs to be added to some of the proofs to make them valid in this generality. See Jacobson [1962], Dixmier [1974], and Humphreys [1972]. Humphreys develops the theory using "toral subalgebras" in place of Cartan subalgebras, at least at first.

Chapter III

The universal enveloping algebra in essence was introduced by Poincaré [1899] and [1900]. The paper [1899] announces a result equivalent with Theorem 3.8, and Poincaré [1900] gives a sketchy proof. Schmid [1982] gives a perspective on this work. Garrett Birkhoff [1937] and Witt [1937] rediscovered Poincaré's theorem and proved it more generally. We have used the proofs as given in Humphreys [1972] and Dixmier [1974].

Cartan [1913] used iterated products of members of the Lie algebra in his work on finite-dimensional representations, but this work did not require the linear independence in the Poincaré-Birkhoff-Witt Theorem. Apart from this kind of use, the first element of order greater than one in a universal enveloping algebra that arose in practice was the Casimir operator (Chapter V), which appeared in Casimir and van der Waerden [1935]. The Casimir operator plays a key role in the proof of the complete reducibility theorem that we give as Theorem 5.29. The universal enveloping algebra did not find further significant application until Gelfand and Harish-Chandra in the 1950s showed its importance in representation theory.

The connection with differential operators (Problems 10–12) is stated by Godement [1952] and is identified on p. 537 of that paper as an unpublished result of L. Schwartz; a published proof is in Harish-Chandra [1956a] as Lemma 13. No generality is gained by adjusting the definition of left-invariant differential operator so as to allow an infinite order operator that is of finite-order on each compact subset of a chart.

For further discussion of the universal enveloping algebra and its properties, see Helgason [1962], 90–92, 97–99, 386, and 391–393. Also see Jacobson [1962], Chapter V, and Dixmier [1974]. The Poincaré-Birkhoff-Witt Theorem is valid over any field.

Symmetrization in §3 is due to Gelfand [1950] and Harish-Chandra [1953].

Chapters IV and V

Historically representations of complex semisimple Lie algebras were considered even before group representations of finite groups were invented, and the connection between the two theories was realized only much later. According to Lie-Engel [1888-90-93], III, 785–788, E. Study proved complete reducibility for the complex-linear finite-dimensional representations of $\mathfrak{sl}(2, \mathbb{C})$, $\mathfrak{sl}(3, \mathbb{C})$, and $\mathfrak{sl}(4, \mathbb{C})$. Lie and Engel then conjectured complete reducibility of representations of $\mathfrak{sl}(n, \mathbb{C})$ for arbitrary n.

Frobenius [1896] is the first paper on the representation theory of finite groups, apart from papers about 1-dimensional representations. Frobenius at first treated the characters of finite groups, coming at the problem by trying to generalize an identity of Dedekind concerning multiplicative characters of finite abelian groups. It was only in later papers that Frobenius introduced matrix representations and related them to his theory of characters. Frobenius credits Molien with independently discovering in 1897 the interpretation of characters in terms of representations. Frobenius also takes note of Molien's 1893 paper realizing certain finite-dimensional semisimple associative algebras as matrices; the theory of semisimple associative algebras has points of contact with the theory of representations of finite groups. Burnside [1904] and then I. Schur [1905] redid the Frobenius theory, taking matrix representations as the primary objects of study and deducing properties of characters as consequences of properties of representations. According to E. Artin [1950], 67, "It was Emmy Noether who made the decisive step. It consisted in replacing the notion of matrix by the notion for which the matrix stood in the first place, namely, a linear transformation of a vector space."

Much of Chapters IV and V stems from work of Cartan and Weyl, especially Cartan [1913] and Weyl [1925–26].

Cartan [1913] contains an algebraic treatment of the complex-linear finite-dimensional representations of complex semisimple Lie algebras, including the Theorem of the Highest Weight essentially as in Theorem 5.5. The paper proves existence by handling fundamental representations case by case and by generating other irreducible representations from highest weight vectors of tensor products. Cartan makes use of iterated products of elements of the Lie algebra, hence implicitly makes use of the universal enveloping algebra. But he does not need to know the linear independence that is the hard part of the Poincaré-Birkhoff-Witt Theorem (Theorem 3.8).

Cartan's paper refers to the underlying transformation groups for his representations, and differentiation leads him to the formalism of representations of Lie algebras on tensor products. But oddly the formal similarity between Cartan's theory for Lie groups and the representation theory of finite groups went unnoticed for many years. Possibly mathematicians at the time were still thinking (with Lie and Engel) that transformation groups were the principal objects of study. Cartan uses language in the 1913 paper to suggest that he regards two group representations in different dimensions as involving different groups, while he regards two Lie algebra representations as involving the same Lie algebra if the bracket relations can be matched.

Cartan [1914] classifies the real forms of complex simple Lie algebras, and one sees by inspection that each complex simple Lie algebra has one and only one compact real form. At this stage the ingredients for a theory of group representations were essentially in place, but it is doubtful that Cartan was aware at the time of any connection between his papers and the theory of group representations.

Having a fruitful theory of representations of compact Lie groups requires having invariant integration, and Hurwitz [1897] had shown how to integrate on $O(n)$ and $U(n)$. I. Schur [1924] put this idea together with his knowledge of the representation theory of finite groups to arrive at a representation theory for $O(n)$ and $U(n)$. Invariant integration on arbitrary Lie groups (defined by differential forms as at the start of §VIII.2) was already known to some mathematicians; it is mentioned in a footnote on the second page of Cartan [1925a]. Weyl was aware of this fact and of Cartan's work, and Weyl [1924] immediately set forth, using analysis, a sweeping representation theory for compact semisimple Lie groups.

Weyl [1925–26] gives the details of this new theory. Kap. I is about $\mathfrak{sl}(n, \mathbb{C})$. After reviewing Cartan's treatment, Weyl points out (footnote in §4) that Cartan implicitly assumed without proof that finite-dimensional representations are completely reducible (Theorem 5.29) when he constructed representations with given highest weights. Weyl then gives a proof of complete reducibility, using his "unitary trick." To push this argument through, he has to lift a representation of $\mathfrak{su}(n)$ to a representation of $SU(n)$. Lie's results give Weyl a locally defined representation, and Weyl observes in a rather condensed argument in §5 that there is no obstruction to extending the locally defined representation to be global if $SU(n)$ is simply connected. Then he proves that $SU(n)$ is indeed simply connected. He goes on in Kap. II to use what is now called the Weyl Integration Formula (Theorem 8.60) to derive formulas for the characters and dimensions of irreducible representations of $SU(n)$. Kap. II treats $\mathfrak{sp}(n, \mathbb{C})$ and $\mathfrak{so}(n, \mathbb{C})$ similarly, taking advantage of $Sp(n)$ and $SO(n)$. The treatment of $SO(n)$ is more subtle than that of $Sp(n)$, because Weyl must consider single-valued and double-valued representations for $SO(n)$. (That is, $SO(n)$ is not simply connected.)

Kap. III begins by redoing briefly some of Cartan [1894] in Weyl's own style. The Killing form is introduced on its own, and Cartan's Criteria are stated and proved. Then Weyl introduces the Weyl group of the root system and derives some of its properties. (The Weyl group appears also in Cartan [1925b].) Finally Weyl proves the existence of a compact real form for any complex semisimple Lie algebra, i.e., a real form on which the Killing form is negative definite. (Later Cartan [1929a] remarks (footnote in §6) that Weyl implicitly assumed without proof that the adjoint group of this compact real form is compact, and then Cartan gives a proof.) Kap. IV begins by proving that every element of a compact semisimple Lie group is conjugate to a member of a maximal torus (Theorem 4.36), the proof being rather similar to the one here.

Next Weyl shows that the universal covering group of a compact semisimple Lie group is compact (Theorem 4.69). He is then free to apply the unitary trick to lift representations from a complex semisimple Lie algebra to a compact simply connected group corresponding to the compact real form and to deduce complete reducibility (Theorem 5.21). The rest of Kap. IV makes use the Weyl Integration Formula (Theorem 8.60). Proceeding along lines that we indicate at the end of Chapter VIII, Weyl quickly derives the Weyl Character Formula (Theorems 5.77 and 5.113) and the Weyl Dimension Formula (Theorem 5.84). To complete the discussion of the Theorem of the Highest Weight, Weyl handles existence by noting on analytic grounds that the irreducible characters are a complete orthogonal set in the space of square-integrable functions constant on conjugacy classes. He dismisses the proof of this assertion as in the spirit of his earlier results; Peter-Weyl [1929] gives a different, more comprehensive argument. Weyl's definition of "integral" is what we have called "algebraically integral"; there does not seem to be a proof that algebraically integral implies analytically integral in the simply connected case (Theorem 5.107).

Let us consider the sections of Chapters IV and V in order. The representations in §1 were known to Cartan [1913]. Representation theory as in §2 began with a theory for finite groups, which has been discussed above. Maschke proved Corollary 4.7 for finite groups, and Loewy and Moore independently proved Proposition 4.6 in this context. Burnside [1904] was the one who saw that Corollary 4.7 follows from Proposition 4.6, and Burnside [1904] proved Proposition 4.8. Although there is a hint of Corollary 4.9 in the earlier work of Burnside, Corollary 4.9 is generally attributed to I. Schur [1905]. Schur [1905] also proved Corollary 4.10. Schur [1924] observed that the results of §2 extend to $O(n)$ and $U(n)$, Hurwitz [1897] having established invariant integration for these groups. Weyl [1925–26] understood that invariant integration existed for all compact semisimple Lie groups, and he derived the Weyl Integration Formula (Theorem 8.60).

Topological groups and covering groups were introduced systematically by Schreier [1926] and [1927], and it was plain that the abstract theory of §2 (and also §3) extends to general compact groups as soon as invariant integration is available. Existence of Haar measure for locally compact groups was proved under a separability assumption in Haar [1933], and uniqueness was established in von Neumann [1934a]. Von Neumann [1934b] gives a quick development of invariant means that handles both existence and uniqueness in the compact case. See Weil [1940] for further historical discussion.

The Peter-Weyl Theorem in §3 originally appeared in Peter-Weyl [1927]. We have followed an argument given in Cartan [1929b]. The text has not discussed infinite-dimensional representations at all, but the Peter-Weyl Theorem has important consequences for this theory that we should mention. A continuous group action by unitary operators in a Hilbert space is called a **unitary representation**. The Hilbert space $L^2(G)$ carries two natural unitary representations, the **left regular representation** $\pi(g)f(x) = f(g^{-1}x)$ and the **right regular representation** $\pi(g)f(x) = f(xg)$. Corollary 4.21 implies that each of these representations decomposes as a Hilbert space orthogonal direct sum of finite-dimensional irreducible representations, each irreducible representation occurring with multiplicity equal to its dimension. It is not hard to derive as a consequence a formula for the orthogonal projection to the sum of the spaces corresponding to a given irreducible representation. In addition, it follows that any unitary representation of a compact group is the Hilbert space orthogonal direct sum of finite-dimensional irreducible representations, and the same kind of formula is valid for the orthogonal projection to the sum of the spaces corresponding to a given irreducible representation. See §I.5 of Knapp [1986] for details.

The results of §4 are influenced by §II.6 of Helgason [1978]. Because of Corollary 4.22, Theorem 4.29 is really a theorem about matrix groups. Goto [1948] proved that a semisimple matrix group is a closed subgroup of matrices, and the proof of Theorem 4.29 makes use of some of Goto's ideas.

The first key result in §5 is Theorem 4.36, which appears as Weyl [1925–26], Kap. IV, Satz 1. We have followed Varadarajan [1974], and the proof is not too different from Weyl's. Other proofs are possible. Adams [1969] gives a proof due to Weil [1935] that uses the Lefschetz Fixed-Point Theorem. Helgason [1978] gives a proof due to Cartan that is based on Riemannian geometry. Serre [1955] discusses both these proofs. For variations, see the proofs in Hochschild [1965] and Wallach [1973].

Theorem 4.34 is a consequence of the proofs by Weyl or Weil of Theorem 4.36; the quick proof that we give is from Hunt [1955]. Theorem 4.50 and its corollaries are due to Hopf [1940–41] and [1942–43]; we have followed Helgason [1978] and ultimately Serre [1955].

The results of §§6–7 are implicit or explicit in Weyl [1925–26]. In connection with §8, see Weyl [1925–26], Kap. IV, Satz 2, for Theorem 4.69. Helgason [1978], 154, discusses a number of other proofs. Our proof of Lemma 4.70 is taken from Varadarajan [1974], 343, and ultimately from Cartier [1955b].

The results of §§1–2 of Chapter V are due to Cartan [1913]. In proving the existence of a representation with a given highest weight, Cartan did not give a general argument. Instead he made explicit computations to produce each fundamental representation (Problems 36–41) and used Cartan composition (Problem 15) to generate the other irreducible representations. Chevalley [1948a] and [1948b] and Harish-Chandra [1951] gave the first general arguments to prove existence. Harish-Chandra [1951] constructs semisimple Lie algebras $\mathfrak{g}$ and their representations together. The paper works with an infinite-dimensional associative algebra $\mathfrak{A}$, Verma-like modules for it (Lemma 12), and quotients of such modules. Then the paper obtains $\mathfrak{g}$ as a Lie subalgebra of $\mathfrak{A}$, and the modules of $\mathfrak{A}$ yield representations of $\mathfrak{g}$. A construction of modules closer to the Verma modules of §3 appears in Harish-Chandra [1954–55],

IV, §§1–6, expecially Lemmas 2, 5, and 16. Cartier [1955a], item (2) on p. 3, constructs Verma modules explicitly and establishes properties of highest weight modules. The name "Verma modules" seems to have been introduced in lectures and discussions by Kostant in the late 1960s, in recognition of the work Verma [1968] that establishes some structure theory for these modules. Verma proved that the space of $U(\mathfrak{g})$ maps between two of these modules is at most 1-dimensional and that any nonzero such map is one-one. Bernstein-Gelfand-Gelfand [1971] developed further properties of these modules. Early publications in which the name "Verma modules" appears are Dixmier [1974] and Kostant [1975].

Complete reducibility (§4) was first proved by Weyl [1925–26], using analytic methods. Casimir and van der Waerden [1935] gave an algebraic proof. Other algebraic proofs were found by Brauer [1936] and Whitehead [1937]. For Proposition 5.19, see §2.6 of Dixmier [1974]. Proposition 5.32 is from Harish-Chandra [1951].

The Harish-Chandra isomorphism in §5 (Theorem 5.44) is a fundamental result in infinite-dimensional representation theory and first appeared in Harish-Chandra [1951]. Most proofs of this theorem make use of a result of Chevalley about invariants in the symmetric algebra. See Humphreys [1972] or Dixmier [1974] or Varadarajan [1974], for example, for this proof. We have chosen to bypass the symmetric algebra and reproduce the more direct argument that is in Knapp-Vogan [1995].

Weyl [1925–26] gives analytic proofs of the Weyl Character Formula and Weyl Dimension Formula of §6. For the algebraic proof of the Weyl Character Formula in this section, we have followed Dixmier [1974]. The proof comes ultimately from Bernstein-Gelfand-Gelfand [1971] and proves the Kostant Multiplicity Formula (Corollary 5.83) (Kostant [1959]) at the same time.

The name "Borel subalgebra" in §7 has come to be standard because of the systematic treatment in Borel [1956] of the corresponding groups in the theory of algebraic groups.

In §8 Theorems 5.110 and 5.117 are due to Weyl [1925–26].

Problem 19 is taken from Humphreys [1972] and ultimately from Steinberg [1961], and Problems 21–23 are from Parthsarathy and Ranga Rao and Varadarajan [1967]. The spin representations of Problems 24–35 are recalled by Weyl [1924]. Chevalley [1946] gives a concrete discussion, Cartan [1938b] has a more abstract book-length development, and Lawson-Michelsohn [1989] gives a more recent treatment.

There are several books with substantial sections devoted to the representation theory of compact Lie groups and/or complex semisimple Lie algebras. Among these are the ones by Adams [1969], Dixmier [1974], Freudenthal and de Vries [1969], Fulton and Harris [1991], Helgason [1984], Humphreys [1972], Jacobson [1962], Knapp [1986], Lichtenberg [1970], Séminaire "Sophus Lie" [1955], Serre [1966], Varadarajan [1974], Wallach [1973], Warner

[1972a], Weyl [1946], Wigner [1959], and Želobenko [1973]. To this list one can add the books by Helgason [1962] and [1978] and Hochschild [1965] as including extensive structure theory of compact Lie groups and complex semisimple Lie algebras, though essentially no representation theory.

Chapter VI

After 1914 Cartan turned his attention to differential geometry and did not return to Lie groups until 1925. His interest in geometry led him eventually to introduce and study Riemannian symmetric spaces, and he found that classifying these spaces was closely tied to the classification of simple real Lie algebras, which he had carried out in 1914. (Many symmetric spaces turn out to be of the form G/K with G semisimple.) He began to study the corresponding Lie groups, bringing to bear all his knowledge and intuition about geometry, and soon the beginnings of a structure theory for semisimple Lie groups were in place. The treatment of structure theory in Helgason [1978] follows Cartan's geometric approach.

In this book the approach is more Lie-theoretic. The existence of a compact real form for any complex semisimple Lie algebra (Theorem 6.11) is proved case by case in Cartan [1914], and Weyl [1925–26] gives a proof independent of classification. Lemmas 6.2 through 6.4 and Theorem 6.6 appear in Weyl's treatment. But Weyl's proof uses more information about the constants $C_{\alpha,\beta}$ of §1 than appears in these results, enough in fact to deduce the Isomorphism Theorem (Theorem 2.108). The proof of the Isomorphism Theorem in Helgason [1978], 173, follows the lines of Weyl's argument. We have used the Serre relations to obtain the Isomorphism Theorem, and the result is a simpler proof of the existence of a compact real form.

Having known all the simple real Lie algebras for many years, Cartan could see many results case by case before he could give general proofs. Cartan [1927a], 122, effectively gives the Cartan decomposition on the Lie algebra level; comments in Cartan [1929a], 14, more clearly give the decomposition and refer back to the spot in the paper [1927a]. Cartan [1929a] gives a general argument for the existence of a Cartan involution and for uniqueness of the Cartan involution up to conjugacy. Essentially this argument appears in Helgason [1978]. In §2 we have followed an approach in lectures by Helgason, which is built around the variant Theorem 6.16 of Cartan's results; this variant is due to Berger [1957].

The global Cartan decomposition of §3 appears in Cartan [1927b], proceeding by a general argument that uses the case-by-case construction of the involution of the Lie algebra. The group-theoretic approach we have followed is due to Mostow [1949]. The computation of the differential in connection with (6.36) is taken from Helgason [1978], 254–255.

One result about structure theory that we have omitted in §3, having no Lie-theoretic proof, is the theorem of Cartan [1929a] that any compact subgroup of G is conjugate to a subgroup of K.

Cartan [1927b] shows that there is a Euclidean subgroup A of G such that any element of G/K can be reached from the identity coset by applying a member of A and then a member of K. This is the subgroup A of §4, and the geometric result establishes Theorem 6.51 in §5 and the KAK decomposition in Theorem 7.39. Cartan [1927b] introduces restricted roots. The introduction of N in §4 is due to Iwasawa [1949], and the decomposition given as Theorem 6.46 appears in the same paper. Lemma 6.44 came after Iwasawa's original proof and appears as Lemma 26 of Harish-Chandra [1953]. Cartan [1927b] uses the group $W(G, A)$ of §5, and Theorem 6.57 is implicit in that paper.

It was apparent from the work of Harish-Chandra and Gelfand-Graev in the early 1950s that Cartan subalgebras would play an important role in harmonic analysis on semisimple Lie groups. The results of §6 appear in Kostant [1955] and Harish-Chandra [1956a]. Kostant [1955] announces the existence of a classification of Cartan subalgebras up to conjugacy, but the appearance of Harish-Chandra [1956a] blocked the appearance of proofs for the results of that paper. Sugiura [1959] states and proves the classification.

In effect Cayley transforms as in §7 appear in Harish-Chandra [1957], §2. For further information, see the Notes for §VII.9.

In §8 the name "Vogan diagram" is new. In the case that $\mathfrak{a}_0 = 0$, the idea of adapting a system of positive roots to given data was present in the late 1960s and early 1970s in the work of Schmid on discrete series representations (see Schmid [1975], for example), and a Vogan diagram could capture this idea in a picture. Vogan used the same idea in the mid 1970s for general maximally compact Cartan subalgebras. He introduced the notion of a θ stable parabolic subalgebra of $\mathfrak{g}$ to handle representation-theoretic data and used the diagrams to help in understanding these subalgebras. The paper [1979] contains initial results from this investigation but no diagrams.

Because of Theorem 6.74 Vogan diagrams provide control in the problem of classifying simple real Lie algebras. This theorem was perhaps understood for a long time to be true, but Knapp [1996] gives a proof. Theorem 6.88 is due to Vogan.

The results of §9 were already recognized in Cartan [1914]. The classification in §10, as was said earlier, is in Cartan [1914]; it is the result of a remarkable computation made before the discovery of the Cartan involution. Lie algebras of each complex type are to be classified in that paper, and the signature of the Killing form is the key invariant. The classification is recalled in Cartan [1927a], and $\mathfrak{k}_0$ is identified in each case. In this paper Cartan provided a numbering for the noncomplex noncompact simple real Lie algebras. This numbering has been retained by Helgason [1978], and we use the same numbering for the exceptional cases in Figures 6.2 and 6.3.

Cartan [1927b] improves the classification by relating Lie algebras and geometry. This paper contains tables giving more extensive information about the exceptional Lie algebras. Gantmacher [1939a] and [1939b] approached classification as a problem in classifying automorphisms and then succeeded in simplifying the proof of classification. This method was further simplified by Murakami [1965] and Wallach [1966] and [1968] independently. Murakami and Wallach made use of the Borel and de Siebenthal Theorem (Borel and de Siebenthal [1949]), which is similar to Theorem 6.96 but slightly different. The original purpose of the theorem was to find a standard form for automorphisms, and Murakami and Wallach both use the theorem that way. Helgason [1978] gives a proof of classification that is based on classifying automorphisms in a different way. The paper Knapp [1996] gives the quick proof of Theorem 6.96 and then deduces the classification as a consequence of Theorem 6.74; no additional consideration of automorphisms is needed.

The above approaches to classification make use of a maximally compact Cartan subalgebra. An alternate line of attack starts from a maximally noncompact Cartan subalgebra and is the subject of Araki [1962]. The classification is stated in terms of "Satake diagrams," which are described by Helgason [1978], 531. Problem 7 at the end of Chapter VI establishes the facts due to Satake [1960] needed to justify the definition of a Satake diagram.

The information in (6.107) and (6.108) appears in Cartan [1927b]. Appendix C shows how this information can be obtained from Vogan diagrams.

Chapter VII

§1. The essence of Theorem 7.8 is already in Cartan [1925b]. Goto [1948] and Mostow [1950] investigated conditions that ensure that an analytic subgroup is closed. The circle of ideas in this direction in §1 is based ultimately on Goto's work. The unitary trick is due to Weyl [1925–26] and consists of two parts—the existence of compact real forms and the comparison of $\mathfrak{g}$ and $\mathfrak{u}_0$.

§2. The necessity for considering reductive groups emerged from the work of Harish-Chandra, who for a semisimple group G was led to form a series of infinite-dimensional representations constructed from the M of each cuspidal parabolic subgroup. The subgroup M is not necessarily semisimple, however, and it was helpful to have a class of groups that would include a rich supply of semisimple groups G and would have the property that the M of each cuspidal parabolic subgroup of G is again in the class. Various classes have been proposed for this purpose. The Harish-Chandra class is the class defined by axioms in §3 of Harish-Chandra [1975], and its properties are developed in the first part of that paper. We have used axioms from Knapp-Vogan [1995], based on Vogan [1981]. These axioms, though more complicated to state than Harish-Chandra's axioms, have the advantage of being easier to check. The present axioms yield a slightly larger class of groups than Harish-Chandra's, according to Problem 2 at the end of the chapter.

§3. The existence of the KAK decomposition is in Cartan [1927b]. See the Notes for Chapter VI.

§4. The Bruhat decomposition was announced for complex classical groups and their real forms in Bruhat [1954a] and [1954b]. Harish-Chandra [1954], citing Bruhat, announced a proof valid for all simple Lie groups, and Harish-Chandra [1956b] gives the proof. Bruhat [1956] repeats Harish-Chandra's proof.

§5. The group M does not seem to appear in Cartan's work, but it appears throughout Harish-Chandra's work. Some of its properties are developed in Harish-Chandra [1958], Satake [1960], and Moore [1964a]. A version of Theorem 7.53 appears in Satake [1960], Lemma 9, and Moore [1964a], Lemmas 1 and 3. See also Knapp-Zuckerman [1982], §2. Theorem 7.55 seems to have been discovered in the late 1960s. See Loos [1969b], Theorems 3.4a and 3.6 on pp. 75–77, for the key step that $2\{H \in i\mathfrak{a}_0 \mid \exp H \in K\}$ is contained in the lattice generated by the vectors $4\pi i|\beta|^{-2}H_\beta$; this step comes out of the work in Cartan [1927b]. The proof that we give, based on Theorem 5.107, is new.

§6. Real-rank-one subgroups appear in Araki [1962]. Gindikin-Karpelevič [1962] shows that integrals $\int_{N^-} e^{-(\lambda+\rho)H(\bar{n})}\,d\bar{n}$, where $x = \kappa(x)e^{H(x)}n$ is the Iwasawa decomposition of x and ρ is half the sum of the positive restricted roots, can be computed in terms of integrals for the real-rank-one subgroups. Theorem 7.66 was known case by case at least as in the early 1950s. The proof here, independent of classification, is from Knapp [1975].

§7. Parabolic subgroups, particularly cuspidal parabolic subgroups, play an important role in the work of Harish-Chandra on harmonic analysis on semisimple Lie groups. For some information about parabolic subgroups, see Satake [1960] and Moore [1964a]. Much of the material of this section appears in Harish-Chandra [1975]. Harish-Chandra was the person to introduce the name "Langlands decomposition" for parabolic subgroups. For Proposition 7.110 and Corollary 7.111, see Knapp-Zuckerman [1982], §2.

§8. Most of the material deriving properties of Cartan subgroups from Cartan subalgebras is based on Harish-Chandra [1956a]. That paper contains an error that is noted in the *Collected Papers*, but the error can be accommodated. Also that paper uses a definition of Cartan subgroup that Harish-Chandra modified later. We use the later definition here.

§9. Spaces G/K for which G/K embeds as a bounded domain in some $\mathbb{C}^n$ with G operating holomorphically were studied and classified by Cartan [1935]. The classification is summarized in (7.147). Hua [1963] develops at length properties of the domains of this kind corresponding to classical groups G. The proof of Theorem 7.117 is based on material in Helgason [1962], 354 and 304–322, and Knapp [1972], as well as a suggestion of J.-i. Hano. Theorem 7.129 and the accompanying lemmas are due to Harish-Chandra [1955–56].

Problem 41 is based on the proof of a conjecture of Bott and Korányi by Moore [1964b]. The hint for the solution of Problem 42 shows why the Cayley transforms of Chapter VI are so named. For more about Problems 42–44, see Korányi-Wolf [1965] and Wolf-Korányi [1965]. For further discussion, see Helgason [1994], §V.4.

Chapter VIII

§1. The development of integration of differential forms is taken from Chevalley [1946] and Helgason [1962]. The proof of Sard's Theorem is taken from Sternberg [1964], 47–49.

§2. Invariant integration on Lie groups, defined in terms of differential forms, is already mentioned in a footnote on the second page of Cartan [1925a]. Existence of a left-invariant measure on a general locally compact group was proved under a separability assumption by Haar [1933], and uniqueness was proved by von Neumann [1934a]. See Weil [1940], Loomis [1953], Hewitt-Ross [1963], and Nachbin [1965] for later developments and refinements.

§3. Theorem 8.32 and its proof are from Bourbaki [1963], 66. The proof of Lemma 8.35 is taken from Helgason [1962]. Knapp-Vogan [1995], 661–663, explains how the natural objects to integrate over G/H are functions on G that are "densities" relative to H. The condition that $\Delta_G|_H = \Delta_H$ forces densities to be right invariant under H, and then they descend to functions on G/H.

§4. This section follows the lines of §V.6 of Knapp [1986]. Proposition 8.46 is due to Harish-Chandra [1958], 287. The technique of proof given here occurs in Kunze-Stein [1967], Lemma 13. Use of densities (see above) makes this proof look more natural; see Knapp-Vogan [1995], 663. Helgason's Theorem is from Helgason [1970], §III.3. Warner [1972a], 210, calls the result the "Cartan-Helgason Theorem." Possibly the mention of Cartan is a reference to Cartan [1929b], 238–241, but Cartan's treatment is flawed. Cartan's work was redone by Harish-Chandra and Sugiura. Harish-Chandra [1958], §2, proved that if ν is the highest restricted weight of an irreducible finite-dimensional representation with a K fixed vector, then $\langle \nu, \beta\rangle/|\beta|^2$ is an integer ≥ 0 for every positive restricted root. Sugiura [1962] proved conversely that any ν such that $\langle \nu, \beta\rangle/|\beta|^2$ is an integer ≥ 0 for every positive restricted root is the highest restricted weight of some irreducible finite-dimensional representation with a K fixed vector. See also Wallach [1971] and [1972] and Lepowsky-Wallach [1973]. For further discussion, see Helgason [1994], §II.4.

§5. The Weyl Integration Formula in the compact case is due to Weyl [1925–26]. The proof is rather rapid. One may find a proof also in Adams [1969]. The formula in the noncompact case is due to Harish-Chandra [1965], Lemma 41, and [1966], Lemma 91; the proof is omitted, being similar to the proof in the compact case.

Appendix A

The material in §§1–3 of this appendix is taken from Knapp [1988], Chapter II.

Appendix B

Representations on vector spaces over $\mathbb{R}$ were considered by Cartan but not by Weyl, and the question of their complete reducibility does not seem to have been addressed. We have taken Lemma B.1 and its proof from Helgason [1984], 601–602, and Helgason [1978] in turn quotes Freudenthal and de Vries [1969] on this point.

The decomposition now called the Levi decomposition (Theorem B.2) was announced by Killing [1888–89–90]. In one of the announcements preceding Cartan [1894], Cartan notes errors in Killing's argument but affirms that the result is true. The first published correct proof appears to be the one in Levi [1905], valid over $\mathbb{C}$. Whitehead [1936] gives a proof valid over $\mathbb{R}$ as well; see Jacobson [1962] for an exposition. The semisimple subalgebra is unique up to conjugacy, according to Malcev [1945]. The proof we have given for Theorem B.2 is from Bourbaki [1960], 89–90, and is reproduced in Fulton-Harris [1991]. A proof of the Malcev theorem also appears in Bourbaki [1960].

The global form of Lie's third theorem in Theorem B.7 is in Cartan [1930]. The proof here is taken from lectures by Kostant.

Lie believed but could not prove that every finite-dimensional Lie algebra (over $\mathbb{C}$) can be realized as a Lie algebra of matrices. Ado [1935] and [1947] finally proved that Lie's conjecture was correct. **Ado's Theorem**, as it is called, is valid over any field of characteristic 0. Other proofs were given by Cartan [1938a] and Harish-Chandra [1949].

Another topic omitted from Chapter I is the **Campbell-Hausdorff formula**. This formula expresses $\exp X \exp Y$ in the form $\exp W$ for X and Y in a suitably small neighborhood of 0 and may be regarded as a sharpening of Lemma 1.92a. We have omitted the result since its known uses with reductive Lie groups are quite limited. References to the work on this formula by Campbell, Baker, Hausdorff, and Dynkin may be found in Bourbaki [1972], 301–302. For a treatment of the result and its proof, see Hochschild [1965].

Appendix C

The information in §§1–2 was all known to Cartan, most of it as early as 1913. For tables giving this information and more, see Bourbaki [1968].

Much of the information in §§3–4 appears in Cartan [1927b]. Tables in that paper give $\mathfrak{k}_0$, a set of simple roots for $\mathfrak{k}_0$, the real rank, the system of restricted roots, and the multiplicities of each restricted root. Cartan's way of obtaining simple roots for $\mathfrak{k}_0$ is different from what has been used here; see Murakami [1965] for an exposition. In addition, Cartan [1927b] tells the order of the center of a simply connected group with each Lie algebra.

Wolf [1965] classified those simple real Lie algebras $\mathfrak{g}_0$ for which G/K has a reasonable quaternionic structure. In §4 a notation is made which $\mathfrak{g}_0$'s have this property. For further discussion of this matter, see Aleksseevskii [1968], Sudbery [1979], and Besse [1987].

Further Topics

Realizations of representations of compact Lie groups. Borel and Weil (see Serre [1954]) and Tits [1955], 112–113, independently discovered an explicit construction of the irreducible representations of compact connected Lie groups. The realizations are geometric ones, in terms of spaces of holomorphic sections of holomorphic line bundles, and the result goes under the name **Borel-Weil Theorem**. An exposition appears in Knapp [1986], §V.7.

At about the same time as the work of Borel and Weil and Tits, Harish-Chandra [1955–56] independently introduced "holomorphic discrete series" representations of semisimple Lie groups G as generalizations of some known representations of $SL(2, \mathbb{R})$. Harish-Chandra's construction works under the assumption of §VII.9 that $Z_{\mathfrak{g}}(\mathfrak{c}) = \mathfrak{k}$, which is valid in particular in the special case that G is compact. In this special case Harish-Chandra's construction reduces to the Borel-Weil Theorem.

Bott [1957] generalized the construction in the Borel-Weil Theorem to allow other realizations in spaces of sheaf cohomology sections (or equivalently Dolbeault cohomology sections). This generalization goes under the name **Bott-Borel-Weil Theorem**, and an exposition appears in Baston-Eastwood [1989]. This theorem is more or less equivalent with an algebraic theorem of Kostant's (see Kostant [1961] and Cartier [1961]). See Knapp-Vogan [1995] for an exposition of Kostant's Theorem and for further discussion.

Linear algebraic groups. The possibility of defining matrix groups over fields other than $\mathbb{R}$ and $\mathbb{C}$ has led to a large theory of linear algebraic groups. Some books on this subject are Chevalley [1951] and [1955], Borel [1969], and Springer [1981].

Representations of reductive Lie groups. The theory in this book leads naturally to the infinite-dimensional representation theory of reductive Lie groups. For orientation, see Knapp [1986]. The first book on the subject was Gelfand-Naimark [1950]. Other books in this field are Warner [1972a] and

[1972b], Vogan [1981], Wallach [1988] and [1992], and Knapp-Vogan [1995]. A book giving a sense of ongoing research is Vogan [1987].

Analysis on symmetric spaces and related spaces. The theory in this book leads naturally also to a field of analysis in settings that involve semisimple or reductive groups. Some of this work, but not all, makes use of some infinite-dimensional representation theory. Some books on the subject are Wallach [1973], Helgason [1984] and [1994], Schlichtkrull [1984], and Varadarajan [1989].

REFERENCES

In the text of this book, a name followed by a bracketed year points to the list that follows. The date is followed by a letter in case of ambiguity.

Adams, J. F., *Lectures on Lie Groups*, W. A. Benjamin, New York, 1969. Reprinted edition: University of Chicago Press, Chicago, 1982.

Ado, I. D., Note on the representation of finite continuous groups by means of linear substitutions, *Bull. Phys. Math. Soc. Kazan* 7 (1935), 3–43 (Russian).

Ado, I. D., The representation of Lie algebras by matrices, *Uspekhi Mat. Nauk* 2 (1947), No. 6 (22), 159-173 (Russian). English translation: *Amer. Math. Soc. Translations* (1) 9 (1962), 308-327.

Aleksseevskii, D. V., Compact quaternionic spaces, *Funktsionalnyi Analiz i Ego Prilozheniya* 2 (1968), No. 2, 11–20 (Russian). English translation: *Functional Anal. and Its Appl.* 2 (1968), 106–114.

Araki, S., On root systems and an infinitesimal classification of irreducible symmetric spaces, *J. of Math., Osaka City Univ.* 13 (1962), 1–34.

Artin, E., The influence of J. H. M. Wedderburn on the development of modern algebra, *Bull. Amer. Math. Soc.* 56 (1950), 65–72. (= *Collected Papers*, Springer-Verlag, New York, 1965, 526–533.)

Artin, E., *Geometric Algebra*, Interscience, New York, 1957.

Artin, M., *Algebra*, Prentice-Hall, Englewood Cliffs, N.J., 1991.

Baston, R. J., and M. G. Eastwood, *The Penrose Transform: Its Interaction with Representation Theory*, Oxford University Press, Oxford, 1989.

Berger, M., Les espaces symétriques non compacts, *Annales Sci. École Norm. Sup.* 74 (1957), 85-177.

Bernstein, I. N., I. M. Gelfand, and S. I. Gelfand, Structure of representations generated by highest weight vectors, *Funktsionalnyi Analiz i Ego Prilozheniya* 5 (1971), No. 1, 1–9 (Russian). English translation: *Functional Analysis and Its Applications* 5 (1971), 1–8.

Besse, A. L., *Einstein Manifolds*, Springer-Verlag, New York, 1987.

Birkhoff, Garrett, Representability of Lie algebras and Lie groups by matrices, *Annals of Math.* 38 (1937), 526–533.

Borel, A., *Linear Algebraic Groups*, W. A. Benjamin, New York, 1969.

Borel, A., Groupes linéaires algébriques, *Annals of Math.* 64 (1956), 20–82. (= *Oeuvres*, Vol. I, 490–552.)

Borel, A., and J. de Siebenthal, Les sous-groupes fermés de rang maximum des groupes de Lie clos, *Comment. Math. Helv.* 23 (1949), 200–221. (= Borel, *Oeuvres*, Vol. I, 11–32.)

Borel, A., *Oeuvres, Collected Papers*, I, II, III, Springer-Verlag, Berlin, 1983.

Bott, R., Homogeneous vector bundles, *Annals of Math.* 66 (1957), 203–248.

Bourbaki, N., *Eléments de Mathématique, Livre II, Algèbre: Chapitre 8, Modules et Anneaux Semi-Simples*, Actualités scientifiques et industrielles 1261, Hermann, Paris, 1958.

Bourbaki, N., *Eléments de Mathématique, Groupes et Algèbres de Lie: Chapitre 1*, Actualités scientifiques et industrielles 1285, Hermann, Paris, 1960.

Bourbaki, N., *Eléments de Mathématique, Livre VI, Intégration: Chapitre 7, Mesur de Haar, Chapitre 8, Convolution et Représentations*, Hermann, Paris, 1963.

Bourbaki, N., *Eléments de Mathématique, Groupes et Algèbres de Lie: Chapitres 4, 5 et 6*, Actualités scientifiques et industrielles 1337, Hermann, Paris, 1968.

Bourbaki, N., *Eléments de Mathématique, Groupes et Algèbres de Lie: Chapitres II et III*, Hermann, Paris, 1972.

Brauer, R., Eine Bedingung für vollständige Reduzibilität von Darstellungen gewöhnlicher und infinitesimaler Gruppen, *Math. Zeitschrift* 41 (1936), 330–339. (= *Collected Papers*, Vol. III, MIT Press, Cambridge, Mass., 1980, pp. 462–471.)

Bremmer, M. R., R. V. Moody, and J. Patera, *Tables of Dominant Weight Multiplicities for Representations of Simple Lie Algebras*, Marcel Dekker, New York, 1985.

Bruhat, F., Représentations induites des groupes de Lie semi-simples complexes, *C. R. Acad. Sci. Paris* 238 (1954a), 437–439.

Bruhat, F., Représentations induites des groupes de Lie semi-simples réels, *C. R. Acad. Sci. Paris* 238 (1954b), 550–553.

Bruhat, F., Sur les représentations induites des groupes de Lie, *Bull. Soc. Math. France* 84 (1956), 97–205.

Burnside, W., On the representations of a group of finite order as an irreducible group of linear substitutions and the direct establishment of the relations between the group-characteristics, *Proc. London Math. Soc.* (2) 1 (1904), 117–123.

Cartan, É., Sur la structure des groupes de transformations finis et continus, *Thèse*, Nony, Paris, 1894. Second edition: Vuibert, 1933. (= *Œuvres Complètes*, I, 137–287.)

Cartan, É., Les groupes projectifs qui ne laissent invariante aucune multiplicité plane, *Bull. Soc. Math. France* 41 (1913), 53–96. (= *Œuvres Complètes*, I, 355–398.)

Cartan, É., Les groupes réels simples finis et continus, *Annales Sci. École Norm. Sup.* 31 (1914), 263–355. (= *Œuvres Complètes*, I, 399–491.)

Cartan, É., Les tenseurs irréductibles et les groupes linéaires simples et semi-simples, *Bull. des Sci. Math.* 49 (1925a), 130–152. (= *Œuvres Complètes*, I, 531–553.)

Cartan, É., Le principe de dualité et la théorie des groupes simples et semi-simples, *Bull. des Sci. Math.* 49 (1925b), 361–374. (= *Œuvres Complètes*, I, 555–568.)

Cartan, É., Sur une classe remarquable d'espaces de Riemann, *Bull. Soc. Math. France* 55 (1927a), 114–134. (= *Œuvres Complètes*, I, 639–659.)

Cartan, É., Sur certaines formes riemanniennes remarquables des géométries à groupe fondamental simple, *Annales Sci. École Norm. Sup.* 44 (1927b), 345–467. (= *Œuvres Complètes*, I, 867–989.)

Cartan, É., Groupes simples clos et ouverts et géométrie riemannienne, *J. de Math. Pures et Appliquées* 8 (1929a), 1–33. (= *Œuvres Complètes*, I, 1011–1043.)

Cartan, É., Sur la détermination d'un système orthogonal complet dans un espace de Riemann symétrique clos, *Rend. Circ. Mat. Palermo* 53 (1929b), 217–252. (= *Œuvres Complètes*, I, 1045–1080.)

Cartan, É., Le troisième théorème fondamental de Lie, *C. R. Acad. Sci. Paris* 190 (1930), 914–916 and 1005–1007. (= *Œuvres Complètes*, I, 1143–1148.)

Cartan, É., Sur les domaines bornés homogènes de l'espace de n variables complexes, *Abhandlungen aus dem Mathematischen Seminar der Hamburgischen Universität* 11 (1935), 116–162. (= *Œuvres Complètes*, I, 1259–1305.)

Cartan, É., Les représentations linéaires des groupes de Lie, *J. de Math. Pures et Appliquées* 17 (1938a), 1–12; correction 17 (1938), 438. (= *Œuvres Complètes*, I, 1339–1351.)

Cartan, É., *Leçons sur la Théorie des Spineurs* I, II, Hermann, Paris, 1938b. English translation: *The Theory of Spinors*, Hermann, Paris, 1966; reprinted by Dover, New York, 1981.

Cartan, É., *Œuvres Complètes*, I, II, III, Gauthiers-Villars, Paris, 1952. Reprinted: Centre National de la Recherche Scientifique, Paris, 1984.

Cartier, P., Réprésentations linéaires des algèbres de Lie semi-simples, Exposé 17, Séminaire "Sophus Lie," *Théorie des algèbres de Lie, topologie des groupes de Lie, 1954–55*, École Normale Supérieure, Paris, 1955a.

Cartier, P., Structure topologique des groupes de Lie généraux, Exposé 22, Séminaire "Sophus Lie," *Théorie des algèbres de Lie, topologie des groupes de Lie, 1954–55*, École Normale Supérieure, Paris, 1955b.

Cartier, P., Remarks on "Lie algebra cohomology and the generalized Borel-Weil theorem," by B. Kostant, *Annals of Math.* 74 (1961), 388–390.

Casimir, H., and B. L. van der Waerden, Algebraischer Beweis der vollständigen Reduzibilität der Darstellungen halbeinfacher Liescher Gruppen, *Math. Annalen* 111 (1935), 1–12.

Chevalley, C., *Theory of Lie Groups I*, Princeton University Press, Princeton, 1946.

Chevalley, C., Sur la classification des algèbres de Lie simples et de leurs représentations, *C. R. Acad. Sci. Paris* 227 (1948a), 1136–1138.

Chevalley, C., Sur les représentations des algèbres de Lie simples, *C. R. Acad. Sci. Paris* 227 (1948b), 1197.

Chevalley, C., *Théorie des Groupes de Lie, Tome II, Groupes Algébriques*, Actualités scientifiques et industrielles 1152, Hermann, Paris, 1951.

Chevalley, C., *Théorie des Groupes de Lie, Tome III, Théorèmes généraux sur les algèbres de Lie*, Actualités scientifiques et industrielles 1226, Hermann, Paris, 1955.

Chevalley, C., Invariants of finite groups generated by reflections, *Amer. J. Math.* 77 (1955), 778–782.

Cohn, P. M., *Lie Groups*, Cambridge University Press, London, 1957.

Coxeter, H. S. M., Discrete groups generated by reflections, *Annals of Math.* 35 (1934), 588–621.

Dixmier, J., *Algèbres Enveloppantes*, Gauthier-Villars Editeur, Paris, 1974. English translation: *Enveloping Algebras*, North-Holland Publishing Co., Amsterdam, 1977.

Dynkin, E., Classification of the simple Lie groups, *Mat. Sbornik* 18 (60) (1946), 347–352 (Russian with English summary).

Dynkin, E. B., The structure of semisimple algebras, *Uspekhi Mat. Nauk* 2 (1947), No. 4 (20), 59–127 (Russian). English translation: *Amer. Math. Soc. Translations* (1) 9 (1962), 328–469.

Engel, F., Sur un groupe simple à quatorze paramètres, *C. R. Acad. Sci. Paris* 116 (1893), 786–788.

Freudenthal, H., and H. de Vries, *Linear Lie Groups*, Academic Press, New York, 1969.

Frobenius, F. G., Über Gruppencharaktere, *Sitzungsberichte der Königlich Preussischen Akademie der Wissenschaften zu Berlin* (1896), 985–1021. (= *Gesammelte Abhandlungen*, Vol. III, Springer-Verlag, Berlin, 1968, pp. 1–37.)

Fulton, W., and J. Harris, *Representation Theory, A First Course*, Springer-Verlag, New York, 1991.

Gantmacher, F., Canonical representation of automorphisms of a complex semi-simple Lie group, *Mat. Sbornik* 5 (47) (1939a), 101–146 (English with Russian summary).

Gantmacher, F., On the classification of real simple Lie groups, *Mat. Sbornik* 5 (47) (1939b), 217–250 (English with Russian summary).

Gelfand, I. M., The center of an infinitesimal group ring, *Mat. Sbornik N.S.* 26 (68) (1950), 103–112 (Russian).

Gelfand, I. M., and M. A. Naimark, *Unitary Representations of the Classical Groups*, Trudy Mat. Inst. Steklov 36, Moskow-Leningrad, 1950 (Russian). German translation: Akademie-Verlag, Berlin, 1957.

Gindikin, S. G., and F. I. Karpelivič, Plancherel measure for Riemann symmetric spaces of nonpositive curvature, *Doklady Akademiia Nauk SSSR* 145 (1962), 252–255 (Russian). English translation: *Soviet Math. Doklady* 3 (1962), 962–965.

Godement, R., A theory of spherical functions I, *Trans. Amer. Math. Soc.* 73 (1952), 496–556.

Goto, M., Faithful representations of Lie groups I, *Math. Japonicae* 1 (1948), 107–119.

Haar, A., Der Maassbegriff in der Theorie der Kontinuierlichen Gruppen, *Annals of Math.* 34 (1933), 147–169.

Harish-Chandra, Faithful representations of Lie algebras, *Annals of Math.* 50 (1949), 68–76. (= *Collected Papers*, Vol. I, 222-230.)

Harish-Chandra, On some applications of the universal enveloping algebra of a semisimple Lie algebra, *Trans. Amer. Math. Soc.* 70 (1951), 28–96. (= *Collected Papers*, Vol. I, 292-360.)

Harish-Chandra, Representations of a semisimple Lie group on a Banach space I, *Trans. Amer. Math. Soc.* 75 (1953), 185–243. (= *Collected Papers*, Vol. I, 391-449.)

Harish-Chandra, Representations of semisimple Lie groups V, *Proc. Nat. Acad. Sci. USA* 40 (1954), 1076–1077. (= *Collected Papers*, Vol. I, 561-563.)

Harish-Chandra, Representations of semisimple Lie groups IV, V, VI *Amer. J. Math.* 77 (1955), 743–777; 78 (1956), 1–41; 78 (1956), 564–628. (= *Collected Papers*, Vol. II, 13-154.)

Harish-Chandra, The characters of semisimple Lie groups, *Trans. Amer. Math. Soc.* 83 (1956a), 98–163. (= *Collected Papers*, Vol. II, 156-221.)

Harish-Chandra, On a lemma of F. Bruhat, *J. de Math. Pures et Appliquées* 35 (1956b), 203–210. (= *Collected Papers*, Vol. II, 223-330.)

Harish-Chandra, Fourier transforms on a semisimple Lie algebra II, *Amer. J. Math.* 79 (1957), 653–686. (= *Collected Papers*, Vol. II, 343-376.)

Harish-Chandra, Spherical functions on a semisimple Lie group I, *Amer. J. Math.* 80 (1958), 241–310. (= *Collected Papers*, Vol. II, 409-478.)

Harish-Chandra, Invariant eigendistributions on a semisimple Lie group, *Trans. Amer. Math. Soc.* 119 (1965), 457–508. (= *Collected Papers*, Vol. III, 351-402.)

Harish-Chandra, Discrete series for semisimple Lie groups II, *Acta Math.* 116 (1966), 1–111. (= *Collected Papers*, Vol. III, 537-647.)

Harish-Chandra, Harmonic analysis on real reductive groups I, *J. Func. Anal.* 19 (1975), 104–204. (= *Collected Papers*, Vol. IV, 102-202.)

Harish-Chandra, *Collected Papers*, I, II, III, IV, Springer-Verlag, New York, 1984.

Helgason, S., *Differential Geometry and Symmetric Spaces*, Academic Press, New York, 1962.

Helgason, S., A duality for symmetric spaces with applications to group representations, *Advances in Math.* 5 (1970), 1–154.

Helgason, S., *Differential Geometry, Lie Groups, and Symmetric Spaces*, Academic Press, New York, 1978.

Helgason, S., *Groups and Geometric Analysis, Integral Geometry, Invariant Differential Operators, and Spherical Functions*, Academic Press, Orlando, Fla., 1984.

Helgason, S., *Geometric Analysis on Symmetric Spaces*, Mathematical Surveys and Monographs, Vol. 39, American Mathematical Society, Providence, R. I., 1994.

Hewitt, E., and K. A. Ross, *Abstract Harmonic Analysis*, Vol. I, Springer-Verlag, Berlin, 1963. Second edition: Springer-Verlag, New York, 1979.

Hochschild, G., *The Structure of Lie Groups*, Holden-Day, San Francisco, 1965.

Hoffman, K., and R. Kunze, *Linear Algebra*, Prentice-Hall, Englewood Cliffs, N.J., 1961. Second edition: Prentice-Hall, 1971.

Hopf, H., Über den Rang geschlossener Liescher Gruppen, *Comment. Math. Helv.* 13 (1940–41), 119–143.

Hopf, H., Maximale Toroide und singuläre Elemente in geschlossenen Lieschen Gruppen, *Comment. Math. Helv.* 15 (1942–43), 59–70.

Hua, L.-K., *Harmonic Analysis of Functions of Several Complex Variables in the Classical Domains*, American Mathematical Society, Providence, R.I., 1963; revised 1979.

Humphreys, J. E., *Introduction to Lie Algebras and Representation Theory*, Springer-Verlag, New York, 1972.

Humphreys, J. E., *Reflection Groups and Coxeter Groups*, Cambridge University Press, Cambridge, 1990.

Hunt, G. A., A theorem of Élie Cartan, *Proc. Amer. Math. Soc.* 7 (1956), 307–308.

Hurwitz, A., Ueber die Erzeugung der Invarianten durch Integration, *Nachrichten von der Königlichen Gesellschaft der Wissenschaften zu Göttingen* (1897), 71–90. (= *Mathematische Werke*, Vol. II, Emil Birkhäuser, Basel, 1933, pp. 546–564.)

Iwasawa, K., On some types of topological groups, *Annals of Math.* 50 (1949), 507–558.

Jacobson, N., *Lie Algebras*, Interscience Publishers, New York, 1962.

Killing, W., Die Zusammensetzung der stetigen endlichen Transformationsgruppen I, II, III, IV, *Math. Annalen* 31 (1888), 252–290; 33 (1889), 1–48; 34 (1889), 57–122; 36 (1890), 161–189.

Knapp, A. W., Bounded symmetric domains and holomorphic discrete series, *Symmetric Spaces*, W. M. Boothby and G. L. Weiss (ed.), Marcel Dekker, New York, 1972, 211–246.

Knapp, A. W., Weyl group of a cuspidal parabolic, *Annales Sci. École Norm. Sup.* 8 (1975), 275–294.

Knapp, A. W., *Representation Theory of Semisimple Groups: An Overview Based on Examples*, Princeton University Press, Princeton, 1986.

Knapp, A. W., *Lie Groups, Lie Algebras, and Cohomology*, Princeton University Press, Princeton, 1988.

Knapp, A. W., A quick proof of the classification of simple real Lie algebras, *Proc. Amer. Math. Soc.*124 (1996), in press for October.

Knapp, A. W., and D. A. Vogan, *Cohomological Induction and Unitary Representations*, Princeton University Press, Princeton, 1995.

Knapp, A. W., and G. J. Zuckerman, Classification of irreducible tempered representations of semisimple groups, *Annals of Math.* 116 (1982), 389–501; typesetter's correction 119 (1984), 639.

Koecher, M., An elementary approach to bounded symmetric domains, duplicated lecture notes, Rice University, Houston, 1969.

Korányi, A., and J. A. Wolf, Realization of Hermitian symmetric spaces as generalized half-planes, *Annals of Math.* 81 (1965), 265–288.

Kostant, B., On the conjugacy of real Cartan subalgebras I, *Proc. Nat. Acad. Sci. USA* 41 (1955), 967–970.

Kostant, B., A formula for the multiplicity of a weight, *Trans. Amer. Math. Soc.* 93 (1959), 53–73.

Kostant, B., Lie algebra cohomology and the generalized Borel-Weil theorem, *Annals of Math.* 74 (1961), 329–387.

Kostant, B., Verma modules and the existence of quasi-invariant differential operators, *Non-Commutative Harmonic Analysis*, Lecture Notes in Mathematics, Vol. 466, Springer-Verlag, Berlin, 1975, pp. 101–128.

Kunze, R. A., and E. M. Stein, Uniformly bounded representations III, *Amer. J. Math.* 89 (1967), 385–442.

Lawson, H. B., and M.-L. Michelsohn, *Spin Geometry*, Princeton University Press, Princeton, 1989.

Lepowsky, J., and N. R. Wallach, Finite- and infinite-dimensional representations of linear semisimple groups, *Trans. Amer. Math. Soc.* 184 (1973), 223–246.

Levi, E. E., Sulla struttura dei gruppi finiti e continui, *Atti della Reale Accademia delle Scienze di Torino* 40 (1905), 423–437 (551–565).

Lichtenberg, D. B., *Unitary Symmetry and Elementary Particles*, Academic Press, New York, 1970; 2nd ed., 1978.

Lie, S., with the help of F. Engel, *Theorie der Transformationsgruppen I, II, III*, B. G. Teubner, Leipzig, 1888, 1890, 1893. (Engel's role is listed as "Unter Mitwirkung von Dr. Friedrich Engel.")

Loomis, L. H., *An Introduction to Abstract Harmonic Analysis*, D. Van Nostrand, New York, 1953.

Loos, O., *Symmetric Spaces*, Vol. I, W. A. Benjamin Inc., New York, 1969a.

Loos, O., *Symmetric Spaces*, Vol. II, W. A. Benjamin Inc., New York, 1969b.

Malcev, A., On the theory of Lie groups in the large, *Mat. Sbornik* 16 (58) (1945), 163–190 (English with Russian summary).

McKay, W. G., and J. Patera, *Tables of Dimensions, Indices, and Branching Rules for Representations of Simple Lie Algebras*, Marcel Dekker, New York, 1981.

Montgomery, D., and L. Zippin, *Topological Transformation Groups*, Interscience, New York, 1955.

Moore, C. C., Compactifications of symmetric spaces, *Amer. J. Math.* 86 (1964a), 201–218.

Moore, C. C., Compactifications of symmetric spaces II, *Amer. J. Math.* 86 (1964b), 358–378.

Mostow, G. D., A new proof of E. Cartan's theorem on the topology of semi-simple groups, *Bull. Amer. Math. Soc.* 55 (1949), 969–980.

Mostow, G. D., The extensibility of local Lie groups of transformations and groups on surfaces, *Annals of Math.* 52 (1950), 606–636.

Mostow, G. D., Self-adjoint groups, *Annals of Math.* 62 (1955), 44–55.

Murakami, S., Sur la classification des algèbres de Lie réelles et simples, *Osaka J. Math.* 2 (1965), 291–307.

Nachbin, L., *The Haar Integral*, D. Van Nostrand, Princeton, 1965.

Parthasarathy, K. R., R. Ranga Rao, and V. S. Varadarajan, Representations of complex semi-simple Lie groups and Lie algebras, *Annals of Math.* 85 (1967), 383–429.

Peter, F., and H. Weyl, Die Vollständigkeit der primitiven Darstellungen einer geschlossenen kontinuierlichen Gruppe, *Math. Annalen* 97 (1927), 737–755. (= Weyl, *Gesammelte Abhandlungen*, Vol. III, 58–75.)

Poincaré, H., Sur les groupes continus, *C. R. Acad. Sci. Paris* 128 (1899), 1065–1069.

Poincaré, H., Sur les groupes continus, *Trans. Cambridge Philosophical Soc.* 18 (1900), 220–255.

Pontrjagin, L., *Topological Groups*, Princeton University Press, Princeton, 1939. Second edition: Gordon and Breach, New York, 1966.

Riesz, F., and B. Sz. Nagy, *Functional Analysis*, Frederick Ungar Publishing Co., New York, 1955.

Satake, I., On representations and compactifications of symmetric Riemannian spaces, *Annals of Math.* 71 (1960), 77–110.

Schlichtkrull, H., *Hyperfunctions and Harmonic Analysis on Symmetric Spaces*, Birkhäuser, Boston, 1984.

Schmid, W., On the characters of discrete series: the Hermitian symmetric case, *Invent. Math.* 30 (1975), 47–144.

Schmid, W., Poincaré and Lie groups, *Bull. Amer. Math. Soc.* 6 (1982), 175–186.

Schreier, O., Abstrakte kontinuierliche Gruppen, *Abhandlungen aus dem Mathematischen Seminar der Hamburgischen Universität* 4 (1926), 15–32.

Schreier, O., Die Verwandtschaft stetiger Gruppen im grossen, *Abhandlungen aus dem Mathematischen Seminar der Hamburgischen Universität* 5 (1927), 233–244.

Schur, F., Zur Theorie der endlichen Transformationsgruppen, *Math. Annalen* 38 (1891), 263–286.

Schur, F., Ueber den analytischen Charakter der eine endliche continuierliche Transformationsgruppe darstellenden Functionen, *Math. Annalen* 41 (1893), 509–538.

Schur, I., Neue Begründung der Theorie der Gruppencharaktere, *Sitzungsberichte der Königlich Preussischen Akademie der Wissenschaften* (1905), 406–432. (= *Gesammelte Abhandlungen*, Vol. I, 143–169.)

Schur, I., Neue Anwendungen der Integralrechnung auf Probleme der Invariantentheorie, *Sitzungsberichte der Preussischen Akademie der Wissenschaften* (1924), 189–208. (= *Gesammelte Abhandlungen*, Vol. II, 440–459.)

Schur, I., *Gesammelte Abhandlungen*, I, II, III, Springer-Verlag, Berlin, 1973.

Séminaire "Sophus Lie," *Théorie des algèbres de Lie, topologie des groupes de Lie, 1954–55*, École Normale Supérieure, Paris, 1955.

Serre, J.-P., Représentations linéaires et espaces homogènes Kählérians des groupes de Lie compacts, Exposé 100, *Séminaire Bourbaki, 6ᵉ année, 1953/54*, Inst. Henri Poincaré, Paris, 1954. Reprinted with corrections: 1965.

Serre, J.-P., Tores maximaux des groupes de Lie compacts, Exposé 23, Séminaire "Sophus Lie," *Théorie des algèbres de Lie, topologie des groupes de Lie, 1954–55*, École Normale Supérieure, Paris, 1955.

Serre, J.-P., *Lie Algebras and Lie Groups*, W. A. Benjamin, New York, 1965.

Serre, J.-P., *Algèbres de Lie Semi-Simples Complexes*, W. A. Benjamin, New York, 1966. English translation: *Complex Semisimple Lie Algebras*, Springer-Verlag, New York, 1987.

Spivak, M., *A Comprehensive Introduction to Differential Geometry*, Vol. 1, Publish or Perish, Houston, TX, 1970.

Springer, T. A., *Linear Algebraic Groups*, Birkhäuser, Boston, 1981.

Steinberg, R., A general Clebsch-Gordon theorem, *Bull. Amer. Math. Soc.* 67 (1961), 406–407.

Sternberg, S., *Lectures on Differential Geometry*, Prentice-Hall, Englewood Cliffs, N.J., 1964.

Sudbery, A., Quaternionic analysis, *Math. Proc. Cambridge Phil. Soc.* 85 (1979), 199–225.

Sugiura, M., Representations of compact groups realized by spherical functions on symmetric spaces, *Proc. Japan Acad.* 38 (1962), 111–113.

Sugiura, M., Conjugate classes of Cartan subalgebras in real semisimple Lie algebras, *J. Math. Soc. Japan* 11 (1959), 374–434; correction 23 (1971), 379–383.

Tits, J., Sur certaines classes d'espaces homogènes de groupes de Lie, Acad. Roy. Belg. Cl. Sci. Mém. Coll. 29 (1955), No. 3.

Tits, J., Classification of algebraic semisimple groups, *Algebraic Groups and Discontinuous Subgroups*, Proceedings Symposia in Pure Mathematics, Vol. 9, American Mathematical Society, Providence, R.I., 1966, pp. 33–62.

Umlauf, K. A., Über die Zusammensetzung der enlichen continuierlichen Transformationsgruppen, insbesondere der Gruppen vom Range Null, thesis, Leipzig, 1891.

van der Waerden, B. L., Die Klassifikation der einfachen Lieschen Gruppen, *Math. Zeitschrift* 37 (1933), 446–462.

Varadarajan, V. S., *Lie Groups, Lie Algebras, and Their Representations*, Prentice-Hall, Englewood Cliffs, N.J., 1974. Second edition: Springer-Verlag, New York, 1984.

Varadarajan, V. S., *An Introduction to Harmonic Analysis on Semisimple Lie Groups*, Cambridge University Press, Cambridge, 1989.

Verma, D.-N., Structure of certain induced representations of complex semisimple Lie algebras, *Bull. Amer. Math. Soc.* 74 (1968), 160–166; correction 74 (1968), 628.

Vogan, D. A., The algebraic structure of the representation of semisimple Lie groups I, *Annals of Math.* 109 (1979), 1–60.

Vogan, D. A., *Representations of Real Reductive Lie Groups*, Birkhäuser, Boston, 1981a.

Vogan, D. A., *Unitary Representations of Reductive Lie Groups*, Princeton University Press, Princeton, 1987.

von Neumann, J., Zum Haarschen Mass in topologischen Gruppen, *Compositio Math.* 1 (1934a), 106–114. (= *Collected Works*, Vol. II, 445–453.)

von Neumann, J., Almost periodic functions in a group I, *Trans. Amer. Math. Soc.* 36 (1934b), 445–492. (= *Collected Works*, Vol. II, 454–501.)

von Neumann, J., *Collected Works*, I to VI, Pergamon Press, Oxford, 1961.

Wallach, N. R., A classification of real simple Lie algebras, thesis, Washington University, 1966.

Wallach, N. R., On maximal subsystems of root systems, *Canad. J. Math.* 20 (1968), 555–574.

Wallach, N. R., Cyclic vectors and irreducibility for principal series representations, *Trans. Amer. Math. Soc.* 158 (1971), 107–113.

Wallach, N. R., Cyclic vectors and irreducibility for principal series representations II, *Trans. Amer. Math. Soc.* 164 (1972), 389–396.

Wallach, N. R., *Harmonic Analysis on Homogenous Spaces*, Marcel Dekker, New York, 1973.

Wallach, N., *Real Reductive Groups I*, Academic Press, San Diego, 1988.

Wallach, N., *Real Reductive Groups II*, Academic Press, San Diego, 1992.

Warner, F. W., *Foundations of Differentiable Manifolds and Lie Groups*, Scott Foresman, Glenview, Ill., 1971. Second edition: Springer-Verlag, New York, 1982.

Warner, G., *Harmonic Analysis on Semi-Simple Lie Groups I*, Springer-Verlag, New York, 1972a.

Warner, G., *Harmonic Analysis on Semi-Simple Lie Groups II*, Springer-Verlag, New York, 1972b.

Weil, A., Démonstration topologique d'un théorème fondamental de Cartan, *C. R. Acad. Sci. Paris* 200 (1935), 518–520.

Weil, A., *L'intégration dans les groupes topologiques et ses applications*, Actualités scientifiques et industrielles 869, Hermann, Paris, 1940. Second edition: Actualités scientifiques et industrielles 869–1145, 1951.

Wells, R. O., *Differential Analysis on Complex Manifolds*, Prentice-Hall, Englewood Cliffs, N.J., 1973. Second edition: Springer-Verlag, New York, 1980.

Weyl, H., Zur Theorie der Darstellung der einfachen kontinuierlichen Gruppen (Aus einem Schreiben an Herrn I. Schur), *Sitzungsberichte der Preussischen Akademie der Wissenschaften* (1924), 338–345. (= *Gesammelte Abhandlungen*, Vol. II, 453–460.)

Weyl, H., Theorie der Darstellung kontinuierlicher halbeinfacher Gruppen durch lineare Transformationen I, II, III, and Nachtrag, *Math. Zeitschrift* 23 (1925), 271–309; 24 (1926), 328–376; 24 (1926), 377–395; 24 (1926), 789–791. (= *Gesammelte Abhandlungen*, Vol. II, 543–645.)

Weyl, H., *The Classical Groups, Their Invariants and Representations*, 2nd ed., Princeton University Press, Princeton, 1946.

Weyl, H., *Gesammelte Abhandlungen*, I, II, III, IV, Springer-Verlag, Berlin, 1968.

Whitehead, J. H. C., On the decomposition of an infinitesimal group, *Proc. Cambridge Phil. Soc.* 32 (1936), 229–237. (= *Mathematical Works*, Vol. I, Macmillan, New York, 1963, 281–289.)

Whitehead, J. H. C., Certain equations in the algebra of a semi-simple infinitesimal group, *Quart. J. Math.* 8 (1937), 220–237. (= *Mathematical Works*, Vol. I, Macmillan, New York, 1963, 291–308.)

Wigner, E. P., *Group Theory and Its Applications to the Quantum Mechanics of Atomic Spectra*, Academic Press, New York, 1959.

Witt, E., Treue Darstellung Liescher Ringe, *J. Reine Angew. Math.* 177 (1937), 152–160.

Witt, E., Spiegelungsgruppen und Aufzählung halbeinfacher Liescher Ringe, *Abhandlungen aus dem Mathematischen Seminar der Hansischen Universität* 14 (1941), 289–322.

Wolf, J. A., Complex homogeneous contact manifolds and quaternionic symmetric spaces, *J. Math. and Mech.* 14 (1965), 1033–1047.

Wolf, J. A., and A. Korányi, Generalized Cayley transformations of bounded symmetric domains, *Amer. J. Math.* 87 (1965), 899–939.

Yang, C. T., Hilbert's fifth problem and related problems on transformation groups, *Mathematical Developments Arising from Hilbert Problems*, Proceedings Symposia in Pure Mathematics, Vol. 28, Part 1, American Mathematical Society, Providence, R.I., 1976, pp. 142–146.

Zariski, O., and P. Samuel, *Commutative Algebra*, Vol. I, D. Van Nostrand, Princeton, 1958.

Zariski, O., and P. Samuel, *Commutative Algebra*, Vol. II, D. Van Nostrand, Princeton, 1960.

Želobenko, D. P., *Compact Lie Groups and Their Representations*, Translations of Mathematical Monographs, Vol. 40, American Mathematical Society, Providence, R.I., 1973 (English translation of 1970 edition in Russian).

INDEX OF NOTATION

See also the list of Standard Notation on page xv. In the list below, Latin, German, and script letters appear together and are followed by Greek symbols, special superscripts, and non-letters.

INDEX

Progress in Mathematics

Edited by:

Hyman Bass
Dept. of Mathematics
Columbia University
New York, NY 10010
USA

J. Oesterlé
Dépt . de Mathématiques
Université de Paris VI
4, Place Jussieu
75230 Paris Cedex 05, France

A. Weinstein
Department of Mathematics
University of California
Berkeley, CA 94720
U.S.A.

Progress in Mathematics is a series of books intended for professional mathematicians and scientists, encompassing all areas of pure mathematics. This distinguished series, which began in 1979, includes authored monographs and edited collections of papers on important research developments as well as expositions of particular subject areas.

We encourage preparation of manuscripts in some form of TeX for delivery in camera-ready copy which leads to rapid publication, or in electronic form for interfacing with laser printers or typesetters.

Proposals should be sent directly to the editors or to: Birkhäuser Boston, 675 Massachusetts Avenue, Cambridge, MA 02139, U. S. A.

21 KATOK. Ergodic Theory and Dynamical Systems II
22 BERTIN. Séminaire de Théorie des Nombres, Paris 1980-81
23 WEIL. Adeles and Algebraic Groups
24 LE BARZ/HERVIER. Enumerative Geometry and Classical Algebraic Geometry
25 GRIFFITHS. Exterior Differential Systems and the Calculus of Variations
26 KOBLITZ. Number Theory Related to Fermat's Last Theorem
27 BROCKETT/MILLMAN/SUSSMAN. Differential Geometric Control Theory
28 MUMFORD. Tata Lectures on Theta I
29 FRIEDMAN/MORRISON. Birational Geometry of Degenrations
30 YANO/KON. CR Submanifolds of Kaehlerian and Sasakian Manifolds
31 BERTRAND/WALDSCHMIDT. Approximations Diophantiennes et Nombres Transcendants
32 BOOKS/GRAY/REINHART. Differential Geometry
33 ZUILY. Uniqueness and Non-Uniqueness in the Cauchy Problem
34 KASHIWARA. Systems of Microdifferential Equations
35 ARTIN/TATE. Arithmetic and Geometry: Papers Dedicated to I. R. Shafarevich on the Occasion of His Sixtieth Birthday. Vol. 1
36 ARTIN/TATE. Arithmetic and Geometry: Papers Dedicated to I. R. Shafarevich on the Occasion of His Sixtieth Birthday. Vol. 1I
37 DE MONVEL. Mathématique et Physique
38 BERTIN. Séminaire de Théorie des Nombres, Paris 1981-82
39 UENO. Classification of Algebraic and Analytic Manifolds
40 TROMBI. Representation Theory of Reductive Groups
41 STANLEY. Combinatorics and Commutative Algebra; 1st ed. 1984 2nd ed. 1996
42 JOUANOLOU. Théorèmes de Bertini et Applications
43 MUMFORD. Tata Lectures on Theta II
44 KAC. Infitine Dimensional Lie Algebras
45 BISMUT. Large deviations and the Malliavin Calculus
46 SATAKE/MORITA. Automorphic Forms of Several Variables, Taniguchi Symposium, Katata, 1983
47 TATE. Les Conjectures de Stark sur les Fonctions L d'Artin en $s = 0$
48 FRÖLICH. Classgroups and Hermitian Modules
49 SCHLICHTKRULL. Hyperfunctions and Harmonic Analysis on Symmetric Spaces
50 BOREL ET AL. Intersection Cohomology
51 BERTIN/GOLDSTEIN. Séminaire de Théorie des Nombres, Paris 1982-83
52 GASQUI/GOLDSCHMIDT. Déformations Infinitesimales des Structures Conformes Plates
53 LAURENT. Théorie de la Deuxième Microlocalisation dans le Domaine Complexe
54 VERDIER/LE POTIER. Module des Fibres Stables sur les Courbes Algébriques: Notes de l'Ecole Normale Supérieure, Printemps, 1983
55 EICHLER/ZAGIER. The Theory of Jacobi Forms
56 SHIFFMAN /SOMMESE. Vanishing Theorems on Complex Manifolds
57 RIESEL. Prime Numbers and Computer Methods for Factorization
58 HELFFER/NOURRIGAT. Hypoellipticité Maximale pour des Opérateurs Polynomes de Champs de Vecteurs
59 GOLDSTEIN. Séminaire de Théorie des Nombres, Paris 1983–84
60 PROCESI. Geometry Today: Giornate Di Geometria, Roma. 1984
61 BALLMANN/GROMOV/SCHROEDER. Manifolds of Nonpositive Curvature
62 GUILLOU/MARIN. A la Recherche de la Topologie Perdue
63 GOLDSTEIN. Séminaire de Théorie des Nombres, Paris 1984–85
64 MYUNG. Malcev-Admissible Algebras
65 GRUBB. Functional Calculus of Pseudo-Differential Boundary Problems; 2nd ed.

66 Cassou-Nogues/Taylor. Elliptic Functions and Rings and Integers
67 Howe. Discrete Groups in Geometry and Analysis: Papers in Honor of G.D. Mostow on His Sixtieth Birthday
68 Robert. Autour de L'Approximation Semi-Classique
69 Faraut/Harzallah. Deux Cours d'Analyse Harmonique
70 Adolphson/Conrey/Ghosh/Yager. Analytic Number Theory and Diophantine Problems: Proceedings of a Conference at Oklahoma State University
71 Goldstein. Séminaire de Théorie des Nombres, Paris 1985–86
72 Vaisman. Symplectic Geometry and Secondary Characteristic Classes
73 Molino. Riemannian Foliations
74 Henkin/Leiterer. Andreotti-Grauert Theory by Integral Formulas
75 Goldstein. Séminaire de Théorie des Nombres, Paris 1986–87
76 Cossec/Dolgachev. Enriques Surfaces I
77 Reyssat. Quelques Aspects des Surfaces de Riemann
78 Borho/Brylinski/MacPherson. Nilpotent Orbits, Primitive Ideals, and Characteristic Classes
79 McKenzie/Valeriote. The Structure of Decidable Locally Finite Varieties
80 Kraft/Petrie/Schwarz. Topological Methods in Algebraic Transformation Groups
81 Goldstein. Séminaire de Théorie des Nombres, Paris 1987–88
82 Duflo/Pedersen/Vergne. The Orbit Method in Representation Theory: Proceedings of a Conference held in Copenhagen, August to September 1988
83 Ghys/de la Harpe. Sur les Groupes Hyperboliques d'après Mikhael Gromov
84 Araki/Kadison. Mappings of Operator Algebras: Proceedings of the Japan-U.S. Joint Seminar, University of Pennsylvania, Philadelphia, Pennsylvania, 1988
85 Berndt/Diamond/Halberstam/Hildebrand. Analytic Number Theory: Proceedings of a Conference in Honor of Paul T. Bateman
86 Cartier/Illusie/Katz/Laumon/Manin/Ribet. The Grothendieck Festschrift: A Collection of Articles Written in Honor of the 60th Birth day of Alexander Grothendieck., Vol. I
87 Cartier/Illusie/Katz/Laumon/Manin/Ribet. The Grothendieck Festschrift: A Collection of Articles Written in Honor of the 60th Birth day of Alexander Grothendieck., Vol. II
88 Cartier/Illusie/Katz/Laumon/Manin/Ribet. The Grothendieck Festschrift: A Collection of Articles Written in Honor of the 60th Birth day of Alexander Grothendieck., Vol. III
89 Van der Geer/Oort/Steenbrink. Arithmetic Algebraic Geometry
90 Srinivas. Algebraic K-Theory
91 Goldstein. Séminaire de Théorie des Nombres, Paris 1988–89
92 Connes/Duflo/Joseph/Rentschler. Operator Algebras, Unitary Representations, Enveloping Algebras, and Invariant Theory. A Collection of Articles in Honor of the 65th Birthday of Jacques Dixmier
93 Audin. The Topology of Torus Actions on Symplectic Manifolds
94 Mora/Traverso (eds.) Effective Methods in Algebraic Geometry
95 Michler/Ringel (eds.) Representation Theory of Finite Groups and Finite Dimensional Algebras
96 Malgrange. Equations Différentielles à Coefficients Polynomiaux
97 Mumford/Nori/Norman. Tata Lectures on Theta III
98 Godbillon. Feuilletages, Etudes géométriques
99 Donato/Duval/Elhadad/Tuynman. Symplectic Geometry and Mathematical Physics. A Collection of Articles in Honor of J.-M. Souriau

100 TAYLOR. Pseudodifferential Operators and Nonlinear PDE
101 BARKER/SALLY. Harmonic Analysis on Reductive Groups
102 DAVID. Séminaire de Théorie des Nombres, Paris 1989-90
103 ANGER /PORTENIER. Radon Integrals
104 ADAMS /BARBASCH/VOGAN. The Langlands Classification and Irreducible Characters for Real Reductive Groups
105 TIRAO/WALLACH. New Developments in Lie Theory and Their Applications
106 BUSER. Geometry and Spectra of Compact Riemann Surfaces
108 BRYLINSKI. Loop Spaces, Characteristic Classes and Geometric Quantization
108 DAVID. Séminaire de Théorie des Nombres, Paris 1990-91
109 EYSSETTE/GALLIGO. Computational Algebraic Geometry
110 LUSZTIG. Introduction to Quantum Groups
111 SCHWARZ. Morse Homology
112 DONG/LEPOWSKY. Generalized Vertex Algebras and Relative Vertex Operators
113 MOEGLIN/WALDSPURGER. Décomposition spectrale et séries d'Eisenstein
114 BERENSTEIN/GAY/VIDRAS/YGER. Residue Currents and Bezout Identities
115 BABELON/CARTIER/KOSMANN-SCHWARZBACH. Integrable Systems, The Verdier Memorial Conference: Actes du Colloque International de Luminy
116 DAVID. Séminaire de Théorie des Nombres, Paris 1991-92
117 AUDIN/LAFONTAINE (eds). Holomorphic Curves in Symplectic Geometry
118 VAISMAN. Lectures on the Geometry of Poisson Manifolds
119 JOSEPH/ MEURAT/MIGNON/PRUM/ RENTSCHLER (eds). First European Congress of Mathematics, July, 1992, Vol. I
120 JOSEPH/ MEURAT/MIGNON/PRUM/ RENTSCHLER (eds). First European Congress of Mathematics, July, 1992, Vol. II
121 JOSEPH/ MEURAT/MIGNON/PRUM/ RENTSCHLER (eds). First European Congress of Mathematics, July, 1992, Vol. III (Round Tables)
122 GUILLEMIN. Moment Maps and Combinatorial Invariants of T^n-spaces
123 BRYLINSKI/BRYLINSKI/GUILLEMIN/KAC. Lie Theory and Geometry: In Honor of Bertram Kostant
124 AEBISCHER/BORER/KALIN/LEUENBERGER/ REIMANN. Symplectic Geometry
125 LUBOTZKY. Discrete Groups, Expanding Graphs and Invariant Measures
126 RIESEL. Prime Numbers and Computer Methods for Factorization
127 HORMANDER. Notions of Convexity
128 SCHMIDT. Dynamical Systems of Algebraic Origin
129 DIJGRAAF/FABER/VAN DER GEER. The Moduli Space of Curves
130 DUISTERMAAT. Fourier Integral Operators
131 GINDIKIN/LEPOWSKY/WILSON (eds). Functional Analysis on the Eve of the 21st Century. In Honor of the Eightieth Birthday of I. M. Gelfand, Vol. 1
132 GINDIKIN/LEPOWSKY/WILSON (eds.) Functional Analysis on the Eve of the 21st Century. In Honor of the Eightieth Birthday of I. M. Gelfand, Vol. 2
133 HOFER/TAUBES/WEINSTEIN/ZEHNDER. The Floer Memorial Volume
134 CAMPILLO LOPEZ/NARVAEZ MACARRO (eds.) Algebraic Geometry and Singularities
135 AMREIN/BOUTET DE MONVEL/GEORGESCU. C_0-Groups, Commutator Methods and Spectral Theory of N-Body Hamiltonians
136 BROTO/CASACUBERTA/MISLIN (eds). Algebraic Topology: New Trends in Localization and Periodicity
137 VIGNERAS. Représentations l-modulaires d'un groupe réductif p-adique avec $l \neq p$
138 BERNDT/DIAMOND/HILDEBRAND (eds.) Analytic Number Theory. Vol. 1 In Honor of Heini Halberstam
139 BERNDT/DIAMOND/HILDEBRAND (eds.) Analytic Number Theory, Vol. 2 In Honor of Heini Halberstam
140 KNAPP. Lie Groups Beyond an Introduction
141 CABANES. Finite Reductive Groups: Related Structures and Representations

www.ingramcontent.com/pod-product-compliance
Ingram Content Group UK Ltd.
Pitfield, Milton Keynes, MK11 3LW, UK
UKHW012140240726
13966UKWH00001B/84